Springer Monographs in Mathematics

Springer
*Berlin
Heidelberg
New York
Barcelona
Hong Kong
London
Milan
Paris
Singapore
Tokyo*

Franz Lemmermeyer

Reciprocity Laws

From Euler to Eisenstein

Springer

Franz Lemmermeyer
http://www.rzuser.uni-heidelberg.de/~hb3/
e-mail: hb3@ix.urz.uni-heidelberg.de

Library of Congress Cataloging-in-Publication Data
Lemmermeyer, Franz, 1962 -
 Reciprocity laws : from Euler to Eisenstein / Franz Lemmermeyer.
 p. cm. -- (Springer monographs in mathematics)
 Includes bibliographical references and index.
 ISBN 3540669574 (hc : alk. paper)
 1. Reciprocity theorems I. Title II. Series

QA241 .L56 2000   00-026905
512'74--dc21

Mathematics Subject Classification (1991): 11A15

ISBN 3-540-66957-4 Springer-Verlag Berlin Heidelberg New York

This work is subject to copyright. All rights are reserved, whether the whole or part of the material is concerned, specifically the rights of translation, reprinting, reuse of illustrations, recitation, broadcasting, reproduction on microfilm or in any other way, and storage in data banks. Duplication of this publication or parts thereof is permitted only under the provisions of the German Copyright Law of September 9, 1965, in its current version, and permission for use must always be obtained from Springer-Verlag. Violations are liable for prosecution under the German Copyright Law.

Springer-Verlag is a company in the BertelsmannSpringer publishing group.
© Springer-Verlag Berlin Heidelberg 2000
Printed in Germany

The use of general descriptive names, registered names, trademarks etc. in this publication does not imply, even in the absence of a specific statement, that such names are exempt from the relevant protective laws and regulations and therefore free for general use.

Cover design: *Erich Kirchner, Heidelberg*
Typesetting by the author using a Springer TEX macro package
Printed on acid-free paper   SPIN 10734033   41/3143AT-5 4 3 2 1 0

Preface

The history of reciprocity laws is a history of algebraic number theory. This is a book on reciprocity laws, and our introductory remark is placed at the beginning as a warning: in fact a reader who is acquainted with little more than a course in elementary number theory may be surprised to learn that quadratic reciprocity does – in a sense that we will explain – belong to the realm of algebraic number theory. Hecke ([348, p. 59]) has formulated this as follows:

> Von der Entdeckung des Reziprozitätsgesetzes kann man die moderne Zahlentheorie datieren. Seiner Form nach gehört es noch der Theorie der rationalen Zahlen an, es läßt sich aussprechen als eine Beziehung lediglich zwischen rationalen Zahlen; jedoch weist es seinem Inhalt nach über den Bereich der rationalen Zahlen hinaus. [...] Die Entwicklung der algebraischen Zahlentheorie hat nun wirklich gezeigt, daß der Inhalt des quadratischen Reziprozitätsgesetzes erst verständlich wird, wenn man zu den allgemeinen algebraischen Zahlen übergeht, und daß ein dem Wesen des Problems angemessener Beweis sich am besten mit diesen höheren Hilfsmitteln führen läßt, während man von den elementaren Beweisen sagen muß, daß sie vielmehr den Charakter einer nachträglichen Verifikation besitzen.[1]

Naturally, along with these higher methods came generalizations of the reciprocity law itself. It is no exaggeration to say that this generalization changed our way of looking at the reciprocity law dramatically; Emma Lehmer ([509, p. 467]) writes

> It is well known that the famous Legendre law of quadratic reciprocity, of which over 150 proofs[2] are in print, has been generalized

---
[1] Modern number theory dates from the discovery of the reciprocity law. By its form it still belongs to the theory of rational numbers, as it can be formulated entirely as a simple relation between rational numbers; however its contents points beyond the domain of rational numbers. [...] The development of algebraic number theory has now actually shown that the content of the quadratic reciprocity law only becomes understandable if one passes to general algebraic numbers and that a proof appropriate to the nature of the problem can be best carried out with these higher methods.

[2] She is apparently referring to Gerstenhaber's article [305]; in his email [306], he writes "The origin of the title was not a list but a statement of André Weil in a

over the years to algebraic fields by a number of famous mathematicians from Gauss to Artin to the extent that it has become virtually unrecognizable.

These quotations show that a thorough command of the techniques of algebraic number theory is indispensable for understanding reciprocity laws; given the wealth of well written introductions to algebraic number theory, it seemed reasonable to assume that the readers are familiar with the basic arithmetic of number fields, in particular that of quadratic and cyclotomic fields. Galois theory is also a conditio sine qua non, and occasionally p-adic numbers or, e.g. for Chapter 8, elliptic functions are needed.

So what is a reciprocity law, anyway? Euler's way of looking at it was the following: the quadratic character of a mod p only depends on the residue class of p mod $4a$. For Legendre (who coined the term reciprocity), the reciprocity law was a statement to the effect that an odd prime p is a quadratic residue modulo another odd prime q if and only if q is a quadratic residue modulo p, except when $p \equiv q \equiv 3$ mod 4; more exactly, Legendre defined, for odd primes q, a symbol (p/q) with values in $\{-1, +1\}$ by demanding $(p/q) \equiv p^{(q-1)/2}$ mod q and announced the

Quadratic Reciprocity Law. *Let $p, q \in \mathbb{N}$ be different odd primes; then*

$$\left(\frac{p}{q}\right) = (-1)^{\frac{p-1}{2} \cdot \frac{q-1}{2}} \left(\frac{q}{p}\right).$$

Moreover, we have

$$\left(\frac{-1}{p}\right) = (-1)^{\frac{p-1}{2}} \quad \text{and} \quad \left(\frac{2}{p}\right) = (-1)^{\frac{p^2-1}{8}};$$

these are called the first and the second supplementary law, respectively.

This is where our story begins; as we have noted above, the most transparent proofs of the quadratic reciprocity law are embedded into the theory of algebraic number fields, and already Gauss noticed that in order to formulate the fundamental theorem of biquadratic residues, one needs an "infinite enlargement" of the integers, namely the ring $\mathbb{Z}[i]$ of Gaussian integers. In fact, the biquadratic residue symbol $[\pi/\lambda]$, where $\pi, \lambda \in \mathbb{Z}[i]$ are primes not dividing 2, is the unique element in $\{\pm 1, \pm i\}$ such that the congruence $[\pi/\lambda] \equiv \pi^{(N\lambda-1)/4}$ mod λ holds. The reciprocity law discovered by Gauss then reads

Seminar at the Institute for Advanced Study in Princeton: he said that he knew 50 proofs of the law, and that for each he had seen there were two he had not. So that made 150 proofs. Then he called my attention to Kubota's, which would have been the 151st. So mine had to be the 152nd!"

At about the same time, Hasse [342, p. 100] wrote that there were more than 50 proofs, with some of them differing only marginally from each other. The tables in Appendix B suggest that Weil's estimate was pretty good.

Quartic Reciprocity Law. *Let $\pi, \lambda \in \mathbb{Z}[i]$ be different primary primes, i.e. assume that $\pi \equiv \lambda \equiv 1 \bmod (2 + 2i)$; then*

$$\left[\frac{\pi}{\lambda}\right] = (-1)^{\frac{N\pi-1}{4} \cdot \frac{N\lambda-1}{4}} \left[\frac{\lambda}{\pi}\right].$$

There are also analogues of the first and second supplementary laws; see Chapter 6 for details. A similar (though simpler) formula holds for the cubic residue symbol and 'primary' primes in $\mathbb{Z}[\rho]$, where ρ is a primitive cube root of unity.

The first complete proofs for the cubic and quartic laws were published 1844 by Eisenstein, who also gave the corresponding supplementary laws; Jacobi, however, had given proofs as early as 1837 in his Königsberg lectures [402]. Jacobi was working on a generalization of the cubic and quartic reciprocity law using cyclotomy, but it turned out that the failure of unique factorization was a major stumbling block. Only after Kummer had introduced his ideal numbers (with the intention of applying the theory to find a general reciprocity law) did it become possible to do arithmetic in cyclotomic fields $\mathbb{Q}(\zeta_p)$. Eisenstein, who had before favored the language of forms, quickly acknowledged the superiority of Kummer's approach and succeeded in finding a special case of the general reciprocity law called

Eisenstein's Reciprocity Law. *Let ℓ be an odd prime and suppose that $\alpha \in \mathbb{Z}[\zeta_\ell]$ is primary, i.e. congruent to a rational integer modulo $(1 - \zeta_\ell)^2$. Then*

$$\left(\frac{\alpha}{a}\right)_\ell = \left(\frac{a}{\alpha}\right)_\ell$$

for all integers $a \in \mathbb{Z}$ prime to ℓ.

Here the ℓth power residue symbol $(\alpha/\mathfrak{p})_\ell$ is the unique ℓ^{th} root of unity such that $(\alpha/\mathfrak{p})_\ell \equiv \alpha^{(N\mathfrak{p}-1)/\ell} \bmod \mathfrak{p}$.

The quintic case had to wait for Kummer, who created the theory of *ideal numbers* along the way and finally produced a reciprocity theorem valid in all regular cyclotomic fields.

In order to define a residue symbols for ideals coprime to ℓ in the case of regular primes ℓ we observe that $\mathfrak{a}^h = \alpha \mathcal{O}_K$ is principal. Kummer showed that we can choose α primary, that is, in such a way that the congruences

$$\alpha \overline{\alpha} \equiv a \bmod \ell, \qquad \alpha \equiv b \bmod (1 - \zeta_\ell)^2$$

hold for some integers a and b. Moreover he proved that the residue symbol $\left(\frac{a}{\mathfrak{b}}\right)_\ell$ does not depend on the choice of α, as long as α is primary. Provided that $(\ell, h) = 1$, we can now define the residue symbol $\left(\frac{a}{\mathfrak{b}}\right)_\ell$ by

$$\left(\frac{\mathfrak{a}}{\mathfrak{b}}\right)_\ell^h = \left(\frac{\alpha}{\mathfrak{b}}\right)_\ell.$$

Kummer's Reciprocity Law. *Let $K = \mathbb{Q}(\zeta_\ell)$ and suppose that ℓ is regular, i.e., that ℓ does not divide the class number h of K. Then*

$$\left(\frac{\mathfrak{a}}{\mathfrak{b}}\right)_\ell = \left(\frac{\mathfrak{b}}{\mathfrak{a}}\right)_\ell,$$

where $\mathfrak{a}$ and $\mathfrak{b}$ are relatively prime integral ideals prime to ℓ.

Kummer also gave explicit formulas for the supplementary laws that look rather complicated at first sight.

Hilbert did the next step forward by returning to the quadratic case: he discovered that there is a quadratic reciprocity law in every number field with odd class number, and he outlined how to include fields with even class number as well. Moreover, Hilbert showed that the quadratic reciprocity laws in algebraic number fields could be given a very simple form by using the norm residue symbol, and conjectured the following generalization:

Hilbert's Reciprocity Law. *Let k be an algebraic number field containing the m-th roots of unity; then for all $\mu, \nu \in k^\times$, we have*

$$\prod_\mathfrak{p} \left(\frac{\mu, \nu}{\mathfrak{p}}\right) = 1.$$

Here $\left(\frac{\cdot, \cdot}{\mathfrak{p}}\right)$ is Hilbert's m-th power norm residue symbol mod $\mathfrak{p}$, and the product is extended over all prime places $\mathfrak{p}$ of k.

In this formulation of a reciprocity law, the power residue symbol does not even occur: in order to derive the classical formulation from Hilbert's one essentially has to compute certain Hilbert symbols which is rather straightforward though extremely technical. Actually, even the definition of the norm residue symbol is far from being obvious when $m > 2$.

Hilbert's conjectured reciprocity law was a part of the program he devised when he formulated the ninth problem in his famous address at the Congress of Mathematicians in Paris (1900):[3]

> **Beweis des allgemeinsten Reziprozitätsgesetzes.** *Für einen beliebigen Zahlkörper soll das Reziprozitätsgesetz der ℓ-ten Potenzreste bewiesen werden, wenn ℓ eine ungerade Primzahl bedeutet, und ferner, wenn ℓ eine Potenz von 2 oder eine Potenz einer ungeraden Primzahl ist. Die Aufstellung des Gesetzes, sowie die wesentlichen*

---
[3] See e.g. Faddeev [227], Kantor [416] or Tate [792].

Hilfsmittel zum Beweis desselben werden sich, wie ich glaube, ergeben, wenn man die von mir entwickelte Theorie des Körpers der ℓ-ten Einheitswurzeln und meine Theorie des relativ-quadratischen Körpers in gehöriger Weise verallgemeinert.[4]

The first sentence of this problem asks for a generalization of Kummer's reciprocity law to fields $\mathbb{Q}(\zeta_\ell)$ for *irregular* primes ℓ. This was accomplished by Furtwängler, who succeeded in showing the existence of the Hilbert class field and used it to prove a quite general reciprocity law. Takagi created a sensation when he found that Furtwängler's results were just a special case of what we call class field theory today. As an application of his theory, Takagi derived a reciprocity law for ℓ-th powers in $\mathbb{Q}(\zeta_\ell)$ that contained Kummer's results for regular primes ℓ as a special case.

Between 1923 and 1926, Artin and Hasse were looking for simpler (and more general) formulations of Takagi's reciprocity law in the hope that this would help them finish Hilbert's quest for the "most general reciprocity law" in number fields. One of the less complicated formulas they found is the following:

The Weak Reciprocity Law of Hasse. *Let ℓ be an odd prime, $K = \mathbb{Q}(\zeta_\ell)$, and suppose that $\alpha, \beta \in \mathcal{O}_K$ satisfy $(\alpha, \beta) = 1$, $\alpha \equiv 1 \bmod \ell$, and $\beta \equiv 1 \bmod \lambda$. Let Tr denote the trace for $K/\mathbb{Q}$. Then*

$$\left(\frac{\alpha}{\beta}\right)_\ell \left(\frac{\beta}{\alpha}\right)_\ell^{-1} = \zeta^{\mathrm{Tr}\left(\frac{\alpha-1}{\ell} \cdot \frac{\beta-1}{\lambda}\right)}.$$

Hasse considered this as an approximation to the full reciprocity law since the assumption that $\alpha \equiv 1 \bmod \ell$ is quite strong (in particular, it doesn't contain Kummer's law). Artin and Hasse succeeded in giving more exact formulations, but the price they had to pay was the introduction of ℓ-adic logarithms into their formulas.

The next break-through was Artin's discovery that all the reciprocity laws of Gauss, Kummer, Hilbert, and Takagi could be subsumed into a *general reciprocity law*. The connection between these laws is, as is demonstrated by E. Lehmer's quote cited above, not of the kind that springs to one's eye at first glance. Using the idèle class group C_k of a number field k, Artin's reciprocity law takes the following simple form:

Artin's Reciprocity Law. *Let k be an algebraic number field and let K/k be a finite extension. Then the global norm residue symbol $\left(\frac{K/k}{\cdot}\right)$ induces an*

---

[4] **Proof of the most general reciprocity law.** For an arbitrary number field, to prove the most general reciprocity law for ℓ-th power residues, when ℓ denotes an odd prime, and moreover, when ℓ is a power of 2 or a power of an odd prime. The formulation of the law, as well as the essential means for proving it will, I believe, result through a proper generalization of the theory of the field of ℓ-th roots of unity that I have developed, and of my theory of relative quadratic fields.

isomorphism
$$C_k/N_{K/k}C_K \simeq \operatorname{Gal}(K/k)^{\mathrm{ab}},$$
where $G^{\mathrm{ab}} = G/G'$ is G made abelian.

In its ideal theoretic formulation, Artin's reciprocity law basically states that the power residue symbol $\left(\frac{\alpha}{\mathfrak{p}}\right)_m$ only depends on the residue class of α modulo some multiple of $\mathfrak{p}$; in the case $m = 2$, this is basically Euler's formulation of the quadratic reciprocity law, while for prime values of m already Eisenstein had shown how to derive the reciprocity law from such a statement.

Immediately after Artin had proved his own four-year old conjecture in 1927 (using methods of Chebotarev), Hasse devoted the second part of his Zahlbericht to the derivation of the known explicit reciprocity laws from Artin's. Moreover, Artin's reciprocity law allowed Hasse to define a norm residue symbol $\left(\frac{\mu, K/k}{\mathfrak{p}}\right)$ for any number field k and an abelian extension K/k, not only for those k containing the appropriate roots of unity; moreover he noticed that a product formula similar to Hilbert's holds. Finally, in the special case $\zeta_m \in k$ and $K = k(\sqrt[m]{\nu})$, Hasse found $\left(\frac{\mu, K/k}{\mathfrak{p}}\right) = \left(\frac{\mu, \nu}{\mathfrak{p}}\right)$.

Hasse's investigation of the norm residue symbol (which is of a central importance in the second part of his Bericht) eventually suggested the existence of a "local class field theory", that is a theory of abelian extensions of local fields. This allowed him to find the local counterpart of Artin's reciprocity law and prove it by deducing it from the global result:

Artin's Reciprocity Law for Local Fields. *Let k be a finite extension of the field $\mathbb{Q}_p$ of p-adic numbers and let K/k be a finite extension. Then the local norm residue symbol induces an isomorphism*
$$k^\times/N_{K/k}K^\times \simeq \operatorname{Gal}(K/k)^{\mathrm{ab}}.$$

Hasse immediately suggested that to look for direct proofs for the local case and build global class field theory on the simpler local one. This program was carried out essentially by him, F.K. Schmidt and Chevalley.

The classical formulation of class field theory in terms of ideal groups was abandoned by Chevalley who introduced idèles in order to describe the class field theory of infinite extensions. It soon became clear that idèles could also be used to reverse the classical approach and to deduce the global class field theory from the (easier) local one. Another revolution was the cohomological formulation of class field theory; using Tate's cohomology groups, the reciprocity law takes the following form:

Tate's Formulation of Artin's Reciprocity Law. *Let K/k be a normal extension, and let $\mathrm{u}_{K/k} \in \mathrm{H}^2(\operatorname{Gal}(K/k), C_K)$ be the fundamental class of*

K/k. Then the cup product with $u_{K/k}$ induces, for every $q \in \mathbb{Z}$, an isomorphism

$$u_{K/k} \smile : H^q(\mathrm{Gal}\,(K/k), \mathbb{Z}) \longrightarrow H^{q+2}(\mathrm{Gal}\,(K/k), C_K).$$

The background necessary for understanding Tate's formulation will be given in Part II; here we only note that the special case $q = -2$ is nothing but Artin's reciprocity law, since $H^{-2}(\mathrm{Gal}\,(K/k), \mathbb{Z}) \simeq \mathrm{Gal}\,(K/k)^{\mathrm{ab}}$ and $H^0(\mathrm{Gal}\,(K/k), C_K) = C_k/N_{K/k}C_K$.

If, at this point, you have the feeling that we've come a long way, you might be surprised to hear that Weil [828] claimed that there was hardly any progress at all from Gauss to Artin:

> on peut dire que *tout* ce qui a été fait en arithmétique depuis Gauss jusqu'à ces dernières années consiste en variations sur la loi de réciprocité: on est parti de celle de Gauss; on aboutit, couronnement de tous les travaux de Kummer, Dedekind, Hilbert, à celle d'Artin, et *c'est la même*.[5]

Although Artin's reciprocity law (the decomposition law for abelian extensions) is not very far away from the quadratic reciprocity law (viewed as the decomposition law for quadratic extensions of $\mathbb{Q}$), unless when measured in terms of the technical difficulties involved in their proofs, I feel that Weil is being too modest here.

In a way, Artin's reciprocity law closed the subject (except for the subsequent work on explicit formulas, not to mention the dramatic progress into non-abelian class field theory that is connected in particular with the names of Shimura and Langlands or the recent generalization of class field theory to "higher dimensional" local fields), and the decline of interest in the classical reciprocity laws was a natural consequence. Nevertheless, two of the papers that helped shape the research in number theory during the second half of this century directly referred to Gauss's work on biquadratic residues: first, there's Weil's paper from 1949 ([We3] in Chapter 10) on equations over finite fields in which he announced the Weil Conjectures and which was inspired directly by reading Gauss:

> In 1947, in Chicago, I felt bored and depressed, and, not knowing what to do, I started reading Gauss's two memoirs on biquadratic residues, which I had never read before. The Gaussian integers occur in the second paper. The first one deals essentially with the number of solutions $ax^4 - by^4 = 1$ in the prime field modulo p, and with the connection between these and certain Gaussian sums; actually the method is exactly the same that is applied in the last section

---

[5] it can be said that *everything* which has been done in arithmetic from Gauss to these last years consists of variations on the law of reciprocity: one started with Gauss's law and arrived, thereby crowning all the works of Kummer, Dedekind and Hilbert, at Artin's reciprocity law, and *it is the same*.

of the Disquisitiones to the Gaussian sums of order 3 and the equations $ax^3 - by^3 = 1$. Then I noticed that similar principles can be applied to all equations of the form $ax^m + by^n + cz^r + \ldots = 0$, and that this implies the truth of the "Riemann hypothesis" [...] for all curves $ax^n + by^n + cz^n = 0$ over finite fields, and also a "generalized Riemann hypothesis" for varieties in projective space with a "diagonal" equation $\sum a_i x_i^n = 0$. This led me in turn to conjectures about varieties over finite fields, ...

namely the Weil Conjectures, now Deligne's theorem (see Chapter 10).

The other central theme in number theory during the last few decades came into being in two papers by Birch & Swinnerton-Dyer: while studying the elliptic curves $y^2 = x^3 - Dx$ they were led to an amazing conjecture that linked local and global data of elliptic curves via their Hasse-Weil L-function; in these papers, the quartic reciprocity plays a central role in checking some instances of their conjectures – for the relation between $y^2 = x^3 - Dx$ and quartic residues, see Chapter 10 again. As a matter of fact, even the explicit formulas of Artin-Hasse were resurrected (and generalized) by Iwasawa, Coates and Wiles in order to make progress on the Birch–Swinnerton-Dyer conjecture.

After reciprocity had disappeared from textbooks[6] in number theory around 1950 (excepting, of course, the ubiquitous quadratic reciprocity law), the renaissance of reciprocity laws began with their inclusion in the influential book [386] of Ireland & Rosen. Gauss and Jacobi sums (not to mention Eisenstein sums) shared a similar fate; in [616], Neumann writes

> H. Weber räumte den Gaußschen Summen in seinem "Lehrbuch der Algebra" noch einen beträchtlichen Platz ein, während in unserm Jahrhundert dieser Teil der Kreisteilungstheorie in den Hintergrund gedrängt wurde.[7]

One of the reasons why Gauss sums came back with a vengeance was the growing interest in finite fields in general due to their applications in primality testing, cryptography and coding theory. Also, Jacobi sums provide a simple means of counting solutions of certain congruences, thus giving a well-motivated introduction to problems around the Weil conjectures (see

---

[6] Compare the role of reciprocity in the books of Bachmann [*Die Lehre von der Kreistheilung*, 1872; *Niedere Zahlentheorie I.*, 1902], Sommer [*Vorlesungen über Zahlentheorie*, 1907], Hecke [*Vorlesungen über die Theorie der algebraischen Zahlen*, 1923], Fueter [*Synthetische Zahlentheorie*, 1917] or Landau [*Vorlesungen über Zahlentheorie*, 1927], with those that appeared in the second part of this century, in particular Hardy & Wright, [*An introduction to the theory of numbers*; 1938] or Borevich & Shafarevich [*Number Theory*; 1964], to mention only two of the best known books.

[7] H. Weber devoted a considerable part of his textbook "Lehrbuch der Algebra" to Gauss and Jacobi sums, whereas this part of cyclotomy was thrust into the background in our century.

Chapter 10). Finally, Kolyvagin's Euler systems of Gauss sums make sure that they're here to stay.

Apart from the reciprocity laws given above, which have been of central importance for the development of algebraic number theory up to the 1930s, so-called *rational reciprocity laws* (linking the power residue character of some algebraic number modulo a rational prime p to the representation of p by binary quadratic forms) have been studied extensively by many mathematicians; the subject started with Euler's conjectures on the cubic and quartic residuacity of 2, and was largely neglected after Gauss, Dirichlet and Jacobi had supplied proofs. Only recently there has been a revival of interest in rational reciprocity. Parts of the sometimes confusing history of discoveries and rediscoveries of results on rational reciprocity can be found in the Notes to Chapter 5. Which brings us to the question of what this book is all about.

To begin with I should mention that it was conceived as a one-volume text on the development of reciprocity laws from Euler to Artin, but eventually it seemed more reasonable to split it into two. This first part deals with reciprocity laws from Euler to Eisenstein (including topics such as rational reciprocity that originated with Dirichlet but had a renaissance during the 1970s), while the second will discuss the contributions of Kummer, Hilbert, Furtwängler, Takagi, Artin and Hasse, in other words: it will present the connection between reciprocity laws and class field theory, and in particular with explicit reciprocity laws. Whether the writing of the second part can be completed will however depend on my being in a position to do so.

Although this book is intended to serve as a source of information on the history of reciprocity laws, it cannot claim to be a substitute for Vol. 4 of Dickson's trilogy on the history of number theory;[8] such a fourth volume (on quadratic reciprocity) had actually been planned, as Dickson himself writes on page 3 of his third volume:

> Euler stated many special empirical theorems on the representability of primes by $x^2 \pm Ny^2$, where x and y are relatively prime (in connection with empirical theorems on the linear forms of the prime divisors of $x^2 \pm Ny^2$, to be quoted und the quadratic reciprocity law in vol IV).

For more in this connection, see I. Kaplansky's letter [425] and D. Fenster's article [231]. It seems that A. Cooper's thesis [136] on the history of quadratic residues was meant to be a part of it. Apparently, Cooper also planned to publish a history of quadratic residues and reciprocity law, but this never

---

[8] In particular, this is not a book on the history of mathematics: I do not hesitate to use the language of finite fields when explaining results of Fermat, and in the presentation Eisenstein's proofs of reciprocity laws using elliptic functions I do *not* follow the original proof line by line but rather present his ideas in a modern setting. The advantage of this approach should become clear upon comparing e.g. Eisenstein's 84pp article on the division of the lemniscate with our 4 pages in Chapter 9.

happened. From what I gleaned from [231], at least parts of the manuscript seem to have survived in an archive at the University of Texas – unfortunately I have not yet had the chance to read them.

This book is likewise not meant to be a textbook (courses on reciprocity laws are a rare breed anyway), but it do hope that it contains plenty of stuff with which to pep up lectures on number theory: Section 5.1 contains an approach to rational reciprocity laws using nothing beyond the quadratic reciprocity law, introductions to algebraic number theory may be seasoned with the genus theory presented in Sections 2.2 and 2.3, and those lecturing on elliptic functions may find parts of Chapter 8 attractive. I also found it appropriate to include a number of exercises. One purpose that they serve is to present material that didn't fit in the text; the main reason for having exercises, however, is that the rocky road to research is paved with interesting questions and problems. It is my hope that readers will find some problems in this book that make them get out pencil and paper. In a similar vein, the many bibliographical references were provided not only as a service to those who are interested in studying reciprocity laws and their history, but also in the hope that they may entice readers to take the dusty volumes of the collected works of Abel, Eisenstein, Kummer etc. from the shelves and start browsing through them.

The proofs up to Chapter 9 are essentially complete, although occasionally exercises are invoked to fill in some details. In Chapters 10 and 11, however, I try to tie up the material with topics that cannot be presented in detail here, and in these places the exposition acquires the character of a survey. In general, I have indicated the lack of a proof by closing the theorem with a box $\Box$.

One problem when dealing with different power residue symbols in number fields is the choice of notation; even in a case as simple as the biquadratic reciprocity law we have to distinguish four different symbols, namely the quadratic and the biquadratic residue symbols in $\mathbb{Z}[i]$, the Legendre symbol in $\mathbb{Z}$, and the rational quartic residue symbol in $\mathbb{Z}$. In order to keep the notation as simple as possible I did *not* invent a globally consistent notation but instead tried to make sure that the meaning of the symbols employed is locally constant. The same remark applies to the notion of primary and semi-primary integers: the sheer multitude of definitions prohibits the introduction of a globally consistent notion of primariness.

Appendix B contains a table of references to published proofs of the quadratic reciprocity law. I tried to make the list as complete as possible, not counting, of course, the innumerable standard proofs given in textbooks. A closer examination of these proofs will reveal that not all of them can be counted as different; on the other hand, a thorough classification (continuing the work of Baumgart [38] and Bachmann [26] from more than a century ago) would have required a book of its own (not to mention the time for writing it).⁹

The bibliography at the end of this book contains lots of references to books and articles connected with reciprocity laws; citations by acronyms like [Has] refer to additional references given at the end of each chapter which treat reciprocity only marginally or not at all. The separation of these two groups of references is of course not always clear, but I felt it was desirable to have a list of references on reciprocity that was not sparkled with entries about other issues. Occasionally, references contain a URL; these have the habit of becoming obsolete very fast – in such a case you better use a search engine or check whether the Number Theory Web maintained by Keith Matthews on
    http://www.maths.uq.oz.au:80/~krm/web_aust.html
contains a link that works.

Finally I would like to thank all the people without whose help this book could not have been written. First of all, thanks go to my teachers Ulrich Felgner in Tübingen, who introduced me to the world of reciprocity laws, and to Albrecht Brandis, Sigrid Böge and Peter Roquette in Heidelberg. Irving Kaplansky kindly sent me Cooper's thesis on quadratic residues, and Roger Cuculière provided me with a copy of [142]. Keith Dennis made Cooke's Lecture Notes [134, 135] available to me, and Toyokazu Hiramatsu sent me a copy of his book [371] on higher reciprocity laws in Japanese. Finally, I gained access to Jacobi's Königsberg lectures [402] through the help of Herbert Pieper. A special thank you goes to Jacques Martinet, Stéphane Louboutin, Richard Mollin and Raimund Seidel for support when I needed it most. I also thank the people who helped to reduce the number of mathematical and typographical errors (mainly in the first seven chapters), namely Robin Chapman and Marinus van den Heuvel, as well as Achava Nakhash, Jim Propp, Udaï Venedem, Stefanie Vögeli-Fandel, and Felipe Zaldivar. Last not least I thank the Deutsche Forschungsgemeinschaft for their financial support over the last four years without which this book would never have seen the light of day.

I close with a quote of Kummer (letter to Kronecker, Oct. 10, 1845):

> Nicht allein in der Absicht, daß Sie meine Arbeiten kennen lernen um sie wo es sein kann bei Ihren eigenen zu benutzen, sondern besonders auch um meiner selbst Willen, um für mich etwas Ordnung und Klarheit hineinzubringen, schreibe ich Ihnen ... davon.¹⁰

                     Heidelberg, Bordeaux, Saarbrücken, Bonn, 1993–1999

---

⁹ If this sounds like an invitation, it's because it is one.
¹⁰ I write ... to you not only with the intention that you get to know my work so that you can use it in your own work wherever it is possible, but also for my own sake, in order to bring in some order and clarity.

Contents

Preface v

1. **The Genesis of Quadratic Reciprocity** 1
   1.1 P. Fermat 1
   1.2 L. Euler 3
   1.3 A.-M. Legendre 6
   1.4 C.-F. Gauss 9

2. **Quadratic Number Fields** 43
   2.1 Quadratic Fields 43
   2.2 Genus Theory 47
   2.3 Genus Characters 52
   2.4 The Lucas-Lehmer Test 56
   2.5 Hilbert Symbols and K_2 60

3. **Cyclotomic Number Fields** 79
   3.1 Cyclotomic Fields 79
   3.2 Primality Tests 85
   3.3 Quadratic Gauss Sums 93
   3.4 Cyclotomic Units 97

4. **Power Residues and Gauss Sums** 111
   4.1 Residue Symbols in Number Fields 111
   4.2 Gauss's Lemma 114
   4.3 Discriminants 116
   4.4 Kummer Extensions 119
   4.5 Characters of Abelian Groups 121
   4.6 Sums of Gauss, Jacobi and Eisenstein 126

5. **Rational Reciprocity Laws** 153
   5.1 L. Dirichlet 154
   5.2 A. Scholz 160
   5.3 E. Lehmer 161
   5.4 Rational Quartic Reciprocity 164
   5.5 Residue Characters of Quadratic Units 168

6. Quartic Reciprocity 185
  6.1 Splitting of Primes 186
  6.2 Quartic Gauss and Jacobsthal Sums 190
  6.3 The Quartic Reciprocity Law 194
  6.4 Applications 197
  6.5 Quartic Reciprocity in some Quartic Fields 198

7. Cubic Reciprocity 209
  7.1 Splitting of Primes 209
  7.2 The Cubic Reciprocity Law 212
  7.3 Sextic Reciprocity 217
  7.4 Cubic Reciprocity in some Quartic Fields 220

8. Eisenstein's Analytic Proofs 235
  8.1 Quadratic Reciprocity 236
  8.2 Abel's Construction of Elliptic Functions 239
  8.3 Elliptic Functions 247
  8.4 Quartic and Cubic Reciprocity 251
  8.5 Quadratic Reciprocity in Quadratic Fields 256
  8.6 Kronecker's Jugendtraum 260
  8.7 The Determination of Gauss Sums 265

9. Octic Reciprocity 289
  9.1 The Rational Octic Reciprocity Law 289
  9.2 Eisenstein's Reciprocity Law 292
  9.3 Elliptic Gauss Sums 299
  9.4 The Octic Reciprocity Law 304
  9.5 Scholz's Octic Reciprocity Law 307

10. Gauss's Last Entry 317
  10.1 Connections with Quartic Reciprocity 317
  10.2 Counting Points with Cyclotomic Numbers 321
  10.3 Counting Points with Jacobi Sums 326
  10.4 The Classical Zeta Functions 328
  10.5 Counting Points with Zeta Functions 335

11. Eisenstein Reciprocity 357
  11.1 Factorization of Gauss Sums 357
  11.2 Eisenstein Reciprocity for ℓ-th Powers 361
  11.3 The Stickelberger Congruence 366
  11.4 Class Groups of Abelian Number Fields 371

A. Dramatis Personae 411

B. Chronology of Proofs 413

C. Some Open Problems... 418

References.. 419

Author Index.. 472

Subject Index 484

1. The Genesis of Quadratic Reciprocity

In this first chapter we will present the fathers of the quadratic reciprocity law. Although some results on quadratic residues modulo 10 have been found very early on (see [Ene]) – in connection with the problem of characterizing perfect squares – the history of modern number theory starts with the editions of the books of Diophantus, in particular with the commented edition by Bachet in 1621.

A word on notation: When we say "prime $p \equiv 1 \bmod 4$" we assume tacitly that $p > 0$ unless stated otherwise.

1.1 P. Fermat

The mathematician who started studying reciprocity questions was Pierre de Fermat. There were no mathematical journals in Fermat's time, and what we know about his results is contained in his letters to other mathematicians (or on the margins of some books he read). The first result related with quadratic reciprocity was stated in a letter to Mersenne [232, II, 212–217]:

> Tout nombre premiere, qui surpasse de l'unité un multiple du quaternaire, est une seul fois la somme de deux carrés. [1]

This claim that every prime $p \equiv 1 \bmod 4$ is the sum of two squares first appeared (without proof) in a book of S. Stevin (see Hofmann [Hof] and Cuculière [Cuc]), and a general criterion for a number to be the sum of two squares is credited to Girard by Grosswald [Gro]. The fact that no number $\equiv 3 \bmod 4$ is the sum of two squares was already known to Diophantus of Alexandria (as is suggested by Problem 9 in his Book V), and its proof is trivial as soon as one knows about congruences. In fact, suppose that $p = x^2 + y^2$ is the sum of two squares; then either $p = 2$ and $x = y = 1$ (up to sign), or p is odd, and then we may assume that x is odd and y is even. But this implies $x^2 \equiv 1 \bmod 4$ and $y^2 \equiv 0 \bmod 4$, hence $p = x^2 + y^2 \equiv 1 \bmod 4$.

We can show even more: if $p = x^2 + y^2$, then certainly neither x nor y is divisible by p. Since $\mathbb{Z}/p\mathbb{Z} \simeq \mathbb{F}_p$ is a finite field, there exists $y' \in \mathbb{Z}$

---

[1] Every prime which is one more than a multiple of 4 is a sum of two squares in one and only one way.

such that $yy' \equiv 1 \bmod p$ (this is Euclid's algorithm at work: p is prime, thus $p \nmid y$ implies $(y,p) = 1$, and Euclid's algorithm guarantees the existence of $y', p' \in \mathbb{Z}$ such that $yy' + pp' = 1$; this gives the desired congruence). Now $x^2 + y^2 \equiv 0 \bmod p$ implies that $(xy')^2 \equiv -1 \bmod p$, i.e., the congruence $X^2 \equiv -1 \bmod p$ is solvable.

We have seen: if a prime p is the sum of two squares, then each of the following claims holds:

$$p = 2 \text{ or } p \equiv 1 \bmod 4, \tag{1.1}$$

the congruence $X^2 \equiv -1 \bmod p$ is solvable. $\tag{1.2}$

Is the converse also true? Yes, and it turns out that the conditions (1.1) and (1.2) are in fact equivalent: this is just the assertion of the first supplementary law of quadratic reciprocity. Its proof is quite easy and uses Fermat's "little theorem":

Proposition 1.1. *If p is a prime not dividing $a \in \mathbb{N}$, then $a^{p-1} \equiv 1 \bmod p$.*

Let us prove that (1.1) and (1.2) are equivalent. The case $p = 2$ is trivial, and we may suppose that p is odd. Assume that the congruence $X^2 \equiv -1 \bmod p$ is solvable; raising this to the $\frac{p-1}{2}$th power yields $X^{p-1} \equiv (-1)^{(p-1)/2} \bmod p$. But $p \nmid X$, hence Fermat's little theorem gives $X^{p-1} \equiv 1 \bmod p$, and we conclude that $\frac{p-1}{2}$ must be even, i.e., that $p \equiv 1 \bmod 4$.

Next suppose that $p \equiv 1 \bmod 4$; since $(\mathbb{Z}/p\mathbb{Z})^\times \simeq \mathbb{F}_p^\times$ is isomorphic to the multiplicative group of a finite field, $(\mathbb{Z}/p\mathbb{Z})^\times$ is cyclic, and any $g \in \mathbb{Z}$ such that $g \bmod p$ generates $(\mathbb{Z}/p\mathbb{Z})^\times$ is called a *primitive root* mod p. Put $x = g^{(p-1)/4}$; then $x^4 \equiv 1 \bmod p$ implies $x^2 \equiv \pm 1 \bmod p$, and since g is a primitive root, we must have $x^2 \equiv -1 \bmod p$. This shows that (1.1) and (1.2) are equivalent.

Proposition 1.2. *If p is an odd prime, then $p \equiv 1 \bmod 4 \iff X^2 \equiv -1 \bmod p$ is solvable $\iff p$ is a sum of two squares.*

For proving that every prime $p \equiv 1 \bmod 4$ is the sum of two squares Fermat used a technique called "descente infinie", which he had invented for showing that certain diophantine equations ($X^4 + Y^4 = Z^2$, for example) have no solution in integers. His line of reasoning was the following: assume that there is a solution $(x, y, z) \in \mathbb{Z}^3$ such that $xyz \neq 0$. Then prove that there must be a solution (x', y', z') with $x'y'z' \neq 0$ which is 'smaller' than (x, y, z) (the notion of 'smaller' may depend on the problem; possible 'heights' are $|x|$, $|x| + |y| + |z|$, $|xyz|$, etc.). This yields a contradiction, because these heights cannot become smaller than 1 for integers $x, y, z > 0$.

Later Fermat found that this technique could also be used to prove 'positive' statements: suppose for example that there is a prime $p \equiv 1 \bmod 4$ which is not a sum of two squares. If we can prove that there must be a prime $q \equiv 1 \bmod 4$ with $q < p$ which is not a sum of two squares, then we will eventually get a contradiction since $5 = 1^2 + 2^2$ *is* a sum of squares.

Fermat could prove analogous theorems for primes of the form $x^2 + ny^2$, where $n = \pm 2, \pm 3, -5$; for example a prime p has the form $p = x^2 + 2y^2$ if and only if $p = 2$ or $p \equiv 1, 3 \bmod 8$, and this in turn is equivalent to the solvability of the congruence $x^2 \equiv -2 \bmod p$ (see Exercise 1.4).

The fact that makes Fermat's classification of primes $p = x^2 + ny^2$ work in the case $n = 1$ is the following: if p is a primitive divisor of $x^2 + ny^2$ (this means that $p \mid x^2 + ny^2$ and $p \nmid xy$) then p itself can be represented as $a^2 + nb^2$ for some $a, b \in \mathbb{N}$. It turns out that this remains true for $n = 2$ and $n = 3$ (at least for primes $p \neq 2$; the connoisseur can see the conductor of the ring $\mathbb{Z}[\sqrt{-3}]$ lurking in the background), but fails to hold for $n = 5$: in fact, $3 \cdot 7 = 1^2 + 5 \cdot 2^2$, although neither 3 nor 7 can be written in the form $a^2 + 5b^2$. For modern readers it is easy to recognize the connection with the fact that $\mathbb{Z}[\sqrt{-5}]$ is not a unique factorization domain – note, however, that the concept of unique factorization was quite foreign to Fermat and Euler.

In a letter to Digby written in 1658 or earlier (see Fermat [232, II, p. 402–408]), Fermat claimed that the product of two primes with last digits 3 or 7 can be written as $x^2 + 5y^2$. In fact, after computing some more examples it is quite easy to guess that, more exactly

$$\left.\begin{array}{r} p \equiv 1, 9 \bmod 20 \Longrightarrow p = x^2 + 5y^2 \\ p, q \equiv 3, 7 \bmod 20 \Longrightarrow pq = x^2 + 5y^2 \end{array}\right\} \quad (1.3)$$

for primes $p \neq 2, 5$. As far as I know, the first part of this conjecture is due to Euler, but neither he nor Fermat succeeded in giving proofs – this was left for Lagrange to do. Anyway, the quadratic character of $-1, \pm 2$ and ± 3 can be deduced easily from results known to Fermat, and the quadratic character of -5 follows from (1.3).

1.2 L. Euler

It seems that Euler began to read Fermat's work seriously soon after he and Christian Goldbach started their correspondence (see [218]). In one of his first letters to Euler (no. 2 in [218], from Dec. 1, 1729), Goldbach asked

> P.S. Notane Tibi est Fermatii observatio omnes numeros hujus formulae $2^{2^{x-1}} + 1$, nempe 3, 5, 17, etc. esse primos, quam tamen ipse fatebatur se demonstrare non posse et post eum nemo, quod sciam, demonstravit.[2]

---

[2] P.S. Is Fermat's observation known to you that all numbers of the form $2^{2^{x-1}} + 1$, namely 3, 5, 17, etc. are primes, which he himself could not prove and which no one after him, to the best of my knowledge, has ever proved.

4   1. The Genesis of Quadratic Reciprocity

His subsequent work on divisors of Fermat numbers $2^{2^n} + 1$ and Mersenne numbers $2^q - 1$ or, more generally, on divisors of binary quadratic forms $nx^2 + my^2$, eventually led Euler to the quadratic reciprocity law, although his progress was slow. Using the binomial theorem he could give a proof for Fermat's little theorem, which he stated in the form [212]

> Significante p numerum primum formula $a^{p-1} - 1$ semper per p dividi poterit, nisi a per p divide queat. [3]

His first result directly connected with quadratic reciprocity was (see [213, 214])

Proposition 1.3. (Euler's criterion) *For integers a and odd primes p such that $p \nmid a$ we have*

$$a^{\frac{p-1}{2}} \equiv \begin{cases} +1 \bmod p, & \text{if } a \text{ is a quadratic residue mod } p, \\ -1 \bmod p, & \text{if } a \text{ is a quadratic non-residue mod } p. \end{cases}$$

The usual proofs of Euler's criterion use the fact that $(\mathbb{Z}/p\mathbb{Z})^\times$ is cyclic, a result conjectured by Lambert [La] which Euler could not prove until 1772. For a beautiful combinatorial argument due to Dirichlet [164], see Brown [89].

Euler first stated a theorem equivalent to the quadratic reciprocity law in his paper [211, p. 217] in 1744 (see Kronecker's remarks in [439]; in modern terminology, Euler observed that the quadratic residue character of p modulo primes of the form $4pn \pm s$ with $0 < s < 4p$ and $(s, 2p) = 1$ does not depend on s). Euler had announced a special case of this theorem already in 1742 in his letter to Goldbach [219]. The research of Lagrange [474] in the years 1773/75 made him take up the subject again, and this time he discovered the complete quadratic reciprocity law (it was finally published in 1783 [215], after Euler's death) without, however, being able to produce a proof:

1. Si divisor primus fuerit formae $4ns + (2x+1)^2$, existente s numero primo, tum in residuis occurent numeri $+s$ et $-s$.
2. Si divisor primus fuerit formae $4ns - (2x+1)^2$, existente s numero primo, tum in residuis occuret numerus $+s$, at $-s$ in non-residuis.
3. Si divisor primus fuerit formae $4ns - 4z - 1$ excludendo omnes valores in forma $4ns - (2x+1)^2$ contentos, existente s numero primo, tum in residuis occuret numerus $-s$; at s erit non-residuum.
4. Si divisor primus fuerit formae $4ns + 4z + 1$ excludendo omnes valores in forma $4ns + (2x+1)^2$ contentos, existente s numero primo, tum tam $+s$ quam $-s$ in non-residuis occuret.

Translated into modern notation, this reads

1. If $p \equiv 1 \bmod 4$ is prime and $p \equiv x^2 \bmod s$ for some prime s, then $\pm s \equiv y^2 \bmod p$. In other words: if $p \equiv 1 \bmod 4$ then $\left(\frac{p}{s}\right) = +1 \Rightarrow \left(\frac{\pm s}{p}\right) = +1$.

---
[3] If p denotes a prime, then the number $a^{p-1} - 1$ is always divisible by p, unless a is divisible by p.

2. If $p \equiv 3 \bmod 4$ is prime and $-p \equiv x^2 \bmod s$ for some prime s, then $s \equiv y^2 \bmod p$ and $-s \not\equiv y^2 \bmod p$. In other words: if $p \equiv 3 \bmod 4$ then $(\frac{-p}{s}) = +1 \Rightarrow (\frac{s}{p}) = +1, (\frac{-s}{p}) = -1$.
3. If $p \equiv 3 \bmod 4$ is prime and $-p \not\equiv x^2 \bmod s$ for some prime s, then $-s \equiv y^2 \bmod p$ and $s \not\equiv y^2 \bmod p$. In other words: if $p \equiv 3 \bmod 4$ then $(\frac{-p}{s}) = -1 \Rightarrow (\frac{-s}{p}) = +1, (\frac{s}{p}) = -1$.
4. If $p \equiv 1 \bmod 4$ is prime and $p \not\equiv x^2 \bmod s$ for some prime s, then $\pm s \not\equiv y^2 \bmod p$. In other words: if $p \equiv 1 \bmod 4$ then $(\frac{p}{s}) = -1 \Rightarrow (\frac{\pm s}{p}) = -1$.

This is equivalent to the version of the quadratic reciprocity law that is best known today and that was formulated by Legendre and Gauss:

Theorem 1.4. *Let p and q be different odd primes; then p is a quadratic residue modulo q if and only if q is a quadratic residue modulo p, unless $p \equiv q \equiv 3 \bmod 4$, where p is a quadratic residue modulo q if and only if q is a quadratic nonresidue modulo p.*

There are many references in Euler's papers to problems related to quadratic reciprocity, but the task of listing and discussing them is made difficult by the extent of Euler's work and the fact that he wrote in Latin. For a small list of relevant papers, see Pieper's [662]; here we will single out one of his results that was later refined by Legendre:

Proposition 1.5. *If $f, g, h \in \mathbb{Z}$ are coprime squarefree integers such that the equation $fx^2 + gy^2 = hz^2$ has solutions in integers, then f (resp. g or h) is a quadratic residue modulo each divisor of gh (resp. fh or fg).*

This necessary criterion is implicitly contained in [216]; more interesting for us is the following conjecture that Euler gives as Theorem 3:

Theorem 1.6. *Let $f, g \in \mathbb{Z}$ be coprime squarefree integers that are not both negative. If h and h' are positive primes such that $h \equiv h' \bmod 4fg$, then the equation $fx^2 + gy^2 = sz^2$ is solvable for $s = h$ if if and only if it is solvable for $s = h'$.*

To see the connection with quadratic reciprocity, put $g = -a$ and $f = 1$. If $x^2 - ay^2 = qz^2$ is solvable, then Euler's conjecture 1.6 says that so is $x^2 - ay^2 = q'z^2$ if only $q \equiv q' \bmod 4a$. It will follow from Legendre's theorem 1.7 below that solvability of $x^2 - ay^2 = qz^2$ is equivalent to $(a/q) = 1$ and $(q/p) = 1$ for all $p \mid a$; since $(q/p) = 1$ and $q \equiv q' \bmod 4a$ imply $(q'/p) = 1$, Euler's conjecture essentially boils down to the statement that, under certain additional assumptions, $(a/q) = 1$ and $q \equiv q' \bmod 4a$ imply $(a/q') = 1$: but this is the quadratic reciprocity law (see Theorem 2.28.iv).) As Legendre was to discover, it is not at all clear how to extract the reciprocity law from such observations, although quadratic reciprocity easily implies Euler's conjecture.

1.3 A.-M. Legendre

The quadratic reciprocity law was published in a form that is more familiar to us in 1788 by Legendre [489] (this paper was presented to the Academy in 1785). On pp. 516–517, Legendre considers primes $a, A \equiv 1 \bmod 4$ and $b, B \equiv 3 \bmod 4$ and states

Théorème I.	Si $b^{\frac{a-1}{2}} \equiv +1$, il s'ensuit $a^{\frac{b-1}{2}} \equiv +1$.
Théorème II.	Si $a^{\frac{b-1}{2}} \equiv -1$, il s'ensuit $b^{\frac{a-1}{2}} \equiv -1$.
Théorème III.	Si $a^{\frac{A-1}{2}} \equiv +1$, il s'ensuit $A^{\frac{a-1}{2}} \equiv +1$.
Théorème IV.	Si $a^{\frac{A-1}{2}} \equiv -1$, il s'ensuit $A^{\frac{a-1}{2}} \equiv -1$.
Théorème V.	Si $a^{\frac{b-1}{2}} \equiv +1$, il s'ensuit $b^{\frac{a-1}{2}} \equiv +1$.
Théorème VI.	Si $b^{\frac{a-1}{2}} \equiv -1$, il s'ensuit $a^{\frac{b-1}{2}} \equiv -1$.
Théorème VII.	Si $b^{\frac{B-1}{2}} \equiv +1$, il s'ensuit $B^{\frac{b-1}{2}} \equiv -1$.
Théorème VIII.	Si $b^{\frac{B-1}{2}} \equiv -1$, il s'ensuit $B^{\frac{b-1}{2}} \equiv +1$.

(Actually, Legendre wrote = instead of $\equiv$ in order to denote equality up to certain multiples; the notion of congruence, and in particular the symbol $\equiv$, was introduced by Gauss in [263], who referred to Legendre on this occasion.) In 1798, Legendre [490] announced the quadratic reciprocity law in its final form; on p. 186, he introduces the "Legendre symbol":

> Comme les quantités analogues $N^{(c-1)/2}$ se rencontreront fréquemment dans le cours de nos recherches, nous emploierons le caractère abrégé $\left(\frac{N}{c}\right)$ pour exprimer le reste que donne $N^{(c-1)/2}$ divisé par c, reste qui suivant ce qu'on vient de voir ne peut être que $+1$ ou -1.[4]

On p. 214, he continues

> Quels que soient les nombres premiers m et n, s'ils ne sont pas tous deux de la forme $4x - 1$, on aura toujours $\left(\frac{n}{m}\right) = \left(\frac{m}{n}\right)$; et s'ils sont les deux de la forme $4x - 1$, on aura $\left(\frac{n}{m}\right) = -\left(\frac{m}{n}\right)$. Ces deux cas généraux sont compris dans la formule
> $$\left(\frac{n}{m}\right) = (-1)^{\frac{n-1}{2} \cdot \frac{m-1}{2}} \left(\frac{m}{n}\right).\text{[5]}$$

---
[4] Since the analogous quantities $N^{(c-1)/2}$ will occur often in our researches, we shall employ the abbreviation $\left(\frac{N}{c}\right)$ for expressing the residue that $N^{(c-1)/2}$ gives upon division by c, and which, according to what we just have seen, only assumes the values $+1$ or -1.

[5] Whatever the prime numbers m and n are, if they are not both of the form $4x - 1$, one always has $\left(\frac{n}{m}\right) = \left(\frac{m}{n}\right)$; and if both are of the form $4x - 1$, one has $\left(\frac{n}{m}\right) = -\left(\frac{m}{n}\right)$. These two general cases are contained in the formula
$$\left(\frac{n}{m}\right) = (-1)^{\frac{n-1}{2} \cdot \frac{m-1}{2}} \left(\frac{m}{n}\right).$$

Although Legendre's reciprocity law was announced only for prime values of m and n, the Legendre symbol $\left(\frac{a}{p}\right)$ makes sense for composite values of a as well, and one of the simplest properties of $\left(\frac{a}{p}\right)$ is the fact that it is multiplicative in its numerator.

How did Legendre attempt to prove the quadratic reciprocity law? His starting point was the following theorem ([489, p. 513]):

Theorem 1.7. *Assume that $a, b, c \in \mathbb{Z}$ satisfy the following conditions:*

1. *$(a,b) = (b,c) = (c,a) = 1$;*
2. *at least one of ab, bc, ca is negative;*
3. *the following congruences are solvable:*

$$u^2 \equiv -bc \bmod a, \quad v^2 \equiv -ca \bmod b, \quad w^2 \equiv -ab \bmod c.$$

Then the diophantine equation

$$ax^2 + by^2 + cz^2 = 0 \tag{1.4}$$

has non-trivial solutions in $\mathbb{Z}$.

Proof. See Exercise 1.8 or Exercise 2.36. □

Using this result, Legendre could prove Théorème I as follows: let $a \equiv 1 \bmod 4$ and $b \equiv 3 \bmod 4$ be primes such that $\left(\frac{b}{a}\right) = 1$, and assume for contradiction that $\left(\frac{a}{b}\right) = -1$. Then $\left(\frac{-a}{b}\right) = 1$ because $b \equiv 3 \bmod 4$, and the equation $x^2 + ay^2 - bz^2 = 0$ has non-trivial solutions by Theorem 1.7. Canceling common divisors of x, y and z we may assume that at least one of x, y or z is odd. But $0 = x^2 + ay^2 - bz^2 \equiv x^2 + y^2 + z^2 \bmod 4$ implies $x \equiv y \equiv z \equiv 0 \bmod 2$, and this contradiction proves the claim.

Théorème II is a formal consequence of Théorème I, and the proof of Théorème VII proceeds along similar lines: assume that $b \equiv B \equiv 3 \bmod 4$ are primes such that $(b/B) = +1$ and $(B/b) = +1$; then $Bx^2 + by^2 - z^2 = 0$ is solvable in coprime integers x, y, z, and reducing a solution modulo 4 yields a contradiction exactly as above.

Legendre's attack on Théorème VIII went like this: assume that $b \equiv B \equiv 3 \bmod 4$ are primes such that $(b/B) = (B/b) = -1$; assume moreover that we can find an auxiliary prime $p \equiv 1 \bmod 4$ such that $(p/B) = (p/b) = -1$. Then, by Théorème II, we must have $(B/p) = (b/p) = -1$, and therefore $Bx^2 + by^2 - pz^2 = 0$ has a nontrivial solution. As above, this leads to a contradiction modulo 4.

Thus Legendre's proofs of the cases I, II and VII were complete, while the proof of case VIII depended on the existence of a certain prime p. The conditions $p \equiv 1 \bmod 4$ and $(p/B) = (p/b) = -1$ are satisfied if p is a prime in a suitable residue class modulo $4bB$; in order to guarantee its existence, Legendre announced the following conjecture later proved by Dirichlet:

Theorem 1.8. *Let a and b be positive integers; if $\gcd(a,b) = 1$, then there exist infinitely many primes $\equiv a \bmod b$.* □

As we will see in the Notes, the role of Dirichlet's theorem in Legendre's attempted proof of the quadratic reciprocity law led to some confusion. As already Gauss has shown in his Disquisitiones [263, Art. 296, 297], Legendre's "proof" of 1785/1788 as given in his Recherches [489] can be formulated in such a way that quadratic reciprocity *does* become a corollary of Theorem 1.8.

Let us start with Théorème III: suppose that we have $\left(\frac{a}{A}\right) = 1$; in order to prove that $\left(\frac{A}{a}\right) = 1$, let us assume that $\left(\frac{A}{a}\right) = -1$ and try to derive a contradiction. Legendre assumes that there is a prime $\beta \equiv 3 \bmod 4$ such that $\left(\frac{\beta}{a}\right) = -1$ (this again would follow from Theorem 1.8). Then $\left(\frac{a}{\beta}\right) = -1$ by II, hence $\left(\frac{-a}{\beta}\right) = +1$ by the first supplementary law, and the assumption $\left(\frac{A}{a}\right) = -1$ shows that $A\beta$ is a quadratic residue mod a. Thus the equation $x^2 + ay^2 - A\beta z^2 = 0$ is solvable, and we derive a contradiction exactly as above by looking at it modulo 4.

Case IV is again a formal consequence of Théorème III, and similarly, VI follows from V. It is therefore sufficient to prove V. So assume that $\left(\frac{a}{b}\right) = +1$ and $\left(\frac{b}{a}\right) = -1$; by Dirichlet's Theorem 1.8, there is a prime $\alpha \equiv 1 \bmod 4$ such that $\left(\frac{\alpha}{a}\right) = \left(\frac{\alpha}{b}\right) = -1$. By Théorème IV, this implies $\left(\frac{a}{\alpha}\right) = -1$, and then $\alpha x^2 + ay^2 - bz^2 = 0$ is solvable, leading to the now familiar contradiction modulo 4.

The proof above is contained in Legendre's Recherches [489] from 1788, alongside with other proofs that assume the existence of auxiliary primes that do *not* follow from Dirichlet's theorem. In later editions of his "Essai" [490], Legendre proved the cases VII and VIII using the theory of Pell's equation and replaced the different auxiliary primes by only one assumption:

Lemma 1.9 (Legendre's Lemma). *For each prime $a \equiv 1 \bmod 4$ there exists a prime $\beta \equiv 3 \bmod 4$ such that $(a/\beta) = -1$.*

Legendre could not rigorously prove the existence of such a prime β, but has clearly seen that this claim needed a proof; Cuculière [142] calls his excuse of not giving one *ce que l'on pourrait appeler vulgairement "le coup de Fermat"*. Legendre stuck to his approach to the proof of the reciprocity law in later editions of his book and each time made fresh attempts to close the gap, but he did not succeed in giving complete proofs for his theorems III – VI (the numbering chosen here is from the first edition).

It should also be remarked that Legendre's Lemma does *not* follow from Dirichlet's theorem 1.8 without assuming parts of the quadratic reciprocity law, contrary to many claims made throughout the mathematical literature. We shall deal with this problem in detail in the Notes below.

1.4 C.-F. Gauss

As we have already remarked, it was Gauss who found the first complete proof of the quadratic reciprocity law. His desire to find similar theorems for reciprocity of higher powers made him look for proofs which would generalize, and by 1818 he had published six proofs; two more were found in his unpublished papers.

His first proof used induction; it is remarkable that, like Legendre, he needed the existence of a certain auxiliary prime in his proof (see Exercise 1.40). His second proof used the genus theory of binary quadratic forms and will be discussed in Chapter 2. Proofs 4 and 6 use quadratic Gauss sums, which we will introduce in Chapter 3. Like most of the simplest proofs of the quadratic reciprocity law, proofs 3 and 5 rest on what we now call Gauss's Lemma; before we state it, we have to introduce a *half system* mod p: this is a set of numbers $\{a_1, \ldots, a_{(p-1)/2}\}$ such that every $a \not\equiv 0$ mod p is congruent to exactly one of the $\pm a_j$. In other words: it is a set of representatives for the cosets of $(\mathbb{Z}/p\mathbb{Z})^\times / \{\pm 1\}$.

Lemma 1.10. *Let $p = 2k + 1$ be prime, and suppose that $a \in \mathbb{N}$ is not divisible by p; moreover, assume that $\{a_1, \ldots, a_k\}$ is a half-system mod p. Then $\left(\frac{a}{p}\right) = (-1)^\mu$, where μ is the number of integers $aa_1, \ldots, aa_k$ which are congruent mod p to one of $-a_1, \ldots, -a_k$.*

Proof. For every $1 \leq i \leq k$, write $aa_i \equiv \pm a_{\sigma(i)}$ mod p. Then $\sigma(i) \neq \sigma(j)$ for all $i \neq j$, because $aa_i \equiv \pm aa_j$ mod p implies $a_i \equiv \pm a_j$ mod p contradicting our assumption that $\{a_1, \ldots, a_k\}$ forms a half-system mod p. Hence σ is a permutation of $\{1, 2, \ldots, k\}$, and multiplying the k congruences $aa_i \equiv \pm a_{\sigma(i)}$ mod p we get

$$a^{\frac{p-1}{2}} \prod_{i=1}^{k} a_i \equiv (-1)^\mu \prod_{i=1}^{k} a_{\sigma(i)} = (-1)^\mu \prod_{i=1}^{k} a_i \mod p.$$

Now cancel $\prod a_i$ and recall Euler's criterion. $\square$

As an application, we will compute $\left(\frac{2}{p}\right)$. To this end, we choose a half-system mod p, and in order to keep things as simple as possible we take $A = \{1, 2, \ldots, \frac{p-1}{2}\}$. Then

$$1 < 2a_j \leq \tfrac{p-1}{2} \text{ if } a_j \leq \tfrac{p-1}{4}, \quad \text{and}$$
$$\tfrac{p+1}{2} \leq 2a_j \leq p-1 \text{ if } \tfrac{p-1}{4} < a_j \leq \tfrac{p-1}{2}.$$

Hence there are $\mu = \frac{p-1}{4}$ sign changes if $p \equiv 1$ mod 4, and $\mu = \frac{p+1}{4}$ sign changes if $p \equiv 3$ mod 4. This implies

$$\left(\frac{2}{p}\right) = \begin{cases} (-1)^{\frac{p-1}{4}}, & \text{if } p \equiv 1 \mod 4, \\ (-1)^{\frac{p+1}{4}}, & \text{if } p \equiv 3 \mod 4, \end{cases}$$

which proves the second supplementary law.

For the proof of the general case Gauss introduces the floor function $\lfloor x \rfloor$, which denotes the greatest integer $\leq x$. It is easy to see (see Exercise 1.32) that the integer μ defined in Gauss's lemma can be represented as

$$\mu = \sum_{x=1}^{(p-1)/2} \left\lfloor \frac{ax}{p} \right\rfloor.$$

Note that Gauss's approach was more complicated – what we give here is in fact a simplified version of Gauss's third proof, with substantial contributions e.g. by Eisenstein (see Laubenbacher & Pengelley [479, 480] for a detailed comparison). Next one proves that, for odd integers $p, q \in \mathbb{N}$ (this is due to Eisenstein: see Exercise 1.33)

$$\sum_{x=1}^{(p-1)/2} \left\lfloor \frac{qx}{p} \right\rfloor + \sum_{x=1}^{(q-1)/2} \left\lfloor \frac{px}{q} \right\rfloor = \frac{p-1}{2} \frac{q-1}{2}. \tag{1.5}$$

This obviously implies the quadratic reciprocity law.

But not only did Gauss give the first complete proofs of the quadratic reciprocity law, he also extended it to composite values of p and q in [263, Art. 133]. The corresponding extension of the Legendre symbol, however, by making the denominator multiplicative first appears in Jacobi's paper [400] from 1837. There Jacobi defined the *Jacobi symbol* (a/b) for odd natural integers $b = p_1 \cdots p_n \in \mathbb{N}$ by $(a/b) = (a/p_1) \cdots (a/p_n)$. Clearly $a \equiv x^2$ mod b implies $(a/b) = 1$; the converse, however, is not true. Quite surprisingly, the quadratic reciprocity law also holds for the Jacobi symbol:

Proposition 1.11. *Let $m, n \in \mathbb{N}$ be relatively prime and odd; then*

$$\left(\frac{m}{n}\right) = (-1)^{\frac{m-1}{2}\frac{n-1}{2}} \left(\frac{n}{m}\right), \quad \left(\frac{-1}{n}\right) = (-1)^{\frac{n-1}{2}}, \quad \left(\frac{2}{n}\right) = (-1)^{\frac{n^2-1}{8}}.$$

This was a great improvement on Euler's and Legendre's version of quadratic reciprocity, as far as the computation of residue symbols $\left(\frac{p}{q}\right)$ was concerned: instead of having to factor the residue of p mod q before inverting the occurring Legendre symbols one could simply invert the Jacobi symbols and apply a computationally cheap Euclidean algorithm.

For a deduction of Proposition 1.11 from the quadratic reciprocity law it suffices to note that

$$\begin{aligned}\frac{mn-1}{2} &\equiv \frac{m-1}{2} + \frac{n-1}{2} \mod 2, \\ \frac{m^2n^2-1}{8} &\equiv \frac{m^2-1}{8} + \frac{n^2-1}{8} \mod 2\end{aligned} \tag{1.6}$$

for odd integers $m, n \in \mathbb{Z}$; induction on the number of prime factors of m and n then yields the reciprocity law for the Jacobi symbol.

Gauss's version of Proposition 1.11 looked as follows: he wrote $p = \prod_i p_i^{a_i}$ and defined the integer $\pi(p,q)$ as the sum of all a_i such that $(q/p_i) = -1$. The reciprocity law for the Jacobi symbol is then equivalent to the statement that $\pi(p,q) \equiv \pi(q,p) \bmod 2$ or $\pi(p,q) \equiv \pi(q,p) + 1 \bmod 2$ according as at least one of p or q is congruent to 1 mod 4 or not. Gauss even gave rules for integers p and q with arbitrary signs.

Notes

The prehistory of quadratic residues is addressed in Eneström's article [Ene]. For more details on the lives and works of Fermat, Euler, Legendre and Gauss we refer the reader to the books of Bühler [Buh], Guinot [Gu1, Gu2, Gu3], Mahoney [Mah], Scharlau & Opolka [SO], Vuillemin [Vui], and A. Weil [We2], as well as to the articles of Antropov [An1, An2], Aubry [Au1, Au2], Bachmann [Bm2], D. Cox [140], Burkhardt [Bur], Hofmann [Hof], Laubenbacher & Pengelley [480], and Weil [We1].

Fermat

Amazingly, we don't even know for sure when Fermat was born: most sources give 1601 as the year of Fermat's birth, other guesses range from around 1599 (the entry in the church's books indicate 1601 as the year some Pierre Fermat was baptized) to around 1608 (claiming that the Pierre de Fermat born in 1601 was an elder brother of Fermat the mathematician who died at an early age). Recent investigations of K. Barner suggest 1607 as Fermat's year of birth, in agreement with the epitaph that Fermat's son had inscribed on his father's tombstone, giving 57 as the age he died. On the other hand we are rather sure that Fermat was born in Beaumont-de-Lomagne near Toulouse (see Goldstein [Go2] and Mahoney [Mah]).

The renaissance of number theory in Europe started with the edition of Diophantus' books by Bachet in 1621. A detailed discussion of Diophantus' immense influence on Bachet, Fermat and Euler can be found in Heath [Hea]. For a long time, only six of originally thirteen books were known, and only in 1971 Roshdi Rashed [Ras] discovered four more in an Arabic translation (see Sesiano [Ses]). It was in a copy of Bachet's edition where Fermat stated his conjecture that the 'diophantine equation' $x^n + y^n = z^n$ has only trivial solutions for $n > 2$. Fermat's son Samuel later published this copy, including Fermat's notes and additions (which is how we know about 'Fermat's Last Theorem'). For more on Diophantus' work, as well as contributions of Indian mathematicians such as Bhaskaracharya and Brahmagupta to the theory of Pell's equation (that is, $x^2 - dy^2 = 1$, where $d \in \mathbb{N}$ is not a square) see Joshi [Jos]. It is well known that it was Euler who named this equation after the British mathematician Pell; this is not too bad, since otherwise we would have to call it the Bhaskaracharya-Brahmagupta equation as Joshi does.

Fermat's two-squares theorem, namely the assertion that every prime $p \equiv 1 \bmod 4$ is the sum of two squares, can be proved in a multitude of ways: most of them (for example, the proof using induction sketched in Exercise 1.1) use more or less explicitly the fact that $\mathbb{Z}[i]$ is a Euclidean ring (see e.g. the discussion in Wagon [Wag]); others use continued fractions (Brillhart [Bri]; this method is actually due to Legendre, Serret, and Cornacchia and is related to the approach via the Euclidean algorithm in $\mathbb{Z}[i]$), geometry of numbers (Scharlau & Opolka [SO]), Dirichlet's pigeonhole principle (cf. Exercise 1.3), chessboards (L.C. Larson [Lar]), or "fixed point theorems" (Zagier [Zag]; this proof made it into THE BOOK – see Aigner & Ziegler [AZ]). More details can be found in Dickson [Di2] and Cohen [Coh].

Fermat's little theorem was first stated in a letter to Frenicle (Oct. 18, 1640); later, proofs were found by Leibniz (see Mahnke [Ma, p. 49]) and Euler (letter to Goldbach, 23.02./06.03. 1742); the special case $a = 2$ is already contained in Euler's letter to Goldbach from Nov. 25, 1731. Regarding Leibniz's contribution, Jacobi ([402], II. Vorlesung) writes

> Ein gewisser König, welcher Leibnizische Briefe verfertigte, behauptet, vermuthlich nur um Euler zu kränken, daß Leibniz diesen Satz schon bewiesen habe. [6]

Here he is referring to a footnote in Gauss's [263, Art. 50]. Actually Euler himself saw no reason to doubt König's assertion (see his letter to Goldbach [218, p. 79]).

Counting Fermat among the "fathers" of the quadratic reciprocity law is justified in that some of his results immediately give the quadratic residue character of small primes. Note, however, that Fermat did not talk about quadratic residues modulo p; in fact our discussion in Section 1.1 is filled with anachronisms. For a historical discussion of Fermat's life and work see C. Goldstein's [Go2]).

Fermat described most of the results of Section 1.1 in a letter to Kenelm Digby in 1658 [232, II, p. 402–408; III, 314–319] (see also his letter to Pascal [232, II, p. 310–314]); he gave criteria which determined the primes q such that $-1, 2, \pm 3$, and 5 are quadratic residues mod q. It is often claimed that Fermat did not claim to have a proof for his conjecture that the numbers $2^{2^n} + 1$ are primes, and there are in fact numerous letters to Frenicle and Mersenne from about 1640 in which he admits that he does not possess a proof; in a letter to Carcavi from August 1659 (Œuvres II, 431–436), however, he gives four theorems which he says can be proved by his method of descente infinie:

1. The diophantine equation $x^3 + y^3 = z^3$ has no solution in integers $\neq 0$;
2. The only integral solution of $y^2 + 2 = x^3$ is $y = 5$, $x = 3$;
3. The only integral solutions to $y^2 + 4 = x^3$ are $(x, y) = (2, 2), (5, 11)$;

---

[6] A certain König who manufactured letters of Leibniz claimed, probably only in order to offend Euler, that Leibniz already had proved this theorem.

4. All numbers of the form $2^{2^n} + 1$ are prime.

Although we know nowadays how to prove the first three of these conjectures (namely by using unique factorization in the rings of integers in $\mathbb{Q}(\sqrt{-3})$, $\mathbb{Q}(\sqrt{-2})$ and $\mathbb{Q}(\sqrt{-1})$, respectively), we have no idea about the techniques Fermat could have used – if he had proofs for 1. – 3. at all (I don't think that he did, and so does Bussey [Bus]). Remarkably, the curves in the first three claims are elliptic curves admitting a rational 3-isogeny; this can be used to prove Fermat's first claim, and that the curves $y^2 = x^3 - 2$ and $y^2 = x^3 - 4$ have Mordell-Weil rank 1. This information about rational points can be exploited to find all integral points on these curves (see Gebel, Pethö & Zimmer [GPZ]), but the techniques to do so were out of reach for Fermat. The only other approach for solving the claims 1. – 3. use the arithmetic of quadratic number fields (Euler proved 1. using binary quadratic forms instead of $\mathbb{Q}(\sqrt{-3})$, and, according to the Zentralblatt review, Z. Xu [Xu] proved Fermat's Last Theorem for $n = 3$ without "complex numbers") and were solved – using the strategy suggested by Euler – by Pépin [Pe].

Fermat's conjecture (1.3) first appears in a letter from Frenicle to Fermat (Œuvres, p. 168–170), but the letter indicates that the result was known to Fermat.

Euler

After Goldbach's letter mentioned at the beginning of Section 1.2, Euler tried to find proofs for the many theorems Fermat had claimed; Euler's (incomplete) proofs of Fermat's claims 1. and 2. above are well known, as is his counter example $2^{32} + 1 = 641 \cdot 6\,700\,417$ to claim 4.

As Euler was unable to prove conjecture (1.3), he did what mathematicians to this day do when they find themselves in a similar situation: he generalized the conjecture to the quadratic forms $x^2 + ny^2$ with $n = 6, 7$ (see Comm. Petersb. VI, 221; VIII, 127; XIV). The first proof of (1.3) is due to Lagrange [474].

As Edwards [182] remarked, Euler's statement of quadratic reciprocity is very close to the form in which it comes out as a corollary of Artin's reciprocity law (compare I. Braun [81], Goldschmidt [316], and Steenvoorden [750]). Edwards article also contains the English translation of a large part of Euler's letter to Goldbach containing the statement of the quadratic reciprocity and tells how Euler's investigations on divisors of quadratic forms led him to conjectures on cubic and quartic reciprocity: in [217, §407, p. 250] we find

$$x^3 \equiv 2 \bmod p \text{ is solvable} \iff p = L^2 + 27M^2. \tag{1.7}$$

Similarly, he gave criteria for the cubic residuacity of $3, 5, 6, 7, 10$ (§§408 – 410), as well as for the quartic residuacity of $2, 3$, and 5; for example, on p. 258 of [217] (§456) one finds

$$x^4 \equiv 2 \bmod p \text{ is solvable} \iff p = a^2 + 64b^2.$$

Euler introduced primitive roots in his paper [Eul]. Gauss [263, Art. 56] discovered an error in Euler's proof of their existence, and Dickson ([Di1, p. 63]) calls it defective; the remarks on pp. XXVIII – XXIX of Euler's Opera Omnia, however, show that Euler's paper at least does *contain* a correct proof. In his presentation of the proofs of the quadratic reciprocity law known in 1885, Baumgart [38], p. 4, asserted that Euler's third conjecture (see page 4) contains a mistake, but I cannot find one.

Lagrange

Lagrange's role in the history of quadratic reciprocity is connected with refining Euler's attempts at the creation of a theory of binary quadratic forms, a topic later brought to perfection by Gauss, who succeeded in giving the "correct" definition of the composition of forms. Yet it was Lagrange who succeeded where Euler had failed by proving Fermat's conjecture 1.3; he also solved Pell's equation using continued fractions, and he studied the special case $c = 1$ of Legendre's Theorem 1.7 in [La1]: he noticed that $ax^2 = y^2 - bz^2$ can only be solved if b is a quadratic residue modulo a; starting from an equation $aa_1 = r^2 - bs^2$, he gave an algorithm which either produced a solution to the original equation or showed that there is none.

At the beginning of this paper, Lagrange complained that the part of mathematics studied by Diophantus had been completely neglected, except for the contributions of Euler; he closed it with another reference to the influence of Diophantus:

> nous observons seulement que M. de Fermat pretend, dans ses Remarques sur Diophante, avoir démontré en general ce théorème, que léquation
> $$r^n + s^n = q^n$$
> n'est jamais résoluble d'une manière rationnelle lorsque n surpasse 2; mais ce Savant ne nous a pas laissé sa demonstration, et il ne parait pas que personne l'ait encore trouvée jusquà present. M. Euler a, à la verité, démontré ce théorème dans le cas de $n = 3$ et de $n = 4$, par une analyse particulière et très-ingenieuse, mais qui ne parait pas applicable en général à tous les autres cas; ainsi, ce théorème est un de ceux qui restent encore à démontrer, et qui meritent le plus l'attention des Géomètres. [7]

---

[7] we observe only that Mr. Fermat claims, in his Remarks on Diophantus, to have shown in general the theorem that the equation $r^n + s^n = q^n$ is never solvable in rational numbers when n is larger than 2; but this scientist did not leave us his demonstration, and it seems that no one has found one to this day. Mr. Euler has actually demonstrated this theorem in the case $n = 3$ and $n = 4$ by a special and very ingenious analysis, which, however, does not seem to be applicable in general to all the other cases; thus this theorem is one of those which remain to be proved, and which most merit the attention of the geometers.

Strictly speaking, Fermat's conjecture became the last one to be proved only when the gaps in Euler's "proof" on the integral solutions of e.g. $y^2 + 2 = x^3$ were closed, probably not before 1875 by Pépin [Pe], although the necessary mathematical background was certainly well known to Gauss, Dirichlet or Eisenstein, to name a few.

The role of Fermat's Last Theorem in the development of algebraic number theory is often overrated, probably due to Hensel's (false) story claiming Kummer's first manuscript to have been an incorrect proof of this problem. The crème de la crème of French mathematicians – Lamé, Legendre, Liouville, and Cauchy – tried their luck but didn't really advance algebraic number theory during their work on FLT. Gauss did not value it very highly, but admits that it made him take up his investigations in number theory again: in a letter to Olbers from March 21, 1816, he writes

> Ich gestehe zwar, daß das Fermat'sche Theorem als isolierter Satz für mich wenig Interesse hat, denn es lassen sich eine Menge solcher Sätze leicht aufstellen, die man weder beweisen, noch widerlegen kann. Allein ich bin doch dadurch veranlaßt, einige alte Ideen zu einer *großen* Erweiterung der höheren Arithmetik wieder vorzunehmen. [...] Allein ich bin überzeugt, wenn das *Glück* mehr tun sollte, als ich erwarten darf und mir einige Hauptschritte in jener Theorie glücken, auch der Fermatsche Satz nur als eines der am wenigsten interessanten Corollarien dabei erscheinen wird.[8]

Gauss's last remark clearly indicates that he was at least thinking about the arithmetic in cyclotomic number fields $\mathbb{Q}(\zeta_p)$, even when, in a letter to Bessel a few months later, he reveals that the investigations in question had to do with the part that he eventually would publish, namely the theory of biquadratic residues. Parts of his research were published in 1828 [272] and 1832 [273], and the last paper contains the statement that cubic reciprocity is best described in $\mathbb{Z}[\rho]$, where $\rho^2 + \rho + 1 = 0$, and that, more generally, the study of higher reciprocity laws should be done after adjoining higher roots of unity.

Even Kummer, who is responsible for the greatest advance towards of Fermat's Last Theorem before the recent developments, got the main motivation for studying cyclotomic fields from his desire to find a general reciprocity law (which he called his "main enemy" in [Ku, Feb 25, 1848]). In almost every letter to Kronecker written between 1842 and 1848, Kummer mentions results related to reciprocity; the Fermat equation is mentioned for the first

---

[8] I admit, that Fermat's Theorem, as an isolated result, has little interest for me, since I can easily make a lot of such claims that can be neither proved nor disproved. Nevertheless it made me take up again some old ideas about a *large* extension of the higher arithmetic. [...] Yet I am convinced, if luck should do more than I may expect and if I succeed in making some of the main steps in that theory, then Fermat's Theorem will appear as one of the less interesting corollaries.

16   1. The Genesis of Quadratic Reciprocity

time in [Ku, Apr. 02, 1847]. In [Ku, Sept. 17, 1849] he informs Kronecker about the prize of 3000 Francs that the French Academy had offered to pay for a solution of Fermat's Last Theorem, and in [Ku, Jan. 14, 1850] he writes

> Wenn ich dies erst noch werde ergründet haben, ob es so ist oder nicht, dann werde ich die habsüchtigen Pläne ganz fallen lassen, und wieder nur für die Wissenschaft arbeiten, namentlich für die Reciprocitätsgesetze, für welche ich schon neue Ideen gefaßt habe.[9]

It is therefore safe to say that it was the quest for higher reciprocity laws that made Kummer study abelian extensions of $\mathbb{Q}$, Eisenstein those of $\mathbb{Q}(i)$ and Hilbert those of general number fields. Ironically, it was Hilbert himself who started the rumour that it was FLT that led Kummer to his ideal numbers: in his famous address at the ICM in Paris 1900, he wrote

> durch die Fermatsche Aufgabe angeregt, gelangte Kummer zu der Einführung der idealen Zahlen und der Entdeckung des Satzes von der eindeutigen Zerlegung der Zahlen eines Kreiskörpers in ideale Primfaktoren.[10]

Already in 1910, Hensel talked about "incontestable evidence" (actually it was something that Gundelfinger had heard from H.G. Grassmann) for the existence of a manuscript in which Kummer had claimed to have solved Fermat's Last Theorem, a rumour eventually dismissed by Edwards [Ed1, Ed2] and Neumann [Neu].

Legendre

In his paper [489] from 1785/88, Legendre announced two theorems that later were recognized as the first materializations of two fundamental ideas in number theory: one of these is of course the quadratic reciprocity law, the other is Legendre's Theorem 1.7, which we today view as the first hint at the existence of the Local-Global Principle that Hasse, inspired by Hensel, turned into an indispensable tool for number theorists. A nice proof of Legendre's Theorem 1.7 using Minkowski's geometry of numbers can be found in the book of Scharlau & Opolka [SO]. For other proofs see Davenport & Hall [DH] or Exercise 1.8. Holzer [Hol] showed that there exist nontrivial solutions such that $|x| \leq \sqrt{|bc|}$, $|y| \leq \sqrt{|ac|}$, $|z| \leq \sqrt{|ab|}$, and Mordell [Mo2] gave a simple proof with a small gap filled later by K.S. Williams [Wil]. See also Skolem [Sko] and Cochrane & P. Mitchell [CM]. Since solving Legendre's

---
[9] Once I will have fathomed whether this is so or not, then I will drop the avaricious plans and work again only for the science, especially for the reciprocity laws for which I have already envisaged some ideas.
[10] stimulated by Fermat's problem, Kummer arrived at the introduction of his ideal numbers and discovered the theorem of unique factorization of the integers of cyclotomic fields into ideal prime factors.

equation has become part of modern algorithms, whether you want to perform a simple 2-descent on elliptic curves or determine explicit equations for certain hyperelliptic curves, improvements to classical algorithms have been suggested recently; see e.g. Cremona [Cre] and Rusin [Rus].

As Legendre himself was quite aware, his proof of the quadratic reciprocity law was incomplete because he could not prove the existence of his auxiliary prime β. In the first edition, he used Theorem 1.8 to substantiate his claims; here is how he addressed the problem of giving a proof:

> Remarque. Il serait peut-être nécessaire de démontrer rigoureusement une chose que nous avons supposée dans plusieurs endroits de cet article, savoir, qu'il y a une infinité de nombres premiers compris dans toute progression arithmétique, dont le premier terme & la raison sont premier entr'eux, ou, ce que revient au même, dans la formule $2mx + \mu$, lorsque $2m$ & μ n'ont point de commun diviseur. Cette proposition est assez difficile à démontrer, cependant on peut s'assurer qu'elle est vraie. [...] Je me contente d'indiquer ce moyen de démonstration qu'il seroit trop long de détailler, d'autant plus que ce Mémoire passe déja les bornes ordinaires. [11]

The first criticism of Legendre's proof came from Gauss in [263, Art. 297]; in the preceding article 296, he had presented Legendre's attempted proof in the following way: first, recall that Gauss wrote pRq if p was a quadratic residue modulo q, and pNq otherwise. Then Gauss showed, using Legendre's technique, that the statements bRB and BRb cannot hold at the same time for primes $b \equiv B \equiv 3 \bmod 4$: this is Legendre's Théorème VII. The other cases Gauss considered are 2. bRa and aNb (Legendre I and II), 3. aRA and ANa (Legendre III and IV), 4. aRb and bNa (Legendre V and VI), and 5. bNB and BNb (Legendre VIII). After agreeing that there is "no room for objection" in case 1. and 2, he explained how Legendre's assumptions depend on the theorem on primes in arithmetic progression which "seems very plausible" but whose proof "is far from geometric rigor".

As Legendre's "Essai" didn't reach Gauss, according to his own words in the preface of [263], until "after the greater part of my work was already in the hands of the publishers", he addressed this problem in an additional note [263, p. 465] at the end: after remarking that Legendre did not prove Theorem 1.8 "with geometric rigor", he wrote

---

[11] Remark. It is perhaps necessary to prove rigorously a statement that we have assumed in several parts of this article, namely that there are infinitely many primes in each arithmetic progression whenever the first term and the difference are prime to each other, or, in other words, in the formula $2mx + \mu$, where $2m$ and μ don't have a common divisor. This proposition is quite difficult to prove, but one can convince oneself that it is true. [...] I am content with indicating this means of proof which would be too long to describe in detail, the more so since this memoir already exceeds the usual limits.

> Attamen tunc quoque, quando hoc theorema plene demonstratum erit: suppositio altera supererit (dari numeros primos formae $4n + 3$, quorum non-residuum quadraticum sit numerus primus datus formae $4n + 1$ positive sumtus), quae an *rigorose* demonstrari possit, nisi theorema fundamentale ipsum iam *supponatur*, nescio.

In Clarke's translation (p. 461), this reads

> But even if this theorem were fully demonstrated, the second supposition remains (that there are prime numbers of the form $4n + 3$ for which a given positive prime number of the form $4n + 1$ is a quadratic nonresidue), and I do not know whether this can be proven *rigorously* unless the fundamental theorem is *presumed*.

It is quite clear from the context (and follows unambiguously from the Latin text) that Gauss had doubts whether Legendre's Lemma can be derived from Theorem 1.8. Maser made an error in translating this part (p. 449) of the Disquisitiones into German by writing "denselben" instead of "dieselbe":

> Aber auch dann, wenn dieser Satz vollständig bewiesen sein wird, wird eine andere Annahme übrigbleiben (daß es Primzahlen von der Form $4n + 3$ giebt, deren quadratischer Nichtrest eine gegebene Primzahl von der Form $4n + 1$, positiv genommen, ist) und ich weiß nicht, ob man denselben *streng* beweisen kann, ohne das Fundamentaltheorem schon *vorauszusetzen*.

Kummer made the same mistake earlier when he claimed in [466, p. 19, 20] that Legendre's Lemma would follow from Theorem 1.8. The role of Dirichlet's theorem in Legendre's proof was discussed often (see e.g. Bachmann [Bm1, p. 50 ff.], Hilbert [Hil, p. 30], Patterson [628], Pieper [662]), but seldom as clearly as by Gauss. What we have seen in this Chapter is that Legendre's proof in his "Recherches d'analyse indéterminée" [489] can be completed with Theorem 1.8, while those in [490] cannot. Note, however, that Dirichlet's original proof of Theorem 1.8 used the quadratic reciprocity law – it was Mertens 1897 and Landau who showed how to do without it (see Siegel [Sie, §2.4] or Scharlau & Opolka [SO, p. 144 ff]).

Legendre did try to fill the gap: in the third edition of his book ([490]), he includes Jacobi's proof of the quadratic reciprocity law but also tries to prove the existence of his auxiliary prime by using binary quadratic forms; his (erroneous) proof is given in Exercise 1.38.

Moreover, in the same edition he writes in §411

> Toute progression arithmétique dont le premier terme et la raison sont premiers entre eux, contient une infinité de nombres premiers.[12]

---

[12] Each arithmetic progression whose first term and difference are coprime contains an infinity of prime numbers.

In §413, he adds

> Soit ψ le plus grand nombre premier contenu dans $\sqrt{2n-4}-1$; je dis que parmi les ψ nombres impairs qui suivent immédiatement $2n-1$, il y aura au moins un nombre premier.
> Remarque: Si on donnait à n des valeurs très-petites, on trouverait que ce théorème est sujet à quelques exceptions; mais comme on a supposé que ψ est un terme de la suite $3, 5, 7, 11, \ldots$ etc., il faut que $\sqrt{2n-4}-1$ soit plus grand que 3, ainsi on doit faire $n > 10$, et alors il n'y aura aucune exception.[13]

Thus Legendre thought he had proved Theorem 1.8. As is well known, the first correct proof of this result was provided by Dirichlet [Dir], who remarked

> Legendre macht den zu beweisenden Satz von der Aufgabe abhängig, die größte Anzahl auf einander folgender Glieder einer arithmetischen Reihe zu finden, welche durch gegebene Primzahlen theilbar sein können, löst aber diese Aufgabe nur durch Induction. Versucht man, die auf diese Weise von ihm gefundene, durch die Einfachkeit der Form des Resultats höchst merkwürdige Auflösung der Maximumsaufgabe zu beweisen, so stößt man auf grosse Schwierigkeiten, deren Überwindung mir nicht hat gelingen wollen.[14]

Whether Dirichlet would have ever studied the problem of primes in arithmetic progressions had not Legendre given an erroneous proof is an open question, although Kummer ([468]) goes as far as calling Dirichlet's theorem an offspring of the study of quadratic reciprocity:

> Um diesen Mangel des Legendreschen Beweises zu heben, hat später Hr. Dirichlet diese Eigenschaft der arithmetischen Reihen streng bewiesen, und zwar durch die neuen, überaus fruchtbaren Methoden, durch deren Hülfe er auch die Klassenzahl der quadratischen Formen gefunden hat. Diese berühmten Arbeiten des Hrn. Dirichlet können daher als solche betrachtet werden, welche der Beschäftigung mit den Reciprocitätsgesetzen ihre Entstehung verdanken. [15]

---

[13] Let ψ be the largest prime smaller than $\sqrt{2n-4}-1$; I say that among the ψ odd integers that immediately follow $2n-1$ there is at least one prime.
   Remark: If one gives n some very small values, one finds that this theorem admits a couple of counter examples; but since we have assumed that ψ is a term of the sequence $3, 5, 7, 11, \ldots$ etc., we see that $\sqrt{2n-4}-1$ must be greater than 3, hence we must have $n > 10$, and now there is no exception.

[14] Legendre bases the theorem he wants to prove on the problem of finding the largest number of consecutive terms of an arithmetic progression that are divisible by one of a set of given primes; however, he solves this problem only by induction. If one tries to prove the solution of the maximum problem found by him, whose form is most remarkable because of its simplicity, then one comes across some big difficulties which I did not succeed in surmounting.

[15] In order to remedy this deficiency in Legendre's proof, Mr. Dirichlet later proved this property of arithmetic progressions rigorously, through the new and exceed-

In 1858, the Parisian Academy[16] asked for a proof or a correct formulation for Legendre's unproven claim that, when p_n denotes the n-th prime, then for any $k \in \mathbb{N}$, at least one of the odd integers $2k+1, 2k+3, \ldots, 2k+2p_{n-1}+1$ is not divisible by any of $p_1, \ldots, p_n$ (this is the "Maximumsaufgabe" Dirichlet was talking about). Dupré [Dup] then showed that this is correct for $p_n \leq 19$ as well as for $p_n = 23, 37$ and 41, and that it does not hold for $43 \leq p_n \leq 113$ (see also Backlund [Bl], Errera [Err], Moreau [Mor] and Piltz [Pil]; Scheffler [Sce] also investigated the problem and tried to give a proof). Finally, A. Brauer & Zeitz [BZ] showed that Legendre's claim is false for all $p_n \geq 43$. See also T.-H. Chang [Ch1].

In 1893, Zignago [Zig] attempted an elementary proof of Dirichlet's theorem; Minkowski praised this proof in his letter [Mnk] to Hilbert, and both the Fortschritte der Mathematik [**25** (1893/94), p. 269] as well as Dickson accepted the proof as valid. Unfortunately, it isn't: see Exercise 1.39.

For an exposition of Legendre's attempted proof (including a proof of "Legendre's Waterloo" (Brown [89]), the existence of his auxiliary prime, based on the analytic methods of Dirichlet), see Teege [795]; Rogers' paper [686] shows how to close the gap using a result of Selberg according to which about half the primes up to x have $(D/p) = +1$ for a given non-square D. According to the review in Fortschritte der Mathematik, Pépin [645] completed Legendre's proof using a lemma from Gauss's first proof, and Ivanov [394] is said to have proved with elementary means the following result: for every non-square $a \in \mathbb{Z}$ there exist infinitely many primes p such that $(a/p) = -1$. Quite recently, Pieper [662] examined Legendre's efforts to prove the quadratic reciprocity law.

Gauss

Gauss's role in forming our perception of mathematics, and of number theory in particular, is immense. In Fowler's [Fow] words, "we see number theory through Gauss's eyes". And it is well documented how the Disquisitiones influenced the number theorists in the first half of the 19th century, namely Abel, Dirichlet, Eisenstein, and Jacobi. Among the many instances where the Disquisitiones depart from tradition, Fowler singles out two: the fundamental theorem of arithmetic, which does not appear in the work of Gauss's immediate predecessors (but see C. Goldstein [Go1] and Agargün & Fletcher [AF]; for unique factorization in Euclid's elements, see Knorr [Kn]), and the theory of continued fractions, which are all over the place in the publications of Euler, Lagrange and Legendre on Pell's equation and related topics, but are hardly mentioned by Gauss.

---

ingly fertile methods that also allowed him to find the class number of quadratic forms. These celebrated papers of Mr. Dirichlet may therefore also be said to owe their existence to the occupation with reciprocity laws.

[16] *Rapport sur le concours pour le grand prix de sciences mathématiques*, C. R. Acad. Sci. Paris **48** (1859), 487–488.

Gauss's first proof [264] of the quadratic reciprocity law, which was similar to Legendre's attempted proof in that Gauss also needed an 'auxiliary prime' (see Lemma 2.32, as well as D.H. Lehmer [493] and H. Shapiro [Shp] for an exposition in English, and E. Brown [89] for a discussion of Gauss's rather ingenious argument; the proof of its existence was found on April 8, 1796, according to the second entry in Gauss's diary [Gra]), has never been regarded as being particularly elegant (Smith [743] calls it "repulsive"), and did not receive much attention, even though Dirichlet [170, 171] managed to simplify it considerably. This attitude changed a little when Tate noticed that Gauss's proof could be used for the computation of $K_2(\mathbb{Q})$ (see Section 2.5, as well as Birch [54] and the book [Ros] of Rosenberg, where the structure of $K_2(\mathbb{Q})$ is given in Theorem 4.4.9, and where Corollary 4.4.10 is the quadratic reciprocity law). For more connections between quadratic reciprocity and K-theory, see Bass [Bas], Hurrelbrink [385], Milnor [586], and Scharlau [705, 706].

Gauss's second proof [265] used the genus theory of binary quadratic forms which he developed in his Disquisitiones Arithmeticae (see Springer [747]); the heart of the proof consists in the verification of two inequalities on the number of genera: subsequent generalizations by Kummer, Hilbert, Weber and Takagi turned these into the first and second inequality of class field theory. For a modern version of Gauss's second proof, see Chapter 2.

In [263, Art. 151], Gauss promises a third proof of the quadratic reciprocity law which, however, cannot be found there. The simple explanation is that Gauss shortened the original manuscript before publication; in the original article 366, he wrote

> Haec igitur est tertia theorematis fundamentalis Capitis IV completa demonstratio [...] At ex eodem hoc fonte, sed via opposita quartam deducamus.[17]

These third and fourth proofs are built on the theory of the quadratic period equation and are counted (according to the order of publication) as proofs number 7 and 8. A similar proof discovered by Lebesgue [487] before Gauss's Nachlaß was published. See Dickson [157] for comments on Gauss's seventh proof.

The idea behind the third [266] and fifth [268] proof (i.e. Gauss's Lemma) was given in this Chapter, for proofs no. 4 [267] and 6 [269] using the quadratic Gauss sum (with resp. without the determination of its sign) see Chapter 3.

Some of the eight proofs of Gauss have been discussed in detail: see in particular Dirichlet [170] and [174, §51] as well as Brown [89] for the first, Frobenius [242], Kronecker [446, 447, 448], Laubenbacher & Pengelley [479], Lerch [527], Schering [709], Tafelmacher [787], Voigt [815] and Zeller [882]

---

[17] This is therefore the third complete proof of the fundamental theorem of Chapter IV [...] From the same source, but going in the opposite direction, we deduce the fourth.

22   1. The Genesis of Quadratic Reciprocity

for the third, Dedekind [174, §115], Jenkins [411], and Kronecker [443] for the fourth, and Lerch [531, 532], Ludwig [561], Rédei [673] and Zeller [882] for Gauss's fifth proof. The fact that Gauss gave eight different proofs of the quadratic reciprocity law has impressed many mathematicians; the following quote is from an interview with Yuri Manin [Man]:

> When I was very young I was extremely interested in the fact that Gauß found seven or eight proofs of the quadratic reciprocity law. What bothered me was why he needed seven or eight proofs. Every time I gained some more understanding of number theory I better understood Gauß' mind. Of course he was not looking for more convincing arguments – one proof is sufficiently convincing. The point is, that proving is the way we are discovering new territories, new features of the mathematical landscape.

For the history of Gauss's progress in number theory, an invaluable source is his own diary: for a commented version thereof see Gray [Gra]. There are also German [BSW] and French [EL] translations of Gauss's diary, which was written in Latin. The fact that Gauss essentially proved the reciprocity for Jacobi symbols in [263, Art. 133] was noticed by Lebesgue [488]. Another nice description of Gauss's contributions to number theory is Rieger's article [Rie]; Fuchs [Fu] discusses Gauss's introduction of congruences. A more prosaic description of Gauss's achievements in his Disquisitiones is contained in Weil's letter [828] to his sister Simone Weil during his imprisonment in the military prison at Rouen in the spring of 1940. Neumann's articles [614, 615] also deal with Gauss's contributions to number theory.

Apparently neither Legendre nor Gauss knew about Euler's work on reciprocity when they started working on it, although Gauss credited Euler (referring to his paper of 1742) in [263, Art. 151] for the discovery of special cases of the "fundamental theorem". In his comments on his own third proof [Ga2], when Gauss discusses the contributions of Lagrange and Euler, he writes

> Es ist indess ein merkwürdiges Spiel des Zufalls, daß beide Geometer durch Induktion nicht auf das allgemeine Fundamental-Theorem gekommen sind, das einer so einfachen Darstellung fähig ist. Dies ist zuerst, obwohl in einer etwas andern Gestalt, von Legendre vorgetragen.[18]

It seems that even in 1858 Euler's second paper on the quadratic reciprocity law was not commonly known, because in that year Kummer, in his historical survey [468] on reciprocity laws, only cited Euler's first paper. The first to refer to Euler's Opuscula were Chebyshev [Che] and Smith [743], but it seems that it was Kronecker's article [439] that made Euler's contributions known to a wide audience (see also Pépin's reactions in [647]).

---

[18] It is, however, a remarkable coincidence that both geometers did not find by induction the general fundamental theorem which can be expressed in such a simple way. This was first done, in a slightly different form, by Legendre.

Quadratic Residues and Gauss's Lemma

Some expositions of the elementary theory of quadratic residues or of elementary problems can be found in Aubry [20, 23, 24], Cox [139], Delezoide [155], Froemke & Grossmann [FG], Goldschmidt [316], Johnson [Joh], Pickert [659], and Rameswar [668], as well as in dozens of other sources – quadratic residues are covered in virtually every book on number theory. Also, Cooper's thesis [136] contains lots of references on the more elementary aspects of quadratic residues.

Problems concerning quadratic residues and their applications are too numerous to be discussed here in detail; we merely mention lower bounds for the the least quadratic nonresidue or the least prime quadratic residue, the nonexistence of squares in second order recurring sequences, quadratic residue codes and analogues for higher powers, quadratic residue covers, or Wagstaff's idea for watermarking files using quadratic residues.[19]

Gauss's Lemma for $\left(\frac{a}{m}\right)$, published for prime values of m by Gauss [266] in 1808, was generalized to composite values of m by Jenkins [411] in 1867 (see also Smith's report [743]), and independently by Schering [710] in 1876; the latter stated (without proof) that $\left(\frac{a}{m}\right) = (-1)^\mu$, where μ is the number of negative residues in $\{a, 2a, \ldots, \frac{m-1}{2}a\}$ (this is Lemma 1.10 with the half-system $\{1, 2, \ldots, \frac{m-1}{2}\}$). Kronecker [440] claimed to have discovered this generalization in his lectures 1869/70, and Schering [714] finally published a proof in 1882. See also Genocchi [303].

More proofs of Gauss's Lemma and its generalizations have been given by Aigner [4], Aubry [21], Brenner [82], Chua [130], Eichenberg [183], Gazzaniga [282], Gegenbauer [288], Lipschitz [551], Mandl [562, 563], Pépin [646], Reichardt [677], Scholz [724] (he uses Gauss's lemma for composite values for first deducing Euler's observation that $(a/m) = (a/m+4a)$ and then deriving the quadratic reciprocity law for the Jacobi symbol; Davenport has adopted this approach in his book [Dav]) and Weber [825, p. 296–298]; see also the Notes in Chapter 4. There is also a 'modern' interpretation of Gauss's Lemma: as was noticed by Cartier [Crt] and Waterhouse [824], $\left(\frac{\cdot}{p}\right)$ can be identified with the group-theoretical transfer of $(\mathbb{Z}/p\mathbb{Z})^\times$ to its subgroup $\{-1, +1\}$. For generalizations to abelian groups, see Delsarte [Del]. The article [56] of van der Blij contains an account of classical results including quadratic reciprocity in a modern language.

The greatest integer function (or floor function) introduced by Gauss in connection with Gauss's Lemma was investigated e.g. by Ahlborn [Ahl], Berndt [46], Berndt & Dieter [49], Bouniakowski [68, 71, 72], Busche [108, 110], Carlitz [118], Friedlander & K. H. Rosen [228], Kronecker [449], Kuipers [459, 460], Popovici [Pop], and Sylvester [780]. Zolotarev [883] found a neat way to express the content of Gauss's Lemma by using permutations (cf. Exercise 1.36); see also Brenner [83], Cartier [119] (for a generalization to

---
[19] See http://www.cs.purdue.edu/faculty/ssw/.

number fields), Dressler & Shult [180], Lerch [528, 529, 530], Morton [591], the exposition of Riesz [684] already mentioned, Roberts [685], Rousseau [690] and Slavutskij [742].

The Quadratic Reciprocity Law

> Der Beweis dieses Satzes hat den ersten Mathematikern die größten Schwierigkeiten gemacht. Jetzt hat man deren sehr viele und sehr leichte, und, wie man es auch anfängt, man kommt immer auf einen Beweis.[20]

This is Jacobi's opinion on proofs of the quadratic reciprocity law, taken from [402]. Almost all the essential ideas used in the many elementary proofs of the quadratic reciprocity law (see the Appendix for a list of published proofs) were known to Jacobi. With a few exceptions such as Eisenstein's idea to use periodic functions or Lebesgue's approach via the number of solutions of $x_1^2 + \ldots + x_q^2 \equiv 0 \bmod p$, one might even say that the main ingredients can already be found in Gauss's proofs.

There are quite a few reports dealing with elementary proofs of the quadratic reciprocity law; see e.g. Ostwald's reprint [262] of Gauss's eight proofs, Baumgart's dissertation [38] (comparing the proofs published before 1884), Bachmann's book [26] (giving a list of ca. 50 proofs up to 1897, including two proofs called "forgotten" by Mel'nikov & Slavutskij [576]), the booklet of Pieper [661] (developing many elementary proofs and expanding Riesz's exposition in [684]), the theses of Banderier [28], Dirks [175], Kirchhoff [429], and Schmitt [720] and the surveys of Cululière [141, 142]. For another detailed and particularly illuminating presentation of the theory of quadratic reciprocity see Hasse [342].

From today's point of view it seems quite natural to make a map defined on prime ideals into a homomorphism by extending it multiplicatively to all nonzero ideals. One can get an impression of how far removed from the thinking of the 1830's this must have been by looking at the paper [162] of Dirichlet: starting from an equation (p. 203) $t^2 + au^2 = pq$, where p, q and $a \equiv 1 \bmod 8$ are primes, he proves $\left(\frac{t}{a}\right) = \left(\frac{t}{p}\right)\left(\frac{t}{q}\right)$ for odd t by factoring $t = \prod g_j$ and deducing

$$\left(\frac{t}{a}\right) = \prod \left(\frac{g_j}{a}\right) = \prod \left(\frac{a}{g_j}\right) = \prod \left(\frac{p}{g_j}\right) \prod \left(\frac{q}{g_j}\right)$$
$$= \prod \left(\frac{g_j}{p}\right) \prod \left(\frac{g_j}{q}\right) = \left(\frac{t}{p}\right)\left(\frac{t}{q}\right).$$

He repeats this argument explicitly six times (on pp. 204, 205, 211 and 213) without realizing the shortcut offered by introducing the Jacobi sym-

---

[20] The proof of this theorem presented considerable difficulties to the first mathematicians. Now we have a lot of simple proofs, and no matter how you start, you always end up with one.

bol. Dirichlet started using the Jacobi symbol in 1839 [168], and later (1854) showed that it can be used to simplify Gauss's first proof of the quadratic reciprocity law.

A 'chronology' of proofs of the quadratic reciprocity law has been compiled in Appendix B. Different entries do not necessarily imply different proofs: some of the proofs are merely reformulations of others. A thorough classification of the published proofs of the quadratic reciprocity law is something that remains to be done.

The following papers deal with simple computations of the quadratic character of small integers: Aubry [22], Bouniakowski [69], who gives the following formulas for $(2/p)$ and $(3/p)$:

$$\left(\frac{2}{p}\right) = (-1)^{\lfloor \frac{p+1}{4} \rfloor}, \quad \left(\frac{3}{p}\right) = (-1)^{\lfloor \frac{p+1}{6} \rfloor},$$

Bricard [85], Frattini [237], v. Grosschmid [319], Lagrange [474], Lebesgue [482], Libri [539, 540, 541], Matrot [570], Nielsen [619], Oltramare [623], K.H. Rosen [687], Schiappa-Monteiro [715], Schönemann [721], Stern [753, 754, 755], Stieltjes [761, 762, 765, 766], Storchi [768], and K.S. Williams [858]. Humble [384] proves the reciprocity law for Jacobi symbols by induction, assuming the reciprocity law for Legendre symbols.

Legendre symbols also appear in the theory of continued fractions, as Schinzel [Scz] has shown, proving some conjectures of P. Chowla & S. Chowla [CC]. Friesen [Fri] and van der Poorten & Walsh [vPW] prove enough about continued fractions to yield $(p/q) = -(q/p)$ for primes $p \equiv q \equiv 1 \bmod 4$; the basic ingredient is the solvability of $px^2 - qy^2 = \pm 1$, so one should not expect a proof of the full quadratic reciprocity law along these lines. See also Grytczuk [Gry]; for a recent survey, see Mollin [Mol].

Analogues of the quadratic reciprocity law in the polynomial ring $\mathbb{F}_q[t]$ of one variable over a finite field were studied by Dedekind [Ded] (see W. Scharlau's articles [Sch1, Sch2]), Artin [Art], Bach [Ba], Feng & Ying [FY], Hellegouarch [Hel], Kühne [Kue], Loewy [Loe], Mantel [Mnt], Merrill & Walling [MW], Ore [Ore], Pocklington [Po2], F.K. Schmidt [Scm], Schwarz [Scw], Vaidyanathaswamy [Vai], and Whiteman [Whm]. See also Fouche [Fou].

For proofs of the quadratic reciprocity law in certain axiom systems, see Cornaros [Cor], Naur [Nau], Smith [Smi], Russinoff [692], and Ryan [Rya].

There are several articles and theses dedicated to the development of reciprocity laws apart from those already mentioned: First, there's Collison's thesis [132] and its summary in [133], dealing mostly with the cubic and quartic laws; Frei's paper [238] on basically the same topic as this book, as well as Schwermer [730] and Esrafilian & Sangani-Monfared [208] who tell the story up to Langlands' program.

Computational Problems

Algorithms for computing Legendre and Jacobi symbols were devised by Dirichlet [174], (compare Sylvester [783, 784] as well as Dedekind's remarks [150]), Eisenstein [186], Gauss [Ga1], Gegenbauer [284, 286, 287, 290, 291], Heinitz [350, 351], Kronecker [443], Lebesgue [488, Leb, 486], McClintock [McC], Nasimoff [611], Pexider [658], Saalschütz [693], Scheibner [707, 708], Schering [713], and Zeller [882]. Since fast algorithms for computing Jacobi symbols have become increasingly important for many applications, this subject is being studied again: see e.g. the papers of Collins & Loos [CL], Lenstra [Len] (who deals with the more general problem of residue symbols in number fields), Meyer & Sorenson [MS], Shallit [Shl], and Shallit & Sorenson [SS].

Attempts at redefining the Jacobi symbol for negative values of the denominator (according to our definition, $(\frac{a}{b}) = (\frac{a}{-b})$) have been made regularly, in particular in connection with algorithmic problems: see e.g. Weber [Web], Estes & Pall [210], L. Taylor [793], or Cohen [Coh].

Once it is known that $(a/p) = +1$ for some $a \in \mathbb{Z}$, there is the problem of computing the square root of a mod p. In some special cases, this was already solved by Lagrange (Exercise 1.16) and Legendre (Exercise 1.17); Vandiver [Va3] gave a general algorithm. An analogous question can be asked in any finite field; see Adleman, Manders & Miller [AMM], Bugaieff [Bug], Caris [Car], Chang [Ch2], Cipolla [Cip], D.H. Lehmer [LeD], Pocklington [Po1], Schönheim [Sch], Shanks [Sha], Speckmann [Spe], Stankowitsch [Sta], Tamarkin & Friedmann [TF], and Tonelli [Ton]. For the present state of the art, see Cohen's book [Coh], as well as Schoof [Scf].

Distribution of Residues

Sequences of integers with prescribed residue character have received quite a lot of attention; Stern [754] (see also C. von Staudt [749]) introduced the symbol RR to denote the number of pairs $(k, k+1)$ of integers $1 \leq k \leq p-2$ such that k and $k+1$ are quadratic residues mod p. The symbols NN, NR, or RRR etc. are defined similarly. The number of sequences of length 2 will be determined in Exercise 1.22 below; check the dissertation of Jacobsthal [405] and Chapter 10 (where these results will be generalized) for more information. Other results have been obtained by Aladov [Ala], Aubry [20], Bennett [Ben], Brauer [Br], Burde [Bu1, Bu2, Bu3, Bu4], Dörge [Doe] (he showed that $RRRR + NNNN > 0$ except when $p = 2, 3, 5, 7, 11, 17$; this seems actually to be due to Sterneck), Giudici & Jatem [GJ], Götting [Gö], von Grosschmid [vGr], Gupta [Gup], Hopf [Hop], Jänichen [409, 410], Koutsky [Ko1, Ko2], D.N. Lehmer [491], Linden [Lin], Mathieu [567], Monzingo [589, Mon], Phong & Jones [PJ], Rédei [Red], Reich [Rei], Rocci [Roc], Singh [Sin], Stern [757], Sterneck [St1, St2], Torelli [Tor], Vandiver [810, Va2], and Whyburn [Wy1, Wy2]. For more details, see Cooper's thesis [136]; a modern presentation of the material can be found in Hasse [342], and in Chapter 10 we will discuss how seemingly innocent questions about the distribution of quadratic residues led to substantial research in algebraic geometry.

Analytic questions such as "How many quadratic residues are there in the interval $[N, N + h]$" are becoming increasingly important in algorithmic number theory. Naively, one would expect about half the values to be quadratic residues, hence the expected number N_p of solutions to $y^2 \equiv x \bmod p$ for $x \in [N, N+h]$ should be about h. In fact, it can be proved quite easily using quadratic Gauss sums that $|N_p - h|$ is bounded by $c\sqrt{p}\log p$ for some constant $c > 0$ (see Stepanov [Ste, Chap. 2] or Exercise 3.20).

Carlitz [Crl] defined polynomials $g_{rs} = \prod(x - a)$ (where the product is over all $a = 2, \ldots, p-1$ such that $\left(\frac{a-1}{p}\right) = r$, $\left(\frac{a}{p}\right) = s$, with $r, s \in \{\pm 1\}$) and showed that the g_{rs} are congruent modulo p to Chebyshev polynomials.

Analogous questions for n-th power residues with $n \geq 3$ were studied by Bierstedt & Mills [BM], Brillhart, D. Lehmer & E. Lehmer [BLL], D. Lehmer, E. Lehmer & Mills [LLM], E. Lehmer [LeE], Mills [Mil] and Unger [806]. See also the 'Handbook of number theory' [MSC, Chapter XV] by Mitrinovic, Sandor & Crstici.

Exercises

1.1 Use induction to prove that, for primes p, the solvability of $x^2 \equiv -1 \bmod p$ implies that p is a sum of two squares. (Hint: Put $a_0 = x$, $b_0 = 1$, and start with $a_0^2 + b_0^2 = pm$ for some integer m; show that one can always choose x in such a way that $m < p$. If $m > 1$, then there is a prime q dividing m; deduce from the induction hypothesis that $q = A^2 + B^2$ is a sum of two squares. Show that you can choose the sign of B to make $A - Bx \equiv 0 \bmod q$, and prove that $x^2 + 1 = q(a_1^2 + b_1^2)$. This gives $a_1^2 + b_1^2 = pm_1$ for some $m_1 < m$. Repeat.)

1.2 Let e, f, m be positive integers such that $e, f \leq m < ef$. Then for every integer r coprime to m there exist integers $0 < x < e$ and $0 < y < f$ such that $yr \equiv \pm x \bmod m$. (Hint: look at the set of integers $\{v + rw : 0 \leq v < e, 0 \leq w < f\}$. It has cardinality $ef > m$; now use Dirichlet's pigeonhole principle).
This result has an interesting history: it is usually credited to Thue in its most simple form where $e = f$, the general statement being ascribed to Scholz. But as A. Brauer & Reynolds [BR] have shown, the special case was already known to Aubry, a generalization of it to Vinogradov. The article [Va1] of Vandiver, however, begins with the sentence "In 1903, Professor G.D. Birkhoff communicated to me the following theorem" and then gives the special case $m = p$ a prime, $e = f = \sqrt{p}$, of "Thue's theorem".

1.3 Use Thue's result to prove Proposition 1.2. (Hint: let $e = f$ be the smallest integer such that $e^2 > p$; if $z^2 \equiv -1 \bmod p$, show that there exist $x, y < \sqrt{p}$ with $yz \equiv x \bmod p$).

1.4 Prove the analogues of Proposition 1.2 for primes $p = x^2 + ny^2$, where $n = \pm 2, \pm 3, -5$. Deduce that $\left(\frac{2}{p}\right) = 1 \iff p \equiv \pm 1 \bmod 8$, and derive similar criteria for $-2, \pm 3$ and 5.

28   1. The Genesis of Quadratic Reciprocity

1.5 Try to prove Fermat's conjecture (1.3) on the representation of primes in the form $x^2 + 5y^2$.

1.6 From (1.3), deduce that -5 is a quadratic residue modulo primes $p \neq 2, 5$ if and only if $p \equiv 1, 3, 7, 9 \bmod 20$.

1.7 (Lagrange [La2]) Show that $x^2 + y^2 + 1 = 0$ is solvable in $\mathbb{F}_p$ for every prime p. (Hint: suppose not; then the polynomial $1 + (-1 - x^2)^{(p-1)/2}$ has p roots in $\mathbb{F}_p$ and degree $p - 1$. This proof seems to be due to Minding [Min]).

1.8 Prove Legendre's Theorem 1.7, i.e. show that $ax^2 + by^2 = cz^2$ ($ab < 0$ or $ac > 0$) is solvable if the congruences $x^2 \equiv bc \bmod a$, $x^2 \equiv ca \bmod b$ and $x^2 \equiv -ab \bmod c$ are.
  1. Show that it is sufficient to prove the claim for square-free $a, b, c \in \mathbb{Z}$; show moreover that we may assume that $a, b, c \in \mathbb{N}$.
  2. Let $p \mid c$ be an odd prime; choose a solution $x = \alpha, y = \beta$ of $ax^2 + by^2 = 0$ in $\mathbb{F}_p$ and deduce that $ax^2 + by^2 \equiv a(x + \alpha\beta^{-1}y)(x - \alpha\beta^{-1}y) \bmod p$. In particular, show that, for every odd prime $p \mid abc$, there exist linear forms $l_p = rx + sy = tz$, $m_p = ux + vy + wz$ such that $ax^2 + by^2 - cz^2 \equiv l_p m_p \bmod p$.
  3. Use the congruence $ax^2 + by^2 - cz^2 \equiv (ax + by + cz)^2 \bmod 2$ to prove the existence of the linear forms l_2 and m_2.
  4. Use the Chinese Remainder Theorem to construct linear forms $l(x, y, z)$, $m(x, y, z)$ such that $l(x, y, z) \equiv l_p(x, y, z) \bmod p$ for all $p \mid abc$ (similarly for m). Conclude that $ax^2 + by^2 - cz^2 \equiv l(x, y, z)m(x, y, z) \bmod abc$.
  5. Let x, y, z run through the integers
  $$0 \leq x < \sqrt{bc}, \quad 0 \leq y < \sqrt{ca}, \quad 0 \leq z < \sqrt{ab}. \qquad (*)$$
  If $a = b = c = 1$ the claim is trivial, otherwise not all of $\sqrt{ab}, \sqrt{bc}, \sqrt{ca}$ can be integers, hence there are more than $\sqrt{ab}\sqrt{bc}\sqrt{ca} = abc$ triples (x, y, z). In particular, there must exist $(x_1, y_1, z_1), (x_2, y_2, z_2)$ such that $l(x_1, y_1, z_1) \equiv l(x_2, y_2, z_2) \bmod abc$. Thus $l(x_0, y_0, z_0) \equiv 0 \bmod abc$, where $x_0 = x_2 - x_1$, $y_0 = y_2 - y_1$ and $z_0 = z_2 - z_1$.
  6. Show that $ax_0^2 + by_0^2 - cz_0^2$ is an integer divisible by abc and strictly between $-abc$ and $2abc$. Use the following identity to conclude the proof:
  $(z^2 + ab)(ax^2 + by^2 - cz^2 - abc) = a(xz + by)^2 + b(yz - ax)^2 - c(z^2 + ab)^2$.

1.9 Mordell [Mo1] has observed that the following assertion is equivalent to the quadratic reciprocity law: Let a, b, c be integers and p be a prime dividing abc. If $ax^2 + by^2 + cz^2 \equiv 0 \bmod 4abc/p$ has nontrivial solutions, then so does $ax^2 + by^2 + cz^2 \equiv 0 \bmod 4abc$. Prove this equivalence.

1.10 Complete Legendre's proof of the quadratic reciprocity law, assuming the existence of the auxiliary prime β.

1.11 Prove $\left(\frac{-3}{p}\right) = \left(\frac{p}{3}\right)$ by imitating the computation of $\left(\frac{2}{p}\right)$ in Sect. 1.4.

1.12 Derive the quadratic reciprocity law for the Jacobi symbol from one for the Legendre symbol. In particular, prove the congruences (1.6).

1.13 (Euler, Op. anal. I, 276) If $p \equiv 1 \bmod 4$ is prime, then $\frac{p-1}{4} - n(n+1)$ is a quadratic residue modulo p for every integer n.

1.14 (Euler, Op. anal. I, 281) If $q \equiv 3 \bmod 4$ is prime, then $\frac{q+1}{4} + n(n+1)$ is a quadratic residue modulo q for every integer n.

1.15 (Bork [67]) If q and $p = q + 4$ are prime, then $(p/q) = 1$.

1.16 (Lagrange [La1, p. 500]) If $p \equiv 3 \bmod 4$ is prime and $x^2 \equiv a \bmod p$ is solvable, then then $x \equiv \pm a^{(p+1)/4} \bmod p$ are its solutions.

1.17 (Legendre) Similarly, if $p \equiv 5 \bmod 8$ is prime, the solutions of $x^2 \equiv a \bmod p$ are $x \equiv \pm a^{(p+3)/8} \bmod p$, if a is a quartic residue modulo p, and $x \equiv \pm a^{(p+3)/8} 2^{(p-1)/4} \bmod p$ if not.

1.18 (Bickmore [52]) Since $S_p = 2^{2p} + 1$ is the sum of two squares, so is each of its factors. Verify that, for $p = 2m+1$, $S_p = A_p B_p$ for $A_p = 2^p - 2^{m+1} + 1$ and $B_p = 2^p + 2^{m+1} + 1$, and write A_p and B_p as a sum of two squares. Use the quadratic reciprocity law to prove that $5 \mid A_p \iff (2/p) = -1$ and $5 \mid B_p \iff (2/p) = +1$.

1.19 (Bickmore [52]) Similarly we have $3^{6m+3} + 1 = K_m L_m M_m$, where $K_m = 3^{2m+1} + 1$, $L_m = 3^{2m+1} - 3^{m+1} + 1$, $M_m = 3^{2m+1} + 3^{m+1} + 1$. Write each factor K_m, L_m, M_m in the form $a^2 + 3b^2$.
Show that $7 \mid K_m \iff p \equiv 3 \bmod 6$; $7 \mid L_m \iff p \equiv \pm 5 \bmod 12$; and $7 \mid M_m \iff p \equiv \pm 1 \bmod 12$.

1.20 (Bickmore [52]) Show that, for odd $u \in \mathbb{N}$, $(3^{3u} - 1)/(3^u - 1)$ is a sum of two squares. Show the same for $(5^{5u} + 1)/(5^u + 1)$. Generalize.

1.21 (Sierpinski [735]) Let q be a prime and $m > 1$ an integer. Show that each $x \in \mathbb{Z}$ is a q-th power modulo m if and only if m is not divisible by any prime $\equiv 1 \bmod q$.

1.22 Let p be an odd prime, and let α_{ij} denote the cardinality of the set

$$A_{ij} = \left\{ k \in \{1, 2, \ldots, p-1\} : \left(\frac{k}{p}\right) = (-1)^i, \left(\frac{k+1}{p}\right) = (-1)^j \right\}.$$

Show that
  i) $p - 2 = \alpha_{00} + \alpha_{01} + \alpha_{10} + \alpha_{11}$;
  ii) $-1 - \left(\frac{-1}{p}\right) = \sum_{k=1}^{p-2} \left\{ \left(\frac{k}{p}\right) + \left(\frac{k+1}{p}\right) \right\} = 2\alpha_{00} - 2\alpha_{11}$;
  iii) $+1 - \left(\frac{-1}{p}\right) = \sum_{k=1}^{p-2} \left\{ \left(\frac{k}{p}\right) - \left(\frac{k+1}{p}\right) \right\} = 2\alpha_{01} - 2\alpha_{10}$;
  iv) $-1 = \sum_{k=1}^{p-2} \left(\frac{k}{p}\right)\left(\frac{k+1}{p}\right) = \alpha_{00} - \alpha_{01} - \alpha_{10} + \alpha_{11}$.

Deduce that $\alpha_{00} = \frac{p-5}{4}, \alpha_{01} = \alpha_{10} = \alpha_{11} = \frac{p-1}{4}$ if $p \equiv 1 \bmod 4$, and $\alpha_{01} = \frac{p+1}{4}, \alpha_{00} = \alpha_{10} = \alpha_{11} = \frac{p-3}{4}$ if $p \equiv 3 \bmod 4$.

1.23 Let $N(x^2 + y^2 = 1)$ denote the number of solutions $(x, y) \in \mathbb{F}_p \times \mathbb{F}_p$ of the equation $x^2 + y^2 = 1$; similarly, let $N(x^2 = a)$ denote the number of solutions of $x^2 = a$ in $\mathbb{F}_p$. Show that $N(x^2 + y^2 = 1) = p - (-1/p)$ by computing the sum

$$\sum_{a+b=1} N(x^2 = a) N(x^2 = b) = \sum_{a+b=1} \left(1 + \left(\frac{a}{p}\right)\right)\left(1 + \left(\frac{b}{p}\right)\right).$$

1.24 Let p be an odd prime; show that each solution (x,y) of $x^2 + y^2 = 1$ in $\mathbb{F}_p$ gives rise to exactly eight solutions $(\pm x, \pm y)$ and $(\pm y, \pm x)$ unless either $x = 0$, $y = 0$ or $x = y$. Count the solutions in the last three cases and derive the second supplementary law of quadratic reciprocity from the formula in Exercise 1.23. This idea of Lebesgue was resdcovered by Williams [858].

1.25 Let $p = 2m+1$ and $q = 2n+1$ be primes; consider the set $S_{p,q}$ of solutions in $\mathbb{F}_p$ of the equation $x_1^2 + \ldots x_q^2 = 1$, and show that $\# S_{p,q} = p^{q-1} + (-1)^{mn} p^n$ (this generalizes the result of Exercise 1.23). Derive the quadratic reciprocity law as follows: consider the map $\phi : S_{p,q} \longrightarrow S_{p,q}$ defined by $(x_1, \ldots, x_q) \longmapsto (x_2, \ldots, x_q, x_1)$. Show that ϕ has a fixed point if and only if $(q/p) = 1$, verify that $(q/p) = -1 \iff \# S_{p,q} \equiv 0 \bmod q$ by counting the orbits of ϕ, and then use the formula for $\# S_{p,q}$ to derive the quadratic reciprocity law.
Remark: This is based on Lebesgue's proof of the quadratic reciprocity law in [485]. See Ireland & Rosen [386, p. 101–102], Ely [205] and Lemmermeyer [515] for related proofs.

1.26 Let p be an odd prime and let $f \in \mathbb{Z}[x]$ be a polynomial of degree ≥ 1. Show that the number $N_p(f)$ of solutions $(x,y) \in \mathbb{F}_p \times \mathbb{F}_p$ of the congruence $y^2 = f(x)$ can be expressed as $N_p(f) = p + S_p(f)$ with $S_p(f) = \sum_{x=0}^{p-1} (\frac{f(x)}{p})$.

1.27 (continued) Show that $N_p(f) = p + (-1)^{(p-1)/2}$ for $f(x) = 1 - x^2$, and that $N_p(f) = p - (\frac{a}{p})$ for $f(x) = x^3 + ax^2$.

1.28 For some $k \in \mathbb{Z} \setminus \{0\}$, put $f(x) = (x+k)(x+k^2)$; show that $S_p(f) = (\frac{k}{p}) T_p(f)$, where $T_p(f) = 1 + \sum_{x=1}^{(p-1)/2} \{(\frac{1+x}{p}) + (\frac{1-x}{p})\}(\frac{1+x^2}{p})$ does not depend on k. Show that $T \bmod 4$ only depends on $p \bmod 8$.

1.29 (E. Lehmer, Jacobsthal) For a fixed prime p and any integer $k \geq 2$, let $N_k(D) = \#\{(u,w) \in \mathbb{Z} \times \mathbb{Z} : u^k + D \equiv w^2 \bmod p\}$. Show that

$$N_k(D) = \sum_{u=0}^{p-1} \left(1 + \left(\frac{u^k + D}{p}\right)\right) = p + \left(\frac{D}{p}\right) + \psi_k(D),$$

where $\psi_k(D) = \sum_{u=1}^{p-1} (\frac{u^k + D}{p})$. Define the Jacobsthal sum

$$\Phi_k(D) = \sum_{u=1}^{p-1} \left(\frac{u}{p}\right)\left(\frac{u^k + D}{p}\right).$$

If the integer D' is defined by $DD' \equiv 1 \bmod p$ and if k is odd, show that $\psi_k(D) = (\frac{D}{p}) \Phi_k(D')$.

1.30 Let RRR denote the cardinality of the set $\{a \in \mathbb{Z}/p\mathbb{Z} : (\frac{a}{p}) = (\frac{a+1}{p}) = (\frac{a+2}{p}) = 1\}$. Show that

$$8RRR = \sum_{n=1}^{p-3} \left(1 + \left(\frac{a}{p}\right)\right)\left(1 + \left(\frac{a+1}{p}\right)\right)\left(1 + \left(\frac{a+2}{p}\right)\right)$$

and relate the sum on the right hand side to the Jacobsthal sum $\Phi_2(D)$. Compute RRR for primes $p \equiv 3 \bmod 4$.

1.31 (Jacobi, Cauchy) Let $d = -\ell$ be the discriminant of a complex quadratic number field, where $\ell \equiv 3 \mod 4$ is a prime > 3, and let
$$R = \sum_{(d/a)=1} a, \qquad N = \sum_{(d/b)=-1} b.$$
Then $R + N = \frac{1}{2}\ell(\ell-1)$ and $R - N = \sum_{1 \le a \le d}(\frac{d}{a})a$. Show that $R \equiv N \equiv 0 \mod d$ and define $h = \frac{1}{|d|}(N - R)$. Put $\chi(a) = (d/a)$ and prove
$$h = \frac{1}{2 - \chi(2)} \sum_{1 \le a \le d/2} \chi(a).$$
Finally, prove that h is odd.

1.32 (Gauss) Let p and q be different odd primes and put $m = \frac{1}{2}(q-1)$; prove
$$\left(\frac{p}{q}\right) = (-1)^\mu, \qquad \mu = \sum_{a=1}^{m} \left\lfloor \frac{ap}{q} \right\rfloor.$$
(Hint: Let $A = \{1, 2, \ldots, m\}$ be a half system modulo q. For $a \in A$ write $pa = \lfloor pa/q \rfloor q + t_a$ for some $1 \le t_a \le q - 1$. Sum over all $a \in A$ and find
$$p\frac{q^2 - 1}{8} = q \sum_{a=1}^{m} \left\lfloor \frac{ap}{q} \right\rfloor + \sum_{a=1}^{m} t_a. \tag{1.8}$$
On the other hand, write $pa \equiv (-1)^{\varepsilon_a} r_a \mod q$ for some $r_a \in A$. Then the number of a such that $\varepsilon_a = -1$ equals μ (as defined in Lemma 1.10), and one finds $\sum t_a = \mu q + \frac{1}{8}(q^2 - 1)$. Plug this into Equation (1.8) and reduce modulo 2.)

1.33 Eisenstein [192] found an ingenious proof for (1.5). Consider the lattice $\mathbb{Z}^2$ spanned by $(1, 0)$ and $(0, 1)$; the lattice points in the rectangle R with edges $(1, 1), (1, \frac{q-1}{2}), (\frac{p-1}{2}, 1)$ and $(\frac{p-1}{2}, \frac{q-1}{2})$ are partitioned into two sets by the line L going through $(0, 0)$ and $(\frac{p}{2}, \frac{q}{2})$. Show that there are no lattice points in R lying on L, and prove the above formula by counting the lattice points in R lying above and below L, respectively.

1.34 (Kronecker's proof) Show that, for odd primes p and q and $1 \le k \le \frac{p-1}{2}$,
$$(-1)^{\lfloor qk/p \rfloor} = \operatorname{sgn} \prod_{i=1}^{\frac{q-1}{2}} \left(\frac{i}{q} - \frac{k}{p}\right).$$
Then use Gauss's Lemma to prove
$$\left(\frac{p}{q}\right) = \operatorname{sgn} \prod_{i=1}^{\frac{q-1}{2}} \prod_{k=1}^{\frac{p-1}{2}} \left(\frac{k}{p} - \frac{i}{q}\right).$$
This implies at once the quadratic reciprocity law.

1.35 (Rousseau's [689] proof of the quadratic reciprocity law) Let p and q denote different odd primes, and consider the group $\Gamma = (\mathbb{Z}/p\mathbb{Z})^\times \times (\mathbb{Z}/q\mathbb{Z})^\times$, its subgroup U of order 2 generated by $(-1, -1)$, and the factor group $G = \Gamma/U$. We will prove the quadratic reciprocity law by computing the product π of all elements in G in two different ways:

i) Choose $\{(i,j) : 1 \leq i \leq p-1, 1 \leq j \leq \frac{q-1}{2}\}$ as a system of representatives and show that $\pi = ((p-1)!^{\frac{q-1}{2}}, (-1)^{\frac{p-1}{2}\frac{q-1}{2}}(q-1)!^{\frac{p-1}{2}})U$.
ii) Choose $\{(k \bmod p, k \bmod q) : 1 \leq k \leq \frac{pq-1}{2}, (k,pq) = 1\}$ as a system of representatives and prove $\pi = \left((p-1)!^{\frac{q-1}{2}}\left(\frac{q}{p}\right), (q-1)!^{\frac{p-1}{2}}\left(\frac{p}{q}\right)\right)U$.
iii) Compare the two expressions for π and deduce that

$$\left(1, (-1)^{\frac{p-1}{2}\frac{q-1}{2}}\right)U = \left(\left(\frac{q}{p}\right), \left(\frac{p}{q}\right)\right)U.$$

iv) Derive the law of quadratic reciprocity.
Rousseau credits H. Schmidt [719] for the original idea. This proof made it into Stillwell's book [Sti].

1.36 The following result is the basis for Zolotarev's [883] proof of the quadratic reciprocity law. For $a \in (\mathbb{Z}/p\mathbb{Z})^\times$, define a permutation σ_a of $(\mathbb{Z}/p\mathbb{Z})^\times$ by mapping $t \pmod{p}$ to $ta \pmod{p}$. Let f denote the order of a in $(\mathbb{Z}/p\mathbb{Z})^\times$.
i)   Show that σ_a is the product of g cycles of length f, where $fg = p-1$;
ii) derive that $\mathrm{sgn}\,\sigma_a = (-1)^{(f-1)g} = (-1)^{(p-1)/f}$;
iii) show that $(p-1)/f$ is even if and only if $(a/p) = +1$;
iv) conclude that $\mathrm{sgn}\,\sigma_a = (a/p)$.
Now show that the permutation

$$\begin{pmatrix} 1 & 2 & \cdots & p-1 & p & p+1 & \cdots & (q-1)p+(p-1) \\ q & 2q & \cdots & (p-1)q & 1 & 1+q & \cdots & (q-1)+(p-1)q \end{pmatrix}$$

has sign $(-1)^{\frac{p-1}{2}\frac{q-1}{2}}$.

1.37 Let $b \equiv B \equiv 3 \bmod 4$ be primes. From a minimal nontrivial solution of $x^2 - bBy^2 = 4$ deduce that there exist $t, u \in \mathbb{N}$ such that $bt^2 - Bu^2 = \pm 1$. Show that the sign coincides with (b/B), and derive Legendre's théorème VII and VIII from this equation.
Check that the corresponding proof for primes $a \equiv 1 \bmod 4$ and $b \equiv 3 \bmod 4$ only works if $a \equiv 5 \bmod 8$ or $b \equiv 3 \bmod 8$, the problem coming from the ramification at 2 in $\mathbb{Q}(\sqrt{ab})$.

1.38 This is Legendre's attempt at filling the gap in his proof: assuming that $a \equiv 1 \bmod 4$ is prime, he had to show that there exists a prime $b \equiv 3 \bmod 4$ such that $(a/b) = -1$:
i)   Show that we may assume that $a \equiv 1 \bmod 8$; (Hint: if $a \equiv 5 \bmod 8$, then $\frac{a+1}{2} \equiv 3 \bmod 4$ is a quadratic nonresidue);
ii) Write $a = 2f^2 - g^2$; prove the claim if f is divisible by some prime $b \equiv 3 \bmod 4$;
iii) Show that it is sufficient to prove that there exist $y, z \in \mathbb{Z}$ such that $2fy^2 + 2gyz + fz^2$ is divisible by some prime $b \equiv 3 \bmod 4$ (Hint: $2fy^2 + 2gyz + fz^2 = (fz+gy)^2 + ay^2)$;
iv) Assume that $2fy^2 + 2gyz + fz^2$ is a sum of two squares for all $y, z \in \mathbb{Z}$; assume that $(y,z) = 1$ and write $2fy^2 + 2gyz + fz^2 = t^2 + u^2$. Show that there exist $A, B, M, N \in \mathbb{N}$ such that $t = Ay + Bz$ and $u = My + Nz$.
v)   From the identity $(Ay+Bz)^2 + (My+Nz)^2 = 2fy^2 + 2gyz + fz^2$, deduce that $A^2 + M^2 = 2f$, $AB + MN = g$, and $B^2 + N^2 = f$.
vi) Show that $a = 2f^2 - g^2 = (AN - MB)^2$; since a is prime, this is a contradiction.

If you have succeeded in proving each step, try to find the error.

1.39 Zignago's [Zig] crucial lemma was the following: "Given integers a and b, put $M(a,b) = \min \{M_y : 1 \leq y \leq a-1\}$, where $M_y = \max \{\gcd(a+by, z) : 1 \leq z \leq a\}$. Then $M_y = \gcd(a, b)$." Find some counterexamples with $\gcd(a, b) = 1$ (you better use a computer).

1.40 Prove the existence of Gauss's auxiliary prime, i.e. show that for every prime $q \equiv 1 \mod 4$ there is a prime $p < q$ such that $\left(\frac{q}{p}\right) = -1$ (you might want to look it up in Brown [89]).

Note that Gauss's assertion is stronger than Legendre's, since Gauss's auxiliary prime p must be $< q$ in order to make the induction proof work; on the other hand, Gauss does not have to demand that $p \equiv 3 \mod 4$.

Additional References

[AMM] L.M. Adleman, K. Manders, G. Miller, *On taking roots in finite fields*, Proc. 18th Annual Symp. Computer Science 1977, pp. 175–178

[AF] A.G. Agargün, C.R. Fletcher, *al-Farisi and the fundamental theorem of arithmetic*, Historia Math. **21** (1994), 162–173

[Ahl] H. Ahlborn, *Über Berechnung von Summen von größten Ganzen auf geometrischem Wege nach der von Eisenstein zuerst angewandten Methode*, Pr. Hamburg (1881), JFM 13.0141.04

[AZ] M. Aigner, G.M. Ziegler, *Proofs from THE BOOK*, Springer Verlag 1998

[Ala] N. S. Aladov, *Sur la distribution des résidus quadratiques et non-quadratiques d'un nombre premier p dans la suite $1, 2, \ldots, p-1$*, Recueil Math. (Moscow Math. Soc.) **18** (1896), 61–75; FdM **27** (1896), 146; FdM **30** (1899), 184; BSMA (2) **21** (1897), 63–64

[An1] A. A. Antropov, *Partitioning of forms into genera and the reciprocity law in papers of L. Euler*, Vopr. Istor. Est. Tekh. (1989), 56–57

[An2] A. A. Antropov, *On Euler's partition of forms into genera*, Historia Math. **22** (1995), 188–193

[Art] E. Artin, *Quadratische Körper im Gebiet der höheren Kongruenzen*, Math. Z. **19** (1924), 153–246

[Au1] A. Aubry, *L'Œuvre arithmétique d'Euler*, L'Ens. Math. **11** (1909), 329–356

[Au2] A. Aubry, *Sur les travaux arithmétiques de Lagrange, de Legendre et de Gauss*, L'Ens. Math. **11** (1909), 430–450

[Ba] E. Bach, *A note on square roots in finite fields*, IEEE Trans. Inf. Theory **36** (1990), 1494–1498

[Bm1] P. Bachmann, *Die analytische Zahlentheorie*, Leipzig 1894

[Bm2] P. Bachmann, *Über Gauß' zahlentheoretische Arbeiten*, Abh. Ges. Wiss. Nachr. Gött., Math.-Phys. Kl. 1911, 455–508; FdM **42** (1911), 201–202

[Bl] R.J. Backlund, *Über die Differenzen zwischen den Zahlen, die zu den n ersten Primzahlen teilerfremd sind*, Comment. in honorem Ernesti Leonardi Lindelöf, Helsinki 1929

[Bas] H. Bass, *Algebraic K-Theory*, New York 1968

[Ben] A.A. Bennett, *Large primes have four consecutive quadratic residues*, Tohôku Math. J. **26** (1925), 53–57

[BSW] K.R. Biermann, E. Schuhmann, H. Wussing, O. Neumann, *Mathematisches Tagebuch 1796–1814 von Carl Friedrich Gauss*, 3rd edition, Ostwalds Klassiker 256, Leipzig 1981

[BM] R.G. Bierstedt, W.H. Mills, *On the bound for a pair of consecutive quartic residues of a prime*, Proc. Am. Math. Soc. **14** (1963), 628–632

[Br] A. Brauer, *Über Sequenzen von Potenzresten*, Sitzungsbericht Preuß. Akad. Wiss. 1928, 9–26

[BZ] A. Brauer, H. Zeitz, *Über eine zahlentheoretische Behauptung von Legendre*, Sitz.ber. Berliner Math. Ges. 1930, 116–124; see also A. Brauer, *Nachruf auf Hermann Zeitz*, ibid. 1933, 2–5

[BR] A. Brauer, R.L. Reynolds, *On a theorem of Aubry-Thue*, Canadian J. Math. **3** (1951), 367–374

[Bri] J. Brillhart, *Note on representing a prime as a sum of two squares*, Math. Comp. **26** (1972), 1011–1013

[BLL] J. Brillhart, D.H. Lehmer, E. Lehmer, *Bound for pairs of consecutive seventh and higher power residues*, Math. Comput. **18** (1964), 397–407

[Buh] W.K. Bühler, *Gauss: A biographical study*, Springer 1981

[Bug] N.V. Bugaieff, *Die Lösung der Congruenzen 2ten Grades in Bezug auf einen Primzahlmodulus*, Mosk. Math. Samml. X (1882); FdM **14** (1882), 126–127

[Bu1] K. Burde, *Verteilungseigenschaften von Potenzresten*, J. Reine Angew. Math. **249** (1971), 133–172

[Bu2] K. Burde, *Sequenzen der Länge 2 von Restklassencharakteren*, J. Reine Angew. Math. **272** (1973), 194–202

[Bu3] K. Burde, *Über allgemeine Sequenzen der Länge 3 von Legendre-Symbolen*, J. Reine Angew. Math. **282** (1975), 203–216

[Bu4] K. Burde, *Eine Verteilungseigenschaft der Legendre-Symbole*, J. Number Theory **12** (1980), 273–277

[Bur] J.J. Burkhardt, *Euler's work on number theory: a concordance for A. Weil's Number Theory*, Historia Math. **13** (1986), 28–35

[Bus] W.H. Bussey, *Fermat's method of infinite descent*, Amer. Math. Monthly **25** (1918), 333–337

[Car] P.A. Caris, *A solution of the quadratic congruence modulo p, $p = 8n + 1$, n odd*, Amer. Math. Monthly **32** (1925), 294–297

[Crl] L. Carlitz, *Quadratic residues and Tchebycheff polynomials*, Portugal. Math. **18** (1959), 193–198

[Crt] P. Cartier, *Sur une generalisation du transfert en theorie des groupes*, Enseign. Math. **16** (1970), 49–57
[Ch1] T.-H. Chang, *Über aufeinanderfolgende Zahlen, von denen jede mindestens einer von n linearen Kongruenzen genügt, deren Moduln die ersten n Primzahlen sind*, Schriften math. Semin. u. Inst. f. angew. Math. Univ. Berlin **4** (1938), Heft 2, 35–55
[Ch2] T.-H. Chang, *Lösung der Kongruenz $x^2 \equiv a \pmod{p}$ nach einem Primzahlmodul $p = 4n + 1$*, Math. Nachr. **22** (1960), 136–142
[Che] P. Chebyshev, *Theorie der Congruenzen* (Russ.), 1849; German Transl. 1889
[CC] P. Chowla, S. Chowla, *Problems on periodic simple continued fractions*, Proc. Natl. Acad. Sci. USA **69** (1972), 37–45
[Cip] M. Cipolla, *Un metodo per la risoluzione della congruenza di secondo grado*, Rend. Accad. Sci. Fis. Mat. Napoli (3) **9** (1903), 153–163
[CM] T. Cochrane, P. Mitchell, *Small solutions of the Legendre Equation*, J. Number Theory **70** (1998), 62–66
[Coh] H. Cohen, *A course in computational algebraic number theory*, GTM 138, Springer-Verlag 1993
[CL] G.E. Collins, R.G.K. Loos, *The Jacobi symbol algorithm*, SIGSAM Bull. **16** (1982), 12–16
[Cor] Ch. Cornaros, *On Grzegorczyk induction*, Ann. Pure Appl. Logic **74** (1995), 1–21
[Cre] J.E. Cremona, *Efficient solution of rational conics*, preprint 1998
[Cuc] R. Cuculière, *Le plus beau théorème de Fermat*, Pour la Science, April 1986
[Dav] H. Davenport, *The Higher Arithmetic. An introduction to the theory of numbers*, 6th ed. Cambridge Univ. Press 1992
[DH] H. Davenport, M. Hall, *On the equation $ax^2 + by^2 + cz^2 = 0$*, Quart. J. Math. **19** (1948), 189–192
[Ded] R. Dedekind, *Abriss einer Theorie der höheren Kongruenzen in Bezug auf einen reellen Primzahl-Modulus*, J. Reine Angew. Math **54** (1857), 1–26; Ges. Math. Werke I, 40–67
[Del] P. Delsarte, *A generalization of the Legendre symbol for finite abelian groups*, Discrete Math. **27** (1979), 187–192
[Di1] L.E. Dickson, *History of the Theory of Numbers*, vol I, Chelsea Publishing, New York 1952
[Di2] L.E. Dickson, *History of the Theory of Numbers*, vol II, Chelsea Publishing, New York 1952
[Dir] L. Dirichlet, *Beweis des Satzes, dass jede unbegrenzte arithmetische Progression, deren erstes Glied und Differenz ganze Zahlen ohne gemeinschaftlichen Faktor sind, unendlich viele Primzahlen enthält*, Abh. Preuss. Akad. Wiss. (1837), 45–81; Werke I (1889), 313–342

[Doe] K. Dörge, *Zur Verteilung der quadratischen Reste*, Jahresb. DMV **38** (1929), 41–49; FdM **55** I (1929), 95–95

[Dup] A. Dupré, *Examen d'un proposition de Legendre relative à la théorie des nombres, ouvrage placé en premier ligne par l'Academie des Sciences dans le concours pour le grand prix de Mathématiques de 1858*, Paris (1859), Mallet-Bachelier

[Ed1] H.M. Edwards, *The background of Kummer's proof of Fermat's last theorem for regular primes*, Arch. Hist. Exact Sci. **14** (1975), 219–236

[Ed2] H.M. Edwards, *Postscript to "The background of Kummer's proof"*, Arch. Hist. Exact Sci. **18** (1977), 381–394

[Ene] G. Eneström, *Über die Geschichte der quadratischen und kubischen Reste im christlichen Mittelalter*, (Anfrage 140), Bibl. Math. (3), **9** (1908), 265–266

[Err] A. Errera, *Sur un théorème de Legendre*, Bull. Sci. Math. (2) **50** (1926), 166–168

[Eul] L. Euler, *Demonstrationes circa residua ex divisione potestatum per numeros primos resultantia*, Novi. Comm. Acad. Sci. Petropol. **18** (1772/74), 85–135; (cf. Opera Omnia vol. 3, pp. 240–281)

[EL] P. Eymard, J.P. Lafon, *Le journal mathématique de Gauss*, Revue d'histoire des sciences **9** (1956), 21–51

[FY] K. Feng, L. Ying, *An elementary proof of quadratic reciprocity law in $F_q[t]$*, J. Sichuan Univ., Nat. Sci. Ed. **26**, Spec. Issue (1989) 36–40

[Fou] W. Fouche, *A reciprocity law for polynomials with Bernoulli coefficients*, Trans. Am. Math. Soc. **288** (1985), 59–67

[Fow] D.H. Fowler, *The mathematics of Plato's Academy. A new reconstruction*, Oxford Univ. Press 1987; revised reprint 1990

[Fri] C. Friesen, *Legendre symbols and continued fractions*, Acta Arith. **59** (1991), 365–379

[FG] J. Froemke, J.M. Grossman, *An algebraic approach to some number-theoretic problems arising from paper-folding regular polygons*, Amer. Math. Monthly **95** (1988), 289–307

[Fu] W. Fuchs, *Zur Lehre von den Kongruenzen bei C. F. Gauss III*, Mitt., Gauss-Ges. Gött. **19** (1982), 63–78

[Ga1] C. F. Gauss, *Algorithmus novus ad decidendum, utrum numerus integer positivus datus numeri primi positivi dati residuum quadraticum sit an non-residuum*, Werke II, p. 59–68

[Ga2] C. F. Gauss, *Anzeige*, Göttingische Gelehrte Anzeigen, 12. Mai 1808, Werke II, p. 151–154

[GPZ] J. Gebel, A. Pethö, H.G. Zimmer, *Computing integral points on elliptic curves*, Acta Arith. **68** (1994), 171–192

[GJ] R.E. Giudici, J. Jatem, *On classes of quadratic residues and some applications*, Sci., Ser. A **3** (1989), 23–32

[Go1] C. Goldstein, *On a seventeenth-century version of the "fundamental theorem of arithmetic"*, Historia Math. **19** (1992), 177–187

[Go2] C. Goldstein, *Un théorème de Fermat et ses lecteurs*, Saint-Denis: Presses Universitaires de Vincennes, 1995

[Gö] Götting, *Über die Verteilung der Reste und Nichtreste einer Primzahl von der Form $4n+3$ innerhalb des Intervalles 1 bis $(p-1)/2$*, J. Reine Angew. Math. **70** (1869), 363–364;

[Gra] J. J. Gray, *A commentary on Gauss's mathematical diary, 1796–1814, with an English translation*, Expositiones Math. **2** (1984), 97–130; Erratum: ibid. **3** (1985), 192

[vGr] L. v. Grosschmid, *Zur Theorie der quadratischen Reste*, J. Reine Angew. Math. **145** (1915), 254–257; FdM **45** (1914/15), 324

[Gro] E. Grosswald, *Representations of integers as sums of squares*, Springer-Verlag 1985

[Gry] A. Grytczuk, *On some connections between Legendre symbols and continued fractions*, Acta Acad. Paedagog. Agriensis, Sect. Mat. (N.S.) **24** (1997), 19–21

[Gu1] M. Guinot, *Les resveries de Fermat*, Aléas, Lyon 1993

[Gu2] M. Guinot, *"Le diable d'homme" d'Euler*, Aléas, Lyon 1994

[Gu3] M. Guinot, *Une époque de transition: Lagrange et Legendre*, Aléas, Lyon 1996

[Gup] H. Gupta, *Chains of quadratic residues*, Math. Comp. **25** (1971), 379–382

[Hea] T.L. Heath, *Diophantus of Alexandria. A study in the history of Greek algebra*, Cambridge Univ. Press 1885; Dover Publications, 1964

[Hel] Y. Hellegouarch, *Loi de réciprocité, critère de primalité dans $\mathbb{F}_q[t]$*, C. R. Math. Acad. Sci., Soc. R. Can. **8** (1986), 291–296

[Hil] D. Hilbert, *Natur und mathematisches Erkennen*, Lecture Notes, Autumn 1919, Göttingen 1989

[Hof] J. Hofmann, *Über die zahlentheoretischen Methoden Fermats und Eulers, ihre Zusammenhänge und ihre Bedeutung*, Archive for History of Exact Sciences **1** (1960/62), 122–159

[Hol] L. Holzer, *Minimal solutions of diophantine equations*, Can. J. Math. **11** (1950), 1–3

[Hop] H. Hopf, *Über die Verteilung quadratischer Reste*, Math. Z. **32** (1930), 222–231

[Joh] K.R. Johnson, *Reciprocity in Ramanujan's sum*, Math. Mag. **59** (1986), 216–222

[Jos] K. Joshi, *Notes on Diophantus*, Current Science **67**, No. 12, December 25, 1994, pp. 957–966

[Kn] W. Knorr, *Problems in the interpretation of Greek number theory: Euclid and the "fundamental theorem of arithmetic"*, Studies in Hist. and Philos. Sci. **7** (1976), 353–368

[Ko1] K. Koutsky, *Eine Bemerkung zum quadratischen Charakter der Zahlen*, (Czech.), Casopis **58** (1929), 42–52; FdM **55-II** (1929), 691

[Ko2] K. Koutsky, *Über den quadratischen Charakter der Zahlen und über die Verallgemeinerung eines Satzes von Lagrange über die Verteilung der quadratischen Reste*, (Czech.) Rozpravy **39** (1930), 21 pp; FdM **56-I** (1930), 158

[Ko3] K. Koutsky, *Über die Verteilung der Potenzreste für einen Primzahlmodul*, Casopis **59** (1930), 65–82 (Czech.); FdM **56-I** (1930), 158

[Ko4] K. Koutsky, *Über die Verteilung der Potenzreste für einen Primzahlmodul. II*, C. R. Congr. Math. Pays Slaves, 119–128; FdM **56-I** (1930), 158

[Kue] H. Kühne, *Eine Wechselbeziehung zwischen Funktionen mehrerer Unbestimmter, die zu Reziprozitätsgesetzen führt*, J. Reine Angew. Math. **124** (1902), 121–133; FdM **32** (1901), 210

[Ku] E. Kummer, *Letters to Kronecker*, Collected Papers vol I, Springer 1975

[La1] J. L. Lagrange, *Problèmes indéterminés du second degré*, Mém. de l'Acad. Royale Berlin **23** (1769); Œuvres II, 377–535

[La2] J. L. Lagrange, *Démonstration d'un théorème d'arithmétique*, Œuvres de Lagrange **3** (1896), 189–201

[La] J.H. Lambert, *Aductata quaedam de numeris eorumque anatomia*, Nova Acta Erudit., Leipzig 1769, 127–128

[Lar] L.C. Larson, *A theorem about primes proved on a chessboard*, Math. Mag. **50** (1977), no. 2, 69–74

[Leb] V. A. Lebesgue, *Extension d'une formule de Gauss*, C. R. Acad. Sci. Paris **59** (1864), 1067–1068

[LeD] D.H. Lehmer, *Computer technology applied to the theory of numbers*, Studies Number Theory (LeVeque, ed.), 117–151 (1969)

[LLM] D.H. Lehmer, E. Lehmer, W.H. Mills, *Pairs of consecutive power residues*, Can. J. Math. **15** (1963), 172–177

[LeE] E. Lehmer, *Patterns of power residues*, J. Number Theory **17** (1983), 37–46

[Len] H.W. Lenstra, *Computing Jacobi symbols in algebraic number fields*, Nieuw Arch. Wiskd. **13** (1995), 421–426

[Lin] R. Linden, *Über die Verteilung quadratischer Reste*, Hausarbeit Göttingen, 1976

[Loe] A. Loewy, *Über die Reduktion algebraischer Gleichungen durch Adjunktion insbesondere reeller Radikale*, Math. Z. **15** (1922),

[Ma] D. Mahnke, *Leibniz auf der Suche nach einer allgemeinen Primzahlgleichung*, Bibl. Math. (3) **13** (1912/13), 26–61

[Mah] M.S. Mahoney, *The mathematical career of Pierre de Fermat, 1601-1665*, 2nd ed. Princeton UP 1994

[Man] Y.I. Manin, *Good proofs are proofs that make us wiser*, DMV-Mitteilungen **2** (1998), 40–44

[Mnt] W. Mantel, *Résidus quadratiques de polynômes*, Nieuw Arch. Wisk. (2) **6** (1905), 374–386; FdM **36** (1905), 270

[McC] E. McClintock, *On the nature and use of the functions employed on the recognition of quadratic residues*, Trans. Amer. Math. Soc. **3** (1908), 92–109; FdM **33** (1902), 204;

[MW] K.D. Merrill, L.H. Walling, *On quadratic reciprocity over function fields*, Pacific J. Math. **173** (1996), 147–150

[MS] S.M. Meyer, J.P. Sorenson, *Efficient algorithms for computing Jacobi symbols*, ANTS II, Proc. Bordeaux 1996, 229–243

[Mil] W.H. Mills, *Bounded consecutive residues and related problems*, Proc. Sympos. Pure Math. **8** (1965), 170–174

[Min] F. Minding, *Anfangsgründe der höheren Arithmetik*, Berlin 1832

[Mnk] H. Minkowski, *Letter to Hilbert*, August 3, 1893; in: *Briefe an David Hilbert, mit Beiträgen und herausgegeben von L. Rüdenberg und H. Zassenhaus*, Springer, 1973

[MSC] D.S. Mitrinovic, J. Sandor, B. Crstici, *Handbook of number theory*, Kluwer 1995

[Mol] R. Mollin, *A survey of Jacobi symbols, ideals, and continued fractions*, Far East J. Math. Sci. **6** (1998), 355–368

[Mon] M.G. Monzingo, *On the distribution of quadratic residues*, The Fibonacci sequence, Collect. Manuscr. (1980), 94–97

[Mo1] L. J. Mordell, *The condition for integer solutions of $ax^2 + by^2 + cz^2 + dt^2 = 0$*, J. Reine Angew. Math. **164** (1931), 40–49

[Mo2] L. J. Mordell, *On the magnitude of the integer solutions of the equation $ax^2 + by^2 + cz^2 = 0$*, J. Number Theory **1** (1968), 1–3

[Mor] Moreau, *Extrait d'une lettre de M. Moreau*, Nouv. Annal. (2) **12** (1873), 323–324

[Nau] P. Naur, *Proof versus formalization*, BIT **34** (1994), 148–164

[Neu] O. Neumann, *Über die Anstöße zu Kummers Schöpfung der "Idealen Complexen Zahlen"*, Math. Persp. K.-R. Biermann 60th Birthday, 179–199 (1981)

[Ore] Ö. Ore, *Contributions to the theory of finite fields*, Trans. Amer. Math. Soc. **36** (1934), 243–274

[Pe] T. Pépin, *Sur certains nombres complexes compris dans la formule $a + b\sqrt{-c}$*, J. Math. Pures Appl. (3) **I** (1875), 317–372

[PJ] B. M. Phong, J. P. Jones, *Quadratic residues and related problems*, Ann. Univ. Sci. Budapest Sect. Comput. **13** (1992), 149–155

[Pil] A. Piltz, *Über die Häufigkeit der Primzahlen in arithmetischen Progressionen und über verwandte Gesetze*, Habilitationsschrift Jena 1884

[Po1] H.C. Pocklington, *The direct solution of the quadratic and cubic binomial congruences with prime moduli*, Proc. Cambridge Phil. Soc. **19** (1917), 57–59

[Po2] H.C. Pocklington, *Quadratic and higher reciprocity of modular polynomials*, Proc. Cambridge Phil. Soc. **40** (1944), 212–214

[vPW] A.J. van der Poorten, P.G. Walsh, *A Note on Jacobi Symbols and Continued Fractions*, Amer. Math. Monthly **106** (1999), 52–56

[Pop] C.P. Popovici, *A relation between the integer part and the Euler indicatrix* (Romanian), An. Univ. București Ser. Şti. Natur. Mat.-Mec. **16** (1967), 49–52

[Ras] R. Rashed, *Les travaux perdus de Diophante. II*, Revue Histoire Sci. **28** (1975), 3–30

[Red] L. Rédei, *Über die Anzahl der Potenzreste mod p im Intervall* $1, \sqrt{p}$, Nieuw Arch. Wiskunde (2) **23** (1950), 150–162

[Rei] K. Reich, *Zur Theorie der quadratischen Reste*, Archiv Math. Phys. (II) **11** (1892), 176–192; FdM **24** (1892), 176–177

[Rie] G.J. Rieger, *Die Zahlentheorie bei C.F. Gauß*, Gauß Gedenkband (ed.: H. Reichardt), Teubner 1957

[Roc] E. Rocci, *Sulla distribuzione dei residui quadratici di un numero primo nella serie naturale*, Giornale di Mat. **65** (1927), 112–134; FdM **53** (1927), 124–125

[Ros] J. Rosenberg, *Algebraic K-theory and its application*, GTM **147**, Springer 1996

[Rus] D. Rusin, *Solution of Legendre's equation*, in preparation

[Rya] W. J. Ryan, *Proof of the quadratic reciprocity law in primitive recursive arithmetic*, Math. Scand. **45** (1979), 177–197

[Sch1] W. Scharlau, *Richard Dedekinds algebraische Arbeiten aus seiner Göttinger Privatdozenten-Zeit 1854–1858*, Abh. Braunschw. Wiss. Ges. **33** (1982), 69–70

[Sch2] W. Scharlau, *Unveröffentlichte algebraische Arbeiten Richard Dedekinds aus seiner Göttinger Zeit 1855–1858*, Arch. Hist. Exact Sci. **27** (1982), 335–367

[SO] W. Scharlau, H. Opolka, *Von Fermat bis Minkowski. Eine Vorlesung über Zahlentheorie und ihre Entwicklung*, Springer 1980; Engl. translation: *From Fermat to Minkowski. Lectures on the theory of numbers and its historical development*, Springer 1985

[Sce] H. Scheffler, *Beleuchtung und Beweis eines Satzes aus Legendre's Zahlentheorie*, Leipzig 1893?

[Scz] A. Schinzel, *On two conjectures of P. Chowla and S. Chowla concerning continued fractions*, Ann. Mat. pura appl. **98** (1974), 111–117

[Scm] F.K. Schmidt, *Allgemeine Körper im Gebiet der höheren Kongruenzen*, Diss. Freiburg 1925

[Sch] I. Schönheim, *Formules pour résoudre la congruence $x^2 \equiv a \pmod{P}$ dans des cas encore inconnus et leur application pour déterminer directement des racines primitives de certains nombres premiers* (Romanian. Russian and French summary), Acad. Republ. Popul. Romine, Fil. Cluj, Studii Cerc. Mat. Fiz. **7** (1956), 51–58

[Scf] R. Schoof, *Elliptic curves over finite fields and the computation of square roots mod p*, Math. Comput. **44** (1985), 483–494

[Scw] S. Schwarz, *A contribution to the arithmetic of finite fields* (Slovak.), Prirodoved. Priloha Techn. Obzoru Sloven. **1** no. 8 (1940), 75–81

[Ses] J. Sesiano, *Books IV to VII of Diophantus' Arithmetica in the Arabic translation attributed to Qusta ibn Luqa*, Springer-Verlag 1982

[Shl] J. Shallit, *On the worst case of three algorithms for computing the Jacobi symbol*, J. Symb. Comp. **10** (1990), 593–610

[SS] J. Shallit, J. Sorenson, *A binary algorithm for the Jacobi symbol*, SIGSAM Bull. **27** (1993), 4–11

[Sha] D. Shanks, *Five number-theoretic algorithms*, Proc. 2nd Manit. Conf. numer. Math., Winnipeg 1972, 51–70 (1973)

[Shp] H. Shapiro, *Introduction to the theory of numbers*, Wiley & Sons 1983

[Sie] C.L. Siegel, *Analytische Zahlentheorie I*, Lecture Notes, Göttingen 1963

[Sin] S. Singh, *Bound for the solutions of a Diophantine equation in prime Galois fields*, Indian J. Pure Appl. Math. **8** (1977), 1428–1430

[Sko] Th. Skolem, *A simple proof of the condition of solvability of the diophantine equation $ax^2 + by^2 + cz^2 = 0$*, Norske Vid. Selsk. Forhdl. **24** (1952), 102–107

[Smi] S.T. Smith, *Quadratic residues and $x^3 + y^3 = z^3$ in models of IE_1 and IE_2*, Notre Dame J. Formal Logic **34** (1993), 420–438

[Spe] G. Speckmann, *Über die Auflösung der Congruenz $x^2 \equiv a \pmod{p}$*, Arch. Math. Phys. **14** (1895), 445–448; ibid. **15** (1896), 335–336

[Sta] B. Stankowitsch, *Zur Theorie der Congruenzen mit einer Veränderlichen in Bezug auf einen Primzahlmodulus*, Mosk. Math. Samml. X (1882); FdM **14** (1882), 124–125

[Ste] S.A. Stepanov, *Arithmetic of Algebraic Curves*, Monographs in Contemporary Mathematics, 1994

[St1] R. V. Sterneck, *Ueber die Verteilung quadratischer Reste und Nichtreste einer Primzahl*, Moscow Math. Samml. **20** (1898), 269–284; FdM **29** (1898), 153–154; RSPM **7** (1898), 142

[St2] R. V. Sterneck, Sitzungsber. Akad. Wiss. Wien 2A **114** (1905), 711–717

[Sti] J. Stillwell, *Numbers and Geometry*, Undergraduate Texts in Mathematics, Springer Verlag 1997

[TF] Y. Tamarkin, A. Friedmann, *Sur les congruences du second degré et les nombres de Bernoulli*, Math. Ann. **62** (1906), 409–412

[Ton] A. Tonelli, *Bemerkung über die Auflösung quadratischer Congruenzen*, Gött. Nachr. 1891, 344–346

[Tor] G. Torelli, *Sulla distribuzione dei resti quadratici di un numero primo*, Rend. Accad. Napoli **15** (1909), 223–230; RSPM **19** (1911), 91

[Vai] R. Vaidyanathaswamy, *The quadratic reciprocity of polynomials modulo p*, J. Indian Math. Soc. **17** (1928), 185–196

[Va1] H.S. Vandiver, *An aspect of the linear congruence with applications to the theory of Fermat's quotient*, Bull. Amer. Math. Soc. **22** (1915), 61–67;

[Va2] H.S. Vandiver, *On the distribution of quadratic and higher residues*, Bull. Amer. Math. Soc. **31**, 346–350; FdM **51** (1925), 129

[Va3] H.S. Vandiver, *Algorithms for the solution of the quadratic congruence*, Amer. Math. Monthly **36** (1929), 83–86; FdM **55** I (1929), 95

[Vui] J. Vuillemin, *La Philosophie de l'algèbre*, Presses Univ. France, 1962

[Wag] S. Wagon, *The Euclidean algorithm strikes again*, Am. Math. Mon. **97** (1990), 125–129

[Web] H. Weber, *Lehrbuch der Algebra 3*, Braunschweig, F. Vieweg und Sohn 1908; see also
http://moa.cit.cornell.edu/MOA/MOA-JOURNALS2/WEBR.html

[We1] A. Weil, *L'Œuvre arithmétique d'Euler*, Gedenkband L. Euler, Birkhäuser 1983

[We2] A. Weil, *Number Theory, An approach through history. From Hammurapi to Legendre*, Birkhäuser. XXI, 375 p. (1984)

[Whm] A. L. Whiteman, *On a theorem of higher reciprocity*, Bull. Amer. Math. Soc. **43** (1937), 567–572

[Wy1] C.T. Whyburn, *The density of power residues and non-residues in subintervals of* $[1, \sqrt{p}]$, Acta Arith **14** (1967/1968), 113–116

[Wy2] C.T. Whyburn, *The distribution of rth powers in a finite field*, J. Reine Angew. Math. **245** (1970), 183–187

[Wil] K.S. Williams, *On the size of a solution of Legendre's equation*, Utilitas Math. **34** (1988), 65–72

[Xu] Z. Xu, *Another proof of Fermat's Last Theorem for cubes*, J. Southwest Teach. Univ., Ser. B (1987), No.1, 20–22; Zbl 743.11017

[Zag] D. Zagier, *A one-sentence proof that every prime* $p \equiv 1 \pmod{4}$ *is a sum of two squares*, Am. Math. Mon. **97** (1990), 144

[Zig] I. Zignago, *Intorno ad un teorema di aritmetica*, Annali di Mat. (2) **21** (1893), 47–55

2. Quadratic Number Fields

The aim of this chapter is to present those proofs of the quadratic reciprocity law which are based on the theory of quadratic number fields. The first proof using such techniques was Gauss's second proof; instead of developing the theory of binary quadratic forms we will give a proof using the ideal theoretic language. After developing the complete genus theory for quadratic number fields, we give some applications to the primality tests of Lucas-Lehmer.

2.1 Quadratic Fields

In the Preface we have remarked that quadratic number fields would play a central role in the theory of quadratic residues, and in this section they will finally enter the picture. We begin by recalling the basic results.

Let $m \in \mathbb{Q} \setminus \mathbb{Q}^2$ be a rational number which is not a square. Then we define the quadratic number field $k = \mathbb{Q}(\sqrt{m})$ by $k = \mathbb{Q}[X]/(X^2 - m)\mathbb{Q}[X]$. We write $\sqrt{m}$ for the residue class $X + (X^2 - m)\mathbb{Q}[X]$ (i.e. for something whose square is m) and identify k with the set

$$k = \mathbb{Q}(\sqrt{m}) = \{x + y\sqrt{m} : x, y, \in \mathbb{Q}\}.$$

Observe that we do not regard $\sqrt{m}$ as an element of $\mathbb{C}$; there are, however, two field isomorphisms (called *embeddings*) $\phi_1, \phi_2 : k \longrightarrow \mathbb{C}$ defined by mapping $X+(X^2-m)\mathbb{Z}[X]$ to the complex numbers $\sqrt{m}$ and $-\sqrt{m}$, respectively. Since a quadratic number field $\mathbb{Q}(\sqrt{m})$ does not change if we replace m by ma^2 for some nonzero $a \in \mathbb{Q}$, we may assume without loss of generality that m is a square-free integer. The quadratic field k is called *imaginary* or *complex* if $m < 0$, and *real* if $m > 0$. The (algebraic) *conjugate* $x - y\sqrt{m}$ of an element $\alpha = x + y\sqrt{m}$ is denoted by α'; $N\alpha = \alpha\alpha'$ and $\text{Tr}\,\alpha = \alpha + \alpha'$ are rational numbers called the *norm* and the *trace* of α.

The set $\mathcal{O}_k$ of algebraic integers contained in k (which is simply the subset of $\alpha \in k$ such that $\text{Tr}\,\alpha, N\alpha \in \mathbb{Z}$) is an integral domain called the *maximal order* of k. It is a free $\mathbb{Z}$-module of rank 2, which is a fancy way of saying that $\mathcal{O}_k = \mathbb{Z} \oplus \omega\mathbb{Z}$ for some $\omega \in \mathcal{O}_k$. The set $\{1, \omega\}$ is called an *integral basis*, and the integer $d = \text{disc}\,k = (\omega - \omega')^2$ the *discriminant* of k. It is an easy exercise to verify the following table:

m	$\mathcal{O}_k$	disc k
$\equiv 2, 3 \bmod 4$	$\mathbb{Z} \oplus \mathbb{Z}\sqrt{m}$	$4m$
$\equiv 1 \bmod 4$	$\mathbb{Z} \oplus \mathbb{Z}\frac{1+\sqrt{m}}{2}$	m

The set $E_k = \mathcal{O}_k^\times$ of invertible elements in $\mathcal{O}_k$ is called the *unit group* of k (by abuse of language: the units of k are $k^\times = k \setminus \{0\}$). For imaginary quadratic fields, the unit group can be determined easily. In fact, if we let ζ_m denote a primitive m-th root of unity, then

$$E_k = \begin{cases} \langle -1 \rangle, & \text{if } m < -4; \\ \langle \zeta_4 \rangle, & \text{if } m = -4; \\ \langle \zeta_6 \rangle, & \text{if } m = -3; \end{cases}$$

If $k = \mathbb{Q}(\sqrt{m})$ is real, however, there are infinitely many units; in fact, if $(t, u) \in \mathbb{N} \times \mathbb{N}$ is the minimal solution of Pell's equation $T^2 - mU^2 = \pm 4$, then $\varepsilon_m = (t + u\sqrt{m})/2$ is a unit in $\mathcal{O}_k$ called the *fundamental unit* of k, since $E_k = \langle -1, \varepsilon_m \rangle$ in this case. If $p \equiv 1 \bmod 4$ is prime and $k = \mathbb{Q}(\sqrt{p})$, then the fundamental unit ε_p is known to have norm -1 (see Exercise 2.2).

Finally we have to discuss the decomposition of primes. Let $p \in \mathbb{N}$ be a rational prime; then p generates the ideal $(p) = p\mathcal{O}_k$ in the Dedekind ring $\mathcal{O}_k$, which – in general – will not be a prime ideal in $\mathcal{O}_k$. In fact, there are three possibilities: if (p) is still prime in $\mathcal{O}_k$, then p is called *inert*; if $(p) = \mathfrak{p}^2$ becomes the square of a prime ideal $\mathfrak{p}$, then p is called *ramified*. Finally, if $(p) = \mathfrak{p}\mathfrak{p}'$ for two different prime ideals $\mathfrak{p} \neq \mathfrak{p}'$, then p is said to *split* in k. The way p behaves in k can be described in terms of Legendre symbols:

Proposition 2.1. *Let $p \in \mathbb{N}$ be an odd prime; then*

$$p \begin{cases} \text{ramifies} & \Longleftrightarrow p \mid d, \\ \text{splits} & \Longleftrightarrow (d/p) = +1, \\ \text{is inert} & \Longleftrightarrow (d/p) = -1. \end{cases}$$

Similarly, $p = 2$
$$\begin{cases} \text{ramifies} & \Longleftrightarrow d \equiv 0 \bmod 4, \\ \text{splits} & \Longleftrightarrow d \equiv 1 \bmod 8, \\ \text{is inert} & \Longleftrightarrow d \equiv 5 \bmod 8. \end{cases}$$

This can be found in almost every book on number theory. The decomposition law can be given a simpler form by using the *Kronecker symbol* $\left(\frac{d}{p}\right)$; this is defined for discriminants d of quadratic number fields $k = \mathbb{Q}(\sqrt{d})$ by

$$\left(\frac{d}{p}\right) = \begin{cases} +1, & \text{if } p \text{ splits in } \mathbb{Q}(\sqrt{d}), \\ 0, & \text{if } p \text{ ramifies in } \mathbb{Q}(\sqrt{d}), \\ -1, & \text{if } p \text{ is inert in } \mathbb{Q}(\sqrt{d}). \end{cases}$$

For odd primes $p \nmid d$, this coincides with the Legendre symbol because of the decomposition law for quadratic number fields.

We will need another simple result which is a little harder to locate, namely the "unique factorization theorem" for discriminants d of quadratic number fields. Such a d is called a *prime discriminant* if d is a prime power (up to sign), that is, if $d \in \{\ldots -19, -11, -8, -7, -4, -3, 5, 8, 13, 17, \ldots\}$.

Proposition 2.2. *Let $d = \operatorname{disc} k$ be the discriminant of a quadratic number field. Then d can be written uniquely (up to order) as a product of prime discriminants.*

Proof. The claim is trivial if d is a prime discriminant. Otherwise, d has at least one odd prime factor p; then $d_1 = p^* = (-1)^{(p-1)/2} p \equiv 1 \bmod 4$ is a prime discriminant. Proceeding by induction on the number of different primes dividing d, it is sufficient to show that $d' = d/d_1$ is a discriminant. This is obvious if $d \equiv 0 \bmod 8$; if d' is odd, then it is clear because $d' \equiv d \equiv 1 \bmod 4$ (d' is square-free because d is). Finally, if $d' \equiv 4 \bmod 8$, then we must have $d'/4 \equiv d/4 \equiv 3 \bmod 4$, and again d' is a discriminant. This proves the existence of the factorization; the uniqueness is just as easy to show. □

Let $\sigma: \sqrt{d} \longrightarrow -\sqrt{d}$ denote the nontrivial automorphism of $k/\mathbb{Q}$; clearly $\alpha^\sigma = \alpha'$. If $\mathfrak{a}$ is an ideal in $\mathcal{O}_k$, then $\mathfrak{a}\mathfrak{a}^\sigma$ is an ideal generated by a rational integer $a \in \mathbb{N}$, and it can be shown that a equals the index $(\mathcal{O}_k : \mathfrak{a})$ of the additive groups $\mathcal{O}_k$ and $\mathfrak{a}$. By a widespread abuse of language we write $a = N\mathfrak{a}$ and call it the *norm* of $\mathfrak{a}$. The norm of a fractional ideal $\mathfrak{a} = \mathfrak{b}/\mathfrak{c}$ is defined by $N\mathfrak{a} = N\mathfrak{b}/N\mathfrak{c}$.

An element $\alpha \in k$ is called *totally positive* (we will write $\alpha \gg 0$) if either k is complex or if k is real and $\alpha > 0$ and $\alpha^\sigma > 0$. More exactly: $\alpha \gg 0$ if and only if $\phi(\alpha) > 0$ for all embeddings $\phi: k \longrightarrow \mathbb{R}$ (there is none if k is complex, and exactly two if k is real; see the beginning of this Section). We call $\alpha \in k$ *totally negative* (and write $\alpha \ll 0$) if $-\alpha \gg 0$. Note that a statement $\alpha \gg 0$ depends on the number field you're working in: clearly $-1 \ll 0$ in $\mathbb{Q}$, but $-1 \gg 0$ in $\mathbb{Q}(i)$. The simple proof of the following lemma is left as an exercise.

Lemma 2.3. *Let k be a quadratic number field and $\alpha \in k$; then $N\alpha > 0$ if and only if $\alpha \gg 0$ or $-\alpha \gg 0$.*

Since $k/\mathbb{Q}$ is a cyclic extension, we have Hilbert's Theorem 90; for its statement we use Hilbert's 'symbolic exponents', that is, we will write $\beta^{1-\sigma}$ for β/β^σ. Equivalently, we can view $1 - \sigma$ as an element of the group ring $\mathbb{Z}[\operatorname{Gal}(k/\mathbb{Q})]$ which we will introduce properly in Chapter 11.

Proposition 2.4. *If $N\alpha = +1$ for some $\alpha \in k = \mathbb{Q}(\sqrt{d})$, then there is a $\beta \in k$ such that $\alpha = \beta^{1-\sigma}$.*

Proof. If $\alpha = -1$ we choose $\beta = \sqrt{d}$; otherwise we put $\beta = 1 + \alpha$ and verify the identity $\alpha = \beta^{1-\sigma}$. □

There is also a 'Satz 90' for ideals:

Proposition 2.5. *Let $\mathfrak{a}$ be a fractional ideal in $\mathcal{O}_k$; if $N\mathfrak{a} = 1$, then there exists an integral ideal $\mathfrak{b}$ such that $\mathfrak{a} = \mathfrak{b}^{1-\sigma}$.*

Proof. The simple proof is left as an Exercise (2.4). □

We will now introduce two different equivalence relations on the semigroup of non-zero integral ideals in $\mathcal{O}_k$: two ideals $\mathfrak{a}$ and $\mathfrak{b}$ are said to be *equivalent* (we write $\mathfrak{a} \sim \mathfrak{b}$) if $\mathfrak{a} = \lambda\mathfrak{b}$ for some $\lambda \in k$. If $\mathfrak{a} = \lambda\mathfrak{b}$ for some $\lambda \gg 0$, then we will say that $\mathfrak{a}$ and $\mathfrak{b}$ are *equivalent in the strict sense* ($\mathfrak{a} \stackrel{+}{\sim} \mathfrak{b}$). The equivalence class of an ideal $\mathfrak{a}$ will be denoted by $[\mathfrak{a}]$ and $[\mathfrak{a}]_+$, respectively. In both cases, the equivalence classes form a finite group called the class group $\mathrm{Cl}\,(k)$ (called class group in the usual or wide sense) and the class group $\mathrm{Cl}^+(k)$ in the strict sense (sometimes also called the narrow class group). Their orders $h(k)$ and $h^+(k)$ differ at most by a factor 2, as can be seen by studying the map $\pi : \mathrm{Cl}^+(k) \longrightarrow \mathrm{Cl}\,(k) : [\mathfrak{a}]_+ \longmapsto [\mathfrak{a}]$. Before we do this, we recall that a sequence

$$A_1 \xrightarrow{\alpha_1} A_2 \xrightarrow{\alpha_2} \ldots \xrightarrow{\alpha_n} A_{n+1}$$

of abelian groups, R-modules, vector spaces etc. is called *exact* if $\ker \alpha_{j+1} = \mathrm{im}\,\alpha_j$ for $j = 1, 2, \ldots, n-1$. We will assume some familiarity with exact sequences in the following.

Proposition 2.6. *Let $\langle \sqrt{d}\,\rangle$ denote the subgroup $\{1, [(\sqrt{d}\,)]_+\}$ of $Cl^+(k)$; then the following sequence of abelian groups is exact:*

$$1 \longrightarrow \langle \sqrt{d}\,\rangle \xrightarrow{\iota} \mathrm{Cl}^+(k) \xrightarrow{\pi} \mathrm{Cl}\,(k) \longrightarrow 1. \qquad (2.1)$$

Proof. It is clear that $\iota : \langle \sqrt{d}\,\rangle \longrightarrow \mathrm{Cl}^+(k)$ is injective, that $\pi : \mathrm{Cl}^+(k) \longrightarrow \mathrm{Cl}\,(k)$ is surjective, and that $\mathrm{im}\,\iota \subseteq \ker \pi$. Assume therefore that $[\mathfrak{a}]_+ \in \ker \pi$. Then $\mathfrak{a} = \alpha\mathcal{O}_k$ for some $\alpha \in k$. If $N\alpha > 0$, then we can choose $\alpha \gg 0$, and we have $[\mathfrak{a}]_+ = 1$. If $N\alpha < 0$, then we can choose $\alpha > 0$, $\alpha^\sigma < 0$, and now $\alpha/\sqrt{d} \gg 0$, i.e. $[\mathfrak{a}]_+ = [(\sqrt{d}\,)]_+$. □

Example. In $k = \mathbb{Q}(\sqrt{3}\,)$, the ideal $(\sqrt{3}\,)$ has order 2 in $\mathrm{Cl}^+(k)$, though of course being principal in the wide sense. The ideal class of $(5 + 2\sqrt{3}\,)$, on the other hand, is trivial in both $\mathrm{Cl}\,(k)$ and $\mathrm{Cl}^+(k)$. In $k = \mathbb{Q}(\sqrt{5}\,)$, the ideal $(\sqrt{5}\,)$ is principal in the narrow sense: a totally positive generator is $\frac{1+\sqrt{5}}{2}\sqrt{5}$. These observations generalize, as the next corollary shows.

Corollary 2.7. *For quadratic number fields k, we have $h^+(k) = 2h(k)$ if $d = \mathrm{disc}\,k > 0$ and the fundamental unit ε has positive norm, and $h^+(k) = h(k)$ otherwise.*

Proof. From Proposition 2.6 we see that $h^+(k) = h(k)$ if and only if the ideal class $[(\sqrt{d})]_+$ is trivial. But $(\sqrt{d})$ is principal in the strict sense if and only if $\eta\sqrt{d} \gg 0$ for some unit $\eta \in \mathcal{O}_k$. For $d < 0$ we can take $\eta = 1$, since $\sqrt{d} \gg 0$ in this case. But if $d > 0$ and $\eta\sqrt{d} \gg 0$, then $N\sqrt{d} < 0$ shows that $N\eta < 0$, and we see that $[(\sqrt{d})]_+ = 1$ if and only if there exists a unit η with negative norm, which happens if and only if $N\varepsilon = -1$. □

2.2 Genus Theory

Below we will develop the genus theory for narrow class groups $\text{Cl}^+(k)$. This will simplify the theory considerably: quite a few results proved below become rather complicated when one replaces $\text{Cl}^+(k)$ with the class group $\text{Cl}(k)$ in the usual sense. An ideal $\mathfrak{a}$ is said to be in the *principal genus* if $N\mathfrak{a} = N(\lambda)$ for some totally positive $\lambda \in k$. More generally, we call ideals $\mathfrak{a}, \mathfrak{b}$ *similar* ($\mathfrak{a} \stackrel{+}{\approx} \mathfrak{b}$) if $N\mathfrak{a} = N(\lambda)N\mathfrak{b}$ for some totally positive $\lambda \in k$. The set of equivalence classes with respect to $\stackrel{+}{\approx}$ forms a finite group called the *genus class group* $\text{Cl}^+_{\text{gen}}(k)$. Our first claim is that the ideal classes in the principal genus are exactly the squares:

Proposition 2.8. *For ideals $\mathfrak{a}, \mathfrak{b}$ in $\mathcal{O}_k$, we have $\mathfrak{a} \stackrel{+}{\approx} \mathfrak{b}$ if and only if $\mathfrak{a} \stackrel{+}{\sim} \mathfrak{b}\mathfrak{c}^2$ for some ideal $\mathfrak{c}$.*

Proof. It is clearly sufficient to show that $\mathfrak{a} \stackrel{+}{\approx} (1)$ if and only if $\mathfrak{a} \stackrel{+}{\sim} \mathfrak{c}^2$. Assume therefore that $N\mathfrak{a} = N(\lambda)$, $\lambda \gg 0$. Then $N(\lambda^{-1}\mathfrak{a}) = (1)$, and Hilbert's Theorem 90 for ideals shows that there is an ideal $\mathfrak{c}$ such that $\lambda^{-1}\mathfrak{a} = \mathfrak{c}^{\sigma-1}$, where σ denotes the non-trivial automorphism of $k/\mathbb{Q}$. Since $\mathfrak{c}^\sigma \stackrel{+}{\sim} \mathfrak{c}^{-1}$, we find that $\mathfrak{a} \stackrel{+}{\sim} \lambda\mathfrak{c}^2$.

Now suppose that $\mathfrak{a} \stackrel{+}{\sim} \mathfrak{c}^2$; then $\mathfrak{a} = \lambda\mathfrak{c}^2$ for some $\lambda \gg 0$, and taking the norm gives $N\mathfrak{a} = N(c\lambda)$, where $c = N\mathfrak{c}$. □

Proposition 2.8 shows that $\text{Cl}^+_{\text{gen}}(k) \simeq C_+/C_+^2$, where $C_+ = \text{Cl}^+(k)$ is the class group in the strict sense. The principal result of genus theory is the determination of the number of genera. Let t denote the number of finite primes ramifying in $k/\mathbb{Q}$; then we will show

- The first inequality: $\# C_+/C_+^2 \leq 2^{t-1}$;
- The second inequality: $\# C_+/C_+^2 \geq 2^{t-1}$.

Gauss used ambiguous binary forms to prove the first, and the theory of ternary quadratic forms to prove the second inequality. He also showed that the quadratic reciprocity law is a corollary of the first inequality.

An ideal $\mathfrak{a}$ in $\mathcal{O}_k$ is called *ambiguous* if $\mathfrak{a}^\sigma = \mathfrak{a}$. Similarly, an ideal class $c \in C_+$ is called *ambiguous* if $c^\sigma = c$. The ambiguous ideal classes of k form a subgroup of C_+ which we will denote by Am^+. From the definition of Am^+ we get the exact sequence

48   2. Quadratic Number Fields

$$1 \longrightarrow \text{Am}^+ \longrightarrow C_+ \xrightarrow{1-\sigma} C_+^{1-\sigma} \longrightarrow 1, \tag{2.2}$$

where the homomorphism $C_+ \longrightarrow C_+^{1-\sigma}$ is defined by $c \longmapsto c^{1-\sigma}$.

Obviously, ideal classes generated by ambiguous ideals are ambiguous; for class groups in the strict sense, the converse is also true:

Proposition 2.9. *The ambiguous ideal classes in $C_+ = \text{Cl}^+(k)$ are exactly those which are generated by ambiguous ideals.*

Proof. Suppose that $c = [\mathfrak{a}]_+$ is ambiguous, i.e. $\mathfrak{a}^\sigma = \lambda\mathfrak{a}$ for some $\lambda \gg 0$. Taking the norm shows that $N\lambda$ must be a unit; since $\lambda \gg 0$ we must have $N\lambda = +1$. Now Hilbert's Theorem 90 shows that $\lambda = \alpha^{1-\sigma}$ for some $\alpha \in k$. From $0 \ll \lambda\alpha^{2\sigma} = N\alpha$ we deduce that $N\alpha > 0$ (if k is complex, the norm is always nonnegative, and if k is real, then an integer in $n \in \mathbb{Z}$ is > 0 if and only if $n \gg 0$ in k). Thus α or $-\alpha$ is totally positive by Lemma 2.3. Replacing α by $-\alpha$ if necessary we find that $\mathfrak{a} \stackrel{+}{\sim} \alpha\mathfrak{a}$. But the ideal $\alpha\mathfrak{a}$ is ambiguous since $(\alpha\mathfrak{a})^\sigma = \alpha^\sigma\mathfrak{a}^\sigma = \alpha^\sigma\lambda\mathfrak{a} = \alpha\mathfrak{a}$. We have shown that c is generated by an ambiguous ideal; since we have already noticed that the other direction is trivial, the proof is complete. □

The corresponding result for class groups in the usual sense is false, as the example $k = \mathbb{Q}(\sqrt{34})$ shows: k has class number 2, but the ambiguous ideals $(2, \sqrt{34}) = (6 + \sqrt{34})$ and $(17, \sqrt{34}) = (17 + 3\sqrt{34})$ are principal. On the other hand, we have $\text{Cl}^+(k) \simeq \mathbb{Z}/4\mathbb{Z}$, and $(17, \sqrt{34}) = (17 + 3\sqrt{34}) \stackrel{+}{\sim} (\sqrt{34})$ is an ambiguous ideal which is not principal in the strict sense (because the fundamental unit $\varepsilon = 35 + 6\sqrt{34}$ has norm $+1$).

We could now prove the first inequality as follows: first observe that σ acts as -1 on the class group, since $\mathbb{Q}$ has class number 1; this implies that $C_+^{1-\sigma} = C_+^2$, hence $(C_+ : C_+^2) = \#\text{Am}^+$ by (2.2). Thus all we need is an upper bound for the number of ambiguous ideal classes. As we have seen in Proposition 2.9, these are generated by ambiguous ideals. Now it is easy to see that an ideal is ambiguous if and only if it is a product of ramified prime ideals and an ideal generated by a rational number (see Exercise 2.7). Let t denote the number of ramified primes. Then there are at most t independent ideal classes generated by ambiguous ideals. The first inequality will follow if we can exhibit a non-trivial relation between these ideal classes.

If k is complex, then $\mathfrak{a} \stackrel{+}{\sim} \mathfrak{b}$ if and only if $\mathfrak{a} \sim \mathfrak{b}$, since any element of k is totally positive. In particular, if $d = -4$, then $(1+i) \stackrel{+}{\sim} 1$ is a nontrivial relation; if $0 > d \neq -4$, we have $(\sqrt{d}) \stackrel{+}{\sim} 1$. If $d > 0$ and $N\varepsilon = -1$, then again $\sqrt{d}\mathcal{O}_k$ is principal (in fact, $\varepsilon\sqrt{d} \gg 0$), and we get our nontrivial relation. Finally, if $N\varepsilon = +1$, then $\varepsilon = \alpha^{1-\sigma}$ by Hilbert's Theorem 90; but then $\alpha\mathcal{O}_k$ is an ambiguous principal ideal which will yield the desired relation (cf. Exercise 2.31).

Here we will choose a different approach and prove the first and second inequality simultaneously. The idea is to look at the group of fractional ambiguous ideals A and the obvious homomorphism $A \longrightarrow \text{Am}^+$ defined by mapping an ambiguous ideal $\mathfrak{a}$ to its class $[\mathfrak{a}]_+$. This homomorphism is surjective by Proposition 2.9, and its kernel is the group

$$H = \{\mathfrak{a} : \mathfrak{a}^\sigma = \mathfrak{a}, \mathfrak{a} = (\alpha) \text{ for some } \alpha \gg 0\}.$$

This group H contains the subgroup I of ideals generated by rational numbers, and applying the Snake Lemma (see Exercise 2.13) to the exact and commutative diagram

$$\begin{array}{ccccccccc}
1 & \longrightarrow & I & \longrightarrow & I & \longrightarrow & & & 1 \\
& & \downarrow & & \downarrow & & \downarrow & & \\
1 & \longrightarrow & H & \longrightarrow & A & \longrightarrow & \text{Am}^+ & \longrightarrow & 1
\end{array}$$

we get the exact sequence of cokernels

$$1 \longrightarrow H/I \longrightarrow A/I \longrightarrow \text{Am}^+ \longrightarrow 1. \qquad (2.3)$$

By Exercise 2.7, $A/I \simeq (\mathbb{Z}/2\mathbb{Z})^t$, where t is the number of ramified prime ideals, so we are left to compute the order of H/I. Now H consists of ambiguous principal ideals in the strong sense; take an element $(\alpha) \in H$ with $\alpha \gg 0$: since (α) is ambiguous, $\varepsilon = \alpha^{1-\sigma}$ must be a totally positive unit. Multiplying α by a totally positive unit η will change ε into $\varepsilon\eta^{1-\sigma}$, hence the map $\rho : H \longrightarrow E/E^{1-\sigma}$ (where E denotes the group of totally positive units in $\mathcal{O}_K$) is a well defined homomorphism, and its kernel is easily seen to be I. Moreover, ρ is surjective: Given $\varepsilon \in E$, Hilbert's Theorem 90 gives $\varepsilon = \alpha^{1-\sigma}$ for some $\alpha \in k^\times$, and since $N\alpha = \varepsilon\alpha^{2\sigma} \gg 0$, we may choose $\alpha \gg 0$ by Lemma 2.3; then $(\alpha) \in H$ and $\rho(\alpha) = \varepsilon E^{1-\sigma}$. Thus $H/I \simeq E/E^{1-\sigma}$, and (2.3) together with $A/I \simeq (\mathbb{Z}/2\mathbb{Z})^t$ implies

Lemma 2.10. *Let k be a quadratic number field, E the group of totally positive units in $\mathcal{O}_k$, and let t denote the number of ramified primes. Then $\#\text{Am}^+ = 2^t/(E : E^{1-\sigma})$.*

The calculation of the index $(E : E^{1-\sigma})$ is easy: if $d = \text{disc } k < 0$, then $E = E_k$ since all nonzero elements are totally positive, hence E is generated by a root of unity ζ. Since $\zeta^{1-\sigma} = \zeta^2$, we get $(E : E^{1-\sigma}) = 2$ in this case. Now assume that $d > 0$ and let ε denote the fundamental unit of k. If $N\varepsilon = +1$, then $E = \langle \varepsilon \rangle$ and $E^{1-\sigma} = \langle \varepsilon^{1-\sigma} \rangle$, and $\varepsilon^{1-\sigma} = \varepsilon^2$ implies that $(E : E^{1-\sigma}) = 2$. If $N\varepsilon = -1$, on the other hand, then $E = \langle \varepsilon^2 \rangle$ and $E^{1-\sigma} = \langle \varepsilon^{2-2\sigma} \rangle = \langle \varepsilon^4 \rangle$, and again we get $(E : E^{1-\sigma}) = 2$. Thus $(E : E^{1-\sigma}) = 2$ in all cases, and in particular $\#\text{Am}^+ = 2^{t-1}$. Since Am^+ is the quotient of an elementary abelian 2-group, we conclude that $\text{Am}^+ \simeq (\mathbb{Z}/2\mathbb{Z})^{t-1}$. Finally, C_+/C_+^2 is also an elementary abelian 2-group, and by (2.2) it has the same order as Am^+; this proves

Theorem 2.11. *Let k be a quadratic number field and let t denote the number of ramified primes. Then $\text{Am}^+ \simeq C_+/C_+^2 \simeq (\mathbb{Z}/2\mathbb{Z})^{t-1}$.*

50   2. Quadratic Number Fields

This concludes the determination of the genus class group $\mathrm{Cl}_{\mathrm{gen}}^+$ in the strict sense. Let us now turn to the relation between $\mathrm{Cl}_{\mathrm{gen}}^+$ and $\mathrm{Cl}_{\mathrm{gen}}$. We call two ideals $\mathfrak{a}, \mathfrak{b}$ *similar* ($\mathfrak{a} \approx \mathfrak{b}$) if $N\mathfrak{a} = N\lambda N\mathfrak{b}$ for some $\lambda \in k^\times$. The group of equivalence classes is called the genus group $\mathrm{Cl}_{\mathrm{gen}}$ in the usual sense. As in Proposition 2.8, it is easy to show that $\mathfrak{a} \approx \mathfrak{b}$ if and only if $\mathfrak{a} \sim \mathfrak{bc}^2$; in particular, we have $\mathrm{Cl}_{\mathrm{gen}} \simeq \mathrm{Cl}(k)/\mathrm{Cl}(k)^2$.

Proposition 2.12. *Let $k = \mathbb{Q}(\sqrt{m})$ be a real quadratic number field and let t denote the number of ramified primes. Then the following assertions are equivalent:*

i) $\mathrm{Cl}_{\mathrm{gen}}^+(k) \simeq \mathrm{Cl}_{\mathrm{gen}}(k)$;

ii) $(\sqrt{m}) \stackrel{+}{\approx} (1)$;

iii) m *is a sum of two squares.*

Moreover, $\mathrm{Cl}_{\mathrm{gen}}(k) \simeq (\mathbb{Z}/2\mathbb{Z})^s$, where $s = t - 1$ if m is a sum of two squares, and $s = t - 2$ otherwise.

Proof. We consider the natural projection $\pi : \mathrm{Cl}_{\mathrm{gen}}^+(k) \longrightarrow \mathrm{Cl}_{\mathrm{gen}}(k)$ that maps the genus of the ideal class $[\mathfrak{a}]_+$ to the genus of $[\mathfrak{a}]$.

Assume first that $\mathrm{Cl}_{\mathrm{gen}}^+(k) \simeq \mathrm{Cl}_{\mathrm{gen}}(k)$; then the fact that $(\sqrt{m}) \approx (1)$ implies that $(\sqrt{m}) \stackrel{+}{\approx} (1)$. Conversely, assume that $(\sqrt{m}) \stackrel{+}{\approx} (1)$ and that the genus of $[\mathfrak{a}]_+$ is in the kernel of π. Then $\mathfrak{a} = \lambda \mathfrak{c}^2$; if $N\lambda > 0$, then $[\mathfrak{a}]_+$ is trivial in $\mathrm{Cl}_{\mathrm{gen}}^+(k)$, and if $N\lambda < 0$, then $[\sqrt{m}\mathfrak{a}]_+$ is. Since $(\sqrt{m}) \stackrel{+}{\approx} (1)$, π is injective, and i) follows. This proves that i) $\iff$ ii).

Now assume that ii) holds. Then $(\sqrt{m}) = \lambda \mathfrak{c}^2$ for some $\lambda = \frac{1}{2}(a + b\sqrt{m}) \gg 0$; taking norms shows that $4m = (a^2 - mb^2)c^2$. Now put $a = mA$ and divide by m; then $mA^2c^2 = 4 + b^2c^2$, and this implies that m is a sum of two squares. Conversely, assume that $m = r^2 + s^2$ and put $\mathfrak{c} = (s, r + \sqrt{m})$. Then $\mathfrak{c}^2 = (r + \sqrt{m})$, hence $\sqrt{m}\mathfrak{c}^2 = (\lambda)$ for $\lambda = m + r\sqrt{m} \gg 0$, and ii) follows.

The last claim is now obvious if m is a sum of two squares, and follows from the observation that $\# \ker \pi \leq 2$ otherwise. □

We will now derive the quadratic reciprocity law from the first inequality of genus theory (of course we have shown both, but we want to stress that for the proof we only need the first). In fact, what we will use is simply the following corollary of the first inequality:

Proposition 2.13. *The quadratic number field k with discriminant d has odd class number h^+ in the strict sense if and only if d is a prime discriminant. It has odd class number in the usual sense if d is a prime discriminant or a product of two negative prime discriminants.*

Proof. If d is prime then $t = 1$, and the first inequality gives $(C_+ : C_+^2) \leq 1$, i.e. $C_+ = C_+^2$. This implies that squaring is an automorphism, hence C_+ must have odd order, and so does $\mathrm{Cl}(k)$. Moreover, if d is the product of two

2.2 Genus Theory   51

negative prime discriminants, then d is not the sum of two squares, and our claim follows from Proposition 2.12.

Conversely, if h^+ is odd, then squaring is an automorphism, and we have $(C_+ : C_+^2) = 1$, that is, $t = 1$. Similarly, if h is odd and $t \neq 1$, then $t = 2$ and $d > 0$ is not a sum of two squares: this implies that d is the product of two negative prime discriminants. The converse is trivial. □

For the proof of the Quadratic Reciprocity Law we will have to consider several cases, and we will start with the first supplementary law:

a) $\left(\frac{-1}{p}\right) = +1 \iff p \equiv 1 \bmod 4$.

If $p \equiv 1 \bmod 4$, then $k = \mathbb{Q}(\sqrt{p})$ has a unit ε with $N\varepsilon = -1$. Writing $\varepsilon = \frac{1}{2}(x + y\sqrt{p})$, we get $x^2 - py^2 = -4$, and this implies $\left(\frac{-1}{p}\right) = +1$. Now assume that $\left(\frac{-1}{p}\right) = +1$; then p splits in the Euclidean field $\mathbb{Q}(\sqrt{-1})$, which implies $p = a^2 + b^2$. Hence, $p \equiv 1 \bmod 4$.

b) If $p \equiv 1 \bmod 4$, then $\left(\frac{p}{q}\right) = +1 \iff \left(\frac{q}{p}\right) = +1$.

First note that $\left(\frac{p}{q}\right) = +1$ implies that q splits in $k = \mathbb{Q}(\sqrt{p})$, i.e. $q\mathcal{O}_k = \mathfrak{q}\mathfrak{q}'$; from Proposition 2.13 we know that $h = h^+$ is odd. Therefore $\mathfrak{q}^h$ is principal, and there exist $x, y \in \mathbb{Z}$ such that $\pm 4q^h = x^2 - py^2$. This yields the congruence $\pm 4q^h \equiv x^2 \bmod p$, and $\left(\frac{-1}{p}\right) = +1$ shows that $\left(\frac{q}{p}\right) = +1$ as claimed.

Now suppose that $\left(\frac{q}{p}\right) = +1$; then $k = \mathbb{Q}(\sqrt{q^*})$ has odd class number, where $q^* = (-1)^{(q-1)/2}q$, and p splits in k. Hence there exist $x, y \in \mathbb{Z}$ such that $\pm 4p^h = x^2 - q^*y^2$, and this implies $\left(\frac{\pm p}{q}\right) = +1$. But since the negative sign can hold only if $q^* \geq 0$, i.e., if $q \equiv 1 \bmod 4$, we get in fact $\left(\frac{p}{q}\right) = +1$.

c) If $p \equiv q \equiv 3 \bmod 4$, then $\left(\frac{p}{q}\right) = +1 \iff \left(\frac{q}{p}\right) = -1$.

Consider $k = \mathbb{Q}(\sqrt{pq})$; by Proposition 2.13, the prime ideal $\mathfrak{p} = (p, \sqrt{pq})$ above p is principal (since the class number h is odd, and $\mathfrak{p}^2 = (p) \sim 1$), say $\mathfrak{p} = \frac{1}{2}(x + y\sqrt{pq})$. Taking the norm gives $\pm 4p = x^2 - pqy^2$, and we must have $x = pz$; dividing through by p we get $\pm 4 = pz^2 - qy^2$. If the positive sign holds, then reducing modulo q and modulo p gives $\left(\frac{p}{q}\right) = 1$ and $\left(\frac{q}{p}\right) = -1$, and if the sign is negative, we get $\left(\frac{p}{q}\right) = -1$ and $\left(\frac{q}{p}\right) = 1$; this proves our claim.

For other proofs of c), see Exercises 1.37 and 2.30.

d) $\left(\frac{2}{p}\right) = +1 \iff p \equiv \pm 1 \bmod 8$.

Put $p^* = (-1)^{(p-1)/2}p$; then $p^* \equiv 1 \bmod 4$, and $k = \mathbb{Q}(\sqrt{p^*})$ has odd class number h. If $p \equiv \pm 1 \bmod 8$, then 2 splits in $k/\mathbb{Q}$, and this implies that $x^2 - p^*y^2 = \pm 4 \cdot 2^h$; we may actually assume that the positive sign holds: if $p \equiv 1 \bmod 4$, the fundamental unit has norm -1, and in case $p \equiv 3 \bmod 4$, we have $x^2 - p^*y^2 > 0$ anyway. Now we get $\left(\frac{2}{p}\right) = +1$.

For the proof of the other direction, assume that $(2/p) = 1$. Then p splits in $\mathbb{Q}(\sqrt{2})$ and we get $\pm p = x^2 - 2y^2 \equiv \pm 1 \bmod 8$, since p is odd.

2.3 Genus Characters

Although the description of C_+/C_+^2 given in the previous section is quite a strong result, it is not as strong as one might wish. Suppose, for example, that we are given an ideal $\mathfrak{a}$; how can we tell whether it lies in the principal genus, i.e. whether the ideal class $[\mathfrak{a}]_+$ is a square or not? The theory of genus characters will allow us to answer such questions.

Another way to put this is to say that we know $C_+/C_+^2 \simeq (\mathbb{Z}/2\mathbb{Z})^{t-1}$ since both groups are elementary abelian 2-groups of the same order: what we would like to have, however, is a canonical isomorphism $X : C_+/C_+^2 \longrightarrow (\mathbb{Z}/2\mathbb{Z})^{t-1}$, because then we could easily determine whether $[\mathfrak{a}]_+$ is a square by computing its image in $(\mathbb{Z}/2\mathbb{Z})^{t-1}$. The genus characters will provide such an isomorphism.

In order to introduce the genus characters, let $d = d_1 \cdots d_t$ be the factorization of d into prime discriminants. To each prime discriminant d_j we will associate a quadratic character $\chi_j : \mathrm{Cl}_{\mathrm{gen}}^+ \longrightarrow \mathbb{Z}/2\mathbb{Z}$; we will start by defining χ_j for prime ideals $\mathfrak{p}$ by putting $d_j' = d/d_j$ and

$$\chi_j(\mathfrak{p}) = \begin{cases} \left(\frac{d_j}{N\mathfrak{p}}\right), & \text{if } (N\mathfrak{p}, d_j) = 1; \\ \left(\frac{d_j'}{N\mathfrak{p}}\right), & \text{if } (N\mathfrak{p}, d_j') = 1. \end{cases}$$

We have to show that the characters χ_j are well defined, i.e. that $\left(\frac{d_j}{N\mathfrak{p}}\right) = \left(\frac{d_j'}{N\mathfrak{p}}\right)$ for all prime ideals $\mathfrak{p} \nmid d$. This is obvious if $N\mathfrak{p}$ is a square; if $N\mathfrak{p} = p$ is prime, however, then p splits, so $1 = (d/p) = (d_j/p)(d_j'/p)$, and the claim follows. Now we extend χ_j multiplicatively to the group of fractional ideals in k and put $\chi_j(\alpha) = \chi_j(\alpha \mathcal{O}_k)$. In order to show that $\chi_j(\mathfrak{a})$ only depends on the class of the ideal in $\mathrm{Cl}_{\mathrm{gen}}^+$ we need the following lemma, which will be proved using the quadratic reciprocity law:

Lemma 2.14. *Let k be a quadratic number field, and suppose that $\lambda \in k$ is totally positive. Then $\chi_j(\lambda) = 1$ for all genus characters χ_j.*

Proof. The proof is a bit technical. Assume first that $(N\lambda, d_j) = 1$; we distinguish several cases:

- $d_j \equiv 1 \bmod 4$: then we write $\lambda = \frac{1}{2}(a + b\sqrt{d})$ and we get $4N\lambda = a^2 - db^2$. Now clearly $d_j \nmid a$, so we find $N\lambda \equiv (a/2)^2 \bmod d_j$, and we conclude that $\chi_j(\lambda) = (d_j/N\lambda) = (N\lambda/d_j) = 1$ by quadratic reciprocity (observe that $N\lambda > 0$ and $d_j \equiv 1 \bmod 4$).

- $d_j = -4$: then $d = 4m$ for some squarefree $m \equiv 3 \bmod 4$. With $\lambda = a + b\sqrt{m}$ we find $N\lambda = a^2 - mb^2 \equiv a^2 + b^2 \bmod 4$. Since the norm was assumed to be odd, this implies that $N\lambda \equiv 1 \bmod 4$. Moreover, $N\lambda > 0$ since λ is totally positive, hence $\chi_j(\lambda) = (-1/N\lambda) = 1$.

- $d_j = \pm 8$: these two cases are done similarly; the details are left as an exercise.

Now take any $\lambda \in O_k$. Then $\lambda \mathcal{O}_k = \mathfrak{p}_1 \cdots \mathfrak{p}_s \mathfrak{b}$, where the $\mathfrak{p}_i$ are (not necessarily distinct) ramified prime ideals and where $\mathfrak{b}$ is an ideal coprime to d. For every $i \leq s$ choose an integral ideal $\mathfrak{a}_i$ in the ideal class $[\mathfrak{p}_i]_+^{-1}$ such that $\mathfrak{a}_i$ is coprime to d; then $\mathfrak{a}_i \mathfrak{p}_i = \alpha \mathcal{O}_k$ for some totally positive $\alpha \in \mathcal{O}_k$. The prime ideal factorization of α contains exactly one ramified prime ideal as a factor, hence α is coprime to either d_j or d_j', and we find $1 = \chi_j(\alpha) = \chi_j(\mathfrak{a}_i\mathfrak{p}_i)$ by what we have proved in the first step. Now $\lambda \mathcal{O}_k = \prod_{i=1}^s \mathfrak{a}_i \mathfrak{p}_i \cdot (\mathfrak{b} \prod_{i=1}^s \mathfrak{a}_i^{-1})$. Since (λ) and the $\mathfrak{a}_i \mathfrak{p}_i$ are principal ideals, so is $\mathfrak{c} = \mathfrak{b} \prod_{i=1}^s \mathfrak{a}_i^{-1}$; moreover, it is coprime to d, hence $\chi_j(\mathfrak{c}) = 1$, and this implies $\chi_j(\lambda) = 1$.

Finally the claim for $\lambda \in k^\times$ follows from the multiplicativity of χ_j. □

Corollary 2.15. *If $\mathfrak{a}$ and $\mathfrak{b}$ are ideals such that $N\mathfrak{a} = N\lambda N\mathfrak{b}$ for some $\lambda \gg 0$, then $\chi_j(\mathfrak{a}) = \chi_j(\mathfrak{b})$. In particular, the character $\chi_j(\mathfrak{a})$ of an ideal $\mathfrak{a}$ only depends on the image of its ideal class in $\mathrm{Cl}^+_{\mathrm{gen}}(k)$.*

Proof. This follows directly from the preceding lemma. □

Proposition 2.16 (Product Formula). *We have $\prod_{j=1}^t \chi_j = \mathbf{1}$.*

Proof. It is sufficient to prove $\prod_{j=1}^t \chi_j(\mathfrak{p}) = 1$ for prime ideals $\mathfrak{p}$. If $\mathfrak{p} \nmid d$ then $\prod_{j=1}^t \chi_j(\mathfrak{p}) = \prod_{j=1}^t (d_j/N\mathfrak{p}) = (d/N\mathfrak{p}) = +1$, because either $N\mathfrak{p} = p^2$ and the assertion is trivial, or $N\mathfrak{p} = p$, and $(d/N\mathfrak{p}) = (d/p) = +1$ follows from the fact that p splits.

Now assume that $\mathfrak{p} \mid d$. Without loss of generality we may take $\mathfrak{p} \mid d_1$. Then we find $\prod_{j=1}^t \chi_j(\mathfrak{p}) = (d_1'/N\mathfrak{p}) \prod_{j=2}^t (d_j/N\mathfrak{p}) = (d_2 \ldots d_t/N\mathfrak{p})^2 = +1$. □

Theorem 2.17. *Let k be a quadratic number field with discriminant d, and let $d = d_1 \cdots d_t$ be its factorization into prime discriminants. The homomorphism $X : \mathrm{Cl}^+(k) \longrightarrow (\mathbb{Z}/2\mathbb{Z})^t : c \longmapsto (\chi_1(c), \ldots, \chi_t(c))$ induces an isomorphism $\mathrm{Cl}^+_{\mathrm{gen}}(k) \longrightarrow (\mathbb{Z}/2\mathbb{Z})^{t-1}$.*

Proof. X is a homomorphism $C_+ \longrightarrow (\mathbb{Z}/2\mathbb{Z})^t$, where $(\mathbb{Z}/2\mathbb{Z})^t$ is written multiplicatively, i.e. its elements are $(\varepsilon_1, \ldots, \varepsilon_t)$ with $\varepsilon_j \in \{-1, +1\}$. The product formula shows that $\mathrm{im}\, X$ is contained in a subgroup of index ≥ 2, or more exactly, that $\mathrm{im}\, X \subseteq \ker \pi$, where $\pi : (\mathbb{Z}/2\mathbb{Z})^t \longrightarrow \mathbb{Z}/2\mathbb{Z} : (\varepsilon_1, \ldots, \varepsilon_t) \longmapsto \prod_{j=1}^t \varepsilon_j$.

Once we have shown that $\ker X = C_+^2$, we can conclude that $\mathrm{im}\, X = \ker \pi$ from the fact that $(C_+ : C_+^2) = 2^{t-1}$. This will show that the sequence

$$1 \longrightarrow C_+^2 \longrightarrow C_+ \xrightarrow{X} (\mathbb{Z}/2\mathbb{Z})^t \xrightarrow{\pi} \mathbb{Z}/2\mathbb{Z} \longrightarrow 1$$

is exact. Now $\ker X \subseteq C_+^2$ is equivalent to the statement that $\chi_1(c) = \ldots = \chi_t(t) = 1$ implies that c is the square of an ideal class (this is often called the *Principal Genus Theorem*, because it describes the ideal classes in the

54   2. Quadratic Number Fields

principal genus). There are several ways to do this; we prefer to use Legendre's Theorem 1.7.

So take an ideal class $c = [\mathfrak{a}]_+$ such that $\chi_1(c) = \ldots = \chi_t(c) = 1$. We want to show that c is a square. By Proposition 2.8 this is equivalent to the existence of a $\lambda \gg 0$ such that $N\mathfrak{a} = N\lambda$. Before we prove this we choose $\mathfrak{a}$ appropriately. First we make sure that $\mathfrak{a}$ and $2d$ are relatively prime (see Exercise 2.8). It is also clear that we may take $\mathfrak{a}$ to be primitive, i.e. not divisible by any rational integer (since dividing $\mathfrak{a}$ by integers does not change its ideal class). Finally, we assume that $\mathfrak{a}$ is square-free ($\mathfrak{a}$ is equivalent to an ideal square if and only if its square-free part is). Now put $a = N\mathfrak{a}$; we have to show that the equation $x^2 - dy^2 = az^2$ has nontrivial solutions (simply put $\lambda = (x + y\sqrt{d})/z$ and observe that $\lambda \gg 0$ if we choose $x, y, z > 0$). Since a is the product of pairwise different odd primes not dividing d and the prime ideal factors of $\mathfrak{a}$ have degree 1, we conclude that $(d/p) = +1$ for all $p \mid a$. On the other hand, we know that $(d_i/a) = \chi_i(\mathfrak{a}) = +1$, and this guarantees that $x^2 \equiv a \bmod d_i$ is solvable. Now the conditions of Legendre's Theorem 1.7 are satisfied, and this completes the proof. □

Theorem 2.17 is a very strong result. For example, Artin's Reciprocity Theorem for the genus class field (Proposition 2.18), Legendre's theorem 1.7 (see Exercise 2.36), or the fact that the Artin symbol for quadratic number fields is onto can all be derived from it.

Proposition 2.18. *Let k be a quadratic number field with discriminant d, and let $d = d_1 \cdot \ldots \cdot d_t$ be the factorization of d into prime discriminants. Then there is a canonical isomorphism $\mathrm{Cl}^+_{\mathrm{gen}}(k) \simeq \mathrm{Gal}(k_{\mathrm{gen}}/k)$, where $k_{\mathrm{gen}} = \mathbb{Q}(\sqrt{d_1}, \ldots, \sqrt{d_t})$. Moreover, a prime ideal $\mathfrak{p}$ generates an ideal class in the principal genus if and only if $\mathfrak{p}$ splits completely in k_{gen}/k.*

Proof. This follows easily from what we have proved. In fact, we just map $c \in \mathrm{Cl}^+_{\mathrm{gen}}$ to the automorphism $\sigma : \sqrt{d_i} \mapsto \chi_i(c)\sqrt{d_i}$ of $k_{\mathrm{gen}}/\mathbb{Q}$. The product formula shows that σ fixes k, hence we get a homomorphism $\mathrm{Cl}^+_{\mathrm{gen}} \longrightarrow \mathrm{Gal}(k_{\mathrm{gen}}/k)$ which is easily seen to be an isomorphism.

Moreover, a prime ideal $\mathfrak{p}$ in $\mathcal{O}_k$ above p splits completely in k_{gen}/k if and only if it splits completely in all quadratic subextensions $k(\sqrt{d_i})/k$. This in turn is true if and only if $\chi_j(\mathfrak{p}) = 1$ for all j, i.e. if $[\mathfrak{p}]_+$ lies in the principal genus. □

Next we take a closer look at the genus class field k_{gen}:

Proposition 2.19. *Let the notation be as in Proposition 2.18. Then the maximal abelian extension k_{gen} of $\mathbb{Q}$ which is unramified above k at all finite primes (k_{gen} is called the genus field of k in the strict sense) is given by $k_{\mathrm{gen}} = \mathbb{Q}(\sqrt{d_1}, \ldots, \sqrt{d_t})$.*

Example 1. First look at $k = \mathbb{Q}(\sqrt{-5})$; k has discriminant $d = \mathrm{disc}\, k = -20$ and class number 2. The inert primes are those $\equiv 11, 13, 17, 19 \bmod$

20, and they generate principal prime ideals in $\mathcal{O}_k$. Now consider primes $\equiv 1, 3, 7, 9 \bmod 20$: here we have $p\mathcal{O}_k = \mathfrak{p}\mathfrak{p}'$. Since $\mathcal{O}_k$ has class number 2, the ideal class generated by $\mathfrak{p}$ is a square in Cl(k) if and only if $\mathfrak{p}$ is principal. Proposition 2.19 says that this is the case if and only if $\left(\frac{-4}{p}\right) = \left(\frac{5}{p}\right) = +1$, i.e., if and only if $p \equiv 1, 9 \bmod 20$. Prime ideals above primes $\equiv 3, 7 \bmod 20$, on the other hand, generate ideal classes of order 2. The fact that $\mathcal{O}_k$ has class number 2 guarantees that the product of two prime ideals $\mathfrak{p}, \mathfrak{q}$ above primes $\equiv 3, 7 \bmod 20$ is always principal. Thus we see that Proposition 2.19 contains Fermat's conjecture (1.3) as a (very) special case.

As a matter of fact, genus theory also implies that the prime ideal of norm 5 is principal, and that the one above 2 is not: from $\left(\frac{-4}{5}\right) = 1$ we conclude that the ideal class of $(5, \sqrt{-5})$ is a square (hence trivial), whereas $\left(\frac{5}{2}\right) = -1$ shows that the ideal class of $(2, 1 + \sqrt{-5})$ is not.

Example 2. Now take $k = \mathbb{Q}(\sqrt{3})$; then $d = 12 = (-3)(-4)$, and k has strict class number $h^+ = 2$. Here the primes $p \equiv \pm 1 \bmod 12$ split. Prime ideals above primes $p \equiv 1 \bmod 12$ are principal in the strict sense, whereas those above primes $p \equiv -1 \bmod 12$ are not. This shows that

$$p \equiv 1 \bmod 12 \implies p = x^2 - 3y^2,$$
$$p \equiv -1 \bmod 12 \implies -p = x^2 - 3y^2.$$

Proof of Proposition 2.19. We first show that $K = \mathbb{Q}(\sqrt{d_1}, \ldots, \sqrt{d_t})$ is the maximal unramified extension of k which is abelian over $\mathbb{Q}$. Clearly $K/\mathbb{Q}$ is abelian. In order to show that K/k is unramified it suffices to prove that $k(\sqrt{d_i})/k$ is unramified. But diff$(k(\sqrt{d_i})/k)$ | diff$(\mathbb{Q}(\sqrt{d_i})/\mathbb{Q}) = d_i$ and diff$(k(\sqrt{d/d_i})/k)$ | d/d_i; since d_i and d/d_i are relatively prime, K/k is unramified, and we have shown that $K \subseteq k_{\text{gen}}$.

Next assume that $F/\mathbb{Q}$ is a cyclic extension of prime power degree ℓ^n which is contained in k_{gen}. Then there is a prime p which is ramified in the unique cyclic extension of $\mathbb{Q}$ contained in F (since $|\text{disc } k| > 1$ by Minkowski's bounds), and Hilbert's theory of ramification in Galois extensions shows that p is completely ramified in $F/\mathbb{Q}$, i.e. p has ramification index $e_p(F/\mathbb{Q}) = \ell^n$. On the other hand, kF/k is unramified, i.e. $e_p(kF/\mathbb{Q}) \mid 2$. This is only possible if $\ell^n \mid 2$. In other words: the genus class field of k is the compositum of quadratic extensions of $\mathbb{Q}$.

Now let $F = \mathbb{Q}(\sqrt{\delta})$ be a quadratic extension with discriminant δ contained in k_{gen}; we have to show that δ is a product of prime discriminants occurring in the factorization of d. This is equivalent to $\delta \mid d$ and $\gcd(\delta, d/\delta) = 1$. Since Fk/k is unramified, we get disc$(kF/\mathbb{Q}) = (\text{disc } k)^2$; on the other hand we have disc$(kF/\mathbb{Q}) = (\text{disc } F)^2 N_{k/\mathbb{Q}}\text{disc}(kF/F)$. This implies that $d^2 = \delta^2 N_{k/\mathbb{Q}}\text{disc}(kF/F)$, i.e. $\delta \mid d$. If there were a prime p dividing δ and $N_{k/\mathbb{Q}}\text{disc}(kF/F)$, then p would ramify in $F/\mathbb{Q}$ and kF/F, which contradicts the fact that F is contained in k_{gen}. □

2.4 The Lucas-Lehmer Test

Let p be an odd prime; then $M_p = 2^p - 1$ is called a *Mersenne number*. The reason why the biggest known prime has almost always been a Mersenne number is that there is a simple test which tells us whether M_p is prime or not:

Proposition 2.20. *Let $S_1 = 4$ and define $S_{n+1} = S_n^2 - 2$ for $n \geq 2$. Then $M_p = 2^p - 1$ is prime if and only if $S_{p-1} \equiv 0 \bmod M_p$.*

Example. Take $p = 5$; then $M_5 = 2^5 - 1 = 31$, and $S_2 = 14$, $S_3 \equiv 8$, $S_4 \equiv 0 \bmod 31$, hence M_5 is prime.

The effectiveness of this test can be gleaned from Table 2.1 that contains the Mersenne primes known at the time of writing. For deriving Lucas' test we need information on the "action of the Frobenius" on elements in $k^\times$:

Proposition 2.21. *Let k be a quadratic number field with discriminant d, and suppose that p is a prime such that $p \nmid d$. Then for every $\alpha \in k$, we have*

$$\alpha^p \equiv \begin{cases} \alpha \bmod p\mathcal{O}_k, & \text{if } \left(\frac{d}{p}\right) = 1, \\ \alpha' \bmod p\mathcal{O}_k, & \text{if } \left(\frac{d}{p}\right) = -1. \end{cases}$$

Here α' denotes the conjugate of α.

Proof. Put $\omega = \sqrt{d}$ if $d \equiv 0 \bmod 4$ and $\omega = \frac{1}{2}(1 + \sqrt{d})$ otherwise. Then $\alpha = a + b\omega$ for some $a, b \in \mathbb{Z}$, and $\alpha^p \equiv a^p + b^p \omega^p \equiv a + b\omega^p$. If $d \equiv 0 \bmod 4$, then p is odd, and $\omega^p = \sqrt{d}^p = d^{(p-1)/2}\sqrt{d} \equiv \left(\frac{d}{p}\right)\sqrt{d} \bmod p$; this proves our claim in this case. If $d \equiv 1 \bmod 4$, the proof is similar if p is odd; assume therefore that $p = 2$. Then $\omega^2 = \frac{d-1}{4} + \omega = \frac{d-1}{4} + \sqrt{d} + \omega'$, and now $\sqrt{d} \equiv 1 \bmod 2$ shows that $\omega^2 \equiv \omega \bmod 2$ if $d \equiv 1 \bmod 8$, and $\omega^2 \equiv \omega' \bmod 2$ if $d \equiv 5 \bmod 8$. □

The appearance of quadratic residue symbols modulo p in the process of raising elements to the p-th power is no coincidence: rather, it lies at the heart of many proofs of reciprocity laws. The term "action of the Frobenius" will be explained in the next Chapter, in connection with more general primality tests.

The proof of Lucas' test is quite simple and follows immediately from the following more general result:

Theorem 2.22. *Let $k = \mathbb{Q}(\sqrt{m})$ be a real quadratic number field, and suppose that ω is a unit in $\mathcal{O}_k$ with $N\omega = \omega\overline{\omega} = +1$. Define a sequence S_n, $n \geq 1$ by $S_n = \omega^{2^{n-1}} + \overline{\omega}^{2^{n-1}}$; then*

i) $S_{n+1} = S_n^2 - 2$ for all $n \geq 1$;
ii) if $q = M_p = 2^p - 1$ is prime and inert in $\mathcal{O}_k$, then $S_{p+1} \equiv 2 \bmod q$; if moreover $S_1 \not\equiv \pm 2 \bmod q$, then there is an $r \leq p - 1$ such that $S_r \equiv 0 \bmod q$, and if $\omega^{(q+1)/2} \equiv -1 \bmod q$ then $r = p - 1$.

Table 2.1. Known Mersenne Primes

no	p	year	discovered by	$S_1 = 4$	$S_1 = 10$
1	2	—	—		
2	3	—	—	+	−
3	5	—	—	+	−
4	7	—	—	−	−
5	13	1456	anonymous	+	+
6	17	1588	Cataldi	−	+
7	19	1588	Cataldi	−	+
8	31	1772	Euler	+	+
9	61	1883	Pervushin	+	+
10	89	1911	Powers	−	+
11	107	1914	Powers	−	+
12	127	1876	Lucas	+	+
13	521	1952	Robinson	−	+
14	607	1952	Robinson	−	−
15	1279	1952	Robinson	−	−
16	2203	1952	Robinson	+	−
17	2281	1952	Robinson	−	+
18	3217	1957	Riesel	−	+
19	4253	1961	Hurwitz	+	+
20	4423	1961	Hurwitz	−	−
21	9689	1963	Gillies	−	+
22	9941	1963	Gillies	+	+
23	11213	1963	Gillies	−	−
24	19937	1971	Tuckerman	+	−
25	21701	1978	Noll & Nickel	−	−
26	23209	1979	Noll	+	−
27	44497	1979	Nelson & Slowinski	−	+
28	86243	1982	Slowinski	+	−
29	110503	1988	Colquitt & Welsh	+	+
30	132049	1983	Slowinski	+	−
31	216091	1985	Slowinski	−	+
32	756839	1992	Slowinski & Gage		
33	859433	1994	Slowinski & Gage		
34	1257787	1996	Slowinski & Gage		
35	1398269	1996	Armengaud & Woltman		
36	2976221	1997	Spence & Woltman		
37	3021377	1998	Clarkson & Woltman		
??	6972593	1999	Hajratwala & Woltman		

iii) *if $S_r \equiv 0 \bmod q$, then every prime divisor $\ell \mid M_p$ satisfies $\ell \equiv \left(\frac{m}{\ell}\right) \bmod 2^{r+1}$. If $2r + 1 \geq p$ then $q = M_p$ is prime.*

Proof. i) $S_n^2 = (\omega^{2^{n-1}} + \overline{\omega}^{2^{n-1}})^2 = S_{n+1} + 2$;
ii) If q is prime and inert in $\mathcal{O}_k$, then we know that $\alpha^{q+1} \equiv N\alpha \bmod q$ from Proposition 2.21; this shows $\omega^{q+1} \equiv 1 \bmod q$, hence $S_{p+1} = \omega^{2^p} + \overline{\omega}^{2^p} = \omega^{q+1} + \overline{\omega}^{q+1} \equiv 2 \bmod q$. If q is prime, then $S_p^2 \equiv S_{p+1} + 2 \bmod q$ implies that $S_p \equiv \pm 2 \bmod q$; if the negative sign holds, then $S_{p-1}^2 \equiv S_p + 2 \bmod q$ shows

that $S_{p-1}^2 \equiv 0 \bmod q$, and we can take $r = p - 1$. If not, we can repeat this process, and we must find some $r \geq 2$ with $S_{r+1} \equiv -2 \bmod q$ because $S_1 \not\equiv \pm 2 \bmod q$. If finally $\omega^{(q+1)/2} \equiv -1 \bmod q$, then $S_p = \omega^{(q+1)/2} + \overline{\omega}^{(q+1)/2} \equiv -2 \bmod q$, and this implies $r = p - 1$.

iii) Suppose that $S_r \equiv 0 \bmod q = M_p$; then $0 \equiv S_r = \omega^{2^{r-1}} + \overline{\omega}^{2^{r-1}} = \overline{\omega}^{2^{r-1}}(\omega^{2^r} + 1) \bmod M_p$ implies that $\omega^{2^r} \equiv -1 \bmod \ell$. Obviously, 2^r is the minimal exponent n such that $\omega^n \equiv -1 \bmod \ell$. Now put $e = \left(\frac{m}{\ell}\right)$; we know from Proposition 2.21 that $\omega^{\ell-e} \equiv 1 \bmod \ell$, and since there exists a power of ω congruent to $-1 \bmod \ell$, we conclude that there is a $j \geq 1$ such that $\omega^{(\ell-e)/2^j} \equiv -1 \bmod \ell$. Now $(\ell - e)/2^j$ must be an odd multiple of 2^r, i.e. we have $\ell - e = u \cdot 2^{r+j}$, where u is an odd integer. This proves our claim (note that $j \geq 1$).

Finally, if $2r + 1 \geq p$ then the result we have just proved shows that every prime divisor of $q = M_p$ is $> \sqrt{q}$: this implies that M_p is prime. □

If $p \geq 3$ is odd, then $q = 2^p - 1 \equiv 7 \bmod 8$ and $q \equiv 1 \bmod 3$; therefore $q \equiv 7 \bmod 24$, and if q is prime then q is inert in $\mathcal{O}_k$, where $k = \mathbb{Q}(\sqrt{3})$. Since $\omega = 2 + \sqrt{3}$ is a unit with positive norm, we get $S_0 = 4$. Moreover, if we put $\mu = 1 + \sqrt{3}$, then $\mu^2 = 2(2 + \sqrt{3})$, and Proposition 2.21 implies that

$$(2 + \sqrt{3})^{(q+1)/2} \equiv \mu^{q+1} 2^{-(q+1)/2} \equiv -\left(\frac{2}{q}\right) \equiv -1 \bmod q.$$

Therefore, Proposition 2.20 is indeed a corollary of Theorem 2.22.

Suppose that $q = M_p$ is prime, and that $S_r \equiv 0 \bmod q$ for some index $r \in \mathbb{N}$; then S_{r-1} is a solution of the congruence $x^2 \equiv 2 \bmod M_p$, and since $x = 2^{(p+1)/2}$ is also a solution, we must have $S_{r-1} \equiv s(p) \cdot 2^{(p+1)/2}$ for some $s(p) \in \{\pm 1\}$. D. Lehmer posed the problem of determining this sign (in the special case $S_1 = 4$ – see Guy's book [Guy, p. 9]). It is of course connected with the residuacity of the unit ω: knowing S_{p-2} modulo $q = M_p$ is essentially equivalent to knowing $\omega^{(q+1)/8} \bmod q$, and this residue can be computed from $\omega^{(q^2-1)/16} \bmod q$; but of course $\omega^{(q^2-1)/16} \equiv (\omega/\mathfrak{q})_{16} \bmod \mathfrak{q}$, where $\mathfrak{q}$ is a prime ideal over q of norm q^2 in the number field $\mathbb{Q}(\zeta_{48})$: note that Mersenne primes do have degree 2 in this field. We will outline a possible solution to this problem in Exercises 9.9 and 9.10, although the actual proof using Artin's reciprocity law will be deferred to Part II.

Table 2.1 gives the currently known Mersenne primes as well as the sign (computed by George Woltman) for both $S_1 = 4$ and $S_1 = 10$ (corresponding to the quadratic number field $\mathbb{Q}(\sqrt{6})$ and its unit $5+2\sqrt{6}$) up to $p = 216,091$. Based on these numerical data, Woltman conjectured that the signs for $S_1 = 4$ and for $S_1 = 10$ are the same if and only if $p \equiv 5, 7 \bmod 8$ with the sole exception of $p = 5$; it is expected that work in progress will verify this conjecture. For $p \leq 127$, the presentations $M_p = a^2 + 3b^2$ are given in Table 2.2; it also includes the last few digits of a and b for some larger p.

If we take $S_1 = 16$ (this corresponds to $m = 7$), then it is easy to see that the primes $q = M_p$ are inert in $\mathcal{O}_k = \mathbb{Z}[\sqrt{7}]$ if and only if $p \equiv 1 \bmod 6$, and

Table 2.2. $2^p - 1 = a^2 + 3b^2$

p	a	b
3	2	1
5	2	3
7	10	3
13	46	45
17	362	3
19	298	381
31	46162	2349
61	1505304098	115329357
89	17376907720394	10279641655875
107	9286834445316902	5033685952807971
127	9328321181472828398	5263826472436016979

Table 2.3 gives some values of $s(p)$, as well as the presentation of M_p as a quadratic form $M_p = a^2 + 7b^2$: these computations seem to suggest that $r \leq p - 3$ for all p (this has been verified up to $p = 44497$ by David M. Einstein). We will show in Exercise 6.32 that this is connected with the observation that all the a's are multiples of 8; for Lenstra's amazingly beautiful proof of this last fact, see Part II or Lenstra & Stevenhagen [516].

Table 2.3. $2^p - 1 = a^2 + 7b^2$

p	a	b	r	s(p)
7	8	3	$p-5$	+
13	48	29	$p-3$	−
19	720	29	$p-3$	+
31	43968	5533	$p-3$	−
61	910810592	459233379	$p-3$	−
127	11435623194218822640	2371488550627875869	$p-3$	−
607	...1293960225982644792	...8930217312597345597	$p-5$	−
1279	...9977221542558664760	...8360229120132082371	$p-5$	−
2203	...2570356917191195528	...7573225381275241667	$p-4$	+
2281	...5059917667099308176	...1315089363216047075	$p-3$	−
3217	...1093403645839124664	...2857863675891212605	$p-5$	−
4423	...5861396100280180920	...8833144922699194051	$p-4$	+

There are also some interesting conjectures about Fermat primes. For example, Lucas shows that $F_n = 2^{2^n} + 1$ is prime, if $S_r \equiv 0 \bmod F_n$ for some $2^{n-1} \leq r \leq 2^n$, where $S_1 = 6$ and $S_{k+1} = S_k^2 - 2$ (see Exercise 2.16). Then he claims (without proof) that in fact Fermat primes F_n divide S_r with $r \leq 2^n - n$ (see the paper of H.C. Williams & Shallit [WS] on the history of factoring), thus making it (slightly) superior to Pépin's test (Exercise 2.15). Actually Pépin's test may be improved in a similar way, but both refinements cannot be proved without explicit reciprocity laws.

2.5 Hilbert Symbols and K_2

Hilbert gave a version of the quadratic reciprocity law that did not involve the Legendre symbol anymore: he used his norm residue symbol instead. The statement of the reciprocity law using Hilbert symbols has the great advantage that it is valid for *all* elements $a, b \in \mathbb{Q}^\times$, whereas the quadratic reciprocity law for the Jacobi symbol, even when extended as far as possible, only makes sense when certain restrictions are applied to a and b. There are several ways to define the quadratic Hilbert symbol; Hasse [341, I, §5] gives a construction based on splitting the inversion factor $(\frac{a}{b})(\frac{a}{b})^{-1}$ of the Jacobi symbol into its "p-adic contributions", which is very instructive but can hardly be called elegant. The definition that we shall present below is used in Borevich & Shafarevich [BS] and has the advantage of being easily transported to completions of arbitrary algebraic number fields.

Let us start by recalling the standard construction of p-adic numbers (see e.g. Cassels [Cas]). For $a \in \mathbb{Q}^\times$ and any prime p, let $v_p(a)$ denote the exponent of p in the prime factorization of a. Then $|a|_p = p^{-v_p(a)}$ defines a valuation on $\mathbb{Q}$; the completion of $\mathbb{Q}$ with respect to $|\cdot|_p$ is denoted by $\mathbb{Q}_p$ and called the field of p-adic numbers. The valuation $|\cdot|_p$ on $\mathbb{Q}$ can be extended uniquely to $\mathbb{Q}_p$, and the set $\mathbb{Z}_p = \{a \in \mathbb{Q}_p : |a|_p \leq 1\}$ is called the ring of p-adic integers. Its unit group $\mathbb{Z}_p^\times$ consists of all $z \in \mathbb{Z}_p$ with $|z|_p = 1$.

One of the most important tools for studying p-adic numbers is Hensel's Lemma: in its most simple form, it states that if the congruence $x^2 \equiv a \bmod p$ is solvable for an odd prime p and an integer a not divisible by p, then so is $x^2 \equiv a \bmod p^s$ for every $s \geq 1$, and this is equivalent to a being a p-adic square. The corresponding result for 2 reads as follows: if a is an odd integer and $x^2 \equiv a \bmod 8$, then $x^2 \equiv a \bmod 2^s$ is solvable for every $s \geq 1$, and this is equivalent to a being a p-adic square.

We will say that a form $ax^2 + by^2 + cz^2$ with $a, b, c \in \mathbb{Q}_p$ represents 0 if there exist $x, y, z \in \mathbb{Q}_p$, not all 0, such that $ax^2 + by^2 + cz^2 = 0$. Now we put

$$\left(\frac{m,n}{p}\right) = \begin{cases} +1 & \text{if } mx^2 + ny^2 - z^2 \text{ represents } 0, \\ -1 & \text{otherwise.} \end{cases}$$

The definition of the Hilbert symbol above also makes sense for the infinite prime: we put $(\frac{m,n}{\infty}) = +1$ if and only if $mx^2 + ny^2 - z^2$ represents 0 in $\mathbb{R}$. In particular, we have $(\frac{m,n}{\infty}) = -1$ if $m < 0$ and $n < 0$, and $(\frac{m,n}{\infty}) = +1$ otherwise.

The properties of the Hilbert symbol that are immediately clear from this definition are the following:

Proposition 2.23. *For all $m, n \in \mathbb{Q}_p^\times$, the Hilbert symbol satisfies the following relations:*

1. $(\frac{m,n}{p}) = (\frac{n,m}{p})$;
2. $(\frac{m,n}{p}) = 1$ *if m or n is a square;*
3. $(\frac{-m,m}{p}) = 1$;
4. $(\frac{1-m,m}{p}) = 1$ *if $m \neq 1$;*

Proof. Property 1. is trivial. If $m = r^2$ is a square, then $mx^2 + ny^2 - z^2$ represents 0: simply take $x = 1$, $y = 0$ and $z = r$. This proves 2. For 3., take $x = y = 1$ and $z = 0$, whereas $x = y = z = 1$ do the job in case 4. □

Observe that $(\frac{\cdot,\cdot}{p})$ is a map $\mathbb{Q}_p^\times / \mathbb{Q}_p^{\times 2} \times \mathbb{Q}_p^\times / \mathbb{Q}_p^{\times 2} \longrightarrow \mathbb{F}_2$, that is: the Hilbert symbol only depends on the square classes of a and b; this allows us to restrict the proofs of the main properties of the Hilbert symbol to $a, b \in \mathbb{Z}_p$ (and even to $a, b \in \mathbb{Z}_p^\times \cup p\mathbb{Z}_p^\times$). One of these main properties is the bilinearity; this will follow from the fact that the subgroup $N_a = \{z \in \mathbb{Q}_p^\times : z = x^2 - ay^2$ for some $x, y \in \mathbb{Z}_p\}$ has index at most 2 in $\mathbb{Q}_p^\times$. The subgroup property is easy to verify: for nonsquares $a \in \mathbb{Q}_p^\times$, the relation $z = x^2 - ay^2$ is equivalent to $z = N(x + y\sqrt{a})$, where N denotes the norm of $\mathbb{Q}_p(\sqrt{a})/\mathbb{Q}_p$, and norms are multiplicative; for squares a, showing the multiplicativity is even more trivial. The computation of $(\mathbb{Q}_p^\times : N_a)$ is done in several stages.

The prime $p = 2$ needs special care; let us start by calculating a few Hilbert symbols:

a	$+1$	-1	$+2$	-2	$+5$	-5	$+10$	-10
$(\frac{a,2}{2})$	$+1$	$+1$	$+1$	$+1$	-1	-1	-1	-1

The values of n come from the fact that $\mathbb{Q}_2^\times / \mathbb{Q}_2^{\times 2}$ is a group of order 8 generated by the classes of -1, 2 and 5. For $a \in \{\pm 1, \pm 2\}$, all we have to do is write down a solution for $ax^2 + 2y^2 = z^2$, which is easy. For $a \in \{\pm 5, \pm 10\}$, we have to show that no such solution exists. Take $a = 5$ and consider $5x^2 + 2y^2 = z^2$. Since the equation is homogeneous, we may assume that $x, y, z \in \mathbb{Z}_2$. Next, if $v_p(x) > 0$, then $v_p(z) > 0$ and $v_p(y) > 0$, and we could divide by 4; therefore we may assume that $v_p(x) = 0$ and $v_p(z) = 0$. But then $x^2 \equiv z^2 \equiv 1 \bmod 8$, and the congruence $5 + 2y^2 \equiv 1 \bmod 8$ has no solution in $\mathbb{Z}_2$. The values $a \in \{-5, \pm 10\}$ are treated similarly.

Since the Hilbert symbol only depends on square classes, we deduce that $\mathbb{Q}_2^\times / N_2$ is a group of order 2 generated by any of the classes $a\mathbb{Q}_2^{\times 2}$ with $a \in \{\pm 5, \pm 10\}$. Similar calculations will reveal that $\mathbb{Q}_2^\times / N_b \simeq \mathbb{Z}/2\mathbb{Z}$ for any nonsquare $b \in \mathbb{Q}_2^\times$ (again, it suffices to consider $b \in \{-1, \pm 2, \pm 5, \pm 10\}$). In particular, we have proved the case $p = 2$ of the following

Lemma 2.24. *We have $N_a = \mathbb{Q}_p^\times$ if and only if $a \in \mathbb{Q}_p^{\times 2}$.*

Proof. We may assume that p is odd. If $a = c^2$ for some $c \in \mathbb{Q}_p^\times$, then every $z \in \mathbb{Q}_p^\times$ can be written as $x^2 - ay^2$: from $z = (x - cy)(x + cy)$ we solve $x - cy = 1$ and $x + cy = z$, giving $x = \frac{z+1}{2}$ and $y = \frac{z-1}{2c}$.

Thus we are left to show that a is a square if $N_a = \mathbb{Q}_p^\times$; assume for contradiction that a is not a square. Let us treat the case $a \in \mathbb{Z}_p^\times$ first. We claim in fact that $(\frac{a, pz}{p}) = -1$ for all $z \in \mathbb{Z}_p^\times$: this implies that in particular that $N_a \neq \mathbb{Q}_p^\times$. For a proof, assume that $pz \in N_a$; by the symmetry of the Hilbert symbol, this is equivalent to $a \in N_{pz}$, hence there exist $x, y \in \mathbb{Q}_p$ such

62   2. Quadratic Number Fields

that $a = x^2 - pzy^2$. But now $0 = v_p(a) = \min\{2v_p(x), 2v_p(y) + 1\}$ implies $v_p(x) = 0$ and $v_p(y) \geq 0$, and reduction modulo p gives $(a/p) = +1$: but since p is odd, this gives a contradiction via Hensel's Lemma.

The case $a \in p\mathbb{Z}_p^\times$ can be reduced to the former: write $a = pz$ with $z \in \mathbb{Z}_p^\times$, pick $b \in \mathbb{Z}_p^\times$ such that $(b/p) = -1$; then, by what we just proved, $\left(\frac{a,b}{p}\right) = \left(\frac{b,pz}{p}\right) = -1$. $\square$

Linearity of the Hilbert symbol in the first argument now follows easily: clearly $\left(\frac{aa',b}{p}\right) = +1$ if $\left(\frac{a,b}{p}\right) = \left(\frac{a',b}{p}\right) = +1$. If $\left(\frac{a,b}{p}\right) = +1$ and $\left(\frac{a',b}{p}\right) = -1$, then $\left(\frac{aa',b}{p}\right) = -1$: for if not, then $\left(\frac{a',b}{p}\right) = \left(\frac{a \cdot aa',b}{p}\right) = +1$. Thus we are left to show that $\left(\frac{aa',b}{p}\right) = +1$ if $\left(\frac{a,b}{p}\right) = \left(\frac{a',b}{p}\right) = -1$. But then b must be a non-square, hence a and a' lie in the same coset of $\mathbb{Q}_p^\times/N_a$. Since this last group has order 2, we must have $aa' = c$ for some $c \in N_b$. But then the claim follows. Since the Hilbert symbol is symmetric, we have proved:

Proposition 2.25. *The Hilbert symbol $\left(\frac{\cdot,\cdot}{p}\right)$ is a symmetric non-degenerate bilinear map $\mathbb{Q}_p^\times/\mathbb{Q}_p^{\times 2} \times \mathbb{Q}_p^\times/\mathbb{Q}_p^{\times 2} \longrightarrow \mathbb{F}_2$.*

Recall that a bilinear map $(\cdot,\cdot): V \times V \longrightarrow k$, where V is a k-vector space, is called non-degenerate if $(a,b) = 0$ for all $b \in V$ implies that $a = 0$. For the Hilbert symbol (where the $\mathbb{F}_2$-vector space $\mathbb{Q}_p^\times/\mathbb{Q}_p^{\times 2}$ is written multiplicatively), this means that $\left(\frac{a,b}{p}\right) = 1$ for all $b \in \mathbb{Q}_p^\times$ implies $a \in \mathbb{Q}_p^{\times 2}$: but this is the content of Lemma 2.24.

For our next result we need to extend the Legendre symbol: for $a \in \mathbb{Z}_p^\times$, we put $\left(\frac{a}{p}\right) = +1$ if $a = x^2$ for some $x \in \mathbb{Z}_p$, and $\left(\frac{a}{p}\right) = -1$ otherwise. For $a \in \mathbb{Z}$, this coincides with the Legendre symbol.

Proposition 2.26. *Let p be an odd prime, and assume that $m, n \in \mathbb{Q}_p^\times$. Put $a = v_p(m)$, $b = v_p(n)$, and write $m^b/n^a = r/s$ for p-adic integers $r, s \in \mathbb{Z}_p$, not both divisible by p. Then*

$$\left(\frac{m,n}{p}\right) = \left(\frac{(-1)^{ab}rs}{p}\right).$$

For odd primes p, we could have used this as a definition of the Hilbert symbol; bilinearity would then have been a rather obvious property.

Proof. First we observe that the relation we want to prove "respects multiplicativity" in the following sense: if $m = m_1 m_2$, and if a_j, r_j and s_j are defined correspondingly, then $r/s = r_1 r_2/s_1 s_2$ and $a = a_1 + a_2$, so if the claimed relation holds for m_1 and m_2, it will also hold for m.

This reduces the proof to the cases where m and n are primes or units. In fact, it is sufficient to verify the following formulas:

1. $\left(\frac{p,p}{p}\right) = \left(\frac{-1}{p}\right)$ for all odd primes p;
2. $\left(\frac{p,n}{p}\right) = \left(\frac{n,p}{p}\right) = \left(\frac{n}{p}\right)$ for all odd primes $p \nmid n$;

3. If $p \nmid mn$, then $\left(\frac{m,n}{p}\right) = 1$

The first claim reduces to the second via $\left(\frac{p,p}{p}\right) = \left(\frac{-p,p}{p}\right)\left(\frac{-1,p}{p}\right) = \left(\frac{-1,p}{p}\right)$. Thus consider the second claim. We have to show that $px^2 + ny^2 - z^2$ has a solution if and only if $(n/p) = +1$. Assume first that there exist $x, y, z \in \mathbb{A}_p^\times$ such that $px^2 + ny^2 = z^2$. Then we may assume that $x, y, z \in \mathbb{Z}_p$ and that $p \nmid y$ (otherwise $p \mid z$ and $p \mid x$, and we could cancel p). Reducing modulo p then shows that $ny^2 \equiv z^2 \bmod p$, hence $\left(\frac{n}{p}\right) = +1$. Conversely, if $\left(\frac{n}{p}\right) = +1$, then $n = z^2$ is a square in $\mathbb{Z}_p$ by Hensel's Lemma, and $x = 0$, $y = 1$ and z satisfy $px^2 + ny^2 = z^2$.

Hensel's Lemma reduces the proof of the last claim to verifying the solution of the congruence $mx^2 + ny^2 \equiv z^2 \bmod p$. But this is easy: the function $m(1 - ny^2)$ takes $\frac{p+1}{2}$ different values modulo p, so at least one of them is a square, say $m(1 - ny^2) \equiv X^2 \bmod p$. Writing $X = mx$ and canceling m does the job. □

Here are the corresponding formulas for the dyadic Hilbert symbol:

Lemma 2.27. *We have*

$$\left(\frac{2,u}{2}\right) = (-1)^{(u^2-1)/8}, \quad \left(\frac{u,v}{2}\right) = (-1)^{(u-1)(v-1)/4}$$

Proof. The first formula follows from $\left(\frac{2,-1}{2}\right) = +1$ and $\left(\frac{2,5}{2}\right) = -1$ since both sides only depend on $u \bmod 8$: for the right hand side, this is obvious, while for the left hand side it is a consequence of the fact that $u \equiv 1 \bmod 8$ implies that u is a square in $\mathbb{Q}_2$.

The same fact shows that we need to prove the second formula only for $u, v \in \{1, 3, 5, 7\}$; the following table gives the values of the dyadic Hilbert symbols $\left(\frac{u,v}{2}\right)$:

	1	3	5	7
1	+1	+1	+1	+1
3	+1	−1	+1	−1
5	+1	+1	+1	+1
7	+1	−1	+1	−1

Using this table, the second formula is easily verified. □

Now we can prove

Theorem 2.28. *The following statements are equivalent:*

i) (Legendre – Gauss) *For odd coprime integers $a, b > 0$, we have*

$$\left(\frac{a}{b}\right) = (-1)^{\frac{a-1}{2}\frac{b-1}{2}} \left(\frac{b}{a}\right).$$

ii) *(Eisenstein) For odd integers $a, a', b, b' > 0$, we have*

$$\left(\frac{a}{b}\right)\left(\frac{b}{a}\right) = \left(\frac{a'}{b'}\right)\left(\frac{b'}{a'}\right)$$

whenever $a \equiv a'$ mod 4, $b \equiv b'$ mod 4 and $(a,b) = (a',b') = 1$.

iii) *(Hilbert's Reciprocity Law) For all $a, b \in \mathbb{Q}^\times$, the product formula*

$$\prod_p \left(\frac{m,n}{p}\right) = 1$$

holds. Here the product is extended over all primes, including $p = \infty$.

iv) *(Euler – Artin) As a function of p, the Legendre symbol $\left(\frac{a}{p}\right)$ for $a > 0$ only depends on the residue class of p mod Δ_a, where $\Delta_a = \operatorname{disc} \mathbb{Q}(\sqrt{a})$.*

For yet more formulations of the quadratic reciprocity law, see Exercises 1.9 and 2.3.

Proof. i) $\Longrightarrow$ ii): This is clear since $\left(\frac{a}{b}\right)\left(\frac{b}{a}\right) = (-1)^{(a-1)(b-1)/4}$, and the right hand side depends only on a mod 4 and b mod 4.

ii) $\Longrightarrow$ i): If $a \equiv 1$ mod 4, then ii) shows that $\left(\frac{a}{b}\right)\left(\frac{b}{a}\right) = \left(\frac{1}{b}\right)\left(\frac{b}{1}\right) = 1$; if $a \equiv b \equiv 3$ mod 4, however, then $\left(\frac{a}{b}\right)\left(\frac{b}{a}\right) = \left(\frac{3}{7}\right)\left(\frac{7}{3}\right) = -1$.

i) $\Longrightarrow$ iii): By the bilinearity of the Hilbert symbol it is sufficient to prove the product formula when a and b are primes or units. Since the formula holds trivially if $a = 1$ or $b = 1$, the only unit we have to consider is -1. Thus we have to prove the following formulas:

1. $\displaystyle\prod_p \left(\frac{-1,-1}{p}\right) = +1,$ 2. $\displaystyle\prod_p \left(\frac{-1,q}{p}\right) = +1,$ 3. $\displaystyle\prod_p \left(\frac{q,q'}{p}\right) = +1,$

where q and q' denote positive primes. The proof of 1. is easy: we have $\left(\frac{-1,-1}{p}\right) = +1$ for odd primes p, and $\left(\frac{-1,-1}{2}\right) = \left(\frac{-1,-1}{\infty}\right) = -1$.

The proof of 2. is hardly more difficult: the only possible nontrivial contributions to the product come from $p = q$ and $p = 2$; if q is odd, then $\left(\frac{-1,q}{q}\right) = \left(\frac{-1}{q}\right)$ and $\left(\frac{-1,q}{2}\right) = (-1)^{(-1-1)(q-1)/4} = (-1)^{(q-1)/2}$, so the product formula holds. If $q = 2$, we have $\left(\frac{-1,2}{2}\right) = +1$.

Finally, contributions to the product in 3. only come from $p \in \{2, q, q'\}$. If $q = q'$ is odd, then $\left(\frac{q,q}{q}\right) = \left(\frac{-q,q}{q}\right)\left(\frac{-1,q}{q}\right) = \left(\frac{-1,q}{q}\right) = \left(\frac{-1}{q}\right)$ and $\left(\frac{q,q}{2}\right) = (-1)^{(q-1)(q-1)/4} = (-1)^{(q-1)/2}$; if $q = q' = 2$, on the other hand, then $\left(\frac{2,2}{2}\right) = +1$, and again the product formula holds. Assume therefore that $q \neq q'$. If both primes are odd, then $\left(\frac{q,q'}{q}\right) = \left(\frac{q'}{q}\right)$, $\left(\frac{q,q'}{q'}\right) = \left(\frac{q}{q'}\right)$, and $\left(\frac{q,q'}{2}\right) = (-1)^{(q-1)(q'-1)/4}$. Thus the product formula follows from the quadratic reciprocity law.

iii) $\Longrightarrow$ i): It is sufficient to prove the reciprocity law for prime values of a and b. Assume therefore that p and q are different positive primes; then the only possibly nontrivial contributions to the product formula for $a = p$ and $b = q$ are $\left(\frac{p,q}{2}\right) = (-1)^{(p-1)(q-1)/4}$, $\left(\frac{p,q}{p}\right) = \left(\frac{p}{q}\right)$ and $\left(\frac{p,q}{q}\right) = \left(\frac{q}{p}\right)$. The product formula then implies that $\left(\frac{p}{q}\right)\left(\frac{q}{p}\right) = (-1)^{(p-1)(q-1)/4}$.

i) $\Longrightarrow$ iv): Assume first that $a \equiv 1 \bmod 4$; then $\left(\frac{a}{b}\right) = \left(\frac{a}{b}\right) = \left(\frac{b+ma}{a}\right) = \left(\frac{a}{b+ma}\right)$. If $a \equiv 3 \bmod 4$, however, then $\left(\frac{a}{b}\right) = \left(\frac{-1}{b}\right)\left(\frac{b}{a}\right) = \left(\frac{-1}{b}\right)\left(\frac{b+4a}{a}\right) = \left(\frac{-1}{b}\right)\left(\frac{-1}{b+4a}\right)\left(\frac{a}{b+4a}\right) = \left(\frac{a}{b+4a}\right)$: in fact, we have $\left(\frac{-1}{b}\right)\left(\frac{-1}{b+4a}\right) = 1$ since $b + 4a \equiv b \bmod 4$.

iv) $\Longrightarrow$ i): First consider the case $a \equiv b \bmod 4$; we may assume that $a > b$ and then put $d = (a - b)/4$. Then $\left(\frac{a}{b}\right) = \left(\frac{a-b}{b}\right) = \left(\frac{d}{b}\right)$ as well as $\left(\frac{b}{a}\right) = \left(\frac{-1}{a}\right)\left(\frac{a-b}{a}\right) = \left(\frac{-1}{a}\right)\left(\frac{d}{a}\right)$. By iv) we know that $\left(\frac{d}{b}\right) = \left(\frac{d}{a}\right)$ since $\Delta_d \mid (a - b)$ and $a \equiv b \bmod (a - b)$, hence $\left(\frac{a}{b}\right)\left(\frac{b}{a}\right) = \left(\frac{-1}{a}\right)$, which is the quadratic reciprocity law if $a \equiv b \bmod 4$. If $a \equiv -b \bmod 4$, we put $d = (a + b)/4$ instead, and the same reasoning as above implies that $\left(\frac{a}{b}\right) = \left(\frac{b}{a}\right)$. $\square$

Steinberg symbols

This subsection does not pretend to be an introduction to algebraic K-theory; the intention is rather to leave the impression that there's something interesting going on here. In fact, K-groups will reappear later in this book every now and then, and the following should help the readers unacquainted with the theory to get a vague glimpse of what $K_2 F$ is all about.

Let R be a commutative ring with 1 and G a multiplicatively written abelian group. A Steinberg symbol is a map $S : R^\times \times R^\times \longrightarrow G$ such that

$$S(aa', b) = S(a, b)S(a', b) \quad \text{for } a, a', b \in R^\times;$$
$$S(a, bb') = S(a, b)S(a, b') \quad \text{for } a, b, b' \in R^\times;$$
$$S(a, 1 - a) = 1 \quad \text{for } a \in R^\times \setminus \{1\}.$$

If $R = F$ is a field, then the K-group $K_2 F$ is defined to be the largest group through which all Steinberg symbols factor, that is: there exists a group $K_2 F$ with elements of the form $\{a, b\}$ such that, given any Steinberg symbol $S : F^\times \times F^\times \longrightarrow G$, there is a unique homomorphism $f : K_2 F \longrightarrow G$ such that $S(a, b) = f(\{a, b\})$. In other words: $K_2 F$ is the free abelian group generated by symbols $\{a, b\}$ with $a, b, \in F^\times$ modulo the subgroup generated by the following relations:

$$\{aa', b\} = \{a, b\}\{a', b\} \quad \text{for } a, a', b \in F^\times;$$
$$\{a, bb'\} = \{a, b\}\{a, b'\} \quad \text{for } a, b, b' \in F^\times;$$
$$\{a, 1 - a\} = 1 \quad \text{for } a \in F^\times \setminus \{1\}.$$

The elements of $K_2 F$ are called symbols. The properties of symbols are reminiscent of those of Hilbert symbols; moreover, the property $\{a, -a\} = 1$ corresponding to $\left(\frac{a, -a}{p}\right) = 1$ always holds: this follows by observing that

66   2. Quadratic Number Fields

$$\{a^{-1}, 1 - a^{-1}\} = \{a, 1-a\}^{-1}\{a, -a\}$$

whenever $a, 1 - a \in F^\times$, and then applying the third axiom.
Similarly we can show that symbols are antisymmetric:

$$\{a, b\} = \{a, b\}\{a, -a\} = \{a, -ab\} = \{a, -ab\}\{ab, -ab\}^{-1}$$
$$= \{b^{-1}, ab\} = \{b^{-1}, ab\}\{b^{-1}, -b^{-1}\} = \{b^{-1}, a\} = \{b, a\}^{-1}.$$

As a warm-up, let us prove

Proposition 2.29. $K_2 F = 0$ for any finite field F.

Proof. We know that $F^\times$ is cyclic, so let γ be a generator. It is sufficient to show that $\{\gamma^m, \gamma^n\} = 1$ for all integers $m, n \in \mathbb{N}$. But $\{\gamma^m, \gamma^n\} = \{\gamma, \gamma\}^{mn}$, and $\{\gamma, \gamma\} = \{\gamma, -\gamma\}\{\gamma, -1\} = \{\gamma, -1\}$. This shows that $K_2 F$ is generated by $\{\gamma, \gamma\} = \{\gamma, -1\}$, and since $\{\gamma, -1\}^2 = 1$, we see that $K_2 F$ is killed by 2. Now consider the equation $\gamma x^2 + \gamma y^2 = 1$; by the proof of Proposition 2.26, it has a solution in F. Hence $1 = \{\gamma x^2, 1 - \gamma x^2\} = \{\gamma x^2, \gamma y^2\} = \{\gamma, \gamma\}\{\gamma, y\}^2\{x, \gamma\}^2\{x, y\}^4 = \{\gamma, \gamma\}$, and therefore $K_2 F = 0$.   □

Our next result is much more interesting: we determine the structure of $K_2\mathbb{Q}$. To this end, we define a symbol $\lambda_p : \mathbb{Q}^\times \times \mathbb{Q}^\times \longrightarrow (\mathbb{Z}/p\mathbb{Z})^\times$ for each prime $p \geq 3$ by

$$\lambda_p : (a, b) \longmapsto (-1)^{v_p(a)v_p(b)} \frac{a^{v_p(b)}}{b^{v_p(a)}}. \tag{2.4}$$

This is a symbol by Exercise 2.43; defining λ_2 by this formula is futile: Exercise 2.44 says it is trivial. For this reason, let us define $\lambda_2 : \mathbb{Q}^\times \times \mathbb{Q}^\times \longrightarrow \mathbb{Z}/2\mathbb{Z}$ by putting $\lambda_2(a, b) := (\frac{a, b}{2})$; a different but equivalent definition is offered in Exercise 2.45.

Now since symbols on $\mathbb{Q}$ factor through $K_2\mathbb{Q}$, the maps λ_p induce homomorphisms $\overline{\lambda}_p : K_2\mathbb{Q} \longrightarrow R_p$, where $R_2 = (\mathbb{Z}/4\mathbb{Z})^\times$ and $R_p = (\mathbb{Z}/p\mathbb{Z})^\times$ for $p \geq 3$.

Theorem 2.30 (Tate). *The homomorphism $\prod_{p\geq 2} \overline{\lambda}_p$ induces an isomorphism $K_2\mathbb{Q} \simeq \prod_{p\geq 2} R_p$.*

Proof. Let Λ_n be the subgroup of $K_2\mathbb{Q}$ generated by symbols $\{a, b\}$ with a and b units or primes $\leq n$. Then $\Lambda_1 \subseteq \Lambda_2 \subseteq \Lambda_3 \subseteq \ldots$, and $K_2\mathbb{Q} = \bigcup_{n\geq 1} \Lambda_n$ (in more fancy terms, $K_2\mathbb{Q}$ is the direct limit of the groups Λ_n). We will show by induction that $\Lambda_p \simeq \prod_{2\leq q\leq p} R_q$. To this end, we first show that $\Lambda_1 = \Lambda_2 \simeq \mathbb{Z}/2\mathbb{Z}$. First, the symbol $\{u, v\}$, where u and v are units in $\mathbb{Z}$, can be nontrivial only when $u = v = -1$ (note that $\{a, 1\}\{a, 1\} = \{a, 1 \cdot 1\}$ implies that $\{a, 1\} = 1$), and $\lambda_2(-1, -1) = -1$ shows that $\{-1, -1\}$ is in fact nontrivial: since its square is trivial, it must have order 2, and we deduce that $\Lambda_1 \simeq \mathbb{Z}/2\mathbb{Z}$. Next, $\Lambda_2 = \Lambda_1$: this follows since $\{2, -2\} = 1$ implies $\{2, 2\} = \{2, -1\}\{2, -2\} = \{2, -1\} = \{2, 1 - 2\} = 1$.

We have to construct a homomorphism $\phi_p : (\mathbb{Z}/p\mathbb{Z})^\times \longrightarrow \Lambda_p/\Lambda_{p-1}$ for all primes $p \geq 3$ and show that this is an isomorphism. Let us put $\phi_p(x + p\mathbb{Z}) = \{x, p\}$ for $1 \leq x < p$; we claim that ϕ_p is a well defined homomorphism. For a proof we have to show that $\{x, p\}\{y, p\} = \{z, p\}$ whenever $0 < x, y, z < p$ and $xy \equiv z \bmod p$. Write $xy = z + kp$; clearly $0 \leq k < p$. If $k = 0$, the relation we want to prove is an axiom, hence true. Thus assume that $k > 0$. Then

$$1 = \left\{\frac{kp}{xy}, 1 - \frac{kp}{xy}\right\} = \left\{\frac{kp}{xy}, \frac{z}{xy}\right\}$$
$$= \{p, z\}\{p, xy\}^{-1}\{k, z\}\{k, xy\}^{-1}\{xy, z\}^{-1}\{xy, xy\}.$$

But the entries of the last four symbols have factorizations into primes $< p$, hence are in Λ_{p-1}, and we get

$$1 \equiv \{p, z\}\{p, xy\}^{-1} \bmod \Lambda_{p-1}$$

as desired.

Next, Λ_p/Λ_{p-1} is generated by the symbols $\{a, p\} = \phi_p(a)$ for $p \nmid a$ since $\{p, p\} = \{-1, p\}$, hence ϕ_p is surjective. Moreover, formula (2.4) implies that $\lambda_p(u, v) = 1$ for $1 \leq u, v < p$, hence λ_p is trivial on Λ_{p-1}. Thus the proof will be complete if we can show that the image of ϕ_p contains an element of order $p - 1$. But this is easy: pick a primitive root g modulo p and observe that $\lambda_p(g, p) = g + p\mathbb{Z}$ has order $p - 1$. Since λ_p factors through $K_2\mathbb{Q}$, we deduce that $\{g, p\}$ has order $p - 1$ in $K_2\mathbb{Q}$. $\square$

The content of Tate's theorem on the structure of $K_2\mathbb{Q}$ can be formulated without any reference on $K_2\mathbb{Q}$ at all:

Corollary 2.31. *For any Steinberg symbol c defined on $\mathbb{Q}$ with values in some abelian group A there exist unique homomorphisms $\psi_p : R_p \longrightarrow A$ such that $c(a, b) = \prod_{p \geq 2} \psi_p \circ \lambda_p(a, b)$.*

The last corollary predicts the existence of homomorphisms $\psi_p : R_p \longrightarrow \mathbb{Z}/2\mathbb{Z}$ such that $\left(\frac{a, b}{\infty}\right) = \prod_{p \geq 2} \psi_p \circ \lambda_p(a, b)$. Now the only homomorphisms $(\mathbb{Z}/p\mathbb{Z})^\times \longrightarrow \mathbb{Z}/2\mathbb{Z}$ for primes $p \geq 3$ are the trivial map and the quadratic residue character defined by $a \longmapsto a^{(p-1)/2} \bmod p$; but Exercise 2.46 tells us that $\lambda_p(a, b)^{(p-1)/2} \equiv \left(\frac{a, b}{p}\right) \bmod p$, hence we must have $\psi_p \circ \lambda_p(a, b) = \left(\frac{a, b}{p}\right)^{a_p}$ for integers a_p taking the values 0 or 1. This shows that

$$\left(\frac{a, b}{\infty}\right) = \prod_{p \geq 2} \left(\frac{a, b}{p}\right)^{a_p}. \tag{2.5}$$

What this means is that the quadratic reciprocity law in Hilbert's formulation will follow from (2.5) if we can show that $a_p = 1$ for all primes $p \geq 2$. Let

68   2. Quadratic Number Fields

us start with $p = 2$: putting $a = b = -1$, we find that $(\frac{-1,-1}{p}) = 1$ for all $p \geq 3$, thus $-1 = (\frac{-1,-1}{\infty}) = (\frac{-1,-1}{2})^{a_2} = (-1)^{a_2}$ implies that $a_2 = 1$. Next put $a = b = p$: if $p \equiv 3 \bmod 4$, then $1 = (\frac{p,p}{\infty}) = (\frac{p,p}{2})(\frac{p,p}{p})^{a_p}$, hence $a_p = 1$ whenever $p \equiv 3 \bmod 4$. Now consider the case $p \equiv 3 \bmod 8$: here $1 = (\frac{2,p}{\infty}) = (\frac{2,p}{2})(\frac{2,p}{p})^{a_p} = -(\frac{2}{p})^{a_p} = (-1)^{a_p+1}$, hence $a_p = 1$ again.

Thus we are left to show that $a_p = 1$ for primes $p \equiv 1 \bmod 8$. This is done by induction: plugging in $a = 3$ and $b = 17$, we see that $a_{17} = 1$ since $1 = (\frac{3,17}{\infty}) = (\frac{3,17}{2})(\frac{3,17}{3})(\frac{3,17}{17})^{a_{17}}$, $(\frac{3,17}{2}) = 1$ and $(\frac{3,17}{3}) = -1$. Now assume that $a_q = 1$ for all odd primes $q < p \equiv 1 \bmod 8$. Then $1 = (\frac{q,p}{\infty}) = (\frac{q,p}{2})(\frac{q,p}{q})(\frac{q,p}{p})^{a_p} = (\frac{p}{q})(\frac{q}{p})^{a_p}$. Thus $a_p = 1$ if we can show that there is an odd prime $q < p$ with $(\frac{p}{q}) = -1$. But this is the main lemma in Gauss's first proof of the quadratic reciprocity law:

Lemma 2.32. *If $p \equiv 1 \bmod 8$ is prime, then there exists an odd prime $q < 2\sqrt{p} + 1$ such that $(\frac{p}{q}) = -1$.*

Proof. Let $p \equiv 1 \bmod 8$ be a prime, put $m = \lfloor \sqrt{p} \rfloor$, and assume that $(\frac{p}{q}) = +1$ for all primes $q \leq 2m + 1$. Then the congruences $x^2 \equiv p \bmod q$ and $x^2 \equiv 1 \bmod 8$ are solvable, and Hensel's Lemma implies that we can solve $x^2 \equiv p \bmod q^s$ for every prime power q^s with $s \geq 1$ and $q \leq 2m + 1$. By the Chinese Remainder Theorem, there exists an integer x such that $x^2 \equiv p \bmod (2m + 1)!$. Since $\binom{x+m}{2m+1} = \frac{(x+m)\cdots(x-m)}{1\cdots(2m+1)}$ is an integer, we see that $(2m+1)! \mid (x+m)\cdots(x-m)$. Hence

$$0 \equiv x(x+m)\cdots(x+1)(x-1)\cdots(x-m)$$
$$\equiv x(x^2 - 1^2)\cdots(x^2 - m^2)$$
$$\equiv x(p - 1^2)\cdots(p - m^2) \bmod (2m+1)!,$$

and since $(x, (2m+1)!) = 1$, this implies that

$$(2m+1)! \mid (p - 1^2)\cdots(p - m^2).$$

On the other hand,

$$(2m+1)! = (m+1+m)\cdots(m+1)\cdots(m+1-m)$$
$$= [(m+1)^2 - m^2]\cdots[(m+1)^2 - 1^2](m+1)$$
$$> (p - m^2)\cdots(p - 1^2)\cdots 1,$$

contradicting the relation $(2m+1)! \mid (p - 1^2)\cdots(p - m^2)$.   □

This completes our proof of the product formula for the Hilbert symbol, and we already know that this is equivalent to the quadratic reciprocity law.

The formula (2.4) immediately generalizes to arbitrary number fields F: for prime ideals $\mathfrak{p}$ in $\mathcal{O}_F$, let $v_\mathfrak{p}(\alpha)$ denote the exponent of $\mathfrak{p}$ in the prime ideal factorization of (α). Then

$$\lambda_{\mathfrak{p}} : (\alpha, \beta) \longmapsto (-1)^{v_{\mathfrak{p}}(\alpha) v_{\mathfrak{p}}(\beta)} \frac{\alpha^{v_{\mathfrak{p}}(\beta)}}{\beta^{v_{\mathfrak{p}}(\alpha)}}$$

defines a symbol $\lambda_{\mathfrak{p}} : F^\times \times F^\times \longrightarrow (\mathcal{O}_F/\mathfrak{p})^\times$, hence induces a homomorphism $\overline{\lambda}_{\mathfrak{p}} : K_2 F \longrightarrow (\mathcal{O}_F/\mathfrak{p})^\times$. The intersection of $\ker \overline{\lambda}_{\mathfrak{p}}$ over all prime ideals $\mathfrak{p}$ in $\mathcal{O}_F$ is called the tame kernel and will be denoted by $K_2 \mathcal{O}_F$. The groups $K_2 \mathcal{O}_F$ can be shown to be finite subgroups of $K_2 F$; for example, $K_2 \mathbb{Z} \simeq \mathbb{Z}/2\mathbb{Z}$ since the kernel of $K_2 F \longrightarrow (\mathbb{Z}/p\mathbb{Z})^\times$ is, by the structure theorem of Tate, isomorphic to $\mathbb{Z}/2\mathbb{Z} \times \prod_{q \neq p} (\mathbb{Z}/q\mathbb{Z})^\times$. The orders of the finite groups $K_n \mathbb{Z}$ for $n \geq 2$ contain interesting arithmetic information: there are connections with Bernoulli numbers and values of zeta functions at negative integers etc.

These results, however, involve K-groups $K_n R$ for arbitrary integers $n \geq 0$, and their definition for $n \geq 3$ requires some effort. For the rings of interest to us, only K_0 and K_1 admit descriptions in more or less known terms: if R is a Dedekind ring, then $K_0 R \simeq \mathbb{Z} \oplus \mathrm{Cl}(R)$. If R is commutative, then $K_1 R \simeq R^\times \oplus SK_1 R$, where $SK_1 R$ is a group that is trivial if R is a Euclidean ring or the ring of integers in a number field. In particular, $K_0 \mathbb{Z} \simeq \mathbb{Z}$ and $K_1 \mathbb{Z} \simeq \mathbb{Z}/2\mathbb{Z}$.

NOTES

Proposition 2.2 on the 'unique factorization for prime discriminants' is probably due to Weber, who gave a proof in his Algebra 3 [Web, p. 322]:

> Jede Stammdiskriminante läßt sich auf eine und nur auf eine Weise in Primdiskriminanten zerlegen.[1]

The proof of the quadratic reciprocity law given in Section 2.2 originated in Gauss's genus theory, and was published in the presented form by Dedekind [174, §154]. Two other proofs along these lines have been given by Kummer [469]. The special case of primes $p \equiv q \equiv 3 \bmod 4$ was given by Arndt [16].

Some parts of Section 2.2 were taken from Zagier's book [Zag], where the principal genus theorem is proved using analytic techniques due to Dirichlet. Though it is well known that Legendre's Theorem follows from the Principal Genus Theorem (see e.g. Cohn [Coh, Theorem 14.57]), the only reference for the converse I could find was the paper of Arndt [Arn]; his proof was presented in a slightly modified way by Dedekind [174, §158], and was rediscovered more recently by Nemenzo & Wada [NW]. Gauss's original proof of the Principal Genus Theorem used ternary quadratic forms, other proofs are due to Kronecker [Kro], Mertens [Me1, Me2] (via binary quadratic forms), de la Vallée Poussin [808], and Speiser [Spe]; Zagier [Zag] uses analytic methods such as Dirichlet's L-functions. For a thorough discussion of ambiguous ideals in orders of quadratic number fields, see Mollin [Mol].

---

[1] Every quadratic field discriminant can be written uniquely as a product of prime discriminants.

The simple derivation of the number of ambiguous ideal classes in quadratic number fields that we have presented in Section 2.2 barely hides its cohomological nature. Compare the treatment in Ono [On].

The genus theory of quadratic number fields that we have presented in Propositions 2.18 and 2.19 can be generalized to cyclic extensions of $\mathbb{Q}$; from the immense literature on the subject we mention Inaba [Ina], Leopoldt [Leo], Frei [Fr] and the monographs of Fröhlich [Fro] and Ishida [Ish]. All this will turn out to be a special case of class field theory (see Part II). There we will see that every abelian extension of a given number field k can be described in terms of certain class groups of k. In the case of genus theory of quadratic number fields k the class group is the genus class group $\mathrm{Cl}_{\mathrm{gen}}(k)$, and the corresponding abelian extension is the genus class field.

The questions raised in the section on the Lucas-Lehmer test are intended to serve as a motivation for some of the work done on the power residue character of quadratic units; moreover, there are lots of unsolved (albeit difficult) open problems in this area, some of which will be presented in the chapters below. A not too difficult problem stemming from this area is discussed by Lenstra & Stevenhagen [516]; Woltman's conjecture is the topic of a recent thesis by Lenstra's student Gebre-Egziabher [GE]. For a complete list of known Mersenne primes, see Table 2.1 and check out http://www.utm.edu/research/primes/mersenne.shtml.

The connections between prime factors of Fermat, Mersenne, or Fibonacci numbers (to name a few) and m-th power reciprocity (especially for $m = 2, 3, 4, 8$) have been examined by Aigner [7], Bickmore [52, 53], Brillhart [Bri], Euler [211], Federighi & Roll [228], Golomb [Gol], E. Lehmer [507], Moree and Stevenhagen [MS], Morehead [Mor], Rédei [674], Robinson [Rob], and Storchi [768, 769, 770] Yorinaga [Yo1, Yo2], Ward [818, 819], and Wilson [Wil]. Wall [Wa] studied the Fibonacci series modulo n; you will find many more articles on this topic in the journal Fibonacci Quarterly.

Hilbert's proof of his product formula can be found in [368]. In [369], Hilbert showed that the same formula holds for totally complex number fields with odd class number, and eventually conjectured that it can be generalized to arbitrary number fields – but this is part of class field theory and will be discussed in Part II. For other expositions of Hilbert's reciprocity law), see Blanchard & Blanchard [55], Borevich & Shafarevich [BS, 6.3, 7.1], and the booklet of Frey [Fre].

For an introduction to K-theory, try Bass [Ba1] (this book predates the introduction of K_2F), Silvester [Sil] (an elementary introduction to K_iR for $0 \leq i \leq 2$) or Milnor [586] (also discusses K_iR for $0 \leq i \leq 2$, but with a strong number theoretic flavor. A better idea for the number theoretically inclined might be to first browse the Lecture Notes [Ba2, LGS] and the articles there. Our proof of Gauss's lemma 2.32 is due to Dirichlet [174, §50] (see also Brown [89]); Gauss's original proof fills up the articles 126–129 of the Disquisitiones. For a slight improvement of Lemma 2.32, see Bock [57].

Exercises

2.1 Suppose that $m \in \mathbb{N}$ is divisible by a prime $p \equiv 3 \bmod 4$; then the fundamental unit ε of $\mathbb{Q}(\sqrt{m})$ has norm $+1$.

2.2 Suppose that $p \equiv 1 \bmod 4$ is prime; show that the fundamental unit of $\mathbb{Q}(\sqrt{p})$ has norm -1. (Hint: Let ε be the smallest unit > 1 with norm $+1$ and write $\varepsilon^3 = x + y\sqrt{p}$ with $x, y \in \mathbb{N}$, x odd, and observe that $(x-1)(x+1) = py^2$. Deduce that $x - 1 = 2a^2$ and $x + 1 = 2pb^2$ or $x - 1 = 2pb^2$ and $x + 1 = 2a^2$. Conclude that $a^2 - pb^2 = \pm 1$, and show that the minus sign holds since ε was minimal).

2.3 Let d be the discriminant of a quadratic number field k. Show that there is exactly one primitive nontrivial real-valued character modulo d (for a definition of primitive characters, see Chapter 4), and that the relation $\chi_d = (d/\cdot)$ is equivalent to the quadratic reciprocity law.
(Hints. Show that this is true for prime discriminants: if $d = p$ is prime, the quadratic residue character modulo p is the desired character; now find the characters corresponding to the discriminants -4, 8 and -8, and use the unique factorization of d into prime discriminants.)

2.4 Prove Hilbert's Theorem 90 for ideals (Proposition 2.5).

2.5 Let k be a quadratic number field. For $\alpha \in k$ such that $N_{k/\mathbb{Q}}\alpha > 0$ show that $\alpha \gg 0$ or $-\alpha \gg 0$.

2.6 Transport the action of $\sigma : \sqrt{d} \longrightarrow -\sqrt{d}$ on K to ideal classes $c = [\mathfrak{a}] \in C = \mathrm{Cl}^+(k)$ by putting $c^\sigma = [\mathfrak{a}^\sigma]$, and show that this is well defined.

2.7 Show that an ideal $\mathfrak{a}$ in a quadratic number field is ambiguous if and only if $\mathfrak{a}$ is the product of a rational number and ramified prime ideals. More exactly, show that $A \simeq (\mathbb{Z}/2\mathbb{Z})^t \oplus I$, where I is the group of ideals generated by rational numbers. Show that this implies that the subgroup of the ideal class group which is generated by ambiguous ideals is actually generated by the ramified prime ideals.

2.8 Let c be an ideal class in a quadratic number field k; given any integral ideal $\mathfrak{m}$, show that there exists an integral ideal $\mathfrak{a}$ coprime to $\mathfrak{m}$ such that $c = [\mathfrak{a}]$. Prove this for both the class group in the wide and the narrow sense.

2.9 Let δ and d be discriminants of quadratic number fields such that $\delta \mid d$. Show that δ is the product of prime discriminants occurring in the factorization of d if and only if $\gcd(\delta, d/\delta) = 1$.

2.10 Given a square-free integer $d \in \mathbb{Z}$ and $\epsilon = \pm 1$, show that there are infinitely many primes p such that $(d/p) = \epsilon$.

2.11 Let k be a quadratic number field; show that an $a \in \mathbb{N}$ is the norm of an ideal if and only if a is the product of split, ramified, and squares of inert primes.

2.12 (Shimura [Shi]) Let $p \equiv q \equiv 3 \bmod 4$ be primes such that $(p/q) = +1$, and put $k = \mathbb{Q}(\sqrt{pq})$, $\mathfrak{p} = (p, \sqrt{pq})$ and $\mathfrak{q} = (q, \sqrt{pq})$. Show that the fundamental unit ε of k satisfies $\varepsilon \equiv 1 \bmod \mathfrak{p}$ and $\varepsilon \equiv -1 \bmod \mathfrak{q}$, and that $\mathfrak{p} = (\pi)$ and $\mathfrak{q} = (\rho)$ are principal with $N\pi < 0$ and $N\rho > 0$. Derive analogous results for $k = \mathbb{Q}(\sqrt{p})$ and $k = \mathbb{Q}(\sqrt{2p})$, where $p \equiv 3 \bmod 4$.

2.13 Prove the Snake Lemma, i.e., show that, given an exact and commuting diagram

$$\begin{array}{ccccccc} A & \xrightarrow{f} & B & \xrightarrow{g} & C & \longrightarrow & 1 \\ {\scriptstyle\alpha}\downarrow & & {\scriptstyle\beta}\downarrow & & {\scriptstyle\gamma}\downarrow & & \\ 1 & \longrightarrow & A' & \xrightarrow{f'} & B' & \xrightarrow{g'} & C' \end{array}$$

of abelian groups, the following sequence is exact:

$$\begin{array}{ccccccccc} 1 & \longrightarrow & \ker f & \longrightarrow & \ker \alpha & \longrightarrow & \ker \beta & \longrightarrow & \ker \gamma \\ & & & & & & & & {\scriptstyle\delta}\downarrow \\ 1 & \longleftarrow & \operatorname{coker} g' & \longleftarrow & \operatorname{coker} \gamma & \longleftarrow & \operatorname{coker} \beta & \longleftarrow & \operatorname{coker} \alpha \end{array}$$

2.14 The version of the snake lemma given in the preceding exercise is often more suitable for applications than the usual version found in textbooks. Here is an example: given two homomorphisms $\alpha : A \longrightarrow B$ and $\beta : B \longrightarrow C$ of abelian groups, show that the sequence

$$\begin{array}{ccccccc} 1 & \longrightarrow & \ker \alpha & \longrightarrow & \ker(\beta \circ \alpha) & \longrightarrow & \ker \beta \\ & & & & & & \downarrow \\ 1 & \longleftarrow & \operatorname{coker} \beta & \longleftarrow & \operatorname{coker}(\beta \circ \alpha) & \longleftarrow & \operatorname{coker} \alpha \end{array}$$

is exact. Hint: apply the snake lemma to the diagram

$$\begin{array}{ccccccc} A & \xrightarrow{\alpha} & B & \longrightarrow & \operatorname{coker} \alpha & \longrightarrow & 1 \\ {\scriptstyle\beta\circ\alpha}\downarrow & & {\scriptstyle\beta}\downarrow & & \downarrow & & \\ 1 & \longrightarrow & C & \xrightarrow{\mathrm{id}} & C & \longrightarrow & 1 \end{array}$$

2.15 (Pépin, cf. [WS]) Let $F_n = 2^{2^n} + 1$ denote the n-th Fermat number. Show that F_n ($n \geq 3$) is prime if and only if $3^{(F_n-1)/2} \equiv -1 \bmod F_n$. (Hint: 3 is a primitive root mod prime F_n). Show that the congruence $(-3)^{2^{2^n-n-1}} \equiv -8 \bmod F_n$ implies that F_n is prime, and verify that this congruence holds for $1 \leq n \leq 4$. Verify that $\left(2^{2^{n-1}} + 1\right) 2^{2^{n-2}+1}$ is a square root of $-8 \bmod F_n$ for all $n \geq 2$.

2.16 (Lucas, cf. [WS]) Put $S_1 = 6$ and define the integers S_i recursively by $S_{i+1} = S_i^2 - 2$. Show that $S_k = \varepsilon^{2^k} + \bar{\varepsilon}^{2^k}$, where $\varepsilon = 1 + \sqrt{2}$ and $\bar{\varepsilon} = 1 - \sqrt{2}$. Prove that if F_n ($n \geq 2$) is prime then $F_n \mid S_k$ for some $k \leq 2^n - 1$. Show that the converse holds if $k \geq 2^{n-1}$.
Next, put $\zeta = \zeta_{2^n+1}$ and $\pi = 1 - 2\zeta$, and show that $N\pi = F_n$. Finally, prove that $F_n \mid S_r$, where $r = 2^n - n$, is equivalent to $\varepsilon^{2r} = \varepsilon^{(F_n-1)/2^{n-1}} \equiv -1 \bmod \pi$, and that $S_{r-1} \equiv \pm 2^{2^{n-2}}(2^{2^{n-1}} - 1) \bmod F_n$ in this case.

2.17 Prove that the congruences in Theorem 2.22.iii) are best possible. (Hint: take $m = 1023$ and $\omega = 32 + \sqrt{1023}$; show that $M_{11} \mid S_2$ and $M_{11} = 23 \cdot 89$.)

2.18 (Lenstra) Put $S_1 = \frac{2}{3}$ and show that $S_{p-1} \equiv 0 \bmod M_p$ if and only if M_p is prime. (Hint: $\frac{2}{3} \equiv \frac{1}{3}(2 + M_p) =: x \bmod M_p$. Put $\eta = \frac{1}{2}(\sqrt{x-2} + \sqrt{x+2})$; then $\varepsilon = \eta^2$ is a quadratic unit.)

2.19 As was already noticed by Fermat and Euler, prime divisors of Fermat numbers lie in certain arithmetic progressions: if $p \mid 2^{2^n} + 1$ and $n \geq 2$, then $p \equiv 1 \bmod 2^{n+2}$. (Hint: $2^{2^n} \equiv -1 \bmod p$, and 2 is a square mod p). Find the factors of F_5.

2.20 A similar result holds for factors of Mersenne numbers: if $q \mid 2^{2k+1} - 1$, then $q \equiv \pm 1 \bmod 8$. If moreover $p = 2k+1$ is prime, then $q \equiv 1 \bmod 2p$ for any $q \mid M_p$.

2.21 (Pellet [632]) If p and $q = 2p+1$ are odd primes, then $q \mid M_p$ if and only if $p \equiv 3 \bmod 4$.

2.22 A prime q is called a *primitive divisor* of M_n if $q \mid M_n$ and $q \nmid M_k$ for all $k < m$. Show that primitive divisors of M_n are $\equiv 1 \bmod n$. If n is odd, show that this implies that $q \equiv 1, 6n+1 \bmod 8n$ if $n \equiv 1 \bmod 4$, and $q \equiv 1, 2n+1 \bmod 8n$ if $n \equiv 3 \bmod 4$.

2.23 If p and $q = 4p+1$ are odd primes, then $q \mid 2^{2p}+1$.

2.24 (Warren & Bray [WB]) If $q^2 \mid M_p$ for primes p, q, then $2^{q-1} \equiv 1 \bmod q^2$. (Hint: first show that $q \mid M_p$ implies $M_q \mid M_{(p-1)/2}$.) Prove the same thing for primes q whose square divides some Fermat number.

2.25 Let $\alpha = \frac{1}{2}(1+\sqrt{5})$ and $\beta = \frac{1}{2}(1-\sqrt{5})$. Define the Fibonacci numbers U_n and the Lucas numbers V_n by Binet's formulas

$$U_n = \frac{\alpha^n - \beta^n}{\alpha - \beta}, \qquad V_n = \alpha^n + \beta^n,$$

and show that the following congruences hold modulo p:

U_{p-1}	U_p	U_{p+1}	V_{p-1}	V_p	V_{p+1}	if
0	1	1	2	1	3	$p \equiv \pm 1 \bmod 5$
1	-1	0	-3	1	-2	$p \equiv \pm 2 \bmod 5$

If, for example, $p \equiv \pm 1 \bmod 5$, then you should prove that $U_p \equiv 1 \bmod p$.

2.26 In the last exercise, the reciprocity relation $(5/p) = (p/5)$ was used to prove that $U_p \equiv (p/5) \bmod p$. Here is an elementary proof of that fact due to A. Nakhash [609] that can be used to prove $(5/p) = (p/5)$ for primes $p \neq 5$. For integers $i \in \mathbb{Z}$, define $S_i^{(n)} = \sum_{j \equiv i \bmod 5} \binom{n}{j}$. Show that $S_i^{(n)} = S_k^{(n)}$ if $i \equiv \pm k \bmod n$ and use the p-divisibility of binomial coefficients to show that $p \mid S_i^{(p)}$ unless $i \equiv 0 \bmod 5$ or $i \equiv p \bmod 5$, when $S_i^p \equiv 1 \bmod p$. The heart of the proof is the identity $U_n = (-1)^n (S_{3n-1}^{(n)} - S_{3n-2}^{(n)})$ that is easily proved by induction. The claimed congruence now follows immediately.

2.27 Show that every factor of U_n, n odd, is $\equiv 1 \bmod 4$.

2.28 With the notation as above, show that the following congruences hold modulo p:

$5U^2_{(p-1)/2}$	$5U^2_{(p+1)/2}$	$V^2_{(p-1)/2}$	$V^2_{(p+1)/2}$	if
-1	-4	-5	0	$p \equiv 3, \ 7 \bmod 20$
4	1	0	5	$p \equiv 11, \ 19 \bmod 20$
-1	0	-1	-4	$p \equiv 13, \ 17 \bmod 20$
0	1	4	1	$p \equiv 1, \ 9 \bmod 20$

2.29 Let h denote the (odd) class number of $\mathbb{Q}(\sqrt{-q})$, where $q \equiv 3 \bmod 4$ is prime. Let $p \equiv 3 \bmod 4$ be a prime such that $(-q/p) = 1$; then $p^h = 4a^2 + qb^2$ for some $a, b \in \mathbb{N}$. If ε_p denotes the fundamental unit of $K = \mathbb{Q}(\sqrt{q})$, then $\varepsilon_q^{(p+1)/2} \equiv (-1)^a \bmod p$. (Hint: The prime ideal above 2 in K is principal, say $2\mathcal{O}_K = \eta_q^2 \mathcal{O}_K$; now compute η_q^{p+1} and $2^{(p+1)/2} \bmod p$.)

2.30 Let $p \equiv q \equiv 3 \bmod 4$ be primes; show that $\mathfrak{p} = (p, \sqrt{pq})$ is principal and give a new proof for the result of Exercise 1.37.

2.31 Let $k = \mathbb{Q}(\sqrt{m})$ be a real quadratic number field, and assume that m is square-free. Show that the fundamental unit ε of $\mathcal{O}_k$ has norm $+1$ if and only if there is a principal ideal $\mathfrak{a}$ which is different from (1) and $(\sqrt{m})$ and which is the product of pairwise distinct ramified prime ideals. (Hint: If $\mathfrak{a} = (\alpha)$, then $\varepsilon = \alpha^2/N\mathfrak{a}$ is a unit (no square) with norm $+1$; conversely, if ε is a unit with norm $+1$, apply Hilbert's theorem 90).

2.32 (continued) Using the notation of the preceding exercise, show that $k(\sqrt{\varepsilon})$ is normal over $\mathbb{Q}$ if and only if $N\varepsilon = +1$, and that $k(\sqrt{\varepsilon}) = k(\sqrt{a})$ in this case, where $a = N_{k/\mathbb{Q}}\alpha$. By observing that you can take $\alpha = \varepsilon + 1$ in the preceding exercise, show that a equals the *square-free kernel* of $N(\varepsilon + 1)$, i.e. show that $N(\varepsilon + 1) = ab^2$ for a square-free and some $b \in \mathbb{N}$.

2.33 (continued) Show that $\chi_i(\mathfrak{a}) = 1$ for the ideal $\mathfrak{a}$ in Exercise 2.31

2.34 (Kaplan) Let d be the discriminant of a real quadratic number field $k = \mathbb{Q}(\sqrt{m})$ with $m \in \mathbb{N}$ squarefree, and assume that $m = a^2 + b^2$ is the sum of two squares with a odd. Assume moreover that the fundamental unit of k has positive norm. Show that the ideal $\mathfrak{b} = (b + \sqrt{m})$ generates an ideal class c of order 2 in $\text{Cl}^+(k)$, and verify that $\mathfrak{a}^2 = \mathfrak{b}$, where $\mathfrak{a} = (a, b + \sqrt{m})$. Use genus theory to conclude that c is a fourth power if and only if $(a/p) = +1$ for all primes p dividing d.

2.35 Let $q \equiv 1 \bmod 4$ be prime and $k = \mathbb{Q}(\sqrt{-q})$; use genus theory to show that

$$h(k) \equiv \begin{cases} 2 \bmod 4 & \text{if } q \equiv 5 \bmod 8; \\ 0 \bmod 4 & \text{if } q \equiv 1 \bmod 8. \end{cases}$$

2.36 Deduce Legendre's Theorem 1.7 from genus theory (cf. Cohn [Coh, Thm. 14.57]). Hint: You have to show that $ax^2 + by^2 = cz^2$ has nontrivial solutions if $-ab$ (resp. bc and ac) is a quadratic residue modulo c (resp. modulo a and b). Show that you may assume that $a, b, c > 0$ and consider the field $k = \mathbb{Q}(\sqrt{-ab})$. Solve $x^2 + ab \equiv 0 \bmod c$ and look at the ideals $\mathfrak{a} = (a, \sqrt{-ab})$, $\mathfrak{b} = (b, \sqrt{-ab})$, and $\mathfrak{c} = (c, x + \sqrt{-ab})$. For the genus characters χ_j of k, show that $\chi_j(\mathfrak{c}) = \chi_j(\mathfrak{a})$, conclude that $\mathfrak{ac}$ is in the principal genus and that $\mathfrak{ac}$ is a norm from k.

2.37 Show that every ambiguous ideal class in a quadratic number field k becomes principal in the genus field k_gen. Discuss both the version in the usual and the narrow sense.

2.38 Let $K/\mathbb{Q}$ be a normal finite extension, and let $G = \text{Gal}(K/\mathbb{Q})$ denote its Galois group. For a prime ideal $\mathfrak{p}$ in $\mathcal{O}_K$, define the *decomposition group* of $\mathfrak{p}$ by $Z_\mathfrak{p}(K/\mathbb{Q}) = \{\sigma \in G : \mathfrak{p}^\sigma = \mathfrak{p}\}$ (this group is also known as the stabilizer of $\mathfrak{p}$). The *inertia group* $T_\mathfrak{p}(K/\mathbb{Q}) = \{\sigma \in G : \alpha^\sigma \equiv \alpha \bmod \mathfrak{P}\}$ is the subgroup of $Z_\mathfrak{p}$ which even fixes the residue classes $\mathcal{O}_k/\mathfrak{p}$. Finally, the n-th *ramification group* is $V_\nu = V_\mathfrak{p}(K/\mathbb{Q})_\nu = \{\sigma \in G : \alpha^\sigma \equiv \alpha \bmod \mathfrak{P}^{\nu+1}\}$. The series of subgroups $G \supseteq Z \supseteq T \supseteq V_1 \supseteq \ldots \supseteq V_\nu \ldots$ is called the Hilbert series. The fixed fields K_Z, K_T, K_{V_ν} of these groups are called the *decomposition field*, *inertia field* and the ν-th *ramification field* of $\mathfrak{p}$. Compute this series for quadratic number fields, and verify the following table:

	K_Z	K_T	K_{V_1}	K_{V_2}	K_{V_3}
$(d/p) = +1$	K	K	K	K	K
$(d/p) = -1$	$\mathbb{Q}$	K	K	K	K
p odd, $p \mid d$	$\mathbb{Q}$	$\mathbb{Q}$	K	K	K
$p = 2,\ 4 \parallel d$	$\mathbb{Q}$	$\mathbb{Q}$	$\mathbb{Q}$	K	K
$p = 2,\ 8 \mid d$	$\mathbb{Q}$	$\mathbb{Q}$	$\mathbb{Q}$	$\mathbb{Q}$	K

2.39 An exact sequence $1 \longrightarrow A \longrightarrow B \longrightarrow C \longrightarrow 1$ of abelian groups is said to *split* if $B \simeq A \oplus C$. Show that the sequence (2.1) splits if and only if $d = \text{disc } k$ cannot be written as a sum of two squares. (Hint: see Cohn [Coh, Thm. 14.17])

2.40 Let k be a real quadratic number field whose discriminant d is not a sum of two squares. Use the preceding exercise to show that $\text{Cl}_\text{gen}^+(k) \simeq \text{Cl}_\text{gen}(k) \oplus \mathbb{Z}/2\mathbb{Z}$, and deduce that the class number h of k is odd if and only if d is a product of two (necessarily negative) prime discriminants.

2.41 Find analogues of Fermat's conjecture (1.3) for $q = 13$ and $q = 37$.

2.42 Let $q \equiv 1 \bmod 8$ be prime and put $k = \mathbb{Q}(\sqrt{-q})$. Show that there exist odd integers $e_1, f_1 \in \mathbb{N}$ such that $q = 2e_1^2 - f_1^2$. Consider the following assertions:
i) For $1 \le i \le m$ there exist integers $e_i, f_i \in \mathbb{Z}$ with $e_0 = 2$ and e_i positive and odd for $i \ge 1$, $f_i \equiv f_{i-1} \bmod e_{i-1}$, and $q = e_0 e_1^2 - f_1^2 = e_1 e_2^2 - f_2^2 = \ldots = e_{m-1} e_m^2 - f_m^2$.
ii) The class group of k contains $\mathbb{Z}/2^{m+2}\mathbb{Z}$ as a subgroup.
Show that i) $\Longrightarrow$ ii); moreover, if $e_m \equiv 3 \bmod 4$, show that actually $\text{Cl}_2(k) \simeq \mathbb{Z}/2^{m+2}\mathbb{Z}$. Verify the following table:

q	$h(k)$	e_1	f_1	e_2	f_2	e_3	f_3
17	4	3	1				
41	8	5	3	3	-2		
257	16	13	9	21	74	39	-178

Also find e_i, f_i for the first two cases with class number 32, namely $q = 521$ and $q = 569$.

Hints: Consider the ideals $\mathfrak{b}_i = (e_i, f_i + \sqrt{-q})$ for $0 \le i \le m$ (with $f_0 = 1$) and show that $N\mathfrak{b}_i = e_i$ and $\mathfrak{b}_{i-1}\mathfrak{b}_i^2 = (f_i + \sqrt{-q})$. This proves the first claim. For

the second, assume that there exists an ideal $\mathfrak{b}_{m+1}$ with norm e_{m+1} and such that $\mathfrak{b}_m \mathfrak{b}_{m+1}^2 = (x + y\sqrt{-q})$. Take the norm and derive $(e_m/q) = (-q/e_m) = +1$ and derive that $e_m \equiv 1 \bmod 4$ from quadratic reciprocity.
Remarks: I conjecture that the converse ii) $\Longrightarrow$ i) is also true, and that a proof is not too difficult.

2.43 Show that the map λ_p defined in (2.4) is a symbol (you have to prove $\lambda_p(a, 1-a) = 1$).

2.44 Show that the map λ_p defined in (2.4) is trivial for $p = 2$: $\lambda_2(a,b) = 1$ for all $a, b, \in \mathbb{Q}^\times$.

2.45 For $a \in \mathbb{Q}^\times$, write $a = (-1)^{i(a)} 2^{j(a)} 5^{k(a)} \frac{m}{n}$, where $m, n \in \mathbb{N}$ are congruent to 1 mod 8. Show that the map $c: \mathbb{Q}^\times \times \mathbb{Q}^\times \longrightarrow \mathbb{Z}/2\mathbb{Z}$ defined by $c(a,b) = (-1)^{i(a)i(b) + j(a)k(b) + j(b)k(a)}$ is a Steinberg symbol. Show that c coincides with the dyadic Hilbert symbol.

2.46 Show that $\lambda_p(a,b)^{(p-1)/2} = (\frac{a,b}{p})$ for all $a,b \in \mathbb{Q}^\times$, where λ_p is defined by (2.4).

2.47 Where in the proof of Lemma 2.32 did we use the assumption that p is a prime? Note that the claim does not hold e.g. when p is a square.

Additional References

[Arn] F. Arndt, *Ueber die Anzahl der Genera der quadratischen Formen*, J. Reine Angew. Math. **56** (1859), 72–78

[Ba1] H. Bass, *Algebraic K-Theory*, New York 1968

[Ba2] H. Bass (ed.), *Algebraic K-theory II - "Classical" algebraic K-theory, and connections with arithmetic*, Lecture Notes in Math. **342**, Springer 1973; 2nd printing 1986

[BS] A.I. Borevich, I.R. Shafarevich, *Number theory*, Academic Press 1966

[Bri] J. Brillhart, *Concerning the numbers $2^{2p} + 1$, p prime*, Math. Comp. **16** (1962), 424–430

[Cas] J. W. S. Cassels, *Local Fields*, Cambridge Univ. Press 1986

[Coh] H. Cohn, *A Classical Introduction to Algebraic Numbers and Class Fields*, 2nd ed., Springer Verlag 1988

[Fr] G. Frei, *On the development of the genus of quadratic forms*, Ann. Sci. Math. Quebec **3** (1979), 5–62

[Fre] G. Frey, *Elementare Zahlentheorie*, Vieweg 1984

[Fro] A. Fröhlich, *Central extensions, Galois groups, and ideal class groups of number fields*, Contemp. Math **24**, Amer. Math. Soc. 1983

[GE] S. Gebre-Egziabher, Ph. D. thesis, UC Berkeley, in preparation

[Gol] S. W. Golomb, *Properties of the sequence $3 \cdot 2^n + 1$*, Math. Comp. **30** (1976), 657–663

[Guy] R.K. Guy, *Unsolved Problems in Number Theory*, Problem Books in Mathematics, Springer Verlag, 2nd ed. 1994

[Ina] E. Inaba, *Über die Struktur der ℓ-Klassengruppe zyklischer Zahlkörper von Primzahlgrad ℓ*, J. Fac. Sci. Tokyo **I 4** (1940), 61–115

[Ish] M. Ishida, *The genus fields of algebraic number fields*, Lecture Notes Math. **555** (1976), Springer Verlag

[Kro] L. Kronecker, *Über den Gebrauch der Dirichlet'schen Methoden in der Theorie der quadratischen Formen*, Monatsber. Akad. Wiss. Berlin 1864, 285–303; Werke IV, 227–244

[Leo] H. W. Leopoldt, *Zur Geschlechtertheorie in abelschen Zahlkörpern*, Math. Nachr. **9** (1953), 351–362

[LGS] E. Lluis-Puebla, J.-L. Loday, H. Gillet, C. Soulé, V. Snaith, *Higher algebraic K-theory: an overview*, Lecture Notes in Math. **1491**, Springer 1992

[Me1] F. Mertens, *Über die Komposition der binären quadratischen Formen*, Wiener Ber. **104** (1895), 103–143; FdM **26** (1895), 225

[Me2] F. Mertens, *Ein Beweis des Satzes, daß jede Klasse von ganzzahligen primitiven binären quadratischen Formen des Hauptgeschlechts durch Duplikation entsteht*, J. Reine Angew. Math. **129** (1905), 181–186; FdM **37** (1906), 245

[Mol] R. Mollin, *Quadratics*, CRC Press 1996

[MS] P. Moree, P. Stevenhagen, *Prime divisors of Lucas sequences*, Acta Arith. **82** (1997), 403–410

[Mor] J. C. Morehead, *More on factors of Fermat's numbers*, Proc. Amer. Math. Soc. **12** (1906), 449–459

[NW] F. Nemenzo, H. Wada, *An elementary proof of Gauss' genus theorem*, Proc. Japan Acad. Sci. **68** (1992), 94–95

[On] T. Ono, *An introduction to algebraic number theory* New York, 1990

[Rob] R. M. Robinson, *A report on primes of the form $k \cdot 2^n + 1$ and factors of Fermat numbers*, Proc. Amer. Trans. Soc. **9** (1958), 673–681

[Shi] G. Shimura, *Class fields over real quadratic fields and Hecke operators*, Ann. Math. **95** (1972), 130–190

[Sil] J.R. Silvester, *Introduction to Algebraic K-theory*, Chapman and Hall 1981

[Spe] A. Speiser, *Über die Komposition binärer quadratischer Formen*, H. Weber Festschrift, 1912, 375–395

[Wa] D.D. Wall, *Fibonacci series modulo m*, Amer. Math. Mon. **67** (1960), 525–532

[WB] L.R.J. Warren, H.G. Bray, *On the square-freeness of Fermat and Mersenne numbers*, Pac. J. Math. **22** (1967), 563–564

[Web] H. Weber, *Lehrbuch der Algebra 3*, Braunschweig, F. Vieweg und Sohn 1908; see also
http://moa.cit.cornell.edu/MOA/MOA-JOURNALS2/WEBR.html

[WS] H. C. Williams, J. O. Shallit, *Factoring integers before computers*, Proc. Symp. Appl. Math. **48** (1994), 481–531

[Wil] N. Y. Wilson, *Conjectures as to a factor of* $2^p + 1$, Edinburgh Math. Notes **39** (1954), 6–9
[Yo1] M. Yorinaga, *On a congruencial property of Fibonacci numbers – numerical experiments*, Math. J. Okayama Univ. **19** (1976), 5–10
[Yo2] M. Yorinaga, *On a congruencial property of Fibonacci numbers – considerations and remarks*, Math. J. Okayama Univ. **19** (1976), 11–17
[Zag] D.B. Zagier, *Zetafunktionen und quadratische Körper*, Springer Verlag 1981

3. Cyclotomic Number Fields

This chapter is devoted to some proofs of the quadratic reciprocity law that make use of the arithmetic of cyclotomic number fields. All of them rest on the fact that $\mathbb{Q}(\zeta_p)$ contains a quadratic number field $\mathbb{Q}(\sqrt{p^*})$; as will become clear in later chapters, these proofs depend more or less explicitly on comparing the splitting of primes in the Kummer extension $\mathbb{Q}(\sqrt{p^*})$ and in the class field $\mathbb{Q}(\zeta_p)$.

3.1 Cyclotomic Fields

Instead of staying inside quadratic number fields to prove the quadratic reciprocity law we can also exploit the fact that every quadratic number field is contained in a suitable cyclotomic field. Before we show how to do this, we will recall a few results about such fields; for proofs, we refer to Ireland & Rosen [386] or Washington [Was]:

Proposition 3.1. *For an integer $m \not\equiv 2 \bmod 4$, let ζ_m denote a primitive m-th root of unity.*

i) $K = \mathbb{Q}(\zeta_m)$ is an extension of degree $(K : \mathbb{Q}) = \varphi(m)$ with ring of integers $\mathcal{O}_K = \mathbb{Z}[\zeta_m]$; here φ denotes Euler's phi[1] function: $\varphi(m) = \#(\mathbb{Z}/m\mathbb{Z})^\times$.

ii) A prime p is ramified in $K/\mathbb{Q}$ if and only if $p \mid m$.
If $p \nmid m$, let f denote the minimal integer ≥ 1 such that $p^f \equiv 1 \bmod m$: then $p\mathcal{O}_K$ is the product of $g = \frac{\varphi(m)}{f}$ different prime ideals with inertia degree f.

iii) $1 - \zeta_m$ is a unit if and only if m is not a prime power; for prime powers $m = \ell^n$, $1 - \zeta_m$ generates the prime ideal above ℓ.

iv) $K/\mathbb{Q}$ is abelian; in fact,

$$\mathrm{Gal}\,(K/\mathbb{Q}) = \{\sigma_a : \zeta_m \longmapsto \zeta_m^a, (a, m) = 1\} \simeq (\mathbb{Z}/m\mathbb{Z})^\times.$$

The decomposition group Z of a prime $p \nmid m$ is generated by σ_p; the isomorphism $\mathrm{Gal}\,(K/\mathbb{Q}) \simeq (\mathbb{Z}/m\mathbb{Z})^\times$ shows that $\#Z = f$, where f is defined as in ii).

---
[1] We will use the symbol φ for Euler's phi function, and ϕ for various homomorphisms, polynomials, etc. occurring in the text.

If $m = p^n$ is a power of an odd prime p, then $K/\mathbb{Q}$ is cyclic, and $\mathrm{Gal}\,(K/\mathbb{Q})$ is generated by the automorphism σ_g, where g is a primitive root modulo p^n. In this case, the quadratic subfield of K is fixed by the automorphisms σ_a, where a ranges through the quadratic residues modulo p^n.
More generally, for every $r \mid p - 1$, the subfield K_r of degree r of K is fixed by the σ_a, where $a \equiv g^{rj} \bmod p$. A prime $q \neq p$ splits completely in K_r if and only if $q^{(p-1)/r} \equiv 1 \bmod p$. □

A very simple proof of the quadratic reciprocity law using the arithmetic of cyclotomic fields is due to R. Swan [779]: let $K_p = \mathbb{Q}(\zeta_p)$; then $1 - \zeta_p$ generates the prime ideal $\mathfrak{p}$ above p, hence $\pi_p = (1 - \zeta_p)(1 - \zeta_p^{-1})$ (this is the norm of $1 - \zeta_p$ to the maximal real subfield K_p^+ of K_p) generates an element of norm p in K_p^+. Define π_q accordingly, where $q \neq p$ is another prime. Then $\eta = \pi_p - \pi_q$ is a unit in the compositum $L = K_p^+ K_q^+$ because every factor in the product $\eta = \zeta_q^{-1}(1 - \zeta_p \zeta_q)(1 - \zeta_p^{-1}\zeta_q)$ is a unit by Proposition 3.1. Let $N = N_{L/\mathbb{Q}}$ denote the absolute norm; then $\eta \equiv \pi_p \bmod \pi_q$ implies $N\eta \equiv N\pi_p \bmod \pi_q$ (cf. Exercise 3.12), and since $N\eta$ and $N\pi_p$ are integers, this yields $N\eta \equiv N\pi_p \bmod q$. But $\pi_p \in K_p^+$, hence

$$N\pi_p = N_{K_p^+/\mathbb{Q}}(\pi_p)^{\frac{q-1}{2}} = p^{\frac{q-1}{2}} \equiv \left(\frac{p}{q}\right) \bmod q.$$

Since $N(\eta) = \pm 1$, we have shown

Proposition 3.2. *For an odd prime p, let $\pi_p = 2 - \zeta_p - \zeta_p^{-1}$; define π_q accordingly, where q is another odd prime. Then $N(\pi_p - \pi_q) = \left(\frac{p}{q}\right)$, where N denotes the norm from $L = \mathbb{Q}(\zeta_p + \zeta_p^{-1}, \zeta_q + \zeta_q^{-1})$ to $\mathbb{Q}$.*

Exchanging p and q gives the quadratic reciprocity law, because

$$N_{L/\mathbb{Q}}(-1) = (-1)^{(L:\mathbb{Q})} = (-1)^{\frac{p-1}{2} \cdot \frac{q-1}{2}}.$$

In Proposition 3.1.iv) we have seen that $\mathbb{Q}(\zeta_p)$ contains a quadratic subfield; in fact it is easy to see that $k = \mathbb{Q}(\sqrt{p^*})$ is contained in $\mathbb{Q}(\zeta_p)$: since $\mathrm{disc}\, K$ is a p-power, so is $\mathrm{disc}\, k$ (cf. Exercise 3.11), and this implies that $k = \mathbb{Q}(\sqrt{p^*})$. The fact that $\mathbb{Q}(\sqrt{p^*}) \subseteq \mathbb{Q}(\zeta_p)$ is a very special case of the following theorem:

(Weak Kronecker-Weber) *The abelian extensions of $\mathbb{Q}$ are exactly the subfields of cyclotomic number fields.*

Actually, proofs of the theorem of Kronecker-Weber show more by giving information on the smallest cyclotomic field containing an abelian extension $K/\mathbb{Q}$. A slightly stronger version of the theorem is the following:

Theorem 3.3. *If $K/\mathbb{Q}$ is an abelian extension, then $K \subseteq \mathbb{Q}(\zeta_n)$ for $n = |\mathrm{disc}\, K|$.* □

3.1 Cyclotomic Fields

The bound $n = |\operatorname{disc} K|$ is quite crude, although it is best possible for quadratic number fields (see Exercise 3.13). For most applications, it is even sufficient to show that one can choose n such that $\mathbb{Q}(\zeta_n)$ is unramified outside the primes ramified in K.

Now we can give another proof of quadratic reciprocity by comparing how an odd prime q splits in $K = \mathbb{Q}(\zeta_p)$ and $\mathbb{Q}(\sqrt{p^*})$. Let $f \geq 1$ be the minimal integer such that $q^f \equiv 1 \bmod p$; then (q) splits into $g = \frac{p-1}{f}$ different prime ideals in K, and the decomposition field K_Z of q in K has degree g over $\mathbb{Q}$. Euler's criterion yields

$$\left(\frac{q}{p}\right) = +1 \iff f \text{ divides } \frac{p-1}{2} = \frac{1}{2}fg$$
$$\iff g \equiv 0 \bmod 2 \iff (K_Z : \mathbb{Q}) \text{ is even}$$
$$\iff K_Z \text{ contains a quadratic subfield}$$
$$\iff K_Z \text{ contains } \mathbb{Q}(\sqrt{p^*}) \text{ (because } \mathbb{Q}(\zeta_p)/\mathbb{Q} \text{ is cyclic)}$$
$$\iff \left(\frac{p^*}{q}\right) = +1.$$

We will have the opportunity to admire this elegant argument quite often in the chapters below. Note that the fact $Z = \langle \sigma_q \rangle$ implies that $(q/p) = 1 \iff q$ is a square modulo $p \iff \sigma_q$ is a square in $\operatorname{Gal}(K/\mathbb{Q}) \iff 2 \mid (G : Z)$.

The Cyclotomic Polynomial

Suppose that a number field K is generated by a root of a monic polynomial $f \in \mathbb{Z}[X]$; if p is a prime not dividing disc f, then, according to Dedekind, p splits in K exactly in the same way as f splits in $\mathbb{F}_p[X]$. This suggests that, in the proof of quadratic reciprocity law above, we may replace the splitting primes by splitting polynomials: this is what we will do here.

For every odd prime p, the *cyclotomic polynomial*

$$\Phi_p(X) = X^{p-1} + \ldots + X + 1 = \frac{X^p - 1}{X - 1} = \prod_{j=1}^{p-1}(X - \zeta_p^j).$$

is the minimal polynomial of the primitive p-th roots of unity ζ_p; it has discriminant disc $\Phi_p = (-1)^{\frac{p-1}{2}} p^{p-2}$. We know that $K = \mathbb{Q}(\zeta_p)$ contains a quadratic subfield $k = \mathbb{Q}(\sqrt{p^*})$, hence by Galois theory $\Phi_p(X)$ must split into two irreducible factors $\phi_j(X) \in \mathcal{O}_k[X]$. In fact $\Phi_p(X) = \phi_1(X)\phi_2(X)$ for

$$\phi_1(X) = \prod_{1 \leq r \leq p-1}(X - \zeta_p^r), \quad \left(\frac{r}{p}\right) = +1, \text{ and}$$

$$\phi_2(X) = \prod_{1 \leq n \leq p-1}(X - \zeta_p^n), \quad \left(\frac{n}{p}\right) = -1.$$

For a proof, observe that the automorphisms σ_r fix ϕ_1 and ϕ_2, whereas the σ_n permute them. The first two factorizations are

$$\Phi_3(X) = \left(X - \tfrac{-1-\sqrt{-3}}{2}\right)\left(X - \tfrac{-1+\sqrt{-3}}{2}\right),$$
$$\Phi_5(X) = \left(X^2 + \tfrac{1-\sqrt{5}}{2}X + 1\right)\left(X^2 + \tfrac{1+\sqrt{5}}{2}X + 1\right).$$

Remark 1. Moreover, we observe that the occurrence of square roots in the expressions for $\phi_j(X)$ for $p = 3$ and $p = 5$ are not accidental: if we had $\phi_1 \in \mathbb{Z}[X]$, then also $\phi_2 \in \mathbb{Z}[X]$, and this contradicts the irreducibility of $\Phi_p(X)$ over $\mathbb{Q}$.

Remark 2. Replacing the primitive p-th root of unity ζ_p by some ζ_p^n, where $(n/p) = -1$, has the effect of exchanging ϕ_1 and ϕ_2. However, if we fix $\zeta_p = e^{2\pi i/p}$, then we can identify the factors of Φ_p unambiguously. Take e.g. $p = 5$ and let $\zeta = \alpha + i\beta$ be the unique primitive fifth root of unity in the first quadrant (i.e. $\alpha, \beta > 0$). Then $2\alpha = \zeta + \zeta^{-1}$ must be positive, and this shows that $\phi_1(X) = X^2 + \tfrac{1-\sqrt{5}}{2}X + 1$. Thus $\alpha = \tfrac{\sqrt{5}-1}{4}$, and from $\alpha^2 + \beta^2 = |\zeta| = 1$ we get $\beta^2 = \tfrac{5+\sqrt{5}}{8}$, hence

$$\zeta = \tfrac{1}{2}\left(\tfrac{\sqrt{5}-1}{2} + i\sqrt{\tfrac{5+\sqrt{5}}{2}}\right). \tag{3.1}$$

Recall that, for primes $q \neq p$, $\Phi_p(X)$ splits into g different factors over $\mathbb{F}_q[X]$ if and only if q splits into g different prime ideals in $\mathbb{Q}(\zeta_p)$ (thanks to the fact that disc Φ_p = disc K; in general, there are problems if q divides disc f/disc K, where K is the field generated by a root of f). If $\left(\tfrac{p^*}{q}\right) = +1$, then we can replace $\sqrt{p^*}$ in the polynomials $\phi_j(X)$ by square roots of p^* mod q and find that $\Phi_p(X) \equiv \phi_1(X)\phi_2(X) \bmod q$ (for polynomials $F, G \in \mathbb{Z}[X]$ we write $F \equiv G \bmod p$ if the coefficients of $F - G$ are divisible by p; in particular, $X^2 + X \not\equiv 0 \bmod 2$ although $x^2 + x \equiv 0 \bmod 2$ for all $x \in \mathbb{Z}$). This shows that $g \equiv 0 \bmod 2$, and we find that $\left(\tfrac{q}{p}\right) = +1$. Of course, this is only a rewording of one direction of the proof given right after Theorem 3.3.

For the converse, assume that $\left(\tfrac{q}{p}\right) = +1$. Then multiplication by q is a permutation of the subgroup R of squares in $\mathbb{F}_q^\times$, hence $\phi_1(X) = \prod(X - \zeta_p^r)$ is fixed by the Frobenius automorphism of $\mathbb{F}_q$ and so is in $\mathbb{F}_q[X]$. By Remark 1 above, this implies that the square root of p^* must be an element of $\mathbb{F}_q$, and this does it.

Let σ be the nontrivial automorphism of $k/\mathbb{Q}$ (thus σ is the restriction of the $\sigma_n : \zeta \longmapsto \zeta_n$ to k, with n being a quadratic nonresidue modulo p); then we have seen that σ exchanges ϕ_1 and ϕ_2. But this shows that $\phi_+ = \phi_1 + \phi_2 \in \mathbb{Z}[X]$, because ϕ_+ is invariant under σ. On the other hand, $\psi_- = \phi_1 - \phi_2$ has the property $\psi_-^\sigma = -\psi_-$, hence we find $\sqrt{p^*}\psi_- \in \mathbb{Z}[X]$. Since the coefficients of $\sqrt{p^*}\psi_-$ are divisible by $\sqrt{p^*}$, they must be divisible by p: thus $\phi_1 - \phi_2 = \sqrt{p^*}\phi_-$ for some $\phi_- \in \mathbb{Z}[X]$, and we have proved

3.1 Cyclotomic Fields

Proposition 3.4. *The quadratic factors $\phi_1(X)$ and $\phi_2(X)$ of the cyclotomic polynomial $\Phi_p(X)$ satisfy $\phi_1 + \phi_2 \in \mathbb{Z}[X]$ and $\phi_1 - \phi_2 \in \sqrt{p^*}\,\mathbb{Z}$.*

The last result allows us to give another proof of the quadratic reciprocity law: on the one hand we have

$$(\phi_1(X) - \phi_2(X))^q \equiv \phi_1(X)^q - \phi_2(X)^q$$
$$\equiv \prod_r (X^q - \zeta^{qr}) - \prod_n (X^q - \zeta^{qn})$$
$$\equiv \left(\tfrac{q}{p}\right)(\phi_1(X^q) - \phi_2(X^q)) \bmod q,$$

on the other hand we see that

$$(\phi_1(X) - \phi_2(X))^q = \sqrt{p^*}^{\,q} \phi_-(X)^q \equiv \left(\tfrac{p^*}{q}\right)\sqrt{p^*}\,\phi_-(X)^q$$
$$\equiv \left(\tfrac{p^*}{q}\right)\sqrt{p^*}\,\phi_-(X^q)$$
$$\equiv \left(\tfrac{p^*}{q}\right)(\phi_1(X^q) - \phi_2(X^q)) \bmod q.$$

Since $\phi_1 \not\equiv \phi_2 \bmod q$ (a factorization $\Phi(X) \equiv \phi_1(X)^2 \bmod q$ would imply that $q \mid \operatorname{disc} \Phi = p^{p-2}$), we must have $\left(\tfrac{p^*}{q}\right) = \left(\tfrac{q}{p}\right)$. $\square$

For proofs of the quadratic reciprocity law, it is usually sufficient to consider subfields of $\mathbb{Q}(\zeta_p)$; for applications of cyclotomy to primality testing, however, we will need to know a few things about cyclotomic polynomials $\Phi_n(X)$ for arbitrary integers $n \geq 1$. First, $\Phi_n(X) \in \mathbb{Z}[X]$ is defined to be the minimal polynomial of ζ_n; clearly $\Phi_n(X) \mid X^n - 1$, and it can also be shown that $\Phi_n(X) = \prod(X - \zeta_n^j)$, where the product if over all $1 \leq j < n$ such that $\gcd(j, n) = 1$.

Proposition 3.5. *We have* $\displaystyle \Phi_n(X) = \frac{X^n - 1}{\gcd\left(X^n - 1, \prod_{1 \leq j < n}(X^j - 1)\right)}.$

Proof. It is sufficient to show that the primitive n-th roots of unity, that is the ζ_n^j with $(j, n) = 1$, are roots of $X^n - 1$ but not of any $\gcd(X^j - 1, X^n - 1)$ for $j < n$. Which is clear. $\square$

Normal Integral Bases

Below we will need to know a few things about integral bases of subfields of cyclotomic fields. Suppose that $K/\mathbb{Q}$ is a finite normal extension with Galois group $G = \operatorname{Gal}(K/\mathbb{Q})$; we say that $\mathcal{O}_K$ (or, by abuse of language, K) has a *normal integral basis* (NIB) if there is an $\alpha \in \mathcal{O}_K$ such that $\{\alpha^\sigma \mid \sigma \in G\}$ forms a $\mathbb{Z}$-basis of $\mathcal{O}_K$. In this case, we say that α *generates* a NIB of K.

Example 1. Let $k = \mathbb{Q}(\sqrt{m}\,)$; if $m \equiv 1 \bmod 4$, then $\frac{1+\sqrt{m}}{2}$ generates a NIB of $\mathcal{O}_k$.

Example 2. If $k = \mathbb{Q}(\sqrt{m}\,)$ for some $m \equiv 2, 3 \bmod 4$, then k does not possess a NIB: in fact, assume that $\eta = a + b\sqrt{m}$ generates one. Then $\operatorname{disc} k =$

84   3. Cyclotomic Number Fields

$(\eta^2 - \eta'^2)^2 = (\eta - \eta')^2(\eta + \eta')^2 = 16a^2b^2m$ contradicts the fact that disc $k = 4m$ and $a, b \in \mathbb{Z}$.

Example 3. Let $k = \mathbb{Q}(\zeta_p)$, where p is an odd prime; then ζ_p generates a NIB of $\mathbb{Z}[\zeta_p]$.

Hilbert noticed that the existence of NIBs in abelian extensions behaves well with respect to forming subfields in that we can get NIBs of these subfields simply by taking the trace:

Proposition 3.6. *Suppose that $K/\mathbb{Q}$ is abelian and that $\mathcal{O}_K$ has a NIB; then $\mathcal{O}_F$ has a NIB for every subfield F of $K/\mathbb{Q}$.*

Proof. Put $G = \mathrm{Gal}\,(K/\mathbb{Q})$ and $H = \mathrm{Gal}\,(K/F)$; we will write $\sigma \in G/H$ for automorphisms $\sigma \in G$ that represent the $(G:H)$ cosets of G/H. Then every automorphism of G has the form $\sigma\tau$ for $\sigma \in G/H$ and $\tau \in H$. Let ν generate the NIB of K, and take any $\alpha \in \mathcal{O}_F$. Since $\mathcal{O}_F \subset \mathcal{O}_K$, there exist integers $a_{\sigma,\tau}$ such that

$$\alpha = \sum_{\sigma\tau \in G} a_{\sigma,\tau} \sigma\tau(\nu) = \sum_{\sigma \in G/H} \sum_{\tau \in H} a_{\sigma,\tau} \sigma\tau(\nu)$$

Applying $\rho \in H$ to this equation and observing that $\rho(\alpha) = \alpha$ and that the $\sigma\tau(\nu)$ are linearly independent over $\mathbb{Q}$, we deduce that $a_{\sigma,\tau} = a_{\sigma,\rho\tau}$ for all $\tau \in H$. Thus the $a_{\sigma,\tau}$ do not depend on τ, so we may write $a_\sigma := a_{\sigma,\tau}$ and get $\alpha = \sum_{\sigma \in G/H} a_\sigma \sigma(\sum_{\tau \in H} \tau(\nu))$. But $\sum_{\tau \in H} \tau(\nu) = \mathrm{Tr}\,_{K/F}\nu$, and since $G/H \simeq \mathrm{Gal}\,(F/\mathbb{Q})$, we conclude that $\mathrm{Tr}\,_{K/F}\nu$ generates a NIB of F. □

Corollary 3.7. *If $4 \mid n$, then $\mathbb{Q}(\zeta_n)$ does not posses a NIB.*

Proof. This follows from Proposition 3.6 and Example 2 since $\mathbb{Q}(\sqrt{-1}) \subseteq \mathbb{Q}(\zeta_n)$. □

If p is an odd prime, then ζ_p generates a normal integral basis of $\mathbb{Z}[\zeta_p]$; if k is an integer dividing $p - 1$, then the subfield F of $K = \mathbb{Q}(\zeta_p)$ of degree k will therefore have a NIB generated by the relative trace $\eta_0 = \mathrm{Tr}\,_{K/F}\zeta_p$. The conjugates η_j ($j = 0, \ldots, k-1$) of η_0 are called the Gaussian periods, and the minimal polynomial of η_0 over $\mathbb{Q}$ is the k-th period equation.

Now a prime $q \neq p$ is a k-th power residue modulo p if and only if $f \mid \frac{p-1}{k}$, where $f > 0$ is the smallest integer such that $q^f \equiv 1 \bmod p$. Next $f \mid \frac{p-1}{k} \iff k \mid \frac{p-1}{f} =: g$, that is, if and only if the number g of prime ideals above q in K is divisible by k. Since $\mathbb{Q}(\zeta_p)/\mathbb{Q}$ is cyclic, this is equivalent to the fact that q splits completely in F. By a classical result of Dedekind, this happens if and only if a polynomial whose root generates F splits completely over $\mathbb{F}_q$, with the possible exception of primes dividing the discriminant. Since F is generated by the roots of the k-th period equation, we have proved

Let p be an odd prime, k an integer dividing $p-1$, and let $F_k(X)$ denote the k-th period equation. Then a prime $q \nmid \operatorname{disc} F_k$ is a k-th power residue modulo p if and only if F_k splits into linear factors modulo q.

In this form the result was given by E. Lehmer and credited to Kummer and Glaisher; she also remarked that F_k splits into linear factors modulo q as soon as it has a single root modulo q: this comes from $F/\mathbb{Q}$ being abelian. Actually the result was known to Gauss around 1797, even in the following stronger form:

Proposition 3.8. *Let p be an odd prime, k an integer dividing $p-1$, and let $F_k(X)$ denote the k-th period equation. Then a prime $q \neq p$ is a k-th power residue modulo p if and only if F_k has a root modulo p.*

Proof. We only have to consider the primes $q \neq p$ dividing $\operatorname{disc} F_k$. It is clearly sufficient to prove the following claim: every prime $q \neq p$ dividing $\operatorname{disc} F_k$ is a k-th power residue modulo p, and F_k has a root modulo q.

Assume therefore that $q \mid \operatorname{disc} F_k$. This discriminant is a product of differences of conjugates of η_0; thus if q divides this product, then some prime ideal $\mathfrak{q}$ above q in $\mathcal{O}_F$ must divide one of the differences, say $\eta_i - \eta_j$. But the η_ν form a (normal) integral basis of $\mathcal{O}_F$, hence any prime ideal dividing $\eta_i - \eta_j$ must have degree 1, in other words: q splits completely in $F/\mathbb{Q}$.

We still have to show that F_k has a root modulo q. Assume it doesn't: then F_k does not have a linear factor over $\mathbb{Q}_q[X]$, hence q will not split completely in $F/\mathbb{Q}$: contradiction. □

Observe that the proof actually shows that primes $q \neq p$ dividing $\operatorname{disc} F_k$ are k-th power residues modulo p.

Put $k = 2$ in Proposition 3.8 to get another proof of the quadratic reciprocity law.

3.2 Primality Tests

In this section, we will discuss several primality tests that can be described in terms of Artin symbols and that encompass classical tests such as those of Pocklington or Lucas-Lehmer. Before we do that, we have to define Frobenius and Artin symbols; we will also derive their basic properties.

The Frobenius Symbol

Frobenius symbols allow us to describe the splitting of primes in normal extensions; we will say more about this in Part II. Actually, we already know some instances of the Artin symbol: for primes $p \neq q$ and the abelian extension $K = \mathbb{Q}(\zeta_p)$, the Artin symbol $\left(\frac{K/\mathbb{Q}}{q}\right)$ coincides with the automorphism $\sigma_q : \zeta_p \longmapsto \zeta_p^q$ introduced above.

A little bit more generally, consider a finite normal extension L/K and a prime ideal $\mathfrak{P}$ in $\mathcal{O}_L$ which is unramified in L/K. The Frobenius automorphism ϕ of $\mathfrak{P}$ is defined to be the unique automorphism in $G = \text{Gal}\,(L/K)$ with the property
$$\phi(\alpha) \equiv \alpha^{N\mathfrak{p}} \bmod \mathfrak{P}$$
for all $\alpha \in \mathcal{O}_L \setminus \mathfrak{P}$, where $N\mathfrak{p}$ is the absolute norm of the prime ideal $\mathfrak{p}$ in $\mathcal{O}_K$ below $\mathfrak{P}$. Hasse introduced the notation
$$\phi = \phi_\mathfrak{P} = \left[\frac{L/K}{\mathfrak{P}}\right]$$
and derived the following basic properties of the Frobenius symbol:

Theorem 3.9. *Let L/K be a normal extension of number fields and put $G = \text{Gal}\,(L/K)$; assume moreover that the prime ideal $\mathfrak{P}$ in $\mathcal{O}_L$ is unramified in L/K. Then the Frobenius symbol has the following properties:*

i) $\left[\frac{L/K}{\mathfrak{P}}\right]$ *generates the decomposition group* $Z(\mathfrak{P}|\mathfrak{p})$; *in particular, the Frobenius automorphism* $\left[\frac{L/K}{\mathfrak{P}}\right]$ *has order* $f = f(\mathfrak{P}|\mathfrak{p})$.

ii) *Let σ be any automorphism of the normal closure $N/\mathbb{Q}$ of $L/\mathbb{Q}$; then $\left[\frac{L^\sigma/K^\sigma}{\mathfrak{P}^\sigma}\right] = \sigma\left[\frac{L/K}{\mathfrak{P}}\right]\sigma^{-1}$, where automorphisms act from left on elements and ideals.*

iii) *Let F/K be a normal subextension of L/K; then the restriction of $\left[\frac{L/K}{\mathfrak{P}}\right]$ to F coincides with $\left[\frac{F/K}{\mathfrak{q}}\right]$, where $\mathfrak{q} = \mathfrak{P} \cap \mathcal{O}_F$ is the prime ideal in $\mathcal{O}_F$ below $\mathfrak{P}$.*

iv) *If F/K is a normal subextension of L/K, and if F is the fixed field of $H \subseteq G$, then $\mathfrak{p}$ splits completely in F/K if and only if $\left[\frac{L/K}{\mathfrak{P}}\right] \in H$.*

v) *If $L = L_1 L_2$ is a compositum of normal extensions L_1/K and L_2/K such that $L_1 \cap L_2 = K$, then $\left[\frac{L/K}{\mathfrak{P}}\right] = \left[\frac{L_1/K}{\mathfrak{P}_1}\right] \cdot \left[\frac{L_2/K}{\mathfrak{P}_2}\right]$, where $\mathfrak{P}_i = \mathfrak{P} \cap \mathcal{O}_{L_i}$ denote the prime ideals below $\mathfrak{P}$ in the extensions L_i, $i = 1, 2$, and where e.g. the $\left[\frac{L_1/K}{\mathfrak{P}_1}\right]$ on the right hand side is the extension of $\left[\frac{L_1/K}{\mathfrak{P}_1}\right] \in \text{Gal}\,(L_1/K)$ to $\text{Gal}\,(L/K)$ that induces the identity on L_2.*

vi) *If F is a subfield of L/K, then $\left[\frac{L/K}{\mathfrak{P}}\right] = \left[\frac{L/F}{\mathfrak{P}}\right]^f$, where f is the inertia degree of $\mathfrak{P}$ in L/F.*

Proof. Property i) is well known (see e.g. Hilbert [368]). For the proof of ii), we first observe that L^σ/K^σ is normal if L/K is, and that $\mathfrak{P}^\sigma$ is unramified in L^σ/K^σ if $\mathfrak{P}$ is unramified in L/K. Now we apply σ to the defining congruence $\left[\frac{L/K}{\mathfrak{P}}\right]\alpha \equiv \alpha^{N\mathfrak{p}} \bmod \mathfrak{P}$ and find $\sigma(\alpha^{N\mathfrak{p}}) \equiv (\sigma \cdot \left[\frac{L/K}{\mathfrak{P}}\right])\alpha \equiv (\sigma \cdot \left[\frac{L/K}{\mathfrak{P}}\right]\sigma^{-1})\sigma(\alpha) \bmod \mathfrak{P}^\sigma$ for all $\sigma(\alpha) \in \mathcal{O}_{L^\sigma} \setminus \mathfrak{P}^\sigma$; this proves the claim.

For iii), let $\phi \in \text{Gal}(F/K)$ be the restriction of $\left[\frac{L/K}{\mathfrak{P}}\right]$ to F; then $\phi(\alpha) \equiv \alpha^{N\mathfrak{p}}$ mod $\mathfrak{P}$ for all $\alpha \in \mathcal{O}_F \subset \mathcal{O}_L$, and since both sides of the congruence are in $\mathcal{O}_F$, it is even valid modulo $\mathfrak{q} = \mathfrak{P} \cap \mathcal{O}_F$. By the definition of $\left[\frac{F/K}{\mathfrak{q}}\right]$, we also have $\left[\frac{F/K}{\mathfrak{q}}\right]\alpha \equiv \alpha^{N\mathfrak{p}}$ mod $\mathfrak{q}$ for all $\alpha \in \mathcal{O}_F$, and iii) follows.

As a corollary of i) and iii), we get iv): we know that $\mathfrak{p}$ splits completely in F/K if and only if $\left[\frac{F/K}{\mathfrak{q}}\right] = 1$, where $\mathfrak{q} = \mathfrak{P} \cap \mathcal{O}_F$. By iii), this holds if and only if the restriction of $\left[\frac{L/K}{\mathfrak{P}}\right]$ is trivial in $\text{Gal}(F/K)$, which by Galois theory is true if and only if $\left[\frac{L/K}{\mathfrak{P}}\right] \in H$.

For v), all we have to do is observe that $\left[\frac{L_j/K}{\mathfrak{P}}\right]$ is the restriction to L_j of $\left[\frac{L/K}{\mathfrak{P}}\right]$; the rest follows from the isomorphism $\text{Gal}(L/K) \simeq \text{Gal}(L_1/K) \times \text{Gal}(L_2/K)$ known from Galois theory.

Finally, vi) follows from the defining congruences: we have $\left[\frac{L/K}{\mathfrak{P}}\right]\alpha \equiv \alpha^{N\mathfrak{p}}$ mod $\mathfrak{P}$ and $\left[\frac{L/F}{\mathfrak{P}}\right]\alpha \equiv \alpha^{N\mathfrak{q}}$ mod $\mathfrak{P}$ for all $\alpha \in \mathcal{O}_L$, where $\mathfrak{q} = \mathfrak{P} \cap F$. Since $\mathfrak{P}$ is unramified in L/K, this implies that $N_{F/K}\mathfrak{q} = \mathfrak{p}^f$; applying $\left[\frac{L/K}{\mathfrak{P}}\right]$ f times to α we end up with $\left[\frac{L/K}{\mathfrak{P}}\right]^f \alpha \equiv \left[\frac{L/F}{\mathfrak{P}}\right]\alpha$ mod $\mathfrak{P}$. □

The Artin Symbol

Now assume in addition that L/K is abelian. Then the Frobenius symbol $\left[\frac{L/K}{\mathfrak{P}}\right]$ does not depend on the choice of the prime $\mathfrak{P} \mid \mathfrak{p}$ for fixed $\mathfrak{p}$, since $\left[\frac{L/K}{\mathfrak{P}^\sigma}\right] = \sigma\left[\frac{L/K}{\mathfrak{P}}\right]\sigma^{-1} = \left[\frac{L/K}{\mathfrak{P}}\right]$. Therefore, we are allowed to put $\left(\frac{L/K}{\mathfrak{p}}\right) = \left[\frac{L/K}{\mathfrak{P}}\right]$ for any $\mathfrak{P} \mid \mathfrak{p}$; the symbol thus defined is called the Artin symbol. Since G is abelian, we can extend the Artin symbol multiplicatively to all ideals $\mathfrak{a} = \prod \mathfrak{p}$ prime to $\text{disc}(L/K)$ by putting $\left(\frac{L/K}{\mathfrak{a}}\right) = \prod \left(\frac{L/K}{\mathfrak{p}}\right)$. The properties of the Frobenius symbol proved above can be transferred immediately to the case of Artin symbols:

Corollary 3.10. *Let L/K be an abelian extension of number fields, and let $\mathfrak{p}$ be a prime ideal in $\mathcal{O}_K$ that is unramified in L/K. Then the Artin symbol has the following properties:*

i) $\left(\frac{L/K}{\mathfrak{p}}\right)$ *generates the decomposition group $Z(\mathfrak{P}|\mathfrak{p})$ for any prime ideal $\mathfrak{P}$ in L above $\mathfrak{p}$; in particular, $\left(\frac{L/K}{\mathfrak{p}}\right)$ has order $f = f(\mathfrak{P}|\mathfrak{p})$.*

ii) *Let σ be any automorphism of the normal closure $N/\mathbb{Q}$ of $L/\mathbb{Q}$; then $\left(\frac{L^\sigma/K^\sigma}{\mathfrak{p}^\sigma}\right) = \sigma\left(\frac{L/K}{\mathfrak{p}}\right)\sigma^{-1}$, where automorphisms act from left on elements and ideals.*

iii) *Let F/K be a normal subextension of L/K; then the restriction of $\left(\frac{L/K}{\mathfrak{p}}\right)$ to F coincides with $\left(\frac{F/K}{\mathfrak{p}}\right)$.*

iv) *If F/K is a normal subextension of L/K, and if F is the fixed field of $H \subseteq G$, then $\mathfrak{p}$ splits completely in F/K if and only if $\left(\frac{L/K}{\mathfrak{p}}\right) \in H$.*

v) If $L = L_1 L_2$ is a compositum of abelian extensions L_1/K and L_2/K such that $L_1 \cap L_2 = K$, then $\left(\frac{L/K}{\mathfrak{p}}\right) = \left(\frac{L_1/K}{\mathfrak{p}}\right) \cdot \left(\frac{L_2/K}{\mathfrak{p}}\right)$, where e.g. the $\left(\frac{L_1/K}{\mathfrak{p}}\right)$ on the right hand side is the extension of $\left(\frac{L_1/K}{\mathfrak{p}}\right) \in \mathrm{Gal}\,(L_1/K)$ to $\mathrm{Gal}\,(L/K)$ that induces the identity on L_2.

vi) If F is a subfield of L/K, then $\left(\frac{L/K}{\mathfrak{p}}\right) = \left(\frac{L/F}{\mathfrak{p}}\right)^f$, where f is the inertia degree of $\mathfrak{P}$ in L/F.

We have already remarked that, for $L = \mathbb{Q}(\zeta_p)$ and primes $q > 0$ different from p, we have $\sigma_q = \left(\frac{L/\mathbb{Q}}{q}\right)$. For quadratic fields $L = \mathbb{Q}(\sqrt{d})$ and primes $p \nmid d$, Proposition 2.21 tells us that

$$\left(\frac{L/\mathbb{Q}}{(q)}\right) = \begin{cases} 1 & \text{if } \left(\frac{d}{q}\right) = +1, \\ \sigma & \text{if } \left(\frac{d}{q}\right) = -1, \end{cases}$$

where σ is the nontrivial automorphism of $L/\mathbb{Q}$. In other words: for quadratic fields, the Artin symbol $\left(\frac{L/\mathbb{Q}}{(q)}\right)$ can be identified with the Kronecker symbol $\left(\frac{d}{q}\right)$ via the action on $\sqrt{d}$, i.e. by $\left(\frac{L/\mathbb{Q}}{(q)}\right)\sqrt{d} = \left(\frac{d}{q}\right)\sqrt{d}$. This relation generalizes to n-th power residue symbols; see Section 4.1. Here we will be content with

Proposition 3.11. *Let p be a prime, and let K denote the subfield of degree $d \mid p-1$ of $\mathbb{Q}(\zeta_p)$. Then a prime $q > 0$ different from p is a d-th power residue modulo p if and only if $\left(\frac{K/\mathbb{Q}}{q}\right) = 1$.*

Proof. Let f denote the inertia degree of a prime ideal $\mathfrak{q}$ in $\mathbb{Q}(\zeta_p)$ above q, and put $fg = p - 1$. Then we find that

$$q \equiv x^d \bmod p \iff q^{(p-1)/d} \equiv 1 \bmod p \iff f \mid \tfrac{p-1}{d}, \text{ i.e. } d \mid g$$
$$\iff q \text{ splits in } K/\mathbb{Q} \iff \left(\tfrac{K/\mathbb{Q}}{q}\right) = +1.$$

This completes the proof. □

As a matter of fact, this Proposition is true for arbitrary integers $q > 0$ not divisible by p, but the proof given here does not seem to work in this generality. Note that the special case $d = 2$ gives the quadratic reciprocity law.

Primality Tests

Primality tests have a long history; the first effective tests used Fermat's little theorem: if n is prime, then $a^n \equiv a \bmod n$, so any integer that doesn't pass this test must be composite. Much later it was realized that Euler's criterion is a better indicator of primes: the equation $a^{(n-1)/2} \equiv \left(\frac{a}{n}\right) \bmod n$ can be shown to fail for most $1 \leq a < n$ if n is composite. This led to the development of probabilistic primality tests ("compositeness test" would be more exact, but we do not want to dwell on linguistic problems here): integers that passed this test for many a were declared to be "probable primes", whatever that means. Later, more sophisticated tests were developed, and not too long ago it was noticed by Lenstra [Len] that a large class of these tests can be described within a quite general framework. The basic observation is that an integer $n > 1$ is prime if and only if every divisor $r > 0$ of n is a power of n.

This trivial remark acquires substance by transferring the condition that $r = n^a$ to some finite group via a suitably defined homomorphism. The general setup is this: pick a subgroup H of $\mathbb{Q}^\times$ containing all integers $r > 0$ dividing n, a finite group G, and a homomorphism $\phi : H \longrightarrow G$. Now if n is prime, then for all positive $r \mid n$ we have $r = n^a$ for some $a \geq 0$, hence $\phi(r) = \phi(n)^a$; in other words: if n is prime, then $\phi(r)$ belongs to the subgroup $\langle\phi(n)\rangle \subseteq G$ generated by $\phi(n)$. Since ϕ is a homomorphism, it is clearly sufficient to check this for prime divisors r of n.

Definition. We say that an integer $n > 1$ passes the ϕ-primality test if $\phi(r) \in \langle\phi(n)\rangle$ for every positive prime divisor r of n.

Clearly, primes pass ϕ-primality tests for any choice of $\phi : H \longrightarrow G$. The big problem is of course this: how do we check whether n passes the ϕ-primality test without knowing its prime divisors beforehand? This is hard to answer in general, but can be done in some concrete cases. Let us therefore choose an abelian extension $K/\mathbb{Q}$ with discriminant prime to n and put $G = \operatorname{Gal}(K/\mathbb{Q})$. Let H denote the subgroup of $\mathbb{Q}^\times$ consisting of all fractions $\frac{a}{b}$ with $(a, \operatorname{disc} K) = (b, \operatorname{disc} K) = 1$; since $(n, \operatorname{disc} K) = 1$ by assumption, we see that all $r \mid n$ are contained in H. Note that the Artin symbol defines a homomorphism $\phi : H \longrightarrow G$ via $\phi(\frac{a}{b}) = \bigl(\frac{K/\mathbb{Q}}{(a)}\bigr)\bigl(\frac{K/\mathbb{Q}}{(b)}\bigr)^{-1}$: indeed, since a and b are prime to $\operatorname{disc} K$, their prime factors are unramified, and the Artin symbols are defined. Given an integer $n > 1$, let F denote the fixed field of $\phi(n)$ and put $\mathcal{O} = \mathcal{O}_F$. Then we claim

Lemma 3.12. *With the notation as above, we have* i) $\Longrightarrow$ ii) $\Longrightarrow$ iii) *for the following assertions:*

i) The integer n is prime.
ii) There exists a ring homomorphism $\mathcal{O} \longrightarrow \mathbb{Z}/n\mathbb{Z}$.
iii) The integer n passes the ϕ-primality test.

Proof. i) $\Longrightarrow$ ii): Assume that n is prime; then the Artin symbol $\phi(n)$ generates the decomposition group of n in $K/\mathbb{Q}$, hence n splits completely in F, and so any prime ideal $\mathfrak{p}$ in $\mathcal{O}$ above n satisfies $\mathcal{O}/\mathfrak{p} \simeq \mathbb{Z}/n\mathbb{Z}$. Thus the reduction modulo $\mathfrak{p}$ induces the desired ring homomorphism $\mathcal{O} \longrightarrow \mathcal{O}/\mathfrak{p} \simeq \mathbb{Z}/n\mathbb{Z}$.

ii) $\Longrightarrow$ iii): Let r be a prime divisor of n, and compose the the ring homomorphism (recall that ring homomorphisms map the identity to the identity) $\mathcal{O} \longrightarrow \mathbb{Z}/n\mathbb{Z}$ with the canonical projection $\mathbb{Z}/n\mathbb{Z} \longrightarrow \mathbb{Z}/r\mathbb{Z}$. This provides us with a ring homomorphism $\mathcal{O} \longrightarrow \mathbb{Z}/r\mathbb{Z}$, whose kernel $\mathfrak{r}$ is a prime ideal since $\mathcal{O}/\mathfrak{r} \simeq \mathbb{Z}/r\mathbb{Z}$ is a field. Since $\mathfrak{r}$ has norm r, we see that r

splits completely in $F/\mathbb{Q}$, hence F must be contained in the decomposition field of r in $K/\mathbb{Q}$. Since F is fixed by $\langle \phi(n) \rangle$ and the decomposition field by $\langle \phi(r) \rangle$, Galois theory implies that $\phi(r) \in \langle \phi(n) \rangle$. □

Observe that condition ii) does not involve the (unknown) prime divisors of n anymore. Next we relate the existence of a ring homomorphism in ii) to a divisibility property:

Lemma 3.13. *Assume that there exists an $\alpha \in \mathcal{O}$ such that the index $m = (\mathcal{O} : \mathbb{Z}[\alpha])$ is finite and relatively prime to n. Let $f \in \mathbb{Z}[X]$ denote the minimal polynomial of α. Then there exists a ring homomorphism $\psi : \mathcal{O} \longrightarrow \mathbb{Z}/n\mathbb{Z}$ if and only if there is an integer $a \in \mathbb{Z}$ such that $n \mid f(a)$.*

Proof. Assume that there is a ring homomorphism $\psi : \mathcal{O} \longrightarrow \mathbb{Z}/n\mathbb{Z}$; we can define a ring homomorphism $\widehat{\psi} : \mathbb{Z}[X]/(f) \longrightarrow \mathbb{Z}/n\mathbb{Z}$ as follows: using the isomorphism $\mathbb{Z}[X]/(f) \simeq \mathbb{Z}[\alpha]$ induced by $g(X) + (f) \longmapsto g(\alpha)$, it is clearly sufficient to define a ring homomorphism $\mathbb{Z}[\alpha] \longrightarrow \mathbb{Z}/n\mathbb{Z}$. But given $\beta \in \mathbb{Z}[\alpha]$, we know that $m\beta \in \mathcal{O}$, and since m is invertible modulo n, we simply map β to $m^{-1}\psi(m\beta)$. Now let $a \in \mathbb{N}$ be an integer such that $a + n\mathbb{Z} = \widehat{\psi}(X + (f))$. Then $\widehat{f(a)} = \widehat{\psi}(f(X) + (f)) = \widehat{\psi}(0) = 0 + n\mathbb{Z}$, and so $n \mid f(a)$ as claimed.

Conversely, given $a \in \mathbb{Z}$ with $n \mid f(a)$, we can define a ring homomorphism $\widehat{\psi} : \mathbb{Z}[X]/(f) \longrightarrow \mathbb{Z}/n\mathbb{Z}$ by mapping $X + (f)$ to $a + n\mathbb{Z}$. Since m is invertible modulo n, this can be extended to a ring homomorphism $\psi : \mathcal{O} \longrightarrow \mathbb{Z}/n\mathbb{Z}$ as before. □

In the special case where $K = \mathbb{Q}(\zeta_s)$ for some $s \mid n - 1$, the congruence $f(a) \equiv 0 \bmod n$ assumes a well known form:

Corollary 3.14. *Let $n > 3$ be an integer, $s > 2$ a divisor of $n - 1$, and put $K = \mathbb{Q}(\zeta_s)$. Then $F = K$ and $\mathcal{O} = \mathbb{Z}[\zeta_s]$, and there exists a ring homomorphism $\mathcal{O} \longrightarrow \mathbb{Z}/n\mathbb{Z}$ if the following congruences are satisfied:*

$$a^s \equiv 1 \bmod n \quad \text{and} \quad \gcd(a^{s/q} - 1, n) = 1 \text{ for every prime } q \mid s. \quad (3.2)$$

Proof. By Proposition 3.5, these congruences imply the existence of an integer a such that $\Phi_s(a) \equiv 0 \bmod n$, where $\Phi_s(X)$ is the minimal polynomial of ζ_s. □

The converse of this corollary will hold if $a^s \equiv 1 \bmod nd$, where $d = \gcd\left(\prod_{j \mid n, j < n} a^j - 1, n \right)$; since primality tests in practice never will discover a prime factor of n, we usually expect to have $d = 1$.

Note that, in order to verify the congruences (3.2), we have to know the prime factorization of s. This brings us to our first application:

Proposition 3.15 (Pocklington's Primality Test). *Let $n > 3$ be an integer, and assume that we know the prime factorization of $s \mid n - 1$. If the congruences (3.2) hold for some integer a, and if $s > \sqrt{n}$, then n is prime.*

Proof. From our general theory we know that $\phi(r) \in \langle\phi(n)\rangle$ for every prime divisor r of n. Since $n \equiv 1 \bmod s$, we have $\phi(n) = 1$, hence we deduce that $\phi(r) = 1$. Now $\phi(r)$ is the automorphism of K induced by $\zeta_s \longmapsto \zeta_s^r$, so $\phi(r) = 1$ implies $r \equiv 1 \bmod s$. But if $s > \sqrt{n}$ and if every prime divisor of n is congruent to 1 mod s, then n must be prime. □

The main problem in applying Pocklington's test is finding the prime factorization of a large enough part of $n-1$. The remark that $s > \sqrt[3]{n}$ is sufficient is therefore a great improvement: in fact, in this case we know as above that every prime divisor r of n satisfies $r \equiv 1 \bmod s$, and $s > \sqrt[3]{n}$ implies that n has at most two prime factors. Assume therefore that $n = (as+1)(bs+1)$ with $a, b > 0$ and $ab < s$. Then we find $0 < a + b \leq s$, so from $a + b \equiv \frac{n-1}{s} \bmod s$ we can deduce the value of $a + b$. But then $n = abs^2 + (a+b)s + 1$ gives us the product ab, hence a and b are the roots of the quadratic equation $x^2 - (a+b)x + ab$ and can easily be computed. If these roots are integers and lead to a prime divisor of n, then n is composite, otherwise n is prime.

Now for an integer n with sufficiently many digits, finding the prime factorization of $s \mid n-1$ for some $s > \sqrt[3]{n}$ may still turn out to be next to impossible. The first idea when thinking about improvements is probably to go and see what happens upon replacing the condition $s > \sqrt[3]{n}$ by something like $s > \sqrt[4]{n}$. A better idea is to use divisors s of $n+1$ instead of $s \mid n-1$ (the Lucas-Lehmer test, for example, exploits the fact that, for Mersenne numbers $n = 2^p - 1$, the factorization of $n+1$ is known), but the bottleneck is the same for both methods. It is quite amazing, however, that these two methods can be combined, that is, we can use factors of $n-1$ and $n+1$ in the same primality test!

In Lenstra's setup, this combination comes quite naturally: it turns out to be sufficient that a) we know the prime factorization of s and b) that there exists a small integer t such that $n^t \equiv 1 \bmod s$. As above, we take $K = \mathbb{Q}(\zeta_s)$, but this time $\phi(n)$ has order t, hence $(K : F) = t$. Instead of constructing a ring homomorphism $\mathcal{O} \longrightarrow \mathbb{Z}/n\mathbb{Z}$, we can get away with constructing one between the ring extension $\mathbb{Z}[\zeta_s]$ of $\mathcal{O}$ and a suitably defined ring extension R of $\mathbb{Z}/n\mathbb{Z}$.

$$\begin{array}{ccc} \mathbb{Z}[\zeta_s] & \longrightarrow & R \\ \big\uparrow & & \big\uparrow \\ \mathcal{O} & \longrightarrow & \mathbb{Z}/n\mathbb{Z} \end{array}$$

Assume that we have a ring homomorphism $\widehat{\psi} : \mathbb{Z}[\zeta_s] \longrightarrow R$. If we know that $\widehat{\psi}(\mathcal{O}) \subseteq \mathbb{Z}/n\mathbb{Z}$, then the restriction of $\widehat{\psi}$ to $\mathcal{O}$ will give us the desired ring homomorphism $\psi : \mathcal{O} \longrightarrow \mathbb{Z}/n\mathbb{Z}$. As above, the existence of a ring homomorphism $\widehat{\psi}$ is seen to be equivalent to the existence of an element $a \in R$ such that $a^s = 1$ and $a^{s/q} \in R^\times$ for each prime $q \mid s$. The condition that $\widehat{\psi}(\mathcal{O}) \subseteq \mathbb{Z}/n\mathbb{Z}$ can be translated into a congruence as follows: first note that, in general, $\mathcal{O}$ is not generated as a ring by the conjugates of $\operatorname{Tr}_{K/F}(\zeta_s)$:

in fact, it is easy to show that this never holds if s is divisible by 4 (see Corollary 3.7). There is, however, a way around this mess:

Lemma 3.16. *Let K be a number field, and assume that $\mathcal{O}_K = \mathbb{Z}[\alpha]$ for some $\alpha \in \mathcal{O}_K$. Then if F is any subfield of K, its ring of integers $\mathcal{O}_F$ is generated as a ring by the coefficients of the minimal polynomial $f \in \mathcal{O}_F[X]$ of α.*

Proof. Let R denote the ring generated by these coefficients. Then $R[\alpha] = \mathbb{Z}[\alpha] = \mathcal{O}_K$, hence $\mathcal{O}_K = R \oplus \alpha R \oplus \ldots \oplus \alpha^{n-1} R$, where $n = \deg f = (K : F)$. But then $\mathcal{O}_F = F \cap \mathcal{O}_K = F \cap R[\alpha] = F \cap (R \oplus \ldots \oplus \alpha^{n-1} R) = (F \cap R) \oplus (F \cap \alpha R) \oplus \ldots \oplus (F \cap \alpha^{n-1} R) = R$ as claimed. □

Applying this lemma to the case at hand shows that $\mathcal{O}_F$ is generated as a ring by the coefficients of the minimal polynomial g of ζ_s over F.

Let us now consider the ring $R = (\mathbb{Z}/n\mathbb{Z})[X]/(h)$ for a suitably chosen irreducible polynomial $h \in (\mathbb{Z}/n\mathbb{Z})[X]$ of degree t. As in the case $R = \mathbb{Z}/n\mathbb{Z}$ above, the existence of an element $a \in R$ such that $\Phi_s(a) = 0$ implies the existence of a ring homomorphism $\widehat{\psi} : \mathbb{Z}[\zeta_s] \longrightarrow R$; the condition that $\widehat{\psi}(\mathcal{O}) \subseteq \mathbb{Z}/n\mathbb{Z}$ will be satisfied if, in addition, we have $g(X) \in (\mathbb{Z}/n\mathbb{Z})[X]$, where g is the minimal polynomial of ζ_s over F. Thus we have proved

Lemma 3.17. *With the notation as above, there exists a ring homomorphism $\psi : \mathcal{O} \longrightarrow \mathbb{Z}/n\mathbb{Z}$ if there is an element $a \in R$ such that $a^s = 1$, $a^{s/q} - 1 \in R^\times$ for all primes $q \mid s$, and $\prod_{j=0}^{t-1}(X - a^{n^j}) \in (\mathbb{Z}/n\mathbb{Z})[X]$.*

If n is an integer that passes this test for some s dividing $n^t - 1$, then every possible prime divisor of n must be congruent to $n^k \bmod s$ for some integer $k \in \{0, 1, \ldots, t-1\}$. This is due to the fact that the existence of ψ implies $\phi(r) \in \langle \phi(n) \rangle$, that is, $\left(\frac{K/\mathbb{Q}}{(r)}\right) = \phi(r) = \phi(n)^k = \left(\frac{K/\mathbb{Q}}{(n)}\right)^k$. But then $r \equiv n^k \bmod s$ since $K = \mathbb{Q}(\zeta_s)$.

Thus if $s > \sqrt{n}$, then it is sufficient to test whether any of the t remainders of $n^r \bmod s$ divides n: if not, then n is prime. There is also a variation of this test that works for $s > \sqrt[3]{n}$; in fact, combining the cyclotomic test here with one based on elliptic curves, it is possible improve this bound to $s > \sqrt[4]{n}$: see Mihailescu & Morain [MM].

Theorem 3.18 (Lenstra's Primality Test). *Let $n > 3$ be an integer, and assume that we know the prime factorization of $s \mid n^t - 1$ for some $t \geq 1$. If there is an $a \in R$ such that the conditions in Lemma 3.17 are satisfied, if $s > \sqrt{n}$, and if none of the integers smallest positive residues of $1, n, n^2, \ldots, n^t - 1 \bmod s$ divide n, then n is prime.*

Let us finally show how the Lucas-Lehmer test fits into this picture: For an odd prime p, consider the Mersenne number $n = 2^p - 1$. Put $s = 2^{p+1}$, $t = 2$, and $R = (\mathbb{Z}/n\mathbb{Z})[X]/(h)$ for $h = X^2 - \sqrt{2}X - 1$, where $\sqrt{2}$ denotes the element of $\mathbb{Z}/n\mathbb{Z}$ such that $\sqrt{2} \equiv 2^{(p+1)/2} \bmod n$. Put $a = X + h(\mathbb{Z}/n\mathbb{Z})[X]$;

then $h(a) = 0$ implies that $h(X) = (X-a)(X-b)$ for some $b \in R$, and we get $a + b = \sqrt{2}$ and $ab = -1$, hence $b = -a^{-1}$. Induction shows that $S_j \equiv a^{2^j} + b^{2^j} \bmod n$ for the integers S_j defined in Proposition 2.20.

Now if n is prime, then $\mathbb{Z}/n\mathbb{Z}$ and R are fields; since a and b are roots of an irreducible polynomial h with coefficients in the ground field, they are conjugated, so by Galois theory we must have $a^n = b$ (this is the Frobenius automorphism at work again). This implies that $a^{2^p} = a^{n+1} = ab = -1$, hence $S_{p-1} \equiv a^{2^{p-1}} + b^{2^{p-1}} = a^{2^{p-1}} + a^{-2^{p-1}} = 0 \bmod n$.

Conversely, assume that $n \mid S_{p-1}$. Then $a^{2^p} = -1$, hence $a^s = a^{2^{p+1}} = 1$ and $a^{s/2} - 1 = -2 \in R^\times$. Moreover, $a^n = a^{2^p - 1} = -a^{-1} = b$ shows that $(X-a)(X-a^n) = (X-a)(X-b) = h \in (\mathbb{Z}/n\mathbb{Z})[X]$. Thus there exists a ring homomorphism $\psi : \mathcal{O} \longrightarrow \mathbb{Z}/n\mathbb{Z}$, which in turn implies that $\phi(r) = \phi(n)^k$ for some $k \in \{0, 1\}$, and we find that each prime divisor r of n is congruent to 1 or n modulo s. But since $s > n$, this shows that n is prime.

3.3 Quadratic Gauss Sums

In Section 3.1 we have seen that the polynomial ψ_- changes its sign when the nontrivial automorphism σ of $\mathbb{Q}(\sqrt{p})$ is applied, and that this behaviour is connected with quadratic reciprocity. But if ψ_- changes its sign then so does every coefficient of ψ_-, and so it is natural to ask if we can simplify the proof by looking at just one coefficient instead of working with the whole polynomial. The first non-vanishing coefficient of $\psi_- = \sum a_j x^j$ is

$$a_{(p-3)/2} = \sum_r \zeta_p^r - \sum_n \zeta_p^n,$$

where r and n run through the quadratic residues and non-residues mod p, respectively. This coefficient can also be written as

$$\tau = \sum_{a=1}^{p-1} \left(\frac{a}{p}\right) \zeta^a.$$

τ is called the *Gauss sum* of the quadratic character $\left(\frac{\cdot}{p}\right)$ or simply quadratic Gauss sum on $\mathbb{F}_p$, and, once it has been suitably generalized, is one of the most far-reaching tools for proving reciprocity laws without class field theory.

As a motivation we will compute the Gauss sum τ for the quadratic character $\chi = \left(\frac{-3}{p}\right)$: let $\rho = \frac{1}{2}(-1 + i\sqrt{3})$; then $\tau = \rho - \rho^2 = i\sqrt{3}$. The quadratic reciprocity law for $q = -3$ is an easy corollary: on the one hand we have $\tau^p = (\rho - \rho^2)^p \equiv \rho^p - \rho^{2p} \equiv \left(\frac{p}{3}\right)\tau \bmod p$, and on the other hand $\tau^p = (i\sqrt{3})^{p-1}\tau = (-3)^{(p-1)/2}\tau \equiv \left(\frac{-3}{p}\right)\tau \bmod p$. Comparing both congruences and observing that τ is relatively prime to 3 in $\mathbb{Q}(\rho)$ yields $\left(\frac{-3}{p}\right) = \left(\frac{p}{3}\right)$.

94   3. Cyclotomic Number Fields

We can treat $\left(\frac{5}{p}\right)$ in a similar manner: let ζ be a primitive fifth root of unity. The splitting of $\Phi_5(X)$ over $\mathbb{Q}(\sqrt{5})$ given in the last section (see (3.1)) yields

$$\tau = \zeta - \zeta^2 - \zeta^3 + \zeta^4 = \frac{\sqrt{5}-1}{2} + \frac{1+\sqrt{5}}{2} = \sqrt{5},$$

and the same technique as above shows $\left(\frac{5}{p}\right) = \left(\frac{p}{5}\right)$.

Arithmetic Properties of Quadratic Gauss Sums

After studying these examples, we will attack the general case. All we need to prove quadratic reciprocity law are the following simple properties of τ:

Proposition 3.19. *Let τ be the Gauss sum on $\mathbb{F}_p$; then*
1. $\tau^2 = (-1)^{(p-1)/2} p = p^*$;
2. $\tau^{q-1} \equiv \left(\frac{q}{p}\right) \bmod q$.

We will use the following lemma in the proof:

Lemma 3.20. *For an odd prime p we have* $\displaystyle S = \sum_{c=1}^{p-1} \left(\frac{c}{p}\right) = 0.$

Proof. Let a be a quadratic non-residue mod p. Then

$$-S = \left(\frac{a}{p}\right) \sum_{c=1}^{p-1} \left(\frac{c}{p}\right) = \sum_{c=1}^{p-1} \left(\frac{ac}{p}\right) = \sum_{b=1}^{p-1} \left(\frac{b}{p}\right) = S,$$

where we have used that $b = ac$ runs through $(\mathbb{Z}/p\mathbb{Z})^\times$ if c does. □

Proof of Proposition 3.19. Let us compute τ^2; we find

$$\tau^2 = \sum_{a=1}^{p-1} \left(\frac{a}{p}\right) \zeta^a \cdot \sum_{b=1}^{p-1} \left(\frac{b}{p}\right) \zeta^b = \sum_{a,b} \left(\frac{ab}{p}\right) \zeta^{a+b}.$$

Now substitute $b = ac$; this yields

$$\tau^2 = \sum_{a,c} \left(\frac{c}{p}\right) \zeta^{a+ac} = \sum_{c=1}^{p-1} \left(\frac{c}{p}\right) \sum_{a=1}^{p-1} (\zeta^{1+c})^a.$$

But if $c \neq -1$, then ζ^{1+c} is a primitive p-th root of unity, and $\sum_{a=1}^{p} \zeta^a = 0$ shows $\sum_{a=1}^{p-1} \zeta^a = -1$, thus

$$\tau^2 = -\sum_{c=1}^{p-2}\left(\frac{c}{p}\right) + \left(\frac{-1}{p}\right)\sum_{a=1}^{p-1} 1 = -\sum_{c=1}^{p-2}\left(\frac{c}{p}\right) + (p-1)\left(\frac{-1}{p}\right)$$

$$= p^* - \sum_{c=1}^{p-1}\left(\frac{c}{p}\right) = p^*.$$

The congruence $\tau^{q-1} \equiv (q/p) \bmod q$ is proved exactly as in the special cases above:

$$\tau^q = \left(\sum_{a=1}^{p-1}\left(\frac{a}{p}\right)\zeta^a\right)^q \equiv \sum_{a=1}^{p-1}\left(\frac{a}{p}\right)\zeta^{aq} = \left(\frac{q}{p}\right)\tau \bmod q.$$

But τ and q are relatively prime (observe that (τ, q) divides $(\tau^2, q) = (p^*, q) = 1$) and our claim follows. $\square$

The proof of the quadratic reciprocity law (this is Gauss's sixth proof) is now almost trivial: it follows from the congruence

$$\left(\frac{p^*}{q}\right) \equiv (p^*)^{\frac{q-1}{2}} = \tau^{q-1} \equiv \left(\frac{q}{p}\right) \bmod q,$$

because $-1 \not\equiv +1 \bmod q$.

But we have proved even more: the equation $\tau^2 = p^*$ shows in a very concrete manner that the quadratic number field $F = \mathbb{Q}(\sqrt{p^*})$ is contained in $K = \mathbb{Q}(\zeta_p)$, a fact which we have already observed in Section 3.1. Conversely, the conceptual proof there can also be used to show that $\tau^2 = p^*$, as Dedekind [149] has shown: by Proposition 3.6, the conjugates of $\operatorname{Tr}_{K/F}\zeta$ form an integral basis of F; the conjugates here are the quadratic periods $\eta = \operatorname{Tr}_{K/F}\zeta = \sum \zeta^r$ and $\eta' = \sum \zeta^n$, where r and n run through the quadratic residues and nonresidues modulo p, respectively. Since η and $\eta' = -1 - \eta$ form an integral basis of F, we conclude that $p^* = \operatorname{disc} F = (\eta - \eta')^2(\eta + \eta')^2 = \tau^2$ "from entirely general principles and without calculation", as Dedekind proudly remarks.

Note that the quadratic periods occur as the coefficients of $x^{(p-3)/2}$ in the polynomials ϕ_1 and ϕ_2 defined in Section 3.1. We have seen that $\eta + \eta' = -1$ and that $\eta - \eta' = \tau$. This gives $4\eta\eta' = (\eta+\eta')^2 - (\eta-\eta')^2 = 1 - p^*$, hence η and η' are roots of the polynomial $\Psi(X) = (X-\eta)(X-\eta') = X^2 + X + \frac{1}{4}(1-p^*) \in \mathbb{Z}[X]$, which is called the *quadratic period equation*. We leave it to the reader to give another proof of the quadratic reciprocity law using $\Psi(X)$.

Sign of Quadratic Gauss Sums

We have shown above that $\tau^2 = p^*$. Thus $\tau = \pm\sqrt{p^*}$, and it is natural to ask whether we can determine the correct sign. We can easily show that the sign depends on the choice of the root of unity ζ in the definition of τ. In fact, put $\tau_p(a) = \sum_{t=0}^{p-1}\left(\frac{t}{p}\right)\zeta_p^{at}$; then

96   3. Cyclotomic Number Fields

$$\tau_p(a) = \left(\frac{a}{p}\right)\sum\left(\frac{at}{p}\right)\zeta_p^{at} = \left(\frac{a}{p}\right)\tau_p(1), \tag{3.3}$$

hence replacing ζ_p in the definition of τ by ζ_p^a for some quadratic non-residue modulo p will change the sign of τ. However, if we fix ζ_p by putting $\zeta_p = \exp(\frac{2\pi i}{p})$, then the question regarding the sign of the quadratic Gauss sum makes sense.

Actually, the problem of determining the correct sign is due to Gauss himself: he needed it for his fourth proof of the quadratic reciprocity law. The result which took him four years to prove (see his letter [Gau] to Olbers) looks quite innocent:

Proposition 3.21. *Let $p > 0$ be an odd prime; then*

$$\tau = \begin{cases} \sqrt{p} \ \text{if } p \equiv 1 \bmod 4, \\ i\sqrt{p} \ \text{if } p \equiv 3 \bmod 4. \end{cases}$$

Proof. See Section 8.7. Note that this result agrees with our computations for $p = 3$ and $p = 5$ at the beginning of this section. □

In order to derive the quadratic reciprocity law from this, introduce the sum

$$\tau_m(a) = \sum_{t=0}^{m-1} \exp\left(\frac{2\pi i a t^2}{m}\right);$$

then $\tau_p(1)$ is the Gauss sum τ defined above (See Exercise 3.6).

Proposition 3.22. *Let $a, m, n \geq 1$ be pairwise coprime odd integers. Then $\tau_{mn}(a) = \tau_m(an)\tau_n(am)$.*

Proof. Let u and v run through a complete system of residues modulo m and n, respectively. Then $t = mv + nu$ runs through a complete system of residues modulo mn, and we get

$$\tau_{mn}(a) = \sum_t \exp\left(\frac{2\pi i a t^2}{mn}\right) = \sum_{u,v} \exp\left(\frac{2\pi i a(mv + nu)^2}{mn}\right).$$

Now $(mv + nu)^2 \equiv m^2v^2 + n^2u^2 \bmod mn$ shows that

$$\tau_{mn}(a) = \sum_{u,v} \exp\left(\frac{2\pi i a u^2}{m}\right)\exp\left(\frac{2\pi i a v^2}{n}\right),$$

and our claim follows. □

Now the quadratic reciprocity law can be proved as follows: for different odd primes p, q we have

$$\tau_{pq}(1) = \tau_p(q)\tau_q(p) = \left(\frac{p}{q}\right)\left(\frac{q}{p}\right)\tau_p(1)\tau_q(1),$$

where we have used (3.3). Plugging in the values of $\tau_{pq}(1)$, $\tau_p(1)$ and $\tau_q(1)$ known from Proposition 3.21, we get the quadratic reciprocity law.

Although Proposition 3.21 can be proved directly in many ways, it is usually proved along with more general results on exponential sums defined for integers $p \in \mathbb{Z}$ and $q \in \mathbb{N}$ by

$$S(p,q) = \sum_{r=0}^{q-1} \exp\left(-\frac{\pi i p r^2}{q}\right).$$

These Gauss sums satisfy the 'reciprocity law'

$$\frac{1}{\sqrt{q}} S(p,q) = \frac{\exp(-\frac{\pi i}{4})}{\sqrt{p}} S(-q,p).$$

Putting $p = 2$ and assuming that q is odd this gives $\tau = \frac{1-i^q}{1-i}\sqrt{q}$, i.e. Proposition 3.21. See the Notes for references.

3.4 Cyclotomic Units

In this section we are going to describe a proof of the quadratic reciprocity law by studying units in cyclotomic number fields. Let p denote an odd prime $\equiv 1 \bmod 4$ and put $Z = \exp\frac{\pi i}{p}$ and $\zeta = Z^2$; then $Z \in \mathbb{Q}(\zeta)$, and $Z^j - Z^{-j}$, where $1 \leq j < p$, generates the prime ideal $\mathfrak{p}$ above p. Now put

$A = \{1 \leq a \leq \frac{p-1}{2} : (\frac{a}{p}) = +1\}$, and $B = \{1 \leq b \leq \frac{p-1}{2} : (\frac{b}{p}) = -1\}$.

Then

$$\eta = \frac{\prod_{b \in B}(Z^b - Z^{-b})}{\prod_{a \in A}(Z^a - Z^{-a})} \tag{3.4}$$

is a unit in $\mathbb{Q}(\zeta)$, because A and B have the same cardinality. We claim that η is in fact a unit in $\mathbb{Q}(\sqrt{p})$. Before we will give a proof of this claim, we transform the formula for η a bit:

Lemma 3.23. *With the notation as above we have*

$$\widetilde{\eta} = \frac{\prod(\zeta^b - \zeta^{-b})}{\prod(\zeta^a - \zeta^{-a})} = \begin{cases} \eta, & \text{if } p \equiv 1 \bmod 8, \\ \eta^{-1}, & \text{if } p \equiv 5 \bmod 8. \end{cases}$$

Proof. First assume that $p \equiv 1 \bmod 8$; then 2 is a quadratic residue mod p, and for $b \in B$ we either have $2b \equiv b' \bmod p$ or $2b \equiv -b' \bmod p$ for some $b' \in B$. In the first case we find $b' = 2b$ and $Z^{2b} - Z^{-2b} = Z^{b'} - Z^{-b'}$; in the second case we get $2b = p - b'$ and therefore $Z^{2b} - Z^{-2b} = Z^{p-b'} - Z^{-p+b'} = Z^{b'} - Z^{-b'}$, since $Z^p = -1$. This clearly shows that

98   3. Cyclotomic Number Fields

$$\prod_{b \in B}(Z^{2b} - Z^{-2b}) = \prod_{b \in B}(Z^b - Z^{-b}), \tag{3.5}$$

and similarly, $\prod_{a \in A}(Z^{2a} - Z^{-2a}) = \prod_{a \in A}(Z^a - Z^{-a})$. This implies

$$\widetilde{\eta} = \frac{\prod(\zeta^b - \zeta^{-b})}{\prod(\zeta^a - \zeta^{-a})} = \frac{\prod(Z^{2b} - Z^{-2b})}{\prod(Z^{2a} - Z^{-2a})} = \frac{\prod(Z^b - Z^{-b})}{\prod(Z^a - Z^{-a})} = \eta.$$

Now suppose that $p \equiv 5 \mod 8$. For the $b \in B$ we find $2b \equiv \pm a$ for some $a \in A$, and the same proof as above yields

$$\prod_{b \in B}(Z^{2b} - Z^{-2b}) = \prod_{a \in A}(Z^a - Z^{-a}), \tag{3.6}$$

and thus

$$\widetilde{\eta} = \frac{\prod(\zeta^b - \zeta^{-b})}{\prod(\zeta^a - \zeta^{-a})} = \frac{\prod(Z^{2b} - Z^{-2b})}{\prod(Z^{2a} - Z^{-2a})} = \frac{\prod(Z^a - Z^{-a})}{\prod(Z^b - Z^{-b})} = \eta^{-1}. \quad \square$$

In order to prove that $\eta \in \mathbb{Q}(\sqrt{p}\,)$, we have to show that the automorphisms $\zeta \longmapsto \zeta^r$, where r ranges through the quadratic residues modulo p, fix η or, equivalently, $\widetilde{\eta}$. This is done as follows: if r is a quadratic residue modulo p, then $ra_i \equiv \pm a_j \mod p$ and $rb_i \equiv \pm b_j \mod p$, where $a_i \in A$ and $b_i \in B$. Let μ_A and μ_B denote the number of negative signs occurring as a and b run through A and B, respectively; since $A \cup B$ is a half-system modulo p, Gauss's Lemma 1.10 says that $(-1)^{\mu_A + \mu_B} = (\frac{r}{p}) = +1$. But

$$\prod(\zeta^{ra} - \zeta^{-ra}) = (-1)^{\mu_A} \prod(\zeta^a - \zeta^{-a}), \text{ and}$$
$$\prod(\zeta^{rb} - \zeta^{-rb}) = (-1)^{\mu_B} \prod(\zeta^b - \zeta^{-b}).$$

Thus $\sigma_r(\eta) = (-1)^{\mu_A + \mu_B}\eta = \eta$, and η is a unit in $\mathbb{Q}(\sqrt{p}\,)$ as claimed. Similarly, it follows that $\sigma_n(\eta) = -\eta^{-1}$ for quadratic non-residues $n \mod p$; this implies $N(\eta) = \eta\sigma_n(\eta) = -1$. But this can only happen if $N\varepsilon = -1$, where $\varepsilon > 1$ is the fundamental unit of $\mathbb{Q}(\sqrt{p}\,)$, and if $\eta = \pm\varepsilon^h$ for some odd integer h. In fact we can even show that $\eta = \varepsilon^h$, because both η and ε are positive. We have proved

Proposition 3.24. *If $p \equiv 1 \mod 4$ is prime, then η is a unit with norm -1; in particular, $\eta = \varepsilon^h$, where ε is the fundamental unit of $\mathbb{Q}(\sqrt{p}\,)$, h is an odd integer, and ε has norm -1.*

Now put $N = \prod(Z^b - Z^{-b})$; since $Z^b - Z^{-b} = Z^{p-b} - Z^{-(p-b)}$ we find that $N^2 = \prod_{b \in B}(Z^b - Z^{-b})(Z^{p-b} - Z^{-(p-b)}) = \prod_n(Z^n - Z^{-n})$, where the product ranges over all quadratic non-residues n modulo p in the interval $[1, p-1]$. Similarly, we have $R^2 = \prod(Z^r - Z^{-r})$, where $R = \prod(Z^a - Z^{-a})$ and r ranges through the quadratic residues modulo p.

Proposition 3.25. *With the above notation, we have $NR = (\frac{2}{p})\sqrt{p}$.*

3.4 Cyclotomic Units

Proof. We first show that $N^2R^2 = p$ and then compute the sign of NR:

$$N^2R^2 = \prod_{k=1}^{p-1}(Z^k - Z^{-k}) = \prod_{k=1}^{p-1} Z^k \cdot \prod_{k=1}^{p-1}(1 - Z^{-2k}) = N_{K/\mathbb{Q}}(1-\zeta) = p.$$

Next observe that $Z^k - Z^{-k} = 2i \cdot \sin\frac{k\pi}{p}$ and that $\sin x$ is positive for $0 < x < \pi$. This gives $\text{sign}(NR) = i^{(p-1)/2} = (-1)^{(p-1)/4} = (2/p)$, and the last equation holds because $p \equiv 1 \bmod 4$. □

Define $S = N^2, T = R^2$ and observe that $(-1)^{(p-1)/4} = (2/p)$ for primes $p \equiv 1 \bmod 4$; if $(q/p) = 1$, we find

$$S^q = \left(\frac{2}{p}\right)\prod(Z^n - Z^{-n})^q \equiv \left(\frac{2}{p}\right)\prod(Z^{qn} - Z^{-qn}) = S \bmod q,$$

as well as $T^q \equiv T \bmod q$. This gives $S^{q-1} \equiv 1 \bmod q$ and $T^{q-1} \equiv 1 \bmod q$. If, however, $(q/p) = -1$, then

$$S^q = \left(\frac{2}{p}\right)\prod(Z^n - Z^{-n})^q \equiv \left(\frac{2}{p}\right)\prod(Z^{qn} - Z^{-qn}) = T \bmod q,$$

hence $S^{q+1} \equiv ST = N^2R^2 = p \bmod q$. We have shown

$$S^{q-(q/p)} \equiv \begin{cases} 1 \bmod q & \text{if } (p/q) = +1, \\ p \bmod q & \text{if } (p/q) = -1. \end{cases} \quad (3.7)$$

A part of the quadratic reciprocity law now follows easily. Recall that we still assume that $p \equiv 1 \bmod 4$; then from $\varepsilon^h = N/R$ we infer that

$$\left(\frac{2}{p}\right)\varepsilon^h\sqrt{p} = N^2 = S. \quad (3.8)$$

1. Assume that $\left(\frac{q}{p}\right) = 1$; then $\varepsilon^{q-1} \equiv 1 \bmod q$, hence

$$\left(\frac{p}{q}\right) \equiv \varepsilon^{q-1}\left(\frac{p}{q}\right) \equiv \varepsilon^{q-1}(\sqrt{p})^{q-1} \equiv S^{q-1} \equiv 1 \bmod q$$

by (3.7), hence we have $\left(\frac{q}{p}\right) = \left(\frac{p}{q}\right)$ in this case.

2. The case $\left(\frac{q}{p}\right) = -1$ is a bit different: recall that ε has negative norm and that $\varepsilon^{q+1} \equiv N\varepsilon = -1 \bmod q$ (see Proposition 2.21), and observe

$$\left(\frac{p}{q}\right)p \equiv -\varepsilon^{q+1}\left(\frac{p}{q}\right)p \equiv -\left(\frac{2}{p}\right)\varepsilon^{q+1}\sqrt{p}^{q+1} \equiv -S^{q+1} \equiv -p \bmod q.$$

Thus $\left(\frac{p}{q}\right) = -1$.

This proves the quadratic reciprocity law if one of the primes is $\equiv 1 \bmod 4$. In the case $p \equiv q \equiv 3 \bmod 4$, one has to work harder (see Exercise 3.19).

The analytic class number formula for real quadratic number fields (see Hasse [Has, p. 14]) tells us

Theorem 3.26. *Let k be a quadratic number field with discriminant d, fundamental unit $\varepsilon = \varepsilon_d$ and class number h. Then*

$$\varepsilon^h = \eta = \frac{\prod(Z^b - Z^{-b})}{\prod(Z^a - Z^{-a})}, \quad \text{where } a \text{ and } b \text{ run through the sets}$$
$$A = \left\{a \in \mathbb{N} : 1 \leq a < \tfrac{d}{2}, (a,d) = 1, \left(\tfrac{d}{a}\right) = +1\right\} \quad \text{and}$$
$$B = \left\{b \in \mathbb{N} : 1 \leq b < \tfrac{d}{2}, (b,d) = 1, \left(\tfrac{d}{b}\right) = -1\right\}. \quad \square$$

This theorem, together with the quadratic reciprocity law, shows that the odd integer h in Proposition 3.24 is nothing but the class number of $\mathbb{Q}(\sqrt{p})$. The connection between Gauss sums, class numbers and reciprocity will become even more transparent in Chapter 11, where we will use the prime ideal factorization of Gauss sums to prove Eisenstein's reciprocity law as well as nontrivial results about class groups of quadratic and cyclotomic number fields.

NOTES

The theory of cyclotomy originated with Gauss's investigation of the solution of $x^n = 1$ in radicals. Euler [Eul] showed that this was possible for $n < 11$ (for these values of n, knowledge about the solution of quadratic, cubic and quartic equations by radicals is sufficient), while the treatment of the case $n = 11$ is due to Vandermonde (see Neumann [615], H. Lebesgue [Lb2] and Loewy [Lo2]).

Gauss succeeded for prime values $n = p$, and found as a byproduct that it is possible to construct a regular p-gon, p an odd prime, if $p = 2^{2^n} + 1$ is a Fermat prime; the cases $p = 3$ and $p = 5$ were already known to the Greeks. Although Gauss claimed that the converse also holds, this was only proved by Wantzel [Wan], with a small gap, as was noticed by Loewy [Lo1, p. 108]. Later, complete proofs were given by Petersen and Pierpont.

Gauss worked out the formulas for $p = 17$ in his Disquisitiones Arithmeticae [263]; for explicit constructions see Erchinger (apparently unpublished; see Bachmann's article [Ba1]), Fontené [Fon], Kirwan [Kir], Lebesgue [Lb1], Paucker [Pau], Pfleiderer [Pfl], Schumacher [Sm], Schröter [Sch], von Staudt [vSt], Strnad [Str] and Tafelmacher [Taf]. Kubota [Kub] discussed the abelian extensions behind the construction of the 17-gon. The case $p = 257$ was treated by Cayley [Cay], Hagge [Hag], Richelot [Ric], and very recently again by Gottlieb [Got], $p = 65537$ by Hermes [Her]. See also Mitzscherling's book [Mit] on the subject. A proof of Gauss's result will be given in Exercise

3.14. More details – especially in connection with the generalization of Gauss result to the lemniscate – can be found in Chapter 8. A more recent book on constructions with ruler and compass is Carrega [Car].

Starting with Gauss's work, cyclotomy acquired the meaning of 'studying cyclotomic numbers'; these are defined as follows: let $n \geq 2$ be an integer and consider primes $p = mn + 1$. Fix a primitive root g of p. Then the *cyclotomic numbers* of order n are defined as

$$(r, s) = \#\{(a, a+1) : 1 \leq a \leq p - 2,$$
$$a \equiv g^{kn+r}, b \equiv g^{ln+s} \bmod p \text{ for some integers } k, l \in \mathbb{N}\}.$$

Choosing a different primitive root will at most permute the cyclotomic numbers. They were determined by Gauss for $m = 3$ in [263] and for $m = 4$ in [272]; we will study a few of their properties in Chapter 10. The parity of (00) contains information on the k-th power character of 2 (see Exercise 3.21). The theory of cyclotomy was covered in the textbooks of Bachmann [27] (see Dedekind's comments [Ded]) and Fueter [246] and in the survey of Rajwade [667].

The proof of the quadratic reciprocity law by comparing the splitting of primes in $\mathbb{Q}(\zeta_p)$ and $\mathbb{Q}(\sqrt{p^*})$ that we gave in Section 3.1 goes back to Hilbert's Zahlbericht [368, §122], where the principal idea is credited to Kronecker [442]. It is however related to the proof of Mathieu [567] who used the quadratic period equation. Incidentally, Mathieu's proof is one of those missing from the lists of both Baumgart [38] and Bachmann [26, 27]; von Staudt's cyclotomic proof shared the same fate, as was noticed by Stengel [751], and the same goes for the proof of Schönemann [722].

The proof in Section 3.1 using the factorization of the cyclotomic polynomial is the 'missing link' between Kronecker's comparison of the splitting of primes and the proof using Gauss sums. The factorization of the cyclotomic polynomial $\Phi_p(X)$ over the quadratic number field $\mathbb{Q}(\sqrt{p^*})$ was discovered by Gauss (Disquisitiones Arithmeticae, Art. 357); he wrote it in the form $4\Phi_p(X) = Y^2 \pm pZ^2$, where the positive sign holds if and only if $p \equiv 1 \bmod 4$. Obviously we have $Y(X) = \phi_1(X) + \phi_2(X)$ and $\sqrt{p^*}Z(X) = \phi_1(X) - \phi_2(X)$. See also Teege [797].

Normal integral bases will reappear in the Notes of Chapter 11. The weak form of Proposition 3.8 on the k-th period equation is can be found in E. Lehmer [498], the stronger form in K.S. Williams [859]. One direction was already proved by Kummer [462], while the validity of the converse was pointed out by Glaisher in a footnote to Smith's report [743, p. 105]. It has not been noticed, however, that the result was already known to Gauss: in a preliminary version of the Disquisitiones Arithmeticae, right after the seventh and eight proof of the quadratic reciprocity law built on the quadratic period equation, he starts discussing more general periods and writes:

> Quamvis ad casum, ubi ν est numerus primus, hic nos restrinxerimus, tamen etiam, si ν est compositus, theremata analoga haud magno

102   3. Cyclotomic Number Fields

negotio determinari possunt, quod fusius exponere brevitatis gratia nunc non licet.

Manifestum est, similes observationes etiam de maiori periodorum multitudine formari posse. Ita si $\frac{\nu-1}{m}$ per 3 dividitur i.e. si p est residuum cubicum numeri primi ν, aequatio, per quam radices aequationis $x^\nu = 1$ in tres periodos distribuuntur quamque in Cap. VI a priori determinandam docuimus, solubilis erit secundum modulum p et vice versa. Ita ex. gr. congruentia $x^3 + xx - 2x - 1 \equiv 0$ secundum modulum primum quemcunque, qui est formae $7n \pm 1$, resolvi potest, si vero aliam formam habeat, non poterit.[2]

For other investigations of connections between e-th power residuacity of primes and the divisors of the e-th period equation, see Pépin [641], Sylvester [781, 782, 785, 786], Huber [376], D.H. Lehmer & E. Lehmer [494] and E. Lehmer [495, 496, 498, 502].

Section 3.2 on primality tests is taken from Lenstra's article [Len]; the proof of Lemma 3.16 is from an email from Sept. 19, 1999. In Exercise 7.24 we show how to subsume a primality test for integers of the form $A3^k - 1$ under Lenstra's test. For an in-depth discussion of the history of primality tests, see H.C. Williams [Wil] and Williams & Shallit [WS].

The section on cyclotomic units has been included as a preparation for E. Lehmer's proof of Scholz's reciprocity law in Chapter 5; unfortunately, a proof of the quadratic reciprocity law along these lines requires a large amount of work in the case where $p \equiv q \equiv 3 \bmod 4$. Related problems are addressed in Skolem [740], who uses the quadratic period equation to prove the quadratic reciprocity law. Connections between the cyclotomic polynomials and power residues are known since Kummer, and have been considered in detail by E. Lehmer (see also Sharifi [731]).

The definition of the quadratic Gauss sum given in the Section 3.3 differs by a factor of -1 from the one given in Chapter 4 below, which is why we have denoted the Gauss sum by τ in this chapter.

Proofs for Proposition 3.21 as well as for vast generalizations thereof have been given by Dirichlet [165, Dir] (using Fourier series), Cauchy [Cau] (using theta functions), and Kronecker [Kr2, Kr3, Kr4] (using complex integration); these can be found in Bachmann's book [Ba1]. See also Genocchi [296], Landau [Lan], Landsberg [Ldb], Lebesgue [Le1, Le2], Lerch [Ler] (who claims

---

[2] Although we restricted us to the case where ν is a prime number, we can determine analogous results also if ν is composite, which – for the sake of brevity – we refrain from discussing in detail.

It is clear that similar observations can also be made about a great number of periods. If, for example, $\frac{\nu-1}{m}$ is divisible by 3, i.e. if p is a cubic residue of the prime number ν, then the equation by which the roots of the equation $x^\nu = 1$ are divided into three periods and which we have shown how to determine a priori in Cap. VI, will have a solution modulo p and vice versa. Thus e.g. the congruence $x^3 + xx - 2x - 1 \equiv 0$ can be solved modulo an arbitrary prime number of the form $7n \pm 1$, but has no solution if the prime has an other form.

that some of Kronecker's proofs are incorrect), Mertens [579] (whose proof also contains a gap), Mordell [Mo1], Petr [Pet] (he uses theta functions), Salvadori [Sal], Schaar [703, 704], Schur [Sr], de Séguier [dSe], Teege [Te1, Te2], and Weber [We2]. For generalizations to number fields, check out Hecke [347] and Siegel [734].

The determination of the sign of the Gauss sum keeps attracting the interest of mathematicians: see the papers of Andrews [An], Bambah & Chowla [BC], Bressoud [Bre], Carlitz [Ca1, Ca3], Estermann [Est], Karamate & Tomic [KT], Mordell [Mo2, Mo3], Shanks [Sha], and Waterhouse [Wat].

The theorem of Kronecker-Weber; was first stated by Kronecker [Kr1] in 1853; the first proofs given by Weber in [We1] contain gaps; for a discussion of Weber's proofs, see Neumann [Neu] and Schappacher [Sca]. The first correct proof was given by Hilbert [Hil, 368]; the simple idea behind it is explained in Exercise 3.13.

Other proofs of the global Kronecker-Weber theorem are due to Chebotarev [Che], Delaunay [Del], M.J. Greenberg [Gre], Lubelski [Lbl], Mertens [Mer], Speiser [Spe], Steinbacher [Ste], Weber [We3], K. Yamamoto & Onuki [YO], and Zassenhaus [Zas]. For proofs of the corresponding local result, namely that abelian extensions of $\mathbb{Q}_p$ are contained in cyclotomic extensions of $\mathbb{Q}_p$, see Kozuka [Koz], Lubin [Lub], Riese [Rie], and M. Rosen [Ros]. The global Kronecker-Weber theorem follows easily from the local result by using the fact that $\mathbb{Q}$ does not admit unramified extensions.

Exercises

3.1 Prove the second supplementary law of quadratic reciprocity with the help of the Gauss sum $G = \zeta + \zeta^{-1}$, where $\zeta = \zeta_8$ is a primitive eighth root of unity. (Hint: use $\zeta + \zeta^{-1} = \sqrt{2}$).

3.2 Use the quadratic period polynomial $\Psi(X)$ to give another proof of the quadratic reciprocity law.

3.3 Show that $\Phi_7(X)$ has the two factors $\phi_1(X) = X^3 + \frac{1-\sqrt{-7}}{2}X^2 - \frac{1+\sqrt{-7}}{2}X - 1$ and $\phi_2(X) = X^3 + \frac{1+\sqrt{-7}}{2}X^2 - \frac{1-\sqrt{-7}}{2}X - 1$ over $\mathbb{Q}(\sqrt{-7})$.

3.4 Let m and n be integers not congruent to 2 mod 4; show that $\mathbb{Q}(\zeta_m) \cap \mathbb{Q}(\zeta_n) = \mathbb{Q}(\zeta_{(m,n)})$.

3.5 Complete the proof of the quadratic reciprocity law given at the end of Section 3.4. (Hint: use the quadratic fields $\mathbb{Q}(\sqrt{p})$, $p \equiv 3 \bmod 4$, or $\mathbb{Q}(\sqrt{pq})$, $p \equiv q \equiv 3 \bmod 4$).

3.6 Show that $\tau_p(1) = \tau$. (Hint: subtract $\sum_{t=0}^{p-1} \zeta^t = 0$ from $\tau_p(1)$.)

3.7 (Carlitz [Ca2]) Let $p > 2$ be prime. Show that the determinant of the matrix $\Delta_p = (a_{rs})_{1 \leq r,s \leq p-1}$, $a_{rs} = (\frac{r-s}{p})$, satisfies $\det \Delta_p = p^{(p-3)/2}$.

3.8 What happens to Theorem 3.9.ii) if we let automorphisms act from right?

104   3. Cyclotomic Number Fields

3.9 Gauss conjectured the correct formula for τ in 1801 (entry 118 in his diary), and found a proof four years later on August 30, 1805 (entry 123). He gave the result in the following form:

a modulo 4	0	1	2	3
$\sum \sin \frac{2\pi n^2}{a}$	$\sqrt{a}$	0	0	$\sqrt{a}$
$\sum \cos \frac{2\pi n^2}{a}$	$\sqrt{a}$	$\sqrt{a}$	0	0

For example, $\sum \sin \frac{2\pi n^2}{a} = \sqrt{a}$ for $a \equiv 0$ or 3 mod 4. Show that this implies Proposition 3.21.

3.10 The following is a version of the proof of quadratic reciprocity given after Theorem 3.3, using some formal properties of the Artin symbol. Let p and q be different odd primes, put $K = \mathbb{Q}(\zeta_p)$ and let $F = \mathbb{Q}(\sqrt{p^*})$ denote the quadratic subfield of K. Show that the subgroup H of $\mathrm{Gal}\,(K/\mathbb{Q}) \simeq (\mathbb{Z}/p\mathbb{Z})^\times$ fixing F consists of squares, and check that the following assertions are equivalent:

$$\left(\frac{p^*}{q}\right) = 1 \iff q \in \mathrm{Spl}\,(F/\mathbb{Q}) \iff \left(\frac{F/\mathbb{Q}}{q}\right) = 1$$
$$\iff \left(\frac{K/\mathbb{Q}}{q}\right) \in H \iff q \text{ is a square modulo } p.$$

3.11 Let $K/\mathbb{Q}$ be a finite extension. Show that, for every subfield $k \subseteq K$, we have $(\mathrm{disc}\,k)^{(K:k)} \mid \mathrm{disc}\,K$.

3.12 Observe that $3 \equiv \sqrt{2} \bmod (3 - \sqrt{2})$ in $\mathbb{Z}[\sqrt{2}]$, but $N(3) = 9 \not\equiv -2 = N(\sqrt{2}) \bmod 7$. Explain why Swan's claim that $\eta \equiv \pi_p \bmod \pi_q$ implies $N\eta \equiv N\pi_p \bmod \pi_q$ is nevertheless correct.

3.13 Let K be a quadratic number field with odd discriminant d. Show that K is contained in the cyclotomic field $\mathbb{Q}(\zeta_{|d|})$ by verifying the following steps:
 • Let p be a prime dividing d, and let k_p be the quadratic subfield of $\mathbb{Q}(\zeta_p)$. Show that the compositum Kk_p contains a quadratic subfield L different from K and k_p in which p does not ramify.
 • Check that K is contained in $\mathbb{Q}(\zeta_{|d|})$ if and only if $L \subseteq \mathbb{Q}(\zeta_{|d/p|})$.
 • Replace K by L and repeat the first step. If K is not contained in a cyclotomic field, you will eventually get a quadratic extension $K/\mathbb{Q}$ which is unramified at all finite primes. This contradicts the Minkowski bounds. Generalize this proof to even discriminants and show that $K \subseteq \mathbb{Q}(\zeta_{|d|})$.

This is a quite complicated proof of a rather trivial assertion; the nice thing is that it generalizes to a complete proof (in this form first given by Hilbert) of the theorem of Kronecker-Weber.

3.14 Here we will prove that the regular p-gon can be constructed by ruler and compass if $p = 2^{2^n} + 1$ is a Fermat prime. See Ireland & Rosen [386] for an approach using Gauss sums instead of Galois theory.
1. A complex number $\alpha \in \mathbb{C}$ is called *constructible*, if there exists a tower of fields $\mathbb{Q} = K_0 \subset K_1 \subset \ldots \subset K_n = \mathbb{Q}(\alpha)$ such that $K_{j+1} = K_j(\sqrt{\alpha_j})$ for some $\alpha_j \in K_j$, $j = 0, \ldots, n-1$.
2. Suppose you are given a circle of radius 1 in the complex plane. Show that $\alpha \in \mathbb{C}$ can be constructed by ruler and compass if and only if α is constructible

(observe that the intersection of circles and lines leads to at most quadratic equations).
3. Show that all $\alpha \in \mathbb{Z}[\zeta_{2^m}]$ are constructible.
4. Let p be an odd prime; show that ζ_p is constructible if and only if $p = 2^{2^n} + 1$ is a Fermat prime.

3.15 In this and the next few exercises we generalize Proposition 3.24 to arbitrary discriminants d of real quadratic number fields. Let χ_d denote the primitive nontrivial quadratic character modulo d (see Exercise 1.3). Put $Z = \exp\frac{\pi i}{d}$, $N = \prod(Z^b - Z^{-b})$ and $R = \prod(Z^a - Z^{-a})$, where $1 \leq a, b < \frac{1}{2}d$ satisfy $\chi_d(a) = +1$ and $\chi_d(b) = -1$. Show that

$$RN = \begin{cases} (\frac{2}{p})\sqrt{p} & \text{if } d = p \text{ is prime,} \\ -\sqrt{2} & \text{if } d = 8, \\ (-1)^{\varphi(d)/4} & \text{otherwise.} \end{cases}$$

As a corollary, note that $\sigma(RN) = -RN$ if d is a prime discriminant and $\sigma(RN) = RN$ otherwise, where σ is the nontrivial automorphism of $K = \mathbb{Q}(\sqrt{d})$.

3.16 (continued) Consider the case $d = 4m$, where $m \equiv 3 \bmod 4$ is squarefree. Show that $N^2 = \sigma(R^2) = (-1)^{\varphi(m)/2} \prod(Z^a + Z^{-a})^2$ and deduce that $\eta = \prod(Z^a + Z^{-a})^2$ with η defined in (3.4). Observe that η is a square in K if and only if $P = \prod(Z^a + Z^{-a}) \in K$. Show that the automorphisms of $\mathbb{Q}(Z)/K$ fix P if and only if m is composite, and that $P \in \mathbb{Q}(\sqrt{q}, \sqrt{2})$ if $m = q$ is prime. Deduce that $\sqrt{\varepsilon} \in \mathbb{Q}(\sqrt{q}, \sqrt{2})$, where ε is the fundamental unit of K.

3.17 (continued) In the case $d = 4m$ with $m = 2k$, show that $\eta = P^2$ with $P = \prod(Z^a + Z^{-a})$ as above, and that $P \in K$ unless $k \equiv 3 \bmod 4$ is prime. Prove that, in this case, $\sqrt{\varepsilon} \in \mathbb{Q}(\sqrt{k}, \sqrt{2})$.

3.18 (continued) Finally assume that $m \equiv 1 \bmod 4$. Here η is a square in K unless $m = pq$, where $p \equiv q \equiv 3 \bmod 4$ are primes, and we have $\sqrt{\eta} = \prod(Z^b - Z^{-b}) \in K$ otherwise.
If $m = pq$ is the product of two primes $p \equiv q \equiv 1 \bmod 4$, show that $N_{K/\mathbb{Q}}\sqrt{\eta} = (p/q)$. Use the class number formula to conclude that the class number in the strict sense of $\mathbb{Q}(\sqrt{m})$ is divisible by 4 if and only if $(p/q) = +1$, and that $N\varepsilon_m = -1$ if $(p/q) = -1$.

3.19 Let q and $p \equiv 3 \bmod 4$ be prime, and let ε denote the fundamental unit in $\mathbb{Q}(\sqrt{p})$. Show that $\varepsilon^q \equiv \varepsilon^{(p/q)} \bmod q$. Assume in addition that $q \equiv 3 \bmod 4$, and show that $q^{(p-1)/2} \equiv (\frac{q}{p})(1+p) + p \bmod 4p$. Now use the results of Exercise 3.16 to prove that $(q/p) = -(p/q)$.

3.20 (Vinogradov, Polya; see Stepanov [Ste, p. 72-73]) Show that the number N_p of solutions of the congruence $y^2 \equiv x \bmod p$ with $x \in [0, h]$ satisfies $|N_p - h| \leq c\sqrt{p}\log p$; here h is a fixed integers as the prime p grows.
(Hints: First show that $N_p = h + \sum_{x=1}^{h}(\frac{x}{p})$. Check that $\sum_{t=1}^{p-1}(\frac{t}{p})\zeta^{xt} = (\frac{x}{p})\tau$ and derive that $\sum_{x=1}^{h}(\frac{x}{p}) = \tau^{-1}\sum_{t=1}^{p-1}(\frac{t}{p})\sum_{x=1}^{h}\zeta^{xt}$. Take absolute values, use the triangle inequality, compute the sum $\left|\sum_{x=1}^{h}\zeta^{xt}\right|$, and show that it is bounded by $(\sin\frac{\pi t}{p})^{-1}$. Finally check that the inequality $\sin\frac{\pi t}{p} \geq \frac{2t}{p}$ holds for our values of t and compare the sum $\sum_{t=1}^{(p-1)/2}\frac{1}{t}$ with $\log t$.)

3.21 Let $k \geq 3$ be an odd integer and $p \equiv 1 \bmod k$ a prime. Show that 2 is a k-th power residue modulo p if and only if the cyclotomic number (00) is odd. (Hint: if $(r, r+1)$ is a pair of consecutive k-th power residues modulo p, then so is $(p-r-1, p-r)$. Thus their number is even unless both pairs coincide.)

Additional References

[An]   G.E. Andrews, *Partitions and the Gaussian sum* The mathematical heritage of C.F. Gauss, Collect. Pap. Mem. C.F. Gauss, 35–42 (1991)
[Ba1] P. Bachmann, *Zahlentheorie II: Die analytische Zahlentheorie*, Abschnitt 7, 146–187, Leipzig 1894; reprint 1968.
[Ba1] P. Bachmann, *Über Gauss' zahlentheoretische Arbeiten*, Opera Omnia $\mathbf{X}_2$, 1–74
[BC]   R. P. Bambah, S. Chowla, *On the sign of the Gaussian sum*, Proc. Nat. Inst. Sci. India **13** (1947), 175–176
[Bre] D. M. Bressoud, *On the value of Gaussian sums*, J. Number Theory **13** (1981), 88–94
[Ca1] L. Carlitz, *A note on Gauss's sum*, Proc. Amer. Math. Soc. **7** (1956), 910–911; Le Matematiche **23** (1968), 147–150
[Ca2] L. Carlitz, *Problem 4179*, Amer. Math. Monthly (1957), 748–749
[Ca3] L. Carlitz, *A note on Gauss sums*, Matematiche **23** (1968), 147–150
[Car] J.-C. Carrega, *Théorie des corps. La règle et le compas*, Paris: Hermann 1981; 2nd ed. 1989
[Cau] A. L. Cauchy, *Méthode simple et nouvelle pour la détermination complete de sommes alternées, formées avec les racines primitives des equations binômes*, C. R. Acad. Sci. Paris **10** (1840), 560–572; J. Math. Pures Appl. **5** (1840), 154–168; Œuvres I 5 (1885), 152–166
[Cay] A. Cayley, *Note sur la solution de l'équation $x^{257} - 1 = 0$*, J. Reine Angew. Math. **41** (1851), 81–83; MAth. Papers I, 564–566
[Che] N.G. Chebotarev, *Proof of the Kronecker-Weber theorem on Abelian fields* (Russ.), Mat. Sbornik **31** (1924), 302–309
[Ded] R. Dedekind, *Anzeige von P. Bachmann, "Die Lehre von der Kreisteilung"*, Z. Math. Phys. **18** (1873), 14–24; Werke III, 408–419
[Del] B.N. Delaunay, *Zur Bestimmung algebraischer Zahlkörper durch Kongruenzen; eine Anwendung auf die Abelschen Gleichungen*, J. Reine Angew. Math. **152** (1923), 120–123
[Dir] G. L. Dirichlet, *Sur l'usage des intégrales définies dans la sommation des séries finies ou infinies*, J. Reine Angew. Math. **17** (1837), 57–67; Werke I, 257–270
[DM]   H. Dym, H.P. McKean, *Fourier series and integrals*, Academic Press 1972
[Est] T. Estermann, *On the sign of the Gaussian sum*, J. London Math. Soc. **20** (1945), 66–67

[Eul] L. Euler, *De extractione radicum ex quantitatibus irrationalibus*, Opera Omnia I **6**, 31–77
[Fon] G. Fontené, *Construction du polygon de dix-sept côtés*, Mathesis **9** (1899), 179–185
[Gau] C.F. Gauss, *Letter to W. Olbers, Sept. 3, 1805*,
[Got] C. Gottlieb, *The simple and straightforward construction of the regular 257-gon*, Math. Intell. **21** (1999), 31–37
[Gre] M.J. Greenberg, *An elementary proof of the Kronecker-Weber theorem*, Am. Math. Mon. **81** (1974), 601–607
[Hag] K. Hagge, *Einfache Behandlung der 257-Teilung des Kreises*, Zeitschrift math.-naturwiss. Unterricht **41** (1910), 448–458
[Has] H. Hasse, *Über die Klassenzahl abelscher Zahlkörper*, 2nd edition, Springer Verlag 1982
[Her] J. Hermes, *Ueber die Teilung des Kreises in 65537 gleiche Teile*, Gött. Nachr. 1894, 170–186; FdM **25** (1893/94), 287
[Hil] D. Hilbert, *Ein neuer Beweis des Kroneckerschen Fundamentalsatzes über Abelsche Zahlkörper*, Gött. Nachr. (1896), 29–39
[Ja1] C. G. J. Jacobi, *Observatio arithmetica de numero classium divisorum quadraticorum formae yy + Azz, designante A numerum primum formae 4n + 3*, J. Reine Angew. Math. **9** (1832), 189–192; Ges. Werke **6**, 240–244, Berlin 1891
[Ja2] C. G. J. Jacobi, *Letter to Gauss*, 31. 01. 1837; Werke VII,
[KT] J. Karamate, M. Tomic, *Sur une inégalité de Kusmin-Landau relative aux sommes trigonométriques et son application à la somme de Gauss*, Acad. Serbe Sci. Publ. Inst. Math. **3** (1950), 207–218
[Kir] F. Kirwan, *Square roots and Seventeengons*, Mathematics Masterclasses (M. Sewell, ed.), Oxford Sci. Publ. 1997, 60–88
[Koz] K. Kozuka, *On the local Kronecker-Weber theorem*, Arch. Math. **54** (1990), 162–163
[Kr1] L. Kronecker, *Über die algebraisch auflösbaren Gleichungen*, Monatsber. Berlin (1853), 365–374; ibid. (1856), 203–215; Werke IV, 1–11; 25–37
[Kr2] L. Kronecker, *Sur une formule de Gauss*, J. Math. Pures Appl. (2) **1** (1856), 392–395; Werke 4 (1929), 171–175
[Kr3] L. Kronecker, *Summirung der Gauss'schen Reihen* $\sum_{h=0}^{n-1} e^{2h^2\pi i/n}$, J. Reine Angew. Math. **105** (1889), 267–268; Werke 4 (1929), 295–300
[Kr4] L. Kronecker, *Ueber die Dirichlet'sche Methode der Wertbestimmung der Gaussschen Reihen*, Festschr. Hamb. Mitt. **2** (1890), 32–36; FdM **22** (1890), 205–206; Werke IV, 301–308
[Kub] T. Kubota, *On the system of equations involved in the cyclotomy* (Japanese), Rep. Fac. Sci. Technol. Meijo Univ. **38** (1998), 31–35
[Lan] E. Landau, *Über das Vorzeichen der Gaußschen Summe*, Gött. Nachr. 1928, 19–20

[Ldb] G. Landsberg, *Zur Theorie der Gauss'schen Summen und der linearen Transformation der Thetafunktionen*, J. Reine Angew. Math. **111** (1893), 234–253; FdM **25** (1893/94), 283–285

[Lb1] H. Lebesgue, *Sur une construction du polygone régulier de 17 côtes, due à André-Marie Ampère, d'après des documents conservés dans les archives de l'academie des sciences*, C. R. Acad. Sci. Paris **204** (1937) 925–928; republished in Enseign. Math. **3** (1957) 31–34

[Lb2] H. Lebesgue, *Leçons sur les constructions géométriques*, Paris 1950

[Le1] V. A. Lebesgue, *Sommation de quelques séries*, J. Math. Pures Appl. **5** (1840),

[Le2] V. A. Lebesgue, *Note sur une formule de M. Cauchy*, J. Math. Pures Appl. **5** (1840), 42–71

[Len] H.W. Lenstra, *Primality testing with Artin symbols*, Number theory related to Fermat's last theorem, Prog. Math. **26** (1982), 341–347

[Ler] M. Lerch, *Zur Theorie der Gaußschen Summen*, Math. Ann. **57** (1903), 554–567; FdM **34** (1903), 228

[Lo1] A. Loewy, *Eine algebraische Behauptung von Gauß*, Jahresber. DMV **26** (1918), 100–109

[Lo2] A. Loewy, *Inwieweit kann Vandermonde als Vorgänger von Gauß bezüglich der algebraischen Auflösung der Kreisteilungsgleichungen $x^n = 1$ angesehen werden?*, Jahresber. DMV **27** (1918), 189–195

[Lbl] S. Lubelski, *Zur Arithmetisierung des Beweises des Minkowskischen Diskriminanten- und Kronecker-Weberschen Einbettungssatzes*, Acta Arith. **3** (1939), 235–254

[Lub] J. Lubin, *The local Kronecker-Weber theorem*, Trans. Am. Math. Soc. **267** (1981), 133–138

[Mer] F. Mertens, *Über zyklische Gleichungen*, J. Reine Angew. Math. **131** (1906), 87–112

[MM] P. Mihailescu, F. Morain, *Dual Elliptic Primes and Cyclotomy Primality Proving*, Winnipeg 1999; see also
http://www.inf.ethz.ch/~mihailes/

[Mit] A. Mitzscherling, *Das Problem der Kreisteilung*, 214 pp, Teubner 1913

[Mo1] L. J. Mordell, *On a simple summation of the series $\sum_{s=0}^{n-1} e^{2s^2\pi i/n}$*, Mess. Math. **48** (1918), 54–56

[Mo2] L. J. Mordell, *The sign of Gaussian sums*, Euclides **30** (1954/55), 293–298

[Mo3] L. J. Mordell, *The sign of the Gaussian sum*, Ill. J. Math. **6** (1962), 177–180

[Neu] O. Neumann, *Two proofs of the Kronecker-Weber theorem "according to Kronecker, and Weber"*, J. Reine Angew. Math. **323** (1981), 105–126

[Pau] M.G. v. Paucker, *Geometrische Verzeichnung dse regelmäßigen 17-Ecks und 257-Ecks in den Kreis*, Jahresverh. Kurl. Ges. Litt. Kunst **2** (1822), 160–219

[Pet] K. Petr, *Über eine Anwendung der elliptischen Funktionen auf die Zahlentheorie*, Prag. Ber. 1907, Nr. 18, 8 S

[Pfl] C.F. v. Pfleiderer, *Manuscript 1802*; cf. Gauss Werke **X–1**, p. 120

[Ric] A. Richelot, *De resolutione algebraica aequationis $x^{257} = 1$, sive de divisione circuli per bisectionem anguli septies repetitam in partes 257 inter se aequales commentatio coronata*, J. Reine Angew. Math. **9** (1832), 1–26, 146–161, 209–230, 337–358

[Rie] U. Riese, *Kronecker-Weber via Kummer*, Expo. Math. **16** (1998), 271–276

[Ros] M. Rosen, *An elementary proof of the local Kronecker-Weber theorem*, Trans. Am. Math. Soc. **265** (1981), 599–605

[Sal] M. Salvadori, *Esposizioni delle teoria delle somme di Gauss et di alcuni teoremi di Eisenstein*, Diss. Freiburg (CH), 116 pp; FdM **35** (1904), 210

[Sca] N. Schappacher, *On the History of Hilbert's Twelfth Problem, I: Paris 1900 – Zürich 1932: The Comedy of Errors*, in: Matériaux pour l'histoire des mathématiques au XX siècle, Actes du colloque à la mémoire de Jean Dieudonné (Nice, 1996), "Séminaires et Congrès" **3** (1998), 243–273.

[Sch] H. Schröter, *Zur v. Staudtschen Construction des regulären Siebzehnecks*, J. Reine Angew. Math. **75**, 13–24

[Sm] J. Schumacher, *Die Auflösung der Gleichung $x^{257} - 1 = 0$*, Arch. Math. Phys. **20** (1912/13), 295–311

[Sr] I. Schur, *Über die Gauß'schen Summen*, Göttinger Nachr. 1921, 147–153; FdM **48** (19121/22), 130–131

[Scz] R. Sczech, *Gaussian sums, Dedekind sums and the Jacobi triple product identity*, Kyushu J. Math. **49** (1995), 233-241

[dSe] J. de Séguier, *Sur les sommes de Gauss*, C. R. Acad. Sci. Paris **123** (1896), 166–168; FdM **27** (1896), 143

[Sha] D. Shanks, *Two theorems of Gauss*, Pac. J. Math. **8** (1958), 609–612

[Spe] A. Speiser, *Die Zerlegungsgruppe*, J. Reine Angew. Math. **149** (1919), 174–188

[vSt] C. von Staudt, *Construction des regulären Siebenzehnecks*, J. Reine Angew. Math. **24** (1842), 251

[Ste] F. Steinbacher, *Abelsche Körper als Kreisteilungskörper*, J. Reine Angew. Math. **139** (1911), 85–100

[Ste] S.A. Stepanov, *Arithmetic of Algebraic Curves*, Monographs in Contemporary Mathematics, 1994

[Str] A. Strnad, *Über das regelmäßige Siebzehneck*, Casopis **33** (1904), 543–558; ibid. **36** (1907), 81–86

[Taf]   A. Tafelmacher, *Die Konstruktion des regelmäßigen 17-Ecks*, Verhandl. d. Deutsch. Wiss. Verein Chile **4** (1900), 205–219; RSPM **9** (1900), 14
[Te1]   H. Teege, Dissertation Univ. Kiel, 1900
[Te2]   H. Teege, *Ein algebraischer Beweis für das Vorzeichen der Gaussschen Summen*, Mitt. Math. Ges. Hamburg **5** (1920), 281–289; FdM **47** (1919–20), 108
[Wan]   P.L. Wantzel, *Recherches sur les moyens de reconnaître si un problème de Géometrie se résoudre avec la règle et le compas*, J. Pures Appl. **2** (1837), 366–372
[Was]   L. Washington, *Introduction to Cyclotomic Fields*, Graduate Texts in Math. **83**, Springer Verlag 1982; 2nd edition 1997
[Wat]   W. Waterhouse, *The sign of the Gaussian sum*, J. Number Theory **2** (1970), 363
[We1]   H. Weber, *Theorie der Abelschen Zahlkörper I, II.*, Acta Math. **8** (1886), 193–263; ibid. **9** (1886/87), 105–130
[We2]   H. Weber, *Über Abel's Summation unendlicher Differenzenreihen*, Acta Math. **27** (1903), 225–233
[We3]   H. Weber, *Über zyklische Zahlkörper*, J. Reine Angew. Math. **132** (1907), 167–188
[Wil]   H.C. Williams, *Edouard Lucas and primality testing*, New York 1998
[WS]   H.C. Williams, J.O. Shallit, *Factoring integers before computers*, Proc. Symp. Appl. Math. **48** (1994), 481–531
[YO]   K. Yamamoto, M. Onuki, *On Kronecker's theorem about Abelian extensions*, Sci. Rep. Tokyo Woman's Chr. Coll. **35–38** (1975), 415–418
[Zas]   H. Zassenhaus, *On a theorem of Kronecker*, Delta **1** (1969), No.3, 1–14

4. Power Residues and Gauss Sums

This chapter is devoted to some technical preliminaries. We start with some general remarks on power residue symbols, discuss a connection between the splitting of primes in a number field K and its discriminant, review Kummer theory and characters of abelian groups, and finally introduce and study character sums, namely those of Gauss, Jacobi and Eisenstein.

4.1 Residue Symbols in Number Fields

In this section we define the general power residue symbol in number fields possessing the necessary roots of unity and prove its basic properties. The discussion of Gauss's Lemma is reserved for Section 4.2.

Let k be a number field, and $n \geq 1$ a natural number. A prime ideal $\mathfrak{p}$ in $\mathcal{O}_k$ is said to be prime to n if $\mathfrak{p} \nmid n\mathcal{O}_k$; it is easy to see that $\mathfrak{p}$ is prime to n if and only if $(q, n) = 1$, where $q = p^f = N\mathfrak{p}$ is the norm of $\mathfrak{p}$. For every $\alpha \in \mathcal{O}_k$ prime to $\mathfrak{p}$, we have

$$\alpha^{q-1} \equiv 1 \bmod \mathfrak{p}. \tag{4.1}$$

If k contains an n-th root of unity ζ_n and $\mathfrak{p}$ is prime to n, then the subgroup of $(\mathcal{O}_k/\mathfrak{p})^\times$ generated by $\zeta_n \bmod \mathfrak{p}$ has order n. This shows that n divides $\#(\mathcal{O}_k/\mathfrak{p})^\times = q - 1$, hence we can define $(\alpha/\mathfrak{p})$ to be the n-th root of unity such that

$$\alpha^{\frac{q-1}{n}} \equiv \left(\frac{\alpha}{\mathfrak{p}}\right) \bmod \mathfrak{p}. \tag{4.2}$$

Then $(\alpha/\mathfrak{p})$ is well defined, and it is called the n-th power residue symbol in k. Whenever we encounter such a symbol below, we will tacitly assume that it exists, i.e., that numerator and denominator are relatively prime, and that the ideal in the denominator is prime to n. Of course we extend the symbol multiplicatively to all ideals $\mathfrak{a} = \prod \mathfrak{p}$ prime to n by setting $(\alpha/\mathfrak{a}) = \prod(\alpha/\mathfrak{p})$. See Exercise 4.1 for examples.

Proposition 4.1. *Let k be a number field containing a primitive n-th root of unity ζ_n, and let $\mathfrak{p}$ be a prime ideal in $\mathcal{O}_k$.*

112   4. Power Residues and Gauss Sums

i) If $\alpha \equiv \beta \bmod \mathfrak{p}$ then $\left(\frac{\alpha}{\mathfrak{p}}\right) = \left(\frac{\beta}{\mathfrak{p}}\right)$;

ii) $\left(\frac{\alpha}{\mathfrak{p}}\right) = 1$ if and only if α is an n-th power residue mod $\mathfrak{p}$, i.e., if and only if $\alpha \equiv \xi^n \bmod \mathfrak{p}$ has solutions $\xi \in \mathcal{O}_k \setminus \mathfrak{p}$;

iii) $a \in \mathbb{Z}$ is an m-th power residue mod a prime $p \equiv 1 \bmod m$ if and only if $\left(\frac{a}{\mathfrak{p}}\right) = 1$, where $\mathfrak{p}$ is any prime ideal above p in $\mathbb{Q}(\zeta_m)$.

Proof. The first property is obvious. For a proof of the second, assume that $\alpha \equiv \xi^n \bmod \mathfrak{p}$ for some $\xi \in \mathcal{O}_k$. Then $\alpha^{\frac{q-1}{n}} \equiv \xi^{q-1} \equiv 1 \bmod \mathfrak{p}$ by (4.1), hence $\left(\frac{\alpha}{\mathfrak{p}}\right) = 1$. To prove the other direction, assume that $\left(\frac{\alpha}{\mathfrak{p}}\right) = 1$ and write $\alpha \equiv \gamma^j \bmod \mathfrak{p}$, where γ is some primitive root mod $\mathfrak{p}$ (recall that $(\mathcal{O}_k/\mathfrak{p})^\times$ is cyclic). Then $1 \equiv \alpha^{\frac{q-1}{n}} \equiv \gamma^{j\frac{q-1}{n}} \bmod \mathfrak{p}$ implies that $n \mid j$; thus α is an n-th power mod $\mathfrak{p}$, and we have proved ii). This in turn immediately implies iii). □

$$\begin{array}{c} K \\ | \\ \zeta \in k \\ | \\ F \end{array}$$

If K/k is a finite extension of number fields and if k contains the n-th roots of unity (compare the diagram on the left), then we can define the n-th power residue symbols $(\cdot/\cdot)_k$ and $(\cdot/\cdot)_K$; if K/F is a normal extension, then $\mathrm{Gal}\,(K/F)$ acts on the power residue symbols $(\cdot/\cdot)_k$ and $(\cdot/\cdot)_K$, and we find

Proposition 4.2. *Let K/F be a normal extension with Galois group G, and let k/F be a subextension such that $\zeta_n \in k$.*

i) *For every $\sigma \in G$, we have $\left(\frac{\alpha}{\mathfrak{a}}\right)_K^\sigma = \left(\frac{\alpha^\sigma}{\mathfrak{a}^\sigma}\right)_K$ for all $\alpha \in K^\times$ and all ideals $\mathfrak{a}$ prime to n.*

ii) *If $\mathfrak{p}$ has inertia degree 1 in K/k, then $\left(\frac{\alpha}{\mathfrak{P}}\right)_K = \left(\frac{\alpha}{\mathfrak{p}}\right)_k$, where $\mathfrak{P}$ is a prime ideal in $\mathcal{O}_K$ above $\mathfrak{p}$ and $\alpha \in \mathcal{O}_k$.*

iii) *If K/k is abelian and $\mathfrak{p}$ is a prime ideal in $\mathcal{O}_k$, then $\left(\frac{\alpha}{\mathfrak{p}}\right)_K = \left(\frac{N\alpha}{\mathfrak{p}}\right)_k$ for all $\alpha \in \mathcal{O}_k$, where $N = N_{K/k}$ denotes the relative norm.*

iv) *Let K/k be a cyclic extension of degree $(K : k) = n$, and suppose that there is a prime ideal $\mathfrak{p} \nmid n\mathcal{O}_k$ such that $\mathfrak{p}\mathcal{O}_K = \mathfrak{P}^n$ is totally ramified in K/k. Then $\left(\frac{N_{K/k}\alpha}{\mathfrak{p}}\right)_k = 1$ for all $\alpha \in \mathcal{O}_K \setminus \mathfrak{P}$.*

Proof. The first two assertions can be proved quite easily: by definition we have

$$\alpha^{\frac{q-1}{n}} \equiv \left(\frac{\alpha}{\mathfrak{p}}\right) \bmod \mathfrak{p} \quad \text{and} \quad (\alpha^\sigma)^{\frac{q-1}{n}} \equiv \left(\frac{\alpha^\sigma}{\mathfrak{p}^\sigma}\right) \bmod \mathfrak{p}^\sigma$$

for every $\sigma \in G$. Applying σ to the first congruence yields $(\alpha^\sigma)^{\frac{q-1}{n}} \equiv \left(\frac{\alpha}{\mathfrak{p}}\right)^\sigma \bmod \mathfrak{p}^\sigma$, and comparing these equations proves our first claim if $\mathfrak{a} = \mathfrak{p}$ is prime. The general case follows from multiplicativity.

If $\mathfrak{P}$ is a prime ideal in $\mathcal{O}_K$ above $\mathfrak{p}$, and if $f(\mathfrak{P}|\mathfrak{p}) = 1$, then $N(\mathfrak{P}) = N(\mathfrak{p}) = q$. Thus

4.1 Residue Symbols in Number Fields

$$\left(\frac{\alpha}{\mathfrak{P}}\right) \equiv \alpha^{\frac{q-1}{n}} \bmod \mathfrak{P} \quad \text{and} \quad \left(\frac{\alpha}{\mathfrak{p}}\right) \equiv \alpha^{\frac{q-1}{n}} \bmod \mathfrak{p}.$$

Our second assertion follows.

The proof of iii) is done in three steps: we show that the claim holds for extensions in which $\mathfrak{p}$ is totally ramified, inert or split; iii) then follows by considering the ramification, inertia and decomposition fields of K.

Assume first that $\mathfrak{p}$ ramifies completely in K/k. Write $\mathfrak{p}\mathcal{O}_K = \mathfrak{P}^{(K:k)}$ and let $G = \operatorname{Gal}(K/k)$ (or, better yet, assume that $k = F$); then

$$\left(\frac{\alpha}{\mathfrak{p}}\right)_K = \left(\frac{\alpha}{\mathfrak{P}}\right)_K^{(K:k)} = \prod_{\sigma \in G}\left(\frac{\alpha}{\mathfrak{P}}\right)_K^{\sigma} = \prod_{\sigma \in G}\left(\frac{\alpha^\sigma}{\mathfrak{P}^\sigma}\right)_K = \left(\frac{N\alpha}{\mathfrak{P}}\right)_K = \left(\frac{N\alpha}{\mathfrak{p}}\right)_k,$$

where we have used ii) in the last step.

Next suppose that $\mathfrak{p}$ is inert in K/k; let $\sigma \in \operatorname{Gal}(K/k)$ be the Frobenius automorphism at $\mathfrak{p}$, and recall that σ is characterized by the congruence $\alpha^\sigma \equiv \alpha^q \bmod \mathfrak{p}\mathcal{O}_K$ for all $\alpha \in \mathcal{O}_K \setminus \mathfrak{p}$; here $q = N\mathfrak{p} = (\mathcal{O}_k : \mathfrak{p}\mathcal{O}_k)$ is the absolute norm of $\mathfrak{p}$ (note that $N_{K/\mathbb{Q}}\mathfrak{p} = q^f$). Now we get

$$\left(\frac{\alpha}{\mathfrak{p}}\right)_K \equiv \alpha^{(q^f-1)/n} = \alpha^{\frac{q-1}{n}(1+q+\ldots+q^{f-1})}$$
$$\equiv \alpha^{\frac{q-1}{n}(1+\sigma+\ldots+\sigma^{f-1})} = (N\alpha)^{\frac{q-1}{n}} \bmod \mathfrak{p},$$

hence $(\alpha/\mathfrak{p})_K = (N\alpha/\mathfrak{p})_k$ as claimed.

Finally suppose that $\mathfrak{p}$ splits completely in K/k. Then $\mathfrak{p}\mathcal{O}_K = \mathfrak{P}_1 \cdot \ldots \cdot \mathfrak{P}_g$, where the $\mathfrak{P}_j$ are permuted transitively by G, and we find

$$\left(\frac{\alpha}{\mathfrak{p}}\right)_K = \left(\frac{\alpha}{\mathfrak{P}_1}\right)_K \cdot \ldots \cdot \left(\frac{\alpha}{\mathfrak{P}_g}\right)_K = \prod_{\sigma \in G}\left(\frac{\alpha}{\mathfrak{P}^\sigma}\right)_K$$
$$= \prod_{\sigma \in G}\left(\frac{\sigma^{-1}\alpha}{\mathfrak{P}}\right)_K^{\sigma} = \prod_{\sigma \in G}\left(\frac{\sigma^{-1}\alpha}{\mathfrak{P}}\right)_K$$
$$= \left(\frac{N_{K/k}\alpha}{\mathfrak{P}}\right)_K = \left(\frac{N\alpha}{\mathfrak{p}}\right)_k \quad \text{(where we used ii) again)}$$

where we have used that σ acts trivially on the power residue symbol since $\zeta_n \in k$. Now we put everything together: let K_T and K_Z be the inertia and the decomposition subfield of $\mathfrak{p}$ in K/k, respectively, and apply the results above to K/K_T, K_T/K_Z, and K_Z/k: this proves iii).

The last claim follows easily from the third: in fact, we have

$$\left(\frac{N_{K/k}\alpha}{\mathfrak{p}}\right)_k = \left(\frac{\alpha}{\mathfrak{p}}\right)_k = \left(\frac{\alpha}{\mathfrak{P}}\right)_k^{(K:k)} = 1.$$

This concludes the proof. □

114   4. Power Residues and Gauss Sums

There is a different (and equivalent) definition of the power residue symbol based on the Artin symbol; it has the advantage that the formal properties of the Artin symbol can be transferred easily to properties of the power residue symbol. Moreover, this connection will become the main source of explicit reciprocity laws in class field theory.

Let k be a number field containing a primitive n-th root of unity ζ_n. For any $\alpha \in k^\times$ put $K = k(\sqrt[n]{\alpha})$ and let $\mathfrak{p}$ denote any prime ideal in $\mathcal{O}_k$ unramified in K/k (clearly, $\mathfrak{p} \nmid n\alpha$ is sufficient). Then we claim that the equation

$$\left(\frac{K/k}{\mathfrak{p}}\right)\sqrt[n]{\alpha} = \left(\frac{\alpha}{\mathfrak{p}}\right)\sqrt[n]{\alpha} \tag{4.3}$$

defines the power residue symbol $(\alpha/\mathfrak{p})$. By the definition of the Artin symbol, the left hand side is congruent to $(\sqrt[n]{\alpha})^{N\mathfrak{p}}$ mod $\mathfrak{p}$, whereas the right hand side is congruent to $\alpha^{(N\mathfrak{p}-1)/n}\sqrt[n]{\alpha}$ mod $\mathfrak{p}$. This clearly proves our claim.

The verification of Proposition 4.2.i) using properties of the Artin symbol goes as follows: put $\phi = (\frac{K/F}{\mathfrak{p}})$; then $(\frac{\alpha}{\mathfrak{p}}) = (\alpha^{1/n})^{\phi-1}$ by (4.3), hence $(\frac{\alpha}{\mathfrak{p}^\sigma}) = (\alpha^{1/n})^{\sigma\phi\sigma^{-1}-1}$ and therefore $(\frac{\alpha^\sigma}{\mathfrak{p}^\sigma}) = (\alpha^{1/n})^{\phi\sigma-\sigma} = (\frac{\alpha}{\mathfrak{a}})^\sigma$.

When it comes to 4.2.ii) and 4.2.iii), however, we experience one of the drawbacks of the Artin symbol for ideals: its inability to include ramified primes. The readers are encouraged, however, to derive ii) and iii) for unramified prime ideals from properties of the Artin symbol.

4.2 Gauss's Lemma

In Section 1.4 we have introduced Gauss's Lemma for quadratic residues; its generalization to general power residues is no problem. Let K be a number field containing the nth roots of unity $\mu_n = \{1, \zeta, \ldots, \zeta^{n-1}\}$. For any ideal $\mathfrak{a}$ coprime to n, the homomorphism $\mu_n \longrightarrow (\mathcal{O}_K/\mathfrak{a})^\times$ induced by $\zeta \longmapsto \zeta$ mod $\mathfrak{a}$ is injective. Any set $A = \{\alpha_1, \ldots, \alpha_m\}$ of representatives in $\mathcal{O}_K$ of the classes in $(\mathcal{O}_F/\mathfrak{a})^\times/\mu_n$ is called a $\frac{1}{n}$-system modulo $\mathfrak{a}$. Obviously, a $\frac{1}{2}$-system modulo p in $\mathbb{Z}$ is what we have called a half-system modulo p in Section 1.4. The general form of Gauss's Lemma is

Lemma 4.3. *Let* $A = \{\alpha_1, \ldots, \alpha_m\}$ *denote a $\frac{1}{n}$-system modulo $\mathfrak{p}$; then we have* $\alpha\alpha_i \equiv \zeta^{a(i)}\alpha_{\pi(i)}$ *mod $\mathfrak{p}$ for some permutation π of $\{1, \ldots, m\}$, and*

$$\left(\frac{\alpha}{\mathfrak{p}}\right) = \zeta^\mu, \text{ where } \mu = \sum_{i=1}^{m} a(i).$$

Proof. Exactly as in the quadratic case (see Lemma 1.10). □

At the heart of Eisenstein's approach to the lower reciprocity laws (see Chapter 8) lies the following special form of Gauss's Lemma:

Lemma 4.4. *Let F be a number field containing a primitive n-th root of unity ζ_n, suppose that $\mathfrak{p} = \pi \mathcal{O}_F$ is a principal prime ideal in $\mathcal{O}_F$ above p such that $\mathfrak{p} \nmid n$, and let $f : F \longrightarrow \mathbb{C}$ be an $\mathcal{O}_F$-periodic function such that*

i) $f(\zeta_n z) = \zeta_n f(z)$ for all $z \in F \setminus \mathcal{O}_F$;
ii) $f(\frac{\alpha}{\pi}) \neq 0$ for all $\alpha \in \mathcal{O}_F \setminus \mathfrak{p}$.

Then, for any $\frac{1}{n}$-system A mod $\mathfrak{p}$,
$$\left(\frac{\gamma}{\mathfrak{p}}\right)_n = \prod_{\alpha \in A} \frac{f(\frac{\gamma\alpha}{\pi})}{f(\frac{\alpha}{\pi})}.$$

Proof. By the definition of a $\frac{1}{n}$-system mod $\mathfrak{p}$, we have $\gamma\alpha \equiv \varepsilon_\alpha \alpha' \mod \mathfrak{p}$ for some n-th root of unity ε_α; the $\mathcal{O}_F$-periodicity of f gives

$$f\left(\frac{\gamma\alpha}{\pi}\right) = f\left(\frac{\varepsilon_\alpha \alpha' + \lambda\pi}{\pi}\right) = f\left(\varepsilon_\alpha \frac{\alpha'}{\pi}\right) = \varepsilon_\alpha f\left(\frac{\alpha'}{\pi}\right),$$

and our claim follows by forming the product $\prod_{\alpha \in A} f(\gamma\alpha/\pi)$. □

The main problem of our version of Gauss's Lemma becomes apparent here: Lemma 4.4 does not work if $\mathfrak{p}$ is not principal. If we want to study reciprocity laws in number fields with class number > 1, we have to evaluate symbols $(\alpha/\beta)_n$, so factoring β into prime ideals does not help. Thus we have to replace the prime ideal $\mathfrak{p}$ in Lemma 4.3 by an integral ideal prime to n; this will allow us to compute $(\alpha/\beta)_n$ without having to decompose β into its prime ideal factors.

Theorem 4.5. *Let $\mathfrak{a}$ be an integral ideal prime to n, and assume that $A = \{\alpha_1, \ldots, \alpha_m\}$ is a $\frac{1}{n}$-system modulo $\mathfrak{a}$. Define integers $a(i)$ mod n by $\alpha\alpha_i \equiv \zeta^{a(i)} \alpha_{i'} \mod \mathfrak{a}$; then $\left(\frac{\alpha}{\mathfrak{a}}\right) = \zeta^\mu$, where $\mu = \sum_{j=1}^{m} a(j)$.*

Proof. The existence of A is easy to prove: let $A_0 = \{0, \beta_1, \ldots, \beta_{N-1}\}$ be a complete residue system modulo $\mathfrak{a}$; here $N = N\mathfrak{a} \equiv 1 \mod n$ is the norm of $\mathfrak{a}$. Choose $\alpha_1 = \beta_1$; clearly the elements $\alpha_1 \zeta^j$, $j = 0, 1, \ldots, n-1$, are congruent modulo $\mathfrak{a}$ to n distinct elements of A_0. Delete them and call the resulting set A_1. Then pick an element from $A_1 \setminus \{0\}$ and repeat this procedure.

Let $A^* = \{\alpha_j \in A : (\alpha_j, \mathfrak{a}) = (1)\}$ be the subset of those elements in A which are coprime to $\mathfrak{a}$. Define two symbols $\{\alpha/\mathfrak{a}\}$ and $\{\alpha/\mathfrak{a}\}^*$ by

$$\left\{\frac{\alpha}{\mathfrak{a}}\right\} = \prod \zeta^{a(j)}, \quad \left\{\frac{\alpha}{\mathfrak{a}}\right\}^* = \prod\nolimits^* \zeta^{a(j)},$$

where $\prod$ is over all $1 \leq j \leq m$, whereas $\prod^*$ only runs through the subset of indices which correspond to an element of A^*. It is easy to see that $\{\alpha/\mathfrak{a}\}^* \equiv \alpha^{\varphi(\mathfrak{a})/n} \mod \mathfrak{a}$, where $\varphi = \varphi_F$ denotes Euler's phi-function (i.e. $\varphi(\mathfrak{a}) = \#(\mathcal{O}_F/\mathfrak{a})^\times$ is the cardinality of a complete prime residue system modulo $\mathfrak{a}$): in fact, just multiply the congruences $\alpha\alpha_i \equiv \zeta^{a(i)} \alpha_{i'} \mod \mathfrak{a}$ and then cancel the factor $\prod \alpha_i = \prod \alpha_{i'}$ on both sides. Since a similar computation for the symbol $\{\alpha/\mathfrak{a}\}$ does not work, we have to make a detour for a proof of $\{\alpha/\mathfrak{a}\} = (\alpha/\mathfrak{a})$, which is what we want to show.

For every ideal $\mathfrak{b} \mid \mathfrak{a}$ consider the set $A_\mathfrak{b} = \{\alpha_j \in A : (\alpha_j, \mathfrak{a}) = \mathfrak{a}\mathfrak{b}^{-1}\}$; clearly $A_\mathfrak{a} = \emptyset$ and $A_{(1)} = A^*$. Moreover we have $A = \bigcup_{\mathfrak{b} \mid \mathfrak{a}} A_\mathfrak{b}$; now we claim

$$\left\{\frac{\alpha}{\mathfrak{a}}\right\} = \prod_{\mathfrak{b} \mid \mathfrak{a}} \left\{\frac{\alpha}{\mathfrak{b}}\right\}^*.$$

To this end, consider the set $A_\mathfrak{b} = \{\alpha_{i_1}, \ldots, \alpha_{i_k}\}$. It is an easy exercise to show that $B_\mathfrak{b} = \{1, \alpha_{i_2}/\alpha_{i_1}, \ldots, \alpha_{i_k}/\alpha_{i_1}\}$ is a prime $\frac{1}{n}$-system modulo $\mathfrak{b}$; since $\alpha\alpha_i/\alpha_1 \equiv \zeta^{b(i)}\alpha_j/\alpha_1 \bmod \mathfrak{a}\mathfrak{b}^{-1}$ is equivalent to $\alpha\alpha_i \equiv \zeta^{b(i)}\alpha_j \bmod \mathfrak{a}$, the proof of Gauss's Lemma shows that $\zeta^{\sum b(i)} = \{\alpha/\mathfrak{b}\}^*$. Now the claimed equality follows from $A = \bigcup_{\mathfrak{b} \mid \mathfrak{a}} A_\mathfrak{b}$.

Next we claim that

$$\left\{\frac{\alpha}{\mathfrak{b}}\right\}^* = \begin{cases} (\alpha/\mathfrak{p}) & \text{if } \mathfrak{b} = \mathfrak{p}^a, a > 0, \\ 1 & \text{otherwise.} \end{cases}$$

In fact, suppose that $\mathfrak{b} = \mathfrak{c}\mathfrak{d}$ for integral coprime ideals $\mathfrak{c}$, $\mathfrak{d}$. Then

$$\alpha^{\varphi(\mathfrak{b})/n} \equiv \left(\alpha^{\varphi(\mathfrak{c})}\right)^{\varphi(\mathfrak{d})/n} \equiv 1 \bmod \mathfrak{c},$$

and similarly we get $\alpha^{\varphi(\mathfrak{b})/n} \equiv 1 \bmod \mathfrak{d}$. This proves the claim if $\mathfrak{b}$ is not a prime ideal power. Assume now that $\mathfrak{b} = \mathfrak{p}^a$; then

$$\left\{\frac{\alpha}{\mathfrak{b}}\right\}^* \equiv \alpha^{\varphi(\mathfrak{b})/n} \equiv \left(\frac{\alpha}{\mathfrak{p}}\right)^{p^{f(a-1)}} = \left(\frac{\alpha}{\mathfrak{p}}\right) \bmod \mathfrak{p},$$

since $p^f \equiv 1 \bmod m$. Now we can put everything together:

$$\left\{\frac{\alpha}{\mathfrak{a}}\right\} = \prod_{\mathfrak{b} \mid \mathfrak{a}} \left\{\frac{\alpha}{\mathfrak{b}}\right\}^* = \prod_{\mathfrak{p}^a \mid \mathfrak{a}} \left\{\frac{\alpha}{\mathfrak{p}^a}\right\}^* = \prod_{\mathfrak{p}^a \mid \mathfrak{a}} \left(\frac{\alpha}{\mathfrak{p}}\right) = \prod_{\mathfrak{p}^a \| \mathfrak{a}} \left(\frac{\alpha}{\mathfrak{p}}\right)^a = \left(\frac{\alpha}{\mathfrak{a}}\right).$$

This is what we wanted to prove. □

4.3 Discriminants

In this section we will say a few things about results which connect quadratic reciprocity with the arithmetic of number fields. In particular we will see that the quadratic residue character of the discriminant of a number field is connected with the splitting of primes in a very explicit way. The story begins with the theorem of Stickelberger and Voronoi due to Pellet:

Proposition 4.6. *Let L be a number field of degree $n = (L : \mathbb{Q})$ with discriminant d. If a prime $p \nmid d$ splits into exactly g different prime ideals in $\mathcal{O}_L$, then $\left(\frac{d}{p}\right) = (-1)^{n-g}$.*

Taking $L = \mathbb{Q}(\zeta_q)$ we find that d is an odd power of q^*, hence Proposition 4.6 gives $\left(\frac{q^*}{p}\right) = (-1)^g$. On the other hand, $g = \frac{p-1}{f}$ is even if and only if $\left(\frac{p}{q}\right) = 1$, (see Section 3.1), and we get the quadratic reciprocity law.

We also remark that Proposition 4.6 has an infinite analogue: let $F(X) \in \mathbb{Z}[X]$ be a polynomial, pick a root α of K, and put $K = \mathbb{Q}(\alpha)$. Assume that F splits into h irreducible factors over $\mathbb{R}$: denoting the number of linear and quadratic factors by r and s, respectively, we know that $n = (K : \mathbb{Q}) = r + 2s$; moreover, the number $g = r + s$ of irreducible factors over $\mathbb{R}$ is just the number of infinite places. Now $s = n - g$ is the number of non-real embeddings of K into $\mathbb{C}$, and the well known $\operatorname{sgn} d = (-1)^{n-g}$ gives the infinite part of Proposition 4.6.

We will prove a generalization of Proposition 4.6 for finite extensions L/K of arbitrary number fields K; in order to keep the proof as simple as possible we will use the basics of p-adic number fields.

Our first step will be to replace the discriminant $d = \operatorname{disc} L$ in Proposition 4.6 by something else. To this end, let E be a finite extension of $\mathbb{Q}_p$ or $\mathbb{Q}$, where $\mathbb{Q}_p$ denotes the field of p-adic numbers. For finite extensions F/E of degree m there exists an $x \in F$ such that $F = E(x)$. Let $x_1 = x, x_2, \ldots, x_m$ denote the conjugates of x in a fixed separable closure of F/E; then the image of

$$\prod_{1 \leq i < j \leq m} (x_i - x_j)^2$$

in $E^\times/E^{\times 2}$ does not depend on the choice of x and is denoted by $\delta_{F/E}$.

Next we provide a result which links the global and local theory:

Proposition 4.7. *Let L/K be a finite extension of number fields, and let $\mathfrak{p}$ be a prime ideal in K not dividing $2\operatorname{disc}(L/K)$ such that $\mathfrak{p}\mathcal{O}_L = \prod_{i=1}^g \mathfrak{P}_i$. Let L_i denote the completion of L at $\mathfrak{P}_i$. Then L_i is a finite extension of the completion $E = K_\mathfrak{p}$ of K at $\mathfrak{p}$, and setting $\delta_i = \delta_{L_i/E}$ we find*

$$\left(\frac{\delta_{L/K}}{\mathfrak{p}}\right) = \prod_{i=1}^g \left(\frac{\delta_i}{\mathfrak{p}_E}\right).$$

Proof. This is immediate from $\delta_{L/K} = \prod \delta_i$, where we view $\delta_{L/K}$ as an element of $K_\mathfrak{p}$ via the embedding $K \hookrightarrow K_\mathfrak{p}$ (see for example [Cas], Chapter 9, Lemma 3.2). □

The basic ingredient for our proof of Proposition 4.6 will be

Proposition 4.8. *If F/E is an unramified extension of local fields of degree $m = (F : E)$, then*

$$\left(\frac{\delta_{F/E}}{\mathfrak{p}_E}\right) = (-1)^{m-1}.$$

Proof. Since F/E is unramified, it is cyclic, and $\mathrm{Gal}\,(F/E)$ is generated by an automorphism σ. Since $\mathfrak{p}_E \nmid \delta_{F/E}$, we have $(\delta_{F/E}/\mathfrak{p}_E) = 1$ if and only if $\prod(x_i - x_j)^2$ is a square in E, that is, if and only if $d = \prod(x_i - x_j) \in E$. Applying σ to d, we see that $d^\sigma = d$ will hold if and only if σ is an even permutation. But σ permutes the x_j cyclically, hence σ has signature $(-1)^{m-1}$, and this secures our claim. □

Now we are ready to prove Proposition 4.6 for general base fields K, namely

Proposition 4.9. *Let L/K be a finite extension of number fields. Put $n = (L:K)$ and let $\mathfrak{p}$ be a prime ideal in $\mathcal{O}_k$ such that $\mathfrak{p} \nmid \mathrm{disc}\,(L/K)$. If $\mathfrak{p}$ splits into exactly g prime ideals in L, then*

$$\left(\frac{\delta_{L/K}}{\mathfrak{p}}\right) = (-1)^{n-g}.$$

In fact, since $\mathfrak{p} \nmid \mathrm{disc}\,(L/K)$, the localization L_i of L at $\mathfrak{P}_i$ is unramified, hence we have $(\delta_i/\mathfrak{p}_E) = (-1)^{f_i-1}$, where $(L_i : K_\mathfrak{p}) = f_i$. Invoking Proposition 4.8 and observing $n = (L:K) = \sum_{i=1}^{g} e_i f_i = \sum_{i=1}^{g} f_i$, we get

$$\left(\frac{\delta_{L/K}}{\mathfrak{p}}\right) = \prod_{i=1}^{g}\left(\frac{\delta_i}{\mathfrak{p}_E}\right) = \prod_{i=1}^{g}(-1)^{f_i-1} = (-1)^{n-g}.$$

Setting $K = \mathbb{Q}$ in Proposition 4.9 gives Proposition 4.6; another not less natural generalization of Proposition 4.6 describes the splitting of ramified primes. Of course, if p divides d to an odd power, then $(d/p) = 0$, so we get no information. However, if all prime ideals above an odd prime p have odd ramification indices, then $d = p^{2t}\Delta$ for some Δ not divisible by p, and we have $(d/p) = (\Delta/p) = \pm 1$ (note that this is in agreement with the definition of the Kronecker symbol (d/p) in Section 2.1).

Proposition 4.10. *Let K be a number field, and suppose that p is a prime which does not ramify wildly in $K/\mathbb{Q}$. Assume moreover that the ramification indices e_i of the prime ideals above p are all odd. Then*

$$\left(\frac{d}{p}\right) = \prod_{i=1}^{g}(-1)^{f_i+1}\left(\frac{p}{e_i}\right)^{f_i}. \tag{4.4}$$

This does generalize Proposition 4.9, since if $p \nmid d$ then $e_i = 1$ and $\sum f_i = n$, and (4.4) gives $\left(\frac{d}{p}\right) = \prod(-1)^{f_i+1} = (-1)^{n+g}$. For a proof of Proposition 4.10 we refer the reader to Barrucand and Laubie [35, 36]; here we only observe the following

Corollary 4.11. *Let k be a number field with discriminant d and odd degree $n = (k : \mathbb{Q})$, and assume that a prime $p \nmid n$ is totally ramified in $k/\mathbb{Q}$. Then $p^{n-1} \| d$, and if we put $d = p^{n-1}\Delta$, then $\left(\frac{\Delta}{p}\right) = \left(\frac{p}{n}\right)$.*

The field $k = \mathbb{Q}(\sqrt[n]{p})$ is generated by a root of the polynomial $f(x) = x^n - p$ with discriminant $(-1)^{(n-1)/2}n^n p^{n-1}$, hence d is an odd power of n times a square. In particular, $\left(\frac{\Delta}{p}\right) = \left(\frac{n^*}{p}\right)$, and Corollary 4.11 yields the quadratic reciprocity law.

As a final treat in this section on discriminants, here is the skeleton of a proof of the quadratic reciprocity law that surfaced in the newsgroup sci.math.research; J. Shallit, who put it into his book [BS] with E. Bach as Exercise 5.45, learned it from J. Tunnell when he was a student: Let p and q be two different odd primes; then the following properties are equivalent (you should have no difficulties filling in the details):

- p is a square modulo q;
- the permutation of $(\mathbb{Z}/q\mathbb{Z})^\times$ induced by $x \longmapsto px$ is even;
- the permutation of the q-th roots of unity μ_q in some algebraic closure of $\mathbb{F}_p$ induced by $\zeta \longmapsto \zeta^p$ is even;
- the discriminant of $X^q - 1$ is a square in $\mathbb{F}_p$;
- $q^* = (-1)^{(q-1)/2}q$ is a square modulo p.

4.4 Kummer Extensions

In Section 3.1 we have proved the quadratic reciprocity law by comparing the splitting of primes in quadratic and cyclotomic extensions. In the general case, these extensions are replaced by Kummer extensions and class fields, respectively. The decomposition law for class fields is provided by class field theory in the form of Artin's reciprocity law, whereas the splitting of primes in Kummer extensions of prime degree is known since Kummer and Hilbert (see Hecke's book [348, 349], for example):

Theorem 4.12. *Let k be a number field containing the ℓ-th roots of unity, where ℓ is a rational prime, and let $K = k(\sqrt[\ell]{\mu})$ be a Kummer extension of degree $\ell = (K : k)$. Then the possible decompositions of a prime ideal $\mathfrak{p}$ in $\mathcal{O}_k$ are the following: either $\mathfrak{p}\mathcal{O}_K = \mathfrak{P}_1 \cdot \ldots \cdot \mathfrak{P}_\ell$, that is, $\mathfrak{p}$ splits into $(K : k)$ different prime ideals, or $\mathfrak{p}\mathcal{O}_K$ remains prime (then $\mathfrak{p}$ is called* inert*), or $\mathfrak{p}\mathcal{O}_K = \mathfrak{P}^\ell$ ramifies completely. This much is true for any cyclic extension of prime degree ℓ; in the special case of Kummer extensions, we can say more:*

a) *If $\mathfrak{p}^a \| \mu$ and $\ell \nmid a$, then $\mathfrak{p}$ ramifies;*
b) *If $\mathfrak{p}^a \| \mu$ and $\ell \mid a$, then we can choose ν in such a way that $k(\sqrt[\ell]{\mu}) = k(\sqrt[\ell]{\nu})$ and $\mathfrak{p} \nmid \nu$; if $\mathfrak{p} \nmid \ell$, then*

$\mathfrak{p}$ is inert if $\left(\frac{\nu}{\mathfrak{p}}\right)_\ell \neq 1$,

$\mathfrak{p}$ splits if $\left(\frac{\nu}{\mathfrak{p}}\right)_\ell = 1$.

c) If $\mathfrak{p} = \mathfrak{l}$, where $\mathfrak{l}^a \parallel (1 - \zeta_\ell)$, and $\mathfrak{l} \nmid \mu$, then

$\mathfrak{p}$ splits if $\mu \equiv \xi^\ell \mod \mathfrak{l}^{a\ell+1}$,
$\mathfrak{p}$ does not ramify if $\mu \equiv \xi^\ell \mod \mathfrak{l}^{a\ell}$, and
$\mathfrak{p}$ ramifies otherwise.

A similarly simple criterion in the case of Kummer extensions of prime power degree is not known, not even for cyclic quartic extensions!

Throughout the rest of this section we make the following assumptions:

1. k/F is abelian with Galois group $G = \mathrm{Gal}\,(k/F)$, and $\zeta_n \in k$;
2. $K = k(\sqrt[n]{\mu})$, and $(K : k) = n$; in particular,
3. $A = \mathrm{Gal}\,(K/k)$ is cyclic of order n.

Proposition 4.13. *Suppose that $\zeta_n \in k$; then $k(\sqrt[n]{\mu}) = k(\sqrt[n]{\nu})$ if and only if $\mu = \nu^a \xi^n$, where $a \in \mathbb{N}$ is prime to n, and $\xi \in k^\times$.*

Proof. The direction "$\Longleftarrow$" is trivial, hence assume that $k(\sqrt[n]{\mu}) = k(\sqrt[n]{\nu})$. If τ is any automorphism of K/k, then $\sqrt[n]{\mu}^{\tau-1}$ and $\sqrt[n]{\nu}^{\tau-1}$ are n-th roots of unity. Since these form a cyclic subgroup of $k^\times$, there must be an $a \in \mathbb{N}$ such that $\sqrt[n]{\mu}^{\tau-1} = (\sqrt[n]{\nu}^a)^{\tau-1}$. This implies that $\sqrt[n]{\mu\nu^a}^{\tau-1} = 1$, hence $\sqrt[n]{\mu\nu^a} \in k^\times$, and we have $\mu = \nu^a \xi^n$ for some $\xi \in k^\times$. $\square$

We begin by asking when K/F will be Galois:

Proposition 4.14. *Let $K/k/F$ be as above; then K/F is normal if and only if for every $\sigma \in G$ there exists an $\alpha_\sigma \in k$ and an $a(\sigma) \in \mathbb{N}$ such that $\mu^{\sigma-a(\sigma)} = \alpha_\sigma^n$; here $a(\sigma)$ is unique mod n, and α_σ is determined up to an n-th root of unity.*

Proof. If K/F is normal, then $\sqrt[n]{\mu}^\sigma \in K$. This implies that $k(\sqrt[n]{\mu}^\sigma) = k(\sqrt[n]{\mu})$, and Proposition 4.13 shows that there exists an $a = a(\sigma) \in \mathbb{N}$ and a $\xi = \alpha_\sigma \in k^\times$ such that $\mu^\sigma = \mu^{a(\sigma)} \alpha_\sigma^n$. The other direction will be proved in Proposition 4.15 below by explicitly writing down the elements of the Galois group. $\square$

Proposition 4.15. *Suppose that K/F is normal; then $\mu^{\sigma-a(\sigma)} = \alpha_\sigma^n$ for every $\sigma \in G$. Define $\omega = \sqrt[n]{\mu}$. Then every $\alpha \in K$ has a unique representation of the form $\alpha = \sum_{\nu=0}^{n-1} a_\nu \omega^\nu$, and the maps*

$$\widetilde{\sigma}_j : K \longrightarrow K : \sum_{\nu=0}^{n-1} a_\nu \omega^\nu \longmapsto \sum_{\nu=0}^{n-1} a_\nu^\sigma \zeta^{j\nu} \alpha_\sigma^\nu \omega^{\nu a(\sigma)}, \quad (0 \leq j \leq n-1)$$

are pairwise different automorphisms of K/F, whose restriction to k coincides with σ.

Proof. The maps $\widetilde{\sigma}_j, 0 \le j \le n-1$, are pairwise different because they act differently on ω; in fact $\widetilde{\sigma}_j(\omega) = \zeta^j \omega^{a(\sigma)}$. Moreover, their restriction to k coincides with σ because $\widetilde{\sigma}_j(a_0) = a_0^\sigma$. In order to prove that the $\widetilde{\sigma}_j$ are homomorphisms, it suffices to observe that $\widetilde{\sigma}_j(\omega^n) = \widetilde{\sigma}_j(\mu) = \mu^{a(\sigma)} \alpha_\sigma^n = (\omega^{a(\sigma)} \alpha_\sigma)^n = \widetilde{\sigma}_j(\omega)^n$. □

An exact sequence $E: 1 \longrightarrow A \longrightarrow \Gamma \longrightarrow G \longrightarrow 1$ of finite groups is called an *extension* of G by A. E is called a *central extension* if $A \subseteq Z(\Gamma)$ is contained in the center of Γ (where we have identified A and its image in Γ). A normal tower $K/k/F$ of fields is called central if the exact sequence $1 \longrightarrow \mathrm{Gal}(K/k) \longrightarrow \mathrm{Gal}(K/F) \longrightarrow \mathrm{Gal}(k/F) \longrightarrow 1$ corresponding to the tower is central.

Proposition 4.16. *Suppose that K/F is normal; then $K/k/F$ is central if and only if $\zeta^\sigma = \zeta^{a(\sigma)}$ for every $\sigma \in G$.*

Proof. $K/k/F$ is central if and only if the automorphisms of K/k commute with the automorphisms of K/F; the automorphisms of K/k are exactly the extensions $\widetilde{\mathrm{id}}_i$ of the identity id on k/F. Now

$$\widetilde{\mathrm{id}}_i \widetilde{\sigma}_j \left(\sum a_\nu \omega^\nu \right) = \widetilde{\mathrm{id}}_i \left(\sum a_\nu^\sigma \zeta^{j\nu} \alpha_\sigma^\nu \omega^{\nu a(\sigma)} \right)$$
$$= \left(\sum a_\nu^\sigma \zeta^{j\nu} \alpha_\sigma^\nu \zeta^{i\nu a(\sigma)} \omega^{\nu a(\sigma)} \right), \text{ and}$$
$$\widetilde{\sigma}_j \widetilde{\mathrm{id}}_i \left(\sum a_\nu \omega^\nu \right) = \widetilde{\sigma}_j \left(\sum a_\nu \zeta^{i\nu} \omega^\nu \right)$$
$$= \left(\sum a_\nu^\sigma \zeta^{i\nu\sigma} \alpha_\sigma^\nu \zeta^{j\nu} \omega^{\nu a(\sigma)} \right),$$

where all sums are over $\nu = 0, \ldots, n-1$. Hence $\widetilde{\mathrm{id}}_i$ and $\widetilde{\sigma}_j$ commute if and only if $\zeta^{i\nu\sigma} = \zeta^{i\nu a(\sigma)}$. □

Now central extensions of cyclic groups are abelian (Exercise 4.10), so

Corollary 4.17. *Suppose that K/F is normal and that k/F is cyclic. Then K/F is abelian if and only if $\zeta^\sigma = \zeta^{a(\sigma)}$ for every $\sigma \in G$.*

Example. Take $F = \mathbb{Q}$, $k = \mathbb{Q}(\sqrt{-3})$, let σ denote the non-trivial automorphism of $k/\mathbb{Q}$, and pick some $\alpha \in k^\times$. Put $\mu_j = \alpha^{j+\sigma}$, $j = 1, 2$. Then both $K_1 = k(\sqrt{\mu_1})$ and $K_2 = k(\sqrt{\mu_2})$ are normal over $\mathbb{Q}$, since $\mu_1^{\sigma-1} = 1$ and $\mu_2^{\sigma-2} = \alpha^{-3}$ are cubes in $k^\times$. Moreover, $\mathrm{Gal}(K_1/\mathbb{Q}) \simeq S_3$ (the symmetric group of order 6) and $\mathrm{Gal}(K_2/\mathbb{Q}) \simeq \mathbb{Z}/6\mathbb{Z}$.

4.5 Characters of Abelian Groups

The Jacobi symbol $\left(\frac{\cdot}{m} \right)$ introduced in Chapter 1 can be viewed as a homomorphism $(\mathbb{Z}/m\mathbb{Z})^\times \longrightarrow \mathbb{C}^\times$; such homomorphisms are called *Dirichlet characters*. Characters of finite abelian groups will play a central role in the

122   4. Power Residues and Gauss Sums

following, be it in connection with character sums (that's what you would expect) or with the description of abelian extensions of $\mathbb{Q}$ (character groups contain all the information about splitting and ramification of primes). Before we can uncover these connections, we have to introduce conductors:

Let $K/\mathbb{Q}$ be an abelian extension of $\mathbb{Q}$; every integer $m \in \mathbb{N}$ with the property	Let χ be a Dirichlet character on $(\mathbb{Z}/n\mathbb{Z})^\times$; every integer $m \in \mathbb{N}$ such that
$K \subseteq \mathbb{Q}(\zeta_m)$	$a \equiv b \bmod m \implies \chi(a) = \chi(b)$

whenever a, b are prime to n

is called a *defining modulus*. If m_1 and m_2 are defining moduli, then so is their greatest common divisor, hence there exists a smallest defining modulus $\mathfrak{f}$, which is called the *conductor*

of the abelian extension $K/\mathbb{Q}$.	of the Dirichlet character χ.

A Dirichlet character χ defined modulo m is called primitive if m is the conductor of χ.

Remark. Actually this is just the finite part of the conductor. It is sometimes useful to multiply $\mathfrak{f}$ by the symbolic factor ∞ if K is complex. An instructive example is given in Hasse's [341, I, §5]: for $a \in \mathbb{Q}^\times$, let $\mathfrak{f}_a$ denote the conductor of $\mathbb{Q}(\sqrt{a}\,)$; then the quadratic reciprocity law for coprime odd integers $a, b \in \mathbb{Z} \setminus \{0\}$ can be stated in the following form: $\left(\frac{a}{b}\right)\left(\frac{b}{a}\right) = 1$ if $(\mathfrak{f}_a, \mathfrak{f}_b) = 1$. Note that $(\mathfrak{f}_3, \mathfrak{f}_7) = 4$ and $(\mathfrak{f}_{-3}, \mathfrak{f}_{-7}) = \infty$.

Let us look at an explicit example by examining Dirichlet characters on $(\mathbb{Z}/15\mathbb{Z})^\times$. The first nontrivial character one might think of is the quadratic residue character $\psi = \left(\frac{\cdot}{15}\right)$. It has defining modulus 15, because $a \equiv b \bmod 15$ implies $\psi(a) = \psi(b)$; in fact, ψ has conductor 15, because neither 3 ($1 \equiv -2 \bmod 3$, but $\psi(1) = -\psi(-2)$) nor 5 ($1 \equiv -4 \bmod 5$, but $\psi(1) = -\psi(-4)$) are defining moduli of ψ.

Next we define a Dirichlet character χ explicitly by giving its values on $(\mathbb{Z}/15\mathbb{Z})^\times$: put $\chi(a) = +1$ for $a \equiv 1, -4 \bmod 15$, $\chi(a) = i$ for $a \equiv 2, -8 \bmod 15$, $\chi(a) = -1$ for $a \equiv 4, -1 \bmod 15$, and $\chi(a) = -i$ for $a \equiv -2, 8 \bmod 15$. It is easily checked that χ is indeed a homomorphism $(\mathbb{Z}/15\mathbb{Z})^\times \longrightarrow \mathbb{C}^\times$. Moreover, χ has order 4 and conductor 5. We can in fact construct the following eight ($= \varphi(15)$) Dirichlet characters mod 15:

- the trivial character of order 1 and conductor 1;
- three quadratic characters: ψ and $\chi^2\psi$ of conductor 15, and χ^2 of conductor 5;
- four characters of order 4: χ and χ^3 of conductor 5, $\chi\psi$ and $\chi^3\psi$ of conductor 15.

We will see in a moment that there are no other Dirichlet characters mod 15. But first we define a character of a finite abelian group G to be a homomorphism $G \longrightarrow \mathbb{C}^\times$: Dirichlet characters are just characters on $G = (\mathbb{Z}/m\mathbb{Z})^\times$. The set of characters $G\hat{\;}$ of G is an abelian group with respect to the multiplication $(\psi\chi)(a) = \psi(a)\chi(a)$. If A and B are finite abelian groups, then we obviously have $(A \oplus B)\hat{\;} \simeq A\hat{\;} \oplus B\hat{\;}$. Now we claim

Proposition 4.18. *If G is a finite abelian group, then $G \simeq G\hat{\;}$ (non-canonically) and $G \simeq G\hat{\;}\hat{\;}$ (canonically).*

Proof. Since G is the direct sum of cyclic groups, and since $(A \oplus B)\hat{\;} \simeq A\hat{\;} \oplus B\hat{\;}$, it is sufficient to prove $G \simeq G\hat{\;}$ for cyclic groups G. Let $G = \langle g \rangle$; then any character $\chi \in G\hat{\;}$ is determined by the value of $\chi(g)$, since we have $\chi(g^a) = \chi(g)^a$. Now $\chi(g)$ must be a $\#G$-th root of unity; there are exactly $\#G$ of them, and they are all powers of a primitive $\#G$-th root of unity. Therefore, each character $\in G\hat{\;}$ is a power of the character χ which maps g to a primitive $\#G$-th root of unity. This shows that $G\hat{\;}$ is a cyclic group of order $\#G$, and in particular, we find $G \simeq G\hat{\;}$.

In order to prove that $G \simeq G\hat{\;}\hat{\;}$ we observe that every $g \in G$ induces a map $\gamma_g : G\hat{\;} \longrightarrow \mathbb{C}^\times : \gamma_g(\chi) = \chi(g)$. The map $\psi : g \longmapsto \gamma_g$ defines a homomorphism $G \longrightarrow G\hat{\;}\hat{\;}$ with $\ker \psi = 1$; it must be onto since $\#G = \#G\hat{\;} = \#G\hat{\;}\hat{\;}$. □

Our next result tells us how dualizing affects short exact sequences:

Proposition 4.19. *If $1 \longrightarrow A \longrightarrow B \longrightarrow C \longrightarrow 1$ is an exact sequence of finite abelian groups, then so is $1 \longrightarrow C\hat{\;} \longrightarrow B\hat{\;} \longrightarrow A\hat{\;} \longrightarrow 1$.*

Proof. A is a subgroup of B; therefore, every character of B restricts to a character of A, and we have a natural map $\hat{\imath} = \mathrm{res} : B\hat{\;} \longrightarrow A\hat{\;}$. If χ is a character on $C \simeq B/A$, define a character $\hat{\pi}\chi \in B\hat{\;}$ by $\hat{\pi}\chi(b) = \chi(bA)$. The homomorphism $\hat{\pi} : C\hat{\;} \longrightarrow B\hat{\;}$ is injective, and it is easy to see that $\ker \hat{\imath} = \mathrm{im}\, \hat{\pi}$. But now $\#\mathrm{im}\,\hat{\imath} = \#B\hat{\;}/\#C\hat{\;} = \#B/\#C = \#A = \#A\hat{\;}$, hence $\hat{\pi}$ is onto. □

For a subgroup H of G put $H^\perp = \{\chi \in G\hat{\;} \mid \chi(h) = 1 \text{ for all } h \in H\}$. Obviously, $H^\perp$ is the kernel of the natural map $\hat{\imath}: G\hat{\;} \longrightarrow H\hat{\;}$ studied above. The dual of the exact sequence $1 \longrightarrow H \longrightarrow G \longrightarrow G/H \longrightarrow 1$ shows that $H^\perp \simeq (G/H)\hat{\;}$, and this in turn implies that $H\hat{\;} \simeq G\hat{\;}/(G/H)\hat{\;} \simeq G\hat{\;}/H^\perp$. Similarly, we can prove that $(H^\perp)^\perp \simeq H$.

An important property of characters are the orthogonality relations:

Proposition 4.20. *Let G be a finite abelian group with character group X. Then*

$$\sum_{x \in G} \chi(x) = \begin{cases} \#G & \text{if } \chi = \mathbb{1} \\ 0 & \text{if } \chi \neq \mathbb{1} \end{cases} \quad \text{and} \quad \sum_{\chi \in X} \chi(x) = \begin{cases} \#G & \text{if } x = \mathbb{1} \\ 0 & \text{if } x \neq \mathbb{1} \end{cases}.$$

Proof. The first assertion is clear if $\chi = \mathbb{1}$. If $\chi \neq \mathbb{1}$, then there must be an $y \in G$ such that $\chi(y) \neq 1$. But now $\chi(y) \sum_{x \in G} \chi(x) = \sum_{x \in G} \chi(xy) = \sum_{x \in G} \chi(x)$, proving our claim. The 'dual' assertion is proved similarly (or reduced to the first case by identifying G and $G\hat{\ }\hat{\ }$). □

Now let $L/\mathbb{Q}$ be an abelian extension with Galois group G. To any subfield K of $L/\mathbb{Q}$ we associate the subgroup $X_K = \text{Gal}\,(L/K)^\perp$ of $G\hat{\ }$; in particular, we have $X_L = G\hat{\ }$ and $X_\mathbb{Q} = \{\mathbb{1}\}$. If we are given a subgroup X of $G\hat{\ }$, then we can identify $X^\perp$ via the isomorphism $G \simeq G\hat{\ }\hat{\ }$ with a subgroup H of G; if we let K be the fixed field of H, then the character group associated to K turns out to be the subgroup X of $G\hat{\ }$ we started with, i.e. we have set up a bijective correspondence between subgroups of $G\hat{\ }$ and subfields of $L/\mathbb{Q}$. Furthermore, if the subgroups X_1 and X_2 of $G\hat{\ }$ correspond to the subfields K_1 and K_2 of L, then $X_1 \subseteq X_2$ if and only if $K_1 \subseteq K_2$, and the compositum $K_1 K_2$ corresponds to the product $X_1 X_2$.

It's example time. Consider $L = \mathbb{Q}(\zeta_n)$, and let X be the character group of $\text{Gal}\,(L/\mathbb{Q}) \simeq (\mathbb{Z}/n\mathbb{Z})^\times$; the maximal real subfield $K = L^+$ of L is the fixed field of complex conjugation $J = (\zeta_n \longmapsto \zeta_n^{-1})$. Thus K corresponds to $X_K = \text{Gal}\,(L/K)^\perp = \{\chi \in X : \chi(-1) = 1\}$. This shows in particular that the field corresponding to a character χ is totally real if and only if $\chi(-1) = 1$; such characters are called *real*, and consequently characters satisfying $\chi(-1) = -1$ are called *odd*.

As a second application, let us describe the quadratic extensions of $\mathbb{Q}$ using characters. To this end, let d be a prime discriminant; the quadratic Gauss sum shows explicitly that $\mathbb{Q}(\sqrt{d}) \subseteq \mathbb{Q}(\zeta_{|d|})$ if $d \equiv 1 \bmod 4$, and for $d = -4, \pm 8$ we can check this directly. Given an arbitrary discriminant d, we can write it as a product of prime discriminants, and it follows easily that $\mathbb{Q}(\sqrt{d}) \subseteq \mathbb{Q}(\zeta_{|d|})$ holds in general (see Exercise 3.13).

$t \bmod 8$	1	3	5	7
$\chi_{-4}(t)$	+1	−1	+1	−1
$\chi_8(t)$	+1	−1	−1	+1
$\chi_{-8}(t)$	+1	+1	−1	−1

The quadratic character corresponding to the quadratic field $\mathbb{Q}(\sqrt{d})$ is easily determined: if d is an odd prime discriminant, then there exists only one quadratic character defined modulo d, namely $\chi_d = (\cdot/d)$; if d is even, there are χ_{-4}, χ_8 and χ_{-8}, defined by the table on the left.

Now assume that $d = d_1 d_2$ is a product of two discriminants, and that $\mathbb{Q}(\sqrt{d_1}) \longleftrightarrow \langle \chi_1 \rangle$ and $\mathbb{Q}(\sqrt{d_2}) \longleftrightarrow \langle \chi_2 \rangle$. Then $\mathbb{Q}(\sqrt{d_1}, \sqrt{d_2}) \longleftrightarrow \langle \chi_1, \chi_2 \rangle$, hence we must have $\mathbb{Q}(\sqrt{d}) \longleftrightarrow \langle \chi_1 \chi_2 \rangle$. Using induction we see that, generally, $\mathbb{Q}(\sqrt{d}) \longleftrightarrow \chi_d$. In particular we find that the conductor of a quadratic character, the conductor of the corresponding quadratic field, and its discriminant are all equal.

We already remarked that character groups tell us all about the splitting and ramification of primes. To see that this is so, let X be the character group of the abelian extension $K/\mathbb{Q}$ and p a rational prime. Now define $Y = \{\chi \in X : \chi(p) \neq 0\}$ and $Z = \{\chi \in X : \chi(p) = 1\}$. Clearly $Z \subseteq Y \subseteq X$ are subgroups of X, and it can be shown (cf. Washington [Was]):

4.5 Characters of Abelian Groups

Theorem 4.21. *With the notation above we have $X/Y \simeq T_p(K/\mathbb{Q})\hat{}$ (the dual of the inertia group) and $X/Z \simeq Z_p(K/\mathbb{Q})\hat{}$ (the dual of the decomposition group – see Exercise 2.38). In particular, $(X:Y) = e$, $(Y:Z) = f$, and $(Z:1) = g$, where $p\mathcal{O}_K = (\mathfrak{p}_1 \cdots \mathfrak{p}_g)^e$ is the prime ideal decomposition of p in K, and where f denotes the degree of the $\mathfrak{p}_j$.*

If we define the conductor $\mathfrak{f}_X$ of a subgroup $X \subseteq G\hat{}$ to be the lowest common multiple of the conductors of all $\chi \in X$, i.e. by $\mathfrak{f}_X = \text{lcm}\{\mathfrak{f}_\chi \mid \chi \in X\}$, then class field theory over $\mathbb{Q}$ shows that the conductors of X and of L coincide. Moreover there is the extremely useful

Theorem 4.22. (Conductor-Discriminant Formula)
Let m be a defining modulus of the abelian extension $L/\mathbb{Q}$; then

$$\mathfrak{f}_L = \text{lcm}\{\mathfrak{f}_\chi : \chi \in X_{L,m}\} \quad \text{and} \quad \text{disc } L = (-1)^u \prod_{\chi \in X_{L,m}} \mathfrak{f}_\chi,$$

where u is the number of odd characters $\chi \in X_{L,m}$. □

This result is due to Dedekind [Ded]. Its simplest proof uses analytic techniques, namely the functional equation of Dirichlet's L-functions – see Chapter 10 for details. Part of the fame of Theorem 4.22 is due to its name in German, where it is often called the Führerdiskriminantenproduktformel (if only to impress foreigners). For a generalization of the Conductor-Discriminant Formula to arbitrary abelian (and even Galois) extensions of number fields, see Part II.

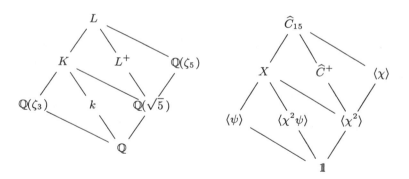

Fig. 4.1. Subfields of $\mathbb{Q}(\zeta_{15})$

This formula can be applied to compute discriminants of abelian extensions: take the field $L = \mathbb{Q}(\zeta_{15})$, for example. We have $\text{Gal}(L/\mathbb{Q}) \simeq C_{15} \simeq C_3 \times C_5$ (here C_n denotes the cyclic group of order n), and this implies that $\widehat{C}_{15} \simeq \widehat{C}_3 \times \widehat{C}_5$. Define characters $\chi \in \widehat{C}_3$ and $\psi \in \widehat{C}_5$ by $\chi(2) = i$ and

$\psi(2) = -1$ (note that 2 is a primitive root mod 3 and mod 5). The subfields of L and the subgroups of $\widehat{C}_{15}$ are displayed in Figure 4.1.

There $K = \mathbb{Q}(\sqrt{-3}, \sqrt{5})$, $k = \mathbb{Q}(\sqrt{-15})$, $L^+ = \mathbb{Q}(\zeta_{15} + \zeta_{15}^{-1})$ (since $\mathbb{Q}(\zeta_5)$ is generated by a square root of $\alpha = \frac{-5+\sqrt{5}}{2}$ over $\mathbb{Q}(\sqrt{5})$, we find that $L^+ = \mathbb{Q}(\sqrt{-3\alpha})$) and $X = \langle \chi^2, \psi \rangle$, $\widehat{C}^+ = \langle \chi^2, \chi\psi \rangle$.

ϕ	1	χ	χ^2	ψ	$\chi\psi$	$\chi^2\psi$
$\mathfrak{f}_\phi$	1	5	5	3	15	15
$\#\langle\phi\rangle$	1	4	2	2	4	2

The table on the left gives the conductors of the characters involved as well as the order of the subgroup of $\widehat{C}_{15}$ that they generate. The sign of the discriminant as given by the conductor-discriminant formula can be checked by the congruence disc $K \equiv 1 \bmod 4$ for odd discriminants, or by the formula sign (disc K) $= (-1)^s$, where s denotes the number of complex primes of K.

K	$\mathbb{Q}(\zeta_3)$	k	$\mathbb{Q}(\sqrt{5})$	K	L^+	$\mathbb{Q}(\zeta_5)$	$\mathbb{Q}(\zeta_{15})$
disc K	-3	-15	5	$3^2 \cdot 5^2$	$3^2 \cdot 5^3$	5^3	$3^4 \cdot 5^6$

4.6 Sums of Gauss, Jacobi and Eisenstein

We have seen in Chapter 3 that the quadratic Gauss sum

$$\tau = \sum_{t=1}^{p-1} \left(\frac{t}{p}\right) \zeta^t$$

allows us to reduce the proof of the quadratic reciprocity law to a mere computation. Of course we would like to have a similar method for proving higher reciprocity laws, so assume we have an m-th power residue character $\chi = (\cdot/\mathfrak{p})$ in $\mathcal{O} = \mathbb{Z}[\zeta_m]$. We clearly should replace the Legendre symbol in the quadratic Gauss sum by χ; since there are no $\mathfrak{p}$-th roots of unity, let us stick with the ζ_p. But then ζ_p^t only makes sense for representatives $t \in \mathbb{Z}$ of $(\mathcal{O}/\mathfrak{p})^\times$.

Before we try to generalize this further, let us assume for the moment that our prime ideal $\mathfrak{p}$ has degree 1; then $\mathcal{O}/\mathfrak{p} \simeq \mathbb{Z}/p\mathbb{Z}$, and if we choose representatives t of $(\mathcal{O}/\mathfrak{p})^\times$ lying in $\mathbb{Z}$, then $\tau = \sum \chi(t)\zeta_p^t$ does make sense. Let us compute an example. Take $m = 4$ and $\mathfrak{p} = (1+2i)$; the values $\chi(t)$ are given in Exercise 4.1, so we find $\tau = \zeta_5 + i\zeta_5^2 - i\zeta_5^3 - \zeta_5^4$. This is an element of $\mathbb{Z}[\zeta_{20}]$, and squaring gives $\tau^2 = -(1+2i)(\zeta_5 - \zeta_5^2 - \zeta_5^3 + \zeta_5^4)$. From the computation of the quadratic Gauss sum over $\mathbb{Z}/5\mathbb{Z}$ we know that $\zeta_5 - \zeta_5^2 - \zeta_5^3 + \zeta_5^4 = \sqrt{5}$ (if $\zeta_5 = \exp(\frac{2\pi i}{5})$), so we find $\tau^2 = -\pi\sqrt{5}$ and $\tau^4 = \pi^3 \overline{\pi}$.

This looks promising, so let us think about what we can do if $\mathfrak{p}$ has degree > 1 (we won't need this for studying cubic and quartic reciprocity, by the way). We start by recalling the role that the ζ_p played in the quadratic Gauss sum: the most important property was of course the fact that ζ_p generates an abelian extension of $\mathbb{Q}$ which contains a quadratic extension $\mathbb{Q}(\sqrt{p^*})$. This is still true if we look at cyclotomic extensions $k(\zeta_p)$ of $k = \mathbb{Q}(\zeta_n)$, because we always have $p \nmid n$ (though the fact that we do not get *all* abelian extensions of k in this way will complicate things when we deal with higher reciprocity laws). Another property of ζ that was needed in the proof of Proposition 3.19 was the fact that $\zeta_p^a \zeta_p^b = \zeta_p^{a+b}$: this is the same as saying that $\psi : \mathbb{F}_p \longrightarrow \mu_p = \{\zeta_p^a : 0 \leq a \leq p-1\} : a \longmapsto \zeta_p^a$ is an additive character on $\mathbb{F}_p = \mathbb{Z}/p\mathbb{Z}$, i.e., that it has the property $\psi(a)\psi(b) = \psi(a+b)$.

Thus the general situation will look like this: we have a finite field $\mathbb{F}_q \simeq \mathcal{O}_k/\mathfrak{p}$ with $q = p^f$ elements, which is a finite extension of its prime field $\mathbb{F}_p$. This extension is normal, the cyclic Galois group $\mathrm{Gal}(\mathbb{F}_q/\mathbb{F}_p)$ being generated by the Frobenius automorphism $\sigma : x \longmapsto x^p$. It is easy to show that the trace $\mathrm{Tr} : \mathbb{F}_q \longrightarrow \mathbb{F}_p$ (mapping an element to the sum of its conjugates) and the norm $N : \mathbb{F}_q^\times \longrightarrow \mathbb{F}_p^\times$ are epimorphisms (see Exercise 4.28). The two ingredients of a Gauss sum will be a *multiplicative character* $\chi : \mathbb{F}_q^\times \longrightarrow \mathbb{C}^\times$ which we will always take to be an m-th power residue symbol via the isomorphism $\mathbb{F}_q \simeq \mathcal{O}_k/\mathfrak{p}$ (consequently, χ will take values in the group μ_m of m-th roots of unity, where $m \mid q-1$ is the order of χ) and an *additive character* $\psi : \mathbb{F}_q \longrightarrow \mathbb{C}^\times$, namely $\psi(t) = \zeta_p^{Tr(t)}$ (this is not a restriction – see Exercise 4.29). Then

$$G_\alpha(\chi) = - \sum_{t \in \mathbb{F}^\times} \chi(t)\psi(\alpha t) \tag{4.5}$$

is called the *Gauss sum* associated to χ (and $\alpha \in \mathbb{F}_q^\times$) on $\mathbb{F} := \mathbb{F}_q$ (strictly speaking we should denote the Gauss sum by $G_\alpha(\chi, \psi)$, but we will always work with a fixed non-trivial additive character ψ). The reason for inserting the minus sign will become clear later on (see e.g. the remarks after Theorem 4.32). The character defined by $\chi(t) = 1$ for all $t \in \mathbb{F}^\times$ is called the unit character and will be denoted by $\chi = \mathbb{1}$.

Observe that $G_\alpha(\chi)$ is an element of $\mathbb{Z}[\zeta_m, \zeta_p]$ (for quadratic characters, we have $m = 2$, thus the quadratic Gauss sum lives in $\mathbb{Z}[\zeta_p]$). The main property of Gauss sums that we are about to prove is the fact that their m-th powers $G_\alpha(\chi)^m$ are elements of $\mathbb{Z}[\zeta_m]$.

Important Remark. It will be convenient to set $\mathbb{1}(0) = 1$, in contrast to $\chi(0) = 0$ for $\chi \neq \mathbb{1}$. If we define multiplication of characters via multiplication of their values, this would imply that $\chi \cdot \chi^{-1} \neq \mathbb{1}$ for characters $\chi \neq \mathbb{1}$ since $\chi(0) = \chi^{-1}(0) = 0$, but $\mathbb{1}(0) = 1$. Therefore we define, in addition to $\mathbb{1}(0) = 1$,

$$(\chi\psi)(a) := \begin{cases} \chi(a)\psi(a) & \text{if } a \neq 0, \\ 0 & \text{if } a = 0 \text{ and } \chi \neq \psi^{-1}, \\ 1 & \text{if } a = 0 \text{ and } \chi = \psi^{-1}. \end{cases}$$

Note that we have $G_1(\mathbb{1}) = 1$ since the sum in (4.5) is over $\mathbb{F}^\times$. This is convenient for the formulation of the Davenport-Hasse relation as well as the Stickelberger congruences.

Put $G(\chi) = G_1(\chi)$; then

$$G_\alpha(\chi) = -\sum_{t \in \mathbb{F}^\times} \chi(t)\psi(\alpha t) = -\chi(\alpha)^{-1} \sum_{t \in \mathbb{F}^\times} \chi(\alpha t)\psi(\alpha t) = \chi(\alpha)^{-1} G(\chi)$$

for any $\alpha \in \mathbb{F}^\times$ since then αt runs through $\mathbb{F}$ if t does. This property generalizes Equation (3.3) and will allow us to extract information about the character χ from the Gauss sums exactly as in the quadratic case.

For the proof of reciprocity laws we need to know how to multiply Gauss sums. So let us take characters $\chi_1 \neq \mathbb{1}$ and $\chi_2 \neq \mathbb{1}$ on $\mathbb{F}^\times$ (these assumptions allow us to extend the summation in (4.5) over all $t \in \mathbb{F}$) and compute the product of their Gauss sums; we will do this as straightforward as possible:

$$G(\chi_1)G(\chi_2) = \sum_{a,b \in \mathbb{F}} \chi_1(a)\chi_2(b)\psi(a+b) = \sum_{a,c \in \mathbb{F}} \chi_1(a)\chi_2(c-a)\psi(c).$$

Here we have substituted $b = c - a$. Next we would like to put $a = ct$: but this is equivalent to introducing the new variable $t = ac^{-1}$, which is only possible if $c \neq 0$. Therefore, we split the sum into two:

$$G(\chi_1)G(\chi_2) = \sum_{a,\, c \neq 0} \chi_1(a)\chi_2(c-a)\psi(c) + \sum_{a} \chi_1(a)\chi_2(-a).$$

Now we evaluate both sums separately. The second sum is easy to compute: if $\chi_1 \neq \chi_2^{-1}$, then $\sum_a \chi_1(a)\chi_2(-a) = 0$ (repeat the proof of Lemma 3.20), and if $\chi_1 = \chi_2^{-1}$, then we find

$$\sum_a \chi_1(a)\chi_2(-a) = \sum_{a \neq 0} \chi_1(-1) = \chi_1(-1) \cdot (q-1).$$

For the computation of the first sum we substitute $a = ct$ as indicated and get

$$\sum_{a, c \neq 0} \chi_1(a)\chi_2(c-a)\psi(c) = \sum_{t,\, c \neq 0} \chi_1(c)\chi_2(c)\psi(c)\chi_1(t)\chi_2(1-t)$$

$$= \sum_{c \neq 0} \chi_1\chi_2(c)\psi(c) \cdot \sum_t \chi_1(t)\chi_2(1-t)$$

$$= G(\chi_1\chi_2)J(\chi_1, \chi_2),$$

where the sum $J(\chi_1, \chi_2) = -\sum_{t \in \mathbb{F}} \chi_1(t)\chi_2(1-t)$, also introduced by Gauss, is called the *Jacobi sum* of χ_1 and χ_2.

In particular, we have the simple relation

4.6 Sums of Gauss, Jacobi and Eisenstein   129

$$G(\chi_1)G(\chi_2) = G(\chi_1\chi_2)J(\chi_1,\chi_2)$$

for all characters such that $\chi_1 \neq \mathbb{1}, \chi_2 \neq \mathbb{1}$, and $\chi_1\chi_2 \neq \mathbb{1}$. This relation is formally similar to the definition of crossed homomorphisms, which is why Jacobi sums are sometimes called the "factor system" of Gauss sums.

If $\chi_1\chi_2 = \mathbb{1}$, however, the first sum contributes

$$G(\mathbb{1})\,J(\chi_1,\chi_1^{-1}) = \chi_1(-1)$$

(we will prove this in a moment), and the second $\chi_1(-1) \cdot (q-1)$. This gives $G(\chi)G(\chi^{-1}) = \chi(-1)q$.

It remains to compute the Jacobi sum $J(\chi,\chi^{-1})$. It is easily checked that the map $\mathbb{F} \setminus \{1\} \longrightarrow \mathbb{F} \setminus \{-1\} : t \longmapsto \frac{t}{1-t}$ is a bijection, hence we get

$$\sum_{t\in\mathbb{F}} \chi(t)\chi(1-t)^{-1} = \sum_{t\neq 1} \chi\left(\frac{t}{1-t}\right) = \sum_{s\neq -1} \chi(s) = -\chi(-1),$$

and this proves our claim.

The following proposition collects the most basic properties of Gauss and Jacobi sums (note that these formulas do not reflect our particular choice of the sign in the Gauss sum (4.5), since these sums come in pairs):

Proposition 4.23. *For all $\alpha \in \mathbb{F}^\times$ and all characters $\chi, \chi_1, \chi_2 \neq \mathbb{1}$ of order dividing m such that $\chi_1\chi_2 \neq \mathbb{1}$ we have*

i) $G_\alpha(\chi) \in \mathbb{Z}[\zeta_p,\zeta_m]$;

ii) $G_\alpha(\chi) = \chi(\alpha)^{-1}G(\chi)$;

iii) $G(\chi_1)G(\chi_2) = G(\chi_1\chi_2)J(\chi_1,\chi_2)$;

iv) $\chi(-1)G(\chi^{-1}) = \overline{G(\chi)}$;

v) $G(\chi)G(\chi^{-1}) = \chi(-1)\cdot q$;

vi) $G(\chi)\overline{G(\chi)} = q$.

Proof. Since vi) follows from iv) and v), only iv) remains to be proved:

$$\overline{G(\chi)} = -\sum \chi(t)^{-1}\psi(t)^{-1} = -\sum \chi(t)^{-1}\psi(-t) = \chi(-1)^{-1}G(\chi^{-1}),$$

where the sum is over all $t \in \mathbb{F}^\times$. The claim follows from $\chi(-1) = \pm 1 = \chi(-1)^{-1}$. □

Note that property vi) implies that $G(\chi)$ is only divisible by prime ideals above p; we will determine its prime ideal decomposition in Chapter 11. Jacobi sums are easier to work with, because they live in the subring $\mathbb{Z}[\zeta_m]$ of $\mathbb{Z}[\zeta_p,\zeta_m]$. An immediate consequence of this observation (proved by repeatedly applying property iii) of Proposition 4.23) is

Corollary 4.24. *If χ is a character of order m on $\mathbb{F}$, then*

$$G(\chi)^r = G(\chi^r) \cdot J(\chi,\chi)J(\chi,\chi^2)\cdot\ldots\cdot J(\chi,\chi^{r-1})$$

for all $1 \leq r \leq m-1$. Moreover,

$$G(\chi)^m = \chi(-1)\cdot q \cdot J(\chi,\chi)J(\chi,\chi^2)\cdot\ldots\cdot J(\chi,\chi^{m-2}) \in \mathbb{Z}[\zeta_m]. \qquad (4.6)$$

130   4. Power Residues and Gauss Sums

We will need a few other properties of Gauss and Jacobi sums in the chapter on Eisenstein's reciprocity law; in particular we will have to strengthen the result of the above Corollary, which says that $G(\chi)^m$ is an element of $\mathbb{Z}[\zeta_m]$:

Proposition 4.25. *Let $\mathfrak{p} \nmid m$ be a prime in $k = \mathbb{Q}(\zeta_m)$; then $\chi = \left(\frac{\cdot}{\mathfrak{p}}\right)_m$ is a multiplicative character of order m on $\mathcal{O}_k/\mathfrak{p} \simeq \mathbb{F}_q$, where $q = p^f$. Let Z denote the decomposition group of $\mathfrak{p}$. Then*

i) $G(\chi)^\sigma = \chi(a(\sigma))^{-1} G(\chi)$ *for all $\sigma \in Z$; here $a(\sigma) \in (\mathbb{Z}/p\mathbb{Z})^\times$ is defined by $\zeta_p^\sigma = \zeta_p^{a(\sigma)}$;*

ii) $G(\chi)^m$ *is contained in the decomposition field of $\mathfrak{p}$ in $\mathbb{Q}(\zeta_{mp})$, i.e., in the subfield of $\mathbb{Q}(\zeta_m)$ fixed by Z.*

Proof. We have $\sigma : \zeta_p \mapsto \zeta_p^{a(\sigma)}$ and $\chi(\alpha)^\sigma = \left(\frac{\alpha^\sigma}{\mathfrak{p}}\right)_m$ (because $\sigma \in Z$); moreover, $\psi(\alpha) = \psi(\alpha^\sigma)$ for $\psi(\alpha) = \zeta^{Tr(\alpha)}$. Now we find

$$\begin{aligned} G(\chi)^\sigma &= -\sum_\alpha \chi(\alpha)^\sigma \psi(\alpha)^\sigma &= -\sum_\alpha \chi(\alpha^\sigma)\psi(\alpha^\sigma)^\sigma \\ &= -\sum_\alpha \chi(\alpha)\psi(\alpha)^\sigma &= -\sum_\alpha \chi(\alpha)\psi(\alpha)^{a(\sigma)} \\ &= -\sum_\alpha \chi(\alpha)\psi(a(\sigma)\alpha) &= \chi(a(\sigma))^{-1} G(\chi). \end{aligned}$$

For the proof of ii) we observe that $G(\chi)^m$ is invariant under $\sigma \in Z$. $\square$

Observe that property 4.25.i) describes the action of the Artin on Gauss sums: for primes $q > 0$ it tells us that $\left(\frac{k/\mathbb{Q}}{(q)}\right) G(\chi) = \chi(q)^{-1} G(\chi)$. We will also need a few relations between Jacobi sums:

Proposition 4.26. *Let χ be a character of order m on $\mathbb{F}_q$; the corresponding Jacobi sums satisfy the following relations:*

i) $J(\chi^r, \chi^s) = J(\chi^s, \chi^r)$;
ii) $J(\chi^r, \chi^s) = \chi(-1)^s J(\chi^{-r-s}, \chi^s)$;
iii) $J(\chi, \chi^s) = \chi(-1) J(\chi, \chi^{m-1-s})$;
iv) $J(\chi, \chi^r) J(\chi, \chi^{r+1}) = J(\chi, \chi) J(\chi^2, \chi^r)$.

Proof. i) $J(\chi^r, \chi^s) = -\sum \chi(t)^r \chi(1-t)^s = -\sum \chi(1-a)^r \chi(a)^s = J(\chi^s, \chi^r)$, where we have substituted $a = 1 - t$.

ii) We find

$$\begin{aligned} -\chi(-1)^s J(\chi^{-r-s}, \chi^s) &= \chi(-1)^s \sum \chi(t)^{-r-s} \chi(1-t)^s \\ &= \sum \chi(t)^{-r-s} \chi(t-1)^s \\ &= \sum \chi(t)^{-r} \chi(1-t^{-1})^s \\ &= \sum \chi(t^{-1})^r \chi(1-t^{-1})^s &= -J(\chi^r, \chi^s). \end{aligned}$$

iii) follows directly from i) and ii): put $r = 1$ in

$$J(\chi^r, \chi^s) = J(\chi^s, \chi^r) = \chi(-1)^r J(\chi^{-r-s}, \chi^r)$$
$$= \chi(-1)^r J(\chi^r, \chi^{-r-s}) = \chi(-1)^r J(\chi^r, \chi^{m-r-s}).$$

iv) Using the fundamental relation $G(\chi)G(\psi) = J(\chi, \psi)G(\chi\psi)$ over and over again we find

$$J(\chi, \chi^r) J(\chi, \chi^{r+1}) = \frac{G(\chi)G(\chi^r)}{G(\chi^{r+1})} \frac{G(\chi)G(\chi^{r+1})}{G(\chi^{r+2})} = \frac{G(\chi)^2 G(\chi^r)}{G(\chi^{r+2})}$$
$$= \frac{J(\chi, \chi)G(\chi^2)G(\chi^r)}{G(\chi^{r+2})} = J(\chi, \chi) J(\chi^2, \chi^r). \quad \square$$

So far, the only connection between Gauss and Jacobi sums on one hand and reciprocity statements on the other is the fact given in Proposition 4.25.i) that the action of the Frobenius automorphism of primes $q > 0$ on Gauss sums over $\mathbb{F}_p$ yields the power residue character of q modulo p. As a foretaste of things to come, here's another little observation:

Proposition 4.27. *Let $p = mn + 1$ be a prime, and χ a character of order m on $\mathbb{F}_p^\times$. Then $\chi(2) \equiv -J(\chi^n, \chi^n) \bmod 2$. In particular, if m is odd then 2 is an m-th power residue modulo p if and only if $J(\chi, \chi) \equiv 1 \bmod 2$.*

Proof. The key observation is that the summation over t in the Jacobi sum $-J(\chi^n, \chi^n) = \sum_{t=2}^{p-1} \chi^n(t)\chi^n(1-t)$ is symmetric with respect to $\frac{p+1}{2}$: since the contributions from t and $p+1-t$ are equal and therefore cancel modulo 2, we find $-J(\chi^n, \chi^n) \equiv \chi^n(\frac{p+1}{2})\chi^n(1 - \frac{p+1}{2}) = \chi^{-2n}(2) = \chi(2) \bmod 2$. Since different m-th roots of unity differ modulo 2 when m is odd, this implies that $\chi(2) = 1$ if and only if $J(\chi^n, \chi^n) \equiv 1 \bmod 2$. But $J(\chi^n, \chi^n) = \sigma_n J(\chi, \chi)$, where $\sigma_n : \zeta_m \longmapsto \zeta_m^n$, and our claim follows. $\square$

In order to apply Gauss and Jacobi sums to power reciprocity laws, one needs their prime ideal factorization. We will show in Chapter 7, for example, that $J(\chi, \chi) = \pi$ for the cubic character $\chi = \left(\frac{\cdot}{\pi}\right)_3$ and primes $\pi = a + b\rho \equiv 1 \bmod 3$ in the ring $\mathbb{Z}[\rho]$ of Eisenstein integers; the result of Proposition 4.27 then implies that 2 is a cubic residue modulo $p = N\pi$ if and only if $2 \mid b$: this is easily seen to be equivalent to the conjecture (1.7) of Euler.

The general solution of this problem is due to Stickelberger and will be discussed in Chapter 11; for now we will be content with what we need for reciprocity laws for small powers, namely the following

Proposition 4.28. *Let $\mathfrak{p}$ be a prime ideal of degree 1 in $K = \mathbb{Q}(\zeta_m)$; its norm is then a prime $p \equiv 1 \bmod m$. Let $\chi = (\cdot/\mathfrak{p})$ be the m-th power character modulo $= f\mathfrak{p}$. Then $J(\chi^a, \chi^b) \equiv 0 \bmod \mathfrak{p}$ for all integers $a, b \geq 1$ such that $a + b \leq m - 1$.*

Proof. We have $\chi(t) \equiv t^{(p-1)/m} \bmod \mathfrak{p}$ for every integer t coprime to p by definition of the power residue symbol. Hence

$$J(\chi^a, \chi^b) \equiv t^{a(p-1)/m}(1-t)^{b(p-1)/m} \bmod \mathfrak{p}.$$

132   4. Power Residues and Gauss Sums

The lemma below shows that the right hand side is divisible by p whenever the degree $(a+b)(p-1)/m$ of the polynomial in t is strictly smaller than $p-1$. Since $\mathfrak{p} \mid p$, our claim follows. □

Lemma 4.29. *Let p be prime; then*

$$\sum_{a=1}^{p-1} a^k \equiv \begin{cases} 0 \bmod p, & \text{if } 0 < k < p-1, \\ -1 \bmod p, & \text{if } k = p-1. \end{cases}$$

Proof. For $k = p-1$ this is obvious, for other k the claim is proved exactly as Lemma 3.20. □

Finally we will present a beautiful result connecting two Jacobi sums:

Proposition 4.30. *Let χ be a character on $\mathbb{F}_q$, q odd, and let ψ denote the quadratic character on $\mathbb{F}_q$. Then $J(\chi, \psi) = \chi(4) J(\chi, \chi)$.*

Proof. The claim is clear if χ has order 2, i.e. if $\chi = \psi$. Assume therefore that χ has order ≥ 3. Then $J(\chi,\chi) = -\sum_t \chi(t(1-t))$; we would like to substitute $s = t - t^2$. How many $t \in \mathbb{F}_q$ correspond to a fixed s? Since $s = t - t^2$ is equivalent to $1-4s = (2t-1)^2$, there are none, one, or two solutions of $s = t-t^2$ if $1-4s$ is a non-square, 0, or a square in $\mathbb{F}_q$, respectively. Thus the number of solutions is $1+\psi(1-4s)$. This shows that $J(\chi,\chi) = -\sum_s \chi(s)(1+\psi(1-4s)) = -\chi(4)^{-1} \sum_s \chi(4s)\psi(1-4s) = -\chi(4)^{-1} \sum_t \chi(t)\psi(1-t) = \chi(4)^{-1} J(\chi,\psi)$, where we have used that $\sum_s \chi(s) = 0$. □

In fact, this result is a special case of a far more general theorem called the Davenport-Hasse-relation:

Theorem 4.31. *Let χ be a character on $\mathbb{F}_q$, and suppose that $m \mid q-1$. Then*

$$\prod G(\chi\psi) = \chi^{-1}(m^m) G(\chi^m) \prod G(\psi),$$

where the products are over all characters ψ on $\mathbb{F}_q$ of order dividing m. □

Putting $m = 2$ in Theorem 4.31 gives $G(\chi)G(\chi\psi) = \chi^{-1}(4)G(\chi^2)G(\psi)$. Dividing through by $G(\chi\psi)G(\chi^2)/G(\chi)$ we get Proposition 4.30. In the same paper [DH] where Davenport and Hasse proved 4.31, they also discovered the following theorem whose importance will become clear only when we start looking at Gauss's Last Entry and the Riemann Hypothesis for Curves in Chapter 10:

Theorem 4.32. *Let E/F be an extension of finite fields, and assume that χ is a character on $F^\times$. Then $\chi' = \chi \circ N_{E/F}$ is a character on $E^\times$, $\psi' = \psi \circ \operatorname{Tr}_{E/F}$ a character on E, and we have $G(\chi', \psi') = G(\chi, \psi)^{(E:F)}$.*

Proof. See Exercise 10.28 □

Without the minus sign in the definition of $G(\chi)$, the Davenport-Hasse theorem would read $G(\chi', \psi') = -(-G(\chi, \psi'))^s$, and this is one of the main reasons for introducing the minus sign.

The Davenport-Hasse theorem for Jacobi sums is an immediate corollary:

Corollary 4.33. *Let χ_j ($j = 1, 2$) be nontrivial characters on $F^\times$, and put $\chi'_j = \chi_j \circ N_{E/F}$. Then $J(\chi'_1, \chi'_2) = J(\chi_1, \chi_2)^{(E:F)}$.*

Proof. Write the Jacobi sum as a quotient of Gauss sums and apply the Davenport-Hasse theorem. □

Eisenstein Sums

The rest of this section is devoted to Eisenstein sums. Let p be an odd prime, and put $q = p^f$. Then $F = \mathbb{F}_{q^n}$ is a finite extension of $K = \mathbb{F}_q$ and $k = \mathbb{F}_p$. We will use the following abbreviations for traces: $\mathrm{T} := \mathrm{Tr}_{F/K}$, $\mathrm{tr} := \mathrm{Tr}_{K/k}$, and $\mathrm{Tr} := \mathrm{Tr}_{F/k} = \mathrm{tr} \circ \mathrm{T}$.

For a character χ of order $m > 1$ on $F^\times$ and an element $\alpha \in K$, we define the Eisenstein sum

$$E(\chi; \alpha) = E_{F/K}(\chi; \alpha) = \sum_{\mathrm{T}(t)=\alpha} \chi(t).$$

Clearly $E(\chi; \alpha) \in \mathbb{Z}[\zeta_m]$. We put $E_a(\chi) = E(\chi; a)$ for $a \in \mathbb{F}_p$ (in particular for $a = 0$ and $a = 2$), $E(\chi) = E_{F/K}(\chi; 1)$, and let χ_K denote the restriction of χ to $K^\times$. The following properties of Eisenstein sums follow immediately from their definition:

Proposition 4.34. *Let χ be a character on $F^\times$, where F is the extension of degree n over $K = \mathbb{F}_q$, $q = p^f$.*

i) $E(\chi; \alpha) = \chi(\alpha) E(\chi)$ for all $\alpha \in K^\times$;
ii) $E(\chi^p; \alpha) = E(\chi, \alpha^p)$;
iii) $G_\alpha(\chi) = E(\chi) G_\alpha(\chi_K) - E_0(\chi)$ for all $\alpha \in K$;
iv) The table below gives some useful relations between Gauss and Eisenstein sums depending on whether $\chi_K = \mathbb{1}$ or not:

	$E_0(\chi)$	$G_\alpha(\chi)$	$E(\chi)E(\chi^{-1})$
$\chi_K \neq \mathbb{1}$	0	$E(\chi) G_\alpha(\chi_K)$	q^{n-1}
$\chi_K = \mathbb{1}$	$(1-q)E(\chi)$	$qE(\chi)$	q^{n-2}

Proof. Let $\alpha \neq 0$ be an element of F. Then

$$E(\chi; \alpha) = \sum_{\mathrm{T}(t)=\alpha} \chi(t) = \chi(\alpha) \sum_{\mathrm{T}(t)=\alpha} \chi(\alpha^{-1} t) = \chi(\alpha) \sum_{\mathrm{T}(s)=1} \chi(s)$$

proves i). The proofs of ii) and iii) are straight forward:

$$E(\chi^p; \alpha) = \sum_{T(t)=\alpha} \chi(t)^p = \sum_{T(t)=\alpha} \chi(t^p),$$

and ii) follows from $T(t^p) = T(t)^p = \alpha^p$. Next

$$-G_\alpha(\chi) = \sum_{t \in F} \chi(t)\zeta^{\mathrm{Tr}\,\alpha t} = \sum_{\beta \in F} \sum_{T(\alpha t)=\beta} \chi(t)\zeta^{\mathrm{tr}\,\alpha\beta}$$
$$= \sum_{\beta \in F} \zeta^{\mathrm{tr}\,\alpha\beta} E(\chi;\beta) = E_0(\chi) + \sum_{\beta \in K^\times} \zeta^{\mathrm{tr}\,\alpha\beta}\chi(\beta)E(\chi)$$
$$= E_0(\chi) - E(\chi)G_\alpha(\chi_K).$$

This proves iii). Now if $\chi_K \ne \mathbb{1}$, then there is an $\alpha \in K^\times$ such that $\chi(\alpha) \ne 1$, and we find

$$\chi(\alpha)E_0(\chi) = \sum_{T(t)=0} \chi(\alpha t) = \sum_{T(s)=0} \chi(s) = E_0(\chi).$$

If $\chi_K = \mathbb{1}$ and $\chi \ne \mathbb{1}$, however, then $G_0(\chi) = 0$ and $G_0(\chi_K) = 1 - q$; now ii) implies the claims in the first column of iv). The second column follows immediately from iii), the first column, and Exercise 4.20; finally, the last column is now a consequence of Proposition 4.23.vi). □

Fourier Analysis on Finite Abelian Groups

The calculations with Gauss and Jacobi sums can be put into a more general framework, that of Fourier analysis on finite abelian groups. It will help us to derive some of our formulas in a less computational way.

Let $\mathbb{F} = \mathbb{F}_q$ be a finite field with $q = p^n$ elements, and let $\psi : \mathbb{F} \longrightarrow \mathbb{C}^\times$ denote the additive character on $\mathbb{F}$ defined by $\psi(t) = \zeta_p^{Tr(t)}$. Consider the space $\mathcal{T}$ of functions $f : \mathbb{F} \longrightarrow \mathbb{C}$; then the map $\mathcal{F}f : \mathbb{F} \longrightarrow \mathbb{C}$ defined by

$$(\mathcal{F}f)(y) = -\sum_{x \in \mathbb{F}} f(x)\psi(xy) \qquad (4.7)$$

is called the Fourier transform of f. The map $\mathcal{F} : \mathcal{T} \longrightarrow \mathcal{T}$ is linear on the $\mathbb{C}$-vector space $\mathcal{T}$. Given two functions $f, g \in \mathcal{T}$, the convolution $f * g$ is defined by

$$(f * g)(y) = -\sum_{x \in \mathbb{F}} f(x)g(y-x). \qquad (4.8)$$

Clearly $f * g = g * f$; you will also notice that for characters χ_1 and χ_2 on $\mathbb{F}^\times$ (extended to functions on $\mathbb{F}$ as in our previous discussions) we have $\chi_1 * \chi_2(1) = J(\chi_1, \chi_2)$. The following results collects the basic properties of Fourier transforms on $\mathbb{F}$:

Theorem 4.35. *The Fourier transform for functions $\mathbb{F} \longrightarrow \mathbb{C}$ has the following properties:*

1. $\mathcal{F}^2 f = qf^-$, that is $(\mathcal{F}^2 f)(z) = qf(-z)$.
2. $\mathcal{F}$ induces a bijection $\mathcal{T} \longrightarrow \mathcal{T}$ with $\widetilde{\mathcal{F}} f(y) = \frac{1}{q} \sum_{x \in \mathbb{F}} f(x)\psi(-xy)$ defining the inverse map.
3. $\mathcal{F}(f * g) = (\mathcal{F}f)(\mathcal{F}g)$.
4. $\mathcal{F}(fg) = \frac{1}{q}(\mathcal{F}f) * (\mathcal{F}g)$.
5. For any character $\chi \neq \mathbf{1}$ on $\mathbb{F}$, we have $\mathcal{F}\chi = G(\chi)\chi^{-1}$, in the sense that the functions on both sides take the same values in $\mathbb{F} \setminus \{0\}$.

Proof. Nothing really new here: since $\psi(xy)\psi(yz) = \psi(x(y+z))$, we get

$$\mathcal{F}^2 f(z) = \sum_x \sum_y f(x)\psi(xy)\psi(yz)$$
$$= \sum_x f(x-z) \sum_y \psi(xy) = qf(-z),$$

and this proves 1. This implies the second assertion as $\widetilde{\mathcal{F}} f = \frac{1}{q}\mathcal{F}f^-$. Next,

$$\mathcal{F}(f * g)(z) = \sum_y (f * g)(y)\psi(yz) = \sum_y \sum_x f(x)g(y-x)\psi(yz);$$

putting $t = y - x$ shows that the last sum is

$$\sum_x f(x)\psi(xz) \sum_t g(t)\psi(tz) = (\mathcal{F}f)(\mathcal{F}g)(z),$$

proving 3. The fourth formula is a formal consequence of earlier results: first observe that by 2. we can write $f = \mathcal{F}f_1$ and $g = \mathcal{F}g_1$; the result now follows from 1. and 3. Finally, $(\mathcal{F}\chi)(y) = \sum_x \chi(x)\psi(xy)$, and by putting $x = ty^{-1}$ we get $(\mathcal{F}\chi)(y) = \sum_t \chi(ty^{-1})\psi(t) = G(\chi)\chi^{-1}(y)$ for all $y \neq 0$. $\square$

Observe that the Fourier transform of a function $\mathbb{F} \longrightarrow \mathbb{C}$ depends crucially on its value at $x = 0$: for example, we have $(\mathbb{F}\mathbf{1})(y) = 1 - \mathbf{1}(0) = 0$ and $(\mathbb{F}\mathbf{1})(0) = 1 - q - \mathbf{1}(0) = -q$ for the principal character $\mathbf{1}$, where we have used our convention that $\mathbf{1}(0) = 0$.

We get a number of immediate corollaries: assume that χ_1 and χ_2 are characters on $\mathbb{F}^\times$ such that $\chi_1, \chi_2, \chi_1\chi_2 \neq \mathbf{1}$. Then

$$\sum_{a \in \mathbb{F}} \chi_1(a)\chi_2(t-a) = \sum_{b \in \mathbb{F}} \chi_1(tb)\chi_2(t-tb) = \chi_1(t)\chi_2(t) \sum_{b \in \mathbb{F}} \chi_1(b)\chi_2(1-b)$$

shows that $\chi_1 * \chi_2(t) = \chi_1(t)\chi_2(t)\chi_1 * \chi_2(1)$. Applying $\mathcal{F}$ to this relation immediately gives $G(\chi_1)G(\chi_2) = G(\chi_1\chi_2)J(\chi_1, \chi_2)$. Similarly, $\mathcal{F}^2\chi = \mathcal{F}(G(\chi)\chi^{-1}) = G(\chi)G(\chi^{-1})\chi$, and comparing with $\mathcal{F}^2 f = qf^-$ proves that $G(\chi)G(\chi^{-1}) = \chi(-1)q$.

The analogy between Fourier analysis on $\mathbb{R}$ and on finite abelian groups extends further: Gauss sums correspond to the Gamma function, Jacobi sums to the Beta function. Jacobi used this analogy to guess (and then prove) properties of Gauss sums!

NOTES

The first residue symbol in number fields other than $\mathbb{Q}$ was introduced by Gauss in his research on biquadratic reciprocity; afterwards Jacobi, Eisenstein and Kummer defined ℓ-th power residue symbols in $\mathbb{Z}[\zeta_\ell]$, and finally Hilbert suggested to study n-th power residue symbols in all number fields containing the n-th roots of unity. Part iii) of Proposition 4.2 seems to be due to K. S. Hilbert [370]. It was extended to normal extensions K/k by Furtwängler [252] and to arbitrary finite extensions by Takagi [789]. For elementary properties of n-th power residues and residue symbols in finite fields see Epstein [206], Giudici [307, 308], Hardman & Jordan [HJ], Longo [554], Mayer [572], Porcelli & Pall [664], the book of Schuh [725], and Whyburn [Wy1, Wy2, Why]. Hensel [361] defined a residue symbol in arbitrary number fields.

An elementary question concerning power residues is the following: suppose that an integer a is an n-th power modulo almost all primes (that is, all primes except for finitely many); does this imply that a is an n-th power? The answer given by Trost [Tro] says that this is almost true: he showed that a is either an n-th power, or that $8 \mid n$ and $a = 2^{n/2} x^n$.

This has been generalized to number fields K: given an $\alpha \in K^\times$ and an integer $n \in \mathbb{N}$, define a set of prime ideals

$$P_K(\alpha, n) = \{\mathfrak{p} \in \mathcal{O}_K : \alpha \equiv \xi^n \bmod \mathfrak{p}\}.$$

The result of Trost can then be expressed as follows:

Proposition 4.36. *Given integers $a \in \mathbb{Z}$ and $n \in \mathbb{N}$, we have $P_\mathbb{Q}(a, n) \stackrel{*}{=} P_\mathbb{Q}(1, n)$ if and only if*

a) $a = x^n$ for some $x \in \mathbb{Z}$; or
b) $8 \mid n$ and $a = 2^{n/2} x^n$ for some $x \in \mathbb{Z}$. □

Here $\stackrel{*}{=}$ means equality up to at most finitely many exceptions. For the necessity of case b), see Exercise 4.33. Results in this direction have been generalized by Flanders [Fla], Gerst [Ger], Schinzel [716, 717, 718], and Schulte [726, 727]. Kraft & Rosen [436] gave a simple proof of Trost's result using Eisenstein's reciprocity law. The problem is discussed in Chapter 10 of the class field theory notes by Artin and Tate [17] in connection with the Grunwald-Wang theorem; in fact, there is a local-global principle behind Trost's result, since its main part can be formulated as follows: a rational number is an n-th power in $\mathbb{Q}$ (with $8 \nmid n$) if and only if it is an n-th power in every completion of $\mathbb{Q}$. T. Ono [On] formulated this in terms of Tate-Shafarevich groups: for simplicity, let K/k be an abelian extension of number fields with Galois group G, and let A be a finite G-module. For each prime ideal $\mathfrak{p}$ in k, let $G_\mathfrak{p}$ denote the decomposition group of $\mathfrak{p}$ (that is, the Galois group of the completions $K_\mathfrak{P}/k_\mathfrak{p}$). Then for each $\mathfrak{p}$ we have a restriction map $\mathrm{res}_\mathfrak{p} : H^1(G, A) \longrightarrow H^1(G_\mathfrak{p}, A)$, and we can define the Tate-Shafarevic group $\mathrm{III}(K/k, A) = \bigcap \ker \mathrm{res}_\mathfrak{p}$, where the intersection is over all primes of k. By putting $K = k(\zeta_8)$, Ono can explain the special role of the exponent 8 in Trost's result as the non-vanishing of the corresponding Tate-Shafarevic group $\text{III}(K/k, A)$ with $A = \langle \zeta_8 \rangle$, thereby correcting an error in Ono & Terasoma [OT].

A related result for real abelian fields was used by Thaine in his work on annihilators of class groups (see Washington [Was, 2nd ed, §15.2]). For generalizations of such and similar problems to elliptic curves and abelian varieties, see Corrales-Rodriganez & Schoof [CRS] and Wong [Won].

Generalizations of Gauss's Lemma 4.3 have been published by Eisenstein, Gmeiner [310] (for fourth powers) and E. Lehmer [508]. The general form 4.5 of Gauss Lemma is due to Reichardt [677], though the main ideas stem from Herglotz [363]. See also Lenstra [Le2] and the Notes in Chapter 1.

The theorem of Stickelberger and Voronoi presented in Section 4.3 is due to Pellet [631] and was rediscovered by Stickelberger [760] and Voronoi [816, 817]; Hasse included it in his famous textbook [341, §26d]. The fact that it has the quadratic reciprocity law as a corollary was noticed by Pellet himself, Lasker [478], Mirimanoff [588], Skolem [739], and Swan [778]; Hensel [360] generalized the result of Stickelberger and Voronoi using his newly created theory of p-adic numbers; parts thereof have been rediscovered by Barrucand & Laubie [35, 36], who generalized the tame part to relative extensions; their results were generalized in turn by Movahhedi & Zahidi [595]. See also Carlitz [116], Dalen [147], Lubelski [557], and Tsvetkov [804].

The results in Section 4.4 were already known to Kummer and Weber; the latter used them in his proof that all abelian extensions of $\mathbb{Q}$ are contained in cyclotomic number fields. The description of Kummer theory we have given is very elementary; there are of course more abstract and powerful formulations of Kummer theory in terms of pairings and cohomology groups. See the nice paper of Waterhouse [Wat] and the references listed there.

Characters first appeared in Gauss's work on genera of binary quadratic forms. Later, Dirichlet introduced characters of residue class groups modulo m in order to separate these classes in his proof of the infinitude of primes in arithmetic progression. The first exposition of characters of finite abelian groups in a textbook seems to be due to Hecke [348]. The consequent utilization of Dirichlet characters in the description of the arithmetic of cyclotomic fields has been propagated mainly by Hasse (see e.g. his book [H2]) and Leopoldt.

Gauss and Jacobi Sums

Gauss sums over $\mathbb{F}_p$ were introduced by Vandermonde and Lagrange [Lag] for the purpose of solving algebraic equations; accordingly, they were called *Lagrange resolvants* for a long time. Gauss used them in [263, Art. 356], and determined the sign of the quadratic Gauss sum in [267]. Proposition 4.23.iii) can be found in his posthumously published [Gau, p. 252]. This last result was also known to Cauchy, Jacobi and Eisenstein.

The prime ideal factorization of Jacobi sums (for primes of the form $p = \ell m + 1$) was essentially discovered by Cauchy [122] and Jacobi [398, 400], of course in a language avoiding ideals; see Exercise 4.24 for another result announced in [398]. Kummer [465] later interpreted Jacobi's and Cauchy's results in terms of his ideal numbers. Stickelberger [759] finally studied Gauss and Jacobi sums over arbitrary finite fields – we will discuss his results in Chapter 11. Stickelberger also was the first to suggest the minus sign in the definition of Gauss sums; in [759], he writes on p. 358:

> besser noch wäre es vielleicht, sowohl die gewöhnliche wie unsere allgemeine Resolvente mit -1 zu multiplizieren.[1]

Actually, the results of Chapter 11 even suggest that one should replace χ by χ^{-1}; according to G. Gras [Gra]

> Outre le signe – utilisé par certains auteurs (Hasse notamment) dans la définition des sommes de Gauss, il semble qu'il serait préférable de définir $G_a(\chi)$ par $G_a(\chi) = -\sum_{t \in \mathbb{F}} \chi^{-1}(t) \zeta_p^{a \operatorname{Tr}(t)}$.[2]

But following Gras' suggestion would not be advisable unless one also makes the corresponding changes in the definition of generalized Bernoulli numbers $B_{n,\chi}$ – these also tend to occur in the form $B_{n,\chi^{-1}}$.

For applications of Gauss and Jacobi sums to other problems, see Joly [414] and Chapter 10. For special problems related to Gauss sums, see Berndt & Chowla [BC] and Berndt & Evans [BE1, BE2, BE3]. The proof of the special case $m = 2$ of the Davenport-Hasse relation (Proposition 4.30) was taken from [BE1] (this is a simplified version of Hasse's own proof in [342]). In [BE2] it was shown that the general case follows by induction from the special cases where m is prime. Huard, Spearman & Williams [HSW] gave a simple proof similar to the one in [DH], but proofs in the spirit of Berndt and Evans have so far not been found. A proof for $m = 3$ was given by Greene & Stanton [GS]; it is, however, much more involved than Berndt's and Evans' proof for $m = 2$. See also McEliece & Rumsey [MR]. A generalization of the Davenport-Hasse theorems is due to Petin [Pet]; for generalizations in a different direction, compare Kubert & Lichtenbaum [KL]. It should also be remarked that a special case of the Davenport-Hasse relation can already be found in Jacobi's article [400].

Proofs of the Davenport-Hasse theorem 4.32 are due to Davenport & Hasse [DH]; different proofs can be found in H.L. Schmid [Scm], Stepanov [Ste], and Weil [Wei] (see also Li [Li] and Exercise 11.9).

The fact that Gauss sums over $\mathcal{O}_k/\mathfrak{p}$ actually are contained in the decomposition field of $\mathfrak{p}$ seems to have been noticed first by G. Gras [Gra];

---

[1] perhaps it would be even better if we multiplied both the usual as well as our general resolvent by -1.

[2] Apart from the sign – used by certain authors (Hasse, in particular) in the definition of Gauss sums, it seems to be preferable to define $G_a(\chi)$ by $G_a(\chi) = -\sum_{t \in \mathbb{F}} \chi^{-1}(t) \zeta_p^{a \operatorname{Tr}(t)}$.

the analogous result for Jacobi sums was rediscovered by Kida & Ono [KO], whereas the corresponding result for the *ideal* generated by a Gauss sum appears as an exercise in the book of Ireland & M. Rosen [386].

For more information about Gauss and Jacobi sums as well as on related sums see the books of Lidl & Niederreiter [LN] (containing the basic material), Li [Li] (going far beyond of what we've been doing) and the book of Berndt, Evans & K.S. Williams [48] (treating character sums and their relation to reciprocity laws in detail). The book on Gauss sums by Salvadori [Sal] was not accessible to me.

Gauss and Jacobi sums are nowadays applied in numerous primality tests: see H.C. Williams [Wil], H.W. Lenstra [Le1] and Cohen & H.W. Lenstra [CL]. These algorithms were developed from ideas in Adleman, Pomerance & Rumely [APR], where power residue symbols in cyclotomic fields were used in a more direct way. For a recent survey on primality testing, see A.K. Lenstra [Le]. Since primality tests deal with integers not (yet) known to be prime, one has to use Gauss sums over $\mathbb{Z}/m\mathbb{Z}$; these are defined in the usual way and share many properties with Gauss sums over finite fields; see e.g. Mihailescu [Mi].

There is an analogy (first noticed by Jacobi) between Gauss and Jacobi sums on one hand, and the Gamma and Beta functions on the other hand; in fact:

$$-G(\chi) = \sum_{t \in \mathbb{F}^\times} \chi(t)\psi(t) \qquad \Gamma(x) = \int_0^\infty t^{x-1} e^{-t} dt$$

$$-J(\chi, \eta) = \sum_{t \in \mathbb{F}} \chi(t)\eta(1-t) \qquad B(x,y) = \int_0^1 t^{x-1}(1-t)^{y-1} dt$$

$$G(\chi)G(\eta) = G(\chi\eta)J(\chi,\eta) \qquad \Gamma(x)\Gamma(y) = \Gamma(x+y)B(x,y)$$

$$G(\chi)G(\chi^{-1}) = \chi(-1) \cdot q \qquad \Gamma(x)\Gamma(1-x) = \frac{\pi}{\sin \pi x}$$

The relations on the left side are valid if the occurring characters are $\neq \mathbb{1}$, whereas those on the right side make sense only if you stay away from the pole $x = 0$ of the Γ-function.

For more on relations between Gauss sums and Fourier analysis on finite groups, see Auslander & Tolimieri [25], Dym & McKean [DM], Lang [La], Langevin [Lan], Pieper [661, p. 90–91] and Terras [Ter].

Finally we mention that Gauss sums have been generalized in various directions: Weber [Web] and Jordan [Jo] considered Gauss sums in many variables, Thakur [Tha] defined Gauss sums to function fields of one variable over a finite field.

Eisenstein Sums

Eisenstein [Eis] was the first to introduce character sums over finite fields of order p^2 (for primes $p \equiv 3, 5, 7 \bmod 8$) and p^3 (for primes $p \equiv 2, 4 \bmod 7$);

140   4. Power Residues and Gauss Sums

these sums were generalized and studied by Stickelberger [759] who called them *Eisenstein sums*. Stickelberger mentions that the first version of his paper, which he presented to the Academy of Science in Göttingen in 1888, was called *Theorie der Eisenstein'schen Summen* ... , and that he was able to eliminate the central role of Eisenstein sums in the second version [759]. Scheffler, unaware of Stickelberger's work, extended Eisenstein's results to primes $p \equiv 1 \bmod 8$ and $p \equiv 1 \bmod 7$ in [Scf]; the rest of his paper is an incoherent attempt at generalizations to primes $p \equiv a \bmod q$ with $(a/q) = 1$. Hahn & Lee [HL] used Stickelberger's relation and the Gross-Koblitz formula to derive general results.

It seems that Eisenstein sums have been largely neglected since then; they make their reappearance in Whiteman [Wh2], who has overlooked Stickelberger's contributions to the theory of Eisenstein sums although he lists his paper in the references; accordingly, he only discusses quadratic Eisenstein sums. The article of Berndt & Evans [BE2] gives the properties of Eisenstein sums in the special case of finite fields of order p^2, and Lidl & Niederreiter [LN] discuss them in a series of exercises. The paper of Williams, Hardy & Spearman [WHS] would be an excellent source if it had not been written in the index-language of the last century. Yamamoto [Yam] rediscovered Eisenstein sums, called them *relative Gauss sums* (although he lists [BE2] in his references) and reproduced many of Eisenstein's results without referring to [Eis]; the result of Exercise 4.37 is due to him. Finally Adler [Adl] presents Eisenstein's computations of Eisenstein sums over $\mathbb{F}_{p^2}$ for primes $p \equiv 3 \bmod 8$ (and falsely claims that Eisenstein used Jacobi sums over $\mathbb{F}_{p^2}$ in [Eis] and that this is why Stickelberger called them Eisenstein sums). It should be remarked that the results of these papers must be compared with great care since they all use different Eisenstein sums: one finds $E_2(\chi)$ in [BE2] and [Adl], $E(\chi)$ in [WHS] and $E(\chi^{-1})$ in [Yam].

Jacobsthal Sums

We have defined Jacobsthal sums $\Phi_k(D)$ in Exercise 1.29 by

$$\Phi_k(D) = \sum_{u=1}^{p-1} \left(\frac{u}{p}\right)\left(\frac{u^k + D}{p}\right) \tag{4.9}$$

The sum $\Phi_2(D)$ was introduced and studied by Jacobsthal himself in [405]. The cubic sum $\Phi_3(D)$ will be discussed in Exercise 7.20, and $\Phi_4(D)$ in the Notes to Chapter 9. Jacobsthal sums were also the subject of investigations by Berndt & Evans [BE1], Chowla [Ch1, Ch2], E. Cohen [CoE], Evans [Ev1, Ev2], Grimson [Gri], Hudson & K.S. Williams [HW1, HW2, 381], Katre [Kat], Katre & Rajwade [KR1, KR2, KR3, KR4, KR5], E. Lehmer [499], Leonard & K.S. Williams [LW], Monzingo [Mon], Postnikov & Stepanov [PS], Whiteman [Wh1, Wh3, Wh4], as well as in the dissertations of Unger [806] and Widmer [Wid]. I haven't seen the dissertation of Walum [Wal]. Jacobsthal and his work are discussed in Selberg's article [Sel]. For generalizations of Jacobsthal sums see S.D. Cohen [CoS] and Rankin [Ran].

Since $\Phi_2(D) = \sum \left(\frac{x}{p}\right)\left(\frac{x^2+D}{p}\right)$ is clearly connected with the number of squares among the values of $x^3 + Dx$, that is, with the number of points on the elliptic curve $y^2 = x^3 + Dx$ over $\mathbb{F}_p$, it is not surprising that there are connections with the theory of elliptic curves over finite fields. Such connections have already been exploited by Poulakis [Pou], Rishi, Parnami & Rajwade [RPR] and Rajwade & Parnami [RP], and recently the case of elliptic curves with complex multiplication by quadratic orders with class number 1 was treated in three papers almost simultaneously (and in different ways): see Joux & Morain [JM], Padma & Venkataraman [PV], and Stark [Sta]. See also Leprevost & Morain [LM].

Jacobsthal sums were employed by Carlitz [Ca1, Ca2] to count the number of solutions of certain equations in finite fields.

Applications of quadratic reciprocity (and in particular of the reciprocity of Gauss sums) to topology and topological quantum field theory are very recent: see e.g. Kronheimer, Larsen & Scherk [451] and Deloup [De1, De2], who proves reciprocity formulas for very general Gauss sums studied by H. Braun [Br], Krazer [Kr] and Turaev [Tur].

Power Residue Characters of Small Primes

As Proposition 4.27 shows, the power residue character of small primes can often be deduced by rather ad hoc methods; in fact, this result was already stated by Pépin [639]. Instead of criteria involving Jacobi sums, most authors preferred to describe the character of primes in terms of solutions of certain diophantine systems; quintic residuacity modulo primes $p \equiv 1 \bmod 5$, for example, is connected with solutions to the system

$$16p = x^2 + 50u^2 + 50v^2 + 125w^2, \qquad xw = v^2 - 4uv - u^2 \qquad (4.10)$$

named after Dickson, but studied before by Burnside [104] and Dickson's student Hull [Hul1, Hul2] in connection with computing the quintic period equation. This polynomial had been studied before by Cayley [Cay], Glashan [Gl], Tanner [Tan], Charlotte Scott [Sco]; innumerable numerical examples (that can already be found in Reuschle [678], as was noticed by Beeger [Bee]) were given by Upadhyaya in [Up].

Results in this direction were obtained by Alderson [10], Evans [220], Glashan [Gla], Karst [426], Katre & Rajwade [427], E. Lehmer [495, 511], Leonard & Williams [519], Rajwade [Raj], K.S. Williams [853, 859, 861], Yoon & Hwang [YH] and Zee [880]. For septimic analogues, see Nashier & Rajwade [NR] and Williams [850].

Explicit criteria for the power residue character of small primes can be found in Alderson [10, 11], Ankeny [15], Bernardi [43, 44, 45], Halter-Koch [325], Helou, Roll & Washington [358], Jakubec [407, 408], E. Lehmer [495],

142   4. Power Residues and Gauss Sums

Leonard & Mortimer [518] & K. S. Williams [517], Llorente [553], Parnami, Agrawal, Pall & Rajwade [625, 627], Pépin [639], Sato [699], K. S. Williams [849, 851, 854, 859], K.S. Williams, Friesen & Howe [869], and Zee [880]. See also Exercise 10.25.

Exercises

4.1 Take $k = \mathbb{Q}(i)$, $\mathfrak{p} = (1 + 2i)$, and let $\chi = [\cdot/\mathfrak{p}]$ be the quartic residue symbol modulo $\mathfrak{p}$ as defined in Section 4.1.
a) show that $\{\pm 1, \pm i\}$ is a set of representatives for $(\mathbb{Z}[i]/\mathfrak{p})^\times$, and show that $\chi(\zeta) = \zeta$ for all $\zeta \in \{\pm 1, \pm i\}$;
b) verify the following table:

t	1	2	3	4
$\chi(t)$	1	i	$-i$	-1

c) compute the values of $\chi(t)$ with Gauss's Lemma.

4.2 Take $k = \mathbb{Q}(i)$, $\mathfrak{p} = (3)$, and let $\chi = [\cdot/\mathfrak{p}]$ be as in Exercise 4.1. Verify the following table:

t	1	2	i	$1+i$	$2+i$	$2i$	$1+2i$	$2+2i$
$\chi(t)$	1	1	-1	$-i$	i	-1	i	$-i$

Check some of these values using Gauss's Lemma.

4.3 Let k be a quadratic number field, and let $(\cdot/\cdot)_2$ denote the quadratic residue symbol in $\mathcal{O}_k$; show that

$$\left(\frac{a}{b}\right)_2 = +1 \quad \text{for all } a, b \in \mathbb{Z}, \ (2a, b) = 1$$

is a special case of Proposition 4.2, and prove it directly.

4.4 Let $p = e^2 - 2f^2 \equiv 1 \mod 8$ be a prime. Put $\pi = e + f\sqrt{2}$, and compute the quadratic residue symbol $[\sqrt{2}/\pi]$ in $\mathbb{Z}[\sqrt{2}]$. (Hint: show that $[f/\pi] = (f/p) = 1$, then use the congruence $f\sqrt{2} \equiv -e \mod \pi$ to deduce that $[\sqrt{2}/\pi] = (e/p)$. The quadratic reciprocity law then gives $[\sqrt{2}/\pi] = (-2/e)$.) Similarly, compute $[\sqrt{-2}/\tau]$ for $\tau = c + d\sqrt{-2}$, where $p = c^2 + 2d^2$. Can you generalize this to e.g. the calculation of $[\sqrt{q}/\pi]$ in $k = \mathbb{Q}(\sqrt{q})$ for odd primes q, at least when k has class number 1?

4.5 (H. Cohen) Let $k = \mathbb{Q}(\sqrt{p^*m})$ be a quadratic number field, where $p^* = \pm p \equiv 1 \mod 4$ is an odd prime and $p \nmid m$. Then $p\mathcal{O}_k = \mathfrak{p}^2$ ramifies in k; choose an element $\pi \in \mathfrak{p} \setminus \mathfrak{p}^2$ and write $\pi^2 = p^*\alpha$: then $\mathfrak{p} \nmid \alpha\mathcal{O}_k$, and the quadratic character $\chi(\mathfrak{p}) = (\frac{\alpha}{\mathfrak{p}})$ does not depend on the choice of π; in fact, prove that $\chi(\mathfrak{p}) = (\frac{m}{p})$. (Hint: Let $K = \mathbb{Q}(\sqrt{p^*}, \sqrt{m})$; then $K = k(\sqrt{m}) = k(\sqrt{p^*}) = k(\sqrt{\alpha})$. Deduce that $(\frac{m}{p}) = (\frac{\alpha}{\mathfrak{p}})$.)

4.6 Let k be a cubic field such that $3\mathcal{O}_k = \mathfrak{p}^2\mathfrak{q}$ for different ideals $\mathfrak{p}$ and $\mathfrak{q}$ of norm 3; then $d = \operatorname{disc} k = -3m$ for some $m \equiv \pm 1 \bmod 3$, and if we choose $\pi \in \mathfrak{p} \setminus \mathfrak{p}^2$ we find $\pi^2 = -3\alpha$ for some $\mathfrak{p}$-integral $\alpha \in k^\times$. Show that $\chi(\mathfrak{p}) = \left(\frac{\alpha}{\mathfrak{p}}\right)$ does not depend on the choice of π, and that we have $\chi(p) = \left(\frac{m}{p}\right)$. In particular, show that α is always a quadratic residue mod $\mathfrak{p}$ if $k = \mathbb{Q}(\sqrt[3]{a})$ is a pure cubic extension for some cube-free $a \equiv \pm 1 \bmod 9$.

4.7 Let K/k be a normal extension with cyclic Galois group $\operatorname{Gal}(K/k) = \langle \sigma \rangle$ of order $m = 2^n$. Show that $L = K(\sqrt{\mu})$ is normal over k if and only if $\mu^{1-\sigma} = \alpha_\sigma^2$ for some $\alpha_\sigma \in K^\times$. Assume that L/k is normal and put $\nu = 1 + \sigma + \ldots + \sigma^{m-1}$ and $\varepsilon_\sigma = \alpha_\sigma^\nu$; then

$$\operatorname{Gal}(L/k) = \begin{cases} C_2 \times C_m & \text{if } \varepsilon_\sigma = +1, \\ C_{2m} & \text{if } \varepsilon_\sigma = -1. \end{cases}$$

This generalizes Lemma 5.14.

4.8 (Furtwängler [Fur]) Let K/k be a normal field extension with $\operatorname{Gal}(K/k) = \langle \sigma, \tau \mid \sigma^2 = \tau^2 = 1, \sigma\tau = \tau\sigma \rangle$ (this is Klein's four group). Then $L = K(\sqrt{\mu})$ is normal over k if and only if $\mu^{1-\sigma} = \alpha_\sigma^2$ and $\mu^{1-\tau} = \alpha_\tau^2$ for some $\alpha_\sigma, \alpha_\tau \in K^\times$. Assume that L/k is normal and write $\eta = \alpha_\tau^{\sigma-1} \alpha_\tau^{1-\sigma}$, $\varepsilon_1 = \alpha_\sigma^{1+\sigma}$, $\varepsilon_2 = \alpha_\tau^{1+\tau}$, and $\varepsilon_3 = \alpha_\sigma^{1+\sigma\tau}$, where $\alpha_{\sigma\tau}^2 = \mu^{1-\sigma\tau}$. Let e denote the number of ε_j which are negative; then $\eta = \varepsilon_1 \varepsilon_2 \varepsilon_3$, $\Gamma = \operatorname{Gal}(L/k)$ is abelian if and only if $\eta = 1$, and

$$\Gamma = \begin{cases} C_2 \times C_2 \times C_2 & \text{if } e = 3, \\ C_2 \times C_4 & \text{if } e = 1, \\ D_4 & \text{if } e = 2, \\ H_8 & \text{if } e = 0. \end{cases}$$

4.9 Here we compute some examples to illustrate Kummer Theory. We start with the cyclic extension $K/\mathbb{Q}$, where $K = \mathbb{Q}(\sqrt{2+\sqrt{2}})$ is the real cyclic quartic subfield of $\mathbb{Q}(\zeta_{16})$. Then $\operatorname{Gal}(K/\mathbb{Q}) = \langle \sigma \rangle \simeq C_4$, where σ is defined by $\sigma(\sqrt{2+\sqrt{2}}) = \sqrt{2-\sqrt{2}}$. Given an extension $L = K(\sqrt{\mu})$, let N denote its normal closure over $\mathbb{Q}$. Now verify the following table:

μ	$\mu^{1+\sigma^2}$	μ^ν	$\operatorname{Gal}(N/\mathbb{Q})$
-1	1	1	$C_2 \times C_4$
$2+\sqrt{2+\sqrt{2}}$	$2-\sqrt{2}$	2	C_8
$1+\sqrt{2}$	$(1+\sqrt{2})^2$	1	$32\Gamma_7 a_1$
$(1+\sqrt{2})(2+\sqrt{2+\sqrt{2}})$	$(1+\sqrt{2})^2(2-\sqrt{2})$	2	$32\Gamma_7 a_2$

Here $32\Gamma_7 a_1$ and $32\Gamma_7 a_2$ denote the groups of order 32 as given in the tables of Hall & Senior [HS], and $\nu = 1 + \sigma + \sigma^2 + \sigma^3$ denotes the 'norm' in $\mathbb{Z}[G]$. Compare Vaughan [Vau] and Waterhouse [Wat].

4.10 Show that central extensions of cyclic groups are abelian.

4.11 Compute the subfields of $\mathbb{Q}(\zeta_{21})$, the corresponding character groups, as well as their conductors. Use the conductor-discriminant formula to calculate the discriminants. Determine how the primes $p < 20$ split in these subfields.

4.12 Show that $K = \mathbb{Q}(\sqrt{m})$ is contained in the cyclotomic field $\mathbb{Q}(\zeta_f)$ if and only if disc K divides f.

4.13 Show that $\left(\frac{a}{m}\right) = \left(\frac{a}{n}\right)$ if $m \equiv n \bmod f$, where $f = \operatorname{disc} \mathbb{Q}(\sqrt{a})$, and prove that no proper divisor of f has this property. Conclude that $|f|$ is the conductor of $\left(\frac{a}{\cdot}\right)$.

4.14 (Spearman & Williams [SW]) Let $D = B^2 + C^2$ be squarefree; show that $K = \mathbb{Q}(\sqrt{A(D + B\sqrt{D})})$ is a cyclic extension of $\mathbb{Q}$ and compute its conductor. (Hint: let d be the conductor of the quadratic subfield k of K; show that there exists a cyclic quartic extension L of conductor d and conclude that the conductors of K and of KL coincide. Then observe that the conductor of KL is the lowest common multiple of the conductors of L and of a quadratic subfield of KL different from k.)

4.15 For abelian extensions $K_i/\mathbb{Q}$ ($i = 1, 2$) with conductors $\operatorname{cond} K_i = f_i$, show that $\operatorname{cond} K_1 K_2 = \operatorname{lcm}(f_1, f_2)$ and $\operatorname{cond} K_1 \cap K_2 = \gcd(f_1, f_2)$.

4.16 Let $K/\mathbb{Q}$ be an abelian extension with conductor $\operatorname{cond} K = f$. Show that a prime p is ramified if and only if $p \mid f$, and that p is tamely ramified if and only if $p \parallel f$.

4.17 Let χ and ψ be characters with conductors f_χ and f_ψ, respectively; if $(f_\chi, f_\psi) = 1$, show that $f_{\chi\psi} = f_\chi f_\psi$.

4.18 For characters $\chi \neq \mathbb{1}$ on $(\mathbb{Z}/m\mathbb{Z})^\times$, show that $\sum_a \chi(a) = 0$, where the sum is over all $a \in (\mathbb{Z}/m\mathbb{Z})^\times$.

4.19 Let G be a finite abelian group. For $a \in G \setminus \{1\}$, show that $\sum_\chi \chi(a) = 0$, where the sum is over all characters $\chi \in \widehat{G}$.

4.20 For all characters $\chi \neq \mathbb{1}$ on $\mathbb{F}_q^\times$, show that $J(\mathbb{1}, \mathbb{1}) = -q$, $J(\chi, \mathbb{1}) = 0$, $J(\chi, \chi^{-1}) = \chi(-1)$, $G_0(\chi) = 0$, $G(\mathbb{1}) = 1$ and $G_0(\mathbb{1}) = 1 - q$. What results would you get if we had defined $\mathbb{1}(0) = 0$?

4.21 Let $m = 8$, $\zeta = \zeta_8$, and $K = \mathbb{Q}(\zeta)$; then $\sqrt{-2} = \zeta + \zeta^3$. Let $\mathfrak{p} = (1 + \sqrt{-2})$ be a prime ideal in $\mathcal{O}_K$ above 3, and put $\chi = \left(\frac{\cdot}{\mathfrak{p}}\right)_8$. Show that $\chi(1 + i) = \zeta^3$, and prove that $G(\chi) = \sqrt{-3}(1 + \sqrt{-2})$.

4.22 Let χ be the cubic residue character modulo $1 - 3\rho$ in $\mathbb{Z}[\rho]$; compute $G(\chi)^3$ by hand.

4.23 Let $\chi \neq \mathbb{1}$ be a character of order $m = 2^k$ on $\mathbb{F}_q$, $q = p^n$, and let ψ be the character of order 2. Then
$$J(\chi^r, \chi^s) \equiv J(\chi^r, \chi^s\psi) \bmod (2 + 2\zeta),$$
where $\zeta = \zeta_m$ is a primitive m-th root of unity. (Hint: compute $J(\chi^r, \chi^s) - J(\chi^r, \chi^s\psi) \bmod (2 + 2\zeta)$, and turn to the proof of Proposition 6.4 for help).

4.24 (Jacobi [398]) Let $p \equiv 1 \bmod 7$ be prime, $\chi : \mathbb{F}_p^\times \longrightarrow \mu_7$ a character of order 7, and $J(\chi, \chi)$ the corresponding Jacobi sum. Show that $p = L^2 + 7M^2$, and that L can be normalized by $L \equiv -1 \bmod 7$. Put $K = \mathbb{Q}(\sqrt{-7}, \sqrt{-3})$ and show that $K(\zeta_7)$ is a Kummer extensions of degree 3 generated by a cube root of $A = \frac{7}{2}(1 + 3\sqrt{-3})$. Let $B = A'$ denote the conjugate of A and show that
$$J(\chi, \chi) = a + b\sqrt{-7} + \sqrt[3]{A}\left(7c + 7e\sqrt{-3} + \sqrt{-7}(d + f\sqrt{-3})\right) + \sqrt[3]{B}\left(7c - 7e\sqrt{-3} + \sqrt{-7}(d - f\sqrt{-3})\right)$$

for integers a, b, c, d, e, f with $a \equiv 1 \bmod 7$, $b \equiv 2M \bmod 7$, and $c + a \equiv b + d \equiv 0 \bmod 3$.

4.25 Let $p = 5n + 1$ be a prime, and let (rs) denote the cyclotomic numbers of order 5 for p. Show that it is sufficient to consider $A = (00)$, $B = (01)$, $C = (02)$, $D = (03)$, $E = (04)$, $Y = (12)$ and $Z = (13)$, and derive the relations $A + B + C + D + E = n - 1$, $2Y + Z + B + E = Y + 2Z + C + D = n$ (turn to Proposition 10.5.iii) for help). Derive $n + 1 = 3(Y + Z) - A$ from these relations.
Now write $J(\chi, \chi) = \sum_{j=1}^{4} a_j \zeta_5^j$, where χ is a character of order 5 on $\mathbb{F}_p$. Write out $p = J(\chi, \chi)\overline{J(\chi, \chi)}$ explicitly, recall $\zeta_5 + \zeta_5^{-1} = \frac{1}{2}(-1 + \sqrt{5})$ and $\zeta_5^2 + \zeta_5^3 = \frac{1}{2}(-1 - \sqrt{5})$, and use the fact that the coefficient of $\sqrt{5}$ must vanish (p is an integer) to deduce that

$$16p = (a_1 + a_2 + a_3 + a_4)^2 + 2(a_1 - a_2 - a_3 + a_4)^2 + 10(a_2 - a_3)^2 + 10(a_1 - a_4)^2.$$

Put $x = -a_1 - a_2 - a_3 - a_4 = 25(Y + Z) - 10n - 4$, $w = Z - Y$, $u = D - C$ and $v = E - B$ and verify that $x \equiv 1 \bmod 5$, $16p = x^2 + 50u^2 = 50v^2 + 125w^2$, and $v^2 - wuv - u^2 = xw$. This is Dickson's diophantine system (4.10). If (x, u, v, w) solves this system, then so do $(x, -u, -v, w)$, $(x, v, -u, -w)$ and $(x, -v, u, -w)$, but up to these permutations the solution is unique (the proofs for uniqueness that I have seen are quite involved). Show that $A = \frac{1}{25}(p - 14 + 3x)$ and use Exercise 3.21 to deduce that $(2/p)_5 = 1$ if and only if $2 \mid x$. Verify the following table of solutions of Dickson's system:

p	x	u	v	w
11	1	0	1	1
31	11	-1	2	1
151	-4	-2	2	-4

4.26 (Stickelberger [759]) Show that Proposition 4.26.ii) can be expressed in the following symmetric form: if a, b, c are integers such that $a + b + c \equiv 0 \bmod m$, then $\chi(-1)^a J(\chi^b, \chi^c) = \chi(-1)^b J(\chi^c, \chi^a) = \chi(-1)^c J(\chi^a, \chi^b)$. Show that this implies $J(\chi_1, \chi_2) = \chi_1(-1)J(\chi_1, \chi_3)$ for characters χ_1, χ_2, χ_3 such that $\chi_1 \chi_2 \chi_3 = \mathbb{1}$.

4.27 Suppose that $m \mid (p + 1)$, and let χ be a character of order m on $\mathbb{F}_{p^2}$. Show that $G(\chi) \in \mathbb{Z}[\zeta_m]$. This result holds more generally when the restriction of a character χ on $\mathbb{F}_q$ to $\mathbb{F}_p$ is the trivial character.

4.28 Let $E = \mathbb{F}_{q^f}$ be a finite extension of a finite field $F = \mathbb{F}_q$; then the norm $N_{E/F} : E^\times \longrightarrow F^\times$ and the trace $\operatorname{Tr}_{E/F} : E \longrightarrow F$ are onto. (Hint: observe that $N_{E/F}\alpha = \alpha \alpha^q \ldots \alpha^{q^{f-1}} = \alpha^m$ with $m = \frac{q^f - 1}{q - 1}$. Then use the fact that $E^\times$ is cyclic to conclude that $N_{E/F}E^\times$ has index m in $E^\times$.)

4.29 (Stickelberger [759]) Let E/F be an extension of finite fields, and let $\Phi : E \longrightarrow F$ be an F-homomorphism. Show that there exists an $\alpha \in E$ such that $\Phi(\omega) = \operatorname{Tr}_{E/F}(\alpha \omega)$ for all $\omega \in E$. Conclude that every additive character $\psi : E \longrightarrow \mu_p$ has the form $\psi(\omega) = \zeta_p^{\operatorname{Tr}(\alpha \omega)}$.

4.30 Generalize the computation of the quadratic Gauss sum to finite fields $\mathbb{F}_q$ with $q = p^n$ by showing that

$$G(\chi) = \begin{cases} (-1)^n \sqrt{q} & \text{if } p \equiv 1 \bmod 4, \\ (-i)^n \sqrt{q} & \text{if } p \equiv 3 \bmod 4, \end{cases}$$

where χ is a quadratic character on $\mathbb{F}_q^\times$ and $G(\chi) = -\sum_{t \in \mathbb{F}^\times} \chi(t) \zeta_p^{\text{Tr}(t)}$ the corresponding Gauss sum.

4.31 (Morain [Mor]) Let $\mathbb{F} = \mathbb{F}_q$ be a finite field with $q = p^n$ elements, and put $N_r = \#\{x \in \mathbb{F} : \text{Tr}_{\mathbb{F}/\mathbb{F}_p}(x^2) = r\}$ for $r \in \{0, 1, 2\}$. Show that $N_0 = \frac{1}{3}(q - G - \overline{G})$, $N_1 = \frac{1}{3}(q - \rho^2 G - \rho \overline{G})$, and $N_2 = \frac{1}{3}(q - \rho G - \rho^2 \overline{G})$, where $G = G(\chi)$ is the quadratic Gauss sum in the preceding exercise.

4.32 (Compare Exercise 1.23) Show that, for primes p,

$$N(x^3 + y^3 = 1) = \begin{cases} p, & \text{if } p \equiv 2 \bmod 3 \\ p - 2 - 2\text{Re } J(\chi, \chi), & \text{if } p \equiv 1 \bmod 3, \end{cases}$$

where χ is a character of order 3 on $\mathbb{F}_p$.

4.33 Show that the congruence $x^8 \equiv 16 \bmod p$ is solvable for every prime p.

4.34 Use the table in Exercise 4.2 to compute the Eisenstein sums $E_0(\chi)$ and $E_1(\chi)$ for the characters χ of order 4 and 8 on $\mathbb{F}_9$.

4.35 (Stickelberger [759]) Let χ be a character of order $m > 1$ on $\mathbb{F}_q^\times$, where $q = p^2$ and $m \mid (p+1)$. Show that $E(\chi) = -(-1)^{(p+1)/m}$.
(Hint: note that $\chi_K = \mathbb{1}$ for $K = \mathbb{F}_p$ and use Proposition 4.34.iv) to reduce the claim to the computation of $E_0(\chi)$. Let γ be a generator of $\mathbb{F}_q^\times$ and show that $\omega = \gamma^{(p+1)/2}$ has trace 0. Then every element of $\mathbb{F}_q$ with vanishing trace has the form $a\omega$ with $a \in \mathbb{F}_p$. Deduce that $E_0(\chi) = (p-1)\chi(\omega)$, and show that $\chi(\omega) = (-1)^{(p+1)/m}$.)

4.36 (Berndt & Evans [BE2, Thm. 2.16]) Let χ and q be as above, but assume that $m \nmid (p+1)$. Show that $E(\chi^{p+1}) = -\chi_K(2) J(\chi_K, \chi_K)$.
(Hint: Using the notation of the preceding exercise, show that elements of trace 2 can be written in the form $1 + b\omega$ with $b \in \mathbb{F}_p$. This gives $E_2(\chi^{p+1}) = \sum_{b=1}^{p-1} \chi_K(1 - b^2\omega^2)$. Show that $g = \omega^2 \in \mathbb{F}_p$ and substitute $1 - gb^2 = n$; this equation has $1 + \left(\frac{g(1-n)}{p}\right) = 1 - \left(\frac{1-n}{p}\right)$ solutions, hence $E_2(\chi^{p+1}) = \sum_{b=1}^{p-1} \chi_K(n)\psi(1-n) = -J(\chi_K, \psi)$, where ψ is the quadratic character on $K^\times$. Now use Proposition 4.30.)

4.37 (Yamamoto [Yam]) Let χ be a character on $F^\times$ such that $\chi_K \neq \mathbb{1}$. Then

$$E(\chi) = \sum_{\substack{\beta \in F^\times/K^\times \\ T(\beta) \neq 0}} \chi_K^{-1}(T(\beta))\chi(\beta).$$

(Hints: Show that every $t \in F^\times$ can be written as $\alpha\beta$ with $\alpha \in K^\times$ and β in some set of representatives of the classes of $F^\times/K^\times$. Then derive

$$E(\chi) = \sum_{T(t)=1} \chi(t) = \sum_{\beta \in F^\times/K^\times} \sum_{\substack{\alpha \in K^\times \\ T(\alpha\beta)=1}} \chi(\alpha)\chi(\beta).$$

The claim follows by replacing α with $T(\beta)^{-1}$.)

4.38 (Yamamoto [Yam]) Let χ be a character of order $m = 2^s \geq 4$ on $K^\times$, where K is a finite extension of $F = \mathbb{F}_q^\times$, and assume that $q \equiv 1 \bmod m$. If γ denotes a generator of $K^\times$, then

$$E(\chi) = \chi_F(2) + 2 \sum_{j=1}^{(q-1)/2} \chi_F(\operatorname{Tr}_{K/F} \gamma^j) \chi(\gamma^j).$$

In particular, $E(\chi) \equiv \chi_F(2) \bmod 2$.

Additional References

[APR] L.M. Adleman, C. Pomerance, R.S. Rumely, *On distinguishing prime numbers from composite numbers*, Ann. of Math. (2) **117** (1983), 173–206

[Adl] A. Adler, *Eisenstein and the Jacobian varieties of Fermat curves*, Rocky Mt. Journal **27** (1997), 1–60

[BS] E. Bach, J. Shallit, *Algorithmic Number Theory*, MIT Press, 1996

[Bee] N.G.W.H. Beeger, *On sexe-section, quinqui-section, etc. treated by P.O. Upadhyaya*, Tohoku Math. Journ. **27** (1926), 211

[BC] B. C. Berndt, S. Chowla, *The reckoning of certain quartic and octic Gauss sums*, Glasgow Math. J. **18** (1977), 153–155

[BE1] B.C. Berndt, R. J. Evans, *Sums of Gauss, Jacobi, and Jacobsthal*, J. Number Theory **11** (1979), 349–398

[BE2] B.C. Berndt, R. J. Evans, *Sums of Gauss, Eisenstein, Jacobi, Jacobsthal, and Brewer*, Ill. J. Math. **23** (1979), 374–437

[BE3] B.C. Berndt, R. J. Evans, *The determination of Gauss sums*, Bull. Amer. Math. Soc. **5 (2)** (1981), 107–129; Corr.: ibid. **7** (1982), 441

[Br] H. Braun, *Geschlechter quadratischer Formen*, J. Reine Angew. Math. **182**, (1940), 32–49

[Ca1] L. Carlitz, *Certain special equations in a finite field*, Monatsh. Math. **58** (1954), 5–12

[Ca2] L. Carlitz, *Pairs of quadratic equations in a finite field*, Amer. J. Math. **76** (1954), 137–154

[Cas] J. W. S. Cassels, *Local Fields*, Cambridge Univ. Press 1986

[Cay] A. Cayley, *The binomial equation $x^p - 1 = 0$; quinquisection*, Proc. London Math. Soc. **11** (1880), 11–14; ibid. **12** (1881), 15–16; ibid. **14** (1883), 61–63

[Ch1] S. Chowla, *On a formula of Jacobsthal*, Norske Vid. Selsk. Forhdl. **34** (1962), 105–106

[Ch2] S. Chowla, *A formula similar to Jacobsthal's for the explicit value of x in $p = x^2 + y^2$ where p is a prime of the form $4k + 1$*, Proc. Lahore Philos. Soc. **7** (1945), 2 pp

[CoE] E. Cohen, *An analogue of a result of Jacobsthal*, Proc. Edinb. Math. Soc. II. Ser. **13** (1962), 139–142

[CL] H. Cohen, H.W. Lenstra, *Primality testing and Jacobi sums*, Math. Comp. **42** (1984), 297–330

[CoS] S.D. Cohen, *Quadratic residues and character sums over fields of square order*, J. Number Theory bf 18 (1984), 391–395

[CRS] C. Corrales-Rodriganez, R. Schoof, *The support problem and its elliptic analogue*, J. Number Theory **64** (1997), 276–290

[DH] H. Davenport, H. Hasse, *Die Nullstellen der Kongruenzzetafunktionen in gewissen zyklischen Fällen*, J. Reine Angew. Math. **172** (1934), 151–182

[Ded] R. Dedekind, *Grundideale von Kreiskörpern*, Gesammelte Werke II, 401–409

[De1] F. Deloup, *Reciprocity for Gauss sums and invariants of links in 3-manifolds*, Diss. Univ. Strasbourg, 1997; see also C. R. Acad. Sci. Paris **326** (1998), 69–73

[De2] F. Deloup, *Linking forms, reciprocity for Gauss sums and invariants of 3-manifolds*, Trans. Amer. Math. Soc. **351** (1999), 1895–1918

[Eis] G. Eisenstein, *Zur Theorie der quadratischen Zerfällung der Primzahlen $8n+3$, $7n+2$ und $7n+4$*, J. Reine Angew. Math. **37** (1848), 97–126; Werke II, 506–535

[Ev1] R. Evans, *Resolution of sign ambiguities in Jacobi and Jacobsthal sums*, Pacific J. Math. **81** (1979), 71–80

[Ev2] R. Evans, *Determinations of Jacobsthal sums*, Pac. J. Math. **110** (1984), 49–58

[Fla] H. Flanders, *Generalization of a theorem of Ankeny and Rogers*, Ann. of Math. **57** (1953), 392–400

[Fur] Ph. Furtwängler, *Über das Verhalten der Ideale des Grundkörpers im Klassenkörper*, Monatsh. Math. Phys. **27** (1916), 1–15

[Gau] C.F. Gauss, *Disquisitionum circa aequationes puras*, Nachlaß, Werke II

[Ger] I. Gerst, *On the theory of n-th power residues and a conjecture of Kronecker*, Acta Arith. **17** (1970), 121-139

[Gl] J.C. Glashan, *Quinquisection of the cyclotomic equation*, Amer. J. **21** (1898), 270–275; ibid. 276–285

[Gra] G. Gras, *Sommes de Gauss sur les corps finis*, Publ. Math. Besançon (1977/78), 71 pp

[GS] J. Greene, D. Stanton, *The triplication formula for Gauss sums*, Aequationes Math. **30** (1986), 134–141

[Gri] R.C. Grimson, *The evaluation of a sum of Jacobsthal*, Norske Vid. Selsk. Skr. 1974, No.4, 6 pp. (1974)

[HL] S. Hahn, D.H. Lee, *Some congruences for binomial coefficients*, preprint 1998

[HS]   M. Hall, J. K. Senior, *The groups of order 2^n ($n \leq 6$)*, Macmillan, New York 1964

[HJ]   N.R. Hardman, J.H. Jordan, *The distribution of quadratic residues in fields of order p^2*, Math. Mag. **42** (1969), 12–17

[H2]   H. Hasse, *Über die Klassenzahl abelscher Zahlkörper*, Akademie-Verlag 1952; 2nd ed. 1985

[HSW]   J.G. Huard, B.K. Spearman, K.S. Williams, *An arithmetic approach to the Davenport-Hasse relation over $GF(p)$*, Rocky Mt. J. Math. **25** (1995), 1341–1350

[HW1]   R.H. Hudson, K.S. Williams, *Resolution of ambiguities in the evaluation of cubic and quartic Jacobsthal sums*, Pacific J. Math. **99** (1982), 379–386

[HW2]   R.H. Hudson, K.S. Williams, *An application of a formula of Western to the evaluation of certain Jacobsthal sums*, Acta Arith. **41** (1982), 261–276

[Hul1] R. Hull, *The numbers of solutions of congruences involving only kth powers*, Trans. Am. Math. Soc. **34** (1932), 908–937

[Hul2] R. Hull, *A determination of all cyclotomic quintic fields*, Ann. of Math., II. Ser. **36** (1935), 366–372; FdM **61-I** (1935), 167–168

[Jo]   C. Jordan, *Sur les sommes de Gauss à plusieurs variables*, C. R. Acad. Sci. Paris **73** 1316–1319

[JM]   A. Joux, F. Morain, *Sur les sommes de caractères liées aux courbes elliptiques à multiplication complexe*, J. Number Theory **55** (1995), 108–128

[Kat]   S.A. Katre, *Jacobsthal sums in terms of quadratic partitions of a prime*, Lect. Notes Math. **1122** (1985), 153–162

[KR1]   S.A. Katre, A.R. Rajwade, *On the Jacobsthal sum $\Phi_4(a)$ and the related sum $\psi_8(a)$*, Math. Stud. **50** (1982), No.1-4, 194–200

[KR2]   S.A. Katre, A.R. Rajwade, *On the Jacobsthal sum $\Phi_4(a)$ and the related sum $\psi_8(a)$*, Ann. Univ. Sci. Budap. Rolando Eötvös, Sect. Math. **29** (1986), 3–7

[KR3]   S.A. Katre, A.R. Rajwade, *Jacobsthal sums of prime order*, Indian J. Pure Appl. Math. **17** (1986), 1345-1362

[KR4]   S.A. Katre, A.R. Rajwade, *Resolution of the sign ambiguity in the determination of the cyclotomic numbers of order 4 and the corresponding Jacobsthal sum*, Math. Scand. 60 (1987), 52–62

[KR5]   S.A. Katre, A.R. Rajwade, *On the Jacobsthal sum $\Phi_9(a)$ and the related sum $\psi_9(a)$*, Math. Scand. **53** (1983), 193–202

[KO]   M. Kida, T. Ono, *A note on Jacobi sums*, Proc. Japan Acad. **69** (1993), 32–34

[Kr]   A. Krazer, *Zur Theorie der mehrfachen Gaußschen Summen*, H. Weber Festschrift, Leipzig 1912, p. 181

[KL] D.S. Kubert, S. Lichtenbaum, *Jacobi-sum Hecke characters and Gauss-sum identities*, Compositio Math. **48** (1983), 55–87

[Lag] J.-L. Lagrange, *Réflexions sur la résolution algébriques des équations*, Nouv. Mém. Acad. Roy. Berlin 1770, 134–215; ibid 1771, 138–254; see also Œuvres **3**, 205–421

[La] S. Lang, *Cyclotomic Fields I and II*, Combined 2nd edition, Graduate Texts in Math. **121**, Springer Verlag 1990

[Lan] P. Langevin, *Les Sommes de Caractères et la Formule de Poisson dans la théorie des codes, des Séquences et des Fonctions Booléennees*, Habilitation Univ. de Toulon et du Var, 1999

[Le] A.K. Lenstra, *Primality testing*, Proc. Sympos. Appl. Math. **42**, Boulder 1989, 13–25

[Le1] H.W. Lenstra, *Primality testing algorithms (after Adleman, Rumely and Williams)*, Bourbaki Seminar 1980/81, pp. 243–257, LNM **901**, Springer 1981.

[Le2] H.W. Lenstra, *Computing Jacobi symbols in algebraic number fields*, Nieuw Arch. Wiskd. **13** (1995), 421–426

[LW] P.A. Leonard, K.S. Williams, *Evaluation of certain Jacobsthal sums*, Boll. Unione Mat. Ital., V. Ser., B **15** (1978), 717–723

[LM] F. Leprevost, F. Morain, *Revêtements de courbes elliptiques à multiplication complexe par des courbes hyperelliptiques et sommes de caractères*, J. Number Theory **64** (1997), 165–182

[Li] W. Li, *Number Theory with applications*, World Scientific 1996

[LN] R. Lidl, H. Niederreiter, *Finite Fields*, Encyclopedia of Mathematics and Its Applications **20**, 2nd ed., Cambridge Univ. Press 1996

[MR] R.J. McEliece, H. Rumsey, *Euler products, cyclotomy, and coding*, J. Number Theory 4 (1972), 302–311

[Mi] P. Mihailescu, *Cyclotomy of Rings & Primality Testing*, Diss. ETH Zurich, 1998

[Mon] M.G. Monzingo, *An elementary evaluation of the Jacobsthal sum*, J. Number Theory **22** (1986), 21–25

[Mor] F. Morain, *Classes d'isomorphismes des courbes elliptiques supersingulières en charactéristique ≥ 3*, Util. Math. **52** (1997), 241–253

[NR] B.S. Nashier, A.R. Rajwade, *Determination of a unique solution of the quadratic partition for primes $p \equiv 1 \bmod 7$*, Pacific J. Math. **72** (1977), 513–521

[On] T. Ono, *A note on Tate-Shafarevich sets for finite groups*, Proc. Japan Acad. **74** (1998), 77–79

[OT] T. Ono, T. Terasoma, *On Hasse principle for $x^n = a$*, Proc. Japan Acad. **73** (1997), no. 7, 143–144

[PV] R. Padma, S. Venkataraman, *Elliptic curves with complex multiplication and a character sum*, J. Number Theory **61** (1996), 274–282

[Pet] B. Petin, *Ein kombinatorisches Beweisverfahren für Produktrelationen zwischen Gauss-Summen über endlichen kommutativen Ringen*, Diss. Bonn, 1990

[PS] A.G. Postnikov, S.A. Stepanov, *On the theory of Jacobsthal sums* (Russ.), Tr. Mat. Inst. Steklov **142** (1976), 208–214. Engl. Transl. Proc. Steklov Inst. Math. **142** (1979), 225–231

[Pou] D. Poulakis, *Evaluation d'une somme cubique de caractères*, J. Number Theory **27** (1987), 41–45

[Raj] A. R. Rajwade, *The period equation for primes congruent to 1 mod 5*, Proc. Cambridge Phil. Soc. **69** (1971), 153–155

[RP] A.R. Rajwade, J.C. Parnami, *A new cubic character sum*, Acta Arith. **40** (1982), 347–356

[Ran] R.A. Rankin, *Generalized Jacobsthal sums and sums of squares*, Acta Arith. **49** (1987), 5–14

[RPR] D.B. Rishi, J.C. Parnami, A.R. Rajwade, *Evaluation of a cubic character sum using the $\sqrt{-19}$ division points of the curve $y^2 = x^3 - 2^3 3 \cdot 19 x + 2 \cdot 19^2$*, J. Number Theory **19** (1984), 184–194

[Sal] M. Salvadori, *Esposizione della teoria delle somme di Gauss*, Pisa 1904

[Scf] H. Scheffler, *Die quadratische Zerfällung der Primzahlen*, Leipzig 1892

[Scm] H.L. Schmid, *Relationen zwischen verallgemeinerten Gauß'schen Summen*, J. Reine Angew. Math. **176** (1937), 189–191

[Sco] Ch.A. Scott, *The binomial equation $x^p - 1 = 0$*, Amer. J. Math. **8** (1886), 261–264

[Sel] S. Selberg, *Ernst Jacobsthal*, Norske Vid. Selsk. Forhdl. **38** (1965), 70–73

[SW] B.K. Spearman, K.S. Williams, *The conductor of a cyclic quartic field using Gauss sums*, Czechoslovak Math. J. **47** (1997), 453–462

[Sta] H.M. Stark, *Counting points on CM elliptic curves*, Rocky Mt. J. Math. **26** (1996), 1115–1138

[Ste] S.A. Stepanov, *Proof of the Davenport-Hasse relations*, Math. Notes **27** (1980), 3–4 transl. from Mat. Zametki **27** (1980), 3–6

[Tan] H.W.L. Tanner, *On the binomial equation $x^p - 1 = 0$: Quinquisection*, Proc. London Math. Soc. **18** (1887), 214–234

[Ter] A. Terras, *Fourier analysis on finite groups and applications*, Cambridge Univ. Press 1999

[Tha] D.S. Thakur, *Gauss sums for $\mathbb{F}_q[T]$*, Invent. Math. **94** (1988), 105–112

[Tro] E. Trost, *Zur Theorie der Potenzreste*, Nieuw Arch. Wiskunde **18** (1934), 58–61

[Tur] V. Turaev, *A reciprocity for Gauss sums on finite abelian groups*, Math. Proc. Cambridge Phil. Soc. **124** (1998), 205–214

[Up] P.O. Upadhyaya, see Tohoku Math. J. **19, 21, 22, 25, 27**, and Ann. Math. **23, 24**.

[Vau] Th.P. Vaughan, *The normal closure of a quadratic extension of a cyclic quartic field*, Can. J. Math. **43** (1991), 1086–1097

[Wal] H. Walum, *Finite Fourier series and Jacobsthal sums*, Diss. Univ. Colorado, Boulder, 1962

[Was] L. Washington, *Introduction to Cyclotomic Fields*, Graduate Texts in Math. **83**, Springer Verlag 1982; 2nd edition 1997

[Wat] W.C. Waterhouse, *The normal closures of certain Kummer extensions*, Can. Math. Bull. **37** (1994), 133–139

[Web] H. Weber, *Über die mehrfachen Gaussischen Summen*, J. Reine Angew. Math. **74** (1871), 14–56

[Wei] A. Weil, *Numbers of solutions of equations in finite fields*, Bull. Amer. Math. Soc. **55** (1949), 497–508

[Wh1] A. L. Whiteman, *Theorems analogous to Jacobsthal's theorem*, Duke Math. J. **16** (1949), 619–626

[Wh2] A. L. Whiteman, *A theorem of Brewer on character sums*, Duke Math. J. **30** (1963), 545–552

[Wh3] A. L. Whiteman, *Theorems on Brewer and Jacobsthal sums. I*, Proc. Sympos. Pure Math. **8** (1965), 44–55

[Wh4] A. L. Whiteman, *Theorems on Brewer and Jacobsthal sums. II*, Mich. Math. J. **12** (1965), 65–80

[Why] C.T. Whyburn, *Note on a method in elementary number theory*, J. Reine Angew. Math. **258** (1973), 153–160

[Wid] A. Widmer, *Über die Anzahl der Lösungen gewisser Kongruenzen nach einem Primzahlmodul*, Diss. Zürich, 1919

[Wil] H.C. Williams, *A class of primality tests for trinomials which includes the Lucas-Lehmer test*, Pac. J. Math. **98** (1982), 477–494

[WHS] K.S. Williams, K. Hardy, B.K. Spearman, *Explicit evaluation of certain Eisenstein sums*, Proc. 1st Conf. Can. Number Theory Assoc., Banff/Alberta 1988, 553–626 (1990)

[Won] S. Wong, *Power residues on Abelian varieties*, Manuscr. Math, to appear

[Yam] K. Yamamoto, *On congruences arising from relative Gauss sums*, Proc. Conf. Number theory and combinatorics Japan 1984, 423–446 (1985)

[YH] D.-S. Yoon, S.-C. Hwang, *On the irreducible quintic polynomial over the rational number field*, Honam Math. J. **20** (1998), no. 1, 21–29.

5. Rational Reciprocity Laws

Rational reciprocity deals with residue symbols which assume only the values ± 1 and which have entries in $\mathbb{Z}$; the first rational reciprocity laws other than quadratic reciprocity were discovered by Dirichlet. His results, however, were soon forgotten and have been rediscovered regularly.

Many reciprocity laws can be rationalized via the isomorphism $\mathcal{O}_k/\mathfrak{p} \simeq \mathbb{Z}/p\mathbb{Z}$ for primes p that split as $p\mathcal{O}_k = \mathfrak{p}\mathfrak{p}'$ in a quadratic number field $k = \mathbb{Q}(\sqrt{m}\,)$: in fact, let j be an integral solution of the congruence $j^2 \equiv m \bmod p$, and choose j in such a way that $\sqrt{m} \equiv j \bmod \mathfrak{p}$. Then

$$\left[\frac{r+s\sqrt{m}}{\mathfrak{p}}\right] = \left(\frac{r+sj}{p}\right),$$

where $[\frac{\cdot}{\cdot}]$ is the quadratic residue symbol in $\mathcal{O}_k$, and where $(\frac{\cdot}{\cdot})$ denotes the Legendre symbol in $\mathbb{Z}$. This equality can be proved easily by noting that $\left[\frac{r+s\sqrt{m}}{\mathfrak{p}}\right] = 1$ if and only if the congruence $r + s\sqrt{m} \equiv \xi^2 \bmod \mathfrak{p}$ has solutions in $k^\times$; using $\sqrt{m} \equiv j \bmod \mathfrak{p}$, this is equivalent to $r + sj \equiv x^2 \bmod \mathfrak{p}$, where x is an integer such that $x \equiv \xi \bmod \mathfrak{p}$. But now both sides of the congruence $r + sj \equiv x^2 \bmod \mathfrak{p}$ are in $\mathbb{Z}$, hence it is valid mod p.

Another way to identify the two residue symbols consists in applying Proposition 4.2.2: in fact, we get

$$\left[\frac{r+s\sqrt{m}}{\mathfrak{p}}\right] = \left[\frac{r+sj}{\mathfrak{p}}\right] = \left(\frac{r+sj}{p}\right).$$

In general, these definitions depend on the choice of the prime ideal $\mathfrak{p}$ over p; clearly they are independent if and only if

$$1 = \left[\frac{r+s\sqrt{m}}{\mathfrak{p}}\right]\left[\frac{r+s\sqrt{m}}{\mathfrak{p}'}\right] = \left[\frac{r+s\sqrt{m}}{\mathfrak{p}}\right]\left[\frac{r-s\sqrt{m}}{\mathfrak{p}}\right]$$

$$= \left[\frac{r^2-ms^2}{\mathfrak{p}}\right] = \left(\frac{r^2-ms^2}{p}\right),$$

i.e. if and only if $r^2 - ms^2$ is a quadratic residue modulo p.

For the discussions in this chapter we will also need a rational n-th power residue symbol which may be defined for integers a and primes $p \equiv 1 \bmod 2n$

by $\left(\frac{a}{p}\right)_n = 1$ if a is an n-th power modulo p, that is if $a^{(p-1)/n} \equiv 1 \bmod p$, and $\left(\frac{a}{p}\right)_n \neq 1$ otherwise. Here we will only deal with a more restrictive case: we will assume that $a^{(p-1)/n} \equiv 1 \bmod p$ and then define $\left(\frac{a}{p}\right)_{2n} = \pm 1$ by $\left(\frac{a}{p}\right)_{2n} \equiv a^{(p-1)/2n} \bmod p$. Needless to say, this rational $2n$-th power residue symbol does not coincide with the symbol introduced at the beginning of Chapter 4; in fact, the symbol just defined equals $\left(\frac{a}{\mathfrak{p}}\right)_{2n}$, where $\mathfrak{p}$ is *any* prime ideal in $\mathbb{Z}[\zeta_{2n}]$ above p. The value of $\left(\frac{a}{p}\right)_{2n} = \pm 1$ can be determined by a variant of Gauss's lemma (see Proposition 5.10).

In this chapter we will first discuss Dirichlet's method of deriving rational reciprocity laws, and then present the results of Scholz and E. Lehmer. The 'universal quartic rational reciprocity law' is the topic in Section 5.4, and we close this chapter with some hints at the connections between rational quartic reciprocity, the power character of quadratic units, and the 2-class groups of quadratic number fields.

5.1 L. Dirichlet

In 1828, Gauss [272] published some of his results on biquadratic residues and remarked that the primes in $\mathbb{Z}[i]$ satisfied a very simple quadratic reciprocity law; in fact, if we let $\left[\frac{\cdot}{\cdot}\right]$ denote the quadratic residue symbol in $\mathbb{Z}[i]$, then

Proposition 5.1. *Let $\pi = a + bi$ and $\lambda = c + di$ be different primes $\equiv 1 \bmod 2$; then $\left[\frac{\pi}{\lambda}\right] = \left[\frac{\lambda}{\pi}\right]$. The supplementary laws are given by $\left[\frac{i}{a+bi}\right] = (-1)^{b/2}$ and $\left[\frac{1+i}{a+bi}\right] = \left(\frac{2}{a+b}\right)$.*

Soon after the publication of Gauss's memoir, Dirichlet [166] noticed that Proposition 5.1 could be deduced very easily from the quadratic reciprocity law in $\mathbb{Z}$; before we reproduce his proof, we record the following facts which we will use quite often in what follows:

Proposition 5.2. *Let $p = a^2 + b^2$ be an odd prime, and suppose that a is odd. Then*

$$\left(\frac{a}{p}\right) = 1, \quad \left(\frac{b}{p}\right) = \left(\frac{2}{p}\right), \quad \text{and} \quad \left(\frac{a+b}{p}\right) = \left(\frac{2}{a+b}\right).$$

Proof. Using the quadratic reciprocity law, we get $\left(\frac{a}{p}\right) = \left(\frac{p}{a}\right) = +1$, because $p = a^2 + b^2 \equiv b^2 \bmod a$. Next the congruence $(a+b)^2 \equiv 2ab \bmod p$ shows that $\left(\frac{a}{p}\right) = \left(\frac{2b}{p}\right)$, and this proves our second claim. Finally $2p = (a+b)^2 + (a-b)^2$ implies that $\left(\frac{a+b}{p}\right) = \left(\frac{p}{a+b}\right) = \left(\frac{2}{a+b}\right)$. □

In order to prove the quadratic reciprocity law in $\mathbb{Z}[i]$, we write $\pi = a + bi, \lambda = c + di$; then $\pi \equiv \lambda \equiv 1 \bmod 2$ implies that $a \equiv c \equiv 1 \bmod 2$ and $b \equiv d \equiv 0 \bmod 2$. If $\pi = p \in \mathbb{Z}$ or $\lambda = \ell \in \mathbb{Z}$, the proof follows directly from the relations $\left[\frac{p}{\lambda}\right] = \left(\frac{p}{N\lambda}\right)$ and $\left[\frac{\pi}{\ell}\right] = \left(\frac{N\pi}{\ell}\right)$, which are special cases of Proposition 4.2. Thus we may assume that $p = N\pi$ and $\ell = N\lambda$ are prime. We find immediately that $ai \equiv b \bmod \pi$ and $ci \equiv d \bmod \lambda$, hence we get

$$\left[\frac{\pi}{\lambda}\right] = \left[\frac{c}{\lambda}\right]\left[\frac{ac+bci}{\lambda}\right] = \left[\frac{c}{\lambda}\right]\left[\frac{ac+bd}{\lambda}\right].$$

Since $c \in \mathbb{Z}$ and $ac + bd \in \mathbb{Z}$, Proposition 4.2 gives

$$\left[\frac{c}{\lambda}\right] = \left(\frac{c}{\ell}\right) \text{ and } \left[\frac{ac+bd}{\lambda}\right] = \left(\frac{ac+bd}{\ell}\right).$$

Using Proposition 5.2, we find

$$\left[\frac{\pi}{\lambda}\right] = \left(\frac{ac+bd}{\ell}\right). \tag{5.1}$$

But now $p\ell = (a^2+b^2)(c^2+d^2) = (ac+bd)^2 + (ad-bc)^2 \equiv (ad-bc)^2 \bmod (ac+bd)$ implies $\left(\frac{\ell}{ac+bd}\right) = \left(\frac{p}{ac+bd}\right)$, and applying the quadratic reciprocity law in $\mathbb{Z}$ twice shows that

$$\left(\frac{ac+bd}{\ell}\right) = \left(\frac{\ell}{ac+bd}\right) = \left(\frac{p}{ac+bd}\right) = \left(\frac{ac+bd}{p}\right).$$

The quadratic reciprocity law in $\mathbb{Z}[i]$ follows by symmetry:

$$\left[\frac{\pi}{\lambda}\right] = \left(\frac{ac+bd}{\ell}\right) = \left(\frac{ac+bd}{p}\right) = \left[\frac{\lambda}{\pi}\right].$$

The supplementary laws follow immediately from (5.1) by putting $a = 0, b = 1$ or $a = b = 1$, and using quadratic reciprocity.

Dirichlet also simplified Gauss's computation of the biquadratic residue character of 2, using quite similar elementary considerations:

Proposition 5.3. *Let $p = a^2 + b^2 \equiv 1 \bmod 4$ be prime, and let i be an integer such that $i \equiv b/a \bmod p$. Then*

$$2^{\frac{p-1}{4}} \equiv i^{\frac{ab}{2}} \bmod p.$$

Proof. We begin with the observation

$$\left(\frac{a+b}{p}\right) \equiv (a+b)^{\frac{p-1}{2}} \equiv (2ab)^{\frac{p-1}{4}} \equiv 2^{\frac{p-1}{4}} i^{\frac{p-1}{4}} a^{\frac{p-1}{2}} \bmod p,$$

where we have used the congruence $(a+b)^2 \equiv 2ab \bmod p$. Next note that $(a/p) = 1$, hence $a^{(p-1)/2} \equiv 1 \bmod p$. Finally

$$\left(\frac{a+b}{p}\right) = \left(\frac{2}{a+b}\right) = (-1)^{((a+b)^2 - 1)/8} \equiv i^{\frac{p-1}{4}} i^{\frac{ab}{2}} \bmod p,$$

because $(a+b)^2 = p + 2ab$ and $i^2 \equiv -1 \bmod p$. □

156   5. Rational Reciprocity Laws

For primes $p \equiv 1 \bmod 8$, this simplifies to $(\frac{2}{p})_4 = 1 \iff p = a^2 + 64b^2$. We can prove similar criteria using other representations of p: let $p = e^2 - 2f^2 \equiv 1 \bmod 8$ (this implies that $f \equiv 0 \bmod 2$) and observe that

$$e^2 \equiv 2f^2 \bmod p \iff e^2 f^2 \equiv 2f^4 \bmod p.$$

Thus $(2/p)_4 = (2f^4/p)_4 = (e^2 f^2/p)_4 = (ef/p)$; writing $f = 2^j n$ for some odd n and keeping in mind that $(2/p) = 1$, we find

$$\left(\frac{e}{p}\right) = \left(\frac{p}{e}\right) = \left(\frac{-2}{e}\right) \text{ and } \left(\frac{f}{p}\right) = \left(\frac{n}{p}\right) = \left(\frac{p}{n}\right) = 1,$$

because $p \equiv e^2 \bmod n$. We have proved that $\left(\frac{2}{p}\right)_4 = \left(\frac{-2}{e}\right)$.

In order to see how this relates to the proposition above, observe that $p = a^2 + b^2 = e^2 - 2f^2$ implies

$$a^2 + 2f^2 \equiv e^2 - b^2 = (e-b)(e+b), \text{ hence } a^2 \equiv -2f^2 \bmod e \pm b.$$

Therefore $\left(\frac{-2}{e \pm b}\right) = 1$; assuming that $e > 0$, we get $e \pm b > 0$, and we see that

$$e \equiv +1, +3 \bmod 8 \iff b \equiv 0 \bmod 8,$$
$$e \equiv -1, -3 \bmod 8 \iff b \equiv 4 \bmod 8.$$

This shows that $\left(\frac{-2}{e}\right) = (-1)^{b/4}$ and gives an independent proof of Proposition 5.3.

Proposition 5.4. *Let $p = a^2 + b^2 = c^2 + 2d^2 = e^2 - 2f^2 = 8n + 1$ be prime, and assume that b is even. Then $\left(\frac{2}{p}\right)_4 = (-1)^{b/4} = \left(\frac{2}{c}\right) = (-1)^{n+d/2} = \left(\frac{-2}{e}\right)$.*

The assertion $\left(\frac{2}{p}\right)_4 = \left(\frac{2}{c}\right)$ may be proved by imitating the reasoning above: start with the congruence $c^2 d^2 \equiv -2d^4 \bmod p$, derive $\left(\frac{-2}{p}\right)_4 = \left(\frac{cd}{p}\right)$, and then show that $\left(\frac{2d}{p}\right) = 1$. The statement of Proposition 5.4 can be extended immediately to products of primes $p \equiv 1 \bmod 8$: see Exercise 5.3.

Using the same approach we can determine the biquadratic residue character modulo primes $p \equiv 1 \bmod 4$ of any prime q^* which is a quadratic residue mod p: let h denote the (odd) class number of $\mathbb{Q}(\sqrt{q^*})$ and write $4p^h = e^2 - q^* f^2$; then we find that $\left(\frac{q^*}{p}\right)_4 = \left(\frac{ef}{p}\right)$. It is however not the best idea to start with the equation $4p^h = e^2 - q^* f^2$ because p and q^* do not enter in a symmetric way (in fact we eventually want to exchange the roles of p and q in order to get reciprocity laws). Therefore, we will use Legendre's equation $\boxed{pf^2 = e^2 - q^* g^2}$ (equations, congruences etc. that are important for the proof will appear in boxes). This equation has non-trivial solutions by Theorem 1.7; writing it as $e^2 = pf^2 + q^* g^2$ makes the symmetry in the case $p \equiv q \equiv 1 \bmod 4$ obvious. Without loss of generality we may assume that at least one of e, f, g is odd; then necessarily $\boxed{e \equiv 1 \bmod 2,}$ since otherwise we would have $0 \equiv e^2 \equiv 1 + 1 \equiv 2 \bmod 4$.

Exactly as above we start with the congruence $q^*g^2 \equiv e^2 \bmod p$ and derive $\boxed{\left(\frac{q^*}{p}\right)_4 = \left(\frac{eg}{p}\right)}$. Since e is odd, we find

$$\left(\frac{e}{p}\right) = \left(\frac{p}{e}\right) = \left(\frac{pf^2}{e}\right) = \left(\frac{pf^2 - e^2}{e}\right) = \left(\frac{-q^*g^2}{e}\right) = \left(\frac{-q^*}{e}\right).$$

If we assume that $\boxed{e > 0}$ (which we may do without loss of generality) we get

$$\left(\frac{-q^*}{e}\right) = \left(\frac{-1}{e}\right)\left(\frac{e}{q}\right).$$

In order to determine $\left(\frac{g}{p}\right)$, we write $g = 2^j u$, where u is odd. Then

$$\left(\frac{u}{p}\right) = \left(\frac{p}{u}\right) = \left(\frac{pf^2}{u}\right) = \left(\frac{e^2 - q^*g^2}{u}\right) = \left(\frac{e^2}{u}\right) = 1.$$

Putting everything together gives

$$\left(\frac{q^*}{p}\right)_4 = \left(\frac{2}{p}\right)^j \left(\frac{-1}{e}\right)\left(\frac{e}{q}\right). \tag{5.2}$$

Our next aim is to show that

$$\left(\frac{2}{p}\right)^j \left(\frac{-1}{e}\right) = \left(\frac{e+bf}{q}\right), \tag{5.3}$$

where b is defined by $\boxed{p = a^2 + b^2,\ b \text{ even.}}$ To this end we first have to make sure that $e + bf$ and q are coprime. From $(e - bf)(e + bf) = e^2 - b^2 f^2 = a^2 f^2 + q^*g^2$ we get $(q, e^2 - b^2 f^2) = (q, a^2 f^2) = (q, a^2)$, i.e., q divides one of $e \pm bf$ if and only if $q \mid a$; on the other hand q cannot divide both $e + bf$ and $e - bf$, hence we may choose the sign of f in such a way that $\boxed{q \nmid (e+bf).}$ Now we put $\boxed{\varepsilon = \text{sign}\,(e+bf)}$ and get

$$\left(\frac{e+bf}{q}\right) = \left(\frac{\varepsilon}{q}\right)\left(\frac{|e+bf|}{q}\right) = \left(\frac{\varepsilon}{q}\right)\left(\frac{q^*}{e+bf}\right) = \left(\frac{\varepsilon}{q}\right)\left(\frac{-1}{e+bf}\right), \tag{5.4}$$

since $q^*g^2 \equiv -a^2 f^2 \bmod (e+bf)$. Now $\boxed{(\varepsilon/q) = \varepsilon{:}}$ this is clear if $\varepsilon = 1$; if $\varepsilon = -1$, then $e + bf < 0$ implies $bf < 0$ (since $e > 0$) and $e - bf > 0$, hence get $0 > e^2 - b^2 f^2 = a^2 f^2 + q^*g^2 > q^*g^2$, that is, $q^* < 0$ and $q \equiv 3 \bmod 4$: thus $(\varepsilon/q) = (-1/q) = -1$ in this case.

Next we claim that

$$\left(\frac{-1}{e+bf}\right) = (-1)^{\frac{bf}{2}}\left(\frac{-1}{e}\right)\varepsilon. \tag{5.5}$$

158   5. Rational Reciprocity Laws

This is due to the fact that $\left(\frac{-1}{m}\right) = \left(\frac{-1}{m+4}\right)$ if $m(m+4) > 0$ and $\left(\frac{-1}{m}\right) = -\left(\frac{-1}{m+4}\right)$ if $m(m+4) < 0$.

On the other hand, $\boxed{(-1)^{bf/2} = (2/p)^j}$: in fact, the left hand side equals -1 if and only if $b \equiv 2 \bmod 4$ (that is, $p \equiv 5 \bmod 8$) and f is odd, i.e. if and only if $(2/p) = -1$ and $1 \equiv e^2 \equiv pf^2 + q^*g^2 \equiv 5 + g^2 \bmod 8$: but the last congruence holds iff $j = 1$. Similarly, $(2/p)^j = -1$ if and only if $(2/p) = -1$ and $j = 1$. This proves our claim that $(-1)^{bf/2} = (2/p)^j$.

Plugging the last box into (5.5), using (5.4) as well as $(\varepsilon/q) = \varepsilon$ we get the desired equality (5.5).

Finally put $\boxed{\sigma \equiv e/f \bmod q;}$ then $\sigma^2 \equiv e^2/f^2 \equiv p \bmod q$, i.e., σ is a square root of $p \bmod q$. From (5.2), (5.3) and $e + bf \equiv f(\sigma + b) \bmod q$ we deduce

$$\left(\frac{q^*}{p}\right)_4 = \left(\frac{e+bf}{q}\right)\left(\frac{e}{q}\right) = \left(\frac{\sigma(b+\sigma)}{q}\right).$$

It is easy to verify that this formula is also valid if we replace σ by $-\sigma$, because

$$\left(\frac{b+\sigma}{q}\right)\left(\frac{b-\sigma}{q}\right) = \left(\frac{b^2-p}{q}\right) = \left(\frac{-a^2}{q}\right) = \left(\frac{-1}{q}\right) = \left(\frac{\sigma}{q}\right)\left(\frac{-\sigma}{q}\right).$$

We have proved

Theorem 5.5. *Let $p = a^2 + b^2$ and q be primes such that $\left(\frac{q}{p}\right) = 1$, and assume that $2 \mid b$. Let σ be a solution of the congruence $\sigma^2 \equiv p \bmod q$; then*

$$\left(\frac{q^*}{p}\right)_4 = \left(\frac{\sigma(b+\sigma)}{q}\right). \tag{5.6}$$

If $q \mid b$ then (5.6) implies that q^* is a biquadratic residue modulo p; if $q \mid a$, then the identity

$$(a+b+\sigma)^2 \equiv 2(a+\sigma)(b+\sigma) \bmod q \tag{5.7}$$

implies $\left(\frac{q^*}{p}\right)_4 = \left(\frac{2\sigma(a+\sigma)}{q}\right) = \left(\frac{2\sigma^2}{q}\right) = \left(\frac{2}{q}\right)$. Now suppose that $q \nmid ab$ and write $\lambda b \equiv \sigma \bmod q$ for some $\lambda \in \mathbb{Z}$; then $\sigma(b+\sigma) \equiv b^2\lambda(\lambda+1) \bmod q$ yields $\left(\frac{q^*}{p}\right)_4 = \left(\frac{\lambda(\lambda+1)}{q}\right)$. If we therefore define $\mu \in \mathbb{Z}$ by $a \equiv \mu b \bmod q$ and assume that $\left(\frac{p}{q}\right) = 1$, then we find $\mu^2 + 1 \equiv p/b^2 \equiv \lambda^2 \bmod q$, and we see that q^* is a biquadratic residue mod p if and only if $\lambda(\lambda + 1)$ is a quadratic residue mod q. This gives the following

Proposition 5.6. *Let $p = a^2 + b^2$ be prime, where $2 \mid b$, and suppose that $\left(\frac{p}{q}\right) = 1$ for another prime q, then*

$$\left(\frac{q^*}{p}\right)_4 = 1 \iff \begin{cases} q \mid b; \text{ or} \\ q \mid a \text{ and } \left(\frac{2}{q}\right) = 1; \text{ or} \\ a \equiv \mu b, \ \mu^2 + 1 \equiv \lambda^2 \bmod q \text{ and } \left(\frac{\lambda(\lambda+1)}{q}\right) = 1. \end{cases}$$

Now we can derive criteria for biquadratic residuacity by computing the numbers μ occuring in Proposition 5.6. As an example, take $q = 13$; the λ satisfying the condition $(\frac{\lambda(\lambda+1)}{q}) = 1$ are $\lambda = 3, 5, 6, 7, 9$, and of these only $\lambda = \pm 6$ give rise to a solution of $\mu^2 \equiv \lambda^2 - 1 \bmod 13$. Thus $\mu \equiv \pm 3 \bmod 13$, and 13 is a biquadratic residue modulo q if and only if $13 \mid b$ or $a \equiv \pm 3 \bmod 13$. The other cases in the following list are derived similarly:

Examples: Let $p = a^2 + b^2$ be a prime and assume that $2 \mid b$; then

-3 is biquadratic residue mod p $\iff$ $3 \mid b$
5 is biquadratic residue mod p $\iff$ $5 \mid b$
-7 is biquadratic residue mod p $\iff$ $7 \mid ab$
-11 is biquadratic residue mod p $\iff$ $11 \mid b$ or $a \equiv \pm 2b \bmod 11$
13 is biquadratic residue mod p $\iff$ $13 \mid b$ or $a \equiv \pm 3b \bmod 13$
17 is biquadratic residue mod p $\iff$ $17 \mid ab$ or $a \equiv \pm b \bmod 17$.

Note that these results have been proved using only the quadratic reciprocity law.

From now on we will assume that $p \equiv q \equiv 1 \bmod 4$, and we will write $p = a^2 + b^2$ and $q = c^2 + d^2$, where b and d are even; then $(\frac{\sigma}{q})$ does not depend on which square root σ of p we choose, and in fact $(\frac{\sigma}{q}) = (\frac{\sqrt{p}}{q}) = (\frac{p}{q})_4$ shows that $(\frac{p}{q})_4 (\frac{q}{p})_4 = (\frac{\beta}{q})$, where $\beta = b + \sqrt{p}$. The identity $(a + b + \sqrt{p})^2 = 2(a + \sqrt{p})(b + \sqrt{p})$ (this is simply another version of (5.7)) shows that $(\frac{\beta}{q}) = (\frac{\alpha}{q})$, where $\alpha = \frac{1}{2}(a + \sqrt{p})$. Similarly $(a + \sqrt{p} + bi)^2 = 2(a + \sqrt{p})(a + bi)$ yields $(\frac{\alpha}{q}) = (\frac{a+bi}{q})$. Finally, from $c(a + bi) \equiv ac + bd \bmod (c + di)$ and Proposition 5.2 we get

$$\left(\frac{a + bi}{q}\right) = \left[\frac{a + bi}{c + di}\right] = \left[\frac{ac + bd}{c + di}\right] = \left(\frac{ac + bd}{q}\right).$$

The following theorem (which I couldn't find in Dirichlet's papers, although it follows easily from Theorem 5.5) collects all these reciprocity laws:

Theorem 5.7. *Let $p = a^2 + b^2$ and $q = c^2 + d^2$ be different primes such that $b \equiv d \equiv 0 \bmod 2$ and $(\frac{p}{q}) = 1$, and define $\alpha = \frac{1}{2}(a + \sqrt{p})$, $\beta = b + \sqrt{p}$, $\gamma = \frac{1}{2}(c + \sqrt{q})$, and $\delta = d + \sqrt{q}$. Then*

$$\left(\frac{p}{q}\right)_4 \left(\frac{q}{p}\right)_4 = \left(\frac{ac + bd}{p}\right) = \left(\frac{c + di}{p}\right) = \left(\frac{\delta}{p}\right) = \left(\frac{\gamma}{p}\right)$$
$$= \left(\frac{ac + bd}{q}\right) = \left(\frac{a + bi}{q}\right) = \left(\frac{\beta}{q}\right) = \left(\frac{\alpha}{q}\right).$$

The special case $(p/q)_4 (q/p)_4 = (ac + bd/p)$ is called Burde's reciprocity law.

5.2 A. Scholz

Let p and q be different primes $\equiv 1 \bmod 4$ such that $\left(\frac{p}{q}\right) = 1$; moreover, let ε_p and ε_q denote the fundamental units in $\mathcal{O}_p$ and $\mathcal{O}_q$, respectively, where $\mathcal{O}_d$ is the ring of integers in $\mathbb{Q}(\sqrt{d})$. It is well known that both ε_p and ε_q have negative norm (see Exercise 2.2 or Proposition 3.24). Our assumptions imply that p splits in $\mathcal{O}_q$, i.e., we have $p\mathcal{O}_q = \mathfrak{p}\mathfrak{p}'$. We also note that the residue symbol $[\varepsilon_q/\mathfrak{p}]$ in $\mathcal{O}_q$ does not depend on the choice of ε_q or $\mathfrak{p}$, because

$$\left[\frac{\varepsilon_q}{\mathfrak{p}}\right]\left[\frac{\varepsilon_q}{\mathfrak{p}'}\right] = \left[\frac{\varepsilon_q}{\mathfrak{p}}\right]\left[\frac{\varepsilon_q'}{\mathfrak{p}}\right] = \left[\frac{-1}{\mathfrak{p}}\right] = \left(\frac{-1}{p}\right) = 1.$$

This shows that we may define $(\varepsilon_q/p) := [\varepsilon_q/\mathfrak{p}]$ without ambiguity (it also shows that $[\varepsilon_q/\mathfrak{p}] = -[\varepsilon_q/\mathfrak{p}']$ for splitting primes $p \equiv 3 \bmod 4$, so our assumption that $p \equiv 1 \bmod 4$ is necessary).

In order to simplify the statement of the supplementary law, we define a symbol $\left(\frac{p}{2}\right)_4$ for primes $p \equiv 1 \bmod 8$ by writing $\left(\frac{p}{2}\right)_4 = +1$ for primes $p \equiv 1 \bmod 16$ and $\left(\frac{p}{2}\right)_4 = -1$ if $p \equiv 9 \bmod 16$ (in other words: $\left(\frac{p}{2}\right)_4 = (-1)^{(p-1)/8}$). The following reciprocity law was discovered by Schönemann [722], and then rediscovered by A. Scholz while he was studying the 2-class group of real quadratic number fields:

Proposition 5.8. *Let p and q be different primes $\equiv 1 \bmod 4$ such that $\left(\frac{p}{q}\right) = 1$; then*

$$\left(\frac{\varepsilon_q}{p}\right) = \left(\frac{p}{q}\right)_4 \left(\frac{q}{p}\right)_4 = \left(\frac{\varepsilon_p}{q}\right).$$

Moreover, if q is a prime $\equiv 1 \bmod 8$, then $\left(\frac{2}{q}\right)_4 \left(\frac{q}{2}\right)_4 = \left(\frac{\varepsilon_2}{q}\right).$

Most proofs of Scholz's reciprocity law are based on the following fact:

Proposition 5.9. *Let $p \equiv 1 \bmod 4$ be prime; then $K = \mathbb{Q}(\sqrt{\varepsilon_p^* \sqrt{p}})$ is the quartic subfield of $\mathbb{Q}(\zeta_p)$, where $\varepsilon_p^* = \left(\frac{2}{p}\right)\varepsilon_p$ and $\varepsilon_p > 1$ is the fundamental unit of $\mathbb{Q}(\sqrt{p})$. Similarly, $K = \mathbb{Q}(\sqrt{2 + \sqrt{2}})$ is the real quartic subfield of $\mathbb{Q}(\zeta_{16})$.*

Proof. In Equation (3.8) we have shown that there is an odd $h \in \mathbb{N}$ such that $\left(\frac{2}{p}\right)\varepsilon^h \sqrt{p} = N^2$ for some $N \in \mathbb{Z}[\zeta_p]$; since $\mathbb{Q}(\zeta_p)$ is cyclic, $\left(\frac{2}{p}\right)\varepsilon^h \sqrt{p}$ must already become a square in the quartic subfield of $\mathbb{Q}(\zeta_p)$.

The assertion about the real quartic subfield of $\mathbb{Q}(\zeta_{16})$ follows from the relation $(\zeta_{16} + \zeta_{16}^{-1})^2 = 2 + \zeta_8 + \zeta_8^{-1} = 2 + \sqrt{2}$. □

Another (quite general) proof of Proposition 5.9 using Galois theory will be given in Section 5.4 below.

Before we will present one of the two proofs of Proposition 5.8 due to Scholz himself, we will make the following observation: if KL/F is a quartic extension of number fields with three quadratic subfields K, L, M, then a prime ideal $\mathfrak{p}$ in $\mathcal{O}_F$ splits in M/F if and only if $\mathfrak{p}$ splits in both K and L or if $\mathfrak{p}$ is inert in both K and L.

For the proof of Scholz's reciprocity law we observe that our assumptions $q \equiv 1 \bmod 4$ and $\left(\frac{p}{q}\right) = 1$ guarantee that q splits completely in $F = \mathbb{Q}(i, \sqrt{p})$. Define quadratic extensions K/F and L/F as in the diagram below, and denote the third quadratic subfield of KL/F by M.

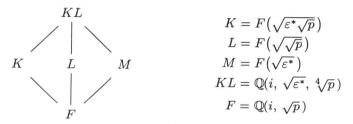

Now K is the compositum of $\mathbb{Q}(i)$ and the quartic cyclic subfield of $\mathbb{Q}(\zeta_p)$, hence a prime $q \equiv 1 \bmod 4$ splits in K/F if and only if $\left(\frac{q}{p}\right)_4 = 1$. Next the decomposition law in Kummer extensions implies that q splits in L/F if and only if $\left(\frac{p}{q}\right)_4 = 1$. This implies that q splits in M/F if and only if $\left(\frac{p}{q}\right)_4 = \left(\frac{q}{p}\right)_4$. On the other hand Kummer theory shows that q splits in M/F if and only if $\left(\frac{\varepsilon^*}{q}\right) = 1$. This completes our proof of Scholz's reciprocity law, since ε^* and ε differ at most by a factor -1. If we replace p by 2 and K by $\mathbb{Q}(\zeta_{16})$, then the proof just given yields the supplementary law.

We close this section with a few remarks about the problem Scholz was interested in when he discovered his reciprocity law. Let p and q be as above; then $\mathbb{Q}(\sqrt{pq})$ has even class number (this follows from genus theory). In fact, one can even show that the class number in the strict sense is divisible by 4; Scholz was interested in the problem when this would hold for the class number in the usual sense, and he found that $4 \mid h_{pq}$ if and only if $(\varepsilon_p/q) = +1$. From symmetry, he concluded that $(\varepsilon_p/q) = (\varepsilon_q/p)$. We will say more about this connection when we discuss explicit reciprocity laws in Part II.

5.3 E. Lehmer

We start by giving Lehmer's variant of Gauss's Lemma:

Proposition 5.10. *Let ℓ and $p = 2mn + 1$ be primes such that $\left(\frac{\ell}{p}\right)_n = 1$. Let $A = \{\alpha_1, \ldots, \alpha_m\}$ be a half-system of n-th power residues. Then $\ell \alpha_j \equiv (-1)^{a(j)} \alpha_{\pi(j)} \bmod p$ for some permutation π of $\{1, \ldots, m\}$, and*

$$\left(\frac{\ell}{p}\right)_{2n} = (-1)^\mu, \text{ where } \mu = \sum_{i=1}^m a(i).$$

Again, the proof mimics the one in the classical quadratic case; we have to define half-systems of n-th power residues, however. To this end, note that $p \equiv 1 \bmod 2n$ guarantees that -1 is a n-th power mod p. This shows that the set of n-th power residues consists of pairs $\pm\alpha_1, \ldots, \pm\alpha_m$, and a half-system is just a set of representatives of n-th power residues mod ± 1. The rest of the proof is left to the reader (see Exercise 5.1).

Now we are ready to give a new proof of Scholz's reciprocity law; it is an adaption of a proof due to E. Lehmer but eventually turned out to be due to Schönemann: both of these authors used the analytic class number formula that we have eliminated from our proof. Recall that in Section 3.4 we have defined elements $R = \prod(Z^a - Z^{-a})$ and $N = \prod(Z^b - Z^{-b})$ in $\mathbb{Z}[\zeta_p]$, and we have shown that $RN = \left(\frac{2}{p}\right)\sqrt{p}$ and $\frac{N}{R} = \varepsilon^h$, where ε is the fundamental unit of $\mathbb{Q}(\sqrt{p})$ and h is an odd integer. This gave

$$\left(\frac{2}{p}\right)\varepsilon^h \sqrt{p} = N^2. \tag{5.8}$$

Now let $q = 4n + 1$ be a prime such that $\left(\frac{q}{p}\right) = 1$; raising (5.8) to the $4n$-th power yielded parts of the quadratic reciprocity law. Taking the $2n$-th power of (5.8) instead we get

$$\left(\frac{p}{q}\right)_4 \left(\frac{\varepsilon}{q}\right) \equiv N^{q-1} \bmod q.$$

In order to deduce Scholz's reciprocity from this relation, we will compute $N^q \bmod q$; if $p \equiv 1 \bmod 8$, then we deduce from Equation (3.5) that

$$N^q = \prod_{b \in B}(Z^b - Z^{-b})^q = \prod(\zeta^b - \zeta^{-b})^q \equiv \prod(\zeta^{bq} - \zeta^{-bq})$$
$$= (-1)^{\mu_B} \prod(\zeta^b - \zeta^{-b}) = (-1)^{\mu_B} N,$$

where μ_B is defined in Lemma 5.12 below. If, on the other hand, $p \equiv 5 \bmod 8$, then a similar reasoning shows that $N^q \equiv (-1)^{\mu_A} N$, with μ_A as in Proposition 5.10.

Using the generalizations of Gauss's Lemma 5.10. and 5.11 due to E. Lehmer ([508]), we find $(-1)^{\mu_A} = \left(\frac{q}{p}\right)_4$ if $p \equiv 5 \bmod 8$ and $(-1)^{\mu_B} = \left(\frac{q}{p}\right)_4$ if $p \equiv 1 \bmod 8$, hence

$$\left(\frac{p}{q}\right)_4 \left(\frac{\varepsilon}{p}\right) \equiv N^{q-1} \equiv \left(\frac{q}{p}\right)_4 \bmod q,$$

and we have given another proof of Scholz's reciprocity law.

Lemma 5.11. *Let $p \equiv 1 \bmod 4$ be prime, and define*
$$B = \left\{b : 1 \leq b \leq \tfrac{p-1}{2}, \left(\tfrac{b}{p}\right) = -1\right\}.$$
If m is a quadratic residue mod p, then for every $b_i \in B$ there is a $b_j \in B$ such that $mb_i \equiv \pm b_j \bmod p$. Let μ_B denote the number of minus signs occuring as b runs through B; then
$$\left(\frac{m}{p}\right)_4 = (-1)^{\mu_B}.$$

The result of E. Lehmer below will be used in Section 5.5 for proving results about the quadratic residuacity of quadratic units:

Proposition 5.12. *Let $p \equiv q \equiv 1 \bmod 4$ be primes such that $\left(\tfrac{p}{q}\right) = 1$, and suppose that $s^2 p^m = c^2 + qd^2$ for some odd $s, m \in \mathbb{N}$ such that $(s,q) = (c,p) = 1$; then*
$$\left(\frac{\varepsilon_q}{p}\right) = \begin{cases} \left(\tfrac{s}{q}\right), & \text{if } q \equiv 1 \bmod 8, \\ (-1)^d \left(\tfrac{s}{q}\right), & \text{if } q \equiv 5 \bmod 8. \end{cases}$$

Proof. Reducing $s^2 p^m = c^2 + qd^2$ first mod p and then mod q, we find
$$\left(\frac{q}{p}\right)_4 = \left(\frac{-1}{p}\right)_4 \left(\frac{cd}{p}\right) = \left(\frac{2cd}{p}\right) \text{ and } \left(\frac{s}{q}\right)\left(\frac{p}{q}\right)_4 = \left(\frac{c}{q}\right).$$
Applying Scholz's reciprocity law gives
$$\left(\frac{\varepsilon_q}{p}\right) = \left(\frac{p}{q}\right)_4 \left(\frac{q}{p}\right)_4 = \left(\frac{sc}{q}\right)\left(\frac{2cd}{p}\right).$$
Now we will distinguish two cases:

1. $c \equiv 1 \bmod 2$: then d is even, and from $qd^2 \equiv s^2 p^m \bmod c$ we get
$$\left(\frac{c}{q}\right) = \left(\frac{q}{c}\right) = \left(\frac{p}{c}\right) = \left(\frac{c}{p}\right), \text{ hence } \left(\frac{c}{q}\right)\left(\frac{c}{p}\right) = 1.$$
Writing $d = 2^j u$ for some $u \equiv 1 \bmod 2$, we find
$$\left(\frac{2d}{p}\right) = \left(\frac{u}{p}\right) = \left(\frac{p}{u}\right) = 1,$$
because $j = 1 \iff p \equiv 5 \bmod 8$. Collecting everything, we find
$$\left(\frac{\varepsilon_q}{p}\right) = \left(\frac{sc}{q}\right)\left(\frac{2cd}{p}\right) = \left(\frac{sc}{q}\right)\left(\frac{c}{q}\right) = \left(\frac{s}{q}\right)$$
as claimed.

2. $c \equiv 0 \bmod 2$: Here we have $pq \equiv 1 \bmod 8$ if and only if $c \equiv 0 \bmod 4$, and this gives $\left(\tfrac{2c}{pq}\right) = 1$. Moreover,
$$\left(\frac{d}{p}\right) = \left(\frac{p}{d}\right) = 1, \text{ and } \left(\frac{\varepsilon_q}{p}\right) = \left(\frac{2s}{q}\right).$$

Taken together, 1. and 2. prove our claim. $\square$

5.4 Rational Quartic Reciprocity

In 1985, K. S. Williams, K. Hardy and C. Friesen discovered that the rational quartic reciprocity laws could be subsumed into one general rational reciprocity law. This section is devoted to a simple proof of their result. The underlying idea is simple: in Chapter 3 we have proved the quadratic reciprocity law by comparing the splitting of primes q in the Kummer extension $\mathbb{Q}(\sqrt{p^*})$ and in the 'class field' $\mathbb{Q}(\zeta_p)$. Here we will assume that $p \equiv 1 \bmod 4$; then $\mathbb{Q}(\zeta_p)$ has a quartic subfield K which we can view as a quadratic extension of $k = \mathbb{Q}(\sqrt{p})$. If we assume in addition that $(q/p) = +1$, then q will split in $k/\mathbb{Q}$, and we can compare the splitting of the primes above q in the extensions K/k and $\mathbb{Q}(\zeta_p)/k$.

Theorem 5.13. *Let $m \equiv 1 \bmod 4$ be a prime, and let A, B, C be integers such that*

$$A^2 = m(B^2 + C^2), \qquad 2 \mid B,$$
$$(A, B) = (B, C) = (C, A) = 1, \qquad A + B \equiv 1 \bmod 4.$$

Then, for every odd prime $q > 0$ such that $(m/q) = +1$,

$$\left(\frac{A + B\sqrt{m}}{q}\right) = \left(\frac{q}{m}\right)_4. \tag{5.9}$$

For the proof of Theorem 5.13, we will need a simple lemma from Galois theory:

Lemma 5.14. *Let k/F and $K = k(\sqrt{\mu})$, $\mu \in k^\times$, be quadratic extensions of fields, and let σ be the non-trivial automorphism of k/F. Then*

$$K/F \text{ is normal} \iff \mu^{1+\sigma} \in k^2$$
$$\mathrm{Gal}\,(K/F) \simeq V_4 \iff \mu^{1+\sigma} \in F^2$$
$$\mathrm{Gal}\,(K/F) \simeq C_4 \iff \mu^{1+\sigma} \in k^2 \setminus F^2$$

Here V_4 and C_4 denote Klein's four group and the cyclic group of order 4, respectively.

Proof. Fix a square root $\sqrt{\mu^\sigma}$; then K/F is normal if and only if $\sqrt{\mu^\sigma} \in K$. Writing $\sqrt{\mu^\sigma} = \alpha + \beta\sqrt{\mu}$ and applying the nontrivial automorphism of K/k we see that $-\sqrt{\mu^\sigma} = \alpha - \beta\sqrt{\mu}$, hence $\alpha = 0$. Thus

$$K/F \text{ is normal} \iff \sqrt{\mu^\sigma} \in K \iff \sqrt{\mu^\sigma} = \beta\sqrt{\mu}$$
$$\iff \sqrt{\mu\mu^\sigma} = \beta\mu \iff \mu^{1+\sigma} \in k^2.$$

Now assume that K/F is normal and extend σ to an automorphism $\widetilde{\sigma}$ of K/F which maps $\sqrt{\mu}$ to $\sqrt{\mu^\sigma}$. Then $\mathrm{Gal}\,(K/F) \simeq V_4$ if and only if $\widetilde{\sigma}^2 = 1$. Now observe that $\widetilde{\sigma}^2 = 1 \iff (\sqrt{\mu\mu^\sigma})^{\widetilde{\sigma}} = \sqrt{\mu\mu^\sigma} \iff \mu\mu^\sigma \in F^2$. □

5.4 Rational Quartic Reciprocity

We will break up the proof of Theorem 5.13 into a few simple steps:

1. $K = \mathbb{Q}(\sqrt{m}, \sqrt{A + B\sqrt{m}})$ is a quartic cyclic extension of $\mathbb{Q}$ containing $k = \mathbb{Q}(\sqrt{m})$.
   This can be verified quickly by noting that $A^2 - mB^2 = mC^2 = (\sqrt{m}\,C)^2$ and $\sqrt{m}\,C \in k \setminus \mathbb{Q}$, and applying Lemma 5.14.

2. $K/\mathbb{Q}$ is unramified outside $m\infty$.
   The identity
   $$2(A + B\sqrt{m})(A + C\sqrt{m}) = (A + B\sqrt{m} + C\sqrt{m})^2 \tag{5.10}$$
   shows that $K = k(\sqrt{2(A + C\sqrt{m})})$, and so the only odd primes that are possibly ramified in K/k are common divisors of $A^2 - mB^2 = mC^2$ and $A^2 - mC^2 = mB^2$. Since B and C are assumed to be prime to each other, only 2 and m can ramify. Now $\sqrt{m} \equiv 1 \bmod 2$ (since $m \equiv 1 \bmod 4$) and $2 \mid B$ imply that $B\sqrt{m} \equiv B \bmod 4$, and we see $A + B\sqrt{m} \equiv A + B \equiv 1 \bmod 4$, which shows that 2 is unramified in K/k (and therefore also in $K/\mathbb{Q}$).

3. K is the quartic subfield of $\mathbb{Q}(\zeta_m)$, the field of m-th roots of unity.
   Both K and the quartic subfield $\widetilde{K}$ of $\mathbb{Q}(\zeta_m)$ are cyclic extensions of $\mathbb{Q}$ unramified outside $m\infty$; therefore the quadratic subfields contained in their compositum must have discriminants dividing m. Since $\mathbb{Q}(\sqrt{m})$ is the only such field, $K\widetilde{K}$ must be a cyclic extension of $\mathbb{Q}$, and this implies that $K = \widetilde{K}$.
   Another way to see this is the following: let $L = K\widetilde{K}$; the inertia subgroup T of $\mathrm{Gal}\,(L/\mathbb{Q})$ fixes a field K_T which is unramified outside ∞, hence we must have $T = \mathrm{Gal}\,(L/\mathbb{Q})$. Next $V_1 = \{1\}$, because $(L : \mathbb{Q})$ is a power of 2 and since m is odd. But we know that T/V_1 is cyclic, thus $\mathrm{Gal}\,(L/\mathbb{Q})$ is cyclic, and this implies that $K = \widetilde{K}$.

The reciprocity formula now follows by comparing the decomposition laws in $K/\mathbb{Q}$ and $\mathbb{Q}(\zeta_m)/\mathbb{Q}$: since $(m/q) = +1$, we know that q splits in $k/\mathbb{Q}$; if $f > 0$ is the smallest natural number such that $q^f \equiv 1 \bmod m$ (here we have to assume that $q > 0$), then q splits into exactly $g = (m-1)/f$ prime ideals in $\mathbb{Q}(\zeta_m)$, and we see

$$\left(\frac{q}{m}\right)_4 = 1 \iff q^{(m-1)/4} \equiv 1 \bmod m$$
$$\iff f \text{ divides } \tfrac{m-1}{4} = \tfrac{fg}{4} \iff g \equiv 0 \bmod 4$$
$$\iff 4 \mid (Z : \mathbb{Q}), \text{ where } Z \text{ is the decomposition field of } q$$
$$\iff Z \text{ contains } K \text{ (because } \mathrm{Gal}\,(\mathbb{Q}(\zeta_m)/\mathbb{Q}) \text{ is cyclic)}$$
$$\iff q \text{ splits completely in } K/\mathbb{Q}$$
$$\iff q \text{ splits in } K/k \text{ (since } q \text{ splits in } k/\mathbb{Q})$$
$$\iff \left(\frac{A + B\sqrt{m}}{q}\right) = 1.$$

This completes the proof of the theorem. □

Letting $m = 2$ and replacing the quartic subfield of $\mathbb{Q}(\zeta_m)$ used above by the cyclic extension $\mathbb{Q}(\sqrt{2+\sqrt{2}})$ contained in $\mathbb{Q}(\zeta_{16})$ yields the equivalence

$$\left(\frac{A+B\sqrt{2}}{q}\right) = 1 \iff q \text{ splits in } \mathbb{Q}\left(\sqrt{2+\sqrt{2}}\right) \iff q \equiv \pm 1 \bmod 16.$$

Formula (5.9) differs from the one given in [870], which reads

$$\left(\frac{A+B\sqrt{m}}{q}\right) = (-1)^{\frac{q-1}{2}\frac{m-1}{4}} \left(\frac{2}{q}\right)\left(\frac{q}{m}\right)_4, \tag{5.11}$$

where $A, B, C > 0$, B is odd, and C is even. Formula (5.10) shows that

$$\left(\frac{A+B\sqrt{m}}{q}\right) = \left(\frac{2}{q}\right)\left(\frac{A+C\sqrt{m}}{q}\right),$$

and so, for B even and C odd, (5.11) is equivalent to

$$\left(\frac{A+B\sqrt{m}}{q}\right) = (-1)^{\frac{q-1}{2}\frac{m-1}{4}} \left(\frac{q}{m}\right)_4. \tag{5.12}$$

Now $A \equiv 1 \bmod 4$ since $A^2 = m(B^2+C^2)$ is the product of $m \equiv 1 \bmod 4$ and of a sum of two relatively prime squares, and we have $A+B \equiv 1 \bmod 4 \iff 4 | B \iff m \equiv 1 \bmod 8$. The sign of B is irrelevant, therefore

$$\left(\frac{-1}{q}\right)^{B/2} = (-1)^{\frac{q-1}{2}\frac{m-1}{4}}.$$

This finally shows that (5.9) is in fact equivalent to (5.11).

Another version of (5.9) which follows directly from (5.12) is

$$\left(\frac{A+B\sqrt{m}}{q}\right) = \left(\frac{q^*}{m}\right)_4, \tag{5.13}$$

where $A, B > 0$ and $q^* = (-1)^{(q-1)/2} p$.

Formula (5.9) is also valid for composite m as long as its prime factors are quadratic residues modulo q (see Lemmermeyer [514]); this corresponds to Jacobi's extension of the quadratic reciprocity law and can be used for deriving generalizations of the known rational reciprocity laws of Burde [99], E. Lehmer [498, 504, 509] and Scholz [723]. These follow from (5.9) by assigning special values to A and B, in other words: they all stem from the observation that the quartic subfield K of $\mathbb{Q}(\zeta_m)$ can be generated by different square roots over $k = \mathbb{Q}(\sqrt{m})$.

Scholz's Reciprocity Law

Let $p \equiv q \equiv 1 \bmod 4$ be primes such that $\left(\frac{p}{q}\right) = 1$, and let ε_p be the fundamental unit of $\mathbb{Q}(\sqrt{p})$. Let $\varepsilon = \varepsilon_p^3$; then $\varepsilon = t + u\sqrt{p}$ for some $t, u \in \mathbb{Z}$, and $t^2 - pu^2 = -1$. Taking $m = p, A = pu, B = t, C = 1$ in Theorem 5.13 gives

$$\left(\frac{q}{p}\right)_4 = \left(\frac{pu + t\sqrt{p}}{q}\right) = \left(\frac{\sqrt{p}}{q}\right)\left(\frac{\varepsilon}{q}\right) = \left(\frac{p}{q}\right)_4 \left(\frac{\varepsilon_p}{q}\right),$$

from which Scholz's reciprocity law follows by symmetry.

Lehmer's Reciprocity Law

The result of Proposition 5.6 due to Dirichlet is usually referred to as Lehmer's (cf. [498]) reciprocity law; it follows from Theorem 5.13 by taking $m = p, A = p, B = b, C = a$ in 5.13 and observing that $b^2 \lambda^2 \equiv a^2 + b^2 \equiv p \bmod q$ if $q \nmid ab$:

$$\left(\frac{q^*}{p}\right)_4 = \left(\frac{p + b\sqrt{p}}{q}\right) = \left(\frac{b^2\lambda^2 + b^2\lambda}{q}\right) = \left(\frac{\lambda(\lambda + 1)}{q}\right).$$

If $q \mid a$, however, we choose $\sqrt{p} \equiv b \bmod q$ and find

$$\left(\frac{q^*}{p}\right)_4 = \left(\frac{p + b\sqrt{p}}{q}\right) = \left(\frac{b^2 + b^2}{q}\right) = \left(\frac{2}{q}\right),$$

whereas in case $q \mid b$ we get

$$\left(\frac{q^*}{p}\right)_4 = \left(\frac{p + b\sqrt{p}}{q}\right) = \left(\frac{a^2}{q}\right) = 1.$$

Burde's Reciprocity Law

The fact that (5.9) is valid for primes $p \mid ABC$ allows us to derive Burde's reciprocity law in a more direct way than was done in [870]: let p and q be primes $\equiv 1 \bmod 4$ such that $p = a^2 + b^2, q = c^2 + d^2, 2|b, 2|d, (p/q) = 1$, and define $A = pq, B = b(c^2 - d^2) + 2acd, C = a(c^2 - d^2) - 2bcd, m = q$. Then $2|B, B \equiv 2d(ac - bd) \bmod q$ (since $c^2 \equiv -d^2 \bmod q$), the sign of A does not matter (since $q \equiv 1 \bmod 4$), and so formula (5.9) yields

$$\left(\frac{q}{p}\right)_4 = \left(\frac{A + B\sqrt{p}}{q}\right) = \left(\frac{B}{q}\right)\left(\frac{p}{q}\right)_4.$$

Now $\left(\frac{2d}{q}\right) = +1$ (see Proposition 5.2) implies Burde's law

$$\left(\frac{p}{q}\right)_4 \left(\frac{q}{p}\right)_4 = \left(\frac{ac - bd}{q}\right). \tag{5.14}$$

5.5 Residue Characters of Quadratic Units

In Exercise 2.25 we have used quadratic reciprocity to show that primes $p \equiv 1, 9 \mod 20$ divide $U_{(p-1)/2}$, the $\frac{p-1}{2}$-th Fibonacci number. In order to characterize the primes dividing $U_{(p-1)/4}$ we need to study the quadratic residue character of the unit $\varepsilon_5 = \frac{1}{2}(1 + \sqrt{5})$. We also have seen in Section 2.4 that the power character of units like $\varepsilon_3 = 2 + \sqrt{3}$ is connected with the output of the Lucas-Lehmer test for Mersenne primes. In this section (and in the Exercises below) we will barely scratch the surface of the subject of the power character of quadratic units; our results will be proved by completely elementary computations which only indicate that there is more to the picture than meets the eye. In fact, the result below will be proved by entirely different means in Part II, and only then will we be able to see exactly *why* the following is true:

Proposition 5.15. *Let $p \equiv q \equiv 1 \mod 4$ be primes such that $\left(\frac{p}{q}\right) = 1$;*

1. *If $q \equiv 5 \mod 8$, then $k = \mathbb{Q}(\sqrt{-q})$ has class number $h = 2m$ for some odd m; in particular there exist $c, d \in \mathbb{N}$ such that $p^m = c^2 + qd^2$, and we have*

$$\left(\frac{\varepsilon_q}{p}\right) = (-1)^d.$$

2. *If $q \equiv 1 \mod 8$, then $k = \mathbb{Q}(\sqrt{-q})$ has class number $h \equiv 0 \mod 4$; let $\mathfrak{p}$ denote a prime ideal in k above p. Then*

$$\left(\frac{\varepsilon_q}{p}\right) = 1 \iff \mathfrak{p}^{\frac{h}{4}} \text{ is principal in } k.$$

3. *If $p = c^2 + 8d^2 \equiv 1 \mod 8$, then*

$$\left(\frac{1 + \sqrt{2}}{p}\right) = 1 \iff d \equiv 0 \mod 2 \iff h(-4p) \equiv 0 \mod 8,$$

where $h(-4p)$ denotes the class number of $k = \mathbb{Q}(\sqrt{-p})$.

Proof. Suppose that $q \equiv 5 \mod 8$; then genus theory (see Exercise 2.35) implies that $\mathbb{Q}(\sqrt{-q})$ has class number $h = 2m$ for some odd $m \in \mathbb{N}$. Moreover, $\left(\frac{p}{q}\right) = 1$ implies that the ideal class of the prime ideal $\mathfrak{p}$ above p is a square, i.e., that $\mathfrak{p}^m = (c + d\sqrt{-q})$ is principal. Taking the norm yields $p^m = c^2 + qd^2$, and applying Proposition 5.12 with $s = p^{(m-1)/2}$ proves our assertion.

The case $q \equiv 1 \mod 8$ is a bit harder: we know that the 2-Sylow subgroup of $\text{Cl}(k)$ is cyclic. Let $\mathfrak{r}$ be a prime ideal above a prime r satisfying $\left(\frac{-1}{r}\right) = \left(\frac{-q}{r}\right) = -1$. Then Proposition 2.19 shows that the ideal class $[\mathfrak{r}]$ is no square, and since $\text{Cl}_2(k)$ is cyclic, it must generate $\text{Cl}_2(k)$.

5.5 Residue Characters of Quadratic Units

Now let $p \equiv 1 \bmod 4$ be a prime such that $(p/q) = 1$; then the ideal class generated by a prime ideal $\mathfrak{p}$ above p is a square, hence there exist $t \in \mathbb{N}$ and an odd $m \in \mathbb{N}$ such that $\mathfrak{r}^{2t}\mathfrak{p}^m = (c + d\sqrt{-q})$ is principal. Applying Proposition 5.12 with $s = r^t p^{(m-1)/2}$ yields

$$\left(\frac{\varepsilon_q}{p}\right) = \left(\frac{r}{q}\right)^t = \left(\frac{q}{r}\right)^t = (-1)^t,$$

because $\left(\frac{q}{r}\right) = -\left(\frac{-q}{r}\right) = -1$. Therefore,

$$\left(\frac{\varepsilon_q}{p}\right) = 1 \iff t = 2u \text{ is even}$$
$$\iff [\mathfrak{p}]^m \text{ is a fourth power}$$
$$\iff [\mathfrak{p}]^{h/4} \text{ is principal in } k.$$

Finally, assume that $p = 8n + 1 = c^2 + 8d^2 = e^2 - 2f^2$ is prime, and that $0 < e \equiv 1 \bmod 4$. The supplementary law to Scholz's reciprocity law and Proposition 5.4 give

$$\left(\frac{1+\sqrt{2}}{p}\right) = \left(\frac{2}{p}\right)_4 \left(\frac{p}{2}\right)_4 = (-1)^{n+d}(-1)^n = (-1)^d.$$

From Exercise 5.21 we know that d is even if and only if $f \equiv 0 \bmod 4$. Write $E = e + 2f$ and $F = e + f$; then $p = 2F^2 - E^2$, and we claim that the ideal $\mathfrak{b} = (F, E + \sqrt{-p})$ of norm F has order 4 in $\mathrm{Cl}\,(k)$. This follows from the factorization $(E + \sqrt{-p}) = (2, 1 + \sqrt{-p})\mathfrak{b}^2$, because $(2, 1 + \sqrt{-p})$ generates the ideal class of order 2 in $\mathrm{Cl}\,(k)$. Now $8 \mid h(-4p)$ if and only if the ideal class of $\mathfrak{b}$ is a square, and by genus theory this is the case if and only if $\left(\frac{-1}{F}\right) = +1$. But now $F = e + f \equiv 1 \bmod 4$ if and only if $f \equiv 0 \bmod 4$, and the proof is complete. □

We note the following two corollaries:

Corollary 5.16. *If $q = 5, 8, 13, 37$, then every prime $p \equiv 1 \bmod 4$ such that $\left(\frac{p}{q}\right) = 1$ has the form $p = c^2 + qd^2$, and*

$$\left(\frac{\varepsilon_q}{p}\right) = (-1)^d.$$

We remark that these values of q are the only sums of two squares such that the biquadratic field $K = \mathbb{Q}(\sqrt{-1}, \sqrt{q})$ has class number 1. In fact, since K has class number 1 and is a quadratic unramified extension of $k = \mathbb{Q}(\sqrt{-q})$, K must contain the *Hilbert class field* of k; this is only possible if k has class number 2, and the complex quadratic fields with small class number have all been determined.

Corollary 5.17. *Let $q \equiv 1 \bmod 8$ be a prime not of the form $x^2 + 32y^2$; assume that $p \equiv 1 \bmod 4$ is prime, $\left(\frac{p}{q}\right) = 1$, and let h denote the class number of $k = \mathbb{Q}(\sqrt{-q}\,)$. Then $h \equiv 4 \bmod 8$, and we have*

$$\left(\frac{\varepsilon_q}{p}\right) = \begin{cases} +1, & \text{if } p^{h/4} = c^2 + qd^2, \\ -1, & \text{if } 2p^{h/4} = c^2 + qd^2. \end{cases}$$

NOTES

Gauss [272] considered even the quadratic reciprocity law in $\mathbb{Z}[i]$ to be quite deep:

> Trotz der großen Einfachheit des Satzes aber unterliegt sein Beweis sehr großen Schwierigkeiten, mit denen wir uns hier jedoch nicht aufhalten, da der Satz selbst nur ein spezieller Fall eines allgemeineren Satzes ist, welcher die ganze Theorie der biquadratischen Reste in sich enthält.[1]

As we already have remarked, it was Dirichlet [166] who showed that it could be derived easily from the quadratic reciprocity law in $\mathbb{Z}$. Other presentations of Dirichlet's proof can be found in Dirichlet's paper [169], in Aubry [23], and in Bianchi's book [51]; see also his article [50] as well as Arwin [19], Busche [109], Scheibner [708], Tschebotaröw [801], Tsunekawa [803], and Western [835].

Dirichlet's ideas were used to generalize the quadratic reciprocity law to other and even to general quadratic number fields in the papers of Bonaventura [66], Dörrie [179], Welmin [834] and Tsunekawa [802]. Proofs relying on genus theory of $\mathbb{Z}[i]$ (first developed by Minnigerode [Mi] and Hilbert [367]; see also Brandt [80]) were given by Smith [744], Brandt [78], and Louboutin [556]. Jacobi derived it from the biquadratic reciprocity law in his lecture [402]; here's Jacobi's ([402], lecture 41) opinion on Dirichlet's paper [166]:

> Im dritten Band des Crelleschen Journals hat Dirichlet zuerst allgemeine Sätze über die biquadratischen Reste mitgetheilt. Sein Beweis, der ziemlich zusammengesetzt, aber eins der scharfsinnigsten Meisterwerke ist, stützt sich hauptsächlich auf eine vielfache und geschickte Anwendung des Gesetzes der Reziprozität, und namentlich auf einen Satz von Legendre, der entscheidet, wann die Gleichung $ax^2 + by^2 + cz^2 = 0$ in ganzen Zahlen aufgelöst werden kann, wenn a, b, c gegebene ganze Zahlen sind.[2]

---

[1] In spite of the great simplicity of the theorem its proof presents considerable difficulties which we will ignore here, since the theorem itself is only a special case of a more general theorem which contains the whole theory of biquadratic residues.

[2] In the third volume of Crelle's Journal, Dirichlet presented for the first time general theorems on biquadratic residues. His proof, which is quite complex but

The simple derivation of the biquadratic character of 2 in Proposition 5.3 is well known and due to Dirichlet; however, in his letter [397] to Gauss, Jacobi not only claimed to possess a proof for this result based on his own methods but also stated a result that is easily seen to be equivalent to Proposition 5.6.

Theorem 5.5 is due to Dirichlet [162] and was rediscovered by Aigner [5] and E. Lehmer [498]. Apart from the fact that Dirichlet used the representation $p^h = e^2 - q^* f^2$ instead of $e^2 = pf^2 + qg^2$, his proof is the same as the one we have presented; a similar technique is used in the book of Venkov [813]. Jacobi [402] derives Dirichlet's theorems from his results on biquadratic residues. Cunningham [144] gave expressions for $(p/q)_4(q/p)_4$ in terms of representations $pm^2 = t^2 \pm qu^2$, but failed to find Burde's reciprocity theorem (see also the followup [145]). Parts of Theorem 5.7 are proved in Burde [101].

There are in fact a few other results in Dirichlet's papers which apparently have been overlooked for a long time. The following is a translation of his results published in [167] into the ideal theoretic language.

Suppose that $q \equiv 1 \bmod 4$ is a prime. According to the theory of genera, $k = \mathbb{Q}(\sqrt{-q})$ has a cyclic 2-class group. Moreover, since the ideal $\mathfrak{b} = (2, 1 + \sqrt{-q})$ generates an ideal class $c = [\mathfrak{b}]$ of order 2, $h(k)$ will be divisible by 4 if and only if c is a square. By genus theory, c is a square if and only if 2 splits in the genus class field $K = k(\sqrt{q})$, i.e. if and only if $q \equiv 1 \bmod 8$. If c is a square, then any prime ideal $\mathfrak{p}$ such that $c = [\mathfrak{p}]$ must have norm $\equiv 1 \bmod 4$, since (again by genus theory) it must split in the genus class field $K = k(\sqrt{-1})$.

From now on we assume that indeed $q \equiv 1 \bmod 8$, and we write $q = a^2 + b^2$, where $a \equiv 1 \bmod 4$. For every ideal class c which is a square Dirichlet shows that the value of $\Delta(c) = \left(\frac{q}{p}\right)_4 \left(\frac{p}{q}\right)_4$ (where $\mathfrak{p} \in c$ is a prime ideal of norm $p \equiv 1 \bmod 4$) is an invariant of the ideal class c and does not depend on the choice of generators $\mathfrak{p}$. Moreover he proves that the class of principal ideals has invariant 1.

Next he calls ideal classes c and c' conjugated if $cc' = [\mathfrak{b}]$, and shows that

$$\Delta(c) = \Delta(c') \quad \Longleftrightarrow \quad \left(\frac{2}{a+b}\right) = 1 \quad \Longleftrightarrow \quad \left(\frac{-4}{q}\right)_8 = 1.$$

Finally Dirichlet studies ideal classes c which coincide with their conjugates, i.e. which satisfy $cc' = [\mathfrak{b}]$ and $c = c'$. It is easy to see that these are exactly the ideal classes of order 4; Dirichlet shows that $\chi_{-4}(c) = 1$ if and only if $a + b \equiv 1 \bmod 8$. Gauss's genus theory implies

5.18. *Let $q = a^2 + b^2 \equiv 1 \bmod 8$ be a prime, and let h denote the class number of $\mathbb{Q}(\sqrt{-q})$. Then*

---

at the same time a shrewd masterpiece, is to a large part based on numerous and clever applications of the reciprocity law, and in particular of a theorem of Legendre which decides when the equation $ax^2 + by^2 + cz^2 = 0$ can be solved in integers, where a, b, c are given integers.

$$h \equiv 0 \bmod 8 \quad \Longleftrightarrow \quad \left(\frac{2}{a+b}\right) = 1 \quad \Longleftrightarrow \quad \left(\frac{-4}{q}\right)_8 = 1.$$

So what Dirichlet actually did was define a rational quartic character on the 2-class group of $\mathbb{Q}(\sqrt{-q})$; later Dirichlet's methods were subsumed into the theory of *spinor genera* by Estes & Pall [209]; see also Halter-Koch [330].

Another special case of his results (which, as Dirichlet noticed, generalizes to all fields $\mathbb{Q}(\sqrt{-q})$ with 2-class group $\simeq C_4$) is

5.19. *Let $p \equiv 1 \bmod 4$ be a prime such that $\left(\frac{17}{p}\right) = 1$; then*

$$p = x^2 + 17y^2 \Longleftrightarrow \left(\frac{17}{p}\right)_4 \left(\frac{p}{17}\right)_4 = +1$$

$$2p = x^2 + 17y^2 \Longleftrightarrow \left(\frac{17}{p}\right)_4 \left(\frac{p}{17}\right)_4 = -1$$

This special case of Corollary 5.17 was rediscovered much later by Brandler [74], and has been generalized to $a = 17, 73, 97$ and 193 by Brown [87] and E. Lehmer [504]. Later Kaplan, K.S. Williams & Yamamoto [424] gave a short proof of these and related results using class field theory.

The explosion of interest in rational reciprocity laws that we have witnessed since the 1970's was caused mainly by the two papers of Barrucand & Cohn [34] and Burde [99]. Actually a rational reciprocity law equivalent to Burde's was already given by T. Gosset [318], who showed that, for primes p and q as above,

$$\left(\frac{q}{p}\right)_4 \equiv \left(\frac{a/b - c/d}{a/b + c/d}\right)^{(q-1)/4} \bmod q. \tag{5.15}$$

Multiplying the numerator and denominator of the term on the right side of (5.15) by $a/b + c/d$ and observing that $c^2/d^2 \equiv -1 \bmod q$ yields

$$\left(\frac{q}{p}\right)_4 \equiv \left(\frac{a^2/b^2 + 1}{q}\right)_4 \left(\frac{a/b + c/d}{q}\right) = \left(\frac{p}{q}\right)_4 \left(\frac{b}{q}\right) \left(\frac{a/b + c/d}{q}\right)$$

$$= \left(\frac{p}{q}\right)_4 \left(\frac{a + bc/d}{q}\right) = \left(\frac{p}{q}\right)_4 \left(\frac{d}{q}\right) \left(\frac{ad + bc}{q}\right) \bmod q,$$

which is Burde's reciprocity law since $\left(\frac{2d}{q}\right) = +1$.

A more explicit form of Burde's reciprocity law for composite values of p and q has been given by L. Rédei [671]; let $\pi_1, \ldots, \pi_r$ denote different primes $\equiv 1 \bmod (2-2i)$ in $\mathbb{Z}[i]$ with norms $p_j = \pi_j \overline{\pi_j}$, and put $\nu = \pi_1 \cdot \ldots \cdot \pi_r = A + Bi$ and $n = p_1 \cdot \ldots \cdot p_r$. For coprime Gaussian integers $\alpha \equiv \beta \equiv 1 \bmod (2-2i)$ we define a symbol

$$(\alpha, \beta)_4 = \left[\frac{N\alpha}{\beta}\right]_4 \left[\frac{N\beta}{\alpha}\right]_4^{-1},$$

where $\left[\frac{\cdot}{\cdot}\right]_4$ denotes the quartic residue symbol which we will study in detail in Chapter 6. Then for any factorization $\nu = \delta \cdot \delta'$ and the corresponding factorization $n = d \cdot d'$, Rédei proves that $(\delta, \delta')_4 = (A/d)$. In the special case where ν is the product of just two primes $\delta_1 = a + bi$ and $\delta_2 = c + di$, we get $p_1 p_2 = A^2 + B^2$ with $A = ac - bd$, and if we assume moreover that $(p_1/p_2) = +1$, then we get (5.14).

Yet another version of Burde's law is due to A. Fröhlich [244]; he showed

$$\left(\frac{p}{q}\right)_4 \left(\frac{q}{p}\right)_4 = \left(\frac{a+bj}{q}\right) = \left(\frac{c+di}{p}\right), \qquad (5.16)$$

where i and j denote rational numbers such that $i^2 \equiv -1 \bmod p$ and $j^2 \equiv -1 \bmod q$. Letting $i = a/b$ and $j = c/d$ and observing that $\left(\frac{a}{p}\right) = \left(\frac{c}{q}\right) = +1$, we find that (5.16) is equivalent to (5.14). Later, K.S. Williams [865], Helou [354] and Sun [774] gave other proofs of Burde's reciprocity law; see Brown [88], Burde [100] and Kaplan [421] for generalizations.

A theorem equivalent to 5.13 has been published by K. S. Williams, K. Hardy & C. Friesen [870] in 1985. A simplified (and generalized) version was later given by R. Evans [223]. The proof we have presented is taken from Lemmermeyer [512]; see [514] for a more general version.

Similarly, the quadratic character of ε_2 (and its connection with Dirichlet's result 5.18) studied by Barrucand & Cohn [34] had already been determined by Aigner & Reichardt [9]; they even proved a result from which the quartic character of ε_2 can be deduced. Moreover, the octic character of ε_2 was known from results of F. Goldscheider [315] on the octic reciprocity law, which we will present in Chapter 9; it is of course also possible to derive this result from the explicit formulas of Hasse (see Part II).

Nevertheless, the papers [34] and [99] were highly influential, as e.g. Emma Lehmer's account of her discovery of Scholz's reciprocity law [509, p. 470] shows:

> In 1969 Barrucand and Cohn [34] proved that $\varepsilon_2 = 1 + \sqrt{2}$ is a quadratic residue of p if and only if $p = c^2 + 32d^2$. Jacob Brandler [73] showed that $\varepsilon_5 = \frac{1+\sqrt{5}}{2}$ and $\varepsilon_{13} = \frac{3+\sqrt{13}}{2}$ are quadratic residues of $p = c^2 + qd^2$ if and only if d is even, and that $\varepsilon_{17} = 4 + \sqrt{17}$ is always a quadratic residue of p. Comparing this with [Proposition 5.12] it was not hard to conjecture that $\left(\frac{\varepsilon_q}{p}\right) = \left(\frac{p}{q}\right)_4 \left(\frac{q}{p}\right)_4$ if $\left(\frac{p}{q}\right) = 1$, from which it would follow from symmetry that $\left(\frac{\varepsilon_q}{p}\right) = \left(\frac{\varepsilon_p}{q}\right)$. I again combed the literature and asked my friends, but nobody knew whether this elegant result was true or false.

In fact, in his article [74] (submitted in May 1971), Brandler calls Scholz's reciprocity law a "theorem due to Mrs. E. Lehmer". The proof given in Section 5.3 is a simplified version of Lehmer's own proof [503]. It should also be noted that V. Ennola's remark in his review (MR 53 # 342) of Brandler's article [75]

In the reviewers opinion there is a point that needs clarification: Since the lemma only holds for a particular choice of μ among its associates, one should show that these choices fit together in the proof of (8) also applies to [74]. A correct proof along Brandler's lines is offered in Exercise 5.9. The element $a + bi$ such that $(a + bi)\varepsilon_m$ is a square constructed there also occurs in Vazzana's construction [Va1] of real quadratic number fields F such that $K_2\mathcal{O}_F$ has trivial 4-rank. For similar connections between power residues and $K_2\mathcal{O}_F$ see Vazzana [Va2] and Hurrelbrink & Kolster [HK].

More proofs of Scholz's reciprocity law were given by Furuta [258], Sun [774] and K. S. Williams [864] (this proof also contains a gap – cf. Exercise 5.10 for the gap and how to close it).

As a matter of fact, the quadratic character of $\varepsilon_5 = \frac{1}{2}(1 + \sqrt{5})$ had been determined previously by Vandiver [809, 810], Ward [818] (implicitly), Rédei [674] (who did not refer to Scholz's work, of which it is a special case) and E. Lehmer [501] (see also Aigner [7], Massell [569], and Ward [819]).

A short history of rational reciprocity can be found in E. Lehmer's (somewhat harsh) answer [509] to Wyman's question "What is a reciprocity law" in [875]. Proofs for the elementary rational reciprocity laws are collected in Vögeli-Fandel [814].

Of all the rational reciprocity laws given in this chapter, Scholz's reciprocity law is probably the most important: not only does it give connections between the splitting of primes, the norm of the fundamental unit in $k_{pq} = \mathbb{Q}(\sqrt{pq})$, and the class number of k_{pq}, it can also be generalized from a law concerning certain quadratic extensions of $\mathbb{Q}$ to one about quadratic extensions of a number field F with odd class number (see Part II). It has not been noticed up to now that Scholz's reciprocity law is much older: as a matter of fact, the statement $(\varepsilon_p/q) = (p/q)_4(q/p)_4$ and its corollary $(\varepsilon_p/q)(\varepsilon_q/p) = 1$ as well as proofs based on Dirichlet's techniques can be found in Schönemann [722].

Finding rational criteria for the n-th power character of small integers was quite popular over the years: Gauss [272] and Dirichlet [172, 173, 163], proved Euler's conjecture on primes for which 2 is a quartic residue. Other elementary calculations of the quartic character of 2 were given e.g. by Halphén [324], Hensel [359], Lebesgue [483], Mordell [590] and Pépin [642, 644]. For more, see the Notes of Chapter 6.7,9

The first part of Proposition 5.15 was found by Leonard & Williams [523], the second part is due to Kramer [437], who thereby disproved a conjecture of Brandler [73, 74]. Corollary 5.16 is due to E. Lehmer [504], but special cases ($q = 5, 13$) have been proved before by Brandler [74]. The third part was, as we have already remarked, proved by Aigner & Reichardt, Barrucand & Cohn, as well as Goldscheider. For yet another proof see Williams [860].

Other papers dealing with the power character of quadratic units are Buell, Leonard, & K. S. Williams [94], Chuman & Ishii [131], Evans [225], Furuta & Kaplan [259], Halter-Koch [329], Halter-Koch & Ishii [332], Hettkamp [366], Ishii [388, 389, 390, 391], E. Lehmer [505, 506, 507], Leonard & K.S. Williams [521, 524], Morton [592], Weinberger [831], and K.S. Williams [866, 867].

Most of these papers are related to class field theory and will be treated more carefully in Part II; the same goes for the many investigations of residuacity and representations of primes by binary quadratic forms, such as Bernardi [43, 44, 45], Halter-Koch [326, 327, 328, 331], Halter-Koch, Kaplan & K.S. Williams [333], Kaplan & K.S. Williams [423], Kaplan, K.S. Williams & Y. Yamamoto [424], Leonard & K.S. Williams [520, 525, 526], and Muskat [602]. These representations are often connected with the decomposition laws of primes in certain non-abelian solvable (especially dihedral) extensions; see e.g. Furuta & Kubota [260], Halter-Koch [330], Kuroda [472], Rédei [672], and Satgé [697, 698].

The problem of determining necessary or sufficient criteria for the fundamental unit of a real quadratic field to have negative (or positive) norm was quite attractive over the years; see e.g. Beach & H. Williams [40], Buell [92], Dirichlet [Dir], Epstein [207], von Lienen [vL], Perott [650], Rédei [671, Re3, Re4], Sansone [695, 696], and Williams & Zarnke [848].

Rational reciprocity laws enjoyed a renaissance recently in connection with the rank of elliptic curves, or rather, their Tate-Shafarevich group. In fact, the method of descent by a 2-isogeny leads to diophantine equations of type $z^2 = ax^4 + bx^2y^2 + cy^4$ first studied by Euler [Eul], Pépin [Pe1] and others (see vol II. of Dickson's History), and then by e.g. Billing [Bil], Brown [Bro], J. Lagrange [Lag], Lind [548], Mordell [590], Pocklington [Poc], Rédei [Re3, Re4], Reichardt [R1], Rose [Ro], and Stroeker & Top [ST]. The solubility of such equations is intimately connected with rational reciprocity and the structure of the 2-class groups of quadratic number fields. The papers of Bölling [Boe], Cassels [Cas], Kramer [Kra], and McGuinness [McG] use reciprocity to construct Tate-Shafarevich groups of elliptic curves with arbitrarily large 2- and 3-ranks.

Exercises

5.1 Prove Proposition 5.10.

5.2 Let $M = \{\mu \in \mathbb{Z}/q\mathbb{Z} \mid \mu^2 + 1 \equiv \lambda^2 \bmod q, \left(\frac{\lambda(\lambda+1)}{q}\right) = 1\}$; show that $1 \in M$ if and only if $q \equiv 1 \bmod 16$.

5.3 Let $P = \prod p_j$ be a product of primes $p_j \equiv 1 \bmod 8$, and write $P = A^2 + B^2 = C^2 + 2D^2 = E^2 - 2F^2 = 8N + 1$ with B even. Define the biquadratic residue symbol $(2/P)_4$ via $(2/P)_4 = \prod (2/p_j)_4$. Show that $(2/P)_4 = (-1)^{B/4} = (2/C) = (-1)^{N+D} = (-2/E)$, independent of the choice of $A, B, \ldots, F$.

5.4 Let $\varepsilon = \frac{1}{2}(1 + \sqrt{5})$ be the fundamental unit of $k = \mathbb{Q}(\sqrt{5})$. the prime ideals above 11 are generated by $\pi = 4 + \sqrt{5}$ and $\pi' = 4 - \sqrt{5}$; show that $[\varepsilon/\pi] = 1$

176   5. Rational Reciprocity Laws

and $[\varepsilon/\pi'] = -1$. Show that the residue symbols for the primes above 29 both equal $+1$.

5.5 Let $p \equiv q \equiv 1 \bmod 4$ be primes such that $\left(\frac{p}{q}\right) = 1$; then $e^2 = pf^2 + qg^2$ has non trivial solutions in $\mathbb{Z}$, and

$$\left(\frac{p}{q}\right)_4 \left(\frac{q}{p}\right)_4 = (-1)^{fg/2} \left(\frac{-1}{e}\right).$$

Derive this reciprocity law from the proof of Theorem 5.5.

5.6 (Brown [86]) Let $p \equiv q \equiv 1 \bmod 4$ be primes such that $p = r^2 + qs^2$ for integers $r, s \in \mathbb{N}$. Show that

$$\left(\frac{p}{q}\right)_4 \left(\frac{q}{p}\right)_4 = \left(\frac{2}{q}\right)^s.$$

5.7 (Lemmermeyer [513]) Let $p \equiv q \equiv 1 \bmod 4$ be primes such that $\left(\frac{p}{q}\right) = 1$; let $(x, y, z) \in \mathbb{Z}^3$ be a primitive solution of $x^2 - py^2 = qz^2$ such that $2 \mid y$, and put $\alpha = x + y\sqrt{p}$ and $k = \mathbb{Q}(\sqrt{pq})$. Replacing x by $-x$ if necessary we can make $k(\sqrt{\alpha})/k$ unramified outside ∞. Show that $\alpha \gg 0$ if and only if $(p/q)_4 (q/p)_4 = 1$. What happens if you replace p and q by arbitrary positive discriminants?

5.8 Let k be a real quadratic number field, and let $d = d_1 d_2 d_3 d_4$ be the factorization of d into prime discriminants; assume that $d_1, d_2 > 0$ and $d_3, d_4 < 0$. Show that $d_1 d_2 X^2 - d_3 d_4 Y^2 = 1$ can only have solutions in integers if $(d_3 d_4/d_1)_4 = (d_3 d_4/d_2)_4$.

5.9 Here is Brandler's [74] proof of Scholz's reciprocity law. Let $m \equiv 1 \bmod 4$ be a square-free integer, and assume that $N\varepsilon_m = -1$.
   1. Let $x, y \in \mathbb{N}$ be integers such that $x^2 - my^2 = -1$. Then there exist $\alpha, \mu \in \mathbb{Z}[i]$ such that $y = \alpha \bar{\alpha}$, $2x = \mu \alpha^2 + \bar{\mu} \bar{\alpha}^2$, $m = \mu \bar{\mu}$, where $\mu \equiv i \bmod 2$. (Hint: $(x + i)(x - i) = x^2 + 1 = my^2$.)
   2. There exist integers $a - 1 \equiv b \equiv 0 \bmod 2$, and $\xi \in \mathcal{O}_K$, $K = \mathbb{Q}(i, \sqrt{m})$ such that $a^2 + b^2 = m$ and $(a + bi)\varepsilon_m = \xi^2$, where $\varepsilon_m = x + y\sqrt{m}$. (Hint: $2\mu\varepsilon = (\mu\alpha + \bar{\alpha}\sqrt{m})^2$.)
   3. Let $q \equiv 1 \bmod 4$ be prime, and write $q = c^2 + d^2$ for $c, d \in \mathbb{Z}, d \equiv 0 \bmod 2$. Then

$$\left(\frac{\varepsilon_m}{q}\right) = \left(\frac{\varepsilon_m}{\mathfrak{q}}\right)_K = \left(\frac{a + bi}{\mathfrak{q}}\right)_K = \left(\frac{a + bi}{q}\right),$$

where $\left(\frac{\cdot}{\cdot}\right)_K$ and $\left[\frac{\cdot}{\cdot}\right]$ denote the quadratic residue symbol in $\mathcal{O}_K$ and $\mathbb{Z}[i]$, respectively, and where $\mathfrak{q}$ is a prime ideal above q in $\mathcal{O}_K$.
Now choose $m = p$ prime and deduce Scholz's reciprocity law from the quadratic reciprocity law in $\mathbb{Z}[i]$.
Remark. Part (1.) can already be found in Epstein's paper [207].

5.10 In the following we will correct K. Williams' [864] proof of Scholz's reciprocity law. Let $p \equiv q \equiv 1 \bmod 4$ be two different primes such that $(p/q) = +1$, let $\varepsilon = \varepsilon_q$ be the fundamental unit of $\mathbb{Q}(\sqrt{q})$, and let T, U be positive integers such that $\varepsilon^3 = T + U\sqrt{q}$. Let h denote the class number of $\mathbb{Q}(\sqrt{q})$; show that there exist integers $u, v \in \mathbb{N}$ such that $p^{3h} = u^2 - 4qv^2$. Now show that

$$\left(\frac{\varepsilon}{p}\right) = \left(\frac{2v}{p}\right)\left(\frac{Uu+2Tv}{p}\right).$$

If we would invert this symbol now (as Williams does) and raise it to the $3h^{th}$ power, then we would get

$$\left(\frac{\varepsilon}{p}\right) = \left(\frac{2v}{p}\right)\left(\frac{u^2-4qv^2}{Uu+2Tv}\right),$$

and now we would like to multiply this with $(U^2/Uu+2Tv)$. Unfortunately, $(U, Uu+2Tv) = (U,v)$ is in general not equal to 1, and then multiplication by $(U^2/Uu+2Tv)$ is multiplication by 0 (this is the gap in Williams' proof). In order to avoid this, write $t = (U,v)$ with $t > 0$, $U = tV$, $v = tw$ and $Uu + 2Tv = tW$. Then show that $t \equiv 1 \bmod 4$, $(t/p) = 1$ and

$$\left(\frac{\varepsilon}{p}\right) = \left(\frac{2v}{p}\right)\left(\frac{t}{p}\right)\left(\frac{W}{p}\right).$$

Proceeding as above show that $(V, W) = 1$ and

$$\left(\frac{\varepsilon}{p}\right) = \left(\frac{2v}{p}\right)\left(\frac{u^2-4qv^2}{W}\right) = \left(\frac{2v}{p}\right)\left(\frac{V^2u^2-4qV^2v^2}{W}\right)$$
$$= \left(\frac{2v}{p}\right)\left(\frac{4T^2w^2-4qU^2w^2}{W}\right) = \left(\frac{2v}{p}\right)\left(\frac{-1}{W}\right)$$

(note that $V^2v^2 = U^2w^2$). Since $tW = Uu + 2Tv$ and $U \equiv 1 \bmod 4$, we get $tW \equiv u \bmod 4$ and $(-1/W) = (-1/t)(-1/u) = (-1/u)$. The rest of the proof is as in Williams' paper.

5.11 (Furuta [258]) Let $m = p_1 \cdots p_t$, suppose that the p_j and q are primes $\equiv 1 \bmod 4$ such that $(p_j/q) = 1$ for $1 \leq j \leq t$, and assume that $N\varepsilon_m = -1$. Deduce from the proof of the preceding exercise that

$$\left(\frac{\varepsilon_m}{q}\right) = \prod_{j=1}^{t}\left(\frac{\varepsilon_j}{q}\right),$$

where ε_j is the fundamental unit in $\mathbb{Q}(\sqrt{p_j})$.

5.12 (H. C. Williams [847]) Let m be a product of primes $p_j \equiv 1 \bmod 4$, and let $q \equiv 3 \bmod 4$ be a prime such that $(p_j/q) = +1$ for all p_j. Moreover, let ε_m denote the fundamental unit of $k = \mathbb{Q}(\sqrt{m})$, and put $\eta = \varepsilon_m\sqrt{m}$.
  1. Show that the quadratic residue symbol (η/q) is well defined (although (ε_m/q) and $(m/q)_4$ are not);
  2. Prove that $\left(\frac{\eta}{q}\right) = \left(\frac{-q}{m}\right)_4$, where the symbol $(\cdot/\cdot)_4$ is the "biquadratic Jacobi symbol".

5.13 Let $p = a^2 + b^2$ and $q = c^2 + d^2$ be primes $\equiv 1 \bmod 4$; show that

$$\left(\frac{ac+bd}{p}\right) = \left(\frac{p}{q}\right)\left(\frac{ac-bd}{p}\right).$$

5.14 Let $p = a^2 + 16b^2 \equiv 1 \bmod 8$ be prime; derive $(2/p)_4 = (-1)^b$ from Burde's reciprocity law.

178   5. Rational Reciprocity Laws

5.15 Let $q \equiv 3 \bmod 4$ be prime; since the prime ideal above 2 in $\mathbb{Z}[\sqrt{q}]$ is principal, there exist $r, s \in \mathbb{Z}$ such that $r^2 - qs^2 = 2\epsilon$, for some $\epsilon = \pm 1$.
   1. Show that $\epsilon = \left(\frac{2}{q}\right)$;
   2. Assume that $p \equiv 1 \bmod 8$ is a prime such that $\left(\frac{p}{q}\right) = 1$; assume moreover that $t^2 p^h = e^2 - qf^2$ for some $t, h, e, f \in \mathbb{N}$, where $(t, rs) = (t, ef) = 1$, and where $t \equiv h \equiv e \equiv 1 \bmod 2$. Then
   $$\left(\frac{r+s\sqrt{q}}{p}\right) = \left(\frac{2\epsilon}{e+f}\right).$$
   3. Write $p = a^2 + b^2$ and $p^k = c^2 + qd^2$, and assume that $a \equiv k \equiv c \equiv 1 \bmod 2$. Then
   $$\left(\frac{2\epsilon}{e+f}\right) = (-1)^{(b+d)/4}.$$
   4. Let ε_q denote the fundamental unit of $\mathbb{Z}[\sqrt{q}]$; show that
   $$\left(\frac{\varepsilon_q}{p}\right)_4 = (-1)^{d/4}.$$

5.16 Suppose that $a^2 - mb^2 = -2$ is solvable in integers for some $m \equiv 3 \bmod 4$; let (a, b) be a solution and define $\eta_m = a + b\sqrt{m}$.
   1. Show that there exist $\mu, \alpha \in \mathbb{Z}[\sqrt{-2}]$ such that $2a = \mu\alpha^2 + \overline{\mu}\overline{\alpha}^2$, $b = \alpha\overline{\alpha}$, $m = \mu\overline{\mu}$;
   2. Verify that $2\mu\eta_m = (\mu\alpha + \overline{\alpha}\sqrt{m})^2$;
   3. Conclude that, for primes $p \equiv 1 \bmod 8$ with $\left(\frac{m}{p}\right) = 1$,
   $$\left(\frac{\eta_m}{p}\right) = \left(\frac{\mu}{p}\right), \quad \text{and} \quad \left(\frac{\varepsilon_m}{p}\right)_4 = \left(\frac{\mu}{p}\right)\left(\frac{2}{p}\right)_4,$$
   where ε_m is the fundamental unit in $\mathbb{Z}[\sqrt{m}]$.

5.17 (continued) Replace m by $2m$ in the last exercise and prove
$$\left(\frac{\varepsilon_{2m}}{p}\right)_4 = \left(\frac{\mu}{p}\right).$$
Conclude that, for primes $m = q \equiv 3 \bmod 8$,
$$\left(\frac{\varepsilon_{2q}}{p}\right)_4 = \left(\frac{2}{p}\right)_4 \left(\frac{\varepsilon_q}{p}\right)_4.$$
Moreover, show that $-2\varepsilon_q \varepsilon_{2q} = \xi^4$ for some $\xi \in \mathbb{Q}(\sqrt{-1}, \sqrt{2}, \sqrt{q})$.

5.18 Suppose that $a^2 - mb^2 = +2$ is solvable in integers for some $m \equiv 3 \bmod 4$; show that the proof of Exercise 16 yields
$$\left(\frac{\varepsilon_m}{p}\right)_4 = \left(\frac{\mu}{p}\right)\left(\frac{2}{p}\right)_4,$$
where $p \equiv 1 \bmod 8$ is a prime such that $\left(\frac{m}{p}\right) = 1$. Similarly, for primes $q \equiv 7 \bmod 8$ one finds that $2\varepsilon_2^2 \varepsilon_q \varepsilon_{2q} = \xi^4$ for some integral $\xi \in \mathbb{Q}(\sqrt{2}, \sqrt{q})$, and

$$\left(\frac{\varepsilon_{2q}}{p}\right)_4 = \left(\frac{-1}{p}\right)_8 \left(\frac{\varepsilon_q}{p}\right)_4.$$

(Hint: use the fact that $\sqrt{\sqrt{2}\varepsilon_2} = \sqrt{2+\sqrt{2}}$ generates the maximal real subfield of $\mathbb{Q}(\zeta_{16})$).

5.19 Let $p = c^2 + 2d^2 \equiv q = e^2 + 2f^2 \equiv 1 \bmod 8$ be primes such that $\left(\frac{p}{q}\right) = 1$. Then there exist η_p and η_q in the maximal orders of $\mathbb{Q}(\sqrt{p})$ and $\mathbb{Q}(\sqrt{q})$, respectively, such that $N(\eta_p) = N(\eta_q) = -2$. Setting $\mu_p = c + d\sqrt{-2}$ and $\mu_q = e + f\sqrt{-2}$, use the preceding exercises to show that

$$\left(\frac{\eta_p}{q}\right) = \left(\frac{\mu_p}{q}\right) = \left(\frac{cf-de}{q}\right) = \left(\frac{cf-de}{p}\right) = \left(\frac{\mu_q}{p}\right) = \left(\frac{\eta_q}{p}\right).$$

Observe the resemblance to Scholz's reciprocity law.

5.20 Let $p \equiv 1 \bmod 4$ and $r \equiv 3 \bmod 4$ be primes such that $(p/r) = +1$. Consider a solution of $x^2 + ry^2 = pz^2$ with $2 \mid y$, $2 \nmid xz$, and put $\pi = x + y\sqrt{-r}$, $\pi' = x - y\sqrt{-r}$. Show that $(\pi/\pi') = (-r/p)_4$.

5.21 Let $\zeta = \frac{1}{2}(\sqrt{2} + i\sqrt{2})$ denote a primitive eighth root of unity, and put $K = \mathbb{Q}(\zeta) = \mathbb{Q}(i, \sqrt{2})$. Then $\{1, \zeta, \zeta^2, \zeta^3\}$ is an integral basis of $\mathcal{O}_K$, and a rational prime p splits completely in $K/\mathbb{Q}$ if and only if $p \equiv 1 \bmod 8$. Suppose that $\pi = A + B\zeta + C\zeta^2 + D\zeta^3$ is an element of norm p in $\mathcal{O}_K$, and show that its relative norms to the three quadratic subfields $k_1 = \mathbb{Q}(i)$, $k_2 = \mathbb{Q}(\sqrt{2})$, and $\mathbb{Q}(i\sqrt{2})$ are

$$a + bi = A^2 - C^2 + 2BD + (B^2 - D^2 - 2AC)i$$
$$c + di\sqrt{2} = A^2 - B^2 + C^2 - D^2 + (AB - BC + CD + DA)i\sqrt{2}$$
$$e + f\sqrt{2} = A^2 + B^2 + C^2 + D^2 + (AB + BC + CD - DA)\sqrt{2}$$

Observe that we always have $e \geq 0$. These formulas were first used by Bickmore [52, 53], then by Barrucand and Cohn [34] to study rational reciprocity. Actually, the idea of considering the field K for arithmetical studies is due to Gauss, (see Bachmann's article [Bac], p. 31–32), who remarked that primes $p \equiv 1 \bmod 8$ are norms of elements $a + bi + c\sqrt{2} + d\sqrt{-2}$. Jacobi [403] also remarked that the different factors of p in the three quadratic fields come from combining the prime factors in $\mathbb{Z}[\zeta_8]$ in different ways; there is also an unpublished article of Jacobi on this subject.

Next show that we can always replace π by an associate in such a way that $\pi \equiv 1 \bmod 2$, i.e, that $A - 1 \equiv B \equiv C \equiv D \equiv 0 \bmod 2$, and that this implies $a \equiv c \equiv e \equiv 1 \bmod 4$, and $b \equiv d \equiv f \equiv 0 \bmod 2$.
Now show that

a) $\quad a + 3 \equiv b \bmod 8 \iff d \equiv 2 \bmod 4$
$\iff f \equiv 2 \bmod 4 \iff B \equiv D + 2 \bmod 4$,

b) $\quad a - 1 \equiv b \bmod 8 \iff d \equiv 0 \bmod 4$
$\iff f \equiv 0 \bmod 4 \iff B \equiv D \bmod 4$,

c) $\quad b \equiv 0 \bmod 8 \iff c \equiv 1 \bmod 8$
$\iff e \equiv 1 \bmod 8 \iff B + C + D \equiv 0 \bmod 4$.

Derive a similar equivalence for $a \equiv 1 \bmod 16$. Generalize these results in the direction of Exercise 5.3.

5.22 Generalize the preceding exercise to $\mathbb{Q}(\sqrt{-1}, \sqrt{3})$ and $\mathbb{Q}(\sqrt{-1}, \sqrt{5})$. Compare your results with those in Exercise 4.4.

5.23 Show that the identity
$$\sum_{\alpha,\beta \in \mathbb{Z}} (-1)^{\alpha+\beta} q^{(4\alpha+1)^2 + 16\beta^2} = \sum_{\gamma,\delta \in \mathbb{Z}} (-1)^{\delta} q^{(4\gamma+1)^2 + 8\delta^2}$$
due to Jacobi generalizes the observation that $d \equiv 0 \bmod 4$ is equivalent to $a - 1 \equiv b \bmod 8$ in Exercise 5.21.b).

5.24 Let $p = c^2 + 2d^2 \equiv 3 \bmod 8$, $c \equiv d \equiv 1 \bmod 2$, and $c, d > 0$. Then
$$2^{\frac{p+1}{4}} \equiv -\left(\frac{-2}{c}\right)\left(\frac{-1}{d}\right)\frac{c}{d} \bmod p.$$
(Hint: $2^{(p+1)/4} = -(-2)^{(p+1)/4} \equiv -(c/d)^{(p+1)/2} \bmod (c + d\sqrt{-2})$).

5.25 Let $p = e^2 - 2f^2 \equiv 7 \bmod 8$, $e \equiv f \equiv 1 \bmod 2$, and $e, f > 0$. Then
$$2^{\frac{p+1}{4}} \equiv \left(\frac{2}{e}\right)\left(\frac{-1}{f}\right)\frac{e}{f} \bmod p.$$

5.26 Let $p = 4a^2 + 3b^2 \equiv 7 \bmod 24$, b odd, and $a, b > 0$. Then
$$3^{\frac{p+1}{4}} \equiv \left(\frac{-3}{a}\right)\left(\frac{-1}{b}\right)\frac{2a}{b} \bmod p.$$

5.27 Let $p = c^2 + 6d^2 \equiv 7 \bmod 24$, $c \equiv d \equiv 1 \bmod 2$, and $c, d > 0$. Then
$$6^{\frac{p+1}{4}} \equiv \left(\frac{-6}{c}\right)\left(\frac{-1}{d}\right)\frac{c}{d} \bmod p.$$

5.28 Let $p = c^2 + 2d^2 \equiv 3 \bmod 8$, $c \equiv d \equiv 1 \bmod 2$, and $c, d > 0$. Then
$$(1 + \sqrt{2})^{\frac{p+1}{2}} \equiv -\left(\frac{-1}{c}\right)\left(\frac{-1}{d}\right)\frac{d}{c}\sqrt{2} \bmod p.$$

5.29 Let $q \equiv 5 \bmod 8$ be a prime, and let ε denote the fundamental unit of $\mathbb{Q}(\sqrt{q})$. Let p be a prime such that $(-1/p) = (q/p) = +1$, and show that there exist $e, f \in \mathbb{N}$ such that $p = e^2 - 4qf^2$ or $4p = e^2 - qf^2$ with e odd. Prove that $\left(\frac{\varepsilon}{p}\right) = \left(\frac{-1}{e-2f}\right)$.
(Hint: replace ε by ε^3 if necessary and write $\varepsilon = t + u\sqrt{q}$ with $t \equiv 2 \bmod 4$ and $u \equiv 1 \bmod 4$. Then verify that $\left(\frac{\varepsilon}{q}\right) = \left(\frac{2f}{q}\right)\left(\frac{2ft-ue}{q}\right)$, and compute the symbol on the right hand side. Note that $\left(\frac{2}{p}\right)\left(\frac{-1}{e}\right) = \left(\frac{-1}{e-2f}\right)$.)

5.30 (Sansone [695, 696], Epstein [207]) Let D be a square-free integer, and suppose that $x^2 - Dy^2 = -1$ has solutions in integers $x, y \in \mathbb{N}$.
1. Write $D = b^2 + d^2$ (cf. Exercise 5.9), with $d \equiv 1 \bmod 2$. Then $X^2 - DY^2 = d$ and $X^2 - DY^2 = -d$ are also solvable in integers;
2. Show that d is a quadratic residue mod D;

3. Show that $x^2 - Dy^2 = -1$ is not solvable if $D = 2p$, $p = a^2 + b^2 \equiv 1 \bmod 8$, $a \equiv 1 \bmod 4$, and $a + 3 \equiv b \bmod 8$;
4. Show that $x^2 - Dy^2 = -1$ is not solvable if $D = pq$, where $p \equiv q \equiv 1 \bmod 4$ are primes such that $\left(\frac{p}{q}\right) = 1$ and $\left(\frac{p}{q}\right)_4 \left(\frac{q}{p}\right)_4 = -1$.

5.31 (Reichardt [R2]) $x^4 - 17y^2 = 2z^2$ has no non-trivial solutions in $\mathbb{Z}$. (Hint: $2z^2$ is a quartic residue mod 17; show that z is a quadratic residue mod 17 and get a contradiction from the fact that $(2/17)_4 = -1$). Generalize.

Remark: This diophantine equation has local solutions everywhere, i.e., it is non-trivially solvable mod p^n for all primes p and all $n \in \mathbb{N}$, and it has solutions in $\mathbb{R}$. This implies that the Tate-Shafarevich group $\mathrm{III}(E/\mathbb{Q})$ of the Jacobian $E : y^2 = x^3 - 68x$ of $x^4 - 17y^2 = 2z^2$ is non-trivial.

5.32 (Pépin [Pe2] and Lemmermeyer [Le]) Show that $z^2 = px^4 - 14y^4$ does not have nontrivial rational solutions for primes $p = 2a^2 + 7b^2$.

5.33 (Rédei [Re1]) Let $p \equiv 1 \bmod 4$ and $q \equiv 3 \bmod 4$ be primes. Then the following assertions are equivalent:
a) $qx^2 + py^2 = z^4$ has a solution with $(x,y) = 1$ and odd z;
b) the class number of $\mathbb{Q}(\sqrt{-qp})$ is divisible by 8;
c) $(-q/p)_4 = +1$.

5.34 (Rédei [Re2]) Let $p \equiv 1 \bmod 8$ be prime; then the following assertions are equivalent:
a) $2x^2 + py^2 = z^4$ has a solution with $(x,y) = 1$;
b) the class number of $\mathbb{Q}(\sqrt{-2p})$ is divisible by 8;
c) $(2/p)_4 = +1$.

5.35 (Continued from Exercise 2.25) Show that the following congruences hold mod p:

	$p \equiv 1, 9 \bmod 40$		$p \equiv 21, 29 \bmod 40$	
	$2 \mid d$	$2 \nmid d$	$2 \mid d$	$2 \nmid d$
$U_{(p-1)/2}$	0	0	0	0
$U_{(p+1)/2}$	1	-1	1	-1
$5U^2_{(p-1)/4}$	0	-4	4	0
$V_{(p-1)/2}$	2	-2	2	-2
$V_{(p+1)/2}$	1	-1	1	-1
$V^2_{(p-1)/4}$	4	0	0	-4

5.36 (Taylor [794]) Let $p = a^2 + b^2 \equiv q \equiv 1 \bmod 4$ be primes such that $\left(\frac{p}{q}\right) = 1$. Call $(x, y, z) \in \mathbb{N}^3$ a *primitive pythagorean triple* (PPT) if $\gcd(x, y, z) = 1$ and $x^2 + y^2 = z^2$.

1. Let $p \equiv 1 \bmod 8$; then the following assertions are equivalent:
   i) $\left(\frac{p}{q}\right)_4 \left(\frac{q}{p}\right)_4 = 1$;
   ii) if (x, y, z) is a PPT, then either $a^2 \equiv x^2 \bmod q$ and $b^2 \equiv y^2 \bmod q$ or $a^2 \equiv -x^2 \bmod q$ and $b^2 \equiv -y^2 \bmod q$, and both possibilities actually occur.

2. Let $p \equiv 1 \bmod 8$; then the following assertions are equivalent:

i) $\left(\dfrac{p}{q}\right)_4 \left(\dfrac{q}{p}\right)_4 = 1$;

ii) if (x,y,z) is a PPT, then either $a^2 \equiv x^2 \bmod 2q$ and $b^2 \equiv y^2 \bmod q$ or $a^2 \equiv q - x^2 \bmod 2q$ and $b^2 \equiv q - y^2 \bmod 2q$, and both possibilities actually occur.

3. Prove the other conjectures in [794].

(Hint: write $\zeta = x+yi$; since $N(\zeta)$ is a square and $(x+yi, x-yi) = (2x, 2y, z) = 1$, conclude that ζ or $i\zeta$ must be a square, too. Now use Burde's reciprocity law, as well as the quadratic reciprocity law in $\mathbb{Z}[i]$.)

Additional References

[Bac] P. Bachmann, *Über Gauss's zahlentheoretische Arbeiten*, Gauss, Opera Omnia **X–2**, 1–74

[Bil] G. Billing, *Beiträge zur arithmetischen Theorie der ebenen kubischen Kurven vom Geschlecht Eins*, Nova Acta Soc. Sci. Upsal., IV. Ser. **11**, No.1 (1938), 1–165

[Boe] R. Bölling, *Die Ordnung der Schafarewitsch-Tate-Gruppe kann beliebig groß werden*, Math. Nachr. **67** (1975), 157–179

[Bro] E. Brown, $x^4 + dx^2y^2 + y^4 = z^2$: *Some cases with only trivial solutions – and a solution Euler missed*, Glasg. Math. J. **31** (1989), 297–307

[Cas] J.W.S. Cassels, *Arithmetic on curves of genus 1. IV: Proof of the Hauptvermutung*, J. Reine Angew. Math. **211** (1962), 95–112

[Dir] L. Dirichlet, *Einige neue Sätze über unbestimmte Gleichungen*, Gesammelte Werke, 219–236.

[Eul] L. Euler, *De casibus quibus formulam $x^4 + mxxyy + y^4$ ad quadratum reducere licet*, Mém. Acad. Sci. St. Pétersbourg **7** (1815/16, 1820), 10–22; Opera Omnia (I) **V**, 35–47

[HK] J. Hurrelbrink, M. Kolster, *Tame kernels under relative quadratic extensions and Hilbert symbols*, J. Reine Angew. Math. **499** (1998), 145–188

[Kra] K. Kramer, *A family of semistable elliptic curves with large Tate-Shafarevitch groups*, Proc. Am. Math. Soc. **89** (1983), 379–386

[Lag] J. Lagrange, *Nombres congruents et courbes elliptiques*, Seminaire Delange-Pisot-Poitou, 1974/75, Exp. 16, 17 pp. (1975)

[Le] F. Lemmermeyer, A Note on Pépin's counter examples to the Hasse principle for curves of genus 1, Abh. Math. Sem. Hamburg **69** (1999), 335–345

[vL] H. von Lienen, *The quadratic form $x^2 - 2py^2$*, J. Number Theory **10** (1978), 10–15

[McG] O. McGuinness, *The Cassels pairing in a family of elliptic curves*, Ph. D. Diss. Brown Univ. 1982

[Mi] C. Minnigerode, *Ueber die Vertheilung der quadratischen Formen mit complexen Coefficienten und Veränderlichen in Geschlechter*, Gött. Nachr. 1873, 160-180

[Pe1] T. Pépin, *Sur certains nombres complexes de la forme $a + b\sqrt{-c}$*, J. Math. Pures Appl. (3) **1** (1875), 317–372

[Pe2] T. Pépin, *Théorèmes d'analyse indéterminée*, C. R. Acad. Sci. Paris **78** (1874), 144–148

[Poc] H.C. Pocklington, *Some diophantine impossibilities*, Proc. Cambridge Phil. Soc. **17** (1914), 108–121

[Re1] L. Rédei, *Aufgabe 175*, Jahresber. DMV **44** (1934), *69*; Solutions by Rédei, Jahresber. DMV **46** (1936), *49–50*, and Scholz, ibid., *80*

[Re2] L. Rédei, *Die Diophantische Gleichung $mx^2 + ny^2 = z^4$*, Monatsh. Math. Phys. **48** (1939), 43–60

[Re3] L. Rédei, *Über den geraden Teil der Ringklassengruppe quadratischer Zahlkörper, die Pellsche Gleichung und die diophantische Gleichung $rx^2 + sy^2 = z^{2^n}$. I, II, III*, Math. Naturwiss. Anz. Ungar. Akad. d. Wiss. **62** (1943), 13–34, 35–47, 48–62

[Re4] L. Rédei, *Die 2-Ringklassengruppe des quadratischen Zahlkörpers und die Theorie der Pellschen Gleichung*, Acta Math. Acad. Sci. Hungaricae **4** (1953) 31–87

[R1] H. Reichardt, *Über die Diophantische Gleichung $ax^4 + bx^2y^2 + cy^4 = ez^2$*, Math. Ann. **117** (1940), 235–276

[R2] H. Reichardt, *Einige im Kleinen überall, im Grossen unlösbare diophantische Gleichungen*, J. Reine Angew. Math. **184** (1942), 12–18

[Ro] H.E. Rose, *On a class of elliptic curves with rank at most two*, Math. Comp. **64** (1995), 1251–1265

[ST] R.J. Stroeker, J. Top, *On the equation $Y^2 = (X+p)(X^2+p^2)$*, Rocky Mt. J. Math. **24** (1994), 1135–1161

[Va1] A. Vazzana, *On the 2-primary part of K_2 of rings of integers in certain quadratic number fields*, Acta Arith. **80** (1997), 225–235

[Va2] A. Vazzana, *Elementary abelian 2-primary parts of $K_2\mathcal{O}$ and related graphs in certain quadratic number fields*, Acta Arith. **81** (1997), 253–264

6. Quartic Reciprocity

In Chapter 5 we have already seen a lot about quartic reciprocity and its applications to rational number theory; these rational laws, however, do not suffice to solve every "rational" problem where quartic reciprocity is involved, as the following example shows.

Let $q \geq 3$ be an odd integer and suppose that $p = 4q + 1$ is prime. Then $S_q = 2^{2q} + 1$ is never prime because of the factorization

$$S_q = A_q \cdot B_q, \quad A_q = 2^q - 2^{\frac{q+1}{2}} + 1, \quad B_q = 2^q + 2^{\frac{q+1}{2}} + 1.$$

Now $p = 4q + 1 \equiv 5 \bmod 8$, hence the quadratic reciprocity law shows that $S_q = 2^{(p-1)/2} + 1 \equiv \left(\frac{2}{p}\right) + 1 \equiv 0 \bmod p$. Thus $p \mid A_q B_q$, and the question (first posed by Brillhart [Bri]) is: which? Since quadratic reciprocity has told us that $p \mid A_q B_q$, we might hope that quartic reciprocity will answer this question. We have already seen that quartic reciprocity is somehow related to the presentation $p = a^2 + b^2$ of p as a sum of two squares, hence we let our computer make a small table:

q	p	*	a	b
3	13	B	3	2
7	29	B	5	2
9	37	A	1	6
13	53	B	7	2
15	61	A	5	6
25	101	A	1	10
27	109	A	3	10

q	p	*	a	b
37	149	A	7	10
39	157	A	11	6
43	173	B	13	2
45	181	A	9	10
49	197	B	1	14
57	229	B	15	2
67	269	A	13	10

These results suggest the following

Conjecture. With the above notation, we have

$$p \mid A_q \iff \frac{b}{2} \equiv \pm 3 \bmod 8, \text{ and } p \mid B_q \iff \frac{b}{2} \equiv \pm 1 \bmod 8.$$

If we want to prove this conjecture, we have to compute $A_q \bmod p$; but $2^q = 2^{(p-1)/4}$ and $\left(\frac{2}{p}\right) = -1$, hence we have to decide whether $2^{(p-1)/4} \equiv i$

or $-i \bmod \pi$, where $\pi = a + bi$. In other words: rational quartic reciprocity does not suffice to solve this problem.

Our first section is devoted to the question how much of the quartic reciprocity law can be proved by comparing the splitting of primes in suitable extensions. Then we will compute quartic Gauss sums and derive the quartic reciprocity law in $\mathbb{Q}(i)$. We conclude this chapter by applying quartic reciprocity to the problem discussed above.

6.1 Splitting of Primes

Among the proofs of quadratic reciprocity that we have studied in Chapters 2 and 3, the most elegant was certainly the proof by comparing the splitting of primes in quadratic and cyclotomic extensions; let us see how this proof generalizes. We take $k = \mathbb{Q}(i)$ as our base field, where $i^2 = -1$. Let $\pi = a + bi$ be a prime in $\mathcal{O}_k$ such that $\pi \equiv 1 \bmod 2 + 2i$. We would like to compare the splitting of primes in K/k, where $K = k(\sqrt[4]{\pi})$, and in a suitable cyclotomic field. Unfortunately this is not possible, because $K/\mathbb{Q}$ is not abelian (it is not even normal). The next best thing to do is to take the maximal abelian subfield K^{ab} of the normal closure N of $K/\mathbb{Q}$. This suggests the following program: compute $G = \mathrm{Gal}\,(N/\mathbb{Q})$, its commutator group G', and then K^{ab} is the fixed field of G'. In order to compute $\mathrm{Gal}\,(N/\mathbb{Q})$, we need a lemma from Galois theory:

Lemma 6.1. *Let F be a field and k/F a quadratic extension. Moreover, let K/k be a cyclic quartic extension and let σ denote an extension of the nontrivial automorphism of k/F to K. If $K \cap K^\sigma = k$, then the normal closure N of K/F has Galois group $\mathrm{Gal}\,(N/F) \simeq C_4 \wr C_2$, the wreath product of C_4 and C_2 (see Exercise 6.1 for the definition of wreath products).*

Proof. K and K^σ are the splitting fields of quartic polynomials $f(X) \in k[X]$ and $f^\sigma(X) \in k[X]$, respectively; since $\sigma^2 = 1$, we have $K^{\sigma^2} = K$. Thus $N = KK^\sigma$ is the normal closure of K/F.

Next let τ generate $\mathrm{Gal}\,(K/k)$. We pick an extension of σ to $\mathrm{Gal}\,(K/F)$ and denote it also by σ. Then $\widetilde{\tau} = \sigma^{-1}\tau\sigma$ generates $\mathrm{Gal}\,(K^\sigma/k)$, and $\mathrm{Gal}\,(N/k) \simeq \langle \tau, \widetilde{\tau} \mid \tau^4 = \widetilde{\tau}^4 = 1,\ [\tau, \widetilde{\tau}] = 1\rangle \simeq C_4 \times C_4$ is a subgroup of index 2 in $\mathrm{Gal}\,(N/F)$. Therefore we have $\mathrm{Gal}\,(N/F) = \langle \sigma, \tau, \widetilde{\tau}\rangle$. We have to compute σ^2 and the commutators $[\sigma, \tau]$ and $[\sigma, \widetilde{\tau}]$. To this end, observe that $\sigma^2 \in \langle \tau, \widetilde{\tau}\rangle$ (since σ^2 fixes k by definition), hence σ^2 commutes with τ and $\widetilde{\tau}$, and we find $\sigma^{-1}\widetilde{\tau}\sigma = \sigma^{-2}\tau\sigma^2 = \tau$. Now assume that $\sigma^2 = \tau^a\widetilde{\tau}^b$ for some $0 \leq a, b < 4$. Then we get $1 = [\sigma, \sigma^2] = [\sigma, \tau^a\widetilde{\tau}^b] = \sigma^{-1}\widetilde{\tau}^{-b}\tau^{-a}\sigma\tau^a\widetilde{\tau}^b = \tau^{-b}\widetilde{\tau}^{-a}\tau^a\widetilde{\tau}^b = \tau^{a-b}\widetilde{\tau}^{b-a}$, that is, we must have $a = b$. Replacing σ by $\tau^{-a}\sigma$ then yields $\sigma^2 = 1$. Hence we may assume without loss of generality that $\sigma^2 = 1$, and we find $\mathrm{Gal}\,(N/F) = \langle \sigma, \tau \mid \sigma^2 = \tau^4 = 1,\ \tau^\sigma = \widetilde{\tau},\ [\tau, \widetilde{\tau}] = 1\rangle$, i.e. $\mathrm{Gal}\,(N/F) \simeq C_4 \wr C_2$. □

Remark. In the special case where F is totally real and k is totally complex (i.e. if k is CM; here CM stands for *complex multiplication* – such fields occur in Shimura's & Taniyama's generalization of the original theory of complex multiplication (see Chapter 8) over imaginary quadratic number fields), the 'normalization' of σ (i.e. replacing it by $\tau^{-1}\sigma$ if necessary) can be avoided by choosing σ as the restriction of complex conjugation to k. In particular, this applies to the case at hand, where we have $F = \mathbb{Q}$ and $k = \mathbb{Q}(i)$.

Now we come back to the fields we are interested in, namely $k = \mathbb{Q}(i)$, $K = k(\sqrt[4]{\pi})$, and the normal closure N of $K/\mathbb{Q}$. Our first aim was to compute the maximal abelian subfield K^{ab} of $N/\mathbb{Q}$; by Galois theory, K^{ab} is the fixed field of the commutator subgroup G' (the group generated by all commutators) of the non-abelian group $G = C_4 \wr C_2$ of order 32. We find $G' = \langle \tau^{-1}\widetilde{\tau} \rangle \simeq C_4$; the abelianization of G is $G/G' \simeq C_4 \times C_2$; we claim that $K^{ab} = \mathbb{Q}(i, \sqrt[4]{\mu})$, where $\mu = \pi^3\overline{\pi} = \pi^2 p$, is the fixed field of the commutator group G'. In order to prove this claim, it is sufficient to show that $K^{ab}/\mathbb{Q}$ is abelian (since K^{ab} has the correct degree $8 = (G : G')$). We observe that

1. K^{ab} contains $\mathbb{Q}(\sqrt{p})$: this is clear, because $\sqrt{\mu} = \pi\sqrt{p}$.
2. σ maps $\sqrt[4]{\pi^3\overline{\pi}}$ to $\sqrt[4]{\overline{\pi}^3\pi}$ in such a way that $\sqrt[4]{\pi^3\overline{\pi}}\sqrt[4]{\overline{\pi}^3\pi} = +p$ (the fact that $\alpha^{1+\sigma} > 0$ follows from σ being complex conjugation).
3. $K^{ab}/\mathbb{Q}$ is normal: this follows immediately from 2.
4. K^{ab} contains the two cyclic extensions $\mathbb{Q}(\sqrt{p}, \sqrt{\alpha})$ and $\mathbb{Q}(\sqrt{p}, \sqrt{-\alpha})$, where $\alpha = \frac{1}{2}(a + \sqrt{p})\sqrt{p}$. This follows directly from
$$\left(\sqrt[4]{\pi^3\overline{\pi}} + \sqrt[4]{\pi\overline{\pi}^3}\right)^2 = \pi\sqrt{p} + \overline{\pi}\sqrt{p} + 2p = 4\alpha.$$

This shows that $\mathrm{Gal}\,(K^{ab}/\mathbb{Q}) \simeq C_4 \times C_2$, and our claim follows. Put $\alpha^* = (-1)^{(p-1)/4}\alpha$; then $K^* = \mathbb{Q}(\sqrt{p}, \sqrt{\alpha^*})$ is the cyclic quartic subfield of $\mathbb{Q}(\zeta_p)$ (see the proof of Theorem 5.13, part 3.).

Let $p = \pi\overline{\pi}$ and $\ell = \lambda\overline{\lambda}$ be primes $\equiv 1 \bmod 4$ such that $(\ell/p) = 1$. Moreover, let $[\,\cdot\,/\,\cdot\,]$ denote the quartic residue symbol in $\mathbb{Z}[i]$. Then

$$\left(\frac{\ell}{p}\right)_4 = 1 \iff \ell^{\frac{p-1}{4}} \equiv 1 \bmod p$$
$$\iff \ell \text{ splits in } K^*/\mathbb{Q} \text{ (Proposition 3.1)}$$
$$\iff \lambda \text{ splits in } K^{ab}/\mathbb{Q}(i)$$
$$\iff \pi^3\overline{\pi} \equiv \xi^4 \bmod \lambda \text{ for some } \xi \in \mathbb{Z}[i]$$
$$\iff \left[\tfrac{\pi}{\lambda}\right]^3 \left[\tfrac{\overline{\pi}}{\lambda}\right] = 1 \iff \left[\tfrac{\overline{\pi}}{\lambda}\right] = \left[\tfrac{\pi}{\lambda}\right].$$

Now $[\alpha/\lambda] \equiv \alpha^{(\ell-1)/4} \bmod \lambda$; write $\pi = a + bi$, where b is even, and compute the binomial expansion of π^n for $n = \frac{\ell-1}{4}$; we find that $\pi^n = g_0 + g_1 i$, where

$$g_0 = \sum_{j \equiv 0(2)} \binom{n}{j}(-1)^{j/2}a^{n-j}b^j, \quad g_1 = \sum_{j \equiv 1(2)} \binom{n}{j}(-1)^{(j-1)/2}a^{n-j}b^j.$$

This shows that $\left[\frac{\overline{\pi}}{\lambda}\right] = \left[\frac{\pi}{\lambda}\right] \iff \ell \mid g_1$ and $\left[\frac{\overline{\pi}}{\lambda}\right] = -\left[\frac{\pi}{\lambda}\right] \iff \ell \mid g_0$.

Remark. We have proved above that $\left(\frac{\ell}{p}\right)_4 = 1 \iff \pi^3 \overline{\pi} \equiv \xi^4 \mod \lambda$. Since $\pi^3 \overline{\pi} = p\pi^2$, this can be written in the form $\left(\frac{\ell}{p}\right)_4 = 1 \iff \left(\frac{p}{\ell}\right)_4 \left[\frac{\pi}{\lambda}\right]_2 = 1$, where $\left[\frac{\cdot}{\cdot}\right]_2$ is the quadratic residue symbol in $\mathbb{Z}[i]$. Using the quadratic reciprocity law in $\mathbb{Z}[i]$, this is at once seen to be equivalent to Burde's rational reciprocity law. This comes of course as no surprise, because we have already seen in Chapter 5 that Burde's law can be deduced by comparing the splitting of primes in certain subfields of the abelian extension K^{ab}.

The proof given above does not carry over easily to the case $\ell \equiv 3 \mod 4$: here $(-\ell/p) = 1$ implies that ℓ splits completely in $K^{\mathrm{ab}}/\mathbb{Q}(i)$, because it splits completely in one of the two cyclic quartic extensions in K^{ab}. This may be considered as an unfortunate accident that complicates investigations; on the other hand, the reader may just as well feel rewarded by the beauty of the Galois theoretic reflections below.

Let K^+ and K^- denote the real and the complex quartic cyclic extension contained in K^{ab}, respectively. We see that $K^* = K^+$ if $p \equiv 1 \mod 8$ and $K^* = K^-$ if $p \equiv 5 \mod 8$. Let

$$G = \mathrm{Gal}\,(N_\pi/\mathbb{Q}) = \langle \sigma, \tau, \widetilde{\tau} \mid \sigma^2 = \tau^4 = \widetilde{\tau}^4 = 1,\ \tau^\sigma = \widetilde{\tau},\ \widetilde{\tau}^\sigma = \tau,\ [\tau, \widetilde{\tau}] = 1 \rangle$$

denote the Galois group of the normal closure N_π of $K_\pi = \mathbb{Q}(i, \sqrt[4]{\pi})$ over $\mathbb{Q}$, where σ denotes complex conjugation. The normalizer of the subgroup $\langle \sigma \rangle$ in G is $R = \langle \sigma, \psi \rangle$, where $\psi = \tau^{-1}\widetilde{\tau}$ generates the commutator subgroup G' of G. The relations $\sigma^{-1}\psi\sigma = \psi^{-1}$ and $\psi^4 = 1$ show that $R \simeq D_4$, where D_4 denotes a dihedral group of order 8. Since K^+ is real and normal over $\mathbb{Q}$, K^+ must be the fixed field of R. Now it is an easy matter to verify the diagrams in Figure 6.1, which give the subfields of K^{ab} and the corresponding Galois groups.

In particular we find that $\mathrm{Gal}\,(N_\pi/K^-) \simeq \langle \sigma\tau^2, \psi \rangle \simeq H_8$, where H_8 denotes the quaternion group of order 8.

Assume first that K^+ is contained in the decomposition field N_Z of a prime ideal $\mathfrak{L}$ in N_π above ℓ; then the decomposition group $Z = Z(\mathfrak{L}|\ell)$ is contained in R. Since $\mathfrak{L}$ is unramified in N_π, we see that the inertia subgroup $T = T(\mathfrak{L}|\ell)$ is trivial. Moreover, ℓ is inert in $\mathbb{Q}(i)/\mathbb{Q}$, hence Z is not a subgroup of $\langle \widetilde{\tau}, \tau \rangle$. By Hilbert's theory of Galois extensions, Z/T is isomorphic to the Galois group of a finite extension of finite fields, hence it is cyclic. This leaves the only possibility $|Z| = 2$. But then ℓ splits completely in $N_\pi/\mathbb{Q}(i)$.

If however K^- is contained in the decomposition field N_Z of $\mathfrak{L}$, then $Z \subseteq \langle \sigma\tau^2, \psi \rangle \simeq H_8$; since H_8 contains exactly one subgroup of order 2, we either have $Z = \langle \psi \rangle$ or $|Z| = 4$. But $Z = \langle \psi \rangle$ again contradicts the fact that ℓ is inert in $\mathbb{Q}(i)/\mathbb{Q}$, hence we have $|Z| = 4$; if ℓ would split completely in $\mathbb{Q}(i, \sqrt[4]{\pi})/\mathbb{Q}(i)$, then it would do so in $\mathbb{Q}(i, \sqrt[4]{\overline{\pi}})/\mathbb{Q}(i)$, because these extensions are conjugate. But this would contradict the fact that $|Z| = 4$, and we have shown:

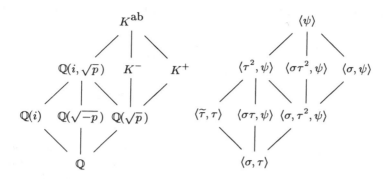

Fig. 6.1. The abelian subfields of K

Theorem 6.2. *In the situation above, a prime $\ell \equiv 3 \bmod 4$ such that $\left(\frac{\ell}{p}\right) = 1$ splits completely in the real cyclic quartic subfield of N_π if and only if ℓ splits completely in $N_\pi/\mathbb{Q}(i)$.*

Now ℓ splits completely in $K^+/\mathbb{Q}$ if and only if either $p \equiv 1 \bmod 8$ and $(\ell/p)_4 = 1$ (in this case we have $K^+ \subseteq \mathbb{Q}(\zeta_p)$) or if $p \equiv 5 \bmod 8$ and $(\ell/p)_4 = -1$ (here $K^- \subseteq \mathbb{Q}(\zeta_p)$). This implies

$$\left(\frac{-\ell}{p}\right)_4 = 1 \iff \ell \text{ splits completely in } K^+/\mathbb{Q}$$
$$\iff \ell \text{ splits completely in } \mathbb{Q}(i, \sqrt[4]{\pi})/\mathbb{Q}(i)$$
$$\iff \left[\frac{\pi}{\ell}\right]_4 = 1.$$

The congruence $\pi^\ell \equiv \overline{\pi} \bmod q$ implies

$$\pi^{(\ell^2-1)/4} \equiv \left(\pi^{\ell-1}\right)^n \equiv \left(\overline{\pi}/\pi\right)^n \bmod \ell, \text{ where } n = \tfrac{\ell+1}{4}.$$

This shows that $\left[\frac{\pi}{\ell}\right]_4 = 1$ if and only if $\pi^n \equiv \overline{\pi}^n \bmod \ell$, and the rest of our proof proceeds exactly as in the case $\ell \equiv 1 \bmod 4$ above.

Proposition 6.3. *Let $\ell = a^2 + b^2$ and $q \neq \ell$ be odd primes, $b \equiv 0 \bmod 2$, and put $n = \frac{1}{4}\left(\ell - \left(\frac{-1}{\ell}\right)\right)$. Define*

$$g_0 = \sum_{j \equiv 0(2)} \binom{n}{j}(-1)^{j/2} a^{n-j} b^j \text{ and } g_1 = \sum_{j \equiv 1(2)} \binom{n}{j}(-1)^{(j-1)/2} a^{n-j} b^j;$$

then

$$\left(\frac{\ell^*}{p}\right)_4 = +1 \iff g_1 \equiv 0 \bmod \ell, \text{ and } \left(\frac{\ell^*}{p}\right)_4 = -1 \iff g_0 \equiv 0 \bmod \ell.$$

Examples. For small primes ℓ, we find that $(\ell/p)_4 = 1$ if and only if the conditions in the following table are satisfied:

ℓ	condition	ℓ	condition
-3	$b \equiv 0 \bmod 3$	11	$b(b^2 - 3a^2) \equiv 0 \bmod 11$
5	$b \equiv 0 \bmod 5$	13	$b(b^2 - 3a^2) \equiv 0 \bmod 13$
-7	$ab \equiv 0 \bmod 7$	17	$ab(b^2 - a^2) \equiv 0 \bmod 17$

Since e.g. $b^2 - 3a^2 \equiv (b-5a)(b+5a) \equiv 5^2(2b+a)(2b-a) \bmod 11$, this agrees with the result derived from Proposition 5.6. In fact, comparing Proposition 6.3 with the corresponding results from Proposition 5.6, we see that the polynomials g_0 and g_1 defined above must split into linear factors over $\mathbb{F}_p$: this does not seem to be obvious. Moreover, comparing the coefficients of the g_j with those of the polynomials derived from Proposition 5.6 we get some curious congruences between solutions of the congruences there and sums of binomial coefficients.

Let's get back to what we said at the beginning of this section. There we wanted to compare the splitting of primes in K/k, where $k = \mathbb{Q}(i)$ and $K = k(\sqrt[4]{\pi})$, and in a suitable cyclotomic field; this was not possible because the extension $K/\mathbb{Q}$ is not normal (let alone abelian). Instead of taking the maximal abelian subfield of the normal closure of $K/\mathbb{Q}$ (as we have done above) we are tempted to ask if there is a "cyclotomic" theory with base field $\mathbb{Q}(i)$. The answer is yes, and the "cyclotomic" theory over $\mathbb{Q}(i)$ is related with Gauss's Last Entry, Kronecker's Jugendtraum, and the division of the lemniscate. We will return to this question in Chapters 8 and 10; comparing the splitting of primes will, however, only give a weak form of the quartic reciprocity law since the distinction between $[\mu/\pi] = i$ and $[\mu/\pi] = -i$ cannot be detected by the splitting behaviour of π.

6.2 Quartic Gauss and Jacobsthal Sums

In the next section we will use Gauss sums to prove the biquadratic reciprocity law; the main ingredient will be the prime factorization of quartic Gauss sums which will be proved now. To this end, let π be a prime of degree 1 in $\mathbb{Z}[i]$ (i.e. $p = N\pi$ is a prime $\equiv 1 \bmod 4$). Then $\mathbb{Z}[i]/(\pi) \simeq \mathbb{F}_p$, hence the quartic residue symbol $[\,\cdot\,/\pi]$ defines a a character of order 4 on $\mathbb{F}_p$ (there are exactly two such characters: the other one is $[\,\cdot\,/\pi]^{-1}$). To any character χ of order dividing 4 on $\mathbb{F}_p$ we have associated quartic Gauss sum $G(\chi^r)$ and quartic Jacobi sums $J(\chi, \chi^r)$ by

$$G(\chi^r) = -\sum_{t=1}^{p-1} \chi(t)^r \zeta^t, \qquad J(\chi, \chi^r) = -\sum_{t=1}^{p-1} \chi(t)\chi(1-t)^r.$$

6.2 Quartic Gauss and Jacobsthal Sums   191

The equations $p = G(\chi)\overline{G(\chi)}$ and $G(\chi)^4 \in \mathbb{Z}[i]$ show that only primes dividing p (i.e. π or $\overline{\pi}$) can occur in the prime factorization of $G(\chi)^4$. We can distinguish π from its associates if we demand that π be *primary*, that is, congruent to $1 \bmod 2 + 2i$. Now the next two propositions show exactly what's going on:

Proposition 6.4. *Let χ be a character of order 4 on $\mathbb{F}_p$, where $p = a^2 + b^2$ is prime; assuming that $a \equiv 1 \bmod 4$, we find*

$$J(\chi, \chi) = a \pm bi, \quad J(\chi, \chi^2) = \chi(-1)J(\chi, \chi), \quad \text{and } G(\chi)^4 = p \cdot J(\chi, \chi)^2.$$

Proof. Write $J(\chi, \chi) = a + bi$ for some $a, b \in \mathbb{Z}$; since χ^2 is a quadratic character, we have $G(\chi^2) = \pm\sqrt{p}$, hence

$$J(\chi, \chi)J(\overline{\chi}, \overline{\chi}) = \frac{G(\chi)^2}{G(\chi^2)} \frac{G(\overline{\chi})^2}{G(\overline{\chi^2})} = \frac{G(\chi)^2}{G(\chi^2)} \frac{\overline{G(\chi)}^2}{\overline{G(\chi^2)}} = \frac{p^2}{p} = p,$$

where we have used Proposition 4.23 iv) and vi). This implies $p = a^2 + b^2$. Similarly,

$$J(\chi, \chi)^2 = \frac{G(\chi)^4}{G(\chi^2)^2} = \chi(-1)J(\chi, \chi)J(\chi, \chi^2)$$

shows that $J(\chi, \chi^2) = \chi(-1)J(\chi, \chi)$, and we find $G(\chi)^4 = p \cdot J(\chi, \chi)^2$.
Next put $p = 2m - 1$; then

$$-J(\chi, \chi) = \sum_{t=1}^{p-1} \chi(t)\chi(1-t)$$

$$= \sum_{t=2}^{m-1} \chi(t)\chi(1-t) + \chi(m)^2 + \sum_{t=m+1}^{2m} \chi(t)\chi(1-t)$$

$$= \chi(m)^2 + 2\sum_{t=2}^{m-1} \chi(t)\chi(1-t).$$

Now $\chi(t) \equiv 1 \bmod (1+i)$, and $-4 = (1+i)^4$ shows that $\chi(m)^2 = \chi(2)^2 = \chi(-1)$. Hence $J(\chi, \chi) \equiv 2(m-2) - \chi(-1) \equiv 2 - \chi(-1) \bmod (2+2i)$, which yields $\chi(-1)J(\chi, \chi) \equiv 2\chi(-1) - 1 \equiv 1 \bmod (2+2i)$.

Equivalently, we can show that $J(\chi, \chi^2) \equiv 1 \bmod (2+2i)$ as follows: first we recall that $\sum \chi(t) = 0$; then we get

$$-J(\chi, \chi^2) = \sum_{t \neq 0, 1} \chi(t)\chi^2(1-t) = \sum_{t \neq 0, 1} \chi(t)\chi^2(1-t) + 1 + \sum_{t \neq 0, 1} \chi(t)$$

$$= 1 + \sum_{t \neq 0, 1} \chi(t)(\chi^2(1-t) + 1).$$

Now $\chi^2(1-t) + 1 \in \{0, 2\}$ and $\chi(t) \in \{\pm i, \pm 1\}$, hence $\chi(t)(\chi^2(1-t) + 1) \equiv \chi^2(1-t) + 1 \bmod (2+2i)$. This shows that we have to count the $t \in \mathbb{F}_p$ such that $\chi^2(1-t) = 1$: there are exactly $\frac{p-1}{2}$ of them, and we find

$$J(\chi,\chi^2) \equiv -1 - \sum_{t\neq 0,1}(\chi^2(1-t)+1) \equiv -1 - 2\cdot\tfrac{p-3}{2}$$
$$= 2 - p \equiv 1 \bmod (2+2i).$$

In particular, $J(\chi,\chi^2)$ is primary. □

Observe that we have proved that every prime $p \equiv 1 \bmod 4$ is the sum of two squares. Since $2 = 1^2 + 1^2$, this implies that *all* prime ideals in $\mathbb{Z}[i]$ are principal. The theorem about unique factorization into prime ideals shows that in $\mathbb{Z}[i]$ *every* ideal is principal. At this point it may seem a bit surprising that the factorizations of Gauss sums of order m have implications for the class group of $\mathbb{Q}(\zeta_m)$; in Section 11.4 we will uncover the deep connections between these two seemingly unrelated subjects.

If we specify the character χ by setting $\chi_\pi = \left[\frac{\cdot}{\pi}\right]$ then we may ask whether $J(\chi_\pi, \chi_\pi^2)$ is associated to π or $\overline{\pi}$. The answer is given by

Proposition 6.5. *If $\pi = a + bi \equiv 1 \bmod (1+i)^3$ is prime, and if $\chi_\pi = \left[\frac{\cdot}{\pi}\right]$ is the biquadratic character mod π, then*

$$J(\chi_\pi,\chi_\pi) = \chi_\pi(-1)\cdot\pi, \quad J(\chi_\pi,\chi_\pi^2) = \pi, \quad \text{and} \quad G(\chi_\pi)^4 = \pi^3\overline{\pi}.$$

Proof. All that remains to show is that $J(\chi_\pi,\chi_\pi) \equiv 0 \bmod \pi$. But this follows by putting $m = 4$ in Proposition 4.28. □

The result that $G(\chi)^4 = \pi^3\overline{\pi}$ shows that $\mathbb{Q}(i, G(\chi))$is just the abelian extension K^{ab} which we encountered in Section 6.1. The fact that both π and $\overline{\pi}$ occur in the prime factorization of $G(\chi)^4$ will complicate our proof of the quartic reciprocity law somewhat.

A pretty corollary of Proposition 6.5 is Gauss's observation that the trace of $\pi = a+bi$ is determined by a binomial coefficient:

Corollary 6.6 (Gauss's congruence). *Let $p = a^2 + b^2 = 4m + 1$ be a prime, and assume $a + bi \equiv 1 \bmod (2+2i)$. Then $2a \equiv (-1)^m \binom{2m}{m} \bmod p$.*

Proof. Put $\pi = a+bi$, $\chi = \chi_\pi$, and write $J(r,s) := J(\chi^r, \chi^s)$; then $J(3,2) = \chi(-1)J(3,3) = \chi(-1)\overline{J(1,1)} = \overline{\pi}$. Then $\chi_\pi(t) \equiv t^m \bmod \pi$, hence

$$-J(3,2) \equiv \sum_{t=0}^{p-1} t^{3m}(1-t)^{2m} \equiv \sum_{t=0}^{p-1} t^{3m}\sum_{j=0}^{m}(-1)^j\binom{2m}{j}t^{2m-j}$$
$$= \sum_{j=0}^{m}(-1)^j\binom{2m}{j}\sum_{t=0}^{p-1}t^{5m-j} \quad \bmod \pi.$$

The inner sum vanishes modulo p except when $j = m$, where it is $\equiv 1 \bmod p$. This shows that $J(3,2) \equiv (-1)^m\binom{2m}{m} \bmod \pi$. Thus $2a = \pi + \overline{\pi} \equiv \overline{\pi} \equiv (-1)^m\binom{2m}{m} \bmod \pi$. Since both sides of the congruence are rational integers, it is valid modulo p, and this proves our claim. □

6.2 Quartic Gauss and Jacobsthal Sums

Remark. Remember that we have inserted the minus sign in the definition of Gauss and Jacobi sums in order to simplify the formulas. Accordingly, Proposition 6.5 says $J(\chi_\pi, \chi_\pi) = \chi_\pi(-1)\pi$ instead of $J(\chi_\pi, \chi_\pi) = -\chi_\pi(-1)\pi$. The annoying occurrence of the factor $\chi_\pi(-1)$ is, however, due to our (and Gauss's) definition of primary integers. Had we heeded Kummer's dictum that 'Jacobi sums should be primary', this factor would disappear: in fact, let us call $\pi = a + bi \in \mathbb{Z}[i]$ *J-primary* if $a \equiv 1 \bmod 4$. Then it is an easy exercise to show that $J(\chi_\pi, \chi_\pi) = \pi$. Moreover, by choosing $a+bi$ in Corollary 6.6 as J-primary we get $\binom{2m}{m} \equiv 2a \bmod p$.

In Exercise 1.29 we introduced the *Jacobsthal sum*

$$\Phi_k(D) = \sum_{u=1}^{p-1} \left(\frac{u}{p}\right)\left(\frac{u^k + D}{p}\right) \tag{6.1}$$

for integers D and odd primes $p \nmid D$. Here we will deal with the sums $\Phi_2(D)$ which were first studied by Jacobsthal. We will show

Theorem 6.7. *The quartic Jacobsthal sum $\Phi_2(D)$ vanishes for primes $p \equiv 3 \bmod 4$; if $p \equiv 1 \bmod 4$, then $\Phi_2(D) \equiv D^{(p-1)/4}\Phi_2(1) \bmod p$. Moreover, if $p = a^2 + b^2$ with $a \equiv 1 \bmod 4$ and b even, then $\Phi_2(1) = 2a$ and $\Phi_2(D) = \pm 2b$ for every quadratic non-residue D modulo p.*

Proof. All these claims follow by using Euler's criterion and the binomial theorem (put $p = 2m + 1$):

$$\Phi_2(D) \equiv \sum_{x=1}^{p-1}(x^3 + Dx)^m \equiv \sum_{x=1}^{p-1}\sum_{j=0}^{m}\binom{m}{j}D^j x^{2j+m} \bmod p.$$

By Lemma 4.29, only exponents divisible by $p-1$ contribute to the sum. The only such exponent occurs for $j = m/2$; thus $\Phi_2(D) \equiv 0 \bmod p$ if m is odd, that is, if $p \equiv 3 \bmod 4$ (since in this case $m/2$ is not integral, and there is no exponent divisible by $p-1$), and $\Phi_2(1) \equiv \binom{2n}{n} \equiv 2a \bmod p$ if $m = 2n$ is even. Here we have used Corollary 6.6 and the Remark above.

Next we observe that $\Phi_2(D) \equiv 0 \bmod 2$: in fact, $\Phi_2(D)$ is a sum of $p-1$ summands ± 1 if $(-D/p) = -1$, and of $p-3$ summands ± 1 if $(-D/p) = +1$. Since $p-1$ and $p-3$ are even, so is $\Phi_2(D)$. Now we clearly have $|\Phi_2(D)| \leq p-1$; since $\Phi_2(D) \equiv 0 \bmod 2p$ if $p \equiv 3 \bmod 4$, we must have $\Phi_2(D) = 0$ in this case. If $p \equiv 1 \bmod 4$, then we have $\Phi_2(1) \equiv 2a \bmod 2p$, and $2|a| < p$ for all $p \geq 5$, so again we must have equality. Finally, from $\Phi_2(D) \equiv D^n\binom{2n}{n} \bmod p$ in this case we get $\Phi_2(D) \equiv D^n \Phi_2(1) \bmod p$ as claimed. Since $D^n \equiv \pm 1 \bmod p$ if $(D/p) = +1$ and $D^n \equiv \pm b/a \bmod p$ if $(D/p) = -1$, we find $\Phi_2(D) = \pm 2a$ in the first and $\Phi_2(D) = \pm 2b$ in the second case (actually we only get congruences modulo p; the equality follows as above). □

By Exercise 1.26, $p + \Phi_2(D)$ is the number of points on the elliptic curve $y^2 = x^3 + Dx$ with coordinates in $\mathbb{F}_p$.

6.3 The Quartic Reciprocity Law

As an immediate corollary of the prime factorization of the quartic Gauss sum we get a special case of the quartic reciprocity theorem:

Proposition 6.8. *If $\alpha \in \mathbb{Z}[i]$ and $a \in \mathbb{Z}$ are relatively prime and primary (i.e., if $\alpha \equiv 1 \bmod 2 + 2i$ and $a \equiv 1 \bmod 4$), then*

$$\left[\frac{\alpha}{a}\right] = \left[\frac{a}{\alpha}\right], \quad \left[\frac{i}{a}\right] = (-1)^{(a-1)/4}, \quad \left[\frac{1+i}{a}\right] = i^{(a-1)/4}.$$

Moreover, we have $\left[\frac{a}{b}\right] = 1$ for all $a, b \in \mathbb{Z}$ which are odd and relatively prime.

Proof. For the proof of $\left[\frac{\alpha}{a}\right] = \left[\frac{a}{\alpha}\right]$ we may assume that α is prime in $\mathbb{Z}[i]$ and that a is prime in $\mathbb{Z}$; we will distinguish the following cases:

1. $a \in \mathbb{Z}, \alpha = q \equiv 3 \bmod 4$; then $\left[\frac{a}{q}\right] = \left[\frac{a}{-q}\right] = 1$.
   In fact $\left[\frac{a}{q}\right] \equiv a^{(q^2-1)/4} \equiv (a^{q-1})^{(q+1)/4} \equiv 1 \bmod q$.
2. $a \in \mathbb{Z}, \alpha = p \equiv 1 \bmod 4$; then $\left[\frac{a}{p}\right] = 1$.
   This follows easily from $\left[\frac{a}{p}\right] = \left[\frac{a}{\pi}\right]\left[\frac{a}{\overline{\pi}}\right] = \left[\frac{a}{\pi}\right]\left[\frac{a}{\pi}\right]^{-1} = 1$.
3. $a = q \equiv 3 \bmod 4, \alpha = \pi$, where $p = \pi\overline{\pi} \equiv 1 \bmod 4$.
   In this case we define $\chi = \chi_\pi = \left[\frac{\cdot}{\pi}\right]$ and find

$$-G(\chi)^q = \Big(\sum_{t=1}^{p-1} \chi(t)\psi(t)\Big)^q \equiv \sum_{t=1}^{p-1} \chi(t)^q \psi(qt)$$

$$= \chi(q)\sum_{t=1}^{p-1} \chi(qt)^3 \psi(qt) = -\chi(q)G(\chi)^3 \bmod q,$$

hence $G(\chi)^q \equiv \left[\frac{q}{\pi}\right]G(\chi^3) \bmod q$. Multiplying through by $G(\chi)$ yields $G(\chi)^{q+1} \equiv \left[\frac{-q}{\pi}\right]p \bmod q$ (use Proposition 4.23.v). On the other hand we have $\pi^q \equiv \overline{\pi} \bmod q$ by Proposition 2.21, hence $p = \pi\overline{\pi} \equiv \pi^{q+1} \bmod q$, and the prime factorization of the quartic Gauss sum yields

$$G(\chi)^{q+1} \equiv (\pi^3\overline{\pi})^{\frac{q+1}{4}} \equiv (\pi^{q+3})^{\frac{q+1}{4}} \equiv \pi^{q+1}\left[\frac{\pi}{q}\right] \equiv p\left[\frac{\pi}{q}\right] \bmod q.$$

Comparing both congruences gives $\left[\frac{-q}{\pi}\right] = \left[\frac{\pi}{q}\right] = \left[\frac{\pi}{-q}\right]$.

4. $a = \ell = \lambda\overline{\lambda} \equiv 1 \bmod 4, \alpha = \pi$, where $\pi\overline{\pi} = p \equiv 1 \bmod 4$. Here $G(\chi)^\ell \equiv \left[\frac{\ell}{\pi}\right]^{-1} G(\chi) \bmod \ell$, and we find

$$\pi^{3(\ell-1)/4}\overline{\pi}^{(\ell-1)/4} \equiv G(\chi)^{\ell-1} \equiv \left[\frac{\ell}{\pi}\right]^{-1} \bmod \ell.$$

This implies $\left[\frac{\pi}{\lambda}\right]^3 \left[\frac{\overline{\pi}}{\lambda}\right] = \left[\frac{\ell}{\pi}\right]^{-1} \bmod \lambda$, hence

$$\left[\frac{\ell}{\pi}\right] = \left[\frac{\pi}{\lambda}\right]\left[\frac{\pi}{\overline{\lambda}}\right]^3 = \left[\frac{\pi}{\lambda}\right]\left[\frac{\pi}{\overline{\lambda}}\right] = \left[\frac{\pi}{\ell}\right].$$

Because of $\frac{m-1}{4} + \frac{n-1}{4} \equiv \frac{mn-1}{4}$ mod 4 for integers $m, n \equiv 1$ mod 4, it is sufficient to prove the supplementary laws only for prime a:

1. $q \equiv 3$ mod 4: then $\left[\frac{i}{q}\right] = (i^{q-1})^{(q+1)/4} = (-1)^{(q+1)/4}$;
2. $p = \pi\overline{\pi} \equiv 1$ mod 4: $\left[\frac{i}{p}\right] = \left[\frac{i}{\pi}\right]\left[\frac{i}{\overline{\pi}}\right] = i^{(p-1)/4}i^{(p-1)/4} = (-1)^{(p-1)/4}$.

The proof of the second supplementary law is hardly more difficult: if $q \equiv 3$ mod 4, then $(1+i)^q \equiv 1 + i^q \equiv 1 - i$ mod q implies that $(1+i)^{q-1} \equiv \frac{1-i}{1+i} = -i$ mod q, hence we find

$$\left[\frac{1+i}{q}\right] \equiv ((1+i)^{q-1})^{\frac{q+1}{4}} \equiv (-i)^{\frac{q+1}{4}} = i^{\frac{-q-1}{4}} \mod q.$$

If $p = \pi\overline{\pi}$ splits, then $\left[\frac{1+i}{p}\right] = \left[\frac{1+i}{\pi}\right]\left[\frac{1+i}{\overline{\pi}}\right] = \left[\frac{1+i}{\pi}\right]\left[\frac{1-i}{\pi}\right]^3 = \left[\frac{i}{\pi}\right]\left[\frac{1-i}{\pi}\right]^4 = \left[\frac{i}{\pi}\right] = i^{(p-1)/4}$, as we have already seen in our proof of the first supplementary law. $\square$

Next we show that the complete quartic reciprocity law stated below can be deduced from Eisenstein's law by the same purely formal computations which we used in Chapter 5 for deriving the quadratic reciprocity law in $\mathbb{Z}[i]$ from that in $\mathbb{Z}$.

Theorem 6.9. (Quartic Reciprocity Law)

1. *The classical formulation (Gauss, Eisenstein):*
   For relatively prime and primary $\alpha = a + bi$ and $\beta = c + di$, we have

$$\left[\frac{\alpha}{\beta}\right]\left[\frac{\beta}{\alpha}\right]^{-1} = (-1)^{\frac{N\alpha-1}{4} \frac{N\beta-1}{4}} = (-1)^{\frac{a-1}{2} \frac{c-1}{2}} = (-1)^{\frac{bd}{4}}.$$

   Moreover there are the supplementary laws

$$\left[\frac{i}{\alpha}\right] = i^{(1-a)/2}, \quad \left[\frac{1+i}{\alpha}\right] = i^{(a-b-b^2-1)/4} \text{ and } \left[\frac{2}{\alpha}\right] = i^{-b/2}.$$

2. *The alternate formulation (Jacobi, Kaplan):*
   If we choose α and β in such a way that $a \equiv c \equiv 1$ mod 4, then

$$\left[\frac{\alpha}{\beta}\right]\left[\frac{\beta}{\alpha}\right]^{-1} = (-1)^{bd/4}.$$

In particular, the characters $\left[\frac{i}{\cdot}\right]^2$, $\left[\frac{1+i}{\cdot}\right]^2$, $\left[\frac{i}{\cdot}\right]$, $\left[\frac{2}{\cdot}\right]$, $\left[\frac{1+i}{\cdot}\right]$, have conductors $\lambda^4, \lambda^5, \lambda^6, \lambda^6$ and λ^7, respectively, where $\lambda = 1 + i$.

Proof. Choose $\alpha = a + bi$ and $\beta = c + di$ in such a way that $a \equiv c \equiv 1$ mod 4, and assume that $(a, b) = (c, d) = 1$; note that we may do this without loss of generality, because we can extract common divisors and invert them with

196   6. Quartic Reciprocity

Eisenstein's law. Now we use the congruences $ci \equiv d \bmod \beta$, $c\alpha \equiv ac + bd \bmod \beta$ and find

$$\left[\frac{\alpha}{\beta}\right] = \left[\frac{c}{\beta}\right]^{-1}\left[\frac{c\alpha}{\beta}\right] = \left[\frac{c}{\beta}\right]^{-1}\left[\frac{ac+bd}{\beta}\right].$$

Similarly we get

$$\left[\frac{\beta}{\alpha}\right] = \left[\frac{a}{\alpha}\right]^{-1}\left[\frac{ac+bd}{\alpha}\right], \text{ i.e., } \left[\frac{\beta}{\alpha}\right]^{-1} = \left[\frac{a}{\alpha}\right]\left[\frac{ac+bd}{\overline{\alpha}}\right].$$

Multiplying both equations yields

$$\left[\frac{\alpha}{\beta}\right]\left[\frac{\beta}{\alpha}\right]^{-1} = \left[\frac{c}{\beta}\right]^{-1}\left[\frac{ac+bd}{\beta}\right]\left[\frac{a}{\alpha}\right]\left[\frac{ac+bd}{\overline{\alpha}}\right].$$

Applying Eisenstein's quartic reciprocity law gives $\left[\frac{c}{\beta}\right] = \left[\frac{\beta}{c}\right] = \left[\frac{di}{c}\right] = \left[\frac{i}{c}\right]$, and correspondingly $\left[\frac{a}{\alpha}\right] = \left[\frac{i}{a}\right]$. Moreover we find

$$\left[\frac{ac+bd}{\overline{\alpha}\beta}\right] = \left[\frac{\overline{\alpha}\beta}{ac+bd}\right] = \left[\frac{ac+bd+(ad-bc)i}{ac+bd}\right] = \left[\frac{i}{ac+bd}\right].$$

Together this shows $\left[\frac{\alpha}{\beta}\right]\left[\frac{\beta}{\alpha}\right]^{-1} = \left[\frac{i}{c}\right]^{-1}\left[\frac{i}{a}\right]\left[\frac{i}{ac+bd}\right]$. The symbols on the right hand side are easily computed using the first supplementary law; in fact $\left[\frac{i}{c}\right]^{-1}\left[\frac{i}{a}\right]\left[\frac{i}{ac+bd}\right] = (-1)^{(a-1-c+1+ac+bd-1)/4} = (-1)^{bd/4}$ because of the congruence $a - c + ac \equiv 1 \bmod 8$.

The first supplementary law for primes $\alpha \equiv 1 \bmod 2+2i$ can be proved very easily:

1. $a \equiv 1, b \equiv 0 \bmod 4$: then $\left[\frac{i}{\alpha}\right] = i^{(a^2+b^2-1)/4} = i^{(a^2-1)/4} = i^{(1-a)/2}$;
2. $a \equiv 3, b \equiv 2 \bmod 4$: then $\left[\frac{i}{\alpha}\right] = i^{(a^2+b^2-1)/4} = i^{(a^2+4-1)/4} = i^{(1-a)/2}$.

The proof of the general case is left as an easy exercise.

Next we show that the quartic reciprocity law holds if we assume α and β to be primary; to this end we consider several cases:

1. $b \equiv d \equiv 0 \bmod 4$: in this case there is nothing to prove because $a \equiv c \equiv 1 \bmod 4$;
2. $b \equiv 2, d \equiv 0 \bmod 4$: then, from what we have proved,

$$\left[\frac{-\alpha}{\beta}\right] = (-1)^{bd/4}\left[\frac{\beta}{-\alpha}\right] = \left[\frac{\beta}{\alpha}\right], \text{ and } \left[\frac{-1}{\beta}\right] = 1.$$

3. $b \equiv d \equiv 2 \bmod 4$: here $\left[\frac{-\alpha}{\beta}\right] = \left[\frac{-\beta}{\alpha}\right]$ and $\left[\frac{-1}{\alpha}\right] = \left[\frac{-1}{\beta}\right] = -1$.

This shows that both formulations are equivalent.

It remains to prove the second supplementary law. Put $\lambda = (1+i)^3 = -2+2i$; then $\left[\frac{1+i}{\alpha}\right] = \left[\frac{\lambda}{\alpha}\right]^3$ shows that it suffices to compute $\left[\frac{\lambda}{\alpha}\right]$. Now $\left[\frac{\lambda}{\alpha}\right] = \left[\frac{\lambda-\alpha}{\alpha}\right] = \left[\frac{-1}{\alpha}\right]\left[\frac{\alpha-\lambda}{\alpha}\right] = \left[\frac{-1}{\alpha}\right]\left[\frac{\lambda}{\alpha-\lambda}\right] = (-1)^{b/2}\left[\frac{\alpha}{\alpha-\lambda}\right]$, where we have used the fact that α or $\alpha - \lambda$ is $\equiv 1 \bmod 4$, and that $\alpha \equiv \lambda \bmod (\lambda - \alpha)$. Writing $\alpha = a + bi$ as $\alpha = c + d\lambda$ (in fact, $c = a+b, d = \frac{b}{2}$) we find

$$\left[\frac{\lambda}{\alpha}\right] = (-1)^d \left[\frac{\lambda}{\alpha - \lambda}\right] = (-1)^{d+(d-1)}\left[\frac{\lambda}{\alpha - 2\lambda}\right] = \ldots = (-1)^{\frac{d(d+1)}{2}}\left[\frac{\lambda}{c}\right].$$

Cubing this equation yields $\left[\frac{1+i}{\alpha}\right] = (-1)^{d(d+1)/2}\left[\frac{1+i}{c}\right]$, and the second supplementary law of Proposition 6.8 gives

$$\left[\frac{1+i}{\alpha}\right] = (-1)^{d(d+1)/2} i^{(c-1)/4} = i^{b(b+2)/4} i^{(a+b-1)/4} = i^{(a-b-b^2-1)/4}$$

as claimed (note that $c \equiv 1 \bmod 4$).

Finally, we have to compute the conductors of some characters. Take $\chi(\alpha) = \left[\frac{1+i}{\alpha}\right]$, for example. Since $\lambda^7 = 8(1+i)$, we have $\chi(\alpha + \mu\lambda^7) = \chi(\alpha)$ for all $\mu \in \mathbb{Z}[i]$; in order to see this, write $\alpha = a + bi$ and $\alpha + \mu\lambda^7 = a_\mu + b_\mu i$. Then $a \equiv a_\mu, b \equiv b_\mu \bmod 8$, and $a + b \equiv a_\mu + b_\mu \bmod 16$. Together with the second supplementary law this proves that χ has conductor dividing λ^7. But $\lambda^6 = -8i$, hence the fact that $1 = \left[\frac{1+i}{1}\right]$ and $-1 = \left[\frac{1+i}{1+8i}\right]$ shows that λ^6 is not the conductor of χ. The other assertions about conductors are proved in a similar way. □

6.4 Applications

Now we are in a position to prove our conjecture from the beginning of this chapter:

Proposition 6.10. *Let q and u be odd integers, and suppose that $p = 4qu + 1 = a^2 + 4b^2$ is prime. Assume moreover that $p \mid S_q = 2^{2q} + 1$ (this is equivalent to $2^{\frac{p-1}{u}} \equiv 1 \bmod p$, thus it is always true for $u = 1$). Then*

$$S_q = A_q \cdot B_q, \text{ where } A_q = 2^q - 2^{\frac{q+1}{2}} + 1 \text{ and } B_q = 2^q + 2^{\frac{q+1}{2}} + 1,$$

$$\text{and } \begin{cases} b \equiv \pm 3u \bmod 8 \iff A_q \equiv 0 \bmod p, \ B_q \equiv 2(1 + 2^q) \bmod p; \\ b \equiv \pm u \bmod 8 \iff B_q \equiv 0 \bmod p, \ A_q \equiv 2(1 + 2^q) \bmod p. \end{cases}$$

Proof. Suppose that $p \mid 2^{2q} + 1$; then $2^{\frac{p-1}{2u}} \equiv -1 \bmod p$, hence $2^{\frac{p-1}{u}} \equiv 1 \bmod p$. If on the other hand $2^{\frac{p-1}{u}} \equiv 1 \bmod p$, then $2^{\frac{p-1}{2u}} \equiv \pm 1 \bmod p$; if the positive sign would hold, then raising the congruence to the u-th power would yield $2^{\frac{p-1}{2}} \equiv 1 \bmod p$, contradicting the fact that $p \equiv 5 \bmod 8$.

198   6. Quartic Reciprocity

Now suppose that $2^{\frac{p-1}{2u}} \equiv -1 \bmod p$ and write $p = \pi\overline{\pi}$, $\pi = a+2bi$. Then $2^{\frac{p-1}{4u}} \equiv i^k \bmod \pi$ for some odd k, and $u^2 \equiv 1 \bmod 4$ implies that $2^{\frac{p-1}{4u}} \equiv i^k \equiv (i^k)^{u^2} \equiv 2^{\frac{p-1}{4}u} \bmod \pi$. The supplementary laws of quartic reciprocity give $2^{\frac{p-1}{4}} \equiv [2/\pi] = i^{a-1-(2b+4b^2)/2} = i^{-b} \bmod \pi$, because $a \equiv 3 \bmod 4$ and $2b \equiv 2 \bmod 4$ in our case. Thus we get $2^{\frac{p-1}{4u}} \equiv i^{-bu} \bmod \pi$. The same reasoning yields after some computation

$$\frac{2^{\frac{p-1+4u}{8u}}}{1+i} \equiv \left(\frac{2^{\frac{p+3}{8}}}{1+i}\right)^u \left(\frac{2^{\frac{u-1}{2}}}{(1+i)^{u-1}}\right)^u$$

$$\equiv \left((1+i)^{\frac{p-1}{4}} i^{\frac{p+3}{8}}\right)^u \left(i^{\frac{1-u}{2}}\right)^u \equiv i^{\frac{3-bu}{2}} \bmod \pi.$$

Now we distinguish the following cases:

i) $b \equiv u \bmod 8$: then $bu \equiv 1 \bmod 8$, hence

$$B_q \equiv i^{-bu} + (1+i)i^{-(3-bu)/2} + 1 \equiv i^3 + (1+i)i + 1 \equiv 0 \bmod \pi.$$

Since there are rational integers on both sides of these congruences, the result is valid mod p.

ii) $b \equiv 3u \bmod 8$: then $A_q \equiv i - (1+i) + 1 \equiv 0 \bmod \pi$.
iii) $b \equiv -3u \bmod 8$: then $A_q \equiv i^3 - (1+i)i^3 + 1 \equiv 0 \bmod \pi$.
iv) $b \equiv -u \bmod 8$: then $B_q \equiv i - (1+i) + 1 \equiv 0 \bmod \pi$.

The other congruences follow from $A_q + B_q = 2 + 2^{q+1}$. □

The corresponding problem for primes $p \equiv 9 \bmod 16$ will be solved in Chapter 9 using the octic reciprocity law.

6.5 Quartic Reciprocity in some Quartic Fields

In Section 5.1 we have shown that the quadratic reciprocity law in $\mathbb{Z}[i]$ can be easily deduced from the one in $\mathbb{Z}$; it is therefore quite natural to ask if we can transfer the quartic reciprocity law to number fields strictly containing $\mathbb{Q}(i)$, for example to $K = \mathbb{Q}(\zeta_8) = \mathbb{Q}(i, \sqrt{2})$. We start by defining semi-primary numbers. Let $\lambda = 1 + \zeta$ be a generator for the prime ideal of norm 2 in $\mathcal{O}_K$; we seek the greatest power λ^m of λ with the following property: "For every $\alpha \in \mathcal{O}_K$ such that $\lambda \nmid \alpha$ there is a unit $\eta \in \mathcal{O}_K^\times$ such that $\alpha\eta \equiv 1 \bmod \lambda^m$." Then we will call $\alpha \in \mathcal{O}_K$ semi-primary if $\alpha \equiv 1 \bmod \lambda^m$. It is easily seen that $m \in \mathbb{N}$ has this property if and only if there exist units $\eta_1, \eta_2, \ldots, \eta_{m-1}$ such that $\lambda^j \parallel (\eta_j - 1)$. Then we verify that $\eta_1 = \zeta$, $\eta_2 = i$, $\eta_3 = i(1+\sqrt{2}) = \zeta + \zeta^2 + \zeta^3$ and $\eta_4 = -1$ are suitable units. Using the fact that $E_K = \langle \zeta, 1+\sqrt{2} \rangle$ one can check that there is no unit η_5, and we have $m = 5$.

Next let $\alpha = A + B\zeta + C\zeta^2 + D\zeta^3$ be semi-primary, i.e., assume that $\alpha \equiv 1 \bmod 2 + 2\zeta$. This is equivalent to $A - 1 \equiv B \equiv C \equiv D \equiv 0 \bmod 2$ and $A - 1 + B + C + D \equiv 0 \bmod 4$. One can easily check that α or $-(3 + 2\sqrt{2})\alpha$ has $B \equiv 0 \bmod 4$; such an α will be called B-primary.

Now let $\alpha = A + B\zeta + C\zeta^2 + D\zeta^3$ and $\beta = E + F\zeta + G\zeta^2 + H\zeta^3$ be B-primary (that is, $\alpha \equiv \beta \equiv 1 \bmod 2 + 2\zeta$, $B \equiv F \equiv 0 \bmod 4$); the congruence $(E+Gi)\alpha \equiv (E+Gi)(A+Ci) - (B+Di)(F+Hi) \bmod \beta$ can be used to reduce quartic residue symbols in K to symbols in $\mathbb{Q}(i)$; these can be evaluated using the quartic reciprocity law, and after some tedious computations we get

Proposition 6.11. *Let $\left[\frac{\cdot}{\cdot}\right]$ denote the quartic power residue symbol in $K = \mathbb{Q}(\zeta_8)$; if $\alpha, \beta \in \mathcal{O}_K$ are semi-primary, then*

$$\left[\frac{\alpha}{\beta}\right] = (-1)^{\frac{CH+DG}{4}} \left[\frac{\beta}{\alpha}\right].$$

For B-primary $\alpha \in \mathcal{O}_K$ wealso have the following supplementary laws:

$$\left[\frac{\zeta}{\alpha}\right] = (-1)^{\frac{D}{2}} i^{\frac{A^2+3C^2-1}{4}},$$

$$\left[\frac{1+\zeta}{\alpha}\right] = (-1)^{\frac{D}{2}} i^{\frac{C^2+D^2}{4}} i^{\frac{1-A+B-C+D}{4}},$$

$$\left[\frac{1+\sqrt{2}}{\alpha}\right] = (-1)^{\frac{C}{2}} i^{\frac{B+D}{2}},$$

$$\left[\frac{\sqrt{2}}{\alpha}\right] = \begin{cases} i^{\frac{C}{2}}, & \text{if } D \equiv 0 \bmod 4, \\ i^{\frac{2-C}{2}}, & \text{if } D \equiv 2 \bmod 4. \end{cases}$$

This is a corollary of the octic reciprocity law in $\mathbb{Z}[\zeta_8]$ which we will discuss in Chapter 9.

NOTES

Unfortunately we do not know when Gauss really discovered the biquadratic reciprocity law; we are quite sure that this must have happened long before the publication of some of his results in 1828 and 1832. According to the remarks of Bachmann after [275], Gauss rediscovered Euler's conjectures on the cubic and quartic residuacity of small integers between 1805 and 1807; according to his diary, he found the main proofs for his theory of cubic and quartic residues on February 22, 1807. In his letter to Sophie Germain [281] (in April 1807) he challenges her to prove the theorems concerning the cubic and the quartic residue character of 2. The complete reciprocity law is stated in one of Gauss's note books [278] and must have been known to him not much later than 1813. This agrees with what Gauss says in a letter to Dirichlet (May 30, 1828; Werke **II**, p. 516):

6. Quartic Reciprocity

> Die ganze Untersuchung, deren Stoff ich schon seit 23 Jahren vollständig besitze, die Beweise der Haupttheoreme aber (zu welchen das in der ersten Commentatio noch nicht zu rechnen ist) seit etwa 14 Jahren, ... [1]

It is therefore reasonable to assume that Gauss discovered rational criteria for cubic and quartic reciprocity starting in 1805, but found the full reciprocity laws along with their proofs only shortly before 1814. Although parts of his investigations dealing with quartic reciprocity were eventually published in 1828 [272] (where he derived the quartic character of 2) and 1832 [273] (where he gave criteria for the quartic residuacity of small primes as well as for the integers ± 6, stated the quartic reciprocity law, and derived the quartic character of $1 + i$), the promised third part containing the proof of the quartic reciprocity law never appeared. We don't even know which method Gauss used for a proof: in his publication [268] containing his fifth and sixth proof of the quadratic reciprocity law, he mentions that these proofs were motivated by his research on cubic and quartic residues, so we might guess that the proof of his quartic reciprocity law was based on either Gauss's Lemma or on quartic Gauss sums. A proof using Gauss sums was found in Gauss's papers [280] after his death, but Bachmann said that this might have been written *after* the publication of Eisenstein's papers.

Bachmann also mentions Schering's [Sch] remark that Riemann told him that Eisenstein told Riemann that Gauss showed Eisenstein sketches of the cyclotomic proof of the biquadratic reciprocity law in a letter; the letters from Gauss to Eisenstein are, however, lost. I should add that this story is hard to fit into the chronology of events: in June 1844, after having received several of Eisenstein's manuscripts, Gauss invited Eisenstein to visit him in Göttingen, which Eisenstein did on June 14, 1844. At this time, his proof of the cubic reciprocity law using Gauss sums had already appeared, while the manuscript on the quartic law was already written. After his return from Göttingen, Eisenstein takes up his correspondence with M. Stern, half a year later with Gauss. Finally, Eisenstein met Riemann in the summer of 1847.

Jacobi gave a proof of the biquadratic reciprocity law in his Königsberg lectures 1836/37 (see the article by G. Frei [238], and H. Pieper's forthcoming edition of [402]). Jacobi's proof is presented in Exercise 6.23 below. Curiously, Jacobi doesn't mention that his formulation of the biquadratic reciprocity law (he assumes that $\alpha = a + bi$ and $\beta = c + di$ have $a \equiv c \equiv 1 \bmod 4$) differs slightly from the one Gauss gave in his second memoir. Jacobi's version was later rediscovered by Kaplan [419]; actually, it is only a special case of Gauss's and Artin's (unpublished) version [239] (cf. Exercise 6.17) and of Kubota's [453] formulation (cf. Exercise 6.18). See Exercise 8.19 for a very simple formulation.

---

[1] The whole investigation, whose content has been in my possession for 23 years, the proofs of the main theorems however (to which the one in the first Commentatio does not belong) for about 14 years, ...

The first published proofs, however, are due to Eisenstein [190, 191, 197, 198]; see Collison [132] for an exposition of Eisenstein's non-analytic proofs. Most of the proofs published after 1850 are variations of those Eisenstein has given (see, for example, Kubota [454, 456, 457], Petr [655], Shiratani [733], and Watabe [821]); exceptions are Kaplan's proof in [419], which starts by computing the number of solutions of the congruence $x_1^4 + \ldots + x_q^4 \equiv q \bmod p$, and the geometric proof of the biquadratic reciprocity law, sketches of which were found in Gauss's posthumous papers [279]. For a reconstruction of Gauss's proof, see Spies [746]. Already Busche [107, 113, 114] had given geometric proofs of the biquadratic reciprocity law. We would also like to draw the reader's attention to the papers of Collison [133] and Rowe [691] for some more history on the genesis of reciprocity laws.

The published proofs of the second supplementary of the biquadratic reciprocity law (with the exception of the one given by Herglotz [362], which is based on his results on the complex multiplication of elliptic functions) consisted of rather ad hoc computations (this applies to proofs due to Eisenstein up to K.S. Williams [862]). The trick of deriving the second supplementary law of quartic reciprocity by computing the quartic character of an odd power of $1+i$ is inspired by a similar method that Kronecker (see König [432]) used in his lectures for computing $(2/p)$ from the quadratic reciprocity law. Dintzl [159] was the first to employ this technique in the quartic case, but he needed 8 pages of computations. Artin & Hasse used this trick repeatedly and generalized it to ℓ-th power reciprocity for general primes ℓ: see e.g. Hasse's paper [335].

Proposition 6.3 is due to Cunningham and Gosset [146] and has been rediscovered by U. Felgner [229]. Some of the examples at the end of Section 6.1 have been conjectured already by Euler; a more extensive list of criteria can be found in [146], for example.

The quartic period equation was discussed by Lebesgue [481] and Cayley [Cay]; for more on this topic, see Berndt, Evans & K.S. Williams [48].

Adkisson [1] computes the quartic Jacobi and Gauss sums and claims that his proofs are simpler than those given by Smith [743]. Bohnicek [59, 61] generalizes Hilbert's proof of the quadratic reciprocity law to fourth powers and corrects an error of Lietzmann [547]. Burde [100] proves the quadratic and the rational quartic reciprocity laws in $\mathbb{Z}[i]$ by computations in finite fields. Chowla [128] used quartic reciprocity to study difference sets. Gegenbauer [294] and Kantz [417] prove Gauss's Lemma for composite values of m. Genocchi [298] (see also [299] in this connection) proves S. Germain's result that $(-4/p)_4 = +1$ for all primes $p \equiv 1 \bmod 4$ (compare Exercise 4.33). Gosset [318] gives a method for computing $(p/q)_4$ using the quartic reciprocity law but avoiding complex integers and ends up proving something equivalent to Burde's rational quartic reciprocity law. Hudson & K.S. Williams [378, 379] derive criteria for the biquadratic character of 3, 5 and 7 from properties of cyclotomic numbers. Mertens [582] and Scheibner [707, 708] use theta functions to express the quartic residue symbol, and Mertens uses this to give a proof of the quartic reciprocity law. Stieltjes [763] computes the quartic character of $1+i$ using the cyclotomic method that Gauss used for computing $(2/p)_4$ in his commentationes; he explains his results in a letter to Hermite [764]. Stouff [771, 772] uses cubic and quartic reciprocity to study certain groups. See also Nagell [606], and Pellet [631, 632, 633, 634].

Gauss's congruence $2a \equiv (-1)^m \binom{2m}{m} \bmod p$ in Corollary 6.6 was first proved by Gauss in [272]. Other proofs were given by Cauchy [123] and Lebesgue [481, p. 279, 283]; for generalizations, see the Notes to Chapter 9. Connections between this congruence and class numbers of quadratic number fields were first noticed by Gauss in a letter to Dirichlet (May 30, 1828). For primes $p = 4n+1$ he defined integers q, r by $q \equiv n! \bmod p$ and $r \equiv \frac{(2n)!}{n!} \bmod p$, and he proved that $2a \equiv r/q \bmod p$ and $2b \equiv r^2 \bmod p$, where $p = a^2 + b^2$, $a \equiv 1 \bmod 4$, whereas b is only determined up to sign (this can be found in [272]). For primes $p \equiv 5 \bmod 8$ he found that the sign of b is given by $b \equiv h+a-1 \bmod 8$, where h is the class number of $\mathbb{Q}(\sqrt{-p})$. This was later rediscovered (without proof) by Jacobi [Ja1, Ja2]. For a generalization, see the paper of K.S. Williams & Currie [868]. For a different generalization of Gauss's congruence, check the paper [CDE] of Chowla, Dwork & Evans where they prove Beuker's conjecture that

$$\binom{2m}{m} \equiv \left(1 + \frac{2^{p-1}-1}{2}\right)\left(2a - \frac{p}{2a}\right) \bmod p^2,$$

where $p = 4m+1 = a^2 + b^2$ and where the sign of a is determined by the congruence $a \equiv 1 \bmod 4$.

The special case $u = 1$ of Proposition 6.10 is due to Th. Gosset [317], and the biquadratic reciprocity law in $\mathbb{Q}(\zeta_8)$ (Proposition 6.11) can already be found in Goldscheider's paper [315, p. 5, formula 8)].

Exercises

6.1 Let A and B be finite groups. What we have denoted by $A \wr B$ is called the *standard complete wreath product* of A and B by Robinson [Rob], p. 313. The construction is as follows: For each element $\beta \in B$, let A_β be a copy of A and put $\widetilde{A} := \bigoplus_{b \in B} A_b$. For a typical element $\widetilde{a} \in \widetilde{A}$, let $(\widetilde{a})_\beta$ denote the coordinate of $\widetilde{a}$ corresponding to A_β. Then B acts on $\widetilde{A}$ via $(\widetilde{a}^b)_\beta := (\widetilde{a})_{\beta b^{-1}}$ and we define $A \wr B = \widetilde{A} \rtimes B$ (the semidirect product). In particular, $A \wr B$ has order $(\#A)^{\#B} \#B$.

6.2 Let $k = \mathbb{Q}(i)$, and assume that $\mu \in \mathcal{O}_k = \mathbb{Z}[i]$ is square-free. Suppose moreover that $\mu \equiv 1 \bmod 2$ and put $m = N\mu$. Then
   i) $k(\sqrt{\mu})/k$ is unramified above $1+i$ if and only if $\mu \equiv 1 \bmod 4$, i.e., if and only if $m \equiv 1 \bmod 8$;
   ii) $k(\sqrt[4]{\mu})/k$ is unramified above $1+i$ if and only if $m \equiv 1 \bmod 16$.

Deduce that $k(\sqrt[4]{\pi^3\overline{\pi}})/k$ is unramified outside p if $\pi \equiv 1 \bmod 2 + 2i$.

6.3 Let $G = C_4 \wr C_2$ be defined as in Section 6.1 above; show that $\tau^{-1}\widetilde{\tau}$ generates the commutator subgroup G' of G, and that $\tau^{-1}\widetilde{\tau}$ fixes $\sqrt[4]{\pi^3\overline{\pi}}$.

6.4 (Eisenstein [Eis]) Let $p \equiv 1 \bmod 4$ be a prime, and χ a quartic character of $\mathbb{F}_p^\times$. Put $a_k = \#\{r : 1 \leq r \leq p-2, \chi(r)\chi(r+1) = i^k\}$. and show that $(a_0 - a_2)^2 + (a_1 - a_3)^2 = p$. (Hint: consider $\sum \chi(r)\chi(r+1)$).
Remarks: Eisenstein's problems in [Eis] were solved by P. Tardy [Tar].

6.5 (cf. Sharifi [731]) Let $p = 1 + 4x^2$ be a prime; for any odd $a \mid x$, write $a^* = (-1)^{(a-1)/2}a$. Show that a^* is a quartic residue modulo p. Moreover, if x is even, then this holds for any divisor a of x. (Hint: $p = N(1 + 2xi)$.)

6.6 (Jacobi) Let $p = a^2 + b^2 \equiv 1 \bmod 8$ be a prime, and suppose that $a \equiv 1 \bmod 4$. Show that $\left(\frac{b}{p}\right)_4 = 1$ and $\left(\frac{a}{p}\right)_4 = (-1)^{(p-1)/8} = (-1)^{(a-1)/4}$. Deduce that $[\pi/\overline{\pi}] = (-1)^{(a-1)/4}(2/p)_4$ for $\pi = a + bi$, and thus $[\pi/\overline{\pi}] = [\overline{\pi}/\pi]$. What happens when $p \equiv 5 \bmod 8$?

6.7 A more direct approach to the last problem goes as follows: observe that $[\pi/\overline{\pi}] = [2bi/\overline{\pi}] = [2i/\overline{\pi}][b/\overline{\pi}]$ and show that the last symbol is 1.

6.8 In [273], Gauss gave the following criterion for the biquadratic character of 6 modulo primes $p = a^2 + b^2$, where a is odd: $(6/p)_4 = +1$ if and only if $b \equiv 0, \pm 2a \bmod 24$ or $3 \mid a$, $8 \mid (b+4a)$. Prove this. Show also that $(6/p)_4 = -1$ if and only if $b \equiv 12a, \pm 10a \bmod 24$ or $3 \mid a$, $8 \mid b$.

6.9 Let $p = 4n + 1$ be a prime. Use Gauss's congruence and Wilson's theorem $(p-1)! \equiv -1 \bmod p$ to show that $(2/p) = (-1)^n$.

6.10 Let $p = 4m + 1 = a^2 + b^2$ be prime, and suppose that $a + b \equiv 1 \bmod 4$. Compute $\binom{3m}{2m}$ (Hint: Corollary 6.6).

6.11 (E. Lehmer [510]) Let $p = a^2 + 4b^2 = c^2 + 2d^2 \equiv 1 \bmod 8$ be prime. Show that all divisors of $c^4 - pa^2$ and of $d^4 - pb^2$ are quartic residues modulo p.

6.12 Let $F_n = 2^{2^n} + 1$ be the n-th Fermat number. Show that F_n is prime if and only if $(-3)^{(F_n-1)/4} \equiv -2^{2^{n-1}} \bmod F_n$, and prove that this supports the conjecture made in Exercise 2.15. (Hints: Write $F_n = N\pi$ for the primary element $\pi = 1 + 2^{2^{n-1}}i \in \mathbb{Z}[i]$, observe that $(-3)^{(F_n-1)/4} \equiv [-3/\pi] \bmod \pi$, and use the quartic reciprocity law.)

6.13 Let $q = a^2 + 64b^2 = 8k + 1$ be prime, and suppose that $b + k$ is even; then $q \mid M_k = 2^k - 1$.

6.14 Determine the quadratic Jacobsthal sum $\Phi_1(D)$ (cf. Exercise 1.22) by computing it modulo $2p$.

6.15 (Continued from Exercise 1.30) Use the determination of $\Phi_2(D)$ to prove that

$$8RRR = \begin{cases} p - 3 & \text{if } p \equiv 3 \bmod 8, \\ p - 7 & \text{if } p \equiv 7 \bmod 8, \\ p - 15 - 2a & \text{if } p \equiv 1 \bmod 8, \\ p - 7 + 2a & \text{if } p \equiv 5 \bmod 8, \end{cases}$$

where a is determined by $p = a^2 + b^2$ and $a \equiv 1 \bmod 4$.

6.16 For an integer $k \geq 2$ and a character χ of order k on $\mathbb{F}_p$ ($p \equiv 1 \bmod 4$ prime), define the sum $G_4 = \sum_{j=1}^{k-1} G(\chi^j)$. Show that $G_4 = G(\chi) + G(\chi^3) - \sqrt{p}$, and compute G_4^2.

6.17 Artin mentioned the following version of the quartic reciprocity law in a letter to Hasse from July 26, 1927 ([239]): suppose that $\alpha = a + 2bi$ and $\beta = c + 2di$; then
$$\left[\frac{\alpha}{\beta}\right] = (-1)^{bd + \frac{a-1}{2}d + \frac{c-1}{2}b}\left[\frac{\beta}{\alpha}\right], \quad \left[\frac{1+i}{\alpha}\right] = i^{\frac{b(a-3b)}{2} - \frac{a^2-1}{8}}.$$
Look it up in Gauss [278] and derive it from Theorem 6.9.

6.18 Let $\alpha, \beta \in \mathbb{Z}[i]$ be odd, and define fourth roots of unity $\varepsilon_\alpha, \varepsilon_\beta$ by $\varepsilon_\alpha \equiv \alpha \bmod (2+2i)$ and $\varepsilon_\beta \equiv \beta \bmod (2+2i)$. Derive the following reciprocity law from Theorem 6.9:
$$\left[\frac{\alpha}{\beta}\right]\left[\frac{\beta}{\alpha}\right]^{-1} = (-1)^{\frac{N\alpha-1}{4}\frac{N\beta-1}{4}}\varepsilon_\alpha^{\frac{N\beta-1}{4}}\varepsilon_\beta^{\frac{N\alpha-1}{4}}.$$
This general version of the quartic reciprocity law appears in Kubota's paper [453] as well as in the book of Stalker [748].

6.19 Derive Burde's reciprocity law directly from Theorem 6.9; give a second proof using Gauss sums.

6.20 Deduce Rédei's generalized version of Burde's reciprocity law (compare p. 172) from the biquadratic reciprocity law.

6.21 In the notation of Proposition 6.11, show that
$$\left[\frac{1-\zeta}{\alpha}\right] = (-1)^{\frac{D}{2}}i^{\frac{C^2+D^2}{4}}i^{\frac{1-A-B-C-D}{4}}.$$

6.22 Compute the conductors of the characters $[i/\cdot]$, $[2/\cdot]$, and $[\pi/\cdot]$, where $\pi \equiv 1 \bmod 2 + 2i$ is prime.

6.23 (Jacobi's proof of the biquadratic reciprocity law)
Let $a + bi$ and $c + di$ be different primes in $\mathbb{Z}[i]$, and assume that $a \equiv c \equiv 1 \bmod 4$.
1. Using Gauss sums, prove the following two formulas:
$$\left[\frac{a-bi}{c \pm di}\right]\left[\frac{a+bi}{c \pm di}\right]^{-1} = \left[\frac{q}{a-bi}\right]; \tag{6.2}$$
$$\left[\frac{a+bi}{q}\right] = \left[\frac{-q}{a-bi}\right]. \tag{6.3}$$

2. a) Let $f \equiv 1 \bmod 4$ be a rational prime dividing $ac - bd$. Use (6.2) to prove that
$$\left[\frac{f}{a-bi}\right]\left[\frac{f}{c-di}\right] = (-1)^{(f-1)/4}.$$
b) Let $q \equiv 3 \bmod 4$ be a rational prime dividing $ac - bd$. Use (6.3) to prove that
$$\left[\frac{-q}{a-bi}\right]\left[\frac{-q}{c-di}\right] = (-1)^{(q+1)/4}.$$

3. Prove that $\left[\frac{-a}{a-bi}\right] = (-1)^{(a+1)/4}$.
4. Use these last results to show that
$$\left[\frac{ac-bd}{a-bi}\right]\left[\frac{ac-bd}{c-di}\right] = (-1)^{(ac-bd-1)/4}.$$
5. Use the congruences $ac - bd \equiv -a(-c-di) \bmod a - bi$ and $ac - bd \equiv -c(-a-bi) \bmod c - di$ to prove
$$\left[\frac{-c-di}{a-bi}\right]\left[\frac{-a-bi}{c-di}\right] = (-1)^{(ac-bd-1)/4}(-1)^{(a+1)/4}(-1)^{(c+1)/4}.$$
Show that the right hand side equals $(-1)^{bd/4}$, and deduce the quartic reciprocity law.

6.24 Let $p = (4\alpha \pm 1)^2 + (8\beta)^2 \equiv 1 \bmod 8$ be prime; show that $\left(\frac{2}{p}\right)_8 = (-1)^{\alpha+\beta}$. This is due to Reuschle [678] and Bickmore [52]).

6.25 Let $p \equiv 1 \bmod 8$ be prime; determine $(2/p)_8$
a) by using $2i = (1+i)^2$ and the quartic reciprocity law in $\mathbb{Z}[i]$;
b) by using $2 = \sqrt{2}^2$ and the quartic reciprocity law in $\mathbb{Z}[\zeta_8]$.

6.26 Complete the proof of the quartic reciprocity law in $\mathbb{Q}(\zeta_8)$; try to generalize this to the quartic fields $\mathbb{Q}(\zeta_{12})$ and $\mathbb{Q}(i, \sqrt{5})$.

6.27 (Goldscheider [315] Let ζ be a primitive eighth root of unity, and $\alpha = A + B\zeta + C\zeta_2 + D\zeta^3 \equiv 1 \bmod 2 + 2\zeta$ a primary element of $\mathbb{Z}[\zeta]$. Let $(\cdot/\cdot)$ and $[\cdot/\cdot]$ denote the quadratic and octic power residue symbol in $\mathbb{Z}[\zeta]$, and let σ_c denote the automorphism $\zeta \mapsto \zeta^c$. Show that $(\sigma_3\alpha/\alpha) = (-1)^{(B+D)/2}$ and $(\sigma_5\alpha/\alpha) = (-1)^{C/2}$. Show also that $[\sigma_3\alpha/\alpha]^2 = \zeta^{2C+(B^2-D^2)/2}$, and that $[\sigma_5\alpha/\alpha] = [\alpha/\sigma_5\alpha]i^C$.

6.28 In this exercise we determine the octic character of $2 + \sqrt{3}$. Consider the biquadratic number field $K = \mathbb{Q}(\sqrt{-1}, \sqrt{3}) = \mathbb{Q}(\zeta_{12})$. Let $[\cdot/\cdot]_K$ and $[\cdot/\cdot]$ denote the quartic residue symbols in $\mathcal{O}_K$ and $\mathbb{Z}[i]$, respectively.
i) Show that every $\alpha \in \mathcal{O}_K$ with odd norm has an associate of the form $A + Bi + C\sqrt{3} + D\sqrt{-3}$ with $A - 1 \equiv B \equiv C \equiv D \equiv 0 \bmod 2$. Such α will be called primary. (Hint: The unit group of $\mathcal{O}_K$ is $\langle \zeta_{12}, \eta \rangle$, where $\eta = (1+\sqrt{3})/(1-i)$ satisfies $\eta^2 = i(2+\sqrt{3})$. You only need that these are units, not that they generate E_K.)
ii) Let $p \nmid 6$ be a prime, and choose a primary generator $\Pi = A + Bi + C\sqrt{3} + D\sqrt{-3}$ of the prime ideal above p in $\mathcal{O}_K$. Write $X = A + Bi$ and $Y = C + Di$, and let π denote the prime in $\mathbb{Z}[i]$ below Π; then
$$\left[\frac{1+\sqrt{3}}{\Pi}\right]_K = (-1)^{D/2}\left[\frac{-2}{X-Y}\right].$$
(Hint: put $Y = (1+i)^r Y'$ and $i^s Y' \equiv 1 \bmod 2 + 2i$; then show that
$$\left[\frac{1+\sqrt{3}}{\Pi}\right]_K = \left[\frac{1+i}{\pi}\right]^{-r}\left[\frac{i}{\pi}\right]^s\left[\frac{1+i}{X-Y}\right]^{2r}\left[\frac{i}{X-Y}\right]^2\left[\frac{-2}{X-Y}\right].$$
Now evaluate these symbols for $r = 2, 3, 4$ and $r \geq 5$.)

iii) Use the relation $2(2+\sqrt{3}) = (1+\sqrt{3})^2$ to deduce that

$$\left(\frac{2+\sqrt{3}}{p}\right)_8 = (-1)^{\frac{C}{2}+\frac{BC+D}{4}} = (-1)^{n+b+d},$$

where $p = 8n+1 = a^2+16b^2 = c^2+192d^2$ (these conditions guarantee that $(2+\sqrt{3}/p)_4 = 1$).

6.29 Compute $(8+3\sqrt{7}/p)_8$ by imitating the last exercise. Show first that every element of $\mathcal{O}_K$ (where $K = \mathbb{Q}(\sqrt{-1}, \sqrt{7})$) with odd norm has an associate of the form $A + Bi + C\sqrt{7} + D\sqrt{-7}$ with $A-1 \equiv B \equiv C \equiv D \equiv 0 \bmod 2$. Then show that

$$\left[\frac{3+\sqrt{7}}{\Pi}\right]_K = (-1)^{D/2}\left[\frac{-2}{X-3Y}\right]$$

and deduce the result

$$\left(\frac{8+3\sqrt{7}}{p}\right)_8 = (-1)^{\frac{B+C}{2}+\frac{BC+D}{4}} = (-1)^{n+d},$$

where $p = 8n+1 = c^2+192d^2$.

6.30 Let $p = 4a^2+3b^2 \equiv 7 \bmod 12$, b odd, and $a, b > 0$. Then

$$(2+\sqrt{3})^{\frac{p+1}{4}} \equiv \begin{cases} +1 & \bmod p, \text{ if } a \equiv 2 \bmod 4, \\ -1 & \bmod p, \text{ if } a \equiv 0 \bmod 4, \\ \left(\frac{-1}{ab}\right)\frac{b}{a}\sqrt{3} & \bmod p \text{ otherwise} \end{cases}$$

Hints: put $k = \mathbb{Q}(i)$, $K = k(\sqrt{3}) = \mathbb{Q}(\zeta_{12})$, $\pi = 2a + b\sqrt{-3} \in \mathcal{O}_K$, and let $[\cdot/\cdot]_K$ and $[\cdot/\cdot]$ denote the quartic residue symbols in K and k, respectively. Then show

i) $[\frac{b}{\pi}]_K = 1$;   iii) $[\frac{b}{b+2ai}] = \pm 1$;

ii) $[\frac{b+2ai}{p}] = (-1)^a[\frac{p}{b+2ai}]$;   iv) $[\frac{p}{b+2ai}] = [\frac{2}{b+2ai}]$;

Next show that $(1+\sqrt{3})^{(p^2-1)/4} \equiv (-1)^a i^{ab} \bmod \pi$; to this end, compute the quartic symbol $[1+\sqrt{3}/\pi]_K$ by reducing it to a symbol in $\mathbb{Z}[i]$ via the congruence $b\sqrt{3} \equiv 2ai \bmod \pi\mathcal{O}_K$.
Finally prove that $(1+\sqrt{3})^{(p+1)/2} \equiv (\frac{3}{a})(\frac{-6}{c})(\frac{-1}{d})\frac{2d}{c}\sqrt{3} \bmod p$ if a is odd, where $p = c^2 + 6d^2$, c, d odd and positive (write $(1+\sqrt{3})^{(p+1)/2} \equiv \varepsilon \frac{2d}{c}\sqrt{3} \bmod p$ for some $\varepsilon = \pm 1$ and raise this congruence to the $\frac{p-1}{2}$-th power).

6.31 Let $p = c^2 + 6d^2 \equiv 7 \bmod 12$, with $c, d > 0$. Use the technique of the preceding exercise to show that $(5+2\sqrt{6})^{\frac{p+1}{4}} \equiv -\left(\frac{-1}{cd}\right)\frac{d}{c}\sqrt{6} \bmod p$.

6.32 Let $p = 4a^2 + 7b^2 \equiv 11, 15, 23 \bmod 28$, b odd, and $a, b > 0$. Then

$$(8+3\sqrt{7})^{\frac{p+1}{4}} \equiv \begin{cases} +1 & \bmod p, \text{ if } a \equiv 0 \bmod 4, \\ -1 & \bmod p, \text{ if } a \equiv 2 \bmod 4, \\ \left(\frac{-1}{ab}\right)\frac{b}{a}\sqrt{7} & \bmod p \text{ otherwise.} \end{cases}$$

Now let $q \equiv 1 \bmod 6$ be a prime such that $p = 2^q - 1$ is a Mersenne prime. Show that $p = 4a^2 + 7b^2$ with a even and $b \equiv \pm 3 \bmod 8$. Define the series S_n by $S_1 = 16$, $S_{n+1} = S_n^2 - 2$ and show that there exists an $r \leq p - 3$ such that $p \mid S_r$ if and only if $a \equiv 0 \bmod 4$.

Additional References

[Bri] J. Brillhart, *Concerning the numbers $2^{2p}+1$, p prime*, Math. Comp. **16** (1962), 424–430

[Cay] A. Cayley, *On the binomial equation $x^p - 1 = 0$; trisection and quartisection*, Proc. London. Math. Soc. **9** (1880), 4–17

[CDE] S. Chowla, B. Dwork and R. Evans, *On mod p^2 determination of $\binom{(p-1)/2}{(p-1)/4}$*, J. Number Theory **24** (1986), 188–196

[Eis] G. Eisenstein, *Aufgaben und Lehrsätze*, J. Reine Angew. Math. **27** (1844), 281–283; Math. Werke I, 108–110

[Rob] D. Robinson, *A course in the theory of groups*, GTM 80, Springer Verlag 1982

[Sch] E. Schering, Göttinger Nachrichten 1879, p. 384; cf. [280], p. 69

[Tar] P. Tardy, *Sopra alcuni teoremi aritmetici*, Annali di Mat. (2) **3** (1870), 331–338

7. Cubic Reciprocity

The cubic reciprocity law can be studied in the same way as quartic reciprocity; both laws live in imaginary quadratic number fields with a finite number of units. We will proceed exactly as in the quartic case: first we study connections with the splitting of primes in certain cubic cyclic extensions, and then we use cubic Gauss and Jacobi sums to derive the cubic reciprocity law (expressed less euphemistically: this chapter won't contain any new ideas – we just compute).

Applying the technique of Dirichlet, we prove a quadratic reciprocity law in $\mathbb{Q}(\sqrt{-3})$ – Eisenstein's original proof used elliptic functions, the proofs of Herglotz and Hilbert will be discussed in Chapter 8 and in Part II, respectively. We will also give applications to primes of the form $3 \cdot 2^n + 1$, and we will determine the cubic character of the quadratic unit $2 + \sqrt{3}$.

7.1 Splitting of Primes

In this section we will derive criteria for $(q/p)_3 = 1$ in terms of the decomposition $4p = L^2 + 27M^2$ by comparing the splitting of primes in Kummer extensions of $\mathbb{Q}(\sqrt{-3})$ and in cyclotomic extensions. The proofs will be simpler than in the quartic case because the extensions considered here are cyclic.

Let ρ denote a primitive cube root of unity, for example $\rho = \frac{-1+i\sqrt{3}}{2}$, and put $k = \mathbb{Q}(\rho)$; if $(\alpha, 3) = 1$ for some $\alpha \in \mathbb{Z}[\rho]$, then we can always find a unit $\varepsilon \in \mathbb{Z}[\rho]$ such that $\alpha\varepsilon \equiv 1 \bmod 3$. We will call α *primary* if $\alpha \equiv \pm 1 \bmod 3$; this is equivalent to $\alpha = a + b\rho$, $3 \nmid a$, and $3 \mid b$. Primes $p \equiv 1 \bmod 3$ split in $\mathbb{Q}(\rho)$, and we have $p = \pi\overline{\pi}$ for some $\pi = a + b\rho$; moreover we may (and will) assume that $\pi = a + b\rho \equiv \pm 1 \bmod 3$. We can also write π in the form $\pi = \frac{1}{2}(L + 3M\sqrt{-3})$ if we put $L = 2a - b$ and $M = b/3$.

Let K be the cubic subfield of $\mathbb{Q}(\zeta_p)$; since $K/\mathbb{Q}$ is cyclic, so is $K(\rho)/\mathbb{Q}(\rho)$, and Hilbert's Theorem 90 gives us a $\mu \in \mathbb{Z}[\rho]$ such that $K(\rho) = \mathbb{Q}(\rho, \sqrt[3]{\mu})$. Since 3 does not ramify in $K/\mathbb{Q}$, we must have $\mu \equiv \pm 1 \bmod 3$. Now $\mathbb{Z}[\rho]$ is a principal ideal ring, and $K/\mathbb{Q}$ is unramified outside p; canceling third powers dividing μ we see that we can assume without loss of generality that $\mu \in \{\pi, \overline{\pi}, \pi\overline{\pi}, \pi^2\overline{\pi}\}$. Let σ denote complex conjugation; since $K(\rho)/\mathbb{Q}$ is abelian, we get $\mu^\sigma = \xi^3\mu^{-1}$ for some $\xi \in \mathbb{Z}[\rho]$ (here we used Corollary 4.17): this excludes the first three possibilities, and we get $\mu = \pi^2\overline{\pi}$, $\mu^\sigma = \pi\overline{\pi}^2 = p^3\mu^{-1}$.

7. Cubic Reciprocity

Next put $\alpha = \sqrt[3]{\mu} + \sqrt[3]{\overline{\mu}}$, with the roots chosen in such a way that α is real; then $\alpha \in K(\rho) \cap \mathbb{R} = K$, and

$$\alpha^3 = \left(\sqrt[3]{\mu} + \sqrt[3]{\overline{\mu}}\right)^3 = \mu + 3\sqrt[3]{\mu^2\overline{\mu}} + 3\sqrt[3]{\mu\overline{\mu}^2} + \overline{\mu} = pL + 3p\alpha$$

shows that α is a root of the *cubic period equation*

$$f_p(X) = X^3 - 3pX - pL. \tag{7.1}$$

Here we have used that $\mu + \overline{\mu} = p(\pi + \overline{\pi}) = p(2a - b) = pL$; the sign of L is not determined: in fact, if α is a root of f_p, then $-\alpha$ is a root of $X^3 - 3pX + pL$. The connection between (7.1) and the "genuine" cubic period equation is presented in Exercise 7.7.

Now we are in a position to prove rational criteria on cubic residuacity. The simplest case is when ℓ is a prime $\equiv 2 \bmod 3$; then ℓ is inert in $\mathbb{Q}(\rho)$, and we have

$$\left(\frac{\ell}{p}\right)_3 = 1 \iff \ell^{\frac{p-1}{3}} \equiv 1 \bmod p$$
$$\iff \ell \text{ splits in } K/\mathbb{Q}$$
$$\iff \ell \text{ splits in } K(\rho)/\mathbb{Q}(\rho)$$
$$\iff \mu \equiv \xi^3 \bmod \ell \text{ for some } \xi \in \mathbb{Z}[\rho].$$

```
          K(ρ)
         /    \
        /      K
   Q(ρ)
       \      /
          Q
```

Putting $\ell = 2$ this shows that $\left(\frac{2}{p}\right)_3 = 1$ if and only if $\mu \equiv \xi^3 \bmod 2$: but $\Phi(2) = 3$ in $\mathbb{Z}[\rho]$, hence we have $\xi^3 \equiv 1 \bmod 2$ for all ξ prime to 2, and we see $\left(\frac{2}{p}\right)_3 = 1 \iff \mu \equiv 1 \bmod 2$. This in turn is equivalent to $\pi \equiv 1 \bmod 2$ (because $\mu = p\pi$), and we have proved the following conjecture of Euler (1.7):

Proposition 7.1. *2 is a cubic residue mod* $p = \frac{1}{4}(L^2 + 27M^2)$ *if and only if* $L \equiv M \equiv 0 \bmod 2$.

If $\ell > 2$, we observe that $p^{(\ell^2-1)/3} = (p^{\ell-1})^{(\ell+1)/3} \equiv 1 \bmod \ell$, hence $\mu = p\overline{\pi}$ is a cube mod ℓ if and only if $\overline{\pi}$ is. Hence we get

$$\left(\frac{\ell}{p}\right)_3 = 1 \iff \overline{\pi} \equiv \xi^3 \bmod \ell$$
$$\iff \overline{\pi}^{(\ell^2-1)/3} \equiv 1 \bmod \ell \text{ by Euler's Lemma,}$$
$$\iff \pi^{\frac{\ell+1}{3}} \equiv \overline{\pi}^{\frac{\ell+1}{3}} \bmod \ell, \text{ since } \pi^{\ell-1} \equiv \overline{\pi}/\pi \bmod \ell.$$

Writing $\pi = \frac{1}{2}(L + M\sqrt{-3})$ and expanding, we find that the last congruence is equivalent to $g_\ell \equiv 0 \bmod \ell$, where $g_\ell = \sum_{j \equiv 1(2)} 3^j \cdot (-3)^{\frac{j-1}{2}} \binom{n}{j} L^{n-j} M^j$ and $n = \frac{\ell+1}{3}$.

Next suppose that $q = 3$; then $\mathfrak{q} = \sqrt{-3}\,\mathcal{O}_k$ is the prime ideal in $\mathcal{O}_k = \mathbb{Z}[\rho]$ above 3, and proceeding as above we find

$$\left(\frac{3}{p}\right)_3 = 1 \iff 3^{\frac{p-1}{3}} \equiv 1 \bmod p$$
$$\iff 3 \text{ splits in } K/\mathbb{Q}$$
$$\iff \mathfrak{q} \text{ splits in } K(\rho)/\mathbb{Q}(\rho)$$
$$\iff \mu \equiv \xi^3 \bmod 3\mathfrak{q} \text{ for some } \xi \in \mathbb{Z}[\rho],$$

where in the last step we have used the decomposition law for Kummer extensions (Theorem 4.12). Put $\xi = \frac{1}{2}(r + s\sqrt{-3})$; then $\xi^3 = \frac{1}{8}(u + v\sqrt{-3})$, and $v = 3s(r^2 - s^2)$. The fact that $(r, 3) = 1$ shows immediately that $s(r^2 - s^2) \equiv 0 \bmod 3$, and now it is easy to verify that $\alpha \in \mathcal{O}_k \setminus \mathfrak{q}$ is a cube mod $3\mathfrak{q}$ if and only if $\alpha \equiv a \bmod 9$ for some $a \in \mathbb{Z} \setminus 3\mathbb{Z}$. Now $\mu = p\overline{\pi}$, hence μ is a cube mod $3\mathfrak{q}$ if and only if $\overline{\pi}$ is, i.e., if and only if $M \equiv 0 \bmod 3$. We have seen:

Proposition 7.2. *3 is a cubic residue mod* $p = \frac{1}{4}(L^2 + 27M^2)$ *if and only if* $M \equiv 0 \bmod 3$.

Finally we will deal with primes $\ell \equiv 1 \bmod 3$; these ℓ split in $\mathcal{O}_k$ as $\ell\mathcal{O}_k = \lambda\overline{\lambda}$, where λ and $\overline{\lambda}$ may be chosen to be primary. We get

$$\left(\frac{\ell}{p}\right)_3 = 1 \iff \ell^{\frac{p-1}{3}} \equiv 1 \bmod p$$
$$\iff \ell \text{ splits in } K/\mathbb{Q}$$
$$\iff \lambda \text{ splits in } K(\rho)/\mathbb{Q}(\rho)$$
$$\iff \mu = p\overline{\pi} \equiv \xi^3 \bmod \lambda \text{ for some } \xi \in \mathbb{Z}[\rho]$$
$$\iff \mu^{\frac{\ell-1}{3}} \equiv 1 \bmod \lambda$$
$$\iff \overline{\pi}^{\frac{\ell-1}{3}} \equiv (\pi^{-2})^{\frac{\ell-1}{3}} \bmod \lambda$$
$$\iff \overline{\pi}^{\frac{\ell-1}{3}} \equiv \pi^{\frac{\ell-1}{3}} \bmod \lambda.$$

Proceeding exactly as in the case $\ell \equiv 2 \bmod 3$, we find

Proposition 7.3. *Let* $p = \frac{1}{4}(L^2 + 27M^2)$ *and* $\ell > 3$ *be prime, put* $n = \frac{1}{3}\left(\ell - \left(\frac{-3}{\ell}\right)\right)$, *and define*

$$g_\ell = \sum_{j \equiv 1(2)} 3^j \cdot (-3)^{\frac{j-1}{2}} \binom{n}{j} L^{n-j} M^j.$$

Then $\left(\frac{\ell}{p}\right)_3 = 1$ *if and only if* $\ell \mid g_\ell$.
The same result holds if we write $p = a^2 - ab + b^2$ *and define*

$$g_\ell = \sum_{j \equiv 1(3)} \binom{n}{j} a^{n-j} b^j - \sum_{j \equiv 2(3)} \binom{n}{j} a^{n-j} b^j.$$

Examples: Let $p = \frac{1}{4}(L^2 + 27M^2)$ be prime; then

$\left(\frac{5}{p}\right)_3 = 1 \iff LM \equiv 0 \bmod 5;$

$\left(\frac{7}{p}\right)_3 = 1 \iff LM \equiv 0 \bmod 7;$

$\left(\frac{11}{p}\right)_3 = 1 \iff LM(L - 3M)(L + 3M) \equiv 0 \bmod 11;$

$\left(\frac{13}{p}\right)_3 = 1 \iff LM(L - 2M)(L + 2M) \equiv 0 \bmod 13.$

There is a cubic analogue to Proposition 5.6, also due to E. Lehmer [498]:

Proposition 7.4. *Let q and $p = \frac{1}{4}(L^2 + 27M^2)$ be primes; then $(q/p)_3 = 1$ if and only if either $q \mid LM$, or $L \equiv \pm\mu M \bmod q$, where $\mu = \pm\frac{9r}{2u+1} \bmod q$, and where $u \neq 0, 1, -\frac{1}{2}, -\frac{1}{3} \bmod q$ satisfies $3u + 1 \equiv r^2(3u - 3) \bmod q$.*

The proof is left as an Exercise.

7.2 The Cubic Reciprocity Law

Exactly as in the quartic case, the proof of the cubic reciprocity law starts with the computation of the cubic Gauss sum $G(\chi) = -\sum_{t=1}^{p-1} \chi(t)\zeta^t$, or rather of $G(\chi)^3 \in \mathbb{Z}[\rho]$. Here χ is a character of order 3 on $\mathbb{F}_p$, and p is a prime $\equiv 1 \bmod 3$. Special attention will of course be given to the cubic power residue character $\chi = \left[\frac{\cdot}{\pi}\right]$.

Cubic Gauss and Jacobsthal Sums

Proposition 7.5. *Let χ be a cubic character on $\mathbb{F}_p$, where $p \equiv 1 \bmod 3$ is prime. Then*

$$J(\chi, \chi) = a + b\rho, \quad G(\chi)^3 = p \cdot J(\chi, \chi),$$

where $p = a^2 - ab + b^2$ and $a - 1 \equiv b \equiv 0 \bmod 3$. If, in particular, $\chi = \left[\frac{\cdot}{\pi}\right]$ is the cubic residue character modulo some prime $\pi \equiv 1 \bmod 3$, then

$$J(\chi, \chi) = \pi, \quad G(\chi)^3 = \pi^2\overline{\pi}.$$

Proof. In Chapter 4 we have seen that $G(\chi)\overline{G(\chi)} = p$ and that $G(\chi)^3 = p \cdot J(\chi, \chi)$. Therefore we have $|J(\chi, \chi)| = \frac{1}{p}|G(\chi)|^3 = \sqrt{p}$. We have to show that $J(\chi, \chi) \equiv 1 \bmod 3$. To this end we observe that

$$G(\chi)^3 \equiv -\sum_a \chi^3(a)\zeta_p^{3a} = -\sum_{a \neq 0} \zeta_p^a = 1 \bmod 3.$$

Now $\chi(-1) = 1$ gives $J(\chi, \chi) \equiv \chi(-1)pJ(\chi, \chi) = G(\chi)^3 \equiv 1 \bmod 3$. Finally, if $\chi = [\cdot/\pi]$, then $J(\chi, \chi) \equiv 0 \bmod \pi$ by Proposition 4.28. □

7.2 The Cubic Reciprocity Law   213

Obviously, these results imply that $\mathbb{Z}[\rho]$ is a principal ideal domain. We also note that $G(\chi)$ generates the abelian extensions $K^{ab}/\mathbb{Q}$ discussed in Section 7.1.

By considering $J(\chi^2, \chi^2)$ we can prove the following corollary in exactly the same way as Corollary 6.6:

Corollary 7.6. *Let $p = 3m + 1 = \frac{1}{4}(L^2 + 27M^2)$ be prime and assume that $L \equiv 2 \bmod 3$. Then $L \equiv \binom{2m}{m} \bmod p$.*

The cubic Jacobsthal sum $\Phi_3(D)$ is taken care of in Exercise 7.20.

Proof of Eisenstein's Cubic Reciprocity Law

Once the cubic Gauss sum is computed, the proof of Eisenstein's cubic reciprocity law $[\alpha/a] = [a/\alpha]$ becomes a matter of routine. Observe that we may assume $\alpha \equiv a \equiv 1 \bmod 3$ without loss of generality, since $[-1/\alpha] = 1$.

Proposition 7.7. *Suppose that $\alpha \in \mathbb{Z}[\rho]$ and $a \in \mathbb{Z}$ are both $\equiv 1 \bmod 3$; then*

$$\left[\frac{\alpha}{a}\right] = \left[\frac{a}{\alpha}\right], \quad \left[\frac{\rho}{a}\right] = \rho^{\frac{1-a}{3}}, \quad \left[\frac{1-\rho}{a}\right] = \rho^{\frac{a-1}{3}}.$$

Moreover we have $\left[\frac{a}{b}\right] = 1$ for all relatively prime $a, b \in \mathbb{Z}$ such that $3 \nmid b$.

Proof. Since the cubic residue symbol is multiplicative, it is sufficient to prove $\left[\frac{\alpha}{a}\right] = \left[\frac{a}{\alpha}\right]$ for prime values of $\alpha \in \mathbb{Z}[\rho]$ and $a \in \mathbb{Z}$. We will distinguish several cases:

1. Let $\alpha = \pi$ be a prime such that $p = N\pi = \pi\overline{\pi} \equiv 1 \bmod 3$ is a rational prime; furthermore, let $a = q = \lambda\overline{\lambda} \equiv 1 \bmod 3$ be a rational prime. Letting $\chi = \chi_\pi = \left[\frac{\cdot}{\pi}\right]$ we find

$$-G(\chi)^q \equiv \sum_{a=1}^{p-1} \chi(a)^q \zeta_p^{aq} = \sum_{a=1}^{p-1} \chi(a) \zeta_p^{aq} = -\chi(q)^{-1} G(\chi) \bmod q.$$

Multiplying both sides by $G(\overline{\chi})$ and using $G(\chi)G(\overline{\chi}) = p$, we get

$$G(\chi)^{q-1} \equiv \chi(q)^{-1} = \chi^2(q).$$

Plugging in the prime factorization of the Gauss sum, this implies

$$\left[\frac{q}{\pi}\right]^2 \equiv G(\chi)^{q-1} \equiv (\pi^2 \overline{\pi})^{\frac{q-1}{3}} \equiv \left[\frac{\pi^2 \overline{\pi}}{\lambda}\right] \bmod \lambda,$$

and we see that

$$\left[\frac{q}{\pi}\right]^2 = \left[\frac{\pi^2 \overline{\pi}}{\lambda}\right]; \qquad (7.2)$$

note that this equality becomes trivial (and useless) if $p = q$. Next we find

$$\left[\frac{\pi^2\overline{\pi}}{\lambda}\right] = \left[\frac{\pi}{\lambda}\right]^2\left[\frac{\overline{\pi}}{\lambda}\right] = \left[\frac{\pi}{\lambda\overline{\lambda}}\right]^2 = \left[\frac{\pi}{q}\right]^2 \mod \lambda,$$

which yields $\left[\frac{q}{\pi}\right] = \left[\frac{\pi}{q}\right]$ after squaring.

2. Let π be as above, but assume that $a = q \equiv 2 \mod 3$ is prime in $\mathbb{Z}[\rho]$. Then

$$-G(\chi)^q \equiv \sum_{a=1}^{p-1}\chi(a)^q\zeta_p^{aq} = \sum_{a=1}^{p-1}\chi(a)^2\zeta_p^{aq} = -\chi(q)G(\chi^2) \mod q,$$

hence $G(\chi)^{q+1} \equiv \chi(q)G(\overline{\chi})G(\chi) = \chi(q)p \mod q$. Therefore we have $\pi\overline{\pi}\chi_\pi(q) \equiv G(\chi)^{q+1} = (\pi^2\overline{\pi})^{(q+1)/3} \mod q$, and using $\pi^q \equiv \overline{\pi} \mod q$ we find

$$\pi^{q+1}\left[\frac{q}{\pi}\right] \equiv (\pi^{q+2})^{\frac{q+1}{3}} \mod q,$$

which gives

$$\left[\frac{q}{\pi}\right] \equiv \pi^{\frac{q^2-1}{3}} \equiv \left[\frac{\pi}{q}\right] \mod q.$$

3. If $p \equiv q \equiv 2 \mod 3$, cubic reciprocity is trivially true:

$$\left[\frac{q}{p}\right] \equiv p^{\frac{q^2-1}{3}} \equiv (p^{q-1})^{\frac{q+1}{3}} \equiv 1 \mod q, \quad \text{hence} \quad \left[\frac{q}{p}\right] = \left[\frac{p}{q}\right] = 1.$$

This takes care of Eisenstein's reciprocity law for cubic residues. It remains to prove the supplementary laws; if $q \equiv 2 \mod 3$ is an inert prime, then $[\rho/q] \equiv \rho^{(q^2-1)/3} = (\rho^{(q+1)/3})^{q-1} = \rho^{(q+1)/3} \mod q$, while for primes $p = \pi\overline{\pi} \equiv 1 \mod 3$ we find $[\rho/p] = [\rho/\pi][\rho/\overline{\pi}] = [\rho/\pi][\rho^2/\pi]^2 = [\rho/\pi]^2 = \rho^{2(p-1)/3} = \rho^{(1-p)/3}$. Finally $[3/a] = 1$ by case 3. above; thus $[1 - \rho/a]^2 = [-3\rho/a] = [\rho/a]$, hence $[1 - \rho/a] = [\rho/a]^2 = \rho^{(a-1)/3}$. $\square$

The cubic reciprocity law follows from this special case (exactly as in the theory of quartic reciprocity) by purely formal manipulations. Before we give these computations, we will present the well known (and much shorter) proof of cubic reciprocity based on Equation (7.2).

First Proof of the Cubic Reciprocity Law

We start by observing that the cubic reciprocity law has already been proved if one of the entries is a rational prime $\equiv 2 \mod 3$. Assume therefore that $p \equiv q \equiv 1 \mod 3$, and that $p = \pi\overline{\pi}$, $q = \lambda\overline{\lambda}$, where π and λ are primary.

7.2 The Cubic Reciprocity Law

We start with Equation (7.2), applied to both p and q (assuming they are different):

$$\left[\frac{q}{\pi}\right]^2 = \left[\frac{\pi^2\overline{\pi}}{\lambda}\right], \quad \left[\frac{\lambda^2\overline{\lambda}}{\pi}\right]^2 = \left[\frac{p}{\lambda}\right].$$

Multiplying both equations and canceling third powers we get $\left[\frac{\pi}{\lambda}\right] = \left[\frac{\lambda}{\pi}\right]$.

We still have to prove reciprocity for primes with the same norm[1] and show that $\left[\frac{\overline{\pi}}{\pi}\right] = \left[\frac{\pi}{\overline{\pi}}\right]$. This is most easily done with a trick due to Evans (see Mollin [Mol, p. 279]): we have

$$\left[\frac{\overline{\pi}}{\pi}\right] = \left[\frac{\overline{\pi}+\pi}{\pi}\right] = \left[\frac{\pi}{\overline{\pi}+\pi}\right] = \left[\frac{-\overline{\pi}}{\overline{\pi}+\pi}\right] = \left[\frac{\overline{\pi}+\pi}{\overline{\pi}}\right] = \left[\frac{\pi}{\overline{\pi}}\right].$$

The extension to composite values of π and λ is trivial and left to the reader.

Proof of the First Supplementary Law

Next we show how to find $[\rho/\alpha]$ for $\alpha \equiv 1 \bmod 3$: we start with the observation that $[\rho/\alpha] = [\rho^2/\overline{\alpha}]^2 = [\rho/\overline{\alpha}]$ (just note that $\rho^4 = \rho$). Multiplying both sides by $[\rho/\alpha]$ gives

$$\left[\frac{\rho}{\alpha}\right]^2 = \left[\frac{\rho}{N\alpha}\right] = \rho^{(1-N\alpha)/3} = \rho^{(1-a^2+ab-b^2)/3},$$

where the second equality comes from Proposition 7.7. Now $b^2 \equiv 0 \bmod 9$, and from $a \equiv 1 \bmod 3$ we deduce that $a^2 \equiv 2a - 1 \bmod 9$ and $ab \equiv b \bmod 9$; this yields

$$\left[\frac{\rho}{\alpha}\right]^2 = \rho^{(1-2a+1+b)/3} = \rho^{-(1-a-b)/3}.$$

Squaring gives the first supplementary law of cubic reciprocity.

Second Proof of the Cubic Reciprocity Law

Now let $\alpha, \beta \in \mathbb{Z}[\rho]$ be primes such that $(\alpha,\beta) = (\alpha,3) = (\beta,3) = (1)$; for our derivation of the cubic reciprocity law from Eisenstein's law we need the following congruences and identities:

$$a\rho \equiv b - a \quad \bmod \alpha \tag{7.3}$$
$$c\rho \equiv d - c \quad \bmod \beta \tag{7.4}$$
$$\overline{\alpha}\beta = ac + bd - bc + (ad - bc)\rho \tag{7.5}$$
$$\alpha\overline{\beta} = ac + bd - ad + (bc - ad)\rho \tag{7.6}$$
$$ac + bd - bc \equiv (bc - ad)\rho \bmod \overline{\alpha} \tag{7.7}$$
$$ac + bd - ad \equiv (ad - bc)\rho \bmod \alpha \tag{7.8}$$

---
[1] This case is often forgotten; note that our proof fails in this case since Equation (7.2) breaks down here.

The first two congruences are clear, (7.5) and (7.6) can be verified immediately, and (7.7) and (7.8) are consequences of (7.5) and (7.6), respectively.
Using (7.4) we get

$$\left[\frac{\alpha}{\beta}\right] = \left[\frac{c}{\beta}\right]^{-1}\left[\frac{ac+bd-bc}{\beta}\right].$$

Similarly, we find by using (7.3), (7.6) and (7.5), as well as $[\rho/\alpha] = [\rho/\overline{\alpha}]$

$$\left[\frac{\beta}{\alpha}\right]^{-1} = \left[\frac{a}{\alpha}\right]\left[\frac{ac+bd-ad}{\alpha}\right]^{-1} = \left[\frac{a}{\alpha}\right]\left[\frac{(bc-ad)\rho}{\alpha}\right]^{-1}$$

$$= \left[\frac{a}{\alpha}\right]\left[\frac{(bc-ad)\rho^2}{\overline{\alpha}}\right] = \left[\frac{a}{\alpha}\right]\left[\frac{\rho}{\overline{\alpha}}\right]\left[\frac{ac+bd-bc}{\overline{\alpha}}\right].$$

This shows that

$$\varepsilon(\alpha,\beta) := \left[\frac{\alpha}{\beta}\right]\left[\frac{\beta}{\alpha}\right]^{-1} = \left[\frac{c}{\beta}\right]^{-1}\left[\frac{a}{\alpha}\right]\left[\frac{\rho}{\overline{\alpha}}\right]\left[\frac{ac+bd-bc}{\overline{\alpha}\beta}\right].$$

Using Eisenstein reciprocity to invert the last symbol, applying (7.5) once more and using the fact that $[m/n] = 1$ for coprime $m, n \in \mathbb{Z}\setminus 3\mathbb{Z}$ we get

$$\left[\frac{ac+bd-bc}{\overline{\alpha}\beta}\right] = \left[\frac{\overline{\alpha}\beta}{ac+bd-bc}\right] = \left[\frac{(ad-bc)\rho}{ac+bd-bc}\right] = \left[\frac{\rho}{ac+bd-bc}\right].$$

Similarly, we have $\left[\frac{a}{\alpha}\right] = \left[\frac{\alpha}{a}\right] = \left[\frac{\rho}{a}\right]$, $\left[\frac{c}{\beta}\right] = \left[\frac{\beta}{c}\right] = \left[\frac{\rho}{c}\right]$, hence

$$\varepsilon(\alpha,\beta) = \left[\frac{\rho}{c}\right]^{-1}\left[\frac{\rho}{a}\right]\left[\frac{\rho}{\overline{\alpha}}\right]\left[\frac{\rho}{ac+bd-bc}\right].$$

But $[\rho/m]$ only depends on the residue class of m mod 9, so we may replace $ac+bd-bc$ by $ac-bc$, and using the supplementary laws of Proposition 7.7 (assuming $a \equiv 1$ mod 3 for the moment) we find

$$\varepsilon(\alpha,\beta) = \left[\frac{\rho}{a}\right]\left[\frac{\rho}{\overline{\alpha}}\right]\left[\frac{\rho}{a-b}\right] = \rho^{\frac{1-a}{3}}\rho^{\frac{1-a-b}{3}}\rho^{\frac{1-a+b}{3}} = \rho^{1-a} = 1.$$

This concludes our second proof of the cubic reciprocity law; most computations done here will be used in our proof of the sextic reciprocity law below.

Proof of the Second Supplementary Law

Finally we have to derive the second supplementary law. This is easy: write $\alpha = a + 3m\rho$; then

$$\left[\frac{3\rho}{\alpha}\right] = \left[\frac{-3\rho}{\alpha}\right] = \left[\frac{\alpha - 3\rho}{\alpha}\right] = \left[\frac{\alpha}{\alpha - 3\rho}\right] = \left[\frac{3\rho}{\alpha - 3\rho}\right].$$

Repeating this reduction m times and using Proposition 7.7 we get

$$\left[\frac{3\rho}{\alpha}\right] = \left[\frac{3\rho}{a}\right] = \left[\frac{\rho}{a}\right] = \rho^{\frac{1-a}{3}}.$$

But now the supplementary law follows: $\left[\frac{1-\rho}{\alpha}\right] = \left[\frac{1-\rho}{\alpha}\right]^4 = \left[\frac{3\rho}{\alpha}\right]^2 = \rho^{\frac{a-1}{3}}$.
We have proved

Theorem 7.8. *Let $\alpha, \beta \in \mathbb{Z}[\rho]$ be primary; then*

$$\left[\frac{\alpha}{\beta}\right] = \left[\frac{\beta}{\alpha}\right].$$

If moreover $\alpha = a + b\rho$, where $a = 3m + 1$ and $b = 3n$, then

$$\left[\frac{\rho}{\alpha}\right] = \rho^{\frac{1-a-b}{3}} = \rho^{-m-n}, \quad \left[\frac{1-\rho}{\alpha}\right] = \rho^{\frac{a-1}{3}} = \rho^m, \quad \left[\frac{3}{\alpha}\right] = \rho^{\frac{b}{3}} = \rho^n.$$

In particular, the characters $\left[\frac{\rho}{\cdot}\right]$ and $\left[\frac{1-\rho}{\cdot}\right]$ have conductors $(1 - \rho)^3$ and $(1 - \rho)^4$, respectively.

7.3 Sextic Reciprocity

Since $\mathbb{Q}(\rho)$ contains the sixth roots of unity, it is reasonable to ask if there exists a sextic reciprocity law in $\mathbb{Q}(\rho)$. This question has been answered positively by Eisenstein [199], and he did so in a very elegant way. In this section we will call $\alpha = a + b\rho$ E-primary if $b \equiv 0 \bmod 3$ and

$$\begin{aligned}
a + b &\equiv 1 \bmod 4, \text{ if } 2 \mid b, & \text{Type I} \\
b &\equiv 1 \bmod 4, \text{ if } 2 \mid a, & \text{Type II} \\
a &\equiv 3 \bmod 4, \text{ if } 2 \nmid ab, & \text{Type III.}
\end{aligned}$$

This is Eisenstein's definition, and I have no idea how he arrived at it. The following lemma, however, gives at least a hint (see Exercise 7.2):

Lemma 7.9. *An $\alpha \in \mathbb{Z}[\rho]$ with $(\alpha, 6) = 1$ is E-primary if and only if $\alpha^3 = A + B\rho$ for $A, B \in \mathbb{Z}$ such that $6 \mid B$ and $A + B \equiv 1 \bmod 4$.*

Now we have

Theorem 7.10. *If $\alpha, \beta \in \mathbb{Z}[\rho]$ are E-primary and relatively prime, then*

$$\left(\frac{\alpha}{\beta}\right)_6 = (-1)^{\frac{N\alpha - 1}{2} \frac{N\beta - 1}{2}} \left(\frac{\beta}{\alpha}\right)_6.$$

218   7. Cubic Reciprocity

Moreover, we have the supplementary laws

$$\left(\frac{1-\rho}{\alpha}\right)_2 = \left(\frac{a}{3}\right)_{\mathbb{Z}} \quad \text{and} \quad \left(\frac{2}{\alpha}\right) = \left(\frac{2}{N\alpha}\right)_{\mathbb{Z}},$$

where $(\frac{\cdot}{\cdot})_2$ and $(\frac{\cdot}{\cdot})_{\mathbb{Z}}$ denote the quadratic residue symbols in $\mathbb{Z}[\rho]$ and in $\mathbb{Z}$, respectively.

It is obviously sufficient to prove the quadratic reciprocity law in $\mathbb{Z}[\rho]$; we will do this using Dirichlet's technique. To this end we put $A = N\alpha$ and $B = N\beta$; we want to prove

$$\left(\frac{\alpha}{\beta}\right)_2 = (-1)^{\frac{A-1}{2}\frac{B-1}{2}} \left(\frac{\beta}{\alpha}\right)_2. \tag{7.9}$$

First consider the case where $\beta = q^*$, $q \equiv 2 \bmod 3$ and $q^* = (-1)^{(q-1)/2}q$; then $\alpha^{q+1} \equiv N\alpha = A \bmod q$, hence

$$\left(\frac{\alpha}{\beta}\right)_2 \equiv (\alpha^{q+1})^{(q-1)/2} \equiv \left(\frac{A}{q}\right)_{\mathbb{Z}} \bmod q,$$

$$\left(\frac{\beta}{\alpha}\right)_2 \equiv q^{(A-1)/2} \equiv \left(\frac{q}{A}\right)_{\mathbb{Z}} \bmod \alpha,$$

and $B = q^2 \equiv 1 \bmod 8$ shows that Equation (7.9) holds.

Now assume that $\alpha = a + b\rho$ and $\beta = c + d\rho$ are prime and E-primary (with $bd \neq 0$). Since $\left(\frac{\alpha}{\beta}\right)_2 = \left(\frac{\alpha^3}{\beta}\right)_2$ we may replace α and β by their cubes, i.e. we may assume that $\alpha \equiv \beta \equiv 1 \bmod 2$ both have type I. The proof of the quadratic reciprocity law consists in transforming residue symbols in $\mathbb{Z}[\rho]$ to symbols in $\mathbb{Z}$ and applying quadratic reciprocity there. Beginning exactly as in the proof of the cubic reciprocity law, we get

$$\left(\frac{\alpha}{\beta}\right)_2 = \left(\frac{c}{\beta}\right)_2 \left(\frac{ac+bd-bc}{\beta}\right)_2, \quad \left(\frac{\beta}{\alpha}\right)_2 = \left(\frac{a}{\alpha}\right)_2 \left(\frac{ac+bd-ad}{\alpha}\right)_2.$$

Next we use (7.5)–(7.8), $(a/\alpha)_2 = (a/A)_2$ and $(\rho/\alpha)_2 = 1$ to deduce

$$\left(\frac{\beta}{\alpha}\right)_2 = \left(\frac{a}{\alpha}\right)_2 \left(\frac{ac+bd-ad}{\alpha}\right)_2 = \left(\frac{-a}{A}\right)_{\mathbb{Z}} \left(\frac{ac+bd-bc}{\overline{\alpha}}\right)_2.$$

This shows that $\varepsilon(\alpha, \beta) := \left(\frac{\alpha}{\beta}\right)_2 \left(\frac{\beta}{\alpha}\right)_2$ equals

$$\varepsilon(\alpha, \beta) = \left(\frac{-a}{A}\right)_{\mathbb{Z}} \left(\frac{c}{B}\right)_{\mathbb{Z}} \left(\frac{ac+bd-bc}{A}\right)_{\mathbb{Z}} \left(\frac{ac+bd-bc}{B}\right)_{\mathbb{Z}}.$$

But now $\left(\frac{-a}{A}\right) = \left(\frac{A}{a}\right) = 1$, because either $A \equiv 1 \bmod 4$ or $A \equiv 3 \bmod 4$ and $b \equiv 2 \bmod 4$, $a \equiv 3 \bmod 4$. Similarly, we have $\left(\frac{c}{B}\right) = \left(\frac{-1}{B}\right)$. Now $ac+bd-bc \equiv (a+b)c \equiv c \bmod 4$, and by inverting the quadratic residue symbols we get

$$\varepsilon(\alpha,\beta) = \left(\frac{-1}{B}\right)_Z (-1)^{\frac{c-1}{2}\frac{A-1}{2}} (-1)^{\frac{c-1}{2}\frac{B-1}{2}} \left(\frac{AB}{ac+bd-bc}\right)_Z.$$

Since $B = c^2 - cd + d^2 \equiv 1 - d \equiv c \bmod 4$, we see that

$$\left(\frac{-1}{B}\right)_Z (-1)^{\frac{c-1}{2}\frac{A-1}{2}} (-1)^{\frac{c-1}{2}\frac{B-1}{2}} = (-1)^{\frac{A-1}{2}\frac{B-1}{2}},$$

which looks promising. All that is left to do now is showing that $(AB/ac + bd - bc)_Z = 1$; we first transform this into a quadratic residue symbol in $\mathbb{Z}[\rho]$ by observing that

$$\left(\frac{AB}{ac+bd-bc}\right)_Z = \left(\frac{\overline{\alpha}\beta}{ac+bd-bc}\right)_2$$

and then use $\overline{\alpha}\beta \equiv (ad - bc)\rho \bmod (ac + bd - bc)$ together with $\left(\frac{m}{n}\right)_2 = 1$ for integers $m, n \in \mathbb{Z}$ (by case 3. in the proof of Proposition 7.7) to see that $(\overline{\alpha}\beta/ac + bd - bc)_2 = (\rho/ac + bd - bc)_2 = 1$ since ρ is a square in $\mathbb{Z}[\rho]$. This concludes our proof of the quadratic reciprocity law in $\mathbb{Z}[\rho]$.

For the proof of the supplementary laws, assume that $\alpha = a + b\rho$ is E-primary and has norm A. Since $-(1 - \rho)$ has type I, we find

$$\left(\frac{1-\rho}{\alpha}\right)_2 = \left(\frac{-1}{\alpha}\right)_2 \left(\frac{\rho-1}{\alpha}\right)_2 = \left(\frac{-1}{A}\right)_Z (-1)^{\frac{A-1}{2}} \left(\frac{\alpha}{1-\rho}\right)_2 = \left(\frac{a}{3}\right)_Z,$$

where we have used the congruence $\alpha \equiv a \bmod (1-\rho)$. Finally, $\left(\frac{2}{\alpha}\right)_2 = \left(\frac{2}{A}\right)_Z$ by Proposition 4.2.ii).

Applications

Here we will use the sextic reciprocity law to discuss a result related to Proposition 6.10:

Proposition 7.11. *Let $q \equiv 1 \bmod 2$, $u \equiv \pm 1 \bmod 6$, and let $p = \pi\overline{\pi} = 6qu + 1$ be prime, where $\pi = a + b\rho$ is E-primary. If $3^{\frac{p-1}{u}} \equiv 1 \bmod p$, then $3^{3q} + 1 = K_q L_q M_q \equiv 0 \bmod p$, where*

$$K_q = 3^q + 1, \quad L_q = 3^q - 3^{\frac{q+1}{2}} + 1, \quad M_q = 3^q + 3^{\frac{q+1}{2}} + 1,$$

and we have
$$\begin{cases} p \mid K_q \iff 9 \mid b, \\ p \mid L_q \iff (-1)^{\frac{p+1}{4}} \equiv \left(\frac{3}{u}\right) \cdot \frac{b}{3} \bmod 3, \\ p \mid M_q \iff (-1)^{\frac{p+1}{4}} \equiv -\left(\frac{3}{u}\right) \cdot \frac{b}{3} \bmod 3. \end{cases}$$

The following table gives a few examples:

q	p	$\chi(p)$	a	b	K_q	L_q	M_q
1	7	$+1$	-2	-3	4	1	0
3	19	-1	-5	-3	9	0	-1
5	31	$+1$	-5	-6	-4	0	23
7	43	-1	7	6	38	0	33
11	67	-1	7	9	0	8	59
13	79	$+1$	-10	-3	25	-29	0

Here $\chi(p) = (-1)^{(p+1)/4}$, and the values of K_q, L_q and M_q are only given modulo p. We will prove Proposition 7.11 only for the special case $u = 1$; the proof of the general case is completed using the same trick as in the quartic case. We begin with the computation of $3^q \pm 3^{(q+1)/2} + 1 \bmod p$; as a matter of fact, it is sufficient to do this modulo π. Since $q = \frac{p-1}{6}$ and $3 = -\rho^2(1-\rho)^2$, we find

$$3^{\frac{p-1}{6}} = (-1)^{\frac{p-1}{6}} (\rho - \rho^2)^{\frac{p-1}{3}} \equiv \left[\frac{\rho - \rho^2}{\pi}\right] = -\rho^{ab/3} \bmod \pi.$$

Similarly, $\quad 3^{\frac{p+5}{12}} = (-1)^{\frac{p+5}{12}} (\rho - \rho^2)(\rho - \rho^2)^{\frac{p-1}{6}}$

$$\equiv (-1)^{\frac{p+1}{4}} (\rho - \rho^2) \left(\frac{\rho}{\pi}\right)_2 \left(\frac{1-\rho}{\pi}\right)_2 \left[\frac{\rho - \rho^2}{\pi}\right]^2$$

$$\equiv (-1)^{\frac{p+1}{4}} (\rho - \rho^2) \left(\frac{a}{3}\right) \rho^{2ab/3} \bmod \pi.$$

A straightforward computation yields

$p \bmod 8$	$a \bmod 3$	$b \bmod 9$	$L_q \bmod \pi$		
7	1	3	$-\rho - (\rho - \rho^2)\rho^2 + 1$	$\equiv$	0
7	1	6	$-\rho^2 - (\rho - \rho^2)\rho + 1$	$\equiv$	$2 - 2\rho^2$
7	2	3	$-\rho^2 + (\rho - \rho^2)\rho + 1$	$\equiv$	0
7	2	6	$-\rho + (\rho - \rho^2)\rho^2 + 1$	$\equiv$	$2 - 2\rho$
3	1	3	$-\rho + (\rho - \rho^2)\rho^2 + 1$	$\equiv$	$2 - 2\rho$
3	1	6	$-\rho^2 + (\rho - \rho^2)\rho + 1$	$\equiv$	0
3	2	3	$-\rho^2 - (\rho - \rho^2)\rho + 1$	$\equiv$	$2 - 2\rho^2$
3	2	6	$-\rho - (\rho - \rho^2)\rho^2 + 1$	$\equiv$	0

This completes the proof in the case $u = 1$.

7.4 Cubic Reciprocity in some Quartic Fields

Using the same ideas as in Section 6.5, we can transfer the cubic reciprocity law to number fields strictly containing $\mathbb{Q}(\rho)$; as an example, we will give

7.4 Cubic Reciprocity in some Quartic Fields

some details for the field $K = \mathbb{Q}(\sqrt{-1}, \sqrt{-3}) = \mathbb{Q}(\zeta_{12})$. We will call an $\alpha \in \mathcal{O}_K$ *primary* if $\alpha \equiv \pm 1 \bmod 3$. $\mathcal{O}_K$ has a fundamental unit $\varepsilon = \frac{1+\sqrt{3}}{1-i}$, and we have $\varepsilon^2 = i(2 + \sqrt{3})$. It is easy to verify that for every $\alpha \in \mathcal{O}_K$ relatively prime to 3 there is a unit $\eta \in \mathcal{O}_K^\times$ such that $\alpha \eta \equiv 1 \bmod 3$ (see Exercises 7.3 and 7.4).

Theorem 7.12. *Let $(\,\cdot\,/\,\cdot\,)$ denote the cubic residue symbol in $\mathbb{Q}(\zeta_{12})$. For primary $\alpha, \beta \in \mathbb{Z}[\zeta_{12}]$, we have the reciprocity law*

$$\left(\frac{\alpha}{\beta}\right) = \left(\frac{\beta}{\alpha}\right).$$

For $\alpha = A + B\rho + C\sqrt{3} + D\rho\sqrt{3} \equiv 1 \bmod 3$, the supplementary laws have the following form:

$$\left(\frac{i}{\alpha}\right) = 1, \quad \left(\frac{\rho}{\alpha}\right) = \rho^{\frac{A+B-1}{3}}, \quad \left(\frac{1-\rho}{\alpha}\right) = \rho^{\frac{1-A}{3}},$$

$$\left(\frac{2+\sqrt{3}}{\alpha}\right) = \rho^C, \quad \left(\frac{\varepsilon}{\alpha}\right) = \rho^{-C}.$$

In particular, for primes $p \equiv 1 \bmod 12$ we have

$$\left(\frac{2+\sqrt{3}}{p}\right)_3 = 1 \iff p = a^2 + 81b^2,$$

where $(\,\cdot\,/p)_3$ denotes a "rational" cubic residue symbol, while for primes $p \equiv 5 \bmod 12$ we get

$$\left(\frac{2+\sqrt{3}}{p}\right)_3 = 1 \iff p = a^2 + b^2, a^2 \equiv b^2 \bmod 9.$$

Proof. (Sketch) We will use $(\,\cdot\,/\,\cdot\,)$ for the cubic residue symbol in $\mathcal{O}_K = \mathbb{Z}[i, \rho]$ and $[\,\cdot\,/\,\cdot\,]$ for the one in $\mathbb{Z}[\rho]$. Write a primary element $\alpha \in \mathcal{O}_K$ in the form $\alpha = X + Y\sqrt{3}$ with $X = A + B\rho$ and $B = C + D\rho$, where $A - 1 \equiv B \equiv C + D \equiv 0 \bmod 3$. If $Y = 0$, then $\alpha \in \mathbb{Z}[\rho]$, and from Proposition 4.2.ii) one easily deduces that $(\beta/\alpha) = (\alpha/\beta)$. If $Y \neq 0$, we may assume that α divides a prime $p \equiv 1 \bmod 12$; write $\beta = V + W\sqrt{3}$, check $(X, Y) = (V, W) = 1$ and put $\tau = (X, W)$ (we may assume that $\tau \equiv 1 \bmod 3$). Put $X = \tau x$, $W = \tau w$, $U = Vx - 3wY$, $\pi = X^2 - 3Y^2$, and $\lambda = V^2 - 3W^2$. Since $V\sqrt{3} \equiv -3W \bmod \beta$, we get

$$\left(\frac{\alpha}{\beta}\right) = \left(\frac{V^2}{V+W\sqrt{3}}\right)\left(\frac{VX-3WY}{V+W\sqrt{3}}\right) = \left[\frac{V^2}{\lambda}\right]\left[\frac{\tau}{\lambda}\right]\left[\frac{U}{\lambda}\right].$$

After this reduction to residue symbols in $\mathbb{Z}[\rho]$ we can use the cubic reciprocity law there and find

222   7. Cubic Reciprocity

$$\left(\frac{\alpha}{\beta}\right) = \left[\frac{\lambda}{V}\right]^2 \left[\frac{\lambda}{\tau}\right]\left[\frac{\lambda}{U}\right] = \left[\frac{9W}{V}\right]\left[\frac{V}{\tau}\right]^2\left[\frac{\lambda}{U}\right].$$

Using cubic reciprocity to invert $[V/\tau]$ and observing $W = w\tau$ we get

$$\left(\frac{\alpha}{\beta}\right) = \left[\frac{9w}{V}\right]\left[\frac{\lambda}{U}\right].$$

Similarly, we can show that

$$\left(\frac{\beta}{\alpha}\right) = \left[\frac{9Y}{x}\right]\left[\frac{\pi}{U}\right].$$

Now we use $Wx = wX$, $x^2\lambda \equiv -3w^2\pi \bmod U$, and $[x/U] = [U/x] = [3wY/x]$ to get $[\lambda/U] = [x/U][3w^2/U][\pi/U]$ and

$$\left(\frac{\alpha}{\beta}\right)\left(\frac{\beta}{\alpha}\right)^{-1} = \left[\frac{3}{V}\right]^2\left[\frac{w}{V}\right]\left[\frac{x}{U}\right]\left[\frac{3w^2}{U}\right]\left[\frac{3}{x}\right]\left[\frac{Y}{x}\right]^2 = \left[\frac{3}{Vx}\right]^2\left[\frac{3}{U}\right]\left[\frac{w}{Vx}\right]\left[\frac{w}{U}\right]$$

Since $V^2x^2U = V^3x^3 - 3V^2x^2wY \equiv 1 \bmod 9$, we find $[3/Vx]^2[3/u] = 1$ (recall that the character $[3/\cdot]$ has conductor 9). Writing $w = (1-\rho)^r\rho^s\sigma$ for some $\sigma \equiv \pm 1 \bmod 3$ we finally get

$$\left[\frac{w}{Vx}\right]\left[\frac{w}{U}\right]^2 = \left[\frac{\sigma}{Vx}\right]\left[\frac{\sigma}{U}\right]^2 = \left[\frac{VxU^2}{\sigma}\right] = \left[\frac{Vx}{\sigma}\right]^3 = 1,$$

and now the reciprocity law follows.

The supplementary laws for ρ and $1-\rho$ follow immediately from the corresponding results in $\mathbb{Z}[\rho]$ (use Proposition 4.2), the cubic character of $2+\sqrt{3}$ is computed as follows:

$$\left(\frac{2+\sqrt{3}}{X+Y\sqrt{3}}\right) = \left(\frac{X}{\pi}\right)^{-1}\left(\frac{2X-3Y}{\pi}\right) = \left(\frac{9Y}{X}\right)\left(\frac{\pi}{2X-3Y}\right)$$

$$= \left(\frac{9Y}{X}\right)\left(\frac{9}{2X-3Y}\right)\left(\frac{3X^2-9Y^2}{2X-3Y}\right)$$

$$= \left(\frac{9Y}{X}\right)\left(\frac{9}{2X-3Y}\right)\left(\frac{X^2}{2X-3Y}\right)$$

$$= \left(\frac{9Y}{X}\right)\left(\frac{9}{2X-3Y}\right)\left(\frac{3Y}{X}\right)^2$$

$$= \left(\frac{3}{X}\right)\left(\frac{9}{2X-3Y}\right) = \left(\frac{9}{2-3Y}\right) = \rho^{-D} = \rho^C,$$

where we have used $X^2 \equiv 1 \bmod 9$ and $3XY \equiv 3Y \bmod 9$.   □

Notes

The first indication where to look for a theory of cubic residues came from the conjectures of Euler that we already mentioned on page 13, namely the criteria for cubic residuacity of small integers. Here are his results: Let $p \equiv 1 \bmod 3$ be a prime and write $p = a^2 + 3b^2$. Then q is a cubic residue modulo p if and only if one of the following divisibility relations for a and b hold:

q	relations
2	$3 \mid a$
3	$9 \mid a$; $9 \mid (a \pm b)$
5	$15 \mid a$; $3 \mid a, 5 \mid b$; $15 \mid (a \pm b)$; $15 \mid (a \pm 2b)$
6	$9 \mid a$; $9 \mid (2a \pm b)$
7	$21 \mid a$; $3 \mid a, 7 \mid b$; $21 \mid (a \pm b)$; $7 \mid (4a \pm b)$; $7 \mid (a \pm 2b)$

The cases $q = 2, 3, 5$ follow almost immediately from our results in Section 7.1: simply observe that if $\pi \equiv a + b\sqrt{-3}$ with $9 \mid (a \pm b)$, then $A + B\rho = \pi\rho$ (if $3 \mid (a - b)$) or $A + B\rho = \pi\rho^2$ (if $3 \mid (a + b)$) is primary and that the condition $9 \mid (a \pm b)$ is equivalent to $9 \mid B$.

The case $q = 6$ takes a little more effort: clearly $1 = (6/p)_3$ for primes $p \equiv 1 \bmod 3$ implies $1 = (2/p)_3 (3/p)_3$, but criteria for $(2/p)_3 = 1$ and $(3/p)_3 = 1$ do not suffice to derive the cubic residuacity of 6. In fact, 6 is a cubic residue modulo $p = \pi\overline{\pi}$ if and only if one of the following holds: a) $[2/\pi] = [3/\pi] = 1$; b) $[2/\pi] = \rho$, $[3/\pi] = \rho^2$; or c) $[2/\pi] = \rho^2$, $[3/\pi] = \rho$. Using Eisenstein's cubic reciprocity law, Euler's conjecture follows after some calculation (see Exercise 7.13).

Gauss's Disquisitiones [263] only contain a few results connected with higher power residues: the part dealing with binary quadratic forms shows that primes $p \equiv 1 \bmod 3$ can be represented as $4p = L^2 + 27M^2$ (which was already known to Euler and Lagrange). In his section on cyclotomy, he computes the cubic period equation in terms of L and M by computing cubic cyclotomic numbers. In a footnote in article 358, Gauss gives the prime factorization of the cubic Gauss sum as well as the congruence $L(m!)^3 \equiv 1 \bmod p$ for $p = 3m + 1$ (this topic will be discussed below).

Gauss started working on higher power residues in earnest around 1805, and as we have seen in the Notes to Chapter 6, seemed to have derived the "rational" part of cubic and quartic reciprocity around 1807. He returned to these problems between 1814 and 1817 (recall his comments on Fermat's Last Theorem on page 1.4), and this time he seems to have discovered the general reciprocity laws for cubic and quartic residues. Around the same time he published his fifth and sixth proof of the quadratic reciprocity law that he had already found between 1807 and 1808 and mentioned that they were motivated by his research on cubic and quartic residues. His 1832 paper also contains the remark that the formulation of the cubic reciprocity law involves the integers $\mathbb{Z}[\rho]$ with $\rho^2 + \rho + 1 = 0$.

With the insight gained from his articles [272, 273] on quartic reciprocity, the number of solutions of the congruence $ax^3 + by^3 \equiv 1 \bmod p$ can be easily derived from Gauss's early results (see Exercise 4.32 and Chapter 10 for more details). His sketches [275] provide proofs for the cubic character of 2 and 3, but the simple proof for $(3/p)_3$ announced in his diary on Jan. 06, 1809 has apparently not been reconstructed.

In 1827, Jacobi [396] published several theorems on cubic residues without proofs; as a matter of fact, Jacobi did not publish a single proof of any of his theorems on cyclotomy, although he presented them in his lectures [402] of 1837. In his letter to Legendre (Aug. 5, 1827), Jacobi writes

> M. Gauss a présenté à la Societé de Goettingen, il y a environ deux ans, un premier mémoire relatif à la théorie des résidus biquadratiques, laquelle est beaucoup plus facile que celle des résidus cubiques.[2]

(He is referring to Gauss's Commentatio prima [272], which appeared in 1828, but had been presented to the Society on April 5, 1825). Jacobi's claim that the theory of biquadratic residues was much easier than that of cubic residues is a bit strange. In fact, in his second paper [400] on cubic residues, he expresses the opposite opinion.

Anyway, in [396] Jacobi announces the following

Theorem 7.13. *Let $p \equiv q \equiv 1 \bmod 6$ be primes, write $4p = L^2 + 27M^2$, and let x be a solution of the congruence $x^2 \equiv -3 \bmod q$. Then*

$$\left(\frac{q}{p}\right)_3 = 1 \iff \left(\frac{L+3Mx}{2}p\Big/q\right)_3 = 1 \iff \left(\frac{L+3Mx}{L-3Mx}\Big/q\right)_3 = 1.$$

This follows easily from our proof of Proposition 7.3: we have shown that

$$\left(\frac{q}{p}\right)_3 = 1 \iff \mu = p\overline{\pi} \equiv \xi^3 \bmod \lambda \iff \left(\frac{\pi^2\overline{\pi}}{\lambda}\right)_3 = 1.$$

But the congruence $\pi = (L + 3M\sqrt{-3})/2 \equiv (L + 3Mx)/2 \bmod q$ shows that

$$\left(\frac{q}{p}\right)_3 = 1 \iff \left(\frac{\pi^2\overline{\pi}}{\lambda}\right)_3 = 1 \iff \left(\frac{L+3Mx}{2}p\Big/q\right)_3 = 1.$$

The second claim follows similarly by observing that $\left(\frac{\pi^2\overline{\pi}}{\lambda}\right)_3 = \left(\frac{\overline{\pi}/\pi}{\lambda}\right)_3$. In this form, the result was already known to Gauss around 1805, as the papers (e.g. [275]) found in his Nachlass show. Generalizations thereof have been given by Pépin [640]. The Kummer theory in our proof of Proposition 7.3 can be replaced by Cardano's formulas: see Cooke [134].

Apparently Jacobi was not aware of the theorem of cubic reciprocity in 1827; only after Gauss had published the biquadratic reciprocity law in 1832 (the Commentatio secunda, presented to Göttingen's Society on April 15, 1831) it became clear that the conjectures of Euler and the results of Gauss

---

[2] Mr. Gauss presented to the Society of Göttingen two years ago a first memoir on the theory of biquadratic residues, which is much easier than the theory of cubic residues.

and Jacobi on cubic residues could be explained by a reciprocity law. Such a law appeared in print for the first time 1837 in Jacobi's paper [400].

In 1844 Eisenstein, at the age of 21, discovered and published proofs for the cubic and biquadratic reciprocity laws using Gauss sums ([185, 188]); the next year Jacobi had his article from 1837 reprinted in Crelle's Journal, having added the famous footnote (see the Notes of Chapter 8 for further details). Eisenstein's first paper [185] on the cubic reciprocity law contains the statement and proof for the case where both elements have different norm; only in [188] does he prove that $[\pi/\overline{\pi}] = [\overline{\pi}/\pi] = 1$, and this part of the proof of the cubic reciprocity law is often neglected in textbooks. The cyclotomic proofs of the cubic reciprocity law given by Eisenstein (including his derivation of the supplementary laws using Newton's formulas in [188]) are discussed in detail by Collison [132].

Proofs similar to Eisenstein's first one have been given by Skolem [740], Friesen, Spearman & K. S. Williams [241] and Joly [413]. Habicht [321] worked out a geometric proof of the cubic reciprocity law (sketches of which were found in Gauss's Nachlass [276], including Gauss's lemma for cubic residues). Fueter's preface [Fu1] to the fifth volume of Euler's collected works derives Euler's conjectures from Eisenstein's results.

Among the papers dealing with the cubic residuacity of small primes, we mention Chowla [127] ($p = 2$), Dedekind [151] ($p = 2, 3$), Jensen [412], (he uses the ring class fields whose existence Dedekind could only prove in special cases), and Stickelberger [759] ($p = 5$). Stieltjes [763] uses elementary cyclotomic methods for computing the cubic character of $1 + \rho$. Analytic proofs of the cubic reciprocity law are due to Dantscher [148] (he also proves the supplementary laws), Gegenbauer [283] and Petr [655]; see also Koschmieder [434], Lewandowski [537], Sbrana [701], and Watabe [820] (he computes $\left(\frac{1-\rho}{\pi}\right)$ using elliptic functions). Dintzl [160] derived the second supplementary law from the cubic reciprocity law; a proof by an ad-hoc computation is due to K.S. Williams [855]. Hayashi [344] gave a proof of the cubic reciprocity law in the spirit of Kaplan's [419] technique in the quartic case. A modern proof of the cubic reciprocity law is contained in Kubert [452].

For other papers dealing with cubic residues, see Dickson [158, Theorem 7] (he gave a criterion for $(2/p)_3 = 1$ using Gauss's cyclotomic method of computing the number of solutions of $ax^3 + by^3 = 1$ over finite fields), Dittmar [176] (elementary stuff), Dunton [Dun] (is interested in pairs of cubic residues modulo a prime p), Genocchi [299] (deals with questions of cubic residues that were discussed in letters between S. Germain and Gauss), Le Vavasseur [534, 535] (discusses the elementary arithmetic of $\mathbb{Q}(\sqrt{-3})$ and states the cubic reciprocity law), Hübler [383], Ito [393], Park & Kim [624] (derive the supplementary laws and a criterion for $(3/p)_3 = 1$), Pellet [631, 632, 633], Pépin [636, 638], Reshetukha [680, 681], Sato & Karakisawa [700] (they work out explicit criteria for cubic residuacity in terms of $4p = L^2 + 27M^2$), Stern [752] (gives Gauss's derivation of cubic cyclotomic numbers and derives criteria for 2 and 3 to be cubic residues), Tihanyi [800] (computes $(2/p)_3$, $(3/p)_3$ etc.), Vaidyanathaswamy [807] (gives cubic cyclotomy in an algebraic form, computes $N(x^3 + y^3 = 1)$ and applies the results to find $(2/p)_3$ and $(3/p)_3$), K. S. Williams [856, 863], and Zuravskij [884] (gives an arithmetic proof of the cubic reciprocity law), [885] (proves the second supplementary law). Fueter's book [246] contains a detailed discussion of the cubic reciprocity law along with the arithmetic of cyclotomic fields which lies behind its proof.

Eisenstein wrote two papers [193, 194] on the theory of cubic forms (which correspond to cyclic cubic fields in the same way binary quadratic forms correspond to quadratic number fields). There he proves, for primes $p = 3m + 1 = \pi\overline{\pi}$, the identity

$$27(x^{p-1} + \ldots + x + 1) = U^3 + p\pi Y^3 + p\overline{\pi} Z^3 - 3pUYZ =: D, \qquad (7.10)$$

where $Y = V + W\rho$, $Z = V + W\rho^2$, and where $U, V, W \in \mathbb{Z}[X]$. He notes that the polynomials satisfy $U + V + W \equiv 0 \bmod 3$, and claims that one can derive e.g. the cubic character of 3 from the properties of these polynomials. At the end of these papers he claims that it is possible to develop a genus theory for these cubic forms and to deduce the cubic reciprocity law from it; contrary to Frei's claim in [238], however, [193] and [194] do not contain a proof of the cubic reciprocity law.

While Gauss was working on cubic residues, he also seems to have come across the cubic form later studied by Eisenstein; the entry 137 in his diary says that he started working on the theory of cubic forms and the solutions of the equation $x^3 + ny^3 + n^2 z^3 - 3nxyz = 1$. In [277] he characterizes primes dividing $x^3 + ny^3 + n^2 z^3 - 3nxyz$ [with $p \nmid xyz$] as those for which n is a cubic residue.

In his papers [311, 312, 313, 314], Gmeiner gave a very complicated proof (about 70 pages) of the sextic reciprocity law, apparently unaware of the previous work of Eisenstein (whose proof is related to the one we will give in Chapter 8), and of Dörrie [179]. Other proofs of the quadratic reciprocity law in $\mathbb{Z}[\rho]$ are due to Brandt [79], Lewandowski [537], Mertens [582], Peitz & Jordan [629], Petr [655], and Scheibner [708]. The dissertation of Glause [309] gives a geometric proof of the supplementary laws of sextic reciprocity.

Proposition 7.3 is (like its quartic counterpart) due to Cunningham & Gosset [146] and has been rediscovered by Felgner [229]; the examples following it can already be found in Jacobi's paper on cubic residues (or in Gauss's Nachlass [275]), and again we refer to [146] for a more extensive list. Other results contained in Jacobi's paper have been rediscovered in a slightly different form by Sun [776]: put $4q = L'^2 + 27M'^2$; then $(L'/3M')^2 \equiv -3 \bmod q$, hence we may replace x in 7.13 by $L'/3M'$, and we get

$$\left(\frac{q}{p}\right)_3 = 1 \iff \left(\frac{LM' + L'M}{LM' - L'M}\bigg/q\right)_3 = 1.$$

A similar result was given by von Lienen [546]: he defined $\left(\frac{p}{q}\right)_3 = \left[\frac{p}{\lambda}\right]$, where $p = \pi\overline{\pi}$ and $q = \lambda\overline{\lambda}$ are different primes $\equiv 1 \bmod 3$, and found

$$\left(\frac{p}{q}\right)_3 \left(\frac{q}{p}\right)_3 = \left[\frac{p}{\lambda}\right]\left[\frac{q}{\pi}\right] = \left[\frac{\pi}{\lambda}\right]\left[\frac{\overline{\pi}}{\lambda}\right]\left[\frac{\lambda}{\pi}\right]\left[\frac{\overline{\lambda}}{\pi}\right] = \left[\frac{\lambda}{\pi}\right]^2\left[\frac{\overline{\lambda}}{\pi}\right]^3 = \left[\frac{\lambda}{\pi}\right]^2.$$

Now $\lambda = \frac{1}{2}(L' + 3M'\sqrt{-3}) \equiv \frac{L'M + LM'}{2M} \bmod \pi$ shows that this can be written in the form

$$\left(\frac{p}{q}\right)_3 \left(\frac{q}{p}\right)_3 = \left(\frac{L'M + LM'}{2M}\bigg/p\right)_3^2.$$

Von Lienen calls this a rational cubic reciprocity law (because of its similarities with Burde's rational reciprocity law); it is easy to see that it contains the results of Jacobi and Sun as a special case. See Burde [100] for a generalization of von Lienen's result.

Proposition 7.4, like Proposition 5.6, is proved in E. Lehmer's paper [498]. The result in Exercise 7.23 is taken from Golomb's paper [Gol]; he credits unpublished papers of E. Lehmer for the cubic part, but at least the case $m \equiv 0 \bmod 2$ follows from [499, Equation (24)]. The results of Exercise 7.22 can also be found in [Gol], but are actually due to Pellet [632].

Let $p = 3m + 1$ be a prime, and write $4p = L^2 + 27M^2$. Then $L \equiv 1 \bmod 3$ and M are cubic residues modulo p. This story starts with a footnote in Gauss's [263, Art. 358], where the congruence $L(m!)^3 \equiv 1 \bmod p$ is given (see Exercise 7.9). This congruence was rediscovered by Jacobi [396] and proved in his lectures [402]. The next rediscovery is due to Clausen [Cla], who solved a problem posed by Stern; the latter had already noticed in [752] that L and M have the same cubic character modulo $p = (L^2 + 27M^2)/4$, but had failed to state the full result there. The statement that divisors of LM are cubic residues modulo p also appears in a letter from Kummer [470] to Kronecker, in Pellet [634], in Bachmann's book [26] (at least partially), and again in Nagell's papers [605, 606]. The most recent rediscoveries are due to Jakubec [Jak], who derives the congruence $L(m!)^3 \equiv 1 \bmod p$ from a more general congruence involving Gaussian periods, and Sato & Shirai [SS], who write $4p^n = f_n(L, M) + 27g_n(L, M)$ for $n \geq 1$ and polynomials f_n, g_n with integral coefficients and show that, for n not divisible by 3, the prime factors of $g_n(L, M)$ are cubic residues modulo p.

The method to determine the cubic character of $2+\sqrt{3}$ used in the proof of Theorem 7.12 (Aigner [8] used class field theory) can be generalized to other quadratic units, and in fact it would be possible to prove results for certain infinite families of units. The insight gained by such computations is, however, probably not worth the effort, as we will see in Part II that there are other (and more sophisticated) ways to do this (and even more). The first cubic residue character of a quadratic unit was determined by Kronecker through the explicit construction of the Hilbert class field of $\mathbb{Q}(\sqrt{-31})$. The next result in this direction was E. Lehmer's determination of $(\varepsilon_5/p)_3$ for primes $p = s^2 + 15t^2 \equiv 1 \bmod 30$; this conjecture of M. Dunton was announced in [501], and proofs were given later by Aigner [7], Sun [776], E. Lehmer [505, 506], Muskat [601], and Weinberger [831].

The number of solutions of the congruence $x^3 + y^3 + cz^3 \equiv 0 \bmod p$ was studied repeatedly by Chowla, J. Cowles & M. Cowles [CC1, CC2, CC3, CC4] using techniques due to Gauss that have also been used by Libri, Lebesgue and Stieltjes.

The cubic period equation was discussed by Gauss in [263, Art. 358] (based on numerical evidence, he had conjectured it on October 1, 1796, and he found a proof on July 20, 1797; see the entries 39 and 67 in his diary), Cayley [Ca1], Carey [115], Cooke [134], Lebesgue [482], Pellet [Pel], and Sylvester [Syl]. See Lazarus [Laz] and D.H. Lehmer & E. Lehmer [LL] for the sextic case.

Solvability of cubic congruences $f(x) = x^3 + ax^2 + bx + c \equiv 0 \bmod p$ can be determined using the cubic reciprocity law if f is abelian, and using quadratic and cubic reciprocity if f has Galois group S_3. As in the quadratic case, the computation of solutions is quite another problem. H.C. Williams & Zarnke [WZ] have shown how to do that using ideas of Cailler, Shanks and D.H. Lehmer.

Anshel & Goldfeld [AG] use the cubic reciprocity law to compute the parity of the number of points of an elliptic curve over a finite field, Dedekind [151] applied it to characterizing the splitting of primes in pure cubic extensions, Selmer [Sel] used it for studying rational points on elliptic curves, and Liverance [552] needed the sextic reciprocity law for computing the root numbers (that is, the signs occurring in the functional equation of $L(s, E)$) of elliptic curves $E : y^2 = x^3 + D$. Saidi [694] applied the cubic reciprocity law to a problem in combinatorial geometry. Applications of cubic reciprocity to primality tests can be found in Fueter [Fu2], H.C. Williams [Wil], Guthmann [Gu] and in Berrizbeitia & Berry [BB].

Exercises

7.1 Prove Proposition 7.4.

7.2 Prove Lemma 7.9. (Hint: we have $\alpha^3 = A + B\rho$ for $A = a^3 - 3ab^2 + b^3$ and $B = 3ab(a - b)$. Now go through the three cases).

7.3 Let $K = \mathbb{Q}(\sqrt{-1}, \sqrt{-3})$; show that an $\alpha \in \mathcal{O}_K$ is primary if and only if $\alpha = A + B\rho + C\sqrt{3} + D\rho\sqrt{3}$ for some $A, B, C, D \in \mathbb{Z}$ such that $A - 1 \equiv B \equiv C + D \equiv 0 \bmod 3$. Moreover, show that every ideal not divisible by $(\sqrt{-3})$ has a primary generator.

7.4 (continued) Using the fact that $E_K = \langle \zeta_{12}, \varepsilon \rangle$, where $\varepsilon = \frac{1+\sqrt{3}}{1+i}$, show that for every $\alpha \in \mathcal{O}_K$ with $(\alpha, 3) = 1$ there is a unit $\eta \in E_K$ such that $\alpha\eta$ is primary.

7.5 Let K/k be a normal extension with Galois group of type $(2,2)$, and let k_j ($j = 1, 2, 3$) denote its quadratic subextensions. Moreover, let E_j be the unit group of the ring of integers in k_j. Show that $E_K^2 \subseteq E_1 E_2 E_3$ (in other words: the square of any unit in $\mathcal{O}_K$ is a product of units lying in the quadratic subextensions). Use this to compute the fundamental unit of $\mathbb{Q}(\sqrt{-1}, \sqrt{-3})$.

7.6 (Eisenstein [188]) Compute the conductors of the cubic characters $[\rho/\cdot]$, $[1 - \rho/\cdot]$, and $[3/\cdot]$. (Dedekind [151] remarked that all these conductors divide 9).

7.7 Let $p \equiv 1 \bmod 3$ be a prime, $K = \mathbb{Q}(\zeta_p)$, and let F denote the cubic subfield of K. Let $\eta_0 = \mathrm{Tr}_{K/F} \zeta_p$ and its conjugates η_1, η_2 denote the cubic periods. Let π be a primary prime in $\mathbb{Z}[\rho]$ dividing p, and let χ denote the corresponding cubic character. Show that $G(\chi) = \eta_0 + \rho \eta_1 + \rho^2 \eta_2$ (for a suitable choice of π) and $G(\chi^{-1}) = \eta_0 + \rho^2 \eta_1 + \rho \eta_2$. Show that this implies that $\alpha = G(\chi) + G(\chi^{-1}) = 3\eta_0 + 1$ for the $\alpha = \sqrt[3]{\mu} + \sqrt[3]{\overline{\mu}}$ studied in Section 7.1, and use the minimal polynomial of α to deduce that

$$F(X) = x^3 + x^2 + \frac{1-p}{3}x + \frac{1 - pL - 3p}{27}$$

is the minimal polynomial of $\eta_0 = \frac{\alpha - 1}{3}$; note that $L \equiv 1 \bmod 3$ here. Also check that $f(X) = X^3 - 3pX - pL$ has discriminant $\Delta_f = 108p^3 - 27p^2 L^2 = -3^6 p^2 M^2$, while $\Delta_F = p^2 M^2$.

7.8 (continued) Use Proposition 3.8 and the results of the preceding exercise to prove that primes $q \mid LM$ are cubic residues modulo p. In particular, derive criteria for the cubic character of 2 and 3.

7.9 Let $p = 6n + 1 = \frac{1}{4}(L^2 + 27M^2)$ be prime; then $\left(\frac{q}{p}\right)_3 = 1$ for all primes $q \mid LM$. Check that $L \equiv ((2n)!)^{-3} \bmod p$ with $L \equiv 1 \bmod 3$ by putting $m = 2n$ in the congruence of Corollary 7.6, observing $4n \equiv -(2n+1), \ldots, 3n + 1 \equiv -3n \bmod p$ and $((3n)!)^2 \equiv -(-1/p) \bmod p$.

7.10 (continued) Use the cubic reciprocity law in Proposition 7.7 to show that primes dividing LM are cubic residues modulo p. Derive this result also from Jacobi's Theorem 7.13.

7.11 (Martinet [566]) Let $p \equiv q \equiv 1 \bmod 3$ be prime, and write $4pq = L^2 + 27M^2$; then $\left(\frac{L}{p}\right)_3 = \left(\frac{L}{q}\right)_3 = 1 \iff \left(\frac{p}{q}\right)_3 = \left(\frac{q}{p}\right)_3 = 1$.

7.12 (cf. Sharifi [731]) Let $p = 1 + 3x + 9x^2$ be prime. Show that any divisor a of x is a cubic residue modulo p. (Hint: $p = N(1 - 3x\rho)$; now use cubic Eisenstein reciprocity.)

7.13 Prove Euler's conjecture concerning the cubic character of 6.

7.14 (Cassels [120]) Suppose that x, y, z are integers such that $x^3 + y^3 + z^3 = 3$; then $x \equiv y \equiv z \bmod 9$. This was generalized by Qi Sun [Sun] and Craig [Cra].

7.15 (Euler [217], Pellet [632], Bickmore [52]) Let $q = 24k + 7$ be prime; if $q = a^2 + 27b^2$, then $q \mid 2^{4k+1} - 1$.

7.16 Suppose that $M_p = 2^p - 1$ is prime for $p > 3$; then $M_p = 4a^2 + 27b^2$ for odd $a, b \in \mathbb{N}$.

230 7. Cubic Reciprocity

7.17 Let q be an odd integer, and suppose that $p = 12a+1 = a^2 +4b^2 = c^2 +3d^2 \equiv 5 \bmod 8$ is prime. Then $3 \mid d$ implies that $p \mid S_q = 2^{2q} + 1$, and we have

$$p \mid A_q \iff b \equiv \pm 1 \bmod 8,$$
$$p \mid B_q \iff b \equiv \pm 3 \bmod 8.$$

7.18 Show that $[\pi/\overline{\pi}] = 1$ for primes $\pi \equiv 1 \bmod 3$. (Hint: Combine $[\pi/\overline{\pi}] = [\overline{\pi}/\pi]$ with the action of complex conjugation on the residue symbol; alternatively, use the congruence $\overline{\pi} = a + b\rho^2 = a - b - b\rho \equiv 2a - b \bmod \pi$ to show that $[\overline{\pi}/\pi] = [2a - b/\pi]$ and use the reciprocity law to compute this symbol.)

7.19 Consider the ring of integers in $K = \mathbb{Q}(\sqrt{2}, \sqrt{-3}\,)$; show that any $\alpha \in \mathcal{O}_K$ prime to 3 has an associate of the form $\alpha = A + B\rho + C\sqrt{2} + D\rho\sqrt{2}$ with $A - 1 \equiv B \equiv C + D \equiv 0 \bmod 3$; call such α primary. Let $p = c^2 + 6d^2$ be a prime, and let choose a primary generator π for the prime ideal in $\mathcal{O}_K$ above p. Show that

$$\left(\frac{1+\sqrt{2}}{\pi}\right)_3 = \rho^C = \rho^{-d}.$$

In particular, $1 + \sqrt{2}$ is a cubic residue mod p if and only if $p = x^2 + 54y^2$. Do the same for the fundamental units of $\mathbb{Q}(\sqrt{q}\,)$ with $q = 3, 5, 17, 41$.

7.20 Let g be a primitive root modulo a prime $p \equiv 1 \bmod 3$. Consider the cubic Jacobsthal sum $\Phi_3(D) = \sum_{x \bmod p} \left(\frac{x}{p}\right) \left(\frac{x^3+D}{p}\right)$. Now show
  1. $\Phi_3(D)$ only depends on $(D/p)_3$;
  2. $\Phi_3(1)+\Phi_3(g)+\Phi_3(g^2) = -3$ (Hint: compute $\sum_{D=0}^{p-1} \Phi_3(D)$ in two different ways);
  3. $\Phi_3(1)^2 + \Phi_3(g)^2 + \Phi_3(g^2)^2 = 6p+3$ (Hint: compute $\sum_{D=0}^{p-1} \Phi_3(D)^2$ in two different ways);
  4. $\Phi_3(4g^\nu) = 1 - p + 3[(0,\nu) + (1, \nu - 1) + (2, \nu - 2)]$, where (μ, ν) are the cyclotomic numbers of order 3;
  5. if $4p = L^2 + 27M^2$ and $L \equiv 1 \bmod 3$, then $L = 1 + \Phi_3(4)$ and $9M = \Phi_3(4g) - \Phi_3(4g^2)$.

7.21 This exercise strengthens results of Felgner [230]. Let $k = 3a^2 + nb^3$, where $a, b, n \in \mathbb{N}$ have the following properties:
  1. $b \equiv \pm 2, \pm 4 \bmod 9$ is square-free;
  2. $n \mid a$, $3 \nmid a$;
  3. $(b, n) = 1$, $3 \mid b$;
  4. $(n/q)_3 = 1$ for all primes $q \equiv 1 \bmod 3$ dividing (a, b).
Then the diophantine equation $x^3 = y^2 + k$ has no solutions in $\mathbb{Z}$. Hints: assume that $(x, y) \in \mathbb{Z}^2$ is a solution.
  a) Show that $(y, n) = 1$;
  b) deduce that $(y + a\sqrt{-3}, n) = 1$ in $\mathbb{Z}[\rho]$;
  c) suppose that $q \equiv 1 \bmod 3$ divides $y^2 + 3a^2$; write $q = \pi\overline{\pi}$ for primes $\pi, \overline{\pi} \equiv 1 \bmod 3$, and show that $[n/\pi]_3 = 1$ or $v_\pi(y + a\sqrt{-3}) \equiv 0 \bmod 3$. This is done as follows:
    1. $q \mid (y, a)$: if $q \nmid x$, then $n \equiv (x/b)^3 \bmod q$. If $q \mid x$, show that $q \mid (a, b)$ and use the assumption.
    2. $q \nmid (y, a)$: if $q \nmid x$, our claim is proved as above. Assume that $q \mid x$. Prove that $q^3 \mid (y^2 + 3a^2)$, and show that there exists a $\pi \mid q$ such that $\pi^3 \mid (y + a\sqrt{-3})$. Put $y + a\sqrt{-3} = \pi^3(y_1 + a_1\sqrt{-3})$, as well as $x = qx_1$ and $b = qb_1$. Show that $q \nmid (y_1, a_1)$, and repeat step 2.

d) show that $3 \nmid (y^2 + 3a^2)$.
The rest of the proof follows [230]: on the one hand we have

$$\left[\frac{y+a\sqrt{-3}}{n}\right] = \left[\frac{y}{n}\right] = +1$$

from Eisenstein's reciprocity law, on the other hand we can show that this cubic residue symbol is $\neq 1$ as follows: factor $y + a\sqrt{-3}$ as

$$y + a\sqrt{-3} = \pm \rho^c \prod q_i \prod \pi_j^{a_j} \prod \pi_k^{b_k},$$

where $0 \leq c \leq 2$, the q_i are primes $\equiv 2 \bmod 3$, the π_j and π_k are primary prime factors of primes $\mathfrak{p} \equiv 1 \bmod 3$, $a_j \equiv 0 \bmod 3$, and $b_k \not\equiv 0 \bmod 3$. Then

$$\left[\frac{y+a\sqrt{-3}}{q_i}\right] = 1, \quad \left[\frac{y+a\sqrt{-3}}{\pi_j}\right]^{a_j} = 1, \quad \left[\frac{y+a\sqrt{-3}}{\pi_k}\right]^{b_k} = 1,$$

where the last relation follows from property 4. Therefore we have

$$\left[\frac{y+a\sqrt{-3}}{n}\right] = \left[\frac{\rho}{n}\right]^c.$$

But $\left[\frac{\rho}{n}\right] \neq 1$ since $n \not\equiv \pm 1 \bmod 9$, and $c \not\equiv 0 \bmod 3$ since this would imply that $y + a\sqrt{-3} = y + a + 2a\rho$ were primary, i.e. that $3 \mid a$. This contradiction proves our claim.

7.22 (Golomb [Gol]) Consider the numbers $T_n = 3 \cdot 2^n + 1$. Assume that $p = T_n$ is prime, and let $e = e_2(p)$ denote the smallest integer ≥ 1 such that $2^e \equiv 1 \bmod p$.
i) Show that $e_2(p) = 3^j 2^k$, where $0 \leq j \leq 1$ and $0 \leq k \leq n$, and that $k \leq n-1$ for $p > 13$.
ii) Show that p divides some Fermat number $F_r = 2^{2^r} + 1$ if and only if $3 \nmid e_2(p)$.
iii) Conclude that if T_n is prime and $n = 2m$ is even, then T_n does not divide any Fermat number.
Remark. It has been known for quite a while that $T_{41} \mid F_{38}$ and $T_{209} \mid F_{207}$; between 1995 to 1998, J. Young [Fer] discovered that $T_{157\,169} \mid F_{157\,167}$, $T_{213\,321} \mid F_{213\,319}$, and $T_{303\,088} \mid F_{303\,093}$. The most recent result that $T_{382449} \mid F_{382447}$ is due to Cosgrave. There are, of course, several other primes of the form T_{2n+1}.

7.23 (Continued) Let $p = 3 \cdot 2^{2m} + 1$ be prime. Show that

$$2^{\frac{p-1}{3}} \equiv \begin{cases} \frac{-1+3\cdot 2^m}{2} \bmod p, & \text{if } m \equiv 0 \bmod 2, \\ \frac{-1-3\cdot 2^m}{2} \bmod p, & \text{if } m \equiv 1 \bmod 2. \end{cases}$$

Moreover, prove that $2^{\frac{p-1}{12}} \equiv \left(\frac{2}{p}\right)_4 \cdot 3(2^{2m} + (-2)^m) \bmod p$.

7.24 In [BB], Berrizbeitia & Berry devised the following primality test for integers of the form $n = A3^k - 1$ with A even and $0 < A < 8 \cdot 3^k - 2$: choose a prime $p \equiv 1 \bmod 3$ such that $(n/p)_3 \neq 1$. Let $\pi \in \mathbb{Z}[\rho]$ be a primary prime dividing p, and put $\tau = (\overline{\pi}/\pi)^A$. Choose an integer Q_0 such that $Q_0 \equiv \tau + \overline{\tau} \bmod n$, and define $Q_{r+1} = Q_r(Q_r^2 - 3)$. Then n is prime if and only if $Q_{k-1} \equiv -1 \bmod n$.

Show that this is a special case of Lenstra's Primality Test in Theorem 3.18. (Hints: First show that $Q_r \equiv \tau^{3^r} + \bar{\tau}^{3^r} \bmod n$. Next, use the assumption that $(n/p)_3 \neq 1$ and the cubic reciprocity law to deduce that $(\bar{\pi}/\pi)^{(n+1)/3} \equiv \rho \bmod n$ if n is prime (you may have to switch the roles of π and $\bar{\pi}$, and it helps to recall the action of the Frobenius automorphism on π). Then show that Lenstra's test applies with $t = 2$, $s = 3^k$, $h(X) = (X - \tau)(X - \bar{\tau})$.)

7.25 The primality test in the preceding exercise is a simplification of a test introduced by H.C. Williams [Wil]. He put $P_1 \equiv p^{-A/2}(\pi^A + \bar{\pi}^A) \bmod n$ and showed that n is prime if and only if $P_k \equiv \pm 1 \bmod n$. Prove this, and show in addition that the sign is given by the Legendre symbol $-(p/n)$. (Hints: $P_k \equiv \pm 1 \bmod n$ implies that $P_{k+1} \equiv \mp 2 \bmod n$. But $P_{k+1} \equiv \tau^{(p+1)/2} + \bar{\tau}^{(p+1)/2} \bmod n$; now $\tau = \bar{\pi}/\pi \equiv \pi^{p-1} \bmod n$, hence $\tau^{(p+1)/2} \equiv (\tau/n) \bmod n$, where $(\cdot/\cdot)$ is the quadratic residue symbol in $\mathbb{Z}[\rho]$.)

Additional References

[AG] M. Anshel, D. Goldfeld, *Zeta functions, one-way functions, and pseudorandom number generators*, Duke Math. J. **88** (1997), 371–390

[BB] P. Berrizbeitia, T.G. Berry, *Cubic reciprocity and generalised Lucas-Lehmer tests for primality of $A \cdot 3^n \pm 1$*, Proc. Amer. Math. Soc. **127** (1999), 1923–1925

[Ca1] A. Cayley, *On the binomial equation $x^p - 1 = 0$; trisection and quartisection*; Proc. London Math. Soc. **11** (1879), 4–17

[Ca2] A. Cayley, *On the binomial equation $x^p - 1 = 0$; quinquisection*; Proc. London Math. Soc. **12** (1880), 15–16; second note, ibid. **16** (1885), 61–63

[Ch] S. Chowla, *A formula similar to Jacobsthal's for the explicit value of x in $p = x^2 + y^2$ where p is a prime of the form $4k + 1$*, Proc. Lahore Philos. Soc. **7** (1945), 2 p.

[CC1] S. Chowla, J. Cowles, M. Cowles, *On the number of zeros of diagonal cubic forms*, J. Number Theory **9** (1977), 502–506

[CC2] S. Chowla, J. Cowles, M. Cowles, *Congruence properties of the number of solutions of some equations*, J. Reine Angew. Math. **298** (1978), 101–103

[CC3] S. Chowla, J. Cowles, M. Cowles, *The number of zeroes of $x^3 + y^3 + cz^3$ in certain finite fields*, J. Reine Angew. Math. **299/300** (1978), 406–410

[CC4] S. Chowla, J. Cowles, M. Cowles, *On the difference of cubes mod p*, Acta Arith. **37** (1980), 61–65

[Cla] S. Clausen, *Auflösung einiger Aufgaben aus gegenwärtigem Journal*, J. Reine Angew. Math. **8** (1832), 138–141

[Cra] M. Craig, *Integer values of $\sum(x^2/yz)$*, J. Number Theory **10** (1978), 62–63

Additional References 233

[Dun] M. Dunton, *Bounds for pairs of cubic residues*, Proc. Am. Math. Soc. **16** (1965), 330–332

[Fer] *Prime factors of Fermat numbers*, URL: http://ballingerr.xray.ufl.edu/proths/fermat.html

[Fu1] R. Fueter, *Vorwort des Herausgebers*, L. Euler's Opera Omnia (I) **V**, Commentationes Arithmeticae, VIII – XXXVII

[Fu2] R. Fueter, *Über primitive Wurzeln von Primzahlen*, Comment. Math. Helv. **18** (1946), 217–223

[Gol] S. W. Golomb, *Properties of the sequence $3 \cdot 2^n + 1$*, Math. Comp. **30** (1976), 657–663

[Gu] A. Guthmann, *Effective primality tests for integers of the forms $N = k \cdot 3^n + 1$ and $N = k \cdot 2^m 3^n + 1$*, BIT **32** (1992), 529–534

[Jak] S. Jakubec, *The congruence for Gauss period*, J. Number Theory **48** (1994), 36–45

[Laz] A. Lazarus, *The sextic period polynomial*, Bull. Austral. Math. Soc. **49** (1994), 293–304

[LL] D.H. Lehmer, E. Lehmer, *The sextic period polynomial*, Pacific J. Math. **111** (1984), 341–355

[Mol] R. Mollin, *Algebraic Number Theory*, CRC 1999

[Pel] A. E. Pellet, *Mémoire sur la théorie algébrique des équations*, Bull. Soc. Math. France **15** (1887), 61–103

[SS] K. Sato, S. Shirai, *On certain rational expressions whose prime divisors are cubic residues (mod p)*, Applications of Fibonacci numbers, Vol. 6 (1996), 423–429

[SvR] L. von Schrutka, *Ein Beweis für die Zerlegbarkeit der Primzahlen von der Form $6n+1$ in ein einfaches und ein dreifaches Quadrat*, J. Reine Angew. Math. **140** (1911), 252–265

[Sel] E.S. Selmer, *The Diophantine Equation $ax^3 + by^3 + cz^3 = 0$*, Acta Math. **85** (1951), 203–302

[Sun] Q. Sun, *On the diophantine equation $x^3 + y^3 + z^3 = n$*, Kexue Tongbao, Sci. Bull. **33**, No.24 (1988), 2007–2010

[Syl] J. J. Sylvester, *Sur les équations à 3 et à 4 périodes des racines de l'unité*, C. R. Assoc. Franc. Reims 1880, 96–98

[Wil] H.C. Williams, *The primality of $N = 2A3^n - 1$*, Can. Math. Bill. **15** (1972), 585–589

[WZ] H.C. Williams, C.R. Zarnke, *Some algorithms for solving a cubic congruence modulo p*, Utilitas Math. **6** (1974), 285–306

8. Eisenstein's Analytic Proofs

In this chapter we will have a closer look at Eisenstein's analytic proofs for the reciprocity laws for quadratic, cubic and quartic residues. In the historical survey which he put at the beginning of his paper [468], Kummer praises them with the following words:

> Für einen der schönsten Beweise dieses von den ausgezeichnetsten Mathematikern viel bewiesenen Theorems wird aber derjenige mit Recht gehalten, welchen Eisenstein in Crelle's Journal, Bd. 29, pag. 177, gegeben hat. In diesem wird das Legendresche Zeichen $\left(\frac{p}{q}\right)$ durch Kreisfunktionen so ausgedrückt, daß bei der Vertauschung von p und q dieser Ausdruck, bis auf eine leicht zu bestimmende Änderung im Vorzeichen, ungeändert bleibt. [...] Wenn dieser Eisensteinsche Beweis schon wegen seiner vorzüglichen Eleganz beachtenswerth ist, so wird der Werth desselben noch dadurch erhöht, daß er, wie Eisenstein selbst gezeigt hat, ohne besondere Schwierigkeit auch auf die biquadratischen und kubischen Reciprocitätsgesetze angewendet werden kann, wenn anstatt der Kreisfunktionen elliptische Funktionen mit bestimmten Moduln angewendet werden.[1]

In fact Eisenstein's proofs of the cubic and quartic reciprocity laws using elliptic functions are much simpler than those using Gauss sums, at least from a computational point of view. Most proofs of the quadratic reciprocity law given in Chapter 3 using cyclotomic methods probably have their counterparts in the lemniscatic theory; so far only a few of these proofs have been uncovered, however.

Here's a short description of what to expect in this chapter: in Section 8.1 we will discuss Eisenstein's proof of the quadratic reciprocity law via the

---

[1] One of the most beautiful proofs of this theorem that was proved by the most distinguished mathematicians is without doubt the one given by Eisenstein in Crelle's Journal, vol. 29, p. 177. There, Legendre's symbol (p/q) is expressed by circular functions in such a way that switching p and q leaves this expression invariant up to an easily determined change of sign. [...] Although this proof of Eisenstein is already remarkable because of its exquisite elegance, its value is even increased by the fact that, as Eisenstein himself has shown, it can be applied without particular difficulties to biquadratic and cubic reciprocity laws if one uses elliptic functions for certain moduli instead of the circular functions.

sine function; one of the relations occurring here will be used to determine the sign of the quadratic Gauss sum in Section 8.7. In Section 8.2 we will see how Abel constructed elliptic functions – today's approach via Weierstraß' ℘-functions is presented in Section 8.3. These results will then be applied to derive quartic, cubic and quadratic reciprocity laws in imaginary quadratic number fields. Finally, we will show what Kronecker's Jugendtraum has got to do with all this.

8.1 Quadratic Reciprocity

Eisenstein's proof of the quadratic reciprocity law using the sine function is built on Gauss's Lemma 4.4; in order to exploit it we need $\mathbb{Z}$-periodic functions $f : \mathbb{Q} \longmapsto \mathbb{C}$ such that $f(-z) = -f(z)$ for all $z \in \mathbb{Q} \setminus \mathbb{Z}$; the simplest one is $f : z \longmapsto (-1)^{[z]}$, and this function f was the basis for Gauss's third proof as well as for many other elementary proofs of the quadratic reciprocity law. Eisenstein's idea to use the $\mathbb{Z}$-periodic function $\sin 2\pi z$ can hardly be overrated: eventually it led him to the discovery of the simplest proofs for cubic and quartic residuacity one can imagine.

In the sequel we will give three proofs of the quadratic reciprocity law which are all based on Eisenstein's work on trigonometric functions.

First Proof. Here is Eisenstein's original proof: applying Gauss's Lemma 4.4 to $f(z) = \sin 2\pi z$ yields

Proposition 8.1. *Let $A = \{\alpha \in \mathbb{Z} \mid 1 \leq \alpha \leq \frac{p-1}{2}\}$ be a half-system modulo an odd prime p; then*

$$\left(\frac{q}{p}\right) = \prod_{\alpha \in A} \frac{\sin(\frac{2\pi}{p} q\alpha)}{\sin(\frac{2\pi}{p} \alpha)}. \tag{8.1}$$

Now we will give three proofs of the quadratic reciprocity law which are all based on Eisenstein's work on trigonometric functions.

We start by examining the right hand side of (8.1); the addition theorem for trigonometric functions yields $\sin 2\alpha = 2 \sin \alpha \cos \alpha$ and $\sin 3\alpha = \sin \alpha (3 - 4 \sin^2 \alpha)$. Induction readily shows that $\sin q\alpha = \sin \alpha \cdot P(\sin \alpha)$ for all odd $q \geq 1$, where $P \in \mathbb{Z}[X]$ is a polynomial of degree $q-1$ and highest coefficient $(-4)^{\frac{q-1}{2}}$. Thus there exist $a_i \in \mathbb{Z}$ such that

$$\frac{\sin qz}{\sin z} = (-4)^{\frac{q-1}{2}} \left((\sin z)^{q-1} + a_{q-2}(\sin z)^{q-2} + \ldots + a_0\right)$$
$$= (-4)^{\frac{q-1}{2}} \psi(X), \quad \text{where } X = \sin z. \tag{8.2}$$

Since $\phi(z) = \frac{\sin qz}{\sin z}$ is an even function, so is $\psi(X)$, hence $a_{q-2} = \ldots = a_1 = 0$. Now $\phi(z)$ has zeros $\{\pm \frac{2\pi}{q} \beta, 1 \leq \beta \leq \frac{q-1}{2}\}$; actually, we may replace

$\{1, \ldots, \frac{q-1}{2}\}$ by any half system B modulo q. Since ψ is monic of degree $q-1$, the fact that the $q-1$ zeros of ψ are $X_\beta = \pm \sin \frac{2\pi}{q} \beta$ shows that

$$\psi(X) = \prod_{\beta \in B} \left(X \pm \sin \frac{2\pi}{q} \beta \right) = \prod_{\beta \in B} \left(X^2 - \sin^2 \frac{2\pi}{q} \beta \right).$$

Replacing X by $\sin z$, we get

$$\frac{\sin qz}{\sin z} = (-4)^{\frac{q-1}{2}} \prod_{\beta \in B} \left(\sin^2 z - \sin^2 \frac{2\pi}{q} \beta \right). \tag{8.3}$$

Now we can reap the harvest of our work: recalling that A and B denote half-systems mod p and q, respectively, we put $z = \frac{2\pi}{p} \alpha$ and find

$$\begin{aligned}\left(\frac{q}{p}\right) &= \prod_{\alpha \in A} \frac{\sin(\frac{2\pi}{p} q \alpha)}{\sin(\frac{2\pi}{p} \alpha)} = \prod_{\alpha \in A} (-4)^{\frac{q-1}{2}} \prod_{\beta \in B} \left(\sin^2 \tfrac{2\pi}{p} \alpha - \sin^2 \tfrac{2\pi}{q} \beta \right) \\ &= (-4)^{\frac{p-1}{2} \frac{q-1}{2}} \prod_{\alpha \in A} \prod_{\beta \in B} \left(\sin^2 \tfrac{2\pi}{p} \alpha - \sin^2 \tfrac{2\pi}{q} \beta \right).\end{aligned} \tag{8.4}$$

Exchanging p and q on the right hand side of (8.4) gives rise to a factor $(-1)^{\frac{p-1}{2} \frac{q-1}{2}}$; whence $\left(\frac{q}{p}\right) = (-1)^{\frac{p-1}{2} \frac{q-1}{2}} \left(\frac{p}{q}\right)$, which is the quadratic reciprocity law. Observe that in order to compute $\left(\frac{q}{p}\right)$ from formula (8.4) we need only the *signs* of the factors $\sin^2 \frac{2\pi}{p} \alpha - \sin^2 \frac{2\pi}{q} \beta$.

This is essentially Eisenstein's well known proof of the quadratic reciprocity law; the rest of this section is devoted to relating Eisenstein's proof to the one using Gauss sums. Both proofs will make use of the formula

$$p = \prod_{\beta=1}^{\frac{p-1}{2}} \left(4 \sin^2 \tfrac{2\pi}{p} \beta \right), \tag{8.5}$$

which is derived by applying $\lim_{z \to 0}$ to Equation (8.3), using L'Hospital's rule and replacing q by p.

Second Proof. Let's go back to equation (8.2); it is not hard to determine the coefficients $a_j \in \mathbb{Z}$: to this end observe that $e^{iz} = \cos z + i \sin z$ and put $Y = e^{iz}$. Then

$$\frac{\sin qz}{\sin z} = \frac{Y^q - Y^{-q}}{Y - Y^{-1}}. \tag{8.6}$$

If we define $\begin{bmatrix} n \\ k \end{bmatrix} = \frac{n+k}{n} \binom{n}{k}$ and $Z = Y - Y^{-1}$, we find

$$Y^q - Y^{-q} = \begin{bmatrix} q \\ 0 \end{bmatrix} Z^q + \begin{bmatrix} q-1 \\ 1 \end{bmatrix} Z^{q-2} + + \ldots + \begin{bmatrix} (q+1)/2 \\ (q-1)/2 \end{bmatrix} Z \tag{8.7}$$

for all odd $q \in \mathbb{N}$ (this is a simple proof by induction; see Exercise 8.2). Moreover the proof shows that the $\begin{bmatrix} n \\ k \end{bmatrix}$ are integers. This in turn implies that for prime $q \in \mathbb{N}$ and for all $1 \leq k \leq \frac{q-1}{2}$, we have $\begin{bmatrix} q-k \\ k \end{bmatrix} = \frac{q}{q-k}\binom{q-k}{k} \equiv 0 \bmod q$. Now we see that

$$\frac{Y^q - Y^{-q}}{Y - Y^{-1}} = \sum_{j=0}^{(q-1)/2} \begin{bmatrix} q-j \\ j \end{bmatrix} Z^{q-2j-1} \equiv Z^{q-1} \bmod q.$$

This congruence due to Euler [Eu1] can be given the form

$$\sin qz \equiv (\sin z)^q \bmod q \tag{8.8}$$

and should of course be read as 'there exists a polynomial $P(X) = X^q + a_{q-1}X^{q-1} + \ldots + a_1 X + a_0 \in \mathbb{Z}[X]$ with coefficients $a_j \equiv 0 \bmod q$ such that $\sin qz = P(\sin z)$'.

Another interpretation is the following: if we substitute $z = \frac{2\pi i}{p}$ into (8.8), then there are algebraic numbers on both sides of the congruence; thus in this case it can be read as a plain congruence in $\mathbb{Q}(\zeta_p)$ describing the action of the Frobenius automorphism on $\sin z$, and there is a simple direct proof (see Exercise 8.3).

Plugging (8.8) into (8.1) and (8.6) we get

$$\left(\frac{q}{p}\right) = \prod_{\alpha \in A(p)} \frac{\sin qz}{\sin z} = \prod_{\alpha \in A(p)} \frac{Y^q - Y^{-q}}{Y - Y^{-1}} \equiv \prod_{\alpha \in A(p)} Z^{q-1} \bmod q.$$

Letting $z = \frac{2\pi}{p}\alpha$ in $Z^2 = -4\sin^2 z$ and using Equation (8.5) gives

$$\left(\frac{q}{p}\right) \equiv \prod_{\alpha \in A(p)} (-4\sin^2 \tfrac{2\pi}{p}\alpha)^{\frac{q-1}{2}} = (-1)^{\frac{p-1}{2}\frac{q-1}{2}} p^{\frac{q-1}{2}} \bmod q,$$

and from this congruence we can read off the quadratic reciprocity law immediately.

Third Proof. From (8.5) we get

$$p^* = \prod_{\alpha=1}^{\frac{p-1}{2}} \left(-4\sin^2 \tfrac{2\pi}{p}\alpha\right) = \left(\prod_{\alpha=1}^{\frac{p-1}{2}} \left(2i \sin \tfrac{2\pi}{p}\alpha\right)\right)^2 = \left(\prod_{a=1}^{\frac{p-1}{2}} (\zeta_p^a - \zeta_p^{-a})\right)^2.$$

In Chapter 3 we have shown that $\tau^2 = p^*$, where τ is the quadratic Gauss sum; hence we must have

$$\Lambda := \prod_{a=1}^{\frac{p-1}{2}} (\zeta_p^a - \zeta_p^{-a}) = \pm \sum_{a=1}^{p-1} \left(\frac{a}{p}\right)\zeta_p^a = \pm\tau \tag{8.9}$$

for suitably chosen signs (see Section 8.7). Replacing the Gauss sum τ by Λ we get another proof for the quadratic reciprocity law:

$$\Lambda^q = \prod_{a=1}^{\frac{p-1}{2}} \left(\zeta_p^a - \zeta_p^{-a}\right)^q \equiv \prod_{a=1}^{\frac{p-1}{2}} \left(\zeta_p^{aq} - \zeta_p^{-aq}\right) \equiv (-1)^\mu \prod_{a=1}^{\frac{p-1}{2}} \left(\zeta_p^a - \zeta_p^{-a}\right) \bmod q,$$

where μ is the number of a in the half-system $A = \{1, 2, \ldots, \frac{p-1}{2}\}$ such that aq is congruent mod p to an element in $-A$. Now Gauss's Lemma shows $\Lambda^q \equiv \left(\frac{q}{p}\right)\Lambda \bmod p$, and $\left(\frac{q}{p}\right) \equiv \Lambda^{q-1} = (\Lambda^2)^{\frac{q-1}{2}} = (p^*)^{\frac{q-1}{2}} \equiv \left(\frac{p^*}{q}\right) \bmod q$ proves the claim.

As a conclusion we note that the proofs of the quadratic reciprocity law using the sine function are closely related to proofs via Gauss sums; this is of course no big surprise in view of Euler's formula $e^{ix} = \cos x + i \sin x$.

8.2 Abel's Construction of Elliptic Functions

The key to Eisenstein's proof of quartic reciprocity is the study of a certain elliptic function which can already be found in the work Abel (and which had been studied before by Gauss, who chose not to publish most of his results in this area), namely the lemniscatic sine $\sin \operatorname{lemn} z = \operatorname{sl}(z)$; as the name suggests, $\operatorname{sl} z$ is related to the usual sine. We will throw some light on this relation by sketching Abel's construction of $\operatorname{sl} z$, together with the corresponding theory of the usual sine (using only real analysis). In the next section we show how to construct $\operatorname{sl} z$ using complex analysis, and then we will apply our results to the theory of cubic and biquadratic residues.

Let us first say a few words about the lemniscate. Geometrically it can be described as follows: let F_1 and F_2 be points in $\mathbb{R}^2$ with distance $\overline{F_1 F_2} = 2c$; then the lemniscate is the set of points P such that $\overline{PF_1} \cdot \overline{PF_2} = c^2$. If we choose $F_1 = (-c, 0)$ and $F_2 = (c, 0)$, we get the equation $(x^2 + y^2)^2 = 2c^2(x^2 - y^2)$. In polar coordinates, lemniscates are described by $r^2 = a^2 \cos 2\theta$ with $a^2 = 2c^2$, as a straightforward computation shows. We will only consider the lemniscate corresponding to $a = 1$; then, from the equations $r^4 = x^2 - y^2$ and $r^2 = x^2 + y^2$, we get the parameterization $2x^2 = r^2 + r^4$, $2y^2 = r^2 - r^4$.

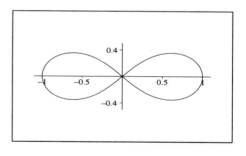

For computing the arclength of the lemniscate (say for the part lying in the first quadrant and corresponding to the interval $r \in [0, 1]$) we need the derivatives $\dot{x} = \frac{dx}{dr}$ and $\dot{y} = \frac{dy}{dr}$. From $x\dot{x} = r + 2r^3$ and $y\dot{y} = r - 2r^3$ we get (after some computing) $\dot{x}^2 + \dot{y}^2 = (1-r^4)^{-1}$; in particular, the arclength of the lemniscate in the first quadrant is given by $\int_0^1 dr/\sqrt{1-r^4}$.

Abel's construction of elliptic functions begins with the integral

$$z = \int_0^w \frac{dx}{\sqrt{1-x^2}}, \qquad z = \int_0^w \frac{dx}{\sqrt{1-x^4}}.$$

Then $z = z(w)$ is a well defined differentiable function on the open interval $(-1, 1)$. The transformation $\phi : [0, 1] \longrightarrow [0, 1]$ given by

$$x \longmapsto \frac{2t}{1+t^2} \qquad x \longmapsto \sqrt{\frac{2t^2}{1+t^4}}$$

shows that

$$\int_0^1 \frac{dx}{\sqrt{1-x^2}} = 2\int_0^1 \frac{dt}{1+t^2} \qquad \int_0^1 \frac{dx}{\sqrt{1-x^4}} = \sqrt{2}\int_0^1 \frac{dt}{\sqrt{1+t^4}}$$

converges. A numerical computation yields

$$\pi = 2\int_0^1 \frac{dx}{\sqrt{1-x^2}} = 3.141592\ldots \qquad \omega = 2\int_0^1 \frac{dx}{\sqrt{1-x^4}} = 2.62205\ldots$$

For a calculation of ω in terms of π and the value $\Gamma(1/4)$ of the Gamma function, see Exercise 8.5. Since the function $z = z(w)$ is strictly increasing on $[-1, 1]$, we can define its inverse function by

$$w = \sin z \qquad\qquad w = \sin \text{lemn } z =: \text{sl } z.$$

Directly from the definition we get

$$\sin 0 = 0, \; \sin \tfrac{\pi}{2} = 1 \qquad\qquad \text{sl } 0 = 0, \; \text{sl } \tfrac{\omega}{2} = 1$$

These functions are differentiable, and a simple computation yields

$$\tfrac{d}{dz}\sin z = \sqrt{1-\sin^2 z} =: \cos z \qquad \tfrac{d}{dz}\text{sl}(z) = \text{f}(z)\text{F}(z),$$

where $\text{f}(z) = \sqrt{1-\text{sl}^2 z}$ and $\text{F}(z) = \sqrt{1+\text{sl}^2 z}$. In all three cases, we choose the plus sign on sufficiently small intervals $[0, \varepsilon]$. Note that $\sin z$ and $\text{sl } z$ must have derivative $+1$ in $z = 0$, because the integrands in their definition have value 1 at $z = 0$. Now straightforward computations show

8.2 Abel's Construction of Elliptic Functions

$\frac{d}{dz}\cos z = -\sin z,$ $\qquad \frac{d}{dz}\mathrm{f}(z) = -\mathrm{sl}(z)\,\mathrm{F}(z),$

$\cos 0 = 1,$ $\qquad \frac{d}{dz}\mathrm{F}(z) = \mathrm{sl}(z)\,\mathrm{f}(z),$

$\qquad\qquad\qquad \mathrm{F}(0) = 0,\ \mathrm{f}(0) = 1,$

$\cos\frac{\pi}{2} = 0.$ $\qquad \mathrm{f}(\frac{\omega}{2}) = 0,\ \mathrm{F}(\frac{\omega}{2}) = \sqrt{2}.$

Next let us derive the addition theorems; to this end we put

$$\widetilde{\omega} = \pi, \qquad\qquad \widetilde{\omega} = \omega,$$

and $I = [-\frac{\widetilde{\omega}}{2}, \frac{\widetilde{\omega}}{2}]$. Then we put $D = \{(\alpha, \beta) \in \mathbb{R}^2 : \alpha, \beta, \alpha+\beta \in I\}$ and define a function $g : D \longrightarrow \mathbb{R}$ by $g(\alpha, \beta) =$

$$\sin\alpha\cos\beta + \cos\alpha\sin\beta, \qquad \frac{\mathrm{sl}(\alpha)\mathrm{f}(\beta)\mathrm{F}(\beta) + \mathrm{sl}(\beta)\mathrm{f}(\alpha)\mathrm{F}(\alpha)}{1 + \mathrm{sl}^2(\alpha)\,\mathrm{sl}^2(\beta)}.$$

Introducing the new variables $\gamma = \frac{\alpha+\beta}{2}$ and $\delta = \frac{\alpha-\beta}{2}$ we find $\partial g/\partial \delta = 0$; this shows that $g(\gamma, \delta)$ does not depend on δ. Evaluating g at $\delta := \gamma$ gives

$$g(\gamma) = \sin 2\gamma, \qquad\qquad g(\gamma) = \mathrm{sl}\, 2\gamma.$$

After reintroducing α and β we find the addition formulae for $\sin x$ and $\mathrm{sl}\,x$; the same technique can be used to verify the corresponding results for $\cos z$, $\mathrm{f}(z)$, and $\mathrm{F}(z)$ (see Table 8.1).

Table 8.1. Addition formulae

$\sin(\alpha+\beta) = \sin\alpha\cos\beta + \cos\alpha\sin\beta,$

$\cos(\alpha+\beta) = \cos\alpha\cos\beta - \sin\alpha\sin\beta,$

$\mathrm{sl}(\alpha+\beta) = \dfrac{\mathrm{sl}(\alpha)\mathrm{f}(\beta)\mathrm{F}(\beta) + \mathrm{sl}(\beta)\mathrm{f}(\alpha)\mathrm{F}(\alpha)}{1 + \mathrm{sl}^2(\alpha)\,\mathrm{sl}^2(\beta)},$

$\mathrm{f}(\alpha+\beta) = \dfrac{\mathrm{f}(\alpha)\mathrm{f}(\beta) - \mathrm{sl}(\alpha)\mathrm{sl}(\beta)\mathrm{F}(\alpha)\mathrm{F}(\beta)}{1 + \mathrm{sl}^2(\alpha)\,\mathrm{sl}^2(\beta)},$

$\mathrm{F}(\alpha+\beta) = \dfrac{\mathrm{F}(\alpha)\mathrm{F}(\beta) + \mathrm{sl}(\alpha)\mathrm{sl}(\beta)\mathrm{f}(\alpha)\mathrm{f}(\beta)}{1 + \mathrm{sl}^2(\alpha)\,\mathrm{sl}^2(\beta)}.$

Putting $\alpha = \beta$ in the addition formula we get the duplication formulae (see Table 8.2).

Substituting $\beta = \frac{1}{2}\widetilde{\omega}$ in the addition formulae gives us a relation that allows us to define the lemniscatic cosine $\mathrm{cl}\,z$:

Table 8.2. Duplication formulae

$\sin(2\alpha) = 2\sin\alpha\cos\alpha,$	$\mathrm{sl}(2\alpha) = \dfrac{2\mathrm{sl}(\alpha)\mathrm{f}(\alpha)\mathrm{F}(\alpha)}{1+\mathrm{sl}(\alpha)^4},$
$\cos(2\alpha) = \cos^2\alpha - \sin^2\alpha,$	$\mathrm{f}(2\alpha) = \dfrac{\mathrm{f}(\alpha)^2 - \mathrm{sl}(\alpha)^2\mathrm{F}(\alpha)^2}{1+\mathrm{sl}(\alpha)^4},$
	$\mathrm{F}(2\alpha) = \dfrac{\mathrm{F}(\alpha)^2 + \mathrm{sl}(\alpha)^2\mathrm{f}(\alpha)^2}{1+\mathrm{sl}(\alpha)^4}.$

$$\sin(\alpha + \tfrac{\pi}{2}) = \cos\alpha, \qquad \mathrm{sl}(\alpha + \tfrac{\widetilde{\omega}}{2}) = \tfrac{\mathrm{f}(\alpha)}{\mathrm{F}(\alpha)} =: \mathrm{cl}\,\alpha.$$

Note that we have proved the addition formulae only for $\alpha \in [-\tfrac{\widetilde{\omega}}{2}, 0]$; for $\alpha \in [0, \tfrac{\widetilde{\omega}}{2}]$, however, we can use these identities to extend the domain of definition for $\sin z$ and $\mathrm{sl}\,z$. Repeating this argument we find that we can extend $\sin z$ and $\mathrm{sl}\,z$ to differentiable functions on $\mathbb{R}$ (actually they even are analytic). It is clear that the addition formulae continue to hold for all $\alpha, \beta \in \mathbb{R}$. Moreover we find that $\sin z$ is 2π-periodic, and that $\mathrm{sl}\,z$ is 2ω-periodic. Differentiation shows that the same is true for $\cos z$ and $\mathrm{f}(z), \mathrm{F}(z)$, respectively. Finally we note the following relations:

$$\sin^2 z + \cos^2 z = 1, \qquad \mathrm{sl}^2 z + \mathrm{cl}^2 z + \mathrm{sl}^2 z\,\mathrm{cl}^2 z = 1.$$

The identity for $\sin z$ and $\cos z$ follows directly from the definition of $\cos z$; in the lemniscatic case we get $\mathrm{sl}^2 z + \mathrm{cl}^2 z + \mathrm{sl}^2 z\,\mathrm{cl}^2 z = (1+\mathrm{sl}^2 z)^{-1}((1+\mathrm{sl}^2 z)\,\mathrm{sl}^2 z + (1-\mathrm{sl}^2 z) + (1-\mathrm{sl}^2 z)\,\mathrm{sl}^2 z) = 1$ as claimed.

We will now evaluate these functions at certain real z; we will need these values later when we compute the zeros and poles of $\mathrm{sl}\,z$:

z	$\sin z$	$\cos z$	$\mathrm{sl}\,z$	$\mathrm{f}(z)$	$\mathrm{F}(z)$
$m\widetilde{\omega}$	0	$(-1)^m$	0	$(-1)^m$	1
$(m+\tfrac{1}{2})\widetilde{\omega}$	$(-1)^m$	0	$(-1)^m$	0	$\sqrt{2}$

Our next goal is to turn the duplication formulas into 'division formulas', i.e. we want to express $\mathrm{sl}(\alpha/2)$, say, in terms of $\mathrm{sl}(\alpha)$, $\mathrm{f}(\alpha)$ and $\mathrm{F}(\alpha)$. To this end we replace α by $\alpha/2$ and define

$$x = \sin\tfrac{\alpha}{2}, \qquad\qquad x = \mathrm{sl}(\tfrac{\alpha}{2}),\ y = \mathrm{f}(\tfrac{\alpha}{2}),$$
$$y = \cos\tfrac{\alpha}{2}, \qquad\qquad z = \mathrm{F}(\tfrac{\alpha}{2}).$$

Then we have

8.2 Abel's Construction of Elliptic Functions

$$y^2 = 1 - x^2, \qquad\qquad y^2 = 1 - x^2, \ z^2 = 1 + x^2$$

and from Table 8.2 we get

$$\cos\alpha = y^2 - x^2 \qquad\qquad f(\alpha) = \tfrac{y^2-x^2z^2}{1+x^4} = \tfrac{1-2x^2-x^4}{1+x^4},$$

$$\sin\alpha = 2xy \qquad\qquad F(\alpha) = \tfrac{z^2+x^2y^2}{1+x^4} = \tfrac{1+2x^2-x^4}{1+x^4},$$

Solving for x and y we find the identities

$$x^2 = \frac{1-\cos\alpha}{2} \qquad\qquad x^2 = \tfrac{1-f(\alpha)}{1+F(\alpha)} = \tfrac{F(\alpha)-1}{1+f(\alpha)},$$

$$y^2 = \frac{1+\cos\alpha}{2} \qquad\qquad y^2 = \tfrac{f(\alpha)+F(\alpha)}{1+F(\alpha)}, \ z^2 = \tfrac{f(\alpha)+F(\alpha)}{1+f(\alpha)}.$$

We leave it as an easy exercise for the reader to show that the expressions on the right hand side are all non-negative. Extracting the square roots finally shows

$$\sin\tfrac{\alpha}{2} = \sqrt{\tfrac{1-\cos\alpha}{2}}, \qquad\qquad \mathrm{sl}\left(\tfrac{\alpha}{2}\right) = \sqrt{\tfrac{1-f(\alpha)}{1+F(\alpha)}},$$

$$f\left(\tfrac{\alpha}{2}\right) = \sqrt{\tfrac{f(\alpha)+F(\alpha)}{1+F(\alpha)}},$$

$$\cos\tfrac{\alpha}{2} = \sqrt{\tfrac{1+\cos\alpha}{2}}, \qquad\qquad F\left(\tfrac{\alpha}{2}\right) = \sqrt{\tfrac{f(\alpha)+F(\alpha)}{1+f(\alpha)}}.$$

Note that the signs of the square roots have to be determined by using that $\mathrm{sl}\,x \geq 0$ for $x \in [0,\omega]$ and $\mathrm{sl}\,x \leq 0$ if $x \in [\omega, 2\omega]$ etc. In particular, we find the following values:

$$\sin(\tfrac{\pi}{4}) = \tfrac{1}{2}\sqrt{2}, \qquad\qquad \mathrm{sl}(\tfrac{\omega}{4}) = \sqrt{\sqrt{2}-1},$$

$$\cos(\tfrac{\pi}{4}) = \tfrac{1}{2}\sqrt{2}. \qquad\qquad f(\tfrac{\omega}{4}) = \sqrt{2-\sqrt{2}}, \ F(\tfrac{\omega}{4}) = \sqrt[4]{2}.$$

Observe that these algebraic numbers generate abelian extensions of $\mathbb{Q}$ and $\mathbb{Q}(i)$, respectively (note also that these can be constructed by straight edge and compass).

Abel's next step is defining $\sin z$ and $\mathrm{sl}\,z$ for $z \in i\mathbb{R}$, that is, on the imaginary axis. This is simple for $\mathrm{sl}\,z$, because replacing z by iz in the definition of $\mathrm{sl}\,z$ shows that we should define $\mathrm{sl}(iz) = i\mathrm{sl}(z)$. In the case of the sine, however, we are led to put $\sin(iz) = i\sinh(z)$, where the real function $w = \sinh(z)$ is defined by inverting the integral $z = \int_0^w \frac{dx}{\sqrt{1+x^2}}$.

Using the addition formula we can extend $\sin z$ and $\mathrm{sl}\,z$ to functions $\mathbb{C} \longrightarrow \mathbb{C} \cup \{\infty\}$ on the entire complex plane (allowing them to take the value '∞' in their poles). Complex functions f defined on an open subset D of $\mathbb{C}$ are called *holomorphic* in D if f is differentiable in every $z \in D$ (alternatively, f can be expanded into a power series around each $z \in D$), and *meromorphic* in D if

244   8. Eisenstein's Analytic Proofs

the set S of poles of f in D is a discrete subset of $\mathbb{C}$ and if f is holomorphic in $D \setminus S$. One checks that $\sin z$ is holomorphic in $\mathbb{C}$, whereas $\operatorname{sl} z$ is meromorphic (see Exercises 8.6 and 8.7). Furthermore, the addition formulae continue to hold for all $\alpha, \beta \in \mathbb{C}$ (note that they are valid per definitionem if $\alpha \in \mathbb{R}$ and $\beta \in i\mathbb{R}$). Replacing z by iz in the definition of f and F shows that $f(iz) = F(z)$ and $F(iz) = f(z)$. We can also compute a small table of values:

Table 8.3.

z	sl z	f(z)	F(z)
$m\omega + n\omega i$	0	$(-1)^m$	$(-1)^n$
$(m + \frac{1}{2})\omega + n\omega i$	$(-1)^{m+n}$	0	$(-1)^n \sqrt{2}$
$m\omega + (n + \frac{1}{2})\omega i$	$(-1)^{m+n} i$	$(-1)^n \sqrt{2}$	0
$(m + \frac{1}{2})\omega + (n + \frac{1}{2})\omega i$	∞	∞	∞

In order to see that sl z has poles at $z = (m + \frac{1}{2})\omega + (n + \frac{1}{2})\omega i$, we observe

$$\operatorname{sl}(\alpha + i\alpha) = \frac{(1+i)\operatorname{sl}(\alpha)\operatorname{f}(\alpha)\operatorname{F}(\alpha)}{\operatorname{f}(\alpha)^2 \operatorname{F}(\alpha)^2} = (1+i)\frac{\operatorname{sl}(\alpha)}{\operatorname{f}(\alpha)\operatorname{F}(\alpha)}. \tag{8.10}$$

Putting $\alpha = \frac{\omega}{2}$ shows that sl z has simple poles at $z = (m + \frac{1}{2})\omega + (n + \frac{1}{2})\omega i$.

Our next step is to show that sl has no zeros and poles except those given in the table above. Suppose that sl $z = 0$ for $z = \alpha + \beta i$, $\alpha, \beta \in \mathbb{R}$. Then the addition formula shows that

$$0 = \operatorname{sl}(z) = \frac{\operatorname{sl}(\alpha)\operatorname{f}(\beta)\operatorname{F}(\beta) + i\operatorname{sl}(\beta)\operatorname{f}(\alpha)\operatorname{F}(\alpha)}{1 - \operatorname{sl}(\alpha)^2 \operatorname{sl}(\beta)^2};$$

now there are only two possibilities:

a) $1 = \operatorname{sl}(\alpha)^2 \operatorname{sl}(\beta)^2$: this implies that $\operatorname{sl}(\alpha)^2 = \operatorname{sl}(\beta)^2 = 1$. This happens if and only if $\alpha = (m + \frac{1}{2})\omega$ and $\beta = (n + \frac{1}{2})\omega$; in this case, we have already shown that sl(z) has a simple pole.
b) $1 \neq \operatorname{sl}(\alpha)^2 \operatorname{sl}(\beta)^2$: then $\operatorname{sl}(\alpha)\operatorname{f}(\beta)\operatorname{F}(\beta) = \operatorname{sl}(\beta)\operatorname{f}(\alpha)\operatorname{F}(\alpha) = 0$, but since $1 \leq F(z) \leq \sqrt{2}$ for $z \in \mathbb{R}$, this is equivalent to $\operatorname{sl}(\alpha)\operatorname{f}(\beta) = \operatorname{sl}(\beta)\operatorname{f}(\alpha) = 0$. Since sl($z$) and f($z$) cannot have common zeros by the definition of f, this implies that either $\operatorname{sl}(\alpha) = \operatorname{sl}(\beta) = 0$ or $\operatorname{f}(\alpha) = \operatorname{f}(\beta) = 0$. From our knowledge of the zeros of sl and f on the real line we conclude that either $\alpha = m\omega$, $\beta = n\omega$ or $\alpha = (m + \frac{1}{2})\omega$ and $\beta = (n + \frac{1}{2})\omega$; we have excluded the second alternative in part a), hence the first possibility must hold.

At this point we know that sl, cl, f and F are meromorphic functions on $\mathbb{C}$ with periods 2ω and $2\omega i$; now we claim that sl z and cl z actually have $(1 + i)\omega$ and $(1 - i)\omega$ as periods. The verification consists in a few simple computations:

8.2 Abel's Construction of Elliptic Functions

$$\mathrm{sl}(\alpha + \omega) = -\mathrm{sl}(\alpha), \ \mathrm{sl}(\alpha + i\omega) = -\mathrm{sl}(\alpha),$$
$$\mathrm{f}(\alpha + \omega) = -\mathrm{f}(\alpha), \quad \mathrm{f}(\alpha + i\omega) = \mathrm{f}(\alpha),$$
$$\mathrm{F}(\alpha + \omega) = \mathrm{F}(\alpha), \ \mathrm{F}(\alpha + i\omega) = -\mathrm{F}(\alpha),$$

and now the addition formula yields $\mathrm{sl}(\alpha + (1+i)\omega) = \mathrm{sl}(\alpha)$ as claimed. Similarly, one shows that $\mathrm{cl}(\alpha + (1+i)\omega) = \mathrm{cl}(\alpha)$, and that f and F change sign under addition of $(1+i)\omega$. Note that $\mathrm{sl}\,z$ is *not* $\frac{1+i}{2}\omega$-periodic by the formulas above. We record a few more computations:

$$\mathrm{sl}\left(\alpha + \tfrac{\omega}{2}\right) = \tfrac{\mathrm{f}(\alpha)}{\mathrm{F}(\alpha)} \qquad \mathrm{sl}\left(\alpha + \tfrac{i\omega}{2}\right) = i\tfrac{\mathrm{f}(\alpha)\mathrm{F}(\alpha)}{1-\mathrm{sl}(\alpha)^2}$$
$$\mathrm{f}\left(\alpha + \tfrac{\omega}{2}\right) = -\sqrt{2}\tfrac{\mathrm{sl}(\alpha)\mathrm{F}(\alpha)}{1+\mathrm{sl}(\alpha)^2} \qquad \mathrm{f}\left(\alpha + \tfrac{i\omega}{2}\right) = \sqrt{2}\tfrac{\mathrm{f}(\alpha)}{1-\mathrm{sl}(\alpha)^2}$$
$$\mathrm{F}\left(\alpha + \tfrac{\omega}{2}\right) = \sqrt{2}\tfrac{\mathrm{F}(\alpha)}{1+\mathrm{sl}(\alpha)^2} \qquad \mathrm{F}\left(\alpha + \tfrac{i\omega}{2}\right) = i\sqrt{2}\tfrac{\mathrm{sl}(\alpha)\mathrm{f}(\alpha)}{1-\mathrm{sl}(\alpha)^2}$$

As a corollary, we note the relations

$$\mathrm{sl}(\alpha)\,\mathrm{sl}\!\left(\alpha + \tfrac{1+i}{2}\omega\right) = -i, \quad \mathrm{sl}(\alpha)\,\mathrm{sl}\!\left(\alpha + \tfrac{1-i}{2}\omega\right) = +i, \qquad (8.11)$$

which show that the zeros of $\mathrm{sl}\,z$ and the poles of $\mathrm{sl}(z + \tfrac{1+i}{2}\omega)$ coincide.

It is often convenient to work with the function $\phi(z) = \mathrm{sl}((1-i)\omega)$ whose period lattice is $\mathbb{Z}[i]$ rather than with the function $\mathrm{sl}\,z$ whose period lattice is $(1+i)\omega\mathbb{Z}[i]$. Our results above show that ϕ is an elliptic function with period lattice $\mathbb{Z}[i]$ such that $\phi(iz) = i\phi(z)$, $\phi(z)\phi(z - \tfrac{1}{2}) = 1$ and $\phi(\tfrac{1+i}{4}) = 1$. It is this function ϕ that we will construct from scratch in the next section, using complex analysis.

Now we come to Abel's main result on the complex multiplication of the lemniscatic sine; its subsequent refinement by Eisenstein will play a central part later on:

Proposition 8.2. *Let $\mu = a + bi \in \mathbb{Z}[i]$ be primary, i.e. assume that $\mu \equiv 1 \bmod 2 + 2i$. Let $m = a^2 + b^2$ denote the norm of μ and put $x = \mathrm{sl}(z)$. Then there exist polynomials $W(X), V(X) \in \mathbb{Z}[i][X]$ of degree $\frac{m-1}{4}$ such that*

$$\mathrm{sl}(\mu z) = x\frac{W(x^4)}{V(x^4)}. \qquad (8.12)$$

Moreover, $W(x^4) = x^{m-1}V(x^{-4})$, i.e. if we write $W(X) = X^{(m-1)/4} + A_{(m-5)/4}X^{(m-5)/4} + \ldots + A_1X + A_0$, then $V(X) = 1 + A_{(m-5)/4}X + \ldots + A_0X^{(m-1)/4}$. Finally, we have $A_0 = \mu$.

Proof. (Sketch) We first claim that, for odd integers $u \in \mathbb{Z}$ such that $u \equiv 1 \bmod 4$, there exist polynomials $W, V, P, Q \in \mathbb{Z}[i][X]$ such that

$$\mathrm{sl}(uz) = \mathrm{sl}(z)\frac{W(x^4)}{V(x^4)},$$
$$\mathrm{f}(uz) = \mathrm{f}(z)\frac{P(x^4) - x^2Q(x^4)}{V(x^4)}, \quad \mathrm{F}(uz) = \mathrm{F}(z)\frac{P(x^4) + x^2Q(x^4)}{V(x^4)}, \qquad (8.13)$$

where, as above, $x = \mathrm{sl}(z)$ (and where the polynomials depend on u, of course). In analogy to the trigonometric case we expect that these formulas can be proved simultaneously by induction for odd $u \in \mathbb{N}$ (going from u to $u + 2$ and using the addition formulas), and then for all odd $u \in \mathbb{Z}$ by using that sl is odd and that f and F are even functions, and this does indeed work. Next one proves the same formulas for all $u \equiv 1 \bmod 2$ in $\mathbb{Z}[i]$ by induction going from u to $u + 2i$ etc. The fact that $V(0) = 1$ is also proved by induction, and then $W(0) = \mu$ follows by applying $\lim_{z \to 0}$ to $\mathrm{sl}(\mu z)/\mathrm{sl}\, z$.

This proves the existence of W and V, but it does not provide us with their degrees (these polynomials might have a common factor, and in fact, they do: see Exercise 8.9). So let us cancel common divisors of W and V in $\mathbb{Z}[i][X]$ and assume that the polynomials are coprime in $\mathbb{Z}[i][X]$.

It clearly suffices to prove our claims with sl replaced by ϕ (simply substitute $(1-i)\omega z$ for z; the polynomials W and V do not change). Since $\phi(\mu z)/\phi(z)$ vanishes for $z = \alpha/\mu$ for any $\alpha \neq 0$ in a complete residue system modulo μ, we find that $W(x^4)$ has the roots $\phi(\alpha/\mu)$, where $\alpha \in A$ and A contains a complete residue system modulo μ except $0 \bmod \mu$. Since none of these zeros is a double root (check the derivative), we get $W(x^4) = \prod_{\alpha \in A}(x - \phi(\alpha/\mu))$. In particular, W is a monic polynomial of degree $(N\mu - 1)/4$.

Similarly, the zeros of V coincide with the poles of $\phi(\mu z)/W(z)$; since these are just the $\frac{1}{2} + \frac{\alpha}{\mu}$ with $\alpha \in A$, we get

$$V(x^4) = c \prod_{\alpha \in A} \left(x - \phi\left(\frac{1}{2} + \frac{\alpha}{\mu}\right)\right)$$

for some constant c. Using $\phi(z)\phi(z - \frac{1}{2}) = 1$, this shows that

$$V(x^4) = c \prod_{\alpha \in A} \left(x - \phi\left(\frac{\alpha}{\mu}\right)^{-1}\right).$$

Multiplying through by $\prod_{\alpha \in A}(-\phi(\frac{\alpha}{\mu})^{-1}))$ we find

$$V(x^4) = c' \prod_{\alpha \in A} \left(1 - \phi\left(\frac{\alpha}{\mu}\right)x\right).$$

Applying $\lim_{z \to 0}$ to

$$\frac{\phi(\mu z)}{\phi(z)} = \frac{\prod_{\alpha \in A}(x - \phi(\frac{\alpha}{\mu}))}{c' \prod_{\alpha \in A}(1 - \phi(\frac{\alpha}{\mu})x)}$$

and using L'Hospital gives $c' = 1$, and our claims are proved. We also remark that $W(x^4) = \prod_{\alpha \in B}(x^4 - \phi(\alpha/\mu)^4)$, where B is a $\frac{1}{4}$-system modulo μ; a similar formula holds for $V(x^4)$. See Exercise 8.11 for examples. □

8.3 Elliptic Functions

In the last section we have seen how Abel constructed elliptic functions (i.e. doubly-periodic meromorphic functions on $\mathbb{C}$). Today's courses on complex analysis build the theory of elliptic functions around Liouville's theorems; here we will review the basic theorems and then show how to derive Abel's results on sl z from them.

Liouville's Theorems

Let $f : D \longrightarrow \mathbb{C}$ be a meromorphic function; if f has no pole in $z_0 \in D$, then f can be written as a Laurent series $f(z) = \sum_{n=-\infty}^{\infty} a_n(z - z_0)^n$. The coefficient a_{-1} is called the residue of f at z_0 and is denoted by $a_{-1} = \operatorname{res}_{z_0} f$.

Remark. The sum $\sum_{n=-\infty}^{-1} a_n(z-z_0)^n$ is called the negative part of the Laurent series. If f is a function with an infinite negative part, it has an *essential singularity* at $z = z_0$ (an example is $f(z) = \exp(1/z)$). The functions we will consider will have Laurent series with finite negative part; their singularities will therefore be 'nice'.

If z is any isolated singularity of f, let C be a small circle around z such that the only pole of f inside C. Then integrating the Laurent series term by term gives $\int_C f(\zeta) d\zeta = 2\pi i a_{-1} = 2\pi i \cdot \operatorname{res}_z f$.

These results will now be applied to elliptic functions with period lattice $\Lambda = \omega_1 \mathbb{Z} \oplus \omega_2 \mathbb{Z}$; these are non-constant meromorphic functions $f : \mathbb{C} \longrightarrow \mathbb{C}$ such that $f(z + \lambda) = f(z)$ for all $\lambda \in \Lambda$, i.e. doubly periodic functions with $\mathbb{R}$-independent periods $\omega_1, \omega_2 \in \mathbb{C}$. We can view f also as a meromorphic function on the torus $\mathbb{C}/\Lambda$; notice that each $z \in \mathbb{C}$ is congruent modulo Λ to exactly one point in the fundamental domain

$$\mathcal{F} = \mathcal{F}_\Lambda = \{z = z_0 + a\omega_1 + b\omega_2 : 0 \leq a, b < 1\}$$

for any choice of $z_0 \in \mathbb{C}$. Since the set of poles and zeros of non-constant meromorphic functions is discrete, we may choose z_0 in such a way that there are no poles or zeros on the boundary of $\mathcal{F}$. Observe also that the number of poles inside $\mathcal{F}$ is finite; since $1/f$ is also Λ-elliptic, the same observation is valid for the zeros of f.

We can define the *divisor* of an elliptic function f as a formal sum $\operatorname{div} f := (f) := \sum_{p \in \mathcal{F}} n_p(p)$ (note that this sum is actually finite), where

$$n_p = \begin{cases} -m, & \text{if } f \text{ has a pole of order } m \text{ at } p \\ +m, & \text{if } f \text{ has a zero of order } m \text{ at } p \end{cases}$$

The order (sometimes also called multiplicity) m of f at p is defined to be the smallest integer n such that $f(p)(z - p)^{-n} \neq 0$. Any finite formal sum

$D = \sum_{p\in \mathbb{C}/\Lambda} n_p(p)$ is called a *divisor* on $\mathbb{C}/\Lambda$; D is called *principal* if there exists an elliptic function f on $\mathbb{C}/\Lambda$ such that $D = (f)$. The *degree* of D is defined to be the integer $\deg(D) = \sum_{p\in \mathbb{C}/\Lambda} n_p$, which can be viewed as the difference between the number of zeros and the number of poles (counted with multiplicity) if $D = (f)$.

Using some basic results of complex analysis we find (Liouville 1847):

Proposition 8.3. *1. Elliptic functions without poles are constant. In particular, if f and g are elliptic functions such that $(f) = (g)$, then $f = c \cdot g$ for some $c \in \mathbb{C} \setminus \{0\}$.*
2. Elliptic functions have only finitely many poles in $\mathcal{F}$, and the sum of their residues is 0. In particular, elliptic functions have at least two poles (counting multiplicities).
3. Elliptic functions assume each value $\mathrm{ord}\,f$ times on $\mathcal{F}$ (counting multiplicities); in particular, $\deg(f) = 0$.

Proof. 1. Liouville's theorem states that the only bounded holomorphic functions on $\mathbb{C}$ are constants. If f is an elliptic function without a pole, then f is bounded on $\mathcal{F}$ and hence on $\mathbb{C}$: therefore it must be constant. Since $(f) = (g)$ implies that f/g is an elliptic function without poles, it must be constant.
2. Integrating f over $\mathcal{F}$ gives $2\pi i \sum \mathrm{res}_z f = 0$. If an elliptic function had exactly one pole z_0 in $\mathcal{F}$, its residue in z_0 would have to be nonzero, which is a contradiction.
3. f' and $g = f'/f$ are elliptic functions with the same period lattice Λ as f; applying 2. to g and observing $\mathrm{ord}_z f = \mathrm{res}_z g$ gives the desired result. $\square$

Weierstraß' Theory

The next step is the construction of Weierstraß' $\wp$-function:[2] to this end, let Λ be a lattice in $\mathbb{C}$ and define

$$\wp(z) = \wp(z;\Lambda) = \frac{1}{z^2} + \sum{}' \Big(\frac{1}{(z-\lambda)^2} - \frac{1}{\lambda^2}\Big), \qquad (8.14)$$

where the prime indicates that the sum is over all $\lambda \in \Lambda \setminus \{0\}$. It is an easy exercise to verify uniform convergence on compacta outside Λ, hence $\wp$ is a meromorphic function on $\mathbb{C}$. Since its derivative $\wp'$ is clearly Λ-periodic, the difference $\wp(z) - \wp(z+\lambda)$ (for $\lambda \in \Lambda$) must be a constant; an evaluation at $\lambda/2$ shows that this constant vanishes, hence $\wp$ is also Λ-periodic. The main properties of the $\wp$-function are collected in

Proposition 8.4. *Weierstraß' $\wp$-function has the following properties:*

---

[2] Professor Jürgen Neukirch manufactured a number of models of famous surfaces and functions, among them the Riemann zeta function and the Weierstraß $\wp$-function. They can be seen at the University of Regensburg as well as on the www: http://www.uni-regensburg.de/Fakultaeten/nat_Fak_I/sammlung.htm

i) ℘ is an elliptic function on Λ, and its singularities are poles of second order in Λ.
ii) ℘ satisfies the differential equation $℘'^2 = 4℘^3 - g_2℘ - g_3$ Moreover, we have $℘''(z) = 6℘(z) - \frac{1}{2}g_2$.
iii) ℘ possesses the addition law

$$℘(z+w) + ℘(z) + ℘(w) = \frac{1}{4}\left(\frac{℘'(z) - ℘'(w)}{℘(z) - ℘(w)}\right)^2,$$

which is valid for all $z, w, z+w \in \mathbb{C} \setminus \Lambda$.

Proof. By developing the summand in (8.14) into a power series it is easy to show (see Exercise 8.8) that

$$℘(z) = \frac{1}{z^2} + \sum_{n=1}^{\infty}(2n+1)G_{2n+2}z^{2n}, \tag{8.15}$$

where the *Eisenstein series* $G_k = G_k(\Lambda)$ are defined for all $k \geq 2$ by

$$G_k(\Lambda) = \sum_{\lambda \in \Lambda \setminus \{0\}} \lambda^{-2k}. \tag{8.16}$$

Taking the derivative of ℘ and putting

$$g_2 = g_2(\Lambda) = 60G_2, \qquad g_3 = g_3(\Lambda) = 140G_3,$$

it is easy to see that $h(z) = ℘'(z)^2 - 4℘(z)^3 + g_2℘(z) + g_3$ is an elliptic function without poles vanishing in $z = 0$; by Liouville's Theorem, $h = 0$.

The addition law follows by taking the difference of the right and left hand side and showing that this elliptic function has no poles. □

Using the addition theorem it is easy to prove that $℘(nz)$ is a rational function of $℘(z)$ for all $n \in \mathbb{Z}$: we say that ℘ has multiplication by $\mathbb{Z}$. If there exist $\alpha \in \mathbb{C} \setminus \mathbb{Z}$ such that $℘(\alpha z)$ is a rational function of $℘(z)$, we say that ℘ has *complex multiplication*, the reason for the nomenclature being that α then necessarily belongs to a complex quadratic number field:

Proposition 8.5. *Let $\Lambda = \mathbb{Z} + \tau\mathbb{Z}$ be a lattice in $\mathbb{C}$, and let ℘ be the corresponding ℘-function. For any $\alpha \in \mathbb{C} \setminus \mathbb{Z}$, the following assertions are equivalent:*

i) *$℘(\alpha z)$ is a rational function of $℘(z)$;*
ii) *$\alpha \Lambda \subseteq \Lambda$;*
iii) *$k = \mathbb{Q}(\tau)$ is an imaginary quadratic number field, the set $\mathcal{O} = \{\beta \in k : \beta\Lambda \subseteq \Lambda\}$ is an order in k containing α, and Λ is a fractional $\mathcal{O}$-ideal.*

If these conditions are satisfied, then $℘(\alpha z) = \frac{A(℘(z))}{B(℘(z))}$ for coprime polynomials $A, B \in \mathbb{C}[X]$ such that $\deg A = 1 + \deg B = (\Lambda : \alpha\Lambda) = N_{k/\mathbb{Q}}\alpha$. In fact, this also holds for $\alpha \in \mathbb{Z}$ with $N_{k/\mathbb{Q}}\alpha$ replaced by α^2.

Next comes the construction of Weierstraß' σ- and ζ-functions. We want the σ-function to be a function with simple zeros in $\lambda \in \Lambda$; the function $z\prod_{\lambda \neq 0}(1-z\lambda^{-1})$, where the product is over all $\lambda \in \Lambda \setminus \{0\}$, would be such a function: unfortunately it does not converge. Weierstraß' idea was to multiply the factors in this product by functions without zeros that would make the product convergent:

$$\sigma(z) = \sigma(z, \Lambda) = z \prod_{\lambda \in \Lambda \setminus \{0\}} \left(1 - \frac{z}{\lambda}\right) \exp\left(\frac{z}{\lambda} + \frac{z^2}{2\lambda^2}\right). \tag{8.17}$$

Weierstraß' ζ-function (which has nothing to do with Riemann's!) is simply the logarithmic derivative of $\sigma(z)$:

$$\zeta(z) = \zeta(z, \Lambda) = \frac{\sigma'(z)}{\sigma(z)} = \frac{1}{z} + \sum_{\lambda \in \Lambda \setminus \{0\}} \left(\frac{1}{z-\lambda} + \frac{1}{\lambda} + \frac{z}{\lambda^2}\right). \tag{8.18}$$

Now it is easy to verify that $\zeta'(z) = -\wp(z)$. In particular, $\zeta(z)$ is 'almost' an elliptic function.

Abel's Theorem

Abel's Theorem is an efficient tool for constructing elliptic functions with prescribed poles and zeros.

Theorem 8.6. *Let $u_1, \ldots, u_r, v_1, \ldots, v_s \in \mathbb{C}/\Lambda$ and integers $m_1, \ldots, m_r$, $n_1, \ldots, n_s \in \mathbb{N}$ be given. Assume that $\sum_{j=1}^r m_j = \sum_{j=1}^s n_j$. Then there exists an elliptic function f with divisor $D = \sum_{j=1}^r m_j(u_j) - \sum_{j=1}^s n_j(v_j)$ on $\mathbb{C}/\Lambda$ if and only if*

$$\sum_{j=1}^r m_j u_j \equiv \sum_{j=1}^s n_j v_j \bmod \Lambda. \tag{8.19}$$

In other words: the divisor D on $\mathbb{C}/\Lambda$ is principal if and only if $\deg(D) = 0$ and if (8.19) is satisfied.

Proof. Let $\{a_1, \ldots, a_n\}$ be the set of zeros (counted with multiplicity), and $\{b_1, \ldots, b_n\}$ the set of poles. The fact that both sets have the same cardinality is due to our assumption that $\sum_{j=1}^r m_j = \sum_{j=1}^s n_j$. From (8.19) we get $\sum a_j - \sum b_j \in \Lambda$, and by changing one of the a_j by a summand in Λ we may assume without loss of generality that $\sum a_j = \sum b_j$. Then the properties of the σ-function show that

$$\frac{\prod_{j=1}^n \sigma(z - a_j)}{\prod_{j=1}^n \sigma(z - b_j)}$$

is an elliptic function with period lattice Λ and divisor (D). $\square$

This theorem allows us to construct the elliptic function sl z at a single stroke: we simply choose the divisor $D = (0) + \left(\frac{1+i}{2}\right) - \left(\frac{1}{2}\right) - \left(\frac{i}{2}\right)$ on the lattice $\Lambda = \mathbb{Z} + \mathbb{Z}i$; since (8.19) is satisfied and $\deg D = 0$, there must be an elliptic function ϕ with $(\phi) = D$. In fact there are infinitely many, all differing by a constant factor $c \neq 0$. By demanding $\phi(\frac{1+i}{4}) = 1$ we get a uniquely determined elliptic function $\phi(z)$ (which we know must coincide with the function $\phi(z) = \text{sl}((1-i)\omega z)$ constructed right before Proposition 8.2). We claim that $\phi(iz) = i\phi(z)$. In fact, since $\phi(iz)$ is an elliptic function with the same divisor as ϕ, the quotient $\phi(iz)/\phi(z) = c$ must be a constant function. Developing $\phi(z) = \sum_{n \geq 0} a_n z^n$ into a Taylor series at $z = 0$ we find $\phi(iz) = \sum_{n \geq 0} a_n i^n z^n$; thus $i^n = c$ whenever $a_n \neq 0$. But since ϕ has a *simple* zero at 0, we must have $a_1 \neq 0$, and we get $c = i$.

Finally we claim that $\phi(z)\phi(z - \frac{1}{2}) = i$. Since $\phi(z)$ and $\phi(z - \frac{1}{2})$ have inverse divisors, their product must be constant. Therefore it is sufficient to evaluate it at $z = \frac{1+i}{4}$; we find $\phi(\frac{1+i}{4})\phi(\frac{-1+i}{4}) = i\phi(\frac{1+i}{4})^2 = i$.

The definition of sl z we have just given has the advantage of leading very quickly to the properties of sl z needed for proving the quartic reciprocity law. On the other hand, showing that $\text{sl}'(z) = \sqrt{1 - \text{sl}(z)^4}$ (which was almost trivial in Abel's approach) will need some serious work (see the Exercises).

8.4 Quartic and Cubic Reciprocity

In the last two sections we have given two different constructions of elliptic functions with properties i) and ii) of the proposition below. Here we will show how to deduce the quartic reciprocity law from these properties.

Proposition 8.7. *The elliptic function $\phi(z) = \text{sl}((1-i)\omega z)$ has the following properties:*

i) *ϕ has period lattice $\mathbb{Z}[i]$ and divisor $(\phi) = (0) + \left(\frac{1+i}{2}\right) - \left(\frac{1}{2}\right) - \left(\frac{i}{2}\right)$; this means that ϕ has simple zeros at $z \equiv 0, \frac{1+i}{2} \mod \mathbb{Z}[i]$ simple poles at $z \equiv \frac{1}{2}, \frac{i}{2} \mod \mathbb{Z}[i]$, and takes finite values everywhere else.*
ii) *$\phi(iz) = i \cdot \phi(z)$, $\phi(\frac{1+i}{4}) = 1$, and $\phi(z)\phi(z - \frac{1}{2}) = i$;*
iii) *Let $\nu \in \mathbb{Z}[i]$, assume that $(2, \nu) = 1$, and define a fourth root of unity ε by $\nu \equiv \varepsilon \mod 2 - 2i$. Then*

$$\phi(\nu z) = \varepsilon \prod_\alpha \phi\left(z - \frac{\alpha}{\nu}\right), \tag{8.20}$$

where α runs through a complete residue system modulo ν.

Proof. Only iii) is left to prove. To this end, observe that the function $\psi(z) = \phi(\nu z)$ has the divisor

$$(\psi) = \sum\left(\frac{\alpha}{\nu}\right) + \sum\left(\frac{\alpha}{\nu} + \frac{1+i}{2}\right) - \sum\left(\frac{\alpha}{\nu} + \frac{1}{2}\right) - \sum\left(\frac{\alpha}{\nu} + \frac{i}{2}\right),$$

252   8. Eisenstein's Analytic Proofs

where the sums are over a complete set of residues α modulo ν (observe that we have used $(\nu, 1+i) = 1$). Moreover, $\phi(z - \frac{\alpha}{\nu})$ has the divisor

$$\left(\frac{\alpha}{\nu}\right) + \left(\frac{\alpha}{\nu} + \frac{1+i}{2}\right) - \left(\frac{\alpha}{\nu} + \frac{1}{2}\right) - \left(\frac{\alpha}{\nu} + \frac{i}{2}\right),$$

hence $\phi(\nu z)$ and $\prod \phi(z - \frac{\alpha}{\nu})$ differ at most by a constant factor δ_ν.

In order to compute δ_ν we put $\gamma = \frac{1+i}{4}$; using property ii) we find

$$\phi\left(\gamma + \frac{\alpha}{\nu}\right)\phi\left(\gamma - \frac{i\alpha}{\nu}\right) = -i\phi\left(\gamma + \frac{\alpha}{\nu}\right)\phi\left(i\gamma + \frac{\alpha}{\nu}\right)$$
$$= -i\phi\left(\gamma + \frac{\alpha}{\nu}\right)\phi\left(\gamma + \frac{\alpha}{\nu} + \frac{1}{2}\right) = 1.$$

Since there exists a 'half-system' M of residue classes such that exactly one of α and $-i\alpha$ is in M, we conclude that $\prod_\alpha \phi(\gamma + \frac{\alpha}{\nu}) = 1$. Thus $\delta_\nu = \phi(\nu\gamma)$. But now $\phi(\nu\gamma) = \phi(\varepsilon_\nu\gamma + (\nu - \varepsilon_\nu)\gamma) = \phi(\varepsilon_\nu\gamma) = \varepsilon_\nu\phi(\gamma) = \varepsilon_\nu$.   □

Using these properties of the elliptic function ϕ, giving a proof of the quartic reciprocity law is child's play. In a second proof, we will dig a little deeper and derive some less obvious arithmetic properties of the division values of ϕ.

First Proof. The quartic reciprocity law in $\mathbb{Z}[i]$ is a simple corollary of Proposition 8.7: let μ and ν be primary primes, and let A and B denote $\frac{1}{4}$-systems mod μ and mod ν, respectively. Applying Gauss's Lemma 4.4 to the $\mathbb{Z}[i]$-periodic function ϕ we get

$$\left[\frac{\nu}{\mu}\right] = \prod_{\alpha \in A} \frac{\phi(\nu\alpha/\mu)}{\phi(\alpha/\mu)}. \qquad (8.21)$$

Now we define $P(x,y) = \prod_{k=1}^{4} \phi(x + i^k y)$; using 8.7.iii) and $P(x,y) = -P(y,x)$ we find

$$\left[\frac{\nu}{\mu}\right] = \prod_{\alpha \in A} \prod_{\beta \in B} P\left(\frac{\alpha}{\mu}, \frac{\beta}{\nu}\right).$$

But the product over all $\alpha \in A$ and $\beta \in B$ has $\frac{N\mu - 1}{4} \frac{N\nu - 1}{4}$ factors, hence

$$\left[\frac{\nu}{\mu}\right] = (-1)^{\frac{N\mu - 1}{4} \cdot \frac{N\nu - 1}{4}} \left[\frac{\mu}{\nu}\right],$$

and this is the quartic reciprocity law in $\mathbb{Z}[i]$.

Using Theorem 4.5 instead of Lemma 4.4 we get the quartic reciprocity law for primary integers.

8.4 Quartic and Cubic Reciprocity 253

Second Proof. Divide (8.20) by $\phi(z)$; assuming that $\nu \equiv 1 \bmod 2 + 2i$ (i.e. that $\varepsilon = 1$) and using L'Hospital's rule we find

$$\nu = \lim_{z \to 0} \frac{\phi(\nu z)}{\phi(z)} = \prod_{\alpha \neq 0} \phi\left(\frac{\alpha}{\nu}\right). \tag{8.22}$$

If ν is prime, then the last product is over a complete set of prime residue classes modulo ν. Let M be a $\frac{1}{4}$-system modulo ν; then

$$\left(\prod_{\alpha \in M} \phi\left(\frac{\alpha}{\nu}\right)\right)^4 = \prod_{\alpha \in M} \left[-\phi\left(\frac{\alpha}{\nu}\right)\phi\left(\frac{i\alpha}{\nu}\right)\phi\left(\frac{-\alpha}{\nu}\right)\phi\left(\frac{-i\alpha}{\nu}\right)\right]$$
$$= (-1)^{(N\nu-1)/4} \nu.$$

If we put $\nu^* = (-1)^{(N\nu-1)/4}\nu$, then $\nu^* = a + bi$ with $a \equiv 1 \bmod 4$. We have seen

Proposition 8.8. *Let $\nu^* = a + bi$ be a prime such that $a \equiv 1 \bmod 4$ and $2 \mid b$. Then*

$$\nu^* = \left(\prod_{\alpha \in M} \phi\left(\frac{\alpha}{\nu}\right)\right)^4. \tag{8.23}$$

This corresponds to Equation (8.9) in the third proof of Section 8.1; before we can imitate the proof given there, we need Eisenstein's refinement of Proposition 8.2:

Theorem 8.9. *Let $\pi \in \mathbb{Z}[i]$ be a primary prime and put $\mu = \pi$ and $m = N\pi$ in Proposition 8.2; then $A_j \equiv 0 \bmod \pi$ for $0 \leq j \leq \frac{N\pi - 5}{4}$.*

Proof. Our first claim is that the splitting field N of the polynomial W is $N = \mathbb{Q}(\phi(1/\pi))$; in particular, N contains all the division values $\phi(\alpha/\pi)$, $\pi \nmid \alpha$. This is easy to see: since $\phi(\alpha/\pi)$ is a $\mathbb{Z}[i]$-rational polynomial in $\phi(1/\pi)$, the claim follows from (8.12) and the fact that the $\phi(\alpha/\pi)$ are roots of the polynomial $W \in \mathbb{Z}[i][X]$.

Next we claim that the roots $\phi(\alpha/\pi)$ of $W(x^4)$ are associated for $\alpha \not\equiv 0 \bmod \pi$. To this end, we recall that we have shown that there exist polynomials $A, B \in \mathbb{Z}[i][X]$ such that

$$\frac{\phi(\pi z)}{\phi(z)} = \frac{\pi + xA(x)}{1 + B(x)}, \quad \text{where } x = \phi(z). \tag{8.24}$$

Let $\mu \in \mathbb{Z}[i]$ be a solution of the congruence $\beta \equiv \mu\alpha \bmod \pi$; then $\phi(\mu\alpha/\pi) = \phi(\beta/\pi)$, and (8.24) implies

$$\frac{\phi(\beta/\pi)}{\phi(\alpha/\pi)} = \mu + \phi\left(\frac{\alpha}{\pi}\right) A\left(\phi\left(\frac{\alpha}{\pi}\right)\right) - \phi\left(\frac{\beta}{\pi}\right) B\left(\phi\left(\frac{\alpha}{\pi}\right)\right).$$

Thus the quotient $\phi(\beta/\pi)/\phi(\alpha/\pi)$ is an algebraic integer; by symmetry, so is its inverse, hence $\phi(\beta/\pi)/\phi(\alpha/\pi)$ is a unit.

Finally, we show that the $\phi(\alpha/\pi)$ generate a prime ideal $\mathfrak{p}$ in N which satisfies $\mathfrak{p}^{N\pi-1} = (\pi)$. Clearly they all generate the same ideal $\mathfrak{p}$. From (8.23) we deduce that $\mathfrak{p}^{p-1} = (\pi)$. Since W has degree $p-1$ and (π) is prime in $\mathbb{Z}[i]$, we conclude that $\mathfrak{p}$ is prime in N and completely ramified in $N/\mathbb{Q}(i)$. □

Eisenstein's Theorem has a number of important corollaries (we continue to assume that π is a primary prime):

Corollary 8.10. *The polynomial $W(x^4)$ is irreducible over $\mathbb{Z}[i]$.*

Proof. Apply Eisenstein's irreducibility criterion.[3] □

While we're at it, let us also prove that the extension $\mathbb{Q}(i, \phi(\frac{1}{\pi}))/\mathbb{Q}(i)$ is unramified outside 2π:

Proposition 8.11. *The discriminant of $W(x^4)$ is $D = 2^{(p-1)^2/2}\pi^{p-2}$.*

Proof. If $x_1, \ldots, x_n$ denote the roots of a polynomial f, then its discriminant is given by $\operatorname{disc} f = \prod_{i \neq j}(x_i - x_j)$. Now put $\phi(z) = \operatorname{sl}((1-i)\omega z)$, $g(z) = f((1-i)\omega z)$ and $G(z) = F((1-i)\omega z)$. Consider the following identity (see Exercise 8.25 for the simple proof):

$$[\phi(\alpha) - \phi(\beta)][\phi(\alpha) + \phi(\beta)][\phi(\alpha+\beta) - \phi(\alpha-\beta)] =$$
$$2\phi(\beta)g(\alpha)G(\alpha)\phi(\alpha+\beta)\phi(\alpha-\beta). \quad (8.25)$$

If we let α and β run through a complete set A of prime residues modulo π and multiply the corresponding equations, we get 0 on both sides: if $\alpha \equiv \beta \bmod \pi$, the first factor on the left hand side vanishes, and if $\alpha \equiv -\beta \bmod \pi$, the second does. For this reason, let $\prod_{\alpha,\beta}$ denote the product over all $(p-1)(p-3)$ pairs $(\alpha, \beta) \in A \times A$ such that $\alpha \not\equiv \pm\beta \bmod \pi$. Then

$$D = \prod_{\alpha,\beta}[\phi(\alpha) - \phi(\beta)] \cdot \prod_{\alpha=-\beta}[\phi(\alpha) - \phi(\beta)]$$
$$= \prod_{\alpha,\beta}[\phi(\alpha) - \phi(\beta)] \cdot \prod_{\alpha \in A} 2\phi(\alpha).$$

Similarly we find

---

[3] Eisenstein's irreducibility criterion was published in 1850 by Eisenstein [202] in order to prove the irreducibility of W; actually this result, along with its proof, also appear in a letter to Gauss from 1847. The criterion for polynomials in $\mathbb{Z}[X]$ was, however, first published in 1846 by Schönemann [Sm1], who claimed priority over Eisenstein in [Sm2]. Eisenstein was aware of Schönemann's work prior to [Sm1], as his references in [202] show. Thanks to Michael Filaseta for pointing this out to me.

8.4 Quartic and Cubic Reciprocity   255

$$D = \prod_{\alpha,\beta}[\phi(\alpha) + \phi(\beta)] \cdot \prod_{\alpha \in A} 2\phi(\alpha), \quad \text{as well as}$$

$$D = \prod_{\alpha,\beta}[\phi(\alpha+\beta) - \phi(\alpha-\beta)] \cdot \prod_{\alpha=0, \beta \in A}[\phi(\alpha+\beta) - \phi(\alpha-\beta)]$$

$$= \prod_{\alpha,\beta}[\phi(\alpha+\beta) - \phi(\alpha-\beta)] \cdot \prod_{\alpha \in A} 2\phi(\alpha).$$

Together with $\prod_{\alpha \in A} 2\phi(\alpha) = 2^{p-1}\pi$ this shows that

$$D^3 = \prod_{\alpha,\beta}[\phi(\alpha) - \phi(\beta)][\phi(\alpha) + \phi(\beta)][\phi(\alpha+\beta) - \phi(\alpha-\beta)] \cdot 2^{3(p-1)}\pi^3.$$

The identity above then gives

$$D^3 = 2^{3(p-1)}\pi^3 2^{(p-1)(p-3)} \prod_{\alpha,\beta} \phi(\beta) g(\alpha) G(\alpha) \phi(\alpha+\beta)\phi(\alpha-\beta).$$

Now $\prod_{\alpha,\beta}\phi(\beta) = \pi^{p-3}$, and Exercise 8.26 gives us $\prod_{\alpha \in A} g(\alpha)G(\alpha) = (1+i)^{p-1}$, thus $\prod_{\alpha,\beta} g(\alpha)G(\alpha) = (1+i)^{(p-1)(p-3)} = 2^{(p-1)(p-3)/2}$ since $(1+i)^2 = 2i$ and $i^{p-1} = 1$. Finally, $\prod_{\alpha,\beta}\phi(\alpha+\beta) = \prod_{\alpha \in A}\phi(\alpha)^{p-3} = \pi^{p-3}$, and collecting everything we get $D^3 = 2^{3(p-1)^2/2}\pi^{3(p-2)}$, hence $D = 2^{(p-1)^2/2}\pi^{p-2}$ as claimed. □

Corollary 8.12. *Let $\lambda \neq \pi$ be a primary prime in $\mathbb{Z}[i]$ above ℓ; then the Frobenius automorphism σ of λ is $\sigma : \phi(\frac{1}{\pi}) \longmapsto \phi(\frac{\lambda}{\pi})$.*

Proof. There exist polynomials $P, Q \in \mathbb{Z}[i][X]$ such that

$$\frac{\phi(\lambda z)}{\phi(z)} = \frac{\lambda P(x) + x^{p-1}}{1 + \lambda Q(x)}, \quad x = \phi(z).$$

In other words: we have Eisenstein's congruence

$$\phi(\lambda z) \equiv \phi(z)^\ell \bmod \lambda, \tag{8.26}$$

which clearly corresponds to Euler's congruence (8.8). In particular, we have $\phi(\frac{1}{\pi})^\ell \equiv \phi(\frac{\lambda}{\pi}) \bmod \lambda$. This implies that $\phi(\frac{1}{\pi})^\sigma \equiv \phi(\frac{\lambda}{\pi}) \bmod \lambda$. But if the difference of two distinct roots of the irreducible polynomial W is divisible by λ, then λ must divide the discriminant of W, which is a product of powers of $1+i$ and π by Proposition 8.11. This contradiction shows that the conjugates $\phi(\frac{1}{\pi})^\sigma$ and $\phi(\frac{\lambda}{\pi})$ must be equal, and this proves our claim. □

Corollary 8.13. *The extension $\mathbb{Q}(i, \phi(\frac{1}{\pi}))/\mathbb{Q}(i)$ is cyclic.*

Proof. Let γ be a primitive root modulo π. The extension $K = \mathbb{Q}(i, \phi(\frac{1}{\pi}))$ has degree $N\pi - 1$ over $\mathbb{Q}$, hence K is the splitting field of $W(x^4)$. Thus there exists an automorphism $\sigma \in \mathrm{Gal}\,(K/\mathbb{Q}(i))$ which maps the root $\phi(1/\pi)$ of $W(x^4)$ to the root $\phi(\gamma/\pi)$. Since σ maps $\phi(\alpha/\pi)$ to $\phi(\alpha\gamma/\pi)$ (use (8.24), σ has order $N\pi - 1$, hence it generates the Galois group of $K/\mathbb{Q}(i)$. □

Now we can complete the second proof of the quartic reciprocity law. From (8.23) we get

$$\left[\frac{\nu^*}{\pi}\right] \equiv (\nu^*)^{\frac{p-1}{4}} = \left(\prod_{\alpha \in M} \phi\left(\frac{\alpha}{\nu}\right)\right)^{p-1} = \left[\frac{\pi}{\nu}\right],$$

where we have used that

$$\left(\prod_{\alpha \in M} \phi\left(\frac{\alpha}{\nu}\right)\right)^p \equiv \prod_{\alpha \in M} \phi\left(\frac{\alpha\pi}{\nu}\right) = \left[\frac{\pi}{\nu}\right] \prod_{\alpha \in M} \phi\left(\frac{\alpha}{\nu}\right) \bmod \pi.$$

This in turn is a simple application of Eisenstein's congruence (8.26) and Gauss's Lemma (8.21).

The formulation $[\nu^*/\pi] = [\pi/\nu]$ of the quartic reciprocity law in $\mathbb{Z}[i]$ is reminiscent of the corresponding version $(q^*/p) = (p/q)$ of the quadratic reciprocity law in $\mathbb{Z}$. The equivalence to the usual version can be verified easily (see Exercise 8.19).

Cubic Reciprocity

Everything we have proved so far can be generalized to the cubic situation. In particular, we have the following analogue of Proposition 8.7:

Proposition 8.14. *There exists a unique elliptic function ψ with the following properties:*

i) *ψ has period lattice $\mathbb{Z}[\rho]$ and divisor $(\psi) = -2(0) + (\gamma) + (-\gamma)$, where $\gamma = \frac{\rho-1}{3}$, and $\rho = \frac{1}{2}(-1 + \sqrt{3})$ denotes a primitive third root of unity.*
ii) *$\psi(\rho z) = \rho \cdot \psi(z)$, $\psi(\frac{1}{3}) = 1$, and $\psi(z)\psi(z+\gamma)\psi(z-\gamma) = 1$;*
iii) *Let $\nu \in \mathbb{Z}[\rho]$, assume that $(3, \nu) = 1$, and define a cube root of unity ε by $\nu \equiv \varepsilon \bmod 3$. Then*

$$\psi(\nu z) = \varepsilon \prod_\alpha \psi\left(z + \tfrac{\alpha}{\nu}\right),$$

where α runs through a complete residue system mod ν.

The derivation of the cubic reciprocity law from this result is closely parallel to the proof of the biquadratic reciprocity law and is therefore left to the reader.

8.5 Quadratic Reciprocity in Quadratic Fields

The first quadratic reciprocity law for a quadratic number field was established by Gauss, who proved the quadratic reciprocity law for $\mathbb{Z}[i]$. Dirichlet's proof (see Chapter 5) can be generalized to arbitrary quadratic number fields (though it becomes quite technical), and Herglotz [363] showed how to derive the quadratic reciprocity law in imaginary quadratic fields from elementary properties of elliptic functions:

8.5 Quadratic Reciprocity in Quadratic Fields

Theorem 8.15. *(Herglotz) Let k be an imaginary quadratic number field with ring of integers $\mathcal{O}_k = \mathbb{Z}\omega_1 \oplus \mathbb{Z}\omega_2$. For any $\nu \in \mathcal{O}_k$ with odd norm define $a, b, c, d \in \mathbb{Z}$ by $\nu\omega_1 = a\omega_1 + b\omega_2$, $\nu\omega_2 = c\omega_1 + d\omega_2$. Then $\chi(\nu) = i^w$, where $w = (b^2 - a + 2)c + (a^2 - b + 2)d + ad$, defines a map defined modulo $4\mathcal{O}_k$, and we have*

$$\left[\frac{\mu}{\nu}\right]\left[\frac{\nu}{\mu}\right] = (-1)^{\frac{m-1}{2}\frac{n-1}{2}} \chi(\mu)^{m-\frac{n-1}{2}} \chi(\nu)^{-n\frac{m-1}{2}} \tag{8.27}$$

for all $\mu, \nu \in \mathcal{O}_k$ with odd norms $m = N\mu$ and $n = N\nu$.

Here we will show that Eisenstein's technique allows us to prove at least some special cases of Theorem 8.15 quite easily. In fact, choose an elliptic function ϕ with divisor $(0) + (\frac{\omega_1+\omega_2}{2}) - (\frac{\omega_1}{2}) - (\frac{\omega_2}{2})$ and normalize it by demanding that $\phi(\omega_1/4) = 1$. Next derive $\phi(-z) = -\phi(z)$ as in the quartic case by considering the Taylor expansion of ϕ at $z = 0$ and conclude that

$$\phi(z)\phi\left(\frac{\omega_1}{2} - z\right) = 1 \tag{8.28}$$

(this product is clearly constant, so it suffices to evaluate it at $z = \omega_1/4$). Now we claim

Proposition 8.16. *Let $k = \mathbb{Q}(\sqrt{-m})$ be an imaginary quadratic number field, m square-free, with discriminant d, and suppose that $\nu \in \mathcal{O}_k$ is odd. Let $\{\omega_1, \omega_2\}$ be an integral basis of $\mathcal{O}_k$, chosen in such a way that $\{\omega_1, \omega_2\}$ is a prime residue system modulo $2\mathcal{O}_k$ if 2 is ramified (for example, one can take $\omega_1 = 1$, $\omega_2 = 1 + \sqrt{-m}$ if $8 \mid d$ and $\omega_1 = 1$, $\omega_2 = \sqrt{-m}$ if $d \equiv 4 \bmod 8$). Finally, if $d \equiv 5 \bmod 8$ assume that $\nu \equiv 1 \bmod 2$. Then*

$$\phi(\nu z) = \xi_\nu \prod_\alpha \phi\left(z + \frac{\alpha}{\nu}\right), \tag{8.29}$$

where $\xi_\nu = \phi(\nu\frac{\omega_1}{4})$ defines a map $(\mathcal{O}_k/4\mathcal{O}_k)^\times \longrightarrow \mu_4 = \langle i \rangle : \nu \longmapsto \xi_\nu$ satisfying $\xi_{\mu\nu} = \xi_\mu \xi_\nu^m$, $m = N\mu$. Moreover, $\xi_\nu = \pm 1$ if $\nu \equiv 1 \bmod 2$, and $\xi_\nu = 1$ if $\nu \equiv 1 \bmod 4$.

Proof. The idea should be clear by now: we first show that both functions in (8.29) have the same divisor. To this end we have to verify that multiplication by ν transforms zeros of the lattice into zeros and poles into poles. If $\nu \equiv 1 \bmod 2$, this is obvious. We consider the case $d \equiv 4 \bmod 8$ first; here ϕ has divisor $(\phi) = (0) + (\frac{\omega_1+\omega_2}{2}) - (\frac{\omega_1}{2}) - (\frac{\omega_2}{2}) = (0) + (\frac{1+\sqrt{-m}}{2}) - (\frac{1}{2}) - (\frac{\sqrt{-m}}{2})$ by choice of ω_1 and ω_2, and it is clear that zeros transform to zeros under multiplication by $\nu \equiv \sqrt{-m} \bmod 2$: in fact, $\sqrt{-m} \cdot \frac{1+\sqrt{-m}}{2} = \frac{1+\sqrt{-m}}{2} - \frac{m+1}{2} \equiv \frac{1+\sqrt{-m}}{2} \bmod 2$. If $d \equiv 0 \bmod 8$, we have $(\phi) = (0) + (\frac{\sqrt{-m}}{2}) - (\frac{1}{2}) - (\frac{1+\sqrt{-m}}{2})$, and again we can check that multiplication by $\nu \equiv 1 + \sqrt{-m} \bmod 2$ maps zeros to zeros.

Now it is an easy matter to see that both functions in (8.29) have the same divisor, hence they can only differ by a constant factor $\xi_\nu \in \mathbb{C}^\times$. Substituting $z = \omega_1/4$ gives

$$\phi\Big(\nu\frac{\omega_1}{4}\Big) = \xi_\nu \phi\Big(\frac{\omega_1}{4}\Big) \prod_{\alpha \neq 0} \phi\Big(\frac{\omega_1}{4} + \frac{\alpha}{\nu}\Big);$$

but $\phi(\omega_1/4) = 1$, and if A denotes a half-system modulo ν, then

$$\prod_{\alpha \neq 0} \phi\Big(\frac{\omega_1}{4} + \frac{\alpha}{\nu}\Big) = \prod_{\alpha \in A} \phi\Big(\frac{\omega_1}{4} + \frac{\alpha}{\nu}\Big) \phi\Big(\frac{\omega_1}{4} - \frac{\alpha}{\nu}\Big)$$

$$= \prod_{\alpha \in A} \phi\Big(\frac{\omega_1}{4} + \frac{\alpha}{\nu}\Big) \phi\Big(\frac{\omega_1}{2} - \Big(\frac{\omega_1}{4} + \frac{\alpha}{\nu}\Big)\Big) = 1.$$

This proves the first claim. Next, ξ_ν only depends on ν mod 4 (this is obvious in view of $\xi_\nu = \phi(\nu\omega_1/4)$ and the $\mathcal{O}_k$-periodicity of ϕ). Now let's give a proof of the relation $\xi_{\mu\nu} = \xi_\mu \xi_\nu^m$. We obviously have

$$\phi(\mu\nu z) = \xi_\mu \prod_\alpha \phi\Big(\nu z + \frac{\alpha}{\mu}\Big), \tag{8.30}$$

where α runs through a residue system modulo μ. Now $\nu z + \frac{\alpha}{\mu} = \nu(z + \frac{\alpha}{\mu\nu})$, hence

$$\phi\Big(\nu z + \frac{\alpha}{\mu}\Big) = \xi_\nu \prod_\beta \phi\Big(z + \frac{\alpha}{\mu\nu} + \frac{\beta}{\nu}\Big),$$

where β runs through a residue system modulo ν. Since there are $m = N\mu$ factors in the product of (8.30), we have

$$\phi(\mu\nu z) = \xi_\mu \xi_\nu^m \prod_{\alpha,\beta} \phi\Big(z + \frac{\alpha + \beta\mu}{\mu\nu}\Big). \tag{8.31}$$

We claim that $\alpha + \beta\mu$ runs through a complete set of residues modulo $\mu\nu$. They are clearly incongruent modulo $\mu\nu$, since $\alpha + \beta\mu \equiv \alpha' + \beta'\mu$ mod $\mu\nu$ implies at once $\alpha \equiv \alpha'$ mod μ (thus $\alpha = \alpha'$), and then $\beta\mu \equiv \beta'\mu$ mod $\mu\nu$ gives $\beta \equiv \beta'$ mod ν, i.e. $\beta = \beta'$. On the other hand, there are $N\mu\nu$ such expressions, and our claim follows.

Thus the product in (8.31) is simply $\prod_\gamma \phi(z + \frac{\gamma}{\mu\nu})$, where γ runs through a complete set of residues modulo $\mu\nu$. Since we also have

$$\phi(\mu\nu z) = \xi_{\mu\nu} \prod_\gamma \phi\Big(z + \frac{\gamma}{\mu\nu}\Big),$$

a comparison gives $\xi_{\mu\nu} = \xi_\mu \xi_\nu^m$.

Now assume that $\nu \equiv 1$ mod 2 and write $\nu = 1 + 2\gamma$ for some $\gamma \in \mathcal{O}_k$; then

8.5 Quadratic Reciprocity in Quadratic Fields

$$\xi_\nu^2 = \phi\Big(\frac{\omega_1}{4} + \gamma\frac{\omega_1}{2}\Big)^2 = \phi\Big(\frac{\omega_1}{4} + \gamma\frac{\omega_1}{2}\Big)\phi\Big(\frac{\omega_1}{2} - \frac{\omega_1}{4} - \gamma\frac{\omega_1}{2}\Big) = 1.$$

Of course we also have $\xi_\nu = 1$ if $\nu \equiv 1 \bmod 4$.

It remains to show that ξ_ν takes values in μ_4. To this end we need to know $\phi(\frac{\omega_2}{4})$ and $\phi(\frac{\omega_1+\omega_2}{4})$. The function $\phi(z)\phi(\frac{\omega_2}{2} - z)$ has trivial divisor and is therefore constant, i.e. $\phi(z)\phi(\frac{\omega_2}{2} - z) = c$. Putting $z = \frac{\omega_1+\omega_2}{4}$ in $\phi(z)\phi(\frac{\omega_1}{2} - z) = 1$ and $\phi(z)\phi(\frac{\omega_2}{2} - z) = c$ shows that $c = -1$, hence

$$\phi(z)\phi\Big(\frac{\omega_2}{2} - z\Big) = -1. \tag{8.32}$$

In particular, we get $\phi(\frac{\omega_2}{4})^2 = -1$, i.e. $\phi(\frac{\omega_2}{4}) \in \mu_4$ (Kubota proved that in fact $\phi(\frac{\omega_2}{4}) = i$; it would be nice if one could prove this with our simple method). Next, $\phi(z)$ and $\phi(z + \frac{\omega_1+\omega_2}{2})$ differ only by a constant; since $\phi(\frac{\omega_1}{4} + \frac{\omega_2}{2}) = -1 = -\phi(\frac{\omega_1}{4})$ we find

$$\phi\Big(z + \frac{\omega_1 + \omega_2}{2}\Big) = -\phi(z). \tag{8.33}$$

Since we obviously have $\xi_\nu \in \mu_4$ if $\nu \equiv 1 \bmod 2$, we only need to prove the claim if $d \equiv 0 \bmod 4$. In these cases, the integral basis was chosen in such a way that $\{\omega_1, \omega_2\}$ is a prime residue system modulo $2\mathcal{O}_k$. If we can show that $\nu \equiv \mu \bmod 2$ implies $\xi_\nu = \pm \xi_\mu$, then it is sufficient to show $\phi(\nu\frac{\omega+1}{4}) \in \mu_4$ only for $\nu = 1$ and $\nu = \omega_2$: this follows from $\phi(z) = \phi(z + \frac{a\omega_1+b\omega_2}{2}) = i^s \phi(z)^t$ for some $s \in \{0, 1, 2, 3\}$ and $t \in \{-1, 1\}$, which in turn is a consequence of (8.28), (8.32) and (8.33).

Since ω_1^2 and $\omega_1\omega_2$ are equivalent to ω_1 and ω_2 modulo $2\mathcal{O}_k$ (or vice versa), we only have to check that $\phi(\frac{\omega_1}{4}) \in \mu_4$ and $\phi(\frac{\omega_2}{4}) \in \mu_4$, which we have already done. □

Now Gauss's Lemma gives

$$\Big[\frac{\mu}{\nu}\Big] = \prod_{\alpha \in A} \frac{\phi(\mu\alpha)}{\phi(\alpha)}, \quad \Big[\frac{\nu}{\mu}\Big] = \prod_{\beta \in B} \frac{\phi(\nu\beta)}{\phi(\beta)},$$

where A and B are half-systems modulo μ and ν, respectively. Using (8.29) we get

$$\Big[\frac{\mu}{\nu}\Big] = \xi_\mu^{(n-1)/2} \prod_{\alpha \in A}\prod_{\beta \in B} \phi(\alpha + \beta)\phi(\alpha - \beta),$$

where the second factor is due to the fact that Proposition 8.16 demands a full system of prime residues. This shows

$$\Big[\frac{\mu}{\nu}\Big]\Big[\frac{\nu}{\mu}\Big] = (-1)^{\frac{n-1}{2}\frac{m-1}{2}} \xi_\mu^{(n-1)/2} \xi_\nu^{-(m-1)/2}. \tag{8.34}$$

The fact that $\xi_\nu = 1$ if $\nu \equiv 1 \bmod 4$ gives the following special case of (8.27):

Proposition 8.17. *If $\mu, \nu \in \mathbb{Z}[i]$ are odd and $\mu \equiv \xi^2$ mod 4 for some $\xi \in \mathcal{O}_k$, then*

$$\left[\frac{\mu}{\nu}\right]\left[\frac{\nu}{\mu}\right] = 1.$$

Proof. Since the residue symbols do not change if we multiply μ or ν by a square, we may assume that $\mu \equiv 1$ mod 4.

Now first suppose that $\nu \equiv 1$ mod 2. In this case, we have $\xi_\mu = 1$, $\xi_\nu = \pm 1$, $m \equiv 1$ mod 4, hence the right hand side of (8.34) is trivial.

If $d \equiv 1$ mod 8, every odd $\nu \in \mathcal{O}_k$ is congruent to 1 mod 2, so there is nothing left to prove here. If $d \equiv 5$ mod 8 and ν is odd, then $\nu^3 \equiv 1$ mod $2\mathcal{O}_k$, and the claim $[\mu/\nu] = [\nu/\mu]$ follows at once from $[\mu/\nu^3] = [\nu^3/\mu]$. If $d \equiv 0$ mod 4, then $\mu \equiv 1$ mod 4 implies $m = N\mu \equiv 1$ mod 8 (in fact, write $\mu = 1 + 4\xi$; then $N\mu \equiv 1 + 4(\xi + \xi')$ mod 8; but since $d = \text{disc } k$ is even, $\xi + \xi'$ must be divisible by 2; thus $\xi_\nu^{(m-1)/2} = 1$, $\xi_\mu = 1$, and the claim follows from (8.34). □

Since ξ_ν depends only on the residue class ν mod 4, (8.34) implies

Proposition 8.18. *If k is an imaginary quadratic number field, then*

$$\left[\frac{\mu}{\nu}\right]\left[\frac{\nu}{\mu}\right] = \left[\frac{\mu'}{\nu'}\right]\left[\frac{\nu'}{\mu'}\right]$$

whenever $\mu \equiv \mu'$ mod 4 and $\nu \equiv \nu'$ mod 4.

This allows us to prove the quadratic reciprocity law of a given imaginary quadratic number field by simply computing $[\mu/\nu][\nu/\mu]$ for a complete set of prime residue classes of μ, ν mod 4. As an example, take $k = \mathbb{Q}(\sqrt{-3})$: as in the proof of Theorem 7.10 we may assume that $\mu \equiv \nu \equiv 1$ mod 2 (replacing μ and ν by their cubes if necessary). Choosing μ and ν primary shows that it is sufficient to prove the quadratic reciprocity law in $\mathbb{Z}[\rho]$ for $\mu, \nu \equiv 1, 3 + 2\rho$ mod 4. But if one of them is $\equiv 1$ mod 4, the inversion factor is trivial by Proposition 8.17, and if both of them are $\equiv 3 + 2\rho$ mod 4, then Proposition 8.18 shows

$$\left[\frac{\mu}{\nu}\right]\left[\frac{\nu}{\mu}\right] = \left[\frac{3 + 2\rho}{-1 + 2\rho}\right]\left[\frac{-1 + 2\rho}{3 + 2\rho}\right] = (+1) \cdot (-1) = -1,$$

in accordance with Theorem 7.10.

8.6 Kronecker's Jugendtraum

Next we will explain how the results of the preceding sections are connected with Kronecker's Jugendtraum, the construction of abelian extensions of imaginary quadratic number fields using elliptic functions (in fact, we have already noticed that the division values sl($\omega/4$), f($\omega/4$) and F($\omega/4$) generate abelian extensions of $\mathbb{Q}(i)$; Kronecker's Jugendtraum for $\mathbb{Q}(i)$ claims[4] cum grano salis that we get *all* abelian extensions of $\mathbb{Q}(i)$ by adjoining the division values of sl, namely the sl(ω/n) for $n \in \mathbb{N}$, to $\mathbb{Q}(i)$). The following theorem due to Takagi [Ta1] is one of the major steps in showing that every abelian extension of $k = \mathbb{Q}(i)$ is contained in the field generated by some $\phi(1/n)$, $n \in \mathbb{N}$. For our formulation of Takagi's results we will need the notion of a ray class field of k modulo $\mathfrak{m}$ (we will denote it by $k\{\mathfrak{m}\}$); for a complete definition, see Part II – for now we will be content with the crude description of $k\{\mathfrak{m}\}$ as the maximal abelian extension of k that is unramified outside $\mathfrak{m}$ and whose ramification at $\mathfrak{m}$ is bounded in some sense (for example, any prime ideal $\mathfrak{p}$ such that $\mathfrak{p}^2 \nmid \mathfrak{m}$ is at most tamely ramified).

Theorem 8.19. *Let $\pi \equiv 1 \bmod (2+2i)$ be a primary prime of norm $p \equiv 1 \bmod 4$, and put $\lambda = 1+i$ and $\nu = \pi^n$ for some $n \geq 1$. Then $L = k\{\lambda^3 \nu\} = \mathbb{Q}\left(i, \phi\left(\frac{1}{\nu}\right)\right)$ is the ray class field $\bmod \lambda^3 \nu$ of $k = \mathbb{Q}(i)$; moreover,*

i) L/k is cyclic of order $\Phi(\nu)$; in fact, $\mathrm{Gal}\,(L/k) \simeq (\mathcal{O}_k/\nu\mathcal{O}_k)^\times$.
ii) L contains $k(\sqrt[4]{\pi^*})$ as a subfield, where $\pi^* = (-1)^{(p-1)/4}\pi$.
iii) L is a quartic extension of the ray class field $k\{\nu\}$, containing $k\{2\nu\}$ as a quadratic subfield.

Proof. The special case $n = 1$ of i) is Corollary 8.13; the general case can be proved similarly as long as $(\mathbb{Z}[i]/\nu\mathbb{Z}[i])^\times$ is cyclic. Claim ii) follows from Proposition 8.8, and iii) is a simple generalization of Proposition 8.11, as it is essentially a computation of the discriminant of the polynomial $W(X)$ corresponding to the division by ν. □

There is a beautiful analogy to the construction of abelian extensions of $\mathbb{Q}$: the exponential function $f(z) = \exp 2\pi i z$ has period lattice $\mathbb{Z}$, and the division points $f(\frac{1}{n})$ generate abelian extensions of $\mathbb{Q}$; similarly, $\phi(z)$ has period lattice $\mathbb{Z} \oplus \mathbb{Z}[i]$, and the division points $\phi(\frac{1}{\nu})$ generate abelian extensions of $\mathbb{Q}(i)$. Actually, the analogy extends much further as you will see by comparing the decomposition of primes in ray class fields (see the proof of Corollary 8.20 below) with the situation in cyclotomic extensions. Moreover, the action of the Frobenius automorphism can be described explicitly in both cases: in the cyclotomic theory, the Frobenius of a prime $0 < \ell \neq p$ maps $\exp \frac{2\pi i}{p}$ to $\exp \frac{2\pi i \ell}{p}$, in the lemniscatic theory the Frobenius of a primary prime $\lambda \in \mathbb{Z}[i]$ maps $\phi(\frac{1}{z})$ to $\phi(\frac{\lambda}{z})$.

The Hasse diagrams in Figure 8.1 display some subfields of $k\{\lambda^3 \pi\}$ for the cases $p \equiv 5 \bmod 8$, $p \equiv 9 \bmod 16$ and $p \equiv 1 \bmod 16$, respectively. We note that these inclusions can be proved easily using class field theory; the computational proof using the theory of complex multiplication is a bit tougher.

The situation for abelian extensions of $\mathbb{Q}$ is completely analogous: for odd n, the extension $\mathbb{Q}(\exp \frac{2\pi i}{n})$ is the ray class field $\mathbb{Q}\{n\infty\}$, and we have

---
[4] See Schappacher's articles [Sc2, Sc3] for a more exact exposition.

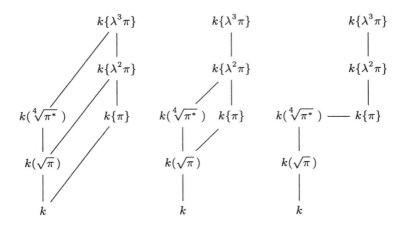

Fig. 8.1. Some ray class fields

$\mathbb{Q}(\sqrt{p}) \subseteq \mathbb{Q}\{p\}$ if $p \equiv 1 \bmod 4$, but $\mathbb{Q}(\sqrt{p^*}) \subseteq \mathbb{Q}\{p\infty\}$ for $p \equiv 3 \bmod 4$. As in the cyclotomic case, we can extract a weak reciprocity law from comparing the splitting of primes in the Kummer extensions $\mathbb{Q}(i, \sqrt[4]{\pi^*})$ and the ray class fields $\mathbb{Q}(i)\{\lambda^3\pi\}$.

Corollary 8.20. *Let $\pi \equiv \kappa \equiv 1 \bmod (2+2i)$ be primary primes in $\mathbb{Z}[i]$ such that $\left(\frac{\pi}{\kappa}\right)_2 = 1$. Then*

$$\left[\frac{\kappa}{\pi}\right] = \left[\frac{\pi^*}{\kappa}\right] = (-1)^{\frac{N\pi-1}{4} \cdot \frac{N\kappa-1}{4}} \left[\frac{\pi}{\kappa}\right].$$

Proof. Before we start we recall the decomposition law for ray class fields: in our case, we let k^1 be the ray class field mod $\lambda^3\pi$. Then a prime ideal $\mathfrak{q} = \kappa\mathcal{O}_k$ splits into g prime ideals of degree f, where $fg = \Phi(\lambda^3\pi) = p - 1$, if f is the smallest integer ≥ 1 such that $\mathfrak{q}^f$ is generated by an element $\equiv 1 \bmod \lambda^3\pi$. Since we may assume that $\kappa \equiv 1 \bmod \lambda^3$ right from the start, this is equivalent to $\kappa^f \equiv 1 \bmod \pi$. Now we see that

$$\left[\frac{\kappa}{\pi}\right] = +1 \iff \kappa^{\frac{p-1}{4}} \equiv 1 \bmod \pi \iff f \mid \frac{p-1}{4}$$
$$\iff g \equiv 0 \bmod 4 \iff K_Z \supseteq k(\sqrt[4]{\pi^*})$$
$$\iff \left[\frac{\pi^*}{\kappa}\right] = +1, \text{ and similarly,}$$

$$\left[\frac{\kappa}{\pi}\right] = -1 \iff \kappa^{\frac{p-1}{4}} \equiv -1 \bmod \pi$$
$$\iff f \mid \frac{p-1}{4} \text{ and } f \nmid \frac{p-1}{2} \iff g \equiv 2 \bmod 4$$
$$\iff K_Z \supseteq k(\sqrt{\pi^*}) \text{ but } K_Z \not\supseteq k(\sqrt[4]{\pi^*})$$
$$\iff \left[\frac{\pi^*}{\kappa}\right] = -1. \qquad \square$$

Complex Multiplication

Before we can say more about complex multiplication, we have to introduce some more complex analytic functions. References for the rest of this section are the books of Cox [139], Knapp [Kna], Koblitz [Kob], and Serre [Ser].

Let $\mathbb{H}$ denote the upper half plane; to any $\tau \in \mathbb{H}$ we can associate a lattice $\Lambda_\tau = \mathbb{Z} + \tau\mathbb{Z}$, and we put $G_k(\tau) = G_k(\Lambda_\tau)$. Thus the Eisenstein series attached to lattices via (8.16) turn into holomorphic functions on $\mathbb{H}$ with the following properties:

Proposition 8.21. *For all $z \in \mathbb{H}$ and every $k \geq 2$, we have*

$$G_k(z+1) = G_k(z); \quad G_k(-\tfrac{1}{z}) = z^{2k} G_k(z); \quad G_k(\infty) = 2\zeta(2k). \quad \square$$

Here $\zeta(s)$ is Riemann's zeta function, and $G_k(\infty)$ is defined as follows: any holomorphic function $f: \mathbb{H} \longmapsto \mathbb{C}$ such that $f(z+1) = f(z)$ for all $z \in \mathbb{H}$ is a function of $q = e^{2\pi i z}$, i.e. $f(z) = f^*(q)$ (note that $|q| < 1$ since $\operatorname{Im} z > 0$ for $z \in \mathbb{H}$). Now put $f(\infty) = f^*(0)$.

We next define the modular group

$$\Gamma = \operatorname{SL}(2, \mathbb{Z}) = \left\{ \begin{pmatrix} a & b \\ c & d \end{pmatrix} : a, b, c, d \in \mathbb{Z}, ad - bc = 1 \right\}.$$

The modular group acts on $\mathbb{H}$ via $\left(\begin{smallmatrix} a & b \\ c & d \end{smallmatrix}\right) z = \frac{az+b}{cz+d}$ (you have to show that the right hand side is again in $\mathbb{H}$). A *modular function* $f: \mathbb{H} \longrightarrow \overline{\mathbb{C}} = \mathbb{C} \cup \{\infty\}$ is required to have the following two properties: first, f behaves nicely under the action of Γ: $f(gz) = (cz+d)^k f(z)$ for all $g = \left(\begin{smallmatrix} a & b \\ c & d \end{smallmatrix}\right) \in \Gamma$. The integer k is called the *weight* of f. Since $\left(\begin{smallmatrix} 1 & 1 \\ 0 & 1 \end{smallmatrix}\right) \in \Gamma$, we see that, in particular, $f(z+1) = f(z)$. Thus $f(z) = f^*(q)$, and the second condition for f to be modular is that f^* be meromorphic on the whole disc $|q| < 1$. This is equivalent to f having an expansion $f(z) = \sum a_n q^n$ where $a_n = 0$ for all $n \leq -N_0$ for some $N_0 \in \mathbb{N}$.

A modular function is called a *modular form* for Γ if f is 'holomorphic at the cusp', i.e. if f has an expansion $f(z) = \sum_{n \geq 0} a_n q^n$. If $a_0 = 0$, we say that f vanishes at the cusp and call f a *cusp form*.

Let us construct a few modular forms. From Proposition 8.21 we know that the Eisenstein series G_k are modular forms of weight $2k$. Now observe that the modular forms of fixed weight $2k$ form a $\mathbb{C}$-vector space. Since the product of modular forms is again a modular form (whose weight is the sum of the weights of the factors), we can construct new modular forms by taking products and then adding up forms of the same weight. Take G_2 and G_3, for example, and recall that $g_2(\Lambda) = 60 G_2(\Lambda)$ and $g_3(\Lambda) = 140 G_3(\Lambda)$. Then both g_2^3 and g_3^2 are modular forms of weight 12, and we find $g_2^3(\infty) = 2^6 3^{-3} \pi^{12}$ and $g_3^2(\infty) = 2^6 3^{-6} \pi^{12}$. Hence $\Delta = g_2^3 - 27 g_3^2$ vanishes at ∞, and we have constructed a cusp form Δ of weight 12:

Theorem 8.22. *The function $\Delta: \mathbb{H} \longrightarrow \mathbb{C}$ has the following properties:*

i) $\Delta(z) \neq 0$ for all $z \in \mathbb{H}$;
ii) $\Delta(z)$ is a cusp form of weight 12;
iii) $\Delta(z) = g_2^3 - 27g_3^2 = (2\pi)^{12} q \prod_{k=1}^{\infty}(1-q^k)^{24}$. □

The fact that $\Delta(\Lambda) \neq 0$ for lattices Λ allows us to define the function

$$j = \frac{1728 g_2^3}{\Delta}.$$

Some analytic properties of this function are collected in

Theorem 8.23. *The "modular invariant" j is a modular form of weight 0; it is holomorphic in $\mathbb{H}$ with a simple pole of residue 1 at ∞, and induces a bijection $\mathbb{H}/\Gamma \longrightarrow \mathbb{C}$.* □

The last assertion is a convenient abbreviation for the fact that $j : \mathbb{H} \longrightarrow \mathbb{C}$ is onto, and that $j(z) = j(z')$ if and only if $z' = gz$ for some $g \in \Gamma$. But the big surprise is the following:

Theorem 8.24. *Write $\mathcal{O}_K = \mathbb{Z} + \tau\mathbb{Z}$ for some $\tau \in \mathcal{O}_K \cap \mathbb{H}$; then $j(\tau)$ is an algebraic integer of degree $h(K)$ over K, and $K(j(\tau))$ is the Hilbert class field of K.* □

The Hilbert class field H of K is the maximal abelian unramified extension of K; its Galois group is isomorphic to the ideal class group of K. In particular, $j(\tau)$ must be a rational integer whenever τ is an element of a quadratic number field with class number 1. For example,

$$\begin{aligned}
j\left(\tfrac{-1+\sqrt{-3}}{2}\right) &= 0, & j(i) &= 12^3, \\
j\left(\tfrac{1+\sqrt{-7}}{2}\right) &= (-15)^3, & j(\sqrt{-2}) &= 20^3, \\
j\left(\tfrac{1+\sqrt{-11}}{2}\right) &= (-32)^3, & j\left(\tfrac{1+\sqrt{-19}}{2}\right) &= (-96)^3, \\
j\left(\tfrac{1+\sqrt{-43}}{2}\right) &= (-960)^3, & j\left(\tfrac{1+\sqrt{-67}}{2}\right) &= (-5280)^3.
\end{aligned}$$

The best known example is, however, $j\left(\tfrac{1+\sqrt{-163}}{2}\right) = (-640\,320)^3$, because it is connected with the impressing decimal expansion

$$e^{\pi\sqrt{163}} = 262\,537\,412\,640\,768\,743.9999999999992\ldots$$

In fact, it is known that $j(\tau) = \tfrac{1}{q} + 744 + 196\,884 q + 21\,493\,760 q^2 + \ldots$, where $q = e^{2\pi i \tau}$. For $\tau = \tfrac{1}{2}(1 + i\sqrt{163})$, the absolute value of $q = -e^{-\pi\sqrt{163}}$ is very small, thus $|j(\tau) - \tfrac{1}{q} - 744|$ is small, and $-e^{\pi\sqrt{163}} = \tfrac{1}{q}$ does not differ much from the integer $j(\tau) - 744 = -262\,537\,412\,640\,768\,744$. In fact, since $j(\tau) \approx e^{\pi\sqrt{163}} - 744$ is a cube, the following expression is even closer to an integer:

$$(e^{\pi\sqrt{163}} - 744)^{1/3} = 640319.99999999999999999999993\ldots$$

Here are some examples of fields with non-trivial class number: the field $K = \mathbb{Q}(\sqrt{-5})$ has class number 2, and $j(\sqrt{-5}) = (50 + 26\sqrt{5})^3$; this gives $K(j(\sqrt{-5})) = K(\sqrt{2})$. Next $K = \mathbb{Q}(\sqrt{-6})$ also has class number 2, and $j(\sqrt{-6}) = 12^3(1 + \sqrt{2})^2(5 + 2\sqrt{2})^3$ shows that its Hilbert class field is $K(\sqrt{2})$. Finally, $\mathbb{Q}(\sqrt{-23})$ has class number 3, and one finds $j\left(\frac{1+\sqrt{-23}}{2}\right) = -5^3 \alpha^{-12}(2\alpha - 1)^3(3\alpha + 2)^3$, where α is the real root of the polynomial $f(x) = x^3 - x - 1$ of discriminant -23.

Observe that some j-values are cubes, while others are not. This is explained by a result of Weber:

Theorem 8.25. *Let K and τ be as in Theorem 8.24, and let H be the Hilbert class field of K. Then*

$$H(\sqrt[3]{j(\tau)}) = \begin{cases} H & \text{if } 3 \nmid \operatorname{disc} K, \\ K\{3\} & \text{if } 3 \mid \operatorname{disc} K, \end{cases}$$

$$H(\sqrt{j(\tau) - 1728}) = \begin{cases} H & \text{if } 2 \nmid \operatorname{disc} K, \\ K\{2\} & \text{if } 2 \mid \operatorname{disc} K. \end{cases} \qquad \square$$

This result has applications to Gauss's class number 1 problem: Let k be an imaginary quadratic number field with discriminant d coprime to 6 and with class number 1. By Theorem 8.25, $j(\tau) = x^3$ and $j(\tau) - 1728 = y^2$ for integers $x, y \in \mathcal{O}_k$; it can be shown that $x \in \mathbb{Z}$ and $y \in \sqrt{-p}\,\mathbb{Z}$, thus to every such number field there is a point $(x, y) \in \mathbb{Z} \times \mathbb{Z}$ on the elliptic curve $py^2 = x^3 + 1728$. This curve still depends on the prime p, but by replacing j with suitable Weber functions Heegner [Hee] succeeded in finding a similar relation with diophantine equations that do not depend on p; by solving these equations Heegner solved Gauss's problem. For a short discussion of his life, see Opolka [Op].

There are a lot of results similar to Theorem 8.25 that have been proved with the aim of reducing the size of the coefficients of generating polynomials of class fields; see Schertz [Sr2] and Gee & Stevenhagen [GS].

8.7 The Determination of Gauss Sums

In this section we will return to the problem of computing the sign of the quadratic Gauss sum τ which we already met in Section 3.3; after giving a proof for Proposition 3.21 we will explain the relevance of elliptic functions in this context.

Quadratic Gauss Sums

Our computation of the sign of the quadratic Gauss sum τ will be based on the formula $\Lambda = \pm \tau$ (Equation (8.9)). If we can show that $\Lambda = \left(\frac{-2}{p}\right)\tau$, then Proposition 3.21 will follow, since $\Lambda = (2i)^{(p-1)/2} \prod_{a=1}^{(p-1)/2} \sin \frac{2\pi a}{p}$ can easily be determined: we already know that $\Lambda = \pm\sqrt{p^*}$, and since the factors $2\sin\frac{2\pi a}{p}$ are all positive, Λ is a positive real number times $i^{(p-1)/2}$. This gives $\Lambda = (2/p)\sqrt{p}$ if $p \equiv 1 \bmod 4$, and $\Lambda = -(2/p)i\sqrt{p}$ if $p \equiv 3 \bmod 4$.

In order to prove that $\Lambda = \left(\frac{-2}{p}\right)\tau$ it is sufficient to show that both sides are congruent modulo $\lambda^{(p+1)/2}$, where $\lambda = \zeta - 1$ generates the prime ideal over p in $\mathbb{Q}(\zeta)$: in fact, since $(\Lambda) = (\tau) = (\lambda)^{(p-1)/2}$ and $\Lambda = \pm\tau$, the congruence modulo $\lambda^{(p+1)/2}$ implies that $\Lambda/\tau \equiv \left(\frac{-2}{p}\right) \bmod \lambda$ in $\mathbb{Z}[i]$, and we must have equality since $1 \not\equiv -1 \bmod \lambda$.

Put $p = 2m + 1$; we start by showing $\Lambda \equiv \left(\frac{2}{p}\right)\lambda^m m! \bmod \lambda^{m+1}$. In fact, we have $\zeta^a - \zeta^{-a} = -\zeta^{-a}(1 - \zeta^{2a}) = -\zeta^{-a}(1 - (1+\lambda)^{2a}) = -\zeta^{-a}(1 - (1 + 2a\lambda)) \bmod \lambda^2$ and $-\zeta^{-a} \equiv -1 \bmod \lambda$, hence $(\zeta^a - \zeta^{-a})/\lambda \equiv 2a \bmod \lambda$. Taking the product over these congruences for $a = 1, \ldots, m$ we get $\Lambda/\lambda^m \equiv 2^m m! \bmod \lambda$. Since $2^m \equiv (2/p) \bmod \lambda$, this proves our claim.

Proposition 3.21 will follow if we can show that $\tau \equiv (-\lambda)^m m! \bmod \lambda^{m+1}$. This is done as follows: we clearly have

$$\tau = \sum_{a=1}^{p-1} \left(\frac{a}{p}\right)\zeta^a \equiv \sum_{a=1}^{p-1} a^m (1+\lambda)^a = \sum_{a=1}^{p-1} a^m \sum_{s=0}^{a} \binom{a}{s}\lambda^s.$$

Now $\binom{a}{s} = 0$ if $s > a$ shows that extending the range of s to all $s \leq m$ does not change the sum, hence

$$\tau \equiv \sum_{s=0}^{m} \lambda^s \sum_{a=1}^{p-1} a^m \binom{a}{s} \bmod \lambda^{m+1}.$$

Next we notice that $\binom{a}{s}$ is a polynomial of degree s for all $s \leq m$. By Lemma 4.29, only $s = m$ will give a contribution modulo p, and we get

$$\tau \equiv \frac{\lambda^m}{m!} \sum_{a=1}^{p-1} a^m a(a-1)\cdots(a-m+1) \equiv -\frac{\lambda^m}{m!} \bmod \lambda^{m+1}. \tag{8.35}$$

From $-1 \equiv (p-1)! = \prod_{a=1}^{m} a(p-a) \equiv (-1)^m (m!)^2 \bmod p$ we get $\frac{1}{m!} \equiv -(-1)^m m! \bmod p$, hence $\tau \equiv (-\lambda)^m m! \bmod \lambda^{m+1}$ as desired.

Quartic Gauss Sums

Now that we have determined quadratic Gauss sums with the help of (8.9), it is natural to ask whether the quartic Gauss sum can similarly be determined by using (8.23); actually, Kubota first posed this question in [454] and [457].

First we observe that the square of the quartic Gauss sum $G = G(\chi)$ for the character $\chi = [\cdot/\pi]$, $\pi = a + bi \equiv 1 \bmod 2 + 2i$ prime, can be determined quite easily: we know $G^2 = J(\chi,\chi)G(\chi^2)$, $J(\chi,\chi) = \chi(-1)\pi$ (see

Proposition 6.5) and $G(\chi^2) = -\tau = -\sqrt{p}$ (see Proposition 3.21, which we just have proved), hence $G^2 = -\chi(-1)\pi\sqrt{p}$.

Let $\wp$ be the Weierstraß $\wp$-function such that $\wp'(z)^2 = 4\wp(z)^3 - \wp(z)$, and let $\theta\mathbb{Z}[i]$ be the corresponding period lattice. With $p = 2N+1$ and

$$P := \prod_{r=1}^{N} \wp'(r\theta/\pi)$$

one finds $P^2 = \chi(-1)\pi^{-3}$ (see Exercise 8.29), hence $(GP\pi)^2 = -\sqrt{p}$. Define $\beta(\pi) = \pm i$ via the congruence $\beta(\pi) \equiv N! \bmod \pi$; then $GP\pi = \pm\beta(\pi)\sqrt[4]{p}$, and computations for primes $p < 5000$ suggested that

$$GP\pi = \beta(\pi)\chi(2i)\sqrt[4]{p}.$$

This conjecture was proved by Matthews [Mt2] (see [Mt1] for the cubic case). Matthew's result gives the following value for the sum G_4 defined in Exercise 6.16:

$$-G_4 = \begin{cases} \sqrt{p} + \left(\frac{|b|}{a}\right)N!(-1)^{(b^2+2|b|)/8}\sqrt{2p+2a\sqrt{p}} & \text{if } p \equiv 1 \bmod 8 \\ \sqrt{p} + \left(\frac{|b|}{a}\right)N!(-1)^{(b^2+2|b|)/8}i\sqrt{2p-2a\sqrt{p}} & \text{if } p \equiv 5 \bmod 8 \end{cases}.$$

NOTES[5]

Elliptic Integrals

Elliptic integrals first occurred in the work of J. Wallis around 1659 in connection with the arclength of ellipses (that's where the 'elliptic' in elliptic functions and elliptic curves comes from). Afterwards, these integrals were studied by Jakob [BJa] and Johann Bernoulli [BJo] (they showed that the arclength of the lemniscate is given by the integral $\int_0^x \frac{dt}{\sqrt{1-t^4}}$), but the first real insight was gained by Fagnano (see Ayoub [Ay]) around 1714, who by and by learned how to divide the lemniscate into 2^m, $3 \cdot 2^m$ and $5 \cdot 2^m$ equal parts. The following explanation of Fagnano's idea is due to Siegel [Sie]: consider the integral $\int_0^c (1-x^4)^{-1/2}dx$ for $c \in [0,1]$. The function $g(t) = 2t^2/(1+t^4)$ is an order-preserving bijection $[0,1] \longrightarrow [0,1]$, and the substitution $x^2 = g(t)$ yields

$$\int_0^c \frac{dx}{\sqrt{1-x^4}} = \sqrt{2}\int_0^b \frac{dt}{\sqrt{1+t^4}},$$

where $b \in [0,1]$ is defined by $c^2 = g(b)$.

Now consider $h(u) = 2u^2/(1-u^4)$; this is an order-preserving bijection $[0, \sqrt{\sqrt{2}-1}] \longrightarrow [0,1]$, and in particular there is an $a \in [0, \sqrt{\sqrt{2}-1}]$ such that $b^2 = h(a)$. Transforming the last integral we get

---
[5] and Footnotes

$$\int_0^b \frac{dt}{\sqrt{1+t^4}} = \sqrt{2} \int_0^a \frac{du}{\sqrt{1-u^4}},$$

and putting everything together we get

$$2 \int_0^a \frac{du}{\sqrt{1-u^4}} = \int_0^c \frac{dx}{\sqrt{1-x^4}}, \quad \text{where } c^2 = \frac{4a^2(1-a^4)}{(1+a^4)^2}.$$

By introducing the elliptic function sl we can write the integral equations above in the form $\text{sl}(w/2) = a$ and $\text{sl}(w) = c$. Plugging this into $c = 2a\sqrt{1-a^4}/(1+a^4)$ (this follows from $a, b, c > 0$) we get a formula for $\text{sl}(w)$ in terms of $\text{sl}(w/2)$: the duplication formula for $\text{sl}(z)$.

In 1750, Fagnano's collected works (Produzioni matematiche; for a new edition, see [Fag]) appeared, and he sent a copy to the Berlin Academy where it was given to Euler for a review on December 23, 1751 (this date was later called the *birthday of the theory of elliptic functions* by Jacobi). Euler immediately started working on a more general theory, to which Lagrange and Legendre made important contributions. Euler [Eu2] falsely claimed that the division of the lemniscate into n equal parts with ruler and compass was possible if and only if n has the form $2^k(2^m + 1)$:

> ... manifestum est hoc modo etiam lemniscaticam in tot partes dividi posse, quarum numerus sit $2^m(1 + 2^n)$. In hac autem formula continentur sequentes numeri 1, 2, 3, 4, 5, 6, 8, 9, 10, 12, 16, 17, 18, 20, 24, 32, 33 etc. [6]

It seems strange that Euler did not realize that the division of the arc of the lemniscate into 33 equal parts implies the division into 11 parts; even more strange is Ayoub's [Ay] question whether Euler could prove his claim or not. Almost impossible to understand, however, is why Euler did not look into the corresponding problem of the division of the circle into n equal parts.

In 1796/97, Gauss started studying the lemniscatic sine, discovered its principal properties, and observed that the theory of cyclotomy had its counterpart in the lemniscatic world: he recorded the discovery that the lemniscate can be divided into 5 equal parts by ruler and compass in his diary on March 21, 1797. Note that the construction must not use the lemniscatic curve: what is constructed using ruler and compass is simply the points on the lemniscate that divide the (invisible) arc into five equal parts. Gauss did not publish on this subject, except for a cryptic remark in his Disquisitiones [263, Art. 335]:

> Ceterum principia theoriae, quam exponere aggredimur, multo latius patent, quam hic extenduntur. Namque non solum ad functiones circulares, sed pari successu ad multas alias functiones transcendentes

---

[6] ... it is clear that the lemniscate can be divided in this way into $2^m(1+2^n)$ parts. This formula contains the following numbers 1, 2, 3, 4, 5, 6, 8, 9, 10, 12, 16, 17, 18, 20, 24, 32, 33 etc.

applicari possunt, e.g. ad eas quae ab integrali $\int_0^c (1-x^4)^{-1/2} dx$ pendent, praetereaque atiam ad varia congruentiarum genera. [7]

Later Gauss published a paper on the arithmetic-geometric mean and its connection with elliptic functions (cf. the book [BB] of J.M. Borwein and P.B. Borwein), but everything else he knew about elliptic functions (see Markuschewitsch [Mar] and Pépin [Pep]) had to be rediscovered by Abel in 1823–1827. Abel mentioned his results on the division of the lemniscate in a letter to Crelle on Dec. 12, 1826; his first article on elliptic functions got lost among Cauchy's papers. In September 1827, the Astronomische Nachrichten contained two letters to the editor by Jacobi where he announced some results on elliptic integrals without proof, and Crelle's Journal featured Abel's first part of his 'Recherches', where he inverted the elliptic integrals for the first time. In November 1827, Jacobi gave proofs for some assertions of his second letter in the Astron. Nachrichten, using the inverse functions of elliptic integrals. It is, however, reasonable to assume that Jacobi discovered this principle independently.

In his great paper from 1827 [Ab1], Abel wrote

> On peut diviser la circonférence entière de la lemniscate en m parties égales *par la règle et le compas seule* si m est de la forme 2^n ou $2^n + 1$, ce dernier nombre étant en même temps premier; ou bien si m est un produit de plusieurs nombres de ces deux formes.[8]

As a matter of fact, M. Libri claimed priority over Abel in connection with the division of the lemniscate; see Libri [Li1, Li2, Li3][9] and Liouville [Lio]. Curiously, Libri is never mentioned in articles or books about the history of this problem (see e.g. Rosen [Ros]). One possible reason may be that his paper does not contain what he claims it does: after discussing Fagnano's result on the 5-division of the lemniscate, he remarks

> En général on voit comment on peut appliquer les mêmes principes aux equations (2.) qui donnent la division de la Lemniscate en $2^n + 1$ parties égales.[10]

In particular, Libri repeats Euler's mistake. Apparently, he didn't even bother to read those of Abel's papers over which he claimed priority.

---

[7] The principles of the theory which we are going to explain actually extend much further than we will indicate. For they can be applied not only to circular functions, but just as well to other transcendental functions, e.g., to those which depend on the integral $\int_0^c (1-x^4)^{-1/2} dx$, and also to various types of congruences.

[8] One can divide the entire circumference of the lemniscate into m equal parts by ruler and compass alone if m has the form 2^n or $2^n + 1$, this last number being at the same time prime; or if m is a product of several numbers of this form.

[9] Thanks to Julio Gonzalez Cabillon for notifying me of this last reference. See also Enneper [Enn], p. 546–547.

[10] In general it can be seen how to apply the same principles to the equation (2.) which then give the division of the lemniscate into $2^n + 1$ equal parts.

The book [Kaw] of Kawada presents the work of Gauss and Abel on elliptic functions to readers who are fluent in Japanese.

For explicit constructions and the corresponding division equations, see Bricard [Bri], Kiepert [Kie], Mathews [Ma2], Meyer [MeP], and Wichert [Wic]. Expositions of Abel's work on the division of the lemniscate can be found in Rosen [Ros], Mel'nikov [Mel], Nagaev [Nag], and Prasolov & Solovyev [PS].

Eisenstein

The mathematical career of G. Eisenstein began in 1844 with the publication of proofs of the quadratic and cubic reciprocity laws in the 27th volume of Crelle's Journal. Though this was the first published proof of the cubic reciprocity law, it was known to Jacobi at least since 1837 when he lectured on number theory and cyclotomy (see [402]) in Königsberg. Jacobi never bothered to publish his proofs, though his results appeared in [400], published in 1837. After Eisenstein's proofs had appeared, he had this article reprinted in 1846 as [401], having added the following footnote:

> Diese aus vielfach verbreiteten Nachschriften der oben erwähnten Vorlesung (an der Königsberger Universität) auch den Herren Dirichlet und Kummer seit mehreren Jahren bekannten Beweise sind neuerdings von Hrn. Dr. Eisenstein im 27ten Bande des Crelleschen Journals auf S. 53 publicirt worden. Der S. 41 des 28ten Bandes von Hrn. Eisenstein gegebene Beweis des quadratischen Reciprocitätsgesetzes ist der nämliche, welchen ich im Jahre 1827 Legendre mitgeteilt und dieser in die 3te Ausgabe seiner Zahlentheorie aufgenommen hat.[11]

Eisenstein defends himself against this attack in vol. 35 of Crelle's Journal and remarks that the proof which Jacobi claims for himself is,

> abgesehen von einer rein äußerlichen Unterschiedenheit kein anderer als der sechste Gaußische Beweis[12]

from 1818. Actually, Cauchy [122] published quite a similar proof in 1829; it was reprinted [123, Note IV] in 1840, together with the remark

> La démonstration que je viens d'en donner, et que j'ai déjà exposée dans le Bulletin de M. Férussac de septembre 1829, est plus

---

[11] These proofs which were known for several years also to Mr. Dirichlet and Mr. Kummer through widely circulated notes of the lecture (at the University of Königsberg) mentioned above have lately been published by Dr. Eisenstein in the 27th volume of Crelle's Journal, p. 53. The proof of the quadratic reciprocity law on p. 41 of vol. 28 given by Mr. Eisenstein is the same which I communicated to Legendre in 1827 and which he included in the 3rd edition of his Number Theory.

[12] apart from a purely superficial difference, no other than Gauss's sixth proof

rigoureuse que celle qu'avait obtenu M. Legendre et plus courte que celles auxquelles M. Gauss était d'abord parvenu[13]

and the following footnote:

Dans la troisième édition de la Théorie des nombres, qui a paru en 1830, M. Legendre présente cette démonstration comme étant la plus simple de toutes et l'attribue à M. Jacobi, sans indiquer aucun Ouvrage òu ce géomètre l'ait publiée, et dont la date soit anterieure au mois de septembre 1829.[14]

In a postscriptum to another article (see Œuvres (I) III, p. 449), however, he is much more cautious:

La note placée au bas de la page 179, et relative à la loi de réciprocité qui existe entre deux nombres premiers, se réduit à cette observation très simple, que la démonstration empruntée par M. Legendre à M. Jacobi ne parait pas avoir été publiée par l'un ou l'autre des ces deux géomètres avant 1830. Je suis loin de vouloir en conclure que cette démonstration n'ait pu être découverte par M. Jacobi à une epoque antérieure.[15]

It has gone unnoticed so far that Borchardt, then editor of Crelle's Journal, seems to have advised Jacobi before he wrote his footnote: in a letter [Bor] to Lipschitz, Borchardt writes

Dirichlet hat sich auf persönliche Reclamationen nie eingelassen, Jacobi dagegen hat z.B. Eisenstein gegenüber die Hand auf sein Eigentum gelegt, und zwar in der von mir angerathenen Form.[16]

Jacobi's accusations seem to have been a severe blow for Eisenstein; this transpires from almost every line on the first three pages of Eisenstein's paper [Eis]:

---

[13] The demonstration I have just given, and which I have already exposed in the Bulletin de M. Férussac in September 1829, is more rigorous than the one obtained by Mr. Legendre and shorter than those at which Mr. Gauss had previously arrived

[14] In the third edition of the 'Theory of numbers', which appeared in 1830, Mr. Legendre presents this proof as the simplest of all and attributes it to Mr. Jacobi without indicating any reference where this geometer has published it, and which appeared before the month of September 1829.

[15] The note placed at the bottom of page 179, referring to the quadratic reciprocity that exists between two prime numbers, is reduced to the very simple observation that the proof attributed by Mr. Legendre to Mr. Jacobi apparently has not been published before 1830 by neither of these two geometers. I am far from wanting to conclude that this demonstration could not have been found by Mr. Jacobi during an earlier period.

[16] Dirichlet never meddled with personal reclamations, Jacobi however laid claim to his property with regard to Eisenstein, and he did so in the form suggested by me.

> Durch die Bemerkung *Jacobi's* über welchen ich mich am Schlusse meiner letzten Abhandlung über Elliptische Functionen bei Gelegenheit neuer Beweise der Reciprocitätsgesetze ausgesprochen habe, wurde ich indessen dergestalt von den Untersuchungen dieser Art abgeschreckt, daß ich dieselben bis auf die neueste Zeit habe liegen lassen ... .[17]

When Dirichlet wanted to explain in 1849 (see Biermann [Bie, p. 39]) why Eisenstein's output of papers had diminished lately he said that "Eisenstein has learned the art of self-criticism in which he had been lacking before"; it seems as if this was only half the truth.

Even if one admits that Jacobi's remarks have damaged Eisenstein's reputation, they do not explain why Eisenstein and his work were forgotten almost as soon as he was dead (see Scharlau's article [Srl] in this connection). Compare e.g. Kummer's remarks on Eisenstein in 1851/52 and 1858: in a letter to Kronecker [Ku] from 1851, he writes about Eisenstein's last paper on the general reciprocity law:

> Nach meiner Vermutung stimmt dieses mit dem Eisensteinschen Resultate im 39. Band von Crelle überein, wo es jedoch weder in seiner einfachsten noch besten Gestalt so verworren dargestellt ist, daß es schwer fällt die Übereinstimmung zu ergründen. Eisenstein hätte, wenn dies wahr ist, durch seine Methode also wirklich mein einfaches Reciprocitätsgesetz finden können, wenn er es geschickter angestellt hätte und wenn er etwas mehr von dem Einflusse der Einheiten auf die complexen Zahlen gewußt hätte.[18]

In his paper [465] published the following year, Kummer repeats these lines in a slightly more diplomatic way and adds

> In der schon oben erwähnten Abhandlung ... hat Herr Eisenstein eine, zwar auf einer unbewiesenen Behauptung beruhende, jedoch sehr sinnreiche Methode entwickelt, um diejenige Potenz von α zu finden,[19] welche dem Verhältnisse $\left(\frac{A}{B}\right) : \left(\frac{B}{A}\right)$ gleich ist, wenn A und B complexe Zahlen sind.[20]

---

[17] By the remark of Jacobi, about whom I have expressed myself on the occasion of new proofs of the reciprocity laws in my treatise on elliptic functions, I was however deterred from investigations of this kind to such a degree that I have left them undone until very recently ...

[18] I guess that this coincides with Eisenstein's results in vol. 39 of Crelle's Journal, where it is presented in a muddled form, neither in its simplest nor in its best form, which makes it difficult to get to the bottom of the correspondence. If this were true, Eisenstein could have found my simple reciprocity law, if he had set to work more cleverly and if he had known more about the influence of the units on the complex numbers.

[19] Kummer used α to denote a primitive ℓth root of unity.

[20] In the paper cited above ... Mr. Eisenstein has developed a method that, although it rests on an unproven claim, nevertheless is very ingenious, and which

However, in his survey [468] from 1858 we already quoted from at the beginning of this chapter, he made a few quite nasty remarks about Eisenstein:

> Als eine Modifikation dieser Gaußischen Beweise ist auch derjenige anzusehen, welchen Eisenstein in Crelle's Journal, Bd. 28, pg. 246, gegeben hat; ... Eisenstein ... hat denselben Beweis im Jahre 1844 in Crelle's Journal, Bd. 28, pag. 41, reproducirt; ... An diesen Beweis von Hrn. Lebesgue schließt sich der von Eisenstein ... gegebene Beweis sehr genau an, obgleich er scheinbar von ganz anderen Prinzipien ausgeht; ... Seine ersten Beweise ... sind zwar nur ganz dieselben, welche Jacobi mehr als zehn Jahre früher gefunden hatte, auch ist der Beweis des biquadratischen Reciprocitätsgesetzes ... zwar scharfsinnig, wie alle Arbeiten Eisensteins, aber doch mehr scheinbar als wirklich originell; ... es ist aber weder ihm selbst noch anderen bisher gelungen, mit Hülfe dieser Principien irgend welche höheren Reciprocitätsgesetze zu beweisen, oder auch nur aufzufinden. Seine eigenen Bemühungen in dieser Beziehung sind schon an den achten Potenzen gescheitert, für deren Reciprocitätsbeziehung er nur specielle Resultate hat gewinnen können; ... Endlich ist noch eine Arbeit von Eisenstein zu erwähnen, ... in welcher er darauf ausgeht, die allgemeinen Reciprocitätsgesetze durch Induktion zu finden. Der Weg, den er dabei einschlägt, hat aber nicht zum Ziele geführt und überhaupt kein Resultat ergeben.[21]

Incidentally, Eisenstein's paper that Kummer mentions here as well as in his letter to Kronecker [Ku] later received praise from Hasse [337], who had a totally different opinion about this paper:

> Die Methode, deren ich mich bedienen werde, ist im Hinblick auf die angeführten früheren Arbeiten ganz neuartig, also in keiner Weise als Verallgemeinerung der dort verwendeten Methoden zu bezeichnen. Sie knüpft vielmehr an die allererste Arbeit zum expliziten

---

he used to find the power of α that equals the ratio $\left(\frac{A}{B}\right) : \left(\frac{B}{A}\right)$, where A and B are complex numbers.

[21] We also have to consider the proof given by Eisenstein in Crelle's Journal, vol. 28, p. 246, as a modification of these Gaussian proofs; ... Eisenstein reproduced the same proof in 1844 in Crelle's Journal, vol. 28, p. 41; ... The proof given by Eisenstein is very close to this proof of Mr. Lebesgue, although it seems to be based upon completely different principles; ... His first proofs ... are the same as those that Jacobi had found more than ten years earlier, and the proof of the biquadratic reciprocity law ... is very clever – as are all of Eisenstein's papers – but more seemingly than really original; ... neither he himself nor others succeeded so far in proving (or only finding) any higher reciprocity laws using these principles. His own efforts in this direction failed even for the eighth powers, whose reciprocity relation he could only find in some special cases; ... Finally we have to mention a paper of Eisenstein, ... in which he tried to find the general reciprocity laws by induction. The path that he chose did however not lead to the goal and did not yield any result at all.

Reziprozitätsgesetz im Kreiskörper an, an den wunderschönen, heute leider nur noch wenig gelesenen Aufsatz von Eisenstein [203].[22]

Of course Hasse was in a better position to judge the importance of Eisenstein's paper: the axiom Eisenstein builds his paper on is a direct consequence of Artin's reciprocity law. In fact, Eisenstein considers the ℓ-th power residue symbol in $\mathbb{Q}(\zeta_\ell)$ for an odd prime ℓ, and he assumes that there exists an integer s with the property that $(\alpha/\beta)(\beta/\alpha)^{-1} = (\alpha'/\beta')(\beta'/\alpha')^{-1}$ whenever $\alpha \equiv \alpha' \bmod (1 - \zeta_\ell)^s$ and $\beta \equiv \beta' \bmod (1 - \zeta_\ell)^s$. Then he shows in a completely elementary way that the mere existence of s implies that one can take $s = \ell + 1$, and then goes on to prove the same for the symbols occurring in the supplementary laws. The same paper also contains the suggestion to formulate reciprocity laws for n-th powers in the form $(\alpha/\beta)(\beta/\alpha)^{-1} = (\alpha'/\beta')(\beta'/\alpha')^{-1}$ whenever $\alpha \equiv \alpha'$ and $\beta \equiv \beta'$ modulo some suitably chosen ideal depending on n (compare Proposition 8.18).

Edwards [181] gives a nice survey of Kummer's and Eisenstein's contributions to number theory. Detailed historical comments can also be found in Neumann's series of articles [614, 615, 616].

In [133], M. Collison writes about Eisenstein's efforts to prove higher reciprocity laws and comments

> Equally unsuccessful were Eisenstein's efforts. The difficulty stems from the fact that the Euclidean algorithm and the Unique Factorization Theorem fail to exist for complex integers composed of most higher roots of unity. It took the theory of ideals by Kummer to restore this aspect of arithmetic.

Actually Eisenstein did use Kummer's ideal numbers in several of his own papers, as Weil [827] acknowledges:

> At first, as we have noted, Eisenstein had been skeptical about Kummer's researches and had rather followed Gauss and Dirichlet by investigating decomposable forms. By 1850, however, he is not only converted to Kummer's ideal theory; he has extended it at any rate to the extensions of $\mathbb{Q}(i)$ generated by the division of the lemniscate, if not further.

As a matter of fact, Eisenstein had to use ideal numbers already for *defining* the n-th power residue symbol (α/β) as the product of the generalized Legendre symbols $(\alpha/\mathfrak{p})$, although he remarked that he knew a different method for doing so that would not involve ideal numbers.

It is surprising that Schönemann rarely gets credited for his irreducibility theorem, given that Dedekind [Werke I, p. 40, 68, 69, 185] mentioned his

---

[22] The method I will employ is, compared with our earlier papers mentioned above, entirely new and cannot be called a generalization of the methods used there. It is rather a continuation of the very first paper on explicit reciprocity laws in cyclotomic fields, the amazingly beautiful and these days unfortunately very rarely read essay of Eisenstein [203].

papers regularly and that Kronecker [Werke III, p. 247] acknowledged his priority. As [DM, p. 33], Krull's introduction to algebra [Kru] or Lüneburg's book [Lun] show, his contributions were never completely forgotten, and recently (see e.g. Panaitopol & Stefanescu [PS]) his name is mentioned quite often.

Analytic Proofs of Reciprocity Laws

Eisenstein's proof of the quadratic reciprocity law using the sine function is taken from [197]; the same article contains proofs (or at least sketches) of the cubic and quartic reciprocity laws using Abel's elliptic functions. The third proof of the quadratic reciprocity law given here is due to Liouville [550]. Abel's construction of elliptic functions can be found in his collected works [Ab2]. We also note that the assertion $W(x^4) = x^{m-1}V(x^{-4})$ of Proposition 8.2 was actually proved by Eisenstein, as well as Jacobi [Jac] and Kronecker [Kr1]. Although the title of this chapter promised "Eisenstein's proofs", the proof we have given proceeds differently. One reason is that Eisenstein's proof is somewhat hard to swallow, and on the other hand there are expositions by Adler [Adl] and Prasolov & Solovyev [PS].

The presentation of quartic reciprocity we have given here is largely drawn from Kubota's excellent papers [454] and [457], as well as from [453]. Note, however, that our approach is simpler than the previous proof in that we do not need any technical calculations to prove the functional equation $\phi(iz) = i\phi(z)$. An exposition of Eisenstein's first proof of quartic reciprocity using elliptic functions can be found in K. Watanabe, Miyagawa, & Higuchi [822], and the introduction to complex analysis by Stalker [748] gives a proof of the biquadratic reciprocity law in its general form

$$\left[\frac{\mu}{\nu}\right]\left[\frac{\nu}{\mu}\right]^{-1} = \frac{\varepsilon(\mu)^{(N\nu-1)/4}}{\varepsilon(\nu)^{(N\mu-1)/4}}(-1)^{(N\mu-1)(N\nu-1)/16},$$

where $\mu, \nu \in \mathbb{Z}[i]$ are primes with odd norm, and where e.g. $\varepsilon(\mu) \in \{\pm i, \pm 1\}$ is defined by $\mu \equiv \varepsilon(\mu) \bmod 2+2i$. Hiramatsu [371] presents Eisenstein's proof based on (8.21), Smith [743, p. 84–86] presents both of Eisenstein's analytic proofs. The development of Eisenstein's proof of the quartic reciprocity law using elliptic functions is nicely portrayed in his letters to Gauss [195, 196].

Propositions 8.4 and 8.7, as well as Theorem 8.6 can be found in virtually every treatise on elliptic functions (cf. e.g. Lang's [Lan]), and Proposition 8.5 is Corollary 10.14 in Cox' wonderful book [140]. Eisenstein's original and completely elementary construction of elliptic functions is discussed in Weil's book [We3] as well as in Quellet [Qu]. Older sources include Hurwitz [Hu1, p. 31], Krusemarck [Kr] (he was one of the students in Eisenstein's 1850 class on elliptic functions), and Scheibner [Sbn].

Proposition 8.8 is formula (9) on p. 305 in Eisenstein's [198]. Eisenstein's original proof of his main theorem 8.9 on the division of the lemniscate is quite complicated (and reproduced in Hurwitz's paper [Hu2]); the simple proof given here is due to Mertens [Mer]. Note that the right hand side of (8.26) only depends on the ideal $\mathfrak{l} = (\lambda)$, whereas the left hand side depends on the choice of the generator λ. The map χ that sends the ideal (λ) to the primary generator $\rho(\lambda)\lambda$, where $\rho(\lambda) \in \mu_4$ is a fourth root of unity making λ primary, defines a *grössencharacter*. Since $\rho(\lambda)\lambda$ is the Jacobi sum of the quartic character modulo λ, this is an example of a Jacobi sum as a grössencharacter. This situation was later studied by Weil [We1, We2]. See also the articles of Kubota [454, 457].

Eisenstein's congruence was generalized by Kronecker [Kr1] to what we today call elliptic curves with complex multiplication; he also observed that the analogous congruence (8.8) was due to Euler [Eu1], but did not mention Eisenstein's work in his article, although he did so in his letter to Dedekind [Kr2] (see Adler's article [Adl]). More proofs of Kronecker's congruence were given by Weber [Web] (for prime ideals of degree 1) and Takagi [Ta2] (who only cites Eisenstein and Weber). Kronecker's congruence was generalized by Shimura & Taniyama to abelian varieties with complex multiplication (see Schappacher [Sc1]).

Eisenstein used his theorem 8.9 for deriving Gauss's congruence (Corollary 6.6) and claimed to have similar proofs for primes $p \equiv 3 \bmod 8$ and $p \equiv 1, 2, 4 \bmod 7$, but those were never published, and Adler [Adl] tried to reconstruct them. Conversely Hurwitz [Hu2] used this result to derive a congruence needed for the proof of a characterization of Hurwitz numbers H_n (these are elliptic analogues of Bernoulli numbers) corresponding to the von Staudt-Clausen theorem. Hurwitz numbers are defined as follows: let $\wp$ be the Weierstraß $\wp$-function satisfying $\wp'^2 = 4\wp^3 - 4\wp$; then

$$\wp(z) = z^{-2} + \sum_{\nu=1}^{\infty} \frac{2^{4n} H_n}{4n} \frac{z^{4n-2}}{(4n-2)!}$$

for rational numbers H_n. The first few values are $H_1 = \frac{1}{10}$, $H_2 = \frac{3}{10}$ and $H_3 = \frac{3^4 \cdot 7}{10 \cdot 13}$; the result corresponding to Euler's summation of Riemann's ζ-function at positive even integers is

$$\sum_{(r,s) \neq (0,0)} \frac{1}{(r+si)^{4n}} = \frac{(2\omega)^{4n}}{(4n)!} H_n,$$

where $\omega = 2 \int_0^1 \frac{dx}{\sqrt{1-x^4}} = 2.62205\ldots$ is half the real period of $\wp(z)$. Generalizations of Hurwitz numbers were given subsequently by Dintzl [Di1] (to $\mathbb{Q}(\sqrt{-2})$), Matter [Mat] (to $\mathbb{Q}(\sqrt{-3})$), Naryskina [Nar] (to imaginary quadratic number fields with class number 1), Dintzl [Di2] and Herglotz [Her] (to general imaginary quadratic number fields). This problem was published as a 'Preisaufgabe' in Math. Ann. **57** (1903), 571–572. See also Niemeyer [620] and Katz [Kat].

The proof of Proposition 8.11 on the discriminant of the lemniscatic division polynomial is taken from Takagi [Ta1] (but I would be surprised if the idea can't be found in Weber [Web]), the proof of Corollary 8.20 is due to R. Fueter [250] and can also be found in Bohnicek's paper [64]. For more on the computation of Hilbert class fields of imaginary quadratic number fields, see Kedlaya's thesis [Ked]. Weber's Theorem 8.25 as well as many other results on the explicit construction of class fields via complex multiplication can be found in Schertz's survey [Sr1].

For more applications of elliptic functions to number theory, see e.g. Bugaieff [Bug], Dantscher [148], Hübler [383], Koschmieder [434, 435], Lewandowski [537], Mel'nikov [575], Mertens [582], Nasimoff [610] (according to the review in the Jahrbuch, he applies elliptic functions for deriving quartic and octic reciprocity laws; see also [611, 612, 613]), Petr [655] (cubic and sextic reciprocity in $\mathbb{Q}(\rho)$ using Weierstraß' $\wp$-function and σ-function, respectively), Sbrana [701], Schwering [729, p. 228], Shiratani [732, 733], Watabe [820]. The books of King [Kin] and Prasolov & Solovyev [PS] both treat the solution of quintic equations using elliptic functions, and the latter also discusses the division of the lemniscate.

The papers of Bohnicek [64], Bonaventura [65], Eisenstein [198, 202], Mathews [Ma2, Ma3], Schwering [Sw1, Sw2], and Takagi [Ta1] contain explicit information about the abelian extensions of $\mathbb{Q}(i)$ constructed via the lemniscatic functions; see Takenouchi [Tak], Lewandowski [537] and Bindschedler [Bin] for the theory over $\mathbb{Q}(\sqrt{-3})$, and Schappacher [Sc2] for historical details.

Proofs of the quadratic reciprocity law in imaginary quadratic number fields using elliptic functions were given by Dintzl [161], Fueter [247, 248, 249], Herglotz [363], Niemeyer [620] (basically an exposition of Herglotz's proof), Bayad [39], and Hajir & Villegas [334]. Our treatment was inspired by the papers of Kubota and Herglotz; Bayad's paper has to be read with care since e.g. his claim that Proposition 8.16 is true without the assumption that $\nu \equiv 1 \bmod 2$ for $d \equiv 5 \bmod 8$ is incorrect. It is possible to prove Herglotz's version of the quadratic reciprocity law using our approach; for the necessary set of tricks see Herglotz [363] and Shiratani [732].

The history of the discovery of elliptic functions and complex multiplication is discussed in quite a few; we mention here the introduction to complex multiplication by Cohn [Coh], the articles of Krazer [Kra] and Masahito [Mas], as well as the books by Alling [All], Enneper [Enn] and Vladut [Vl3]; the latter seems to contain material drawn from [Vl1, Vl2]. Schappacher is preparing a series of notes on Kronecker's Jugendtraum [Sc3].

There are more analogies in the cyclotomic and lemniscatomic theory than those we have already mentioned. Bachmann [Ba] proved the following class number formula for quadratic extensions $K = \mathbb{Q}(\sqrt{\pi})$ of $\mathbb{Q}(i)$, where $\pi = A+Bi$ is a squarefree Gaussian integer. Let T, U be a solution of $T^2 - \pi U^2 = 1$ such that $|T + U\sqrt{\pi}| > 1$ is minimal, and write $\xi + \varepsilon\eta\sqrt{\pi} = \prod_r \mathrm{sl}\frac{r(1+i)\omega}{\pi}$, where the product runs over all quadratic residues r modulo π and where $\varepsilon = (\frac{1+i}{\pi})$ denotes the quadratic character of $1+i$ modulo π. If $\pi \equiv 1 \bmod 8$ is prime, then $\theta = (\xi - \eta\sqrt{\pi})/\sqrt{\pi}$ is a unit in $\mathcal{O}_K$ with relative norm ± 1, hence a power of $T + U\sqrt{\pi}$, up to a root of unity. Bachmann's class number formula states that $|T + U\sqrt{\pi}|^H = |\theta|^2$, where H denotes the class number of K. See Exercise 8.17 for more details.

Bachmann's result was generalized and is nowadays part of the theory of elliptic units; these are units constructed from values of elliptic functions. Examples are the units $\phi(\frac{\alpha}{z})/\phi(\frac{1}{z})$ in the field $\mathbb{Q}(i, \phi(\frac{1}{z}))$ that occurred in our proof of Theorem 8.9. For more, see e.g. the papers of El Hassani & Gillard [EG], Nakamula [Nak], C. Meyer [MeC] and F. Hajir [Ha1, Ha2].

Gauss sums

Discussions of various conjectures about quartic Gauss sums can be found in Barkan [Bar], Ito [It1], and Yamamoto [Yam].

The problem of explicitly determining cubic Gauss sums was posed by Lebesgue [Leb, p. 70] and Kummer [461, 463]. The Kummer sum $G_3 = G(\chi) + G(\overline{\chi})$, where χ is a cubic character on $\mathbb{F}_p$, can be written in the form $G_3 = \sum_{n=1}^{p-1} \exp\left(\frac{2\pi i n^3}{p}\right)$, and is clearly a root of the cubic period equation (7.1). This polynomial has one root in each of the intervals $[-2\sqrt{p}, -\sqrt{p}]$, $[-\sqrt{p}, \sqrt{p}]$ and $[\sqrt{p}, 2\sqrt{p}]$, and the question is which of them contained the root G_3; see Moreno [Mor] for the history of this problem.

The Gauss sum $G(\chi)$ has absolute value $\sqrt{p}$, hence $G(\chi)/\sqrt{p}$ lies on the complex unit circle. The problem of determining whether this quotient is equidistributed on the unit circle was called Kummer's problem. It was finally solved by D. R. Heath-Brown & S. J. Patterson in [HP]. See the papers of Ito [It3], Loxton [Lo1], McGettrick [MG1, MG2], Reshetukha [679, 680, 681, 682, 683] and Vinogradov [Vin] for more details.

Exercises

8.1 Read A. Weil's review [827] of "Mathematische Werke, by Gotthold Eisenstein".

8.2 Prove Equation (8.7). Show that $F_3(Z) = Z^3 + 3Z$, $F_5(Z) = Z^5 + 5Z^3 + 5Z$, and $F_7(Z) = Z^7 + 7Z^5 + 14Z^3 + 7Z$, where $F_q(Z) = Y^q - Y^{-q}$.

8.3 (cf. (8.8)) Let p be an odd prime and ζ a primitive p-th root of unity; show that $\sin \frac{2\pi}{p} = \frac{1}{2i}(\zeta - \zeta^{-1})$, and derive Euler's congruence $\sin \frac{2\pi q}{p} \equiv (\sin \frac{2\pi}{p})^q \bmod q$.

8.4 Observe that $\text{sl}(0) = 0$ and $\text{sl}'(0) = 1$; then use induction to show that $\text{sl}^{(n)}(0) = 0$ if $n \not\equiv 1 \bmod 4$. Deduce that the Taylor series of $\text{sl } z$ has the form $\text{sl } z = z(1 + a_4 z^4 + a_8 z^8 + \ldots)$ with $a_i \in \mathbb{R}$. This also implies that

$\mathrm{sl}(iz) = i\mathrm{sl}(z)$ (and shows incidentally how to find infinitely many analytic functions f satisfying $f(iz) = if(z)$).

8.5 Use the transformation $x^2 = t$ to show that $2\int_0^1 \frac{dx}{\sqrt{1-x^4}} = \int_0^1 \frac{dt}{\sqrt{t-t^3}}$. Then use $t = z^2$ and the definition of the Beta function on p. 139 to show that $2\int_0^1 \frac{dt}{\sqrt{t-t^3}} = B(\frac{1}{4}, \frac{1}{2})$. Show that this gives $4\int_0^1 \frac{dx}{\sqrt{1-x^4}} = \frac{1}{\sqrt{2\pi}}\Gamma(\frac{1}{4})^2$.

8.6 Prove that $\lim_{h \to 0} \frac{1}{h}\mathrm{sl}(h) = 1$. (This is clear from Abel's definition if $h \in \mathbb{R}$; the problem is to show this for all $h \in \mathbb{C}$. Hint: put $h = a + bi$ and use the addition formula to write $\mathrm{sl}(h)$ in the form $u(a,b) + iv(a,b)$, where u and v are real-valued functions. Then verify the Cauchy-Riemann equations).

8.7 Use the preceding exercise and the addition formulas to show that $\mathrm{sl}(z)$ is differentiable away from its poles.

8.8 Verify (8.15) by writing $(z - \omega)^2 = \omega^2(1 - z/\omega)^2$ and using the geometric series.

8.9 Prove Proposition 8.2 by induction and show that it does not yield the correct degrees: in fact, verify that

$$W_{u+2} = W_u V_u (1 - 6x + x^2) + 2(1 - x^2)(P_u^2 - xQ_u^2),$$
$$V_{u+2} = V_u^2(1+x)^2 + 4W_u(x - x^2),$$

and conclude that $\deg W_3 = 2 = \frac{3^2-1}{4}$, $\deg W_5 = 6 = \frac{5^2-1}{4}$, but $\deg W_7 = 13 = \frac{7^2-1}{4} + 1$. Thus W_7 and V_7 have a common factor which apparently cannot be detected by induction.

8.10 Let $\mu \equiv 1 \mod 2 + 2i$ be primary, and put $x = \mathrm{sl}(z)$ and $y = \mathrm{sl}(\mu z)$. Verify the following differential equation used extensively by Eisenstein:

$$\frac{dy}{\sqrt{1-y^4}} = \mu \frac{dx}{\sqrt{1-x^4}}.$$

8.11 Use the addition formula to verify Theorem 8.2 in the following special cases:

π	$W(x)$
$-1+2i$	$x + (-1+2i)$
-3	$x^2 + 6x - 3$
$3+2i$	$x^3 + (-11+10i)x^2 + (7-4i)x + (3+2i)$
$1+4i$	$x^4 + (12-20i)x^3 + (-10+28i)x^2 - (20+12i)x + (1+4i)$

Compute the value $W(1)$ in each case. Conjecture? Proof?

8.12 Let $\wp(z)$ be the Weierstraß $\wp$-function with invariants $g_2 = 4$ and $g_3 = 0$. Write $\wp((1+i)z)$ and $\wp(2z)$ as rational functions of $\wp(z)$.

8.13 Define a holomorphic function f in a real neighborhood of 0 by inverting the integral $z = \int_0^w \frac{dx}{\sqrt{1-x^3}}$; prove Proposition 8.14 using Abel's method outlined in Section 8.2.

8.14 Give a proof for Proposition 8.14 and then derive the cubic reciprocity law from it.

8.15 Let $\pi \in \mathcal{O} = \mathbb{Z}[i]$ be a prime such that $N\pi$ is odd. Put $\lambda = (1+i)$ and define a homomorphism $\psi : E = \{\pm i, \pm 1\} \longrightarrow (\mathcal{O}/\lambda^3\pi\mathcal{O})^\times$ by $\psi(\varepsilon) = \varepsilon \bmod \lambda^3\pi$. Show that
$$(\mathcal{O}/\lambda^3\pi\mathcal{O})^\times \big/ \psi(E) \simeq (\mathcal{O}/\pi\mathcal{O})^\times \simeq \mathbb{Z}/(N\pi - 1)\mathbb{Z}.$$

Hint: Define a homomorphism $\theta : (\mathcal{O}/\lambda^3\pi\mathcal{O})^\times \longrightarrow (\mathcal{O}/\pi\mathcal{O})^\times$ by mapping the residue class defined by $\alpha \equiv \varepsilon \bmod \lambda^3$, $\varepsilon \in E$ and $\alpha \equiv \beta \bmod \pi$ (where β runs through some fixed set of representatives mod π) to the class $\theta(\alpha \bmod \lambda^3\pi) = \varepsilon^{-1}\beta \bmod \pi$ and show that the following sequence is exact:
$$1 \longrightarrow E \longrightarrow (\mathcal{O}/\lambda^3\pi\mathcal{O})^\times \longrightarrow (\mathcal{O}/\pi\mathcal{O})^\times \longrightarrow 1.$$

8.16 Let $\pi \equiv 1 \bmod 4$ be a prime in $\mathbb{Z}[i]$. Show that the fundamental unit of $\mathbb{Q}(i, \sqrt{\pi})$ has norm $\pm i$.

8.17 Let $\pi \in \mathbb{Z}[i]$ be a primary prime; show that $\theta = \prod_r \phi(r/\pi)$ and $\theta' = \prod_n \phi(n/\pi)$ (where r and n run through a complete set of quadratic residues and non-residues modulo π, respectively) can be written in the form $\theta = \frac{1}{2}(\xi + \eta\sqrt{\pi})$ and $\theta' = \frac{1}{2}(\xi - \eta\sqrt{\pi})$. Moreover, show that $\theta\theta' = \pi$, and that $\varepsilon = \theta/\theta'$ is a unit in $\mathbb{Q}(\sqrt{\pi})$. Prove that this unit is ± 1 if $N\pi \equiv 5 \bmod 8$. Can you show that $\varepsilon^4 \neq 1$ if $N\pi \equiv 1 \bmod 8$?

8.18 (Schwering [Sw1]) Let $W(X) = X^n + A_{n-1}X^{n-1} + \ldots + A_1 X + A_0$ (where $m = 4n + 1$) be the polynomial we have defined in Proposition 8.2. Show that $W(z + 1) = z^n + 2ia_{n-1}z^{n-1} + \ldots + (2i)^{n-1}a_1 z + (2i)^n a_0$ for some $a_j \in \mathbb{Z}[i]$. Conclude that if γ is a root of $W(X^4)$, then $\frac{1}{2}(1 - \gamma^4)$ is a unit.

8.19 (Compare Corollary 8.20) Show that the quartic reciprocity law can be written in the form $[\kappa/\pi = [\pi^*/\kappa]$.

8.20 (Niemeyer [620]) Let $\pi = a + b\sqrt{-2}$ be a prime in $\mathbb{Z}[\sqrt{-2}]$, and assume that $a \equiv 1 \bmod 4$. Show that $[\sqrt{-2}/\pi] = (-1)^{b(b+1)/2}(\frac{2}{a})$.

8.21 Try to find as many proofs of the cubic and quartic reciprocity laws as possible, by imitating proofs of the quadratic reciprocity law, replacing cyclotomic extensions of $\mathbb{Q}$ by abelian extensions of $\mathbb{Q}(i)$, and cyclotomic units by elliptic units, etc.

8.22 Find an "elliptic" analogue of Scholz's reciprocity law and prove it.

8.23 (Euler) Show that $\frac{\tan nz}{\tan z} = \frac{w(x)}{v(x)}$ is a rational function of $x = \tan z$; compute $w(x), v(x) \in \mathbb{Z}[x]$ explicitly and, for n odd, verify a congruence similar to Eisenstein's.

8.24 Compare the product expansion of $\sin \pi z$
$$\sin \pi z = \pi z \prod_{n \in \mathbb{Z}\setminus\{0\}} \left(1 - \frac{z}{n}\right) \exp\left(\frac{z}{n}\right)$$
with the definition of Weierstraß' σ-function. Compute the logarithmic derivative of $\sin \pi z$ and compare the result to the definition of $\wp(z)$. Derive the better known product formula

$$\sin \pi z = \pi z \prod_{n \geq 1} \left(1 - \frac{z^2}{n^2}\right),$$

as well as the corresponding formula for Weierstraß' σ-function.

8.25 Use the addition formula to prove the identities

$$\text{sl}(u+v) - \text{sl}(u-v) = 2\frac{\text{sl}(v)\text{f}(u)\text{F}(u)}{1 + \text{sl}(u)^2\text{sl}(v)^2}$$

and

$$\text{sl}(u+v)\text{sl}(u-v) = \frac{\text{sl}(u)^2 - \text{sl}(v)^2}{1 + \text{sl}(u)^2\text{sl}(v)^2}.$$

Use these results to derive the identity (8.25).

8.26 Let $\pi \in \mathbb{Z}[i]$ be a primary prime with norm p. Show that if α runs through a complete set A of prime residues modulo π, then so does $(1+i)\alpha$. Use Equation (8.10) to deduce that $\prod_{\alpha \in A} \text{f}(\alpha)\text{F}(\alpha) = (1+i)^{p-1}$.

8.27 Let $\pi \in \mathbb{Z}[i]$ be a primary prime. Use formula (8.10) to show that $\text{f}((1-i)\omega/\pi)\text{F}((1-i)\omega/\pi)$ is an element of $K = \mathbb{Q}(i, \phi(1/\pi))$ and that it is associated to $1+i$ in $\mathcal{O}_K$. Similarly, use the duplication formula to deduce that $1+\phi(1/\pi)^4$ is associated to $2+2i$ in $\mathcal{O}_K$. This last result is already due to Eisenstein [202].

8.28 Put $\wp(z) := \text{sl}(z)^{-2}$. Check that $\wp$ satisfies the differential equation $\wp'^2 = 4\wp^3 - 4\wp$. Observe that $\text{sl}\, z$ has period lattice Λ generated by $(1+i)\omega$ and $(1-i)\omega$, but that the period lattice of $\wp(z)$ is $\Lambda_0 = \omega\mathbb{Z}[i]$. Compute the divisor of $\wp$ with respect to Λ and Λ_0. Put $\wp(z) = 2\widetilde{\wp}(z)$ and $\wp'(z) = 2\sqrt{2}\widetilde{\wp}'(z)$, and check that $\widetilde{\wp}'(z)^2 = 4\widetilde{\wp}^3(z) - \widetilde{\wp}(z)$.

8.29 (continued) Let $\pi \in \mathbb{Z}[i]$ be a primary prime with prime norm $p \equiv 1 \bmod 4$. Use Exercise 8.28 to show that

$$\prod_{\alpha \neq 0} \wp(\alpha\omega/\pi) = \pi^{-2},$$

where α runs through all $\alpha \not\equiv 0 \bmod \pi$. Explain why the product over $\wp((1-i)\alpha\omega/\pi)$ yields the same result. Use $\wp'(z) = -2\text{sl}'(z)/\text{sl}(z)^3$ to show that

$$\prod_{\alpha \neq 0} \wp'\left(\frac{\alpha\omega}{\pi}\right) = (2-2i)^{p-1}\pi^{-3}.$$

Use Exercise 8.28 to show that $\prod_{\alpha \neq 0} \widetilde{\wp}'(\alpha\omega/\pi) = (-1)^{(p-1)/4}\pi^{-3}$. Conclude that $P^2 = (-1)^{(p-1)/4}\pi^{-3}$ for $P = \prod_{\alpha=1}^{(p-1)/2} \widetilde{\wp}'(\alpha\omega/\pi)$.

8.30 Let $P \in \mathbb{R}[X]$ be a polynomial of degree 3, and suppose that its leading coefficient is positive. Let x denote its maximal real root; then $P(t) > 0$ for all $t > x$, and we can define a positive square root $\sqrt{P(t)}$ in (x, ∞). Show that $f(w) = \int_w^\infty \frac{dt}{\sqrt{P(t)}}$ converges for all $z \geq x$ and that $f'(w) < 0$ for all $w > x$. Deduce that f has an inverse function $\wp(z)$ on $[x, \infty]$, and show that $\wp'(z)^2 = P(\wp(z))$. In particular, conclude that

$$\int_{\wp(u)}^\infty \frac{dt}{\sqrt{P(t)}} = u \tag{8.36}$$

for all $u \in (0, \xi)$, where $\wp(\xi) = x$. (Hint: see Lawden [Law]).

282   8. Eisenstein's Analytic Proofs

8.31 In Exercise 8.28 we have shown how to construct the $\wp$-function for the lattice $\Lambda = (1 - i)\omega\mathbb{Z}[i]$ from the lemniscatic sine sl z. Here we will start with the lattice $L = \mathbb{Z}[i]$ and compute the invariant $g_2(L)$ (clearly $g_3(L) = 0$).
i) Let $\wp(z)$ denote the Weierstraß $\wp$-function attached to the lattice L. Show that $(\wp) = 2(\frac{1+i}{2}) - 2(0)$ and $(\wp') = (\frac{1}{2}) + (\frac{i}{2}) + (\frac{1+i}{2}) - 3(0)$.
ii) Show that $\wp(z) \in \mathbb{R} \cup \{\infty\}$ on the real line (i.e. for $z \in \mathbb{R}$); this implies immediately that $\wp(z)$ is also real on $i\mathbb{R}$ since $\wp(iz) = -\wp(z)$.
iii) We have $\wp'^2 = 4\wp^3 - g_2\wp$; conclude that $g_2 = 4\wp(1/2)^2 > 0$ from the fact that $\wp'(z)$ vanishes for $z = \frac{1}{2}$ (actually, one has $\wp(1/2) = \sqrt{g_2}/2 > 0$, since $\wp$ has no real zeros and $\wp(0) = +\infty$).
iv) write $P(t) = 4t^3 - g_2 t$; use the preceding exercise to show that $\frac{1}{2} = \int_{\sqrt{g_2}/2}^{\infty} dt/\sqrt{P(t)}$. Substitute $t = \frac{1}{2}\sqrt{g_2}/u^2$ to conclude that $g_2 = 4\omega^4$, where $\omega = 2\int_0^1 dt/\sqrt{1-t^4}$.

8.32 Exercise 8.28 shows that Weierstraß' $\wp$-function for $\Lambda = (1+i)\omega\mathbb{Z}[i]$ is 'a square'. Such square roots can be constructed explicitly via Weierstraß' σ-functions. Let $\Lambda = \omega_1\mathbb{Z} \oplus \omega_2\mathbb{Z}$ be a lattice in $\mathbb{C}$ and put $\lambda_1 = \omega_1/2$, $\lambda_2 = -(\omega_1 + \omega_2)/2$, and $\lambda_3 = \omega_3/2$. Then define complex numbers $\eta_j \in \mathbb{C}$ (depending on Λ) by $\zeta(z + 2\lambda_j) = \zeta(z) + \eta_j$ and functions

$$\sigma_j(z) = \frac{\sigma(z+\lambda_j)}{\sigma(\lambda_j)}\exp(-z\eta_j). \qquad (8.37)$$

Now let $\wp(z)$ be Weierstraß' $\wp$-function for Λ, put $e_j = \wp(\lambda_j)$, and observe that $\wp(z) - e_j$ has 'even' divisor. In fact, it has a square root, since

$$\sqrt{\wp(z) - e_j} = \frac{\sigma_j(z)}{\sigma(z)}. \qquad (8.38)$$

These σ-functions can also be used to define Jacobi's elliptic functions sn z, cn z and dn z: put $u = z\sqrt{e_1 - e_3}$ and $k = \sqrt{e_2 - e_3}/\sqrt{e_1 - e_3}$. Then

$$\operatorname{sn}(u,k) = \frac{\sigma_3(z)}{\sigma(z)}, \quad \operatorname{cn}(u,k) = \frac{\sigma_1(z)}{\sigma_3(z)}, \quad \operatorname{dn}(u,k) = \frac{\sigma_2(z)}{\sigma_3(z)}. \qquad (8.39)$$

Observe that $\sqrt{2}\,\operatorname{sl}(z) = \dfrac{\operatorname{sn}(\sqrt{2}z)}{\operatorname{dn}(\sqrt{2}z)}.$

Additional References

[Ab1] N.H. Abel, *Recherches sur les fonctions elliptiques*, J. Reine Angew. Math. **2** (1827), 101–181; ibid. **3** (1828), 160–190; Œuvres I, 263–388
[Ab2] N.H. Abel, *Œuvres complètes de Niels Henrik Abel*, Johnson Reprint Corporation, 1965
[Adl] A. Adler, *Eisenstein and the Jacobian varieties of Fermat curves*, Rocky Mt. Journal **27** (1997), 1–60
[All] N. L. Alling, *Real Elliptic Curves*, Mathematics Studies **54**, North Holland, 1981

[Ay] R. Ayoub, *The lemniscate and Fagnano's contributions to elliptic integrals*, Archive for the History of Exact Sciences **29** (1984), 131–149

[Ba] P. Bachmann, *Ergänzung einer Untersuchung von Dirichlet*, Math. Ann. **16** (1880), 537–550; FdM **12** (1880), 126–127

[Bar] P. Barkan, *Démonstration d'une conjecture de K. Yamamoto sur les sommes de Gauss biquadratiques*, C. R. Acad. Sci. Paris **310** (1990), 69–72

[BJa] Jac. Bernoulli, Acta erudit. 1694, p. 276, 336; Opera, Geneva 1744, p. 601, 608;

[BJo] Joh. Bernoulli, Acta erudit. 1694, p. 394; Opera omnia, Lausanne 1742, p. 119

[Bie] K.-R. Biermann, *Johann Peter Gustav Lejeune Dirichlet – Dokumente für sein Leben und Wirken*, Akademie-Verlag Berlin, 1959

[Bin] C. Bindschedler, *Die Teilungskörper der elliptischen Funktionen im Bereich der dritten Einheitswurzeln*, J. Reine Angew. Math. **152** (1922), 49–75; FdM **48** (1921–22), 189–190

[Bor] C. Borchardt, *Letter to R. Lipschitz*, Dec. 21, 1875; in: Dokumente zur Geschichte der Mathematik, Bd. 2. DMV Vieweg & Sohn, 1986

[BB] J.M. Borwein, P.B. Borwein, *Pi and the AGM. A study in analytic number theory and computational complexity*, Wiley-Interscience 1987

[Bri] R. Bricard, *Sur l'arc de la lemniscate*, Nouv. Ann. (4) **2** (1902), 150

[Bug] Bugaieff, *Quelques applications de la théorie des fonctions elliptiques à la théorie des fonctions discontinues* (Russ.), Moskow 1884; cf. Darboux Bull. (2) **9**, 89–103

[Ca1] J.W.S. Cassels, *On the determination of generalised Gauss sums*, Archiv Math. Brno **5** (1969), 79–84

[Ca2] J.W.S. Cassels, *On Kummer sums*, Proc. London Math. Soc. (3) **21** (1970), 19–27

[Coh] H. Cohn, *Introductory remarks on complex multiplication*, Intern. J. Math. & Math. Sci. **5** (1982), 675–690

[DM] L.E. Dickson, H.H. Mitchell, H.S. Vandiver, G.E. Wahlin, *Report on algebraic numbers*, part I 1923, part II 1928; reprint Chelsea 1967

[Di1] E. Dintzl, *Über die Zahlen im Körper $k(\sqrt{-2})$, welche den Bernoulli'schen Zahlen analog sind*, Wiener Ber. **108** (1899), 196–226

[Di2] E. Dintzl, *Über die Entwicklungskoeffizienten der elliptischen Funktionen, insbesondere im Falle singulärer Moduln*, Monatsh. Math. **25** (1914), 125

[EG] C. El Hassani, R. Gillard, *Unités elliptiques et groupes de classes*, Ann. Inst. Fourier **36** (1986), 29–41

[Eis] G. Eisenstein, *Zur Theorie der quadratischen Zerfällung der Primzahlen $8n+3$, $7n+2$ und $7n+4$*, J. Reine Angew. Math. **37** (1848), 97–126; Werke II, 506–535

[Enn] A. Enneper, *Elliptische Funktionen. Theorie und Geschichte*, Halle 1890

[Eu1] L. Euler, *Introductio in analysis infinitorum, tomi pari; "De multiplicatione ac divisione angulorum"*, Lausanne 1748; Opera Omnia **I** (8), 258–283; German Translation: *Einleitung in die Analysis des Unendlichen*, Springer Verlag 1983

[Eu2] L. Euler, *Observationes de comparatione arcuum irrecitificabilium*, Opera Omnia **I** (20), 80–107

[Fag] G. C. di Fagnano, *Opere matematiche*, vols. I – III, Milano – Roma – Napoli 1911, 1912

[GS] A.C.P. Gee, P. Stevenhagen, *Generating class fields using Shimura reciprocity*, in: Algorithmic Number Theory (J. P. Buhler, ed.), LNCS 1423, Springer Verlag (1998).

[Ha1] F. Hajir, *Elliptic units of cyclic unramified extensions of complex quadratic fields*, Acta Arith. **64** (1993), 69–85

[Ha2] F. Hajir, *On units related to the arithmetic of elliptic curves with complex multiplication*, Arch. Math. **66** (1996), 280–291

[HP] D. R. Heath-Brown, S. J. Patterson, *The distribution of Kummer sums at prime arguments*, J. Reine Angew. Math. **310** (1979), 111–130

[Hee] K. Heegner, *Diophantische Analysis und Modulfunktionen*, Math. Z. **56** (1952), 227–253

[Her] G. Herglotz, *Über die Entwicklungskoeffizienten der Weierstraßschen ℘-Funktion*, Leipziger Ber. **74** (1922), 269–289; Ges. Werke 436–456

[Hu1] A. Hurwitz, *Grundlagen einer independenten Theorie der elliptischen Modulfunctionen und Theorie der Multiplicatorgleichungen erster Stufe*, Diss. Leipzig 1881; Math. Ann. **18** (1881), 528–592; Werke I, 1–66

[Hu2] A. Hurwitz, *Über die Entwicklungskoeffizienten der lemniskatischen Funktionen*, Math. Ann. **51** (1899), 196–226

[It1] H. Ito, *On a problem of Yamamoto concerning biquadratic Gauss sums*, Proc. Japan Acad. **63** (1987), 35–38

[It2] H. Ito, *On a problem of Yamamoto concerning biquadratic Gauss sums. II*, J. Math. Soc. Japan **40** (1988), 647–660

[It3] H. Ito, *On a product related to the cubic Gauss sum*, J. Reine Angew. Math. **395** (1989), 202–213

[Jac] C.G.J. Jacobi, *Suite des notices sur les fonctions elliptiques*, Werke I, 266–275

[Kat] N.M. Katz, *The congruences of Clausen-von Staudt and Kummer for Bernoulli-Hurwitz numbers*, Math. Ann. **216** (1975), 1–4

[Kaw] Y. Kawada, *Gauss and elliptic functions*, Tokyo (1986), 184 pp.

[Ked] K. Kedlaya, *Complex multiplication and explicit class field theory*, Senior thesis, Harvard University, 1996;
URL: http://www.math.princeton.edu/~kkedlaya/math/

[Kie] L. Kiepert, *Siebzehnteilung des Lemniskatenumfangs durch alleinige Anwendung von Lineal und Cirkel*, J. Reine Angew. Math. **75** (1873), 255–263

[Kin] R.B. King, *Beyond the quartic equation*, Birkhäuser 1996

[Kna] A. Knapp, *Elliptic curves*, Princeton University Press 1992

[Kob] N. Koblitz, *Introduction to elliptic curves and modular forms*, Graduate Texts in Math. **97**, Springer-Verlag, 2nd edition 1993

[Kra] A. Krazer, *Zur Geschichte des Umkehrproblems der Integrale*, Jahresber. DMV **18** (1909), 44–75

[Kr1] L. Kronecker, *Zur Theorie elliptischer Funktionen*, 1886; Werke IV, 345–495

[Kr2] L. Kronecker, *Letter to Dedekind*, March 15, 1880; in: Werke **5**, 453–457

[Kru] W. Krull, *Elementare und klassische Algebra vom modernen Standpunkt*, Göschen Band 930, de Gruyter 1952

[Kr] R. Krusemarck, *Zur Theorie der elliptischen Functionen*, J. Reine Angew Math. **46** (1853), 189–233

[Ku] E.E. Kummer, *Letter to Kronecker*, December 4, 1851; Collected Papers I, 89–91

[Lan] S. Lang, *Complex Analysis*, 3rd ed. Springer Verlag 1993

[Law] D.F. Lawden, *Elliptic functions and applications*, Appl. Math. Sci. **80**, Springer-Verlag 1989

[Leb] V. A. Lebesgue, *Sommation de quelques séries*, J. Math. Pures Appl. **5** (1840), 42–71

[Li] J. Li, *Hilbert's twelfth problem - the reciprocity law and Langlands conjecture* (Chinese), Acta Sci. Nat. Univ. Sunyatseni **3** (1981), 104–114

[Li1] M. Libri, *Note de M. Libri sur un théorème de M. Dirichlet*, C. R. Acad. Sci. Paris **10** (1840), 311–314

[Li2] M. Libri, *Résponse de M. Libri aux observations de M. Liouville*, C. R. Acad. Sci. Paris **10** (1840), 345–347

[Li3] M. Libri, *Memoire sur la résolution des équations algébriques dont les racines ont entre elles un rapport donné, et sur l'integration des équations différentielles linéaires dont les integrales particulières peuvent s'exprimer les unes par les autres*, J. Reine Angew. Math. **10** (1833), 167–194

[Lio] J. Liouville, *Observations sur une note de M. Libri*, C. R. Acad. Sci. Paris **10** (1840), 343–345

[Lo1] J. H. Loxton, *Products related to Gauss Sums*, J. Reine Angew. Math. **268/269** (1974), 53–67

[Lo2] J. H. Loxton, *On the determination of Gauss Sums*, Sém. Delange-Pisot-Poitou 1976/77, 12pp

[Lun] H. Lüneburg, *On the rational normal form of endomorphisms*, B.I. Mannheim, 1987

[Mar] A.I. Markuschewitsch, *Die Arbeiten von C.F. Gauß über Funktionentheorie*, Gauß Gedenkband (ed.: H. Reichardt), Teubner 1957

[Mas] T. Masahito, *Three aspects of the theory of complex multiplication*, The intersection of history and mathematics (S. Chikara, ed.) Papers presented at the history of mathematics symposium, Tokyo, Japan, August 31 – September 1, 1990; Hist. Stud. v. **15** (1994), 91–108.

[Ma1] G.B. Mathews, *The complex multiplication of the Weierstrass lemniscatic function*, Quart. J. **32** (1900), 257–265

[Ma2] G.B. Mathews, *Division of the lemniscate into seven equal parts*, Proc. London Math. Soc. (2) **14** (1915), 464–466

[Ma3] G.B. Mathews, *A direct method in the multiplication theory of the lemniscatic function and other elliptic functions*, Proc. London Math. Soc. (2) **14** (1915), 467–475

[Mat] K. Matter, *Die den Bernoulli'schen Zahlen analogen Zahlen im Körper der dritten Einheitswurzeln*; Vierteljahresschrift Naturf. Ges. Zürich (1900), 238–271

[Mt1] C.R. Matthews, *Gauss sums and elliptic functions. I: The Kummer sum*, Invent. Math. **52** (1979), 163–185

[Mt2] C.R. Matthews, *Gauss sums and elliptic functions. II: The quartic sum* Invent. Math. **54** (1979), 23–52

[MG1] A. D. McGettrick, *On Gaussian Sums*, Diss. 1970

[MG2] A. D. McGettrick, *A result in the theory of Weierstrass elliptic functions*, Proc. London Math. Soc. **25** (1972), 41–54

[MG3] A. D. McGettrick, *On the biquadratic Gauss sum*, Proc. Cambridge Phil. Soc. **71** (1972), 79–83

[Mel] I.G. Melnikov, *Das Problem der Lemniskatenteilung. I, II, III* (Russian), Leningr. Gos. Ped. Inst. A.I. Gertsen, Uch. Zap. **197** (1958), 20–83, 39–42; ibid. **218** (1961), 171–177

[Mer] F. Mertens, *Über Abel'sche Gleichungen und den Satz von Kronecker über die Teilungsgleichungen der Lemniskate*, Wiener Ber. **118** (1909), 245–291

[MeC] C. Meyer, *Zur Theorie und Praxis der elliptischen Einheiten*, Ann. Univ. Sarav., Ser. Math. **6**, No. 2 (1995), 215–572

[MeP] P. Meyer, *Über die Siebenteilung der Lemniskate*, Diss. Strassburg, 1900

[Mor] C.J. Moreno, *Sur le problème de Kummer*, Enseignement Math. (2) **20** (1974), 45–51

[Nag] N.D. Nagaev, *The irreducibility of one form of the equations for the division of a lemniscate* (Russian), Modern algebra, No. 5 (1976), 93–101

[Nak] K. Nakamula, *Elliptic units and the class numbers of non-Galois fields* J. Number Theory **31** (1989), 142–166

[Nar] E. Naryskina, *Über Zahlen, die den Bernoullischen Zahlen analog sind und mit den einklassigen quadratischen Körpern von negativer Diskriminante verknüpft sind* (Russ.), Bull. Assoc. Leningrad (6) **19** (1925), 145–176; FdM **51** (1925), 154

[Op] H. Opolka, *Das wissenschaftliche Werk von Kurt Heegner – ein kurzer Bericht*, preprint 1999

[PS] L. Panaitopol, D. Stefanescu, *A resultant condition for the irreducibility of the polynomials*, J. Number Theory **25** (1987), 107–111

[Pep] T. Pépin, *Introduction à la théorie des fonctions elliptiques d'après les oeuvres posthumes de Gauss*, Rom. Acc. Pontif. d. N. L. **9** (1893), 1 ff.

[Pet] K. Petr, *Über eine Anwendung der elliptischen Funktionen auf die Zahlentheorie*, Prager Ber. **18** (1907), 8 pp.; FdM **38** (1907), 260–261

[PS] V. Prasolov, Y. Solovyev, *Elliptic Functions and Elliptic Integrals*, Translations of Mathematical Monographs **170**, AMS 1997

[Qu] J.-P. Quellet, *De la construction de fonctions elliptiques selon une méthode d'Eisenstein*, Diss. Univ. Montreal 1979

[Ros] M. Rosen, *Abel's theorem on the lemniscate*, Amer. Math. Monthly **88** (1981), 387–395

[Sc1] N. Schappacher, *CM motives and the Taniyama group*, in: Motives (Jannsen et al., eds.), Proc. Symp. Pure Math. **55**, part I, AMS 1994, 485–508

[Sc2] N. Schappacher, *Some Milestones of Lemniscatomy*, Algebraic Geometry (Sinan Sertoz, ed.), Lecture Notes in Pure and Applied Mathematics **193**, 1997

[Sc3] N. Schappacher, *On the History of Hilbert's Twelfth Problem, I*, Matériaux pour l'histoire des math. (Nice, 1996), 243–273

[Srl] W. Scharlau, *Zur Entstehung der Algebraischen Zahlentheorie – ein Bericht eines Augenzeugen*, Arch. Hist. Exact. Sci. **36** (1986), 63–74

[Sbn] W. Scheibner, *Über elliptische Doppelsummen*, Leipziger Ber. **42** (1890), 130–152

[Sr1] R. Schertz, *Problèmes de construction en multiplication complexe*, Sém. Théor. Nombres **4** (1992), 239–262

[Sr2] R. Schertz, *Construction of ray class fields by elliptic units*, J. Théor. Nombres Bordx. **9** (1997), 383–394

[Sm1] T. Schönemann, *Von denjenigen Moduln, welche Potenzen von Primzahlen sind*, J. Reine Angew. Math. **32** (1846), 93–105

[Sm2] T. Schönemann, *Notiz*, J. Reine Angew. Math. **40** (1850), 188

[Sw1] K. Schwering, *Zerfällung der lemniskatischen Teilungsgleichung in vier Faktoren*, J. Reine Angew. Math. **110** (1891), 42–72; Zusatz zur Abhandlung: ibidem **112** (1893), 37–38

[Sw2] K. Schwering, *Zur Auflösung der lemniskatischen Theilungsgleichung*, J. Reine Angew. Math. **111** (1893), 170–204

[Ser] J.P. Serre, *A course in arithmetic*, Graduate Texts in Math. **7**, Springer Verlag 1978

[Sie] C.L. Siegel, *Vorlesungen über ausgewählte Kapitel der Funktionentheorie. Teil I*, Mathem. Inst. Univ. Göttingen 1965

[Ta1] T. Takagi, *Über die im Bereich der rationalen komplexen Zahlen Abelschen Zahlkörper*, J. Coll. Sci. Tokyo **19** (1903), 1–42; Coll. Papers, 13–39

[Ta2] T. Takagi, *On a fundamental property of the 'Equation of division' in the theory of complex multiplication*, Proc. Phys.-Math. Soc. Japan **7** (1914), 414–417; Coll. Papers 40–42

[Tak] T. Takenouchi, *On the relatively abelian corpora with respect to the corpus defined by a primitive cube root of unity*, J. Coll. Sci. Imper. Univ. Tokyo **37** (1916), 70 pp.; FdM **46** (1916–18), 247

[Vin] A. I. Vinogradov, *On the cubic Gauss sum*, Math. USSR, Izv. **1** (1967), 113–138

[Vl1] S. G. Vladut, *On the history of complex multiplication; Abel's complex multiplication* (Russ.), Istor. Mat. Issled **26** (1982), 91–114

[Vl2] S. G. Vladut, *On the history of complex multiplication; Gauss, Eisenstein and Kronecker* (Russ.), Istor. Mat. Issled **27** (1983), 190–238

[Vl3] S. G. Vladut, *Kronecker's Jugendtraum and modular functions*, Studies in the Development of Modern Math. **2**, Gordon and Breach, 1991

[Web] H. Weber, *Elliptische Funktionen und algebraische Zahlen*, 2nd edition (Lehrbuch der Algebra III), 1908

[We1] A. Weil, *Jacobi sums as "Grössencharaktere"*, Trans. Am. Math. Soc. **73** (1952), 487–495

[We2] A. Weil, *Sommes de Jacobi et caractères de Hecke*, Nachr. Akad. Wiss. Gött. 1974, 1–14

[We3] A. Weil, *Elliptic functions according to Eisenstein and Kronecker*, Ergebnisse Math. **88**, Springer-Verlag 1976

[Wic] Wichert, *Die Fünf- und Siebzehnteilung der Lemniskate*, Programmabh. Conitz 1846

[Yam] K. Yamamoto, *On Gaussian sums with biquadratic residue character*, J. Reine Angew. Math. **219** (1965), 200–213

9. Octic Reciprocity

The proof of the octic reciprocity law differs considerably from those in the cubic and quartic cases; in fact one needs methods that are much more sophisticated than Gauss sums, and this is why we resurrect elliptic Gauss sums. These were introduced by Eisenstein while he was working on octic reciprocity and the division of the lemniscate; apparently, nobody ever bothered to study these sums that I wouldn't have hesitated to call Eisenstein sums were it not for the fact that this name is already being used for the sums that we have studied in Chapter 4. In contrast to the full octic reciprocity law, rational octic reciprocity laws can be proved quite easily: the octic version of Burde's reciprocity law is presented in Section 9.1, Eisenstein's octic reciprocity law and the formulas of Western are discussed in Section 9.2, and the proof of Scholz's octic reciprocity law is given in Section 9.5.

9.1 The Rational Octic Reciprocity Law

We begin by fixing some notation. Let $\zeta = \zeta_8 = \frac{1}{2}(\sqrt{2} + \sqrt{2}i)$ denote a primitive eighth root of unity. Then $K = \mathbb{Q}(\zeta) = \mathbb{Q}(\sqrt{-1}, \sqrt{2})$ is a biquadratic Euclidean number field with unit group $E_K = \langle \zeta, \varepsilon \rangle$, where $\varepsilon = 1 + \sqrt{2}$. The Galois group of $K/\mathbb{Q}$ contains the four automorphisms σ_j, $j = 1, 3, 5, 7$, defined by $\sigma_j(\zeta) = \zeta^j$. For any $\alpha \in \mathbb{Z}[\zeta]$, we write $\alpha_j = \sigma_j(\alpha)$; in particular, $\alpha_1 = \alpha$. The absolute norm of $\alpha \in K$ is the product $N\alpha = \alpha_1 \alpha_3 \alpha_5 \alpha_7$, and the relative norms to $\mathbb{Q}(\sqrt{-2})$, $\mathbb{Q}(i)$ and $\mathbb{Q}(\sqrt{2})$ are given by $\alpha_1 \alpha_3$, $\alpha_1 \alpha_5$ and $\alpha_1 \alpha_7$, respectively. You should also recall the formulas proved in Exercise 5.21.

Call an element $A + B\zeta + C\zeta^2 + D\zeta^3 \in \mathbb{Z}[\zeta]$ *primary* if it is congruent to 1 mod $(2 + 2\zeta)$. It is an easy exercise to show that every $\alpha \in \mathbb{Z}[\zeta]$ with odd norm is associated to a primary element (and in fact to infinitely many, since the unit $-3 + 2\sqrt{2}$ is primary). Moreover, if α is primary, then so are its conjugates α_j. If η is a primary root of unity, then $\eta = 1$; this applies in particular to the units in $\mathbb{Z}[i]$ and $\mathbb{Z}[\sqrt{-2}]$.

For a primary prime $\Pi \in \mathbb{Z}[\zeta]$ with norm $p = N(\Pi) \equiv 1 \bmod 8$, we let $\chi = [\,\cdot\,/\Pi]$ denote the octic residue character modulo Π. As usual, we will have to compute some character sums before we can attack octic reciprocity.

Octic Gauss and Jacobi Sums

We will start by computing the Jacobi sums for characters of order 8 on $\mathbb{F}_p$; the proof of Exercise 4.23 shows

Proposition 9.1. *Let χ be a character of order 2^m on $\mathbb{F} = \mathbb{F}_q$, and let ψ denote the quadratic character on $\mathbb{F}$. Then $J(\chi, \psi) \equiv 1 \bmod (2 + 2\zeta)$, where ζ is a primitive 2^m-th root of unity.*

We now introduce the abbreviation $J(i,k) = J(\chi^i, \chi^k)$. In Chapter 11 we will see that $J(1,4) = \eta \Pi_1 \Pi_3$ for some root of unity η; the proposition above shows that $J(1,4) \equiv 1 \bmod (2 + 2\zeta)$, and we conclude that $\eta = 1$.

A proof of $J(1,4) = \Pi_1 \Pi_3$ not relying on the results of Chapter 11 goes as follows: let $\sigma \in \text{Gal}(K/\mathbb{Q})$ be the automorphism induced by mapping $\zeta \longrightarrow \zeta^3$. Then $\mathbb{Q}(\sqrt{-2})$ is the fixed field of σ, and we find $J(1,4)^\sigma = J(3,4)$. Proposition 4.26.ii) gives $J(3,4) = \chi(-1)^4 J(1,4) = J(1,4)$. Hence $J(1,4)$ lies in the fixed field of σ, and we must have $J(1,4) = \pm \Pi_1 \Pi_3$ or $J(1,4) = \pm \Pi_5 \Pi_7$. But $\Pi_1 \mid J(1,4)$ by Proposition 4.28, and this excludes the last possibility. Now $J(1,4)$ and $\Pi_1 \Pi_3$ are primary, and the fact that the only primary unit in $\mathbb{Z}[\sqrt{-2}]$ is 1 proves the claim.

Let us find $J(1,1)$ next. Using Proposition 4.30 and the fact that $\chi(4)^2 = \chi(\sqrt{2})^8 = 1$ we find $J(1,1) = \chi(4)^{-1} J(1,4) = \chi(4) \Pi_1 \Pi_3$. Similarly, we have $J(1,3) = \chi(-1) J(1,4) = \chi(-1) \Pi_1 \Pi_3$ from Proposition 4.26. Next we observe the equality $J(1,2) J(1,3) = J(1,1) J(2,2)$, which follows from Proposition 4.26.iv); but $J(2,2) = \Pi_1 \Pi_5$ is the quartic Jacobi sum. Finally we use Proposition 4.26.ii) to compute the Jacobi sums $J(1,5) = J(5,1) = \chi(-1) J(2,1) = \chi(-1) J(1,2)$ and $J(1,6) = J(6,1) = \chi(-1) J(1,1)$; the factorization of $G(\chi)^8$ then follows from Corollary 4.24, and we have proved

Proposition 9.2. *Let $\chi = \left[\frac{\cdot}{\Pi}\right]$ be the residue character of order 8 on $\mathbb{F}_p$, where $\Pi \equiv 1 \bmod (2 + 2\zeta)$ is a primary prime in $\mathbb{Z}[\zeta]$ such that $p = N(\Pi) = \Pi_1 \Pi_3 \Pi_5 \Pi_7 \equiv 1 \bmod 8$. Then*

$$J(1,1) = \chi(4) \Pi_1 \Pi_3, \quad J(1,4) = \Pi_1 \Pi_3,$$
$$J(1,2) = \chi(-4) \Pi_1 \Pi_5, \quad J(1,5) = \chi(4) \Pi_1 \Pi_5,$$
$$J(1,3) = \chi(-1) \Pi_1 \Pi_3, \quad J(1,6) = \chi(-4) \Pi_1 \Pi_3,$$
$$J(2,2) = \Pi_1 \Pi_5, \quad G(\chi)^8 = \Pi_1^7 \Pi_3^5 \Pi_5^3 \Pi_7.$$

Burde's octic reciprocity law

We now will prove the octic counterpart to Burde's rational reciprocity law: to this end let $p \equiv q \equiv 1 \bmod 8$ denote primes such that $\left(\frac{p}{q}\right)_4 = \left(\frac{q}{p}\right)_4 = 1$. If we denote the primary prime factors of p by Π_j ($j = 1, 3, 5, 7$), then $\Pi_1 \Pi_5 = a + bi$ and $\Pi_1 \Pi_3 = c + d\sqrt{-2}$, where $p = a^2 + b^2 = c^2 + 2d^2$ and $b \equiv 2d \equiv 0 \bmod 4$. The factorization of the Gauss sum $G(\chi)$, where $\chi = [\,\cdot\,/\Pi_1]_8$, is

9.1 The Rational Octic Reciprocity Law

$$G(\chi)^8 = \Pi_1^7 \Pi_3^5 \Pi_5^3 \Pi_7 = p(a+bi)^2(c+d\sqrt{-2})^4.$$

This implies the following formula due to Western for the rational octic residue symbol $(q/p)_8 = [q/\Pi]$ which is defined whenever $(q/p)_4 = +1$:

$$\left(\frac{q}{p}\right)_8 \equiv G(\chi)^{q-1} = \left(G(\chi)^8\right)^{\frac{q-1}{8}} = p^{\frac{q-1}{8}}(a+bi)^{\frac{q-1}{4}}(c+d\sqrt{-2})^{\frac{q-1}{2}} \bmod q. \tag{9.1}$$

Of course we have $p^{(q-1)/8} \equiv \left(\frac{q}{p}\right)_8 \bmod q$. Next write $q = A^2+B^2 = C^2+2D^2$; letting $[\,\cdot\,/\,\cdot\,]_2$ and $[\,\cdot\,/\,\cdot\,]_4$ denote the quadratic and quartic residue symbol in $\mathbb{Z}[i]$, we get

$$(a+bi)^{\frac{q-1}{4}} \equiv \left[\frac{a+bi}{A+Bi}\right]_4 \bmod (A+Bi).$$

For the computation of this reciprocity symbol we use (observe that $2AB \equiv (A+B)^2 \bmod (A+Bi)$):

$$\left[\frac{2AB}{A+Bi}\right]_4 = \left[\frac{A+B}{A+Bi}\right]_2 = \left(\frac{A+B}{q}\right) = \left(\frac{A^2+B^2}{A+B}\right) = \left(\frac{2}{A+B}\right),$$

because $A^2 + B^2 - (A+B)(A-B) = 2A^2$. Moreover we see

$$\left[\frac{2i}{A+Bi}\right]_4 = \left[\frac{1+i}{A+Bi}\right]_2 = \left(\frac{2}{A+B}\right), \quad \text{hence} \quad \left[\frac{B}{A+Bi}\right]_4 = \left[\frac{Ai}{A+Bi}\right]_4.$$

But

$$\left[\frac{A}{A+Bi}\right]_4 = i^{\frac{A-1}{4}} = (-1)^{\frac{q-1}{8}} \quad \text{and} \quad \left[\frac{i}{A+Bi}\right]_4 = (-1)^{\frac{q-1}{8}}$$

show that we have

$$\left[\frac{A}{A+Bi}\right]_4 = (-1)^{\frac{q-1}{8}} \quad \text{and} \quad \left[\frac{B}{A+Bi}\right]_4 = 1.$$

This gives

$$\left[\frac{a+bi}{A+Bi}\right]_4 = \left[\frac{B}{A+Bi}\right]_4 \left[\frac{aB-Ab}{A+Bi}\right]_4 = \left(\frac{aB-Ab}{q}\right)_4,$$

as well as

$$\left[\frac{a+bi}{A+Bi}\right]_4 = \left[\frac{A}{A+Bi}\right]_4 \left[\frac{aA+bB}{A+Bi}\right]_4 = (-1)^{\frac{p-1}{8}}\left(\frac{aA+bB}{q}\right)_4.$$

The symbols on the right hand side are well defined by Burde's reciprocity law if we assume that $(p/q)_4 = (q/p)_4 = +1$. A similar computation shows that

292   9. Octic Reciprocity

$$(c+d\sqrt{-2})^{\frac{q-1}{2}} \equiv \left(\frac{c+d\sqrt{-2}}{C+D\sqrt{-2}}\right) = \left(\frac{D}{q}\right)\left(\frac{cD-Cd}{q}\right)$$
$$= \left(\frac{cD-Cd}{q}\right) \bmod (C+D\sqrt{-2}),$$

where $q = C^2 + 2D^2$, because $\left(\frac{D}{q}\right) = 1$ (this follows from $q \equiv 1 \bmod 8$ by extracting the 2-part of D and inverting the symbol). On the other hand, we also have

$$\left(\frac{c+d\sqrt{-2}}{C+D\sqrt{-2}}\right) = \left(\frac{C}{q}\right)\left(\frac{cC+2dD}{q}\right) = \left(\frac{2}{q}\right)_4 \left(\frac{cC+2dD}{q}\right),$$

where we have used

$$\left(\frac{C}{q}\right) = \left(\frac{q}{C}\right) = \left(\frac{2}{C}\right) = \left(\frac{2}{q}\right)_4$$

(see Proposition 5.4). Collecting our results, we find that we have proved

Proposition 9.3. *Let $p \equiv q \equiv 1 \bmod 8$ be primes and assume that $\left(\frac{p}{q}\right)_4 = \left(\frac{q}{p}\right)_4 = 1$; write $p = a^2 + b^2 = c^2 + 2d^2$ and $q = A^2 + B^2 = C^2 + 2D^2$, where $a \equiv A \equiv 1 \bmod 4$. Then*

$$\left(\frac{p}{q}\right)_8 \left(\frac{q}{p}\right)_8 = \left(\frac{aB-Ab}{q}\right)_4 \left(\frac{cD-Cd}{q}\right)$$
$$= (-1)^{\frac{q-1}{8}} \left(\frac{aA+bB}{q}\right)_4 \left(\frac{cD-Cd}{q}\right)$$
$$= \left(\frac{-1}{q}\right)_8 \left(\frac{2}{q}\right)_4 \left(\frac{aA+bB}{q}\right)_4 \left(\frac{cC+2dD}{q}\right).$$

9.2 Eisenstein's Reciprocity Law

Using the factorization of the octic Gauss sum, it is quite easy to prove Eisenstein's reciprocity law for eighth powers. To this end, let $[\,\cdot\,/\,\cdot\,]$ denote the octic residue symbol in $\mathbb{Z}[\zeta]$; we will use $(\,\cdot\,/\,\cdot\,)_8$ for the rational symbol as before.

Proposition 9.4. *(Octic Eisenstein Reciprocity) Let $a \in \mathbb{Z}$ and $\alpha \in \mathbb{Z}[\zeta]$ be primary; then*

$$\left[\frac{\alpha}{a}\right] = \left[\frac{a}{\alpha}\right], \quad \left[\frac{\zeta}{a}\right] = \zeta^{\frac{a^2-1}{4}}, \quad \left[\frac{1+\zeta}{a}\right] = \zeta^{\frac{a^2-1}{8}}, \quad \text{and} \quad \left[\frac{\varepsilon}{a}\right] = 1.$$

9.2 Eisenstein's Reciprocity Law

Actually we can replace the condition that a and α be primary by the weaker assumption that $a \equiv 1 \bmod 4$ and $\alpha \equiv 1$, $1 + \zeta + \zeta^3 \bmod 2$. In fact, for any such α there is a real unit $\eta = \pm\varepsilon^{\pm 1}$ such that $\beta := \alpha\eta$ is primary (see Exercise 9.4), and then we can use the supplementary law $[\frac{\varepsilon}{a}] = 1$ to deduce that $[\frac{\alpha}{a}] = [\frac{\beta}{a}] = [\frac{a}{\beta}] = [\frac{a}{\alpha}]$. We may even weaken the condition on a to $a \equiv 1 \bmod 2$ and write the reciprocity law in the form $[\frac{\alpha}{a}] = [\frac{a^*}{\alpha}]$, where $a^* = (-1)^{(a-1)/2}a$ as usual. In this form, the octic Eisenstein reciprocity law is given by Berndt, Evans & K.S. Williams [48].

Let us start the proof of Proposition 9.4 by recalling the congruence

$$G(\chi)^q \equiv \chi(q)^{-a} G(\chi^a) \bmod q \qquad (9.2)$$

for primes $q \equiv a \bmod 8$. Our first step will be the proof of the reciprocity law $[\Pi/q] = [q/\Pi]$ for primes Π of inertia degree 1. Assume first that $q \equiv 1 \bmod 8$, and write $q\mathcal{O}_K = \Lambda_1 \Lambda_3 \Lambda_5 \Lambda_7$. From (9.2) we get $\chi(q)^{-1} G(\chi) \equiv G(\chi)^q$, i.e.

$$\chi(q)^{-1} \equiv (G(\chi)^8)^{(q-1)/8} = (\Pi_1^7 \Pi_3^5 \Pi_5^3 \Pi_7)^{(q-1)/8} \bmod q,$$

and using $\Pi^{(q-1)/8} \equiv [\Pi/\Lambda_1] \bmod \Lambda_1$ we get

$$\chi(q)^{-1} \equiv \left[\frac{\Pi_1}{\Lambda_1}\right]^7 \left[\frac{\Pi_3}{\Lambda_1}\right]^5 \left[\frac{\Pi_5}{\Lambda_1}\right]^3 \left[\frac{\Pi_7}{\Lambda_1}\right] \bmod \Lambda_1.$$

Now we take inverses, and by using the equality

$$\left[\frac{\alpha_a}{\Lambda_1}\right]^a = \left[\frac{\alpha_1}{\Lambda_a}\right] \qquad (9.3)$$

we get

$$\chi(q) = \left[\frac{\Pi_1}{\Lambda_1}\right] \left[\frac{\Pi_3}{\Lambda_1}\right]^3 \left[\frac{\Pi_5}{\Lambda_1}\right]^5 \left[\frac{\Pi_7}{\Lambda_1}\right]^7 = \left[\frac{\Pi_1}{\Lambda_1}\right] \left[\frac{\Pi_1}{\Lambda_3}\right] \left[\frac{\Pi_1}{\Lambda_5}\right] \left[\frac{\Pi_1}{\Lambda_7}\right] = \left[\frac{\Pi_1}{q}\right].$$

Next consider the case $q \equiv a \bmod 8$ for $a = 3, 5, 7$; for any such a, choose $c \in \{3, 5, 7\}$ such that $a \neq 3$. Then we claim that

$$\chi(q^*) = \left[\frac{\Pi_1}{\Lambda_1}\right] \left[\frac{\Pi_c}{\Lambda_1}\right]^c. \qquad (9.4)$$

By (9.3), this implies at once that $[q^*/\Pi] = [\Pi/q]$.

We first observe that a direct attack on (9.4) is not successful: in fact, (9.2) implies

$$G(\chi)^{q^2} \equiv \chi(q^2)^{-1} G(\chi) \bmod q,$$

hence $G(\chi)^{q^2-1} \equiv \chi(q)^{-2}$ mod q, which is not good enough to determine $\chi(q)$. Actually, we had the same problem in Chapter 6 when we computed $[\alpha/q]_4$ for primes $q \equiv 3$ mod 4. There is, however, a way to circumvent these difficulties. In fact, from (9.2) we get

$$G(\chi)^{q+8-a} \equiv \chi(q)^{-a} G(\chi^a) G(\chi)^{8-a} \bmod q.$$

But $G(\chi)^{8-a} = G(\chi^{8-a}) \prod_{r=1}^{8-a-1} J(1,r)$ by Corollary 4.24, and Proposition 4.23 yields and $G(\chi^a)G(\chi^{8-a}) = \chi(-1) \Pi_1 \Pi_3 \Pi_5 \Pi_7$. Thus

$$G(\chi)^{q+8-a} \equiv \chi(-q)^{-a} \Pi_1 \Pi_3 \Pi_5 \Pi_7 \prod_{r=1}^{8-a-1} J(1,r) \bmod q.$$

In the case $a = 3$ we get $\prod_{r=1}^{8-a-1} J(1,r) = \Pi_1^4 \Pi_3^3 \Pi_5$, i.e.

$$G(\chi)^{q+5} \equiv \chi(-q)^{-3} \Pi_1^5 \Pi_3^4 \Pi_5^2 \Pi_7 \bmod q.$$

Now we use $\Pi_3 \equiv \Pi_1^q$ mod q and $\Pi_7 \equiv \Pi_5^q$ mod q and, after some manipulation, we get

$$\chi(-q)^5 \equiv \Pi_1^{5(q^2-1)/8} \Pi_5^{(q^2-1)/8} \bmod q;$$

raising this to the fifth power yields the claim (9.4) in case $a = 3$. The cases $a = 5$ and $a = 7$ follow just as easily.

Since we need Eisenstein's octic reciprocity only in the case where Π has inertia degree 1, the case where Π has degree 2 will be treated in the Exercises. Observe that we do not need the factorization of the octic Gauss sums in this case; nevertheless, here's how to compute them.

In the following, put $L = \mathbb{F}_{p^2}$ and $F = \mathbb{F}_p$; note that $\mathcal{O}_K/(\Pi) \simeq L$ if $N\Pi = p^2$, and $\mathcal{O}_K/(\Pi) \simeq F$ if $N\Pi = p$ is prime.

If Π is a primary prime dividing some $p \equiv a$ mod 8 for $a = 3, 5, 7$, then $N\Pi = p^2$, and $\chi = [\cdot/\Pi]$ is a character of order 8 on $L^\times$; for primes $p \equiv 1$ mod 8, we can define such a character for every prime $\Pi \mid p$ by $\chi = [\cdot/\Pi] \circ N_{L/F}$ (this χ is the lifted character χ' occurring in Theorem 4.32).

From Proposition 4.34.iii) and iv) we know that $G(\chi) = E(\chi)G(\chi_F)$, where χ_F is χ restricted to $F^\times$, and where $E(\chi)$ denotes the Eisenstein sum defined in Chapter 4. Now $\chi_F = \chi^{a+1}$ for primes $p \equiv a$ mod 8 (see Exercise 9.5), and since $G(\chi_F)^4$ is known from Chapter 6, it is sufficient to determine the sums $E(\chi)$; here are the results:

Proposition 9.5. *Let p be an odd prime, and let χ be a character of order 8 on $\mathbb{F}_{p^2}^\times$. Define integers $a, b, c, d \in \mathbb{Z}$ up to sign by $p = a^2 + 4b^2$ for primes $p \equiv 1$ mod 4 and $p = c^2 + 2d^2$ for $p \equiv 1, 3$ mod 8. Choose $c \equiv 1$ mod 4 if $p \equiv 1$ mod 8, $c \equiv \frac{p-3}{4}$ mod 4 if $p \equiv 3$ mod 8, and $a \equiv 1$ mod 4 if $p \equiv 5$ mod 8. The signs of b and d are determined by the conditions $\Pi \mid a + bi$ and $\Pi \mid c + d\sqrt{-2}$ respectively, where $\chi = [\cdot/\Pi]$ if $p \equiv 3, 5, 7$ mod 8, or where χ is the lift of $[\cdot/\Pi]$ to $\mathbb{F}_{p^2}$ if $p \equiv 1$ mod 8. Then*

9.2 Eisenstein's Reciprocity Law   295

$$E(\chi) = \begin{cases} c + d\sqrt{-2} & \text{if } p \equiv 1,3 \bmod 8, \\ i^{b/2}(a+bi) & \text{if } p \equiv 5 \bmod 8, \text{ and} \\ (-1)^{(p-7)/8} & \text{if } p \equiv 7 \bmod 8. \end{cases}$$

The fact that the factors $e + f\sqrt{2}$ do not appear is easily explained by noting that $|G(\chi)| = p$: whereas $a+bi$ and $c+d\sqrt{-2}$ have absolute value p, the absolute value of $e \pm f\sqrt{2}$ is not even uniquely determined by p, but depends on the choice of e and f since the fundamental unit $1+\sqrt{2}$ has absolute value different from 1; compare Exercise 4.27. This is the reason why we cannot deduce that $\mathbb{Q}(\zeta_8)$ has class number 1: Gauss sums don't say anything about the real subfields of abelian extensions of $\mathbb{Q}$. Note, however, that Proposition 9.5 shows that $\mathbb{Q}(\sqrt{-2})$ has class number 1.

For the following proof we need certain integers associated to characters χ of order m on $\mathbb{F}_q^\times$: put $s_j = \#\{t \in \mathbb{F}_q^\times : \operatorname{Tr}(t) = 2, \chi(t) = \zeta_m^j\}$ and define a polynomial $\varphi(X) = \sum_{j=0}^{m-1} s_j X^j$. From the definition of Eisenstein sums we then get $E_2(\chi^\nu) = \varphi(\zeta^\nu)$. This in turn implies immediately that $8s_0 = \sum_{\nu=0}^{m-1} E_2(\chi^\nu)$.

Proof. Write $p \equiv r \bmod 8$ for $r \in \{1,3,5,7\}$; the easiest case is $r = 7$, since here $\chi_F = 1$, hence $|E(\chi)| = 1$ by Proposition 4.34.vi). In fact, applying the result of Exercise 4.35 with $p \equiv 7 \bmod 8$ yields $E(\chi) = -(-1)^{(p+1)/m}$.

Our next claim is that $E(\chi) \equiv 0 \bmod \Pi$ for $r \leq 5$. In fact, we have

$$E_2(\chi) = \sum_{\operatorname{Tr}(t)=2} \chi(t) \equiv \sum_{\operatorname{Tr}(t)=2} t^{(p^2-1)/8} \bmod \Pi.$$

Now $\mathbb{F}_q = \mathbb{F}_p[i]$ for $r = 3$, and an element $t \in \mathbb{F}_q$ has trace 2 if and only if $t = 1 + iz$ for some $z = 0, 1, \ldots, p-1$. Moreover, $(1+iz)^p = 1-iz$, and since $p^2 - 1 = p^2 - ap + ap - 1 = p(p-a) + (ap-1)$, we get

$$E_2(\chi) = \sum_{z=0}^{p-1}(1+iz)^{(q-1)/8} \equiv \sum_{z=0}^{p-1}(1-iz)^{(p-a)/8}(1+iz)^{(ap-1)/8} \bmod \Pi.$$

But now Lemma 4.29 shows that the last sum is congruent to 0 mod p unless $\frac{p-a}{8} + \frac{ap-1}{8} \geq p-1$, that is, unless $a = 7$. Since $E_2(\chi) = \chi(2)E(\chi) = \pm E(\chi)$, our claim follows.

Next we study the case $r = 3$. Applying Exercise 4.35 to primes $p \equiv 3 \bmod 8$ and characters $\chi^{2\nu}$ we easily get $E_2(\chi^{2\nu}) = E(\chi^{2\nu}) = (-1)^{1+\nu}$ for $\nu = 1,2,3$, whereas $E_2(1) = E(1) = p+1$ follows directly from the definition (the surjectivity of the trace from $\mathbb{F}_{p^2}$ to $\mathbb{F}_p$ shows that there are exactly $p+1$ elements with vanishing trace). Since $E(\chi) = E(\chi^p) = E(\chi^3)$ by Proposition 4.34.ii), we can write $E_2(\chi) = E_2(\chi^3) = c + d\sqrt{-2}$ and $\overline{E_2(\chi)} = E_2(\chi^7) = E_2(\chi^5) = c - d\sqrt{-2}$. Moreover, $c^2 + 2d^2 = p$ by Proposition 4.34.vi). Now $8s_0 = \sum_{\nu=0}^{7} E_2(\chi^\nu) = 4c + p + 1$, hence $c = 2s_0 - \frac{p+1}{4}$. But s_0 is clearly odd, since $2 = \operatorname{Tr}(t) = t + t^p$ shows that elements with trace 2 come in pairs

296   9. Octic Reciprocity

except when $t = t^p = 1$. This shows that $c \equiv 2 - \frac{p+1}{4} \equiv \frac{p+1}{4} \mod 4$. Since $E_2(\chi) = \chi(2)E(\chi) = -E(\chi)$, our claim follows.

Now we consider primes $p \equiv 5 \mod 8$; here we derive $E_2(\chi) = E_2(\chi^5) = x + yi$ with $x^2 + y^2 = p$ as for $r = 3$. It turns out that it is more convenient to determine $E_2(\chi)$. Now $E_2(\chi) = E_2(\chi^5) = x + yi$, $E_2(\chi^3) = E_2(\chi^7) = x - yi$ and $E_2(\chi^6) = \chi^6(2)E(\chi) = -J(\chi_F, \chi_F)$ by Exercise 4.36, hence $E_2(\chi^6) = -\chi_F(-1)\overline{\Pi} = a - bi$ (note that $\Pi \equiv 3 + 2i \mod 4$ here), as well as $E_2(\chi^4) = 1$ and $E_2(\mathbb{1}) = p$. Thus $8s_0 = \sum_{\nu=0}^{7} \varphi(\zeta^\nu) = 4x + 2a + p + 1$, hence $0 \equiv 4x + 6 + 6 \mod 8$, and we find that x is odd. Thus $x = \pm a$, and since s_0 is odd, we get $8 \equiv \pm 4a + 2a + p + 1 \mod 16$. But $p = a^2 + b^2 \equiv a^2 + 4 \mod 16$ implies that $p + 2a + 1 \equiv (a+1)^2 + 4 \equiv 4 \mod 16$, and this shows that we must have $x = -a$, that is, $x \equiv 1 \mod 4$. Summarizing, we have $E_2(\chi) = a + bi$ with $a \equiv 1 \mod 4$, and $8s_0 = 6a + p + 1$.

Finally consider primes $p \equiv 1 \mod 8$. From the Davenport-Hasse Theorem 4.32 we know that $G(\chi)$ is the square of the octic Gauss sum for $[\cdot/\Pi]$, hence we have $G(\chi)^4 = \Pi_1^7 \Pi_3^5 \Pi_5^3 \Pi_7 = p(a + bi)^2(c + d\sqrt{-2})^4$; on the other hand $G(\chi_F)^4 = p(a + bi)^2$, and this gives $E(\chi)^4 = G(\chi)^4/G(\chi_F)^4 = (c + d\sqrt{-2})^4$. Thus $E(\chi) = \varepsilon(c + d\sqrt{-2})$ for some fourth root of unity ε. From Exercise 4.38 we immediately deduce that $\varepsilon = \pm 1$, and a more careful analysis easily shows that $c \equiv 1 \mod 4$. This concludes the proof. □

Western's Formulas

As a corollary of Eisenstein's reciprocity law (Proposition 9.4) we are now going to derive Western's formulas. To this end, let $\Pi \in \mathbb{Z}[\zeta]$ be a primary prime with prime norm $p = \Pi_1 \Pi_3 \Pi_5 \Pi_7 \equiv 1 \mod 8$. Let $q \equiv 1 \mod 8$ be another prime; then we have seen in (9.1) that

$$\left[\frac{q}{\Pi}\right] \equiv p^{\frac{q-1}{8}}(a + bi)^{\frac{q-1}{4}}(c + d\sqrt{-2})^{\frac{q-1}{2}} \mod q.$$

Similar formulas hold for primes $q \equiv 3 \mod 8$: write $q = \Lambda \Lambda_5$ for some $\Lambda \in \mathbb{Z}[\sqrt{-2}]$, and where Λ_5 is the conjugate of Λ. Then Proposition 9.4 yields

$$\left[\frac{q^*}{\Pi}\right] = \left[\frac{\Pi}{q}\right] = \left[\frac{\Pi}{\Lambda}\right]\left[\frac{\Pi}{\Lambda_5}\right] = \left[\frac{\Pi}{\Lambda}\right]\left[\frac{\Pi_5}{\Lambda}\right]^5.$$

Recalling that $\Pi^q \equiv \Pi_3 \mod q$ (this is an obvious generalization of Proposition 2.21) we find

$$\Pi^{(q^2-1)/8} = \Pi^{(q^2-3q)/8}\Pi^{(3q-1)/8} = (\Pi^q)^{(q-3)/8}\Pi^{(q-3)/8}\Pi^{(q+1)/2}$$
$$\equiv (\Pi\Pi_3)^{(q-3)/8}\Pi^{(q+1)/2} \mod q.$$

Treating Π_5 similarly, we get

$$\Pi_5^{5(q^2-1)/8} = \Pi_5^{(q^2-1)/8}\Pi_5^{(q^2-1)/2}$$
$$\equiv (\Pi_5\Pi_7)^{(q-3)/8}\Pi^{(q+1)/2}(\Pi_5\Pi_7)^{(q-1)/2} \mod q.$$

Collecting everything shows
$$\left[\frac{q^*}{\Pi}\right] \equiv p^{\frac{q-3}{8}}(a+bi)^{\frac{q+1}{4}}(c-d\sqrt{-2})^{\frac{q-1}{2}} \bmod \Lambda.$$

The primes $q \equiv 5, 7 \bmod 8$ can be handled in a similar manner, and we have proved

Proposition 9.6. *Let* $p = a^2 + b^2 = c^2 + 2d^2 = e^2 - 2f^2 \equiv 1 \bmod 8$ *and q be odd primes, and let Π and Λ be primary prime divisors of p and q, respectively; then*

$$\left[\frac{q^*}{\Pi}\right] \equiv \begin{cases} p^{\frac{q-1}{8}}(a+bi)^{\frac{q-1}{4}}(c+d\sqrt{-2})^{\frac{q-1}{2}} \bmod \Lambda & \text{if } q \equiv 1 \bmod 8; \\ p^{\frac{q-3}{8}}(a+bi)^{\frac{q+1}{4}}(c-d\sqrt{-2})^{\frac{q-1}{2}} \bmod \Lambda & \text{if } q \equiv 3 \bmod 8; \\ p^{\frac{q-5}{8}}(a-bi)^{\frac{q-1}{4}}(c+d\sqrt{-2})^{\frac{q+1}{2}} \bmod \Lambda & \text{if } q \equiv 5 \bmod 8; \\ p^{\frac{q-7}{8}}(a+bi)^{\frac{q+1}{4}}(c+d\sqrt{-2})^{\frac{q+1}{2}} \bmod \Lambda & \text{if } q \equiv 7 \bmod 8. \end{cases}$$

These formulas can be simplified if we assume that q is a quadratic residue modulo p:

Lemma 9.7. *If q and $p = c^2 + 2d^2 \equiv 1 \bmod 8$ are primes such that $(p/q) = +1$, and if $e = \left(\frac{-2}{q}\right)$, then*

$$(c + d\sqrt{-2})^{\frac{q-e}{2}} \equiv \operatorname{Re}(c + d\sqrt{-2})^{\frac{q-e}{2}} \bmod q,$$

where $\operatorname{Re}\alpha$ *denotes the real part of some* $\alpha \in \mathbb{C}$.

Proof. Suppose first that $e = -1$, i.e. $q \equiv 5, 7 \bmod 8$ and define integers r, s by $(c + d\sqrt{-2})^{(q+1)/2} = r + s\sqrt{-2}$. Squaring yields $p = c^2 + 2d^2 \equiv (c + d\sqrt{-2})^{q+1} = r^2 - 2s^2 + 2rs\sqrt{-2} \bmod q$. This implies $2rs \equiv 0 \bmod q$; if $q \mid r$, then we would have $p \equiv -2s^2 \bmod q$, which contradicts our assumptions that $(p/q) = 1$ and $(-2/q) = -1$. Therefore we must have $q \mid s$, and this shows that the imaginary part of $(c + d\sqrt{-2})^{\frac{q+1}{2}}$ vanishes modulo q: but that's exactly what we wanted to prove.

Next assume that $q \equiv 1, 3 \bmod 8$ and write $q = \Lambda\Lambda'$ for some $\Lambda, \Lambda' \in \mathbb{Z}[\sqrt{-2}]$; then we find

$$1 = \left(\frac{p}{q}\right)_{\mathbb{Z}} = \left(\frac{p}{\Lambda}\right)_2 = \left(\frac{c+d\sqrt{-2}}{\Lambda}\right)_2 \left(\frac{c-d\sqrt{-2}}{\Lambda}\right)_2$$
$$= \left(\frac{c+d\sqrt{-2}}{\Lambda}\right)_2 \left(\frac{c+d\sqrt{-2}}{\Lambda'}\right)_2.$$

Writing $(c + d\sqrt{-2})^{(q-1)/2} = r + s\sqrt{-2}$ as before, this shows that the $r + s\sqrt{-2} \equiv \pm 1 \bmod \Lambda$ and $r + s\sqrt{-2} \equiv \pm 1 \bmod \Lambda'$ (the signs being equal), hence $r + s\sqrt{-2} \equiv \pm 1 \bmod q$. But then $q \mid s$ as claimed. □

A similar simplification is possible for the $(a+bi)$-factor if $(q^*/p)_4 = +1$:

Lemma 9.8. *If q and $p = a^2 + b^2 \equiv 1 \bmod 8$ are primes (of course we assume that b is even) such that $\left(\frac{q^*}{p}\right)_4 = +1$, and if $e = \left(\frac{-1}{q}\right)$, then*

$$(a+bi)^{\frac{q-e}{4}} \equiv \operatorname{Re}(a+bi)^{\frac{q-e}{4}} \bmod q.$$

Proof. As in the proof of Lemma 9.7, write

$$(a+bi)^{(q-e)/4} = r + si. \tag{9.5}$$

Suppose first that $q \equiv 3 \bmod 4$; then $e = -1$, and raising (9.5) to the fourth power we get $p \equiv (a+bi)^{q+1} = r^4 - 6r^2s^2 + s^4 + 4rs(r^2 - s^2)i \bmod q$. This implies at once that $q \mid rs(r^2 - s^2)$.

Suppose that $r^2 \equiv s^2 \bmod q$; then $p \equiv (a+bi)^{q+1} \equiv -4r^2s^2 \bmod q$, and this contradicts our assumption that $\left(\frac{p}{q}\right) = +1$.

Next assume that $q \mid r$; squaring (9.5) yields $(a+bi)^{(q+1)/2} \equiv -s^2 \bmod q$, and this gives

$$(a+bi)^{(q^2-1)/4} \equiv (-s^2)^{(q-1)/2} \equiv -1 \bmod q.$$

On the other hand we find

$$\left(\frac{-q}{p}\right)_4 = \left[\frac{q}{a+bi}\right]_4 = \left[\frac{a+bi}{q}\right]_4 \equiv (a+bi)^{(q^2-1)/4} \bmod q,$$

and this contradiction shows that $q \nmid r$.

Therefore we must have $q \mid s$ as claimed. The proof for $q \equiv 1 \bmod 4$ is left to the reader. □

We have proved

Corollary 9.9. *Let $p = a^2 + b^2 = c^2 + 2d^2 = e^2 - 2f^2 \equiv 1 \bmod 8$ and q be odd primes, and assume that $(q^*/p)_4 = +1$. Then*

$$\left(\frac{q^*}{p}\right)_8 \equiv \begin{cases} p^{\frac{q-1}{8}} \operatorname{Re}(a+bi)^{\frac{q-1}{4}} \operatorname{Re}(c+d\sqrt{-2})^{\frac{q-1}{2}} \bmod q & \text{if } q \equiv 1 \bmod 8; \\ p^{\frac{q-3}{8}} \operatorname{Re}(a+bi)^{\frac{q+1}{4}} \operatorname{Re}(c-d\sqrt{-2})^{\frac{q-1}{2}} \bmod q & \text{if } q \equiv 3 \bmod 8; \\ p^{\frac{q-5}{8}} \operatorname{Re}(a-bi)^{\frac{q-1}{4}} \operatorname{Re}(c+d\sqrt{-2})^{\frac{q+1}{2}} \bmod q & \text{if } q \equiv 5 \bmod 8; \\ p^{\frac{q-7}{8}} \operatorname{Re}(a+bi)^{\frac{q+1}{4}} \operatorname{Re}(c+d\sqrt{-2})^{\frac{q+1}{2}} \bmod q & \text{if } q \equiv 7 \bmod 8. \end{cases}$$

Note that these congruences are valid modulo q (and not only modulo Λ) since both sides are rational integers. As an immediate corollary of Western's formulas we get the octic character of several small primes:

Corollary 9.10. *Let $p = a^2 + b^2 = c^2 + 2d^2 \equiv 1 \bmod 8$ be prime, and suppose that $a \equiv c \equiv 1 \bmod 4$. Then*

$(-3/p)_8 = 1 \iff (-3/p)_4 = 1$ *and* $ac \equiv 1 \bmod 3$;
$(5/p)_8 = 1 \iff (5/p)_4 = 1$ *and* $ac(c^2 - d^2) \equiv 1 \bmod 5$;
$(-7/p)_8 = 1 \iff (-7/p)_4 = 1$ *and*
$$(a^2 - b^2)(c^2 - d^2)(c^2 - 4d^2) \equiv 1 \bmod 7.$$

Proof. Let $q = 3$; then $q^* = -3$, hence $(-3/p)_8 \equiv \mathrm{Re}\,(a + bi)\mathrm{Re}\,(c + di) = ac \bmod q$, and our claim follows. If $q = 5$, then $(5/p)_8 \equiv \mathrm{Re}\,(a + bi)\mathrm{Re}\,(c + di)^2 = ac(c^2 - 6d^2) \equiv ac(c^2 - d^2) \bmod 5$. The case $q = 7$ is left as an exercise. □

Evans [226] gave slightly different criteria; take for example $q = 5$: if $5 \nmid cd$, then $p = c^2 + 2d^2 \equiv \pm 1 \bmod 5$ implies that $c^2 \equiv -d^2 \bmod 5$. Thus
$$ac(c^2 - d^2) \equiv 1 \iff 2ac^3 \equiv 1 \iff c \equiv 2a \bmod 5.$$
Similarly, in the case $5 \mid cd$ we find $5 \mid d$, hence
$$ac(c^2 - d^2) \equiv 1 \iff ac^3 \equiv 1 \iff c \equiv a \bmod 5.$$
This shows that
$$(5/p)_8 = 1 \iff 5 \mid b \text{ and } \begin{cases} 5 \nmid d, & c \equiv 2a \bmod 5 \\ 5 \mid d, & c \equiv a \bmod 5. \end{cases}$$

9.3 Elliptic Gauss Sums

In Chapters 6 and 7 we have deduced the cubic and quartic reciprocity laws by formal computations from the corresponding 'Eisenstein' reciprocity laws, which allow us to invert symbols of the form (a/α) for rational integers a and elements α in the appropriate cyclotomic field. Unfortunately, this does not work here. In fact, write $\alpha, \beta \in \mathbb{Z}[\zeta]$ in the form $\alpha = A + B\zeta$ and $\beta = C + D\zeta$, with $A, B, C, D \in \mathbb{Z}[i]$. Proceeding as in the quartic case, we find
$$\left[\frac{\alpha}{\beta}\right] = \left[\frac{C}{\beta}\right]^{-1}\left[\frac{AC + BC\zeta}{\beta}\right] = \left[\frac{C}{\beta}\right]^{-1}\left[\frac{AC - BDi}{\beta}\right].$$

But Eisenstein reciprocity is not strong enough to invert these symbols: what we need is something slightly stronger, namely a reciprocity law which treats symbols of the form $[a/\alpha]$ with $a \in \mathbb{Z}[i]$. In order to get a reciprocity law stronger than the one that can be proved with Gauss sums, we need tools that are stronger than Gauss sums, namely elliptic Gauss sums. These were introduced by Eisenstein, who started studying octic reciprocity as early as 1848: in his letter to Gauss [201] from March 9, 1848, he gives a few properties of octic elliptic Gauss sums that he never touches on in his publications, although he mentions quartic elliptic Gauss sums in [202, p. 578].

The following sketch of the (essentially non-existent) theory of elliptic Gauss sums deals only with the lemniscatic case. Let sl z denote the lemniscatic sine, $\phi(z) = \mathrm{sl}(1-i)\omega z$ the elliptic function studied in Proposition 8.7, and $\pi \in \mathbb{Z}[i]$ a primary prime with norm $p = 4n+1$. For any integer $n \geq 1$, let ζ_{4n} denote a primitive $4n$-th root of unity, and put $k = \mathbb{Q}(i)$ and $K = \mathbb{Q}(\zeta_{4n})$. Assume that $\mathfrak{p}$ is a prime ideal above π in $\mathcal{O}_K$ with relative inertia degree 1, that is, with $\mathcal{O}_K/\mathfrak{p} \simeq \mathcal{O}_k/(\pi)$. Let $\chi = (\,\cdot\,/\mathfrak{p})_m$ denote the m-th power residue symbol in K, where $m \mid 4n$, and let A denote a complete set of prime residues modulo $\mathfrak{p}$ whose representatives are chosen in $\mathbb{Z}[i]$. Then[1]

$$\mathrm{Eis}(\chi) = -\sum_{t \in A} \chi(t)\phi(\tfrac{t}{\pi})^{(p-1)/m} \tag{9.6}$$

is called the elliptic Gauss sum associated to χ. Since $\phi(\tfrac{t}{\pi})$ generates an abelian extension $M = k\{(1+i)^3\pi\}$ of $k = \mathbb{Q}(i)$, we know that $k(\mathrm{Eis}(\chi))/k$ is an abelian extension contained in the compositum $L = KM$.

Proposition 9.11. *With* $\mathrm{Eis}(\chi)$ *as above, we have* $\mathrm{Eis}(\chi)^m \in \mathbb{Z}[\zeta_{4n}]$.

Proof. Let $\lambda \in \mathbb{Z}[i]$ be a primary prime not dividing p, let σ denote the Frobenius[2] automorphism of λ, and put $p = rn + 1$. We will use σ also to denote the extension of σ to K that fixes ζ_{4n}. It is sufficient to show that $\mathrm{Eis}(\chi)^m$ is invariant under σ. But

$$\mathrm{Eis}(\chi)^\sigma = -\sum_{t \in A} \chi(t)\phi(\tfrac{t}{\pi})^{r\sigma} = -\sum_{t \in A} \chi(t)\phi(\tfrac{\lambda t}{\pi})^r = \chi(\lambda)^{-1}\mathrm{Eis}(\chi),$$

and the claim follows. □

General results on elliptic Gauss sums are rare (the Jacobi sums do not have a counterpart in the elliptic theory; the reason is that the property $\zeta^a \zeta^b = \zeta^{a+b}$ for division values of the exponential function does not carry over to division values of elliptic functions). We will therefore concentrate on quartic and octic character sums.

Quartic elliptic Gauss sums

Consider the case of a quartic character $\chi = [\,\cdot\,/\pi]_4$ in $\mathbb{Z}[i]$ with $\pi = -1 + 2i$; then $r = 1$, $\chi(2) = -i$, hence $\mathrm{Eis}(\chi) = \phi(\tfrac{1}{z}) - i\phi(\tfrac{2}{z}) + i\phi(\tfrac{3}{z}) - \phi(\tfrac{4}{z})$. But $2 \equiv -i \bmod \pi$ implies $\phi(\tfrac{2}{z}) = -i\phi(\tfrac{1}{z})$, and similarly we find $\phi(\tfrac{3}{z}) = i\phi(\tfrac{1}{z})$ and $\phi(\tfrac{4}{z}) = -\phi(\tfrac{1}{z})$. This gives $\mathrm{Eis}(\chi) = \phi(\tfrac{1}{z})[1 + i^2 + i^2 + 1] = 0$.

Worse yet, this is not a coincidence:

---

[1] The exponent $(p-1)/m$ attached to $\phi(t/\pi)$ in (9.6) simplifies the resulting formulas extremely and even helps to produce nonvanishing elliptic Gauss sums. Its introduction is also an idea of Eisenstein.

[2] Strictly speaking this is the Frobenius of a prime ideal in $\mathbb{Q}(i, \phi(1/\pi))$ above λ.

9.3 Elliptic Gauss Sums

Lemma 9.12. *Let $\chi = [\cdot/\pi]_4$ be the quartic residue symbol modulo some primary prime $\pi \in \mathbb{Z}[i]$ with norm $p = 4r + 1$. Put $\psi = \chi^2$ and define*

$$\text{Eis}^+(\chi) = -\sum_{\psi(t)=+1} \chi(t)\phi(\tfrac{t}{\pi})^r, \quad \text{Eis}^-(\chi) = -\sum_{\psi(t)=-1} \chi(t)\phi(\tfrac{t}{\pi})^r.$$

Then $\text{Eis}(\chi) = \text{Eis}^+(\chi) + \text{Eis}^-(\chi)$ and $\text{Eis}^-(\chi) = (-1)^r \text{Eis}^+(\chi)$. In particular, $\text{Eis}(\chi) = 0$ whenever $p \equiv 5 \bmod 8$.

Proof. Since these sums do not depend on the choice of representatives of $t \bmod \pi$, we can substitute $t = is$ in the second sum and find $\psi(is) = (-1)^r \psi(s)$, hence

$$\text{Eis}^-(\chi) = -\sum_{\psi(s)=+1} \chi(is)\phi(\tfrac{is}{\pi})^r = i^r \chi(i) \text{Eis}^+(\chi).$$

Since $\chi(i) = i^r$ we get $\text{Eis}^-(\chi) = (-1)^r \text{Eis}^+(\chi)$ as claimed. □

But all is not lost:

Lemma 9.13. *The sums $\text{Eis}^+(\chi)$ do not vanish for $p \equiv 5 \bmod 8$.*

Proof. We can split the sum $\text{Eis}^+(\chi)$ in two parts, the summation ranging over $1 \leq t < p/2$ and $p/2 < t \leq p-1$, respectively. Since $\phi(-z)^r = (-1)^r \phi(z)$ and $\psi(-1) = 1$, we get $\text{Eis}^+(\chi) = 2\sum \chi(t)\phi(t/\pi)^r$, the summation being over $1 \leq t < p/2$ with $\psi(t) = +1$. This implies $\frac{1}{16}\text{Eis}^+(\chi)^4 \equiv \sum \phi(t/\pi)^{4r} \bmod 2$, and since $\phi(t/\pi)^4 \equiv 1 \bmod 2$ by Exercise 8.27, we deduce that $\frac{1}{16}\text{Eis}^+(\chi)^4 \equiv \frac{p-1}{4} \equiv 1 \bmod 2$. □

The next thing to do is to find the prime factorization of the sums $\text{Eis}^+(\chi)$. To this end we first have to study the Galois action on them. Letting σ denote the Frobenius of a primary prime $\lambda \in \mathbb{Z}[i]$, we easily get

$$\text{Eis}^+(\chi)^\sigma = \chi(\lambda)^{-1} \sum_{\psi(t)=1} \chi(\lambda t)\phi(\tfrac{\lambda t}{\pi})^r.$$

The last sum equals $\text{Eis}^+(\chi)$ if $\psi(\lambda) = +1$, and $\text{Eis}^-(\chi) = -\text{Eis}^+(\chi)$ otherwise. This shows $\text{Eis}^+(\chi)^\sigma = \psi(\lambda)\chi(\lambda)^{-1}\text{Eis}^+(\chi)$, that is, $\text{Eis}^+(\chi)^\sigma = \chi(\lambda)\text{Eis}^+(\chi)$. Now we claim

Proposition 9.14. *Let $\chi = [\cdot/\pi]_4$ be the fourth power residue character modulo some primary prime $\pi \in \mathbb{Z}[i]$ with norm $p \equiv 5 \bmod 8$. Then*

$$\text{Eis}^+(\chi)^4 = \alpha^4 \pi^*$$

for some nonzero $\alpha \in \mathbb{Z}[i]$.

Proof. Observe that the proof that works for ordinary Gauss sums does not carry over since we don't know how to multiply elliptic Gauss sums. Eisenstein's idea to overcome this problem is simple and powerful: let M denote a $\frac{1}{4}$-system modulo π and put $P(\chi) = \prod_{t \in M} \phi(\frac{t}{\pi})$. As above, let σ denote the Frobenius of a primary prime $\lambda \neq \pi$; then $P(\chi)^\sigma = \prod_{t \in M} \phi(\frac{\lambda t}{\pi})$. But $\lambda t \equiv i^{a(t)} t' \bmod \pi$ for some $t' \in M$, hence $\phi(\frac{\lambda t}{\pi}) = i^{a(t)} \phi(\frac{t}{\pi})$. Gauss's Lemma 4.3 shows that $\prod_{t \in M} i^{a(t)} = [\lambda/\pi]_4 = \chi(\lambda)$, and we get $P(\chi)^\sigma = \chi(\lambda) P(\chi)$.

Now assume that $p \equiv 5 \bmod 8$; then $\text{Eis}^+(\chi)^\sigma = \chi(\lambda) \text{Eis}^+(\chi)$, hence $\beta := P(\chi)^3 \text{Eis}^+(\chi)$ is fixed under Frobenius automorphisms and so must lie in $\mathbb{Z}[i]$. But now $P(\chi)^4 = \pi^*$ by Proposition 8.8, hence $\text{Eis}^+(\chi)^4 = \beta^4 / \pi^{*3}$. Since $\text{Eis}^+(\chi)^4 \in \mathbb{Z}[i]$ and π is prime, we conclude that $\beta = \pi^* \alpha$ for some $\alpha \in \mathbb{Z}[i]$, and we get $\text{Eis}^+(\chi)^4 = \alpha^4 \pi^*$ as claimed. □

As an exercise, the reader may want to derive a special case of the quartic reciprocity law from this result, namely that $[\pi^*/\lambda]_4 = [\lambda/\pi]_4$ for primes π with norm $p \equiv 5 \bmod 8$. The remaining case of primes of norm $p \equiv 1 \bmod 8$ seems to be more difficult. If $p \equiv 9 \bmod 16$, a slight modification of the above approach shows that $\text{Eis}^+(\chi) \neq 0$, but I don't know what happens when $p - 1$ is divisible by a high power of 2. Since $\text{Eis}^+(\chi) = \text{Eis}^-(\chi)$ in this case, we find $\text{Eis}^+(\chi)^\sigma = \chi(\lambda)^{-1} \text{Eis}^+(\chi)^\sigma$ for the Frobenius σ of a primary prime $\lambda \in \mathbb{Z}[i]$, and as above it follows that $\text{Eis}^+(\chi) = \alpha^4 \pi^{*3}$ for some $\alpha \in \mathbb{Z}[i]$. If we could show that α is nonzero, the complete quartic reciprocity law would be an immediate corollary.

Octic elliptic Gauss sums

For the rest of this section, we let Π and Λ denote primary primes in $\mathbb{Z}[\zeta]$ such that $\pi = \Pi_1 \Pi_5$ and $\lambda = \Lambda_1 \Lambda_5$ are their relative norms down to $\mathbb{Z}[i]$. Also, $\chi = [\cdot/\Pi]$ is the eighth power character modulo Π.

Eisenstein asserted in his letter to Gauss that the prime factorization of $\text{Eis}(\chi^3)$ is given by $\text{Eis}(\chi^3)^8 = \alpha^8 \Pi_1 \Pi_5^5$. Moreover he claimed that $\alpha^8 \Pi_1^4 \equiv 1 \bmod \Pi_5$, but as he did no repeat this assertion in [202], I don't know what to make of it. As the problems for octic elliptic Gauss sums are the same as in the quartic case, we will do what Eisenstein did when he noticed that elliptic Gauss sums may vanish and replace them by other objects which we can prove to be nonzero.

Let M be a $\frac{1}{4}$-system modulo $\pi \mathbb{Z}[i]$. Let ψ denote the quadratic character modulo $\pi \mathbb{Z}[i]$ and let $M^+ = \{t \in M : \psi(t) = +1\}$ and $M^- = \{t \in M : \psi(t) = -1\}$ be the subsets of M consisting of quadratic residues and nonresidues, respectively. An elementary calculation shows that $\prod_{t \in M^+} t^4 \equiv -1 \bmod \pi$ and $\prod_{t \in M^-} t^4 \equiv +1 \bmod \pi$, hence $\zeta_M = \prod_{t \in M^+} t / \prod_{t \in M^-} t$ is an element of $\mathbb{Z}[i]$ satisfying the congruence $\zeta_M^4 \equiv -1 \bmod \pi$.

The last congruence is also valid modulo Π, and in fact $\zeta_M \equiv \zeta^j \bmod \Pi$ for some $j \in \{1, 3, 5, 7\}$. Since we may replace any $t \in M$ by $t, it, -t$ or $-it$, we can (and will) choose M in such a way that $\zeta_M \equiv \zeta \bmod \Pi$.

9.3 Elliptic Gauss Sums

Now define
$$Q = \prod_{t \in M^+} \phi(\tfrac{t}{\pi}), \quad R = \prod_{t \in M^-} \phi(\tfrac{t}{\pi}).$$

Then $QR = P = \prod_{t \in M} \phi(\tfrac{t}{\pi})$, and $P^4 = \pi^*$ by (8.23). Next put $S = Q + \zeta R$ and $S' = Q - \zeta R$. Before we start, we will have to study the behaviour of these elements under Galois action.

Proposition 9.15. *Let σ denote the Frobenius automorphism of a primary prime $\lambda \in \mathbb{Z}[i]$ as well as its extension to $\mathbb{Q}(\zeta)$ that fixes ζ. Then*

$$Q^\sigma = \begin{cases} \chi(\lambda)Q & \text{if } \psi(\lambda) = +1, \\ \chi(\lambda)\zeta R & \text{if } \psi(\lambda) = -1, \end{cases} \quad R^\sigma = \begin{cases} \chi(\lambda)R & \text{if } \psi(\lambda) = +1, \\ \chi(\lambda)\zeta^{-1}Q & \text{if } \psi(\lambda) = -1. \end{cases}$$

Proof. We clearly have $Q^\sigma = \prod_{t \in M^+} \phi(\tfrac{\lambda t}{\pi})$; we have $\lambda t \equiv i^{a(t)} t' \bmod \pi$, where $t' \in M^+$ if λ is a quadratic residue modulo π and $t' \in M^-$ otherwise. If we denote the quadratic residue character modulo π by ψ, we get $Q^\sigma = i^{\sum a(t)} Q$ if $\psi(\lambda) = +1$, and $Q^\sigma = i^{\sum a(t)} R$ if $\psi(\lambda) = -1$.

On the other hand we have $\prod_{t \in M^+} (\lambda t) \equiv i^{\sum a(t)} \prod_{t \in M^+} t' \bmod \pi$ if $\psi(\lambda) = +1$, hence $\lambda^{(p-1)/8} \equiv i^{\sum a(t)} \bmod \pi$ and therefore $i^{\sum a(t)} = \chi(\lambda)$. If $\psi(\lambda) = -1$, however, then $\prod_{t \in M^+} (\lambda t) \equiv i^{\sum a(t)} \prod_{t \in M^-} t' \bmod \pi$, and we get $i^{\sum a(t)} \equiv \lambda^{(p-1)/8} \zeta_M \bmod \pi$. Since this congruence is also valid modulo Π, we find $i^{\sum a(t)} = \chi(\lambda) \zeta^{-1}$.

Similarly, $R^\sigma = \prod_{t \in M^-} \phi(\tfrac{\lambda t}{\pi})$. If $\psi(\lambda) = 1$, we have $\lambda t \equiv i^{b(t)} t' \bmod \pi$ for some $t' \in M^-$ and thus $i^{\sum b(t)} = \chi(\lambda)$. If $\psi(\lambda) = -1$, however, then $i^{\sum b(t)} = \chi(\lambda)\zeta$. □

Corollary 9.16. *Let the notation be as above. Then*

$$\begin{aligned} S^\sigma &= \chi(\lambda) S, & (Q^2 + iR^2)^\sigma &= \chi(\lambda)^2 (Q^2 + iR^2) \\ S'^\sigma &= \chi(\lambda)^5 S', & (Q^2 - iR^2)^\sigma &= \chi(\lambda)^6 (Q^2 - iR^2). \end{aligned}$$

In particular, $SS'P$ and $(Q^2 + iR^2)P^3$ are elements of $\mathbb{Z}[i]$ divisible by π.

Proof. The first part follows from the Galois action on Q and R that we just have proved. The last claim holds since multiples of P in $\mathbb{Z}[i]$ are divisible by π. □

The last corollary allows us to define $u, v \in \mathbb{Z}[i]$ by putting $(Q^2 + iR^2)P^3 = u\pi$ and $SS'P = v\pi$. Since $S^2 = (Q^2 + iR^2) + 2QR\zeta$, we find $S^2 P^3 = u\pi + 2\pi\zeta$. The $S'^2 P^3 = \pi(u - 2\zeta)$; multiplying them we find $S^2 S'^2 P^6 = \pi^2(u^2 - 4i)$, hence $v^2 \pi^2 = S^2 S'^2 P^2 = \pi(u^2 - 4i)$. Thus we finally get $v^2 \pi = u^2 - 4i$, and in particular we see that $uv \neq 0$. The fact that $u \neq 0$ will also follow from our determination of the residue class $u \bmod \Pi$, which is our next task.

To this end we recall that the elements $\phi(t/\pi)$ are all conjugate in $L = \mathbb{Q}(i, \phi(1/\pi))$, so the quotient Q/R is a unit in $\mathcal{O}_L$. Our basic claim is that

$\phi(t/\pi)/\phi(1/\pi) \equiv t \bmod \phi(1/\pi)$. For a proof, we first observe that we may assume that t is primary: if it has even norm, replace it by $t + \pi$, and if it still isn't primary, multiply by a suitable fourth root of unity. Now equation (8.12) says that $\phi(t/\pi)/\phi(1/\pi)$ is the quotient of polynomials $W(x)$ and $V(x)$ in $x = \phi(1/\pi)$; since the constant terms of $W(x)$ and $V(x)$ are t and 1, respectively, we find $W(x)/V(x) \equiv t \bmod x$, and our claim follows.

The congruence we just proved implies that $Q/R \equiv \prod_{t \in M^+} t / \prod_{t \in M^-} t \equiv \zeta_M \bmod \phi(1/\pi)$, hence $Q/R \equiv \zeta \bmod \mathfrak{P}$, where $\mathfrak{P}$ is the ideal generated by Π and $\phi(1/\pi)$ in $L(\zeta)$. This implies $u = (Q^2 + iR^2)/P = Q/R + iR/Q \equiv \zeta + i\zeta^{-1} = 2\zeta \bmod \mathfrak{P}$, and since both u and 2ζ are elements of $\mathbb{Z}[\zeta]$, we conclude that $u \equiv 2\zeta \bmod \Pi$.

Now let us determine $u \bmod 2$. We start with the congruence $\phi(1/\pi)^4 \equiv -1 \bmod 2 + 2i$ proved in Exercise 8.27. From the definition of Q and R we then get $Q^4 \equiv R^4 \equiv (-1)^{(p-1)/8} \bmod 2 + 2i$, hence $S^4 \equiv Q^4 - R^4 + 2Q^2 R^2 i \equiv -2P^2 \bmod 2 + 2i$. Squaring gives $S^8 \equiv 4P^4 \equiv 4\pi \bmod 4 + 4i$, hence $(1+i)^4 \parallel S^8$. Moreover, the equation $S^8 = \pi(u + 2\zeta)^4$ implies that $u \equiv 1 + i \bmod 2$, and from $v^2\pi = u^2 - 4i$ we then get $v \equiv 1 + i \bmod 2$. Note that this implies that $\frac{1}{2}(u + v\sqrt{\pi})$ is integral and therefore a unit in $\mathbb{Q}(i, \sqrt{\pi})$ with relative norm i; this was already noticed by Eisenstein [202, footnote p. 596].

Define $\alpha \in \mathbb{Z}[i]$ by putting $u + 2\zeta = (1-i)(\alpha + (1+i)\zeta)$. Since the left hand side is exactly divisible by $1 + i$, we must have $\alpha \equiv 1 \bmod 1 + i$, hence $\alpha \equiv 1, i \bmod 2$. In both cases, $\alpha + (1+i)\zeta \equiv \gamma^2 \varepsilon \bmod 2$ for some eighth root of unity γ, and since $\varepsilon = 1 + \sqrt{2} = (1+\zeta)^2/(1-i)$, we conclude that $u + 2\zeta = (1+\zeta)^2 \gamma^2 \beta$ for some $\beta \in \mathbb{Z}[\zeta]$ that is $\equiv 1 \bmod 2$. Since any such β can be made primary by multiplication with some even power of a unit, we conclude that $(u + 2\zeta) = (1+\zeta)^2 \Delta$ for some primary $\Delta \in \mathbb{Z}[\zeta]$.

Forming the norm to $\mathbb{Z}[i]$, this implies $v^2\pi = u^2 - 4i = (1-i)^2 \delta$ for some primary $\delta \in \mathbb{Z}[i]$. Unique factorization in $\mathbb{Z}[i]$ then shows that $\delta = \pi\gamma^2$ for some $\gamma \in \mathbb{Z}[i]$. But since Δ and its conjugate Δ' are coprime, $\Delta\Delta' = \pi\gamma^2$ implies that $\Delta = \Pi_1 \Gamma^2$ or $\Delta = \Pi_5 \Gamma^2$. From the congruence $u \equiv 2\zeta \bmod \Pi$ we then conclude that the second possibility holds, and we have proved that $(u + 2\zeta) = (1+\zeta)^2 \Pi_5 \Gamma^2$ for some $\Gamma \in \mathbb{Z}[\zeta]$. This in turn shows that $S^2 P^3 = \pi(u + 2\zeta) = (1+\zeta)^2 \Pi_1 \Pi_5^2 \Gamma^2$, that is, $S^8 = (S^2 P^3)^4/\pi^3 = (1+\zeta)^8 \Pi_1 \Pi_5^5 \Gamma^8$.

Theorem 9.17. *We have $S^8 = \Gamma^8 (1+\zeta)^8 \Pi_1 \Pi_5^5$ for some $\Gamma \in \mathbb{Z}[\zeta]$ with odd norm.*

9.4 The Octic Reciprocity Law

The reciprocity law necessary for inverting octic power residue symbols with one entry in $\mathbb{Z}[i]$ was first proved by Eisenstein and lies much deeper than Proposition 9.4:

Theorem 9.18. *Let $\alpha \in \mathbb{Z}[i]$ and $\beta \in \mathbb{Z}[\zeta]$ be relatively prime and primary, and let N denote the norm for $\mathbb{Q}(\zeta)/\mathbb{Q}$. Then*

$$\left[\frac{\alpha}{\beta}\right] = (-1)^{\frac{N\alpha-1}{8}\frac{N\beta-1}{8}}\left[\frac{\beta}{\alpha}\right].$$

Proof. We may assume without loss of generality that $\alpha = \Pi$ and $\beta = \Lambda$ are prime in $\mathbb{Z}[\zeta]$. Since the claim is elementary (Proposition 6.11) when both entries are in $\mathbb{Z}[i]$, we may also assume that the norm of Π is a prime $p = 8n + 1$. Let $\chi = [\,\cdot\,/\Pi]_8$ be the octic residue character modulo Π. There are two cases to consider:

1. $N\Lambda = q$ is also a prime $\equiv 1 \bmod 8$. With $\lambda = \Lambda_1\Lambda_5$ we get $S^q \equiv \chi(\lambda)S \bmod \Lambda$, hence $\chi(\Lambda) \equiv (S^8)^{(q-1)/8} \equiv [S^8/\Lambda]_8 \bmod \Lambda$. The prime factorization $S^8 = (1+\zeta)^8 \Pi_1 \Pi_5^5 \Gamma^8$ then implies that $[S^8/\Lambda]_8 = [\Pi_1\Pi_5^5/\Lambda]_8$. But $[\Pi_5^5/\Lambda]_8 = [\Pi_1/\Lambda_5]^5$, and so $[\lambda/\Pi]_8 = \chi(\lambda) = [\Pi_1\Lambda_1]_8[\Pi_1\Lambda_5]_8 = [\Pi_1/\lambda]_8$ as claimed.

2. $N\Lambda = q^2$ for some prime $q = 8m + 5$. Then $S^{q+3} \equiv \Gamma^{q+3}(\Pi_1\Pi_5^5)^{m+1}$ and $S^q \equiv \chi(\lambda)^5 S' \bmod \lambda$ give $S^3 S'\chi(\lambda)^5 \equiv \Gamma^{q+3}(\Pi_1\Pi_5^5)^{m+1} \bmod \lambda$. Since $S^3 S' = \chi(-1)\Gamma^3\Gamma_5\Pi_1\Pi_5^2$, this implies

$$\chi(-1)\Gamma_5\chi(\lambda)^5 \equiv \Gamma^q\Pi_1^m\Pi_5^{5m+3} \bmod \lambda.$$

Using the facts that $\Gamma^q \equiv \Gamma_5 \bmod \lambda$ and $\Pi_5^q \equiv \Pi_5 \bmod \lambda$, we get $\chi(-\lambda) \equiv [\Pi_1/\lambda] \bmod \lambda$, and this implies our claim.

If you've been following the proof carefully, you will have noticed that we have been cheating: since we don't know the prime factors of Γ, we cannot conclude that $[\Gamma/\Lambda]^8 = 1$, since Λ may just as well divide Γ. Here's Eisenstein's way around this last difficulty: assume that Λ is a prime of degree 1 with norm $\ell \neq p$ dividing Γ. Then $S^\ell = \chi(\lambda)S + \Lambda T$ for some integral T in $L(\zeta)$. Write this in the form $S = \chi(\lambda)^{-1}S^\ell - \Lambda T$, replace S^ℓ by $S^8 S^{\ell-8}$ and plug in $S^8 = \Lambda U$. Then $S = \Lambda S_1$, where $S_1 = \chi(\lambda)^{-1}S^{\ell-8}U - T$. Now S_1 is a $\mathbb{Z}[\zeta]$-linear combination of values $\phi(t/\pi)$ with the same properties as S (for example, the Frobenius of a primary prime λ acts via multiplication by $\chi(\lambda)$ etc.). Repeating the proof in the last section shows that $S_1^8 = (1+\zeta)^8 \Pi_1 \Pi_5^5 \Gamma_1^8$ with $\Gamma = \Lambda\Gamma_1$. Repeating this process we eventually get a sum S_n such that Γ_n is not divisible by any prime $\neq \Pi_1, \Pi_5$. This does it. □

Goldscheider gave an elementary (but very technical) proof of Theorem 9.18 and used it to deduce the complete octic reciprocity law using the method indicated by Eisenstein that we have discussed at the beginning of Section 9.3:

Theorem 9.19. *Suppose that $\alpha, \beta \in \mathbb{Z}[\zeta]$ are relatively prime and primary; let N_1, N_2 and N_3 denote the relative norms of the extensions $\mathbb{Q}(\zeta)/\mathbb{Q}(i)$, $\mathbb{Q}(\zeta)/\mathbb{Q}(\sqrt{-2})$ and $\mathbb{Q}(\zeta)/\mathbb{Q}(\sqrt{2})$, respectively, and write $N_1\alpha = a(\alpha)^2 + b(\alpha)^2$, $N_2\alpha = c(\alpha)^2 + 2d(\alpha)^2$, $N_3\alpha = e(\alpha)^2 - 2f(\alpha)^2$ etc.; then*

$$\left[\frac{\alpha}{\beta}\right] = (-1)^{\frac{N\alpha-1}{8}\frac{N\beta-1}{8}}\zeta^{d(\alpha)f(\beta)-d(\beta)f(\alpha)}\left[\frac{\beta}{\alpha}\right].$$

9. Octic Reciprocity

Moreover, we have the following supplementary laws:

$$\left[\tfrac{1-\zeta}{\alpha}\right] = \zeta^{(5a-5+5b+18d+b^2-2bd+d^4/2)/8},$$
$$\left[\tfrac{1+\zeta}{\alpha}\right] = \zeta^{(a-1+b+6d+b^2+2bd+d^4/2)/8},$$
$$\left[\tfrac{\zeta}{\alpha}\right] = \zeta^{\frac{N\alpha-1}{8}} = \zeta^{(a-1+4b+2bd+2d^2)/4},$$
$$\left[\tfrac{1+\zeta+\zeta^2}{\alpha}\right] = \zeta^{(a-1-2b+2d-2d^2)/4},$$
$$\left[\tfrac{\varepsilon}{\alpha}\right] = \zeta^{(d-3b-bd-2d^2)/2}.$$

Using the supplementary law for the real unit $\varepsilon = 1+\sqrt{2}$ of Theorem 9.19 we will prove the octic analogue of Proposition 6.10; recall that $A_q = 2^q - 2^{\frac{q+1}{2}} + 1$ and $B_q = 2^q + 2^{\frac{q+1}{2}} + 1$.

Proposition 9.20. *Let $q, u \in \mathbb{N}$ be odd, and let $p = 8qu+1 = a^2+b^2 = c^2+2d^2$ be prime, where $4 \parallel b$. Assume moreover that $p \mid S_q = 2^{2q}+1$ (this is equivalent to $2^{(p-1)/u} \equiv 1 \bmod p$; if $u=1$, this is trivially true). Then*

$$p \mid A_q \iff d + \tfrac{b}{4}u \equiv \pm 1 \bmod 8,$$
$$p \mid B_q \iff d + \tfrac{b}{4}u \equiv \pm 3 \bmod 8.$$

Proof. $4 \parallel b$ shows that 2 is quadratic residue and biquadratic non-residue mod p; this implies $2^{\frac{p-1}{4}} \equiv 2^{2qu} \equiv -1 \bmod p$. Together with the congruence $2^{(p-1)/u} \equiv 1 \bmod p$ this implies that $2^{(p-1)/4u} \equiv -1 \bmod p$.

In order to determine which of the factors p divides, we observe

$$2 = (1+\zeta)^4(1-\sqrt{2})^2\zeta^6 = (1-\zeta)^4(1+\zeta+\zeta^2)^2\zeta^4;$$

writing $\pi = A+B\zeta+C\zeta^2+D\zeta^3$ and keeping in mind that $N\pi = p \equiv 9 \bmod 16$ and $4 \parallel b$, we find that $C \equiv 2 \bmod 4$ and $B \equiv D \bmod 4$. Now

$$2^q = 2^{\frac{p-1}{8}} \equiv (-1)^{\frac{p-1}{8}} \left[\tfrac{1-\zeta}{\pi}\right]_2 \left[\tfrac{1+\zeta+\zeta^2}{\pi}\right]_4$$
$$= -i^{-(5a-5+5b+18d+b^2-2bd+d^4/2)/4} i^{(a-1-2b+2d-2d^2)/4}$$
$$= -i^{b/4} \bmod \pi,$$

and similarly,

$$2^{\frac{q+1}{2}} = 2^{\frac{p+7}{16}} \equiv (-1)^{\frac{p+7}{16}}(1-\zeta)^2 \varepsilon \left[\tfrac{1-\zeta}{\pi}\right]_4 \left[\tfrac{\varepsilon}{\pi}\right]_8$$
$$\equiv -(-1)^{\frac{p+7}{16}} \zeta(1+i) \left[\tfrac{1-\zeta}{\pi}\right]_4 \left[\tfrac{\varepsilon}{\pi}\right]_8 \bmod \pi.$$

Applying the supplementary laws of octic reciprocity we get, after simplifying, $\left[\tfrac{\zeta}{\pi}\right]_4 \left[\tfrac{1-\zeta}{\pi}\right]_4 \left[\tfrac{\varepsilon}{\pi}\right]_8 = -\zeta^{(3b+4d)/4}$, where we have used that $16 \parallel b^2$ and $-2bd + 2d^2 + d^4/2 \equiv 0 \bmod 32$. Therefore

$$2^{\frac{q+1}{2}} \equiv -\zeta(1+i)i\zeta^{(3b+20d)/4} \equiv \zeta(1+i)i\zeta^{d-\frac{b}{4}} \bmod \pi,$$

and collecting everything we find

$$A_p = 2^q - 2^{\frac{q+1}{2}} + 1 \equiv -i^{b/4} + (1+i)i\zeta^{1-(d+b/4)} + 1 \bmod \pi.$$

Now we distinguish:

$d + \frac{b}{4} \bmod 8$	$\frac{b}{4} \bmod 4$		
$+1$	$+1$	$A_p \equiv$	$-i + (1+i)i + 1 \equiv 0 \bmod \pi$
-1	-1	$A_p \equiv$	$i - (1+i) + 1 \equiv 0 \bmod \pi$
$+3$	-1	$B_p \equiv$	$i - (1+i) + 1 \equiv 0 \bmod \pi$
-3	$+1$	$B_p \equiv$	$-i + (1+i)i + 1 \equiv 0 \bmod \pi$

Our claim follows. □

9.5 Scholz's Octic Reciprocity Law

Scholz's reciprocity law says that $\left(\frac{\varepsilon_p}{q}\right)_2 = \left(\frac{\varepsilon_q}{p}\right)_2$ for primes $p \equiv q \equiv 1 \bmod 4$ such that $\left(\frac{p}{q}\right) = 1$; if we assume that $\left(\frac{\varepsilon_p}{q}\right)_2 = 1$, it is natural to ask if we can determine $\left(\frac{\varepsilon_p}{q}\right)_4$. The following theorem answers this question:

Theorem 9.21. *Let $p \equiv q \equiv 1 \bmod 8$ be primes such that $\left(\frac{p}{q}\right)_4 = \left(\frac{q}{p}\right)_4 = 1$. Then*

$$\left(\frac{p}{q}\right)_8 \left(\frac{q}{p}\right)_8 = \begin{cases} \left(\frac{\varepsilon_p}{q}\right)_4 \left(\frac{\varepsilon_q}{p}\right)_4, & \text{if } N\varepsilon_{pq} = -1; \\ (-1)^{h(pq)/4} \left(\frac{\varepsilon_p}{q}\right)_4 \left(\frac{\varepsilon_q}{p}\right)_4, & \text{if } N\varepsilon_{pq} = +1. \end{cases}$$

Here $h(pq)$ and ε_{pq} are the class number and the fundamental unit of $\mathbb{Q}(\sqrt{pq})$, respectively.

Since the statement involves the class number $h(pq)$ of $\mathbb{Q}(\sqrt{pq})$, it seems reasonable to suspect that we need the class number formula for its proof. Nevertheless we will show that the case $N\varepsilon_{pq} = -1$ can be treated arithmetically; the proof for $N\varepsilon_{pq} = +1$ will use the class number formula.

The main ingredient of the proof is the formula

$$\varepsilon_{pq}^{-h} = \pm N_{\mathbb{Q}(\zeta_{pq})/\mathbb{Q}(\sqrt{pq})}(1 - \zeta_{pq}), \tag{9.7}$$

where $\varepsilon_{pq} > 1$ is the fundamental unit of $\mathbb{Q}(\sqrt{pq})$ and where h is an integer. In fact, the class number formula states that h is the class number (in the usual sense) of $\mathbb{Q}(\sqrt{pq})$, but we will not need this for our proof in the case $N\varepsilon_{pq} = -1$. Equation (9.7) is proved by observing that $1 - \zeta_{pq}$ is a unit since the conductor pq is composite.

308   9. Octic Reciprocity

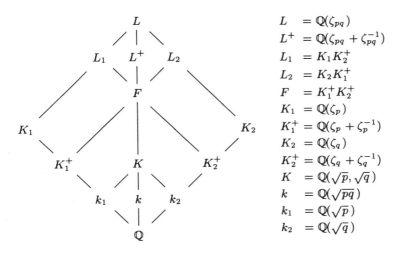

Fig. 9.1. Some subfields of $\mathbb{Q}(\zeta_{pq})$

The proof of Scholz's octic reciprocity law consists basically in a computation of norms in subfields of $\mathbb{Q}(\zeta_f)$. Figure 9.1 shows the Hasse diagram for $L = \mathbb{Q}(\zeta_{pq})$; it contains only the fields that occur in the proof.

In the first part of the proof it is sufficient to assume that $p \equiv q \equiv 1 \bmod 4$ are primes such that $\left(\frac{p}{q}\right) = 1$. Put $\lambda_p = (1 - \zeta_p)(1 - \zeta_p^{-1})$ (this is the norm of $1 - \zeta_p$ to the maximal real subfield of $\mathbb{Q}(\zeta_p)$), and define λ_q similarly. Our first claim is

$$N_{L/k}(1 - \zeta_{pq}) = N_{F/K}(\lambda_p - \lambda_q)^4. \tag{9.8}$$

For a proof, observe first that $\zeta_{pq}^{p+q} = \zeta_p \zeta_q$; since $p + q$ is a quadratic residue modulo p and q, this implies that ζ_{pq} and $\zeta_p \zeta_q$ are conjugate over $\mathbb{Q}(\sqrt{p}, \sqrt{q})$, and we find $N_{L/k}(1 - \zeta_{pq}) = N_{L/k}(1 - \zeta_p \zeta_q)$. Now the identity

$$N_{L/F}(1 - \zeta_p\zeta_q) = (1 - \zeta_p\zeta_q)(1 - \zeta_p\zeta_q^{-1})(1 - \zeta_p^{-1}\zeta_q)(1 - \zeta_p^{-1}\zeta_q^{-1}) = (\lambda_p - \lambda_q)^2$$

shows $N_{L/k}(1 - \zeta_{pq}) = N_{K/k}(N_{F/K}(\lambda_p - \lambda_q)^2)$, thus it is sufficient to prove that $\alpha := N_{L/K}(1 - \zeta_p\zeta_q) \in k$. To this end, let σ and τ denote the non-trivial automorphisms of K/k_1 and K/k_2, respectively. Then $\alpha^{1+\sigma} = N_{K/k_1}\alpha = N_{L/k_1}(1 - \zeta_p\zeta_q) = N_{K_1/k_1}\left(\frac{1-\zeta_p^q}{1-\zeta_p}\right) = 1$, since ζ_p^q and ζ_p are conjugate over k_1. Similarly, $\alpha^{1+\tau} = 1$, and now $\alpha^{\sigma\tau} = (\alpha^{-1})^\tau = \alpha$.

The "class number formula" (9.7), Equation (9.8) and the fact that ε_{pq} is positive now show that

$$\varepsilon_{pq}^{-h/4} = \sigma_{q,p} N_{F/K}(\lambda_p - \lambda_q), \tag{9.9}$$

where $\sigma_{q,p} = \pm 1$ is defined by $\sigma_{q,p} = \text{sign}(N_{F/K}(\lambda_p - \lambda_q))$. Observe that $\sigma_{q,p} = (-1)^{(p-1)(q-1)/16}\sigma_{p,q}$.

Scholz's octic reciprocity law will follow from comparing $N_{F/K}(\lambda_p - \lambda_q)$ modulo $(\sqrt{p})$ and $(\sqrt{q})$, respectively. In fact, we easily see that

$$N_{F/K}(\lambda_q - \lambda_p) \equiv N_{F/K}(\lambda_q) = (N_{K_2^+/k_2}\lambda_q)^{(p-1)/4}$$
$$= (N_{K_2/k_2}(1 - \zeta_q))^{(p-1)/4} \bmod \lambda_p,$$

hence modulo $\sqrt{p}$ (since $(\sqrt{p}) = (\lambda_p) \cap k_2$), and now Proposition 3.24 shows that

$$N_{F/K}(\lambda_p - \lambda_q) \equiv \left(\varepsilon_p^{-h(p)}\sqrt{p}\right)^{(q-1)/4} \bmod \sqrt{q}\mathcal{O}_K, \qquad (9.10)$$

where $h(p)$ is an odd integer (equal to the class number of $\mathbb{Q}(\sqrt{p})$ by the class number formula).

Assume that $p \equiv q \equiv 1 \bmod 8$; then we similarly get the congruence

$$N_{F/K}(\lambda_p - \lambda_q) \equiv \left(\varepsilon_q^{-h(q)}\sqrt{p}\right)^{(p-1)/4} \bmod \sqrt{p}\mathcal{O}_K, \qquad (9.11)$$

and if we assume that $(\varepsilon_p/q) = (p/q)_4 = 1$, then these congruences can be written as

$$N_{F/K}(\lambda_p - \lambda_q) \equiv \begin{cases} \left(\frac{\varepsilon_p}{q}\right)_4 \left(\frac{p}{q}\right)_8 \bmod \sqrt{q}, \\ \left(\frac{\varepsilon_q}{p}\right)_4 \left(\frac{q}{p}\right)_8 \bmod \sqrt{p}. \end{cases} \qquad (9.12)$$

The left hand side in these congruences is – up to sign – a power of $\varepsilon_{pq} = t + u\sqrt{pq}$. Now assume that $N\varepsilon_{pq} = -1$; then $\varepsilon_{pq}^2 \equiv t^2 \equiv -1 \bmod \sqrt{pq}$ shows two things: first, we must have $N_{F/K}(\lambda_q - \lambda_p) = \pm \varepsilon_{pq}^h/4$ for some even h, and second, the right hand sides of the congruences in (9.12) must coincide (since they are residue classes of an even power of ε_{pq} and congruent to $\pm 1 \bmod \sqrt{p}$ and $\sqrt{q}$). This proves Scholz's octic reciprocity law in the case $N\varepsilon_{pq} = -1$.

The proof in the case $N\varepsilon_{pq} = +1$ can be completed easily: first recall that $N\varepsilon_{pq} = +1$ if and only if the ideal $(p, \sqrt{pq})$ is principal, that is, if and only if $px^2 - qy^2 = \alpha = \pm 1$ has solutions in $\mathbb{Z}$. Now $\varepsilon_{pq} \equiv \alpha \bmod \sqrt{p}$ and $\varepsilon_{pq} \equiv -\alpha \bmod \sqrt{q}$ shows that the right hand sides of (9.12) coincide if and only if $h/4$ is even; this proves the claim once we know that the integer $h = h(pq)$ in (9.8) is the class number of $\mathbb{Q}(\sqrt{pq})$. We observe that we do not need the full power of the class number formula: the congruence $h \equiv h(pq) \bmod 8$ would suffice; maybe this can be proved using genus theory.

Remark. Although (9.9) is valid whenever $(p/q) = 1$, this does not imply that $4 \mid h$ in these cases, since ε_{pq} might become a square in K. This is exactly what happens whenever $N\varepsilon_{pq} = +1$.

NOTES

The first results on the octic character of small primes can be found in Reuschle's tables [678] and are partially due to Jacobi. Bickmore [52, 53] conjectured the correct formula for $(2/p)_8$, which was subsequently proved by Western [838] using Eisenstein's Octic Reciprocity Law in Proposition 9.4. Other proofs are due to Perott [650], Beeger [41] and Aigner [3]. For the octic character of small odd primes, see Halter-Koch [325]

In [542], von Lienen claimed that Western's criteria (9.10) for -3 and 5 to be octic powers modulo p were incorrect; this was an error on von Lienen's part who used the normalization $a > 0$, $c > 0$ instead of Western's $a \equiv c \equiv 1 \mod 4$. Later, von Lienen [544] showed that Western's criteria were equivalent to some results obtained by Hudson & K. S. Williams [380]; see also Burde [103], Evans [222, 226], and von Lienen [543]. Similar results for 16-th powers have been obtained by Goldscheider [315], Cunningham [143], Whiteman [843], and much later by Hasse [338, 339]. Other references on 2^r-th powers are the papers of Aigner [5], Aigner & Reichardt [9], Cunningham & Gosset [146], Hudson [377], Hudson & K.S. Williams [378, 382], and E. Lehmer [498]. A. Brauer [Br] used biquadratic and octic reciprocity for solving the problem of Sierpinski whether the sets of prime divisors of $2^r + 1$ and $2^{2s+1} - 1$ each contain infinitely many primes $\equiv 1 \mod 8$.

The rational octic reciprocity law (Proposition 9.3) was first proved by K.S. Williams [857] and Wu [874]. The corresponding result for 16th powers was obtained by Leonard & K. S. Williams [522], and the general case of 2^nth powers was discussed by Evans [223] and Furuta [257]; see also Burde [102] and Helou [355].

Proposition 9.5 was proved by Eisenstein [Eis] (for primes $p \equiv 3 \mod 8$), by Stickelberger [759] (for primes $p \equiv 7 \mod 8$) and by Berndt & Evans [BE] in the other cases.

Eisenstein was well aware that his special case of the octic reciprocity law could be generalized without substantial difficulty to the general octic reciprocity law; in [202, p. 618], he writes

> Man sieht, daß die eben kurz ihrem Wesen nach geschilderte Untersuchung keine principiellen Schwierigkeiten hat, und daß es nur noch auf die Herzählung der dabei sich ergebenden verschiedenen Fälle und deren Betrachtung im Einzelnen ankommt, um das allgemeinste Reciprocitätsgesetz in aller Vollständigkeit und in seiner definitiven Form aufstellen zu können. Dies letztere und die Angabe einiger anderen zu demselben Ziele führenden Methoden wird der Gegenstand einer, wie ich hoffe, bald erscheinenden, den achten Potenzresten insbesondere gewidmeten Abhandlung ausmachen.[3]

---

[3] It can be seen that the investigation whose essentials just have been described does not offer any principal difficulties, and that the formulation of the most general [octic] reciprocity law in its complete and final form boils down to an

It is a great loss that this publication was prevented by Eisenstein's untimely death.

Elliptic Gauss sums were introduced by Eisenstein [202] and instantly forgotten (or, more likely, have never been noticed at all. Even A. Weil who mentioned in his review of Eisenstein's Mathematical Papers that "it is not at all clear that we have even caught up with him", referring to the "p-adic investigation of lemniscatic functions" in [202], does not even mention them). There are only two papers I am aware of that mention elliptic Gauss sums, one by Pinch [Pin] (he explicitly computes some factorizations of elliptic Gauss sums, though not for elliptic functions with complex multiplication by $\mathbb{Z}[i]$), the other by Cassou-Noguès & M.J. Taylor [CT].

The complete octic reciprocity law was first stated and proved by F. Goldscheider [315]. Given that his paper appeared in a rather obscure journal, it is surprising that it has been noticed at all (it is even mentioned in Dickson's History, Vol. II, p. 58). The introduction to his article suggests that even cubic and biquadratic reciprocity laws were widely known outside the circle of university mathematics (Goldscheider was a school teacher):

> Jedermann weiß, daß die Beweise der Reziprozitätsgesetze für die quadratischen, kubischen und biquadratischen Reste mit Leichtigkeit aus der Kreisteilung erhalten werden können. Weniger bekannt scheint dagegen zu sein, daß auch noch für das Gesetz der 8. Reste die Mittel, welche die Kreisteilung bietet, hinreichen, ohne daß es nötig sei, wie Jacobi (vgl. Crelle's J. B. 19 und Monatsber. 1837) vermutete, wesentlich neue Prinzipien zu suchen; es scheint sogar das Gesetz selbst noch nirgends ausgesprochen zu sein, da auch die Eisensteinschen Untersuchungen mittelst der Lemniskatischen Funktionen wohl nicht zum Abschluß gelangt sind.[4]

After recalling the basic arithmetic properties of $\mathbb{Z}[\zeta_8]$, Goldscheider continues:

> Die in der Kreisteilung wurzelnden Untersuchungen ergeben nun als Hauptresultat die Sätze: ...[5]

---

enumeration of the different cases and their detailed consideration. This last problem as well as the presentation of several other methods leading to the same result will be the subject of a treatise that hopefully will appear soon and that will be dedicated especially to eighth power residues.

[4] Everybody knows that the proofs of the reciprocity laws for quadratic, cubic and biquadratic residues can be derived easily from cyclotomy. However it seems to be less known that the methods that cyclotomy offers suffice for the law of eighth powers, and that it is not necessary, as Jacobi (cf. Crelle's J. v. 19 and Monatsber. 1837) conjectured, to search for essentially new principles; in fact it seems that the law itself has not been published anywhere, since Eisenstein's investigations via the lemniscatic functions apparently have not been completed.

[5] The investigations rooting in cyclotomy now yield as our main result the following theorems: ...

... and then he lists, among others, Eisenstein's reciprocity law (Proposition 9.4), the quartic reciprocity law in $\mathbb{Z}[\zeta_8]$ (Proposition 6.11), as well as the supplementary laws of the full octic reciprocity law (Theorem 9.18).

Next Goldscheider derives Eisenstein's octic reciprocity law (Theorem 9.21) using only octic Gauss and Jacobi sums by solving a lot of auxiliary equations of type (1.4), and performing some massive computations (this takes up approximately the first half of his paper); the second half proceeds with the reduction mentioned at the beginning of Section 9.4, and on the final pages he gives a few applications.

Bohnicek [63] gave another proof of Goldscheider's result which was based on his own work [60, 62] on power reciprocity laws in number fields: in fact he deduces it from Hilbert's reciprocity law $\prod_{\mathfrak{p}} \left(\frac{\mu, \nu}{\mathfrak{p}} \right) = 1$ for octic residues by computing the norm residue symbols $\left(\frac{\mu, \nu}{1-\zeta} \right)$. As for Goldscheider's contributions, Bohnicek remarks:

> Er behauptet, man könne diese Sätze mittelst der Kreisteilung erhalten, gibt jedoch für diese Behauptung wie auch für einige weitere in der Arbeit vorkommende Behauptungen keinen Beweis.[6]

Bohnicek's observation that Goldscheider did not give proofs for each claim turned into something completely different in Beeger's 'review' [42]:

> I know that St. Bohnicek has written in a paper that Goldscheider did not prove his results and that he, Bohnicek, gives therefore another proof with the theory of ideals, but in that paper he does not give the demonstrations of the criteria for $(\frac{1-\zeta}{\pi})$, $(\frac{\varepsilon}{\pi})$, etc., and as he uses frequently results from two extensive memoirs of D. Hilbert, his paper is difficult to read.

Apparently Beeger also overlooked that Bohnicek does indeed prove the supplementary laws (he does, however, not show that his results agree with Goldscheider's). Beeger also asked if Goldscheider's results on the supplementary laws could be proved by cyclotomic methods alone. Although we will see in Chapter 10 that this is indeed possible, it would be highly desirable if we had details about Goldscheider's proof, because it is hard to imagine that he followed the quite technical calculations of Section 10.2.

Helou [352, 353] used explicit reciprocity laws to derive, among others, Eisenstein's law 9.18. The octic reciprocity law is also dealt with in Welmin [832, 833].

Scholz's octic reciprocity law was conjectured [95] and proved [96] by Buell & Williams, using a result of Bucher [91]. The proof we have given is essentially due to Stevenhagen [758], but both he and Bucher made use of the analytic class number formula even in the case $N\varepsilon_{pq} = -1$.

---

[6] He claims that these theorems can be proved with cyclotomy, however he does not give proofs neither for these claims nor for several other claims that occur in this paper.

Congruences involving binomial coefficients like those in Corollaries 6.6, 7.6 and Exercise 9.1 have been studied extensively in the last century; the first examples were provided by Gauss, others by Stern [St1, St2], Eisenstein [200, 201, Eis], Goldscheider [315], and Jacobi [396] (see also [398] and [400]). Cauchy [123] proved a congruence for x mod p, where $p \equiv 1$ mod m is prime and $4p^h = x^2 + my^2$, h being the class number of $\mathbb{Q}(\sqrt{-m})$. This was generalized by Stickelberger [759] to primes $p \equiv k$ mod m, where $(-m/k) = +1$.

Among the immense number of more recent papers, we mention those of Brewer [Bre], Buell, Hudson & Williams [93], Hudson & Williams [HW1, HW2], Rajwade [Raj], Yamamoto [Yam], and Whiteman [Wh1]. Here are some of the results: if $p = 8m + 3 = c^2 + 2d^2$ is prime, then $\binom{4m+1}{m} \equiv -2c(-1)^m$ mod p; this was already known to Gauss. A more complex result is Cauchy's congruence $\binom{10m}{m}\binom{10m}{3m} \equiv 4u^2$ mod p, where $p = 20m+1 = u^2 + 5v^2$.

The octic Jacobsthal sum $\Phi_4(D) = \sum_{x=1}^{p-1} \left(\frac{x}{p}\right)\left(\frac{x^4 + D}{p}\right)$ was studied by Whiteman [Wh2] for primes $p = c^2 + 2d^2 \equiv 1$ mod 8 (where c is normalized by $c \equiv 1$ mod 4) and showed that

$$\Phi_4(D) = \begin{cases} -4(-1)^{(p-1)/8}c & \text{if } (D/p)_8 = +1, \\ +4(-1)^{(p-1)/8}c & \text{if } (D/p)_8 = -1, \\ 0 & \text{if } (D/p)_4 = -1, \\ \pm 4d & \text{if } (D/p)_2 = -1. \end{cases}$$

Later, Hudson & Williams [381] removed the ambiguity in the last line by appealing to Western's formula. See also Evans [Eva] and the Notes in Chapter 6.

Burnside [104] and Evans [224] studied the octic period equation, Perott [650] determined the octic cyclotomic numbers.

Exercises

9.1 Let $p = 8m + 1 = a^2 + 4b^2 = c^2 + 2d^2$ be a prime. Show that $\binom{2m}{m}$, $\binom{3m}{m}$ and $\binom{4m}{m}$ are congruent to $\pm 2c$, $\pm 2a$ and $\pm 2c$ mod p, respectively, and determine the correct signs.

9.2 (Aigner [6]) Show that there are infinitely many primes $p \equiv 1$ mod 4 such that 2 has odd order in $(\mathbb{Z}/p\mathbb{Z})^\times$.

9.3 Here we complete the proof of Eisenstein's octic reciprocity law. We have to distinguish the cases $N\Pi = p^2$, $p \equiv 3, 5, 7$ mod 8.
   1. $p \equiv 3$ mod 8
      Here you have to use that $[q/\Pi] \equiv \left(q^{(p-1)/2}\right)^{(p+1)/4} \equiv (q/p)_\mathbb{Z}$ mod Π, where $(q/p)_\mathbb{Z}$ is the Legendre symbol in $\mathbb{Z}$. If $q \equiv 1$ mod 8, write $q = \Lambda_1 \Lambda_3 \Lambda_5 \Lambda_7$ and let Π' denote the prime factor of p conjugate to Π. Now show $[\Pi/q] = [\Pi/\Lambda_1]^4 [\Pi'/\Lambda_1]^4 = [p/\Lambda_1]^4 = (p/q)_\mathbb{Z} = (q/p)_\mathbb{Z} = [q/\Pi]$.
      If $q = \Lambda_1 \Lambda_5 \equiv 3$ mod 8, show that $[\Pi/q] = [p/\Lambda_1][\Pi'/\Lambda_1]^4$, and that $[p/\Lambda_1] = (p/q)$ and $[\Pi'/\Lambda_1]^4 = 1$.
      The cases $q \equiv 5, 7$ mod 8 are treated similarly.

2. $p \equiv 5 \bmod 8$
   First show that we may assume that $\Pi \in \mathbb{Z}[i]$; then prove that $[q/\Pi] = [q/\Pi]_4^3$, where $[\cdot/\cdot]_4$ is the quartic residue symbol in $\mathbb{Z}[i]$, and use the quartic reciprocity law to prove $[q/\Pi] = [\Pi/q]$ by imitating the proof in the case $p \equiv 3 \bmod 8$.
3. $p \equiv 7 \bmod 8$
   We claim that $[\Pi/q] = [q/\Pi] = +1$ in this case. Show first that $\Pi = e + f\sqrt{2}$, f even. Then verify the congruence $(q/\Pi) \equiv \left(q^{(p+1)/8}\right)^{p-1} \equiv 1 \bmod \Pi$. Put $\Pi' = e - f\sqrt{2}$; if $q = \Lambda_1 \Lambda_3 \Lambda_5 \Lambda_7 \equiv 1 \bmod 8$, show that

$$\left[\frac{\Pi}{q}\right] = \left[\frac{\Pi}{\Lambda_1}\right]\left[\frac{\Pi}{\Lambda_3}\right]\left[\frac{\Pi}{\Lambda_5}\right]\left[\frac{\Pi}{\Lambda_7}\right] = \left[\frac{\Pi}{\Lambda_1}\right]\left[\frac{\Pi'}{\Lambda_1}\right]^3\left[\frac{\Pi'}{\Lambda_1}\right]^5\left[\frac{\Pi}{\Lambda_1}\right]^7 = 1.$$

If $q \equiv 3, 5, 7 \bmod 8$, proceed similarly.

Next we want to prove the first supplementary law $[\zeta/a] = \zeta^{(a^2-1)/4}$. The congruence $\frac{a^2b^2-1}{8} \equiv \frac{a^2-1}{8} + \frac{b^2-1}{8} \bmod 8$ shows that it is sufficient to consider primes q. If $q \equiv 1 \bmod 8$, write $q = \Lambda_1 \Lambda_3 \Lambda_5 \Lambda_7$, deduce

$$\left[\frac{\zeta}{q}\right] = \left[\frac{\zeta}{\Lambda_1}\right]\left[\frac{\zeta}{\Lambda_3}\right]\left[\frac{\zeta}{\Lambda_5}\right]\left[\frac{\zeta}{\Lambda_5}\right] = \zeta^{(q-1)/2},$$

and show that this implies the claim. Then consider the other cases $q \equiv 3, 5, 7 \bmod 8$. Finally, prove the second supplementary law by doing similar computations.

9.4 Assume that $\alpha \equiv 1, 1 + \zeta + \zeta^3 \bmod 2$ for some $\alpha \in \mathbb{Z}[\zeta_8]$. Show that there is a real unit $\eta = \pm\varepsilon^{\pm 1}$ such that $\alpha\eta$ is primary.

9.5 Let p be an odd prime, Π a primary prime dividing p in $\mathbb{Z}[\zeta_8]$, $q = p^2$, and put $L = \mathbb{F}_q$ and $F = \mathbb{F}_p$. Let χ be the character of order 8 on $L^\times$ defined by $\chi = [\cdot/\Pi]$ if $p \equiv 3, 5, 7 \bmod 8$ and by $\chi = [\cdot/\Pi] \circ N_{L/F}$ if $p \equiv 1 \bmod 8$. Show that the restriction χ_F of χ to $F^\times$ is given by $\chi_F = \chi^{a+1}$ for primes $p \equiv a \bmod 8$. (Hint: $\chi(t) \equiv t^{(q-1)/8} \bmod \Pi$; now use $q-1 = p(p-a) + ap - 1$).

9.6 Let $\pi \in \mathbb{Z}[i]$ be a primary prime, and let M denote a $\frac{1}{4}$-system modulo π. Show that $\left(\prod_{t \in \mathbb{Z}[i]/(\pi)} t\right)^4 \equiv -1 \bmod 4$ by multiplying over pairs (t, t^{-1}) and observing that the map $t \mapsto t^{-1} \bmod \pi$ has exactly 2 fixed points. Show that this implies $\left(\prod_{t \in M} t\right)^4 \equiv (-1)^{(p-5)/4} \bmod \pi$.
Next assume that $p = N\pi = 8n+1$ is a prime. Define the subsets $M^+, M^- \subset M$ as in the text and show that $\left(\prod_{t \in M^+} t\right)^4 \equiv -1 \bmod \pi$ and $\left(\prod_{t \in M^-} t\right)^4 \equiv +1 \bmod \pi$.

9.7 (Continued from Exercise 6.12) Show that the n-th Fermat $F_n = 2^{2^n} + 1$ is prime if and only if $(-3)^{(F_n-1)/8} \equiv (-8)^{2^{n-2}} \bmod F_n$, and prove that this supports the conjecture made in Exercise 2.15.
(Hints: proceed as in Exercise 6.12 and use the octic reciprocity law from Theorem 9.19.)

9.8 Show that one gets Proposition 6.11 by squaring the octic reciprocity law in Theorem 9.19.

9.9 Let $p = 4a^2 + 3b^2 = c^2 + 6d^2 = e^2 - 2f^2 \equiv 7 \bmod 24$, $a \equiv b \equiv 1 \bmod 2$, and $a, b, c, d, e, f > 0$. Then

$$(2+\sqrt{3})^{\frac{p+1}{8}} \equiv \left(\frac{2}{e}\right)\left(\frac{-1}{f}\right)\left(\frac{2}{a}\right)\frac{e}{2f} + \left(\frac{6}{a}\right)\left(\frac{-6}{c}\right)\left(\frac{-1}{d}\right)\frac{d}{c}\sqrt{3} \bmod p.$$

Actually I'm not really serious here. Of course it is possible to use the octic reciprocity law for proving this congruence, but unless you find a very clever trick which I missed, this will result in pages of horrible calculations (it is by no means impossible that there exist a simple proof; using results from the Exercises in Chapter 6 it can be shown that the congruence above is equivalent to

$$(2+\sqrt{3})^{\frac{p+1}{8}} + (2-\sqrt{3})^{\frac{p+1}{8}} \equiv \left(\frac{2}{a}\right) \cdot 2^{\frac{p+1}{4}} \bmod p,$$

which looks only half as horrible. Nevertheless it may be best to wait for Part II where we will see how to prove such results using class field theory and a computer. It is however easy to show that the congruence above implies the result in Exercise 6.30 for odd values of a.

9.10 Let q be a prime and put $p = 2^q - 1$; define a sequence S_n of integers by $S_1 = 4$ and $S_{n+1} = S_n^2 - 2$. Then p is prime if and only if $S_{q-1} \equiv 0 \bmod 2$. This implies that $S_{q-2} \equiv \pm 2^{(q+1)/2} \bmod p$; show that in fact $S_{q-2} \equiv \left(\frac{2}{a}\right)2^{(q+1)/2} \bmod p$. (Hint: observe that
i) $p = e^2 - 2f^2$ with $e = 2^{(q+1)/2} - 1$ and $f = 2^{(q-1)/2} - 1$;
ii) $S_{q-2} = (2+\sqrt{3})^{(p+1)/8} + (2-\sqrt{3})^{(p+1)/8}$.
Now use the preceding exercise).

9.11 Let $p = 4a^2 + 3b^2 = c^2 + 6d^2 = e^2 - 2f^2 \equiv 7 \bmod 24$, $a \equiv b \equiv 1 \bmod 2$, and $a, b, c, d, e, f > 0$. Then

$$(5+2\sqrt{6})^{\frac{p+1}{8}} \equiv -\left(\frac{2}{de}\right)\left(\frac{-1}{f}\right)\frac{e}{2f} + \left(\frac{6}{c}\right)\left(\frac{2}{d}\right)\left(\frac{-3}{a}\right)\left(\frac{-1}{b}\right)\frac{b}{2a}\sqrt{6} \bmod p.$$

9.12 Let $p = 4a^2 + 7b^2$ be prime, and suppose that $4 \mid a$. Then

$$(8+3\sqrt{7})^{\frac{p+1}{8}} \equiv (-1)^{a/4}\left(\frac{2}{b}\right) \bmod p.$$

Show that $b \equiv \pm 3 \bmod 8$ if p is a Mersenne prime.

9.13 Let $p \equiv 5 \bmod 8$ be prime, and write $p = a^2 + b^2$, where $a \equiv 3$ and $b \equiv 2 \bmod 4$. Then

$$(1+\sqrt{2})^{\frac{p+1}{2}} \equiv \left(\frac{2}{a+b}\right)\frac{a}{b} \bmod p.$$

Additional References

[BE] B.C. Berndt, R. J. Evans, *Sums of Gauss, Eisenstein, Jacobi, Jacobsthal, and Brewer*, Ill. J. Math. **23** (1979), 374–437

[Br] A. Brauer, *A note on a number theoretical paper of Sierpinsky*, Proc. Amer. Math. Soc. **11** (1960), 406–409

[Bre] B. W. Brewer, *On certain character sums*, Trans. Amer. Math. Soc. **99** (1961), 241–245

[CT] Ph. Cassou-Noguès, M.J. Taylor, *Un élément de Stickelberger quadratique*, J. Number Theory 37 (1991), 307–342

[Eis] G. Eisenstein, *Zur Theorie der quadratischen Zerfällung der Primzahlen $8n+3$, $7n+2$ und $7n+4$*, J. Reine Angew. Math. **37** (1848), 97–126; Werke II, 506–535

[Eva] R.J. Evans, *Determinations of Jacobsthal sums*, Pac. J. Math. **110** (1984), 49–58

[HW1] R.H. Hudson, K.S. Williams, *Binomial coefficients and Jacobi sums*, Trans. Amer. Math. Soc. **281** (1984), 431–505

[HW2] R.H. Hudson, K.S. Williams, *Cauchy-type congruences for binomial coefficients*, Proc. Amer. Math. Soc. **85** (1982), 169–174

[Pin] R. Pinch, *Galois module structure of elliptic functions*, Computers in mathematical research, Cardiff 1986, 69–91 (1988)

[Raj] A.R. Rajwade, *Certain classical congruences via elliptic curves*, J. London Math. Soc. (2) **8** (1974), 60–62

[St1] M. Stern, Jahresbücher f. wissensch. Critik (1831), 679

[St2] M. Stern, *Eine Bemerkung zur Zahlentheorie*, J. Reine Angew. Math. **32** (1846), 89–90

[Wh1] A. L. Whiteman, *A theorem of Brewer on character sums*, Duke Math. J. **30** (1963), 545–552

[Wh2] A. L. Whiteman, *Theorems analogous to Jacobsthal's theorem*, Duke Math. J. **16** (1949), 619–626

[Yam] K. Yamamoto, *On congruences arising from relative Gauss sums*, Number Theory and Combinatorics, Japan 1984, 423–446;

10. Gauss's Last Entry

It is well known that Gauss recorded many of his discoveries in a diary; it ends with the 'Last Entry' from July 9, 1814, which reads as follows[1] (see [Kle]):

> Observatio per inductionem facta gravissima theoriam residuorum biquadraticorum cum functionibus lemniscaticis elegantissime nectens. Puta si $a + bi$ est numerus primus, $a - 1 + bi$ per $2 + 2i$ divisibilis, multitudo omnium solutionum congruentiae
> $$1 \equiv xx + yy + xxyy \pmod{a + bi}$$
> inclusis $x = \infty, y = \pm i$; $x = \pm i, y = \infty$ fit $= (a-1)^2 + b^2$.

We will deal with the problem of counting infinite solutions later and remark that an observation by induction is simply a conjecture supported by numerical evidence; in modern notation the Last Entry is

Proposition 10.1. *Let $p = a^2 + b^2 \equiv 1 \bmod 4$ be a prime, and suppose that $a + bi \equiv 1 \bmod 2 + 2i$. Then the congruence $x^2 + y^2 + x^2 y^2 \equiv 1 \bmod p$ has exactly $p - 3 - 2a = (a-1)^2 + b^2 - 4$ solutions.*

At first sight this certainly doesn't look as if it were very interesting or deep. This chapter seeks to correct this first impression by showing how this result is connected to biquadratic reciprocity, elliptic functions, elliptic curves, and the Riemann Hypothesis.

10.1 Connections with Quartic Reciprocity

Let us take our curve

$$C_1 : x^2 + y^2 + x^2 y^2 = 1 \tag{10.1}$$

---

[1] I have made by induction the most important observation that connects the theory of biquadratic residues with the lemniscatic functions. Suppose that $a + bi$ is a prime number, $a - 1 + bi$ divisible by $2 + 2i$, then the number of all solutions to the congruence $1 \equiv x^2 + y^2 + x^2 y^2 \bmod a + bi$ including $x = \infty, y = \pm i$; $x = \pm i, y = \infty$ is $(a-1)^2 + b^2$.

over $\mathbb{F}_p$ and write it as $y^2(x^2 + 1) = 1 - x^2$; multiplying by $1 + x^2$ we find $[(1+x^2)y]^2 = 1-x^4$. Substituting $v = x$ and $w = (1+x^2)y$ we find $w^2 = 1-v^4$. Thus we have transformed (10.1) into a second curve

$$C_2 : w^2 = 1 - v^4. \tag{10.2}$$

The following table suggests a connection between the number of points over $\mathbb{F}_p$ on C_1 and C_2:

	3	5	7	11	13	17	19	23	29	31	37	41
N_1	4	4	8	12	4	12	20	24	36	32	36	28
N_2	4	6	8	12	6	14	20	24	38	32	38	30

Lemma 10.2. *We have* $N_1 = N_2$ *if* $p \equiv 3 \bmod 4$, *and* $N_1 = N_2 - 2$ *if* $p \equiv 1 \bmod 4$.

Proof. Consider the map

$$\Phi : C_1 \longrightarrow C_2 : (x,y) \longmapsto (x, (x^2+1)y).$$

Clearly Φ is injective. Moreover,

$$\Psi : C_2 \longrightarrow C_1 : (v,w) \longmapsto (v, w(v^2+1)^{-1})$$

is easily seen to be a well defined map on those points of C_2 for which $x^2 + 1 \neq 0$. If $p \equiv 3 \bmod 4$, then this expression is always nonzero; since $\Psi \circ \Phi$ and $\Phi \circ \Psi$ are the identity maps on C_1 and C_2, respectively, this implies that $N_1 = N_2$ in this case. If $p \equiv 1 \bmod 4$, on the other hand, we have $x^2 + 1 = 0$ for the two points $(i,0)$ and $(-i,0)$, where i is a solution of the congruence $x^2 \equiv -1 \bmod p$. We find that Φ and Ψ are bijections between C_1 and $C_2 \setminus \{(i,0),(-i,0)\}$, hence we have $N_1 = N_2 - 2$. □

It is therefore sufficient to count the number of points on the curve C_2 over $\mathbb{F}_p$. Let us first consider the case $p \equiv 3 \bmod 4$. Here we have to count the number of solutions of the congruence $w^2 \equiv 1 - v^4 \bmod p$; let us denote this quantity by $N(w^2 = 1 - v^4)$. Since $p \equiv 3 \bmod 4$, an element of $\mathbb{F}_p$ is a fourth power if and only if it is a square. Thus $N(w^2 = 1 - v^4) = N(w^2 = 1 - v^2)$. But the elements of $\mathbb{F}_p$ which can be written as $-v^2$ are just 0 and the quadratic nonresidues.

Let (10) denote the number of quadratic non-residues $a \in \mathbb{F}_p^\times$ such that $a + 1$ is a (nonzero) quadratic residue modulo p. Then we can write $a = -v^2$ (for exactly two different $v \in \mathbb{F}_p^\times$) and $a + 1 = w^2$ (for exactly two different $w \in \mathbb{F}_p^\times$), thus the number of solutions (v,w) of C_2 with $vw \not\equiv 0 \bmod q$ is simply $4 \cdot (10)$. Since $(0, \pm 1)$ and $(\pm 1, 0)$ are also solutions, we find $N(w^2 = 1 - v^4) = 4 \cdot (10) + 4$. In Exercise 2.22 you have shown (or if you didn't: we will prove a more general statement below) that $(10) = \frac{p-3}{4}$, and this gives $N(w^2 = 1 - v^4) = p + 1$.

10.1 Connections with Quartic Reciprocity   319

Let us now compute $N(w^2 = 1 - v^4)$ for primes $p \equiv 1 \bmod 4$. Making the transformation $w \longmapsto iw$ we see that $N(w^2 = 1 - v^4) = N(w^2 = v^4 - 1) = N(w^2 + 1 = v^4)$. Apart from the six solutions $(\pm i, 0)$, $(0, \pm i)$ and $(0, \pm 1)$ (where i is of course a solution of $x^2 = -1$ in $\mathbb{F}_p$) we have to count the number of quadratic residues a such that $a + 1$ is a biquadratic residue modulo p. This is exactly what Gauss did in his memoir on biquadratic residues.

In fact, let $p \equiv 1 \bmod 4$ be a prime, and choose a primary $\pi \in \mathbb{Z}[i]$ such that $p = \pi\bar{\pi}$. Let (r, s) denote the number of $a \in \mathbb{F}_p^\times$ such that $[\frac{a}{\pi}] = i^r$ and $[\frac{a+1}{\pi}] = i^s$. Then clearly $N(w^2 + 1 = v^4) = 8 \cdot (00) + 8 \cdot (20) + 6$. It is quite easy to determine these numbers (rs) using Jacobi sums, as we will see in the next section. Here we will sketch Gauss's own computation.

Let $p = 8n + 1$ be a prime and define subsets $A_r \subset \mathbb{F}_p^\times$ by $A_r = \{a \in \mathbb{F}_p^\times : [a/\pi] = i^r\}$. Then (rs) is the number of solutions of $\alpha_r + 1 = \alpha_s$ over $\mathbb{F}_p$ with $\alpha_j \in A_j$. Multiply this equation through by -1; observing that $\alpha'_j = -\alpha_j \in A_j$ for $\alpha_j \in A_j$ (since $p \equiv 1 \bmod 8$) we find $\alpha'_r = \alpha'_s + 1$. This shows $(rs) = (sr)$. Next we divide $\alpha_r + 1 = \alpha_s$ by α_r; this yields $1 + \alpha'_{-r} = \alpha'_{s-r}$, where the indices are to be read modulo 4, hence we get $(33) = (01)$, $(22) = (02)$, $(11) = (03)$, $(12) = (13) = (23)$.

Let us put $h = (00)$, $j = (01)$, $k = (02)$, $l = (03)$, and $m = (13)$; then we have

$$h + j + k + l = 2n - 1. \qquad (10.3)$$

In fact, A_0 has cardinality $\frac{p-1}{4}$, and one of them is -1. For all $\alpha_0 \in A_0 \setminus \{-1\}$ the element $\alpha_0 + 1$ lies in exactly one of the sets A_0, A_1, A_2, or A_3. This proves the formula. Similarly, we can show

$$j + l + 2m = 2n \qquad (10.4)$$
$$k + m = n \qquad (10.5)$$

Now we need

$$hm + j^2 + kl + lm = jk + lm + km + m^2. \qquad (10.6)$$

Proof. Consider the equation

$$1 + \alpha_0 + \alpha_1 + \alpha_2 = 0 \qquad (10.7)$$

over $\mathbb{F}_p$, where, of course, $\alpha_r \in A_r$. Gauss counts its number of solutions in two different ways, thereby proving the claimed equality.

In fact, consider the case that $1 + \alpha_0 = \beta_0 \in A_0$: this occurs (00) times, and after dividing through by β_0, the equation $\beta_0 + \alpha_1 + \alpha_2 = 0$ becomes $1 + \alpha'_1 + \alpha'_2 = 0$, which has (12) solutions. Thus (10.7) has $(00)(12) = hm$ solutions which satisfy $1 + \alpha_0 \in A_0$.

Similarly, if $1 + \alpha_0 = \beta_1 \in A_1$, we find $\beta_1 + \alpha_1 + \alpha_2 = 0$, which becomes $1 + \alpha'_0 + \alpha'_1 = 0$ after dividing through by β_1, and we find $(01)(01) = j^2$

solutions. Treating the other cases $1 + \alpha_0 = \beta_2$ and $1 + \alpha_0 = \beta_3$ similarly, we find that the total number of solutions of (10.7) equals $hm + j^2 + kl + lm$.

On the other hand we can do the very same counting by considering $1+\alpha_1 = \beta_r$, $r = 0, 1, 2, 3$; in this case we find that (10.7) has $jk+lm+km+m^2$ solutions. This proves our claim. □

Now we are almost done. Subtracting (10.4) from (10.3) we get $h = 2m - k - 1$, and by eliminating h from (10.6) we find

$$0 = (k-m)^2 + j^2 + k(l-j-k) - m.$$

Subtracting (10.5) twice from (10.4) yields $k = \frac{1}{2}(l+j)$, and we get $4j^2 + 4k(l-j-k) = (l-j)^2$, i.e.

$$0 = 4(k-m)^2 + (l-j)^2 - 4m.$$

Now $4m = 2(k+m) - 2(k-m) = 2n - 2(k-m)$ from (10.5); plugging this into our last equation we get

$$2n = 4(k-m)^2 + 2(k-m) + (l-j)^2,$$

which also can be written in the form

$$p = 8n + 1 = [4(k-m) + 1]^2 + 4(l-j)^2.$$

This clearly implies that $a = 4(k-m)+1$ and $b = \pm 2(l-j)$, where $p = a^2 + b^2$ with $a \equiv 1 \bmod 4$.

It remains to determine b. To this end consider $S = \sum_{z=1}^{p-1}(z^4 + 1)^{(p-1)/4}$; we have $S = T - 1$, where $T = \sum_{z=0}^{p-1}(z^4+1)^{(p-1)/4}$. Expanding T using the binomial theorem we get $T \equiv -1 \bmod 2$, hence $S \equiv -2 \bmod p$. On the other hand we clearly have $S \equiv 4(00) + 4i(01) - 4(02) - 4i(03) \bmod \pi$, and we find $-2 \equiv -2a - 2bi - 2 \bmod \pi$. This shows that $\pi = a + bi$.

Similarly, we can compute the numbers (r, s) in the case $p = 8n + 5$; note, however, that here $1 + \alpha_r = \alpha_s$ is equivalent to $\alpha_{2+r} = 1 + \alpha_{2+s}$, hence we have e.g. $(00) = (22)$. Dividing the equation corresponding to (00) by an element in A_2 one also sees that $(00) = (20)$. Eventually, we get

Proposition 10.3. *Let $p = a^2 + b^2$ be prime, where $a \equiv 1 \bmod 4$. If the sets A_r are defined via the character $\chi = [\frac{\cdot}{a+bi}]$, then the values of $(00), \ldots, (13)$ are given in the following table:*

p	$8n+1$	$8n+5$
(00)	$\frac{1}{8}(4n - 3a - 5)$	$\frac{1}{8}(4n + a - 1)$
(01)	$\frac{1}{8}(4n + a - 2b - 1)$	$\frac{1}{8}(4n + a + 2b + 3)$
(02)	$\frac{1}{8}(4n + a - 1)$	$\frac{1}{8}(4n - 3a + 3)$
(03)	$\frac{1}{8}(4n + a + 2b - 1)$	$\frac{1}{8}(4n + a - 2b + 3)$
(13)	$\frac{1}{8}(4n - a + 1)$	$\frac{1}{8}(4n - a + 1)$

Now we can compute $N(w^2 = 1 - v^4)$: if $p = 8n + 1$, we get

$$N(w^2 = 1 - v^4) = 8 \cdot (00) + 8 \cdot (20) + 6 = 8n - 2a = p - 1 - 2a.$$

If $p = 8n + 5$, on the other hand, we find

$$N(w^2 = 1 - v^4) = 8 \cdot (00) + 8 \cdot (20) + 6 = 16 \cdot (00) + 6 = p - 1 + 2a.$$

This coincides with the claim in Proposition 10.1 because we assumed that $a \equiv 1 \bmod 4$ in Proposition 10.3, so we have to replace a by $-a$ if $p \equiv 5 \bmod 8$.

Gauss used the results of Proposition 10.3 to derive the biquadratic residue character of 2 (see Exercise 10.1).

10.2 Counting Points with Cyclotomic Numbers

In this section we aim at a generalization of Gauss's computation of the cyclotomic numbers (r, s) that we discussed for $m = 4$ in Section 10.1 to odd prime values of m. As applications of our results, we will compute the number of solutions of congruences and derive certain supplementary laws for ℓ-th power residues.

The first general results on reciprocity laws that Kummer was able to prove were the supplementary laws for ℓ-th power residues (see [464]). His first approach used cyclotomic methods; Kummer's proofs along these lines have never been published, and the statement of his results is erroneous, but the idea behind his sketch does lead to correct and interesting results. In particular, we will show that the supplementary laws may, at least in principle, be derived by cyclotomic methods alone.

Our aim is to compute the m-the power residue symbol $\chi(1 - \zeta_m)$, where $\chi = (\,\cdot\,/\mathfrak{p})$, and where $\mathfrak{p}$ is a prime ideal of degree 1 in $\mathbb{Q}(\zeta_m)$, that is, $p = N\mathfrak{p} = mn + 1$ for some $n \in \mathbb{N}$. Kummer's idea was to transplant this problem from $\mathbb{Z}[\zeta_m]$ to the much larger ring $\mathbb{Z}[\zeta_{p-1}]$. To this end, choose a prime ideal $\mathfrak{P}$ in $\mathbb{Z}[\zeta_{p-1}]$ above $\mathfrak{p}$ and observe that $\mathbb{Z}[\zeta_m]/\mathfrak{p} \simeq \mathbb{Z}[\zeta_{p-1}]/\mathfrak{P}$. The reason why this simplifies things is that there is a canonical system of representatives for the residue classes modulo $\mathfrak{P}$, namely the group of roots of unity μ_{p-1}.

This implies that, for every $\gamma \in \mathbb{Z}[\zeta_{p-1}] \setminus \mathfrak{P}$, there exists a unique root of unity $\omega(\gamma) \in \mu_{p-1}$ such that $\gamma \equiv \omega(\gamma) \bmod \mathfrak{P}$. Clearly this induces a multiplicative character (actually an isomorphism) $\omega : (\mathbb{Z}[\zeta_{p-1}]/\mathfrak{P})^\times \longrightarrow \mu_{p-1}$ defined modulo $\mathfrak{P}$. For any $\alpha \in \mathbb{Z}[\zeta_m] \setminus \mathfrak{p}$ we have $\chi(\alpha) \equiv \alpha^{(p-1)/m} \equiv \alpha^n \bmod \mathfrak{p}$, and in particular $\chi(\alpha) \equiv \omega^n(\alpha) \bmod \mathfrak{P}$.

Next we factorize $1 - \zeta_m$ in $\mathbb{Z}[\zeta_{p-1}]$ by putting $X = \zeta_{p-1}^i$ in the identity

$$1 - X^n = \prod_{s=0}^{n-1}(1 - \zeta_n^s X), \quad \text{which gives} \quad 1 - \zeta_m^i = \prod_{s=0}^{n-1}(1 - \zeta_{p-1}^{sm+i}),$$

and, in particular,

$$\chi(1 - \zeta_m^i) = \prod_{s=0}^{n-1} \omega^n(1 - \zeta_{p-1}^{sm+i}). \tag{10.8}$$

The $\omega^n(1 - \zeta_{p-1}^{sm+i})$ are roots of unity; more exactly, $\omega^n(1 - \zeta_{p-1}^{sm+i}) = \zeta_{p-1}^{nj}$ for some $j \in \{0, \ldots, p-1\}$. But

$$\omega^n(1 - \zeta_{p-1}^{sm+i}) = \zeta_{p-1}^{nj} \iff \omega(1 - \zeta_{p-1}^{sm+i}) = \zeta_{p-1}^{tm+j}$$
$$\iff 1 - \zeta_{p-1}^{sm+i} \equiv \zeta_{p-1}^{tm+j} \bmod \mathfrak{P}$$

for some t. Since $(\mathbb{Z}[\zeta_{p-1}]/\mathfrak{P})^\times \simeq \mathbb{Z}/p\mathbb{Z}$, there exists a $g \in \mathbb{Z}$ such that $\zeta_{p-1} \equiv g \bmod \mathfrak{P}$; clearly g is a primitive root modulo p, and $\mathfrak{P} = (p, g - \zeta_{p-1})$. This implies

$$\omega^n(1 - \zeta_{p-1}^{sm+i}) = \zeta_{p-1}^{nj} \iff 1 - g^{sm+i} \equiv g^{tm+j} \bmod p.$$

Let $N_{i,j}$ denote the number of $g^{sm+i} \bmod p$ that satisfy this last congruence. Then exactly $N_{i,1}$ factors of the product

$$\prod_{s=0}^{n-1} \omega(1 - \zeta_{p-1}^{sm+i})$$

have the form ζ_{p-1}^{tm+1}, exactly $N_{i,2}$ have the form ζ_{p-1}^{tm+2}, etc. If we put $S_i = N_{i,1} + 2N_{i,2} + \ldots + (m-1)N_{i,m-1}$, then (10.8) gives $\chi(1 - \zeta_m^i) = \zeta_m^{S_i}$.

This reduces the computation of $\chi(1 - \zeta_m^i)$ to counting the number N_{ij} of solutions of the congruence $1 - g^{sm+i} \equiv g^{tm+j} \bmod p$; these integers N_{ij} are called cyclotomic numbers. This will be done using Jacobi sums. Using the isomorphism $\mathbb{Z}/p\mathbb{Z} \simeq \mathbb{Z}[\zeta_{p-1}]/\mathfrak{P}$, we can write

$$-J(\chi^r, \chi^s) = \sum_{b=0}^{p-1} \chi^r(b)\chi^s(1-b) = \delta_r + \sum_{a=0}^{p-2} \omega^{nr}(\zeta_{p-1}^a)\omega^{ns}(1 - \zeta_{p-1}^a),$$

where δ_r denotes the 'Kronecker delta' $\delta_{r,0}$ which is 1 if $r = 0$ and 0 otherwise. But $\omega^{nr}(\zeta_{p-1}^a) = \zeta_{p-1}^{anr}$, and the situation $\omega^n(1 - \zeta_{p-1}^{sm+i}) = \zeta_{p-1}^{nj}$ occurs exactly $N_{i,j}$ times; this shows that

$$-J(\chi^r, \chi^s) = \delta_r + \delta_s + \sum_{i=0}^{m-1} \sum_{j=0}^{m-1} \zeta^{ri+sj} N_{i,j},$$

since the $N_{i,j}$ did not count the contribution $\chi^r(1)\chi^s(0) = \delta_s$. We have proved:

10.2 Counting Points with Cyclotomic Numbers

Proposition 10.4. *Let $p = mn + 1$ be prime, χ a character mod p of order m, and $\zeta = \zeta_m$. Then*

$$-J(\chi^r, \chi^s) = \delta_r + \delta_s + \sum_{i=0}^{m-1}\sum_{j=0}^{m-1} \zeta^{ri+sj} N_{i,j}.$$

The cyclotomic numbers $N_{i,j}$ introduced above satisfy a number of identities; the most obvious ones are

Proposition 10.5. *Suppose that $p = mn + 1$ is prime and let the $N_{i,j}$ be defined as above. Then*

1. $N_{i,j} = N_{j,i}$;
2. $N_{i,j} = N_{-i,j-i}$ *(with indices reduced modulo m) if m is odd;*
3. $\sum_{j=0}^{m-1} N_{i,j}$ *equals n or $n-1$ according as $i \neq 0$ or $i = 0$.*

We also remark that, for $m = 4$, we have $N_{r,s} = (s - r, -r)$, where the (r, s) are the cyclotomic numbers of order 4 defined in the last section.

Instead of expressing the Jacobi sums $J(\chi^r, \chi^s)$ in terms of the $N_{i,j}$, we can likewise compute the $N_{i,j}$ from the $J(\chi^r, \chi^s)$:

Proposition 10.6. *Let $p = mn + 1$ be prime and χ a character of order m on $\mathbb{F}_p$. Then*

$$m^2 \cdot N_{i,j} + m \cdot (\delta_i + \delta_j) = p - \sum_{r,s=1}^{m-1} \zeta^{-ri-sj} J(\chi^r, \chi^s).$$

Proof. Again, this is a straightforward calculation:

$$S := \sum_{r,s=0}^{m-1} \zeta^{-ri-sj} J(\chi^r, \chi^s) = \sum_{a=0}^{p-1}\sum_{s=0}^{m-1} \zeta^{-sj} \chi^s(1-a) \sum_{r=0}^{m-1} \zeta^{-ri} \chi^r(a).$$

The last sum equals 1 if $a = 0$, it vanishes for $a \neq 0$ such that $\chi(a) \neq \zeta^i$, and it equals m for $a \in A_i := \{a \in \mathbb{F}_p^\times : \chi(a) = \zeta^i\}$. Thus

$$S = m \sum_{a \in A_i}\sum_{s=0}^{m-1} \zeta^{-sj} \chi^s(1-a) + \sum_{s=0}^{m-1} \zeta^{-sj} \chi^s(1);$$

the last sum clearly equals $m\delta_j$, and applying the arguments used before for the sum over r we get $-S = m^2 \#A_{ij} + m\delta_i + m\delta_j$, where $A_{ij} = \{a \in \mathbb{F}_p^\times : \chi(a) = \zeta^i, \chi(1-a) = \zeta^j\}$. Since $\#A_{ij} = N_{i,j}$, this almost proves our claim. The only thing still missing is the remark that the sums over $r = 0$ or $s = 0$ only contribute $J(\mathbb{1}, \mathbb{1}) = -p$. □

Almost without knowing, we have determined the number of points on the Fermat curve $F_m : x^m + y^m = 1$ over $\mathbb{F}_p$; see Exercise 10.3.

324   10. Gauss's Last Entry

In Proposition 10.6 we have expressed the cyclotomic numbers $N_{k,h}$ as sums of $(m-1)(m-2)$ Jacobi sums (we regard the $m-1$ trivial sums $J(\chi^r, \chi^{m-r}) = \chi^r(-1)$ as trivial). From now on, let $m = \ell$ be an odd prime; then we can exploit known symmetries of Jacobi sums to express them as sums of only $\ell - 2$ Jacobi sums. We define $S_{i,j} = \ell^2 N_{i,j} + \ell(\delta_i + \delta_j)$; then

$$-S_{i,j} = -p + \sum_{r=1}^{\ell-1}\sum_{s=1}^{\ell-1} \zeta^{-ri-sj} J(\chi^r, \chi^s)$$

Substituting $s = rt$ and introducing the automorphism $\sigma_r : \zeta_m \longrightarrow \zeta^r$ gives

$$-S_{i,j} + p = \sum_{r=1}^{\ell-1}\sum_{t=1}^{\ell-1} \zeta^{-ri-rtj} J(\chi^r, \chi^{rt}) = \sum_{r=1}^{\ell-1} \sigma_r\left(\sum_{t=1}^{\ell-1} \zeta^{-i-tj} J(\chi, \chi^t)\right).$$

Since $J(\chi, \chi^{-1}) = \chi(-1) = 1$, the sum for $t = \ell - 1$ contributes

$$\sum_{r=1}^{\ell-1} \sigma_r(\zeta^{j-i}) = \ell\delta_{j-i} - 1,$$

and we find $\quad -S_{i,j} + p - \ell\delta_{j-i} + 1 = \sum_{r=1}^{\ell-1} \sigma_r\left(\sum_{t=1}^{\ell-2} \zeta^{-i-tj} J(\chi, \chi^t)\right).$

Next we define integers $a_{t,j}$ $(1 \le t \le \ell - 1)$ by writing $J(\chi, \chi^t) = a_{t,0} + a_{t,1}\zeta + \ldots + a_{t,\ell-2}\zeta^{\ell-2}$ and get

$$-S_{i,j} + p - \ell\delta_{j-i} + 1 = \sum_{r=1}^{\ell-1} \sigma_r\left(\zeta^{-i}\sum_{t=1}^{\ell-2} \zeta^{-tj} \sum_{h=0}^{\ell-2} a_{t,h}\zeta^h\right)$$

$$= \sum_{t=1}^{\ell-2}\sum_{h=0}^{\ell-2} a_{t,h} \sum_{r=1}^{\ell-1} \zeta^{r(h-i-tj)}.$$

The inner sum over r yields $\ell - 1$ if $h \equiv i + tj \bmod \ell$, and -1 otherwise; thus we get

$$\sum_{h=0}^{\ell-2} a_{t,h} \sum_{r=1}^{\ell-1} \zeta^{r(h-i-tj)} = (\ell - 1)a_{t,i+tj} - \sum_{h \ne i+tj} a_{t,h} = \ell a_{t,i+tj} - \sum_{h=0}^{\ell-2} a_{t,h},$$

where the index in $a_{t,i+tj}$ has to be reduced modulo ℓ and where $a_{t,\ell-1} = 0$ for $1 \le t \le \ell - 2$. This gives

$$S_{i,j} - p + \ell\delta_{j-i} - 1 = \sum_{t=1}^{\ell-2}\sum_{h=0}^{\ell-2} a_{t,h} - \ell\sum_{t=1}^{\ell-2} a_{t,i+tj}.$$

Thus we have proved the following formula for the cyclotomic integers $N_{i,j}$ in terms of the coefficients of the Jacobi sums $J(\chi, \chi^t)$:

10.2 Counting Points with Cyclotomic Numbers

Proposition 10.7. *Let ℓ and $p = \ell n + 1$ denote odd primes, and define integers $a_{t,j}$ and A by*

$$J(\chi, \chi^t) = \sum_{h=0}^{\ell-2} a_{t,h} \zeta^h, \quad A = \sum_{t=1}^{\ell-2} \sum_{h=0}^{\ell-2} a_{t,h};$$

then $\quad N_{i,j} = \dfrac{1}{\ell}\left(n + \dfrac{A+2}{\ell} - (\delta_i + \delta_j + \delta_{i-j}) - \sum_{t=1}^{\ell-2} a_{t,i+tj}\right).$

The simple proof that $A \equiv -2 \bmod \ell$ is left to the reader as Exercise 10.6.

Kummer's Supplementary Laws

Here we will milk the results of this section and derive an unfortunately not very explicit form of the supplementary laws for ℓ-th power reciprocity in $\mathbb{Q}(\zeta_\ell)$.

From $1 - \zeta = \zeta(\zeta^{\ell-1} - 1)$ we get $\chi(\zeta) = \chi(1-\zeta)\chi^{-1}(1-\zeta^{\ell-1}) = \zeta^{S(1)-S(\ell-1)}$ with

$$S(1) = N_{1,1} + 2N_{1,2} + \ldots + (\ell-1)N_{1,\ell-1},$$
$$S(\ell-1) = N_{\ell-1,1} + 2N_{\ell-1,2} + \ldots + (\ell-1)N_{\ell-1,1}.$$

From Proposition 10.5 we get $N_{\ell-1,i} = N_{1,i+1}$, hence $S(\ell-1) = N_{1,2} + \ldots + (\ell-2)N_{1,\ell-1} + (\ell-1)N_{1,0}$. Since $\ell - 1 \equiv -1 \bmod \ell$ this implies

$$S(1) - S(\ell-1) = N_{1,1} + N_{1,2} + \ldots + N_{1,\ell-1} - (\ell-1)N_{1,0}$$
$$\equiv N_{1,0} + N_{1,1} + \ldots + N_{1,\ell-1} = \tfrac{p-1}{\ell} = n \bmod \ell,$$

which is what we expected since $\chi(\zeta) = \zeta^n$. We get a non-trivial consequence if we do not replace $\ell - 1$ by -1:

$$S(1) - S(\ell-1) = N_{1,1} + N_{1,2} + \ldots + N_{1,\ell-1} - (\ell-1)N_{1,0}$$
$$= \sum_{j=1}^{\ell-1}(N_{1,j} - N_{1,0}) = \frac{1}{\ell}\sum_{j=1}^{\ell-1}\left(\sum_{t=1}^{\ell-2}(a_{t,1+tj} - a_{t,1}) + 1 - \delta_{1-j}\right)$$

by Proposition 10.7. We have proved:

Theorem 10.8 (Supplementary Laws). *Let ℓ and $p = \ell n + 1$ denote odd primes, and let χ be a character of order ℓ on $\mathbb{F}_p^\times$. Using the notation of Proposition 10.7, we have*

$$p - 1 \equiv \ell \cdot (S(1) - S(-1)) \equiv \sum_{h=1}^{l-2}\left(1 - \delta_{1-h} + \sum_{t=1}^{l-2}(a_{t,1+th} - a_{t,1})\right) \bmod \ell^2,$$

In particular, $\chi(\zeta) = \zeta^{\frac{p-1}{\ell}} = \zeta^{S(1)-S(-1)}$ and $\chi(1-\zeta) = \zeta^{S(1)}$ are determined by the coefficients of the Jacobi sums $J(\chi, \chi^t)$.

Example. Take $\ell = 3$; then $J(\chi,\chi) = a + b\rho \equiv 1 \bmod 3$ $J(\chi,\chi^2) = 1$, $A = a+b$, and e.g. $9N_{0,0} + 6 = p - (a+b\rho) - (a+b\rho^2) - 2 = p - 2a + b - 2$ in agreement with Proposition 10.7, which gives $N_{0,0} = \frac{1}{3}(\frac{p-1}{3} + \frac{a+b+2}{3} - 3 - a) = \frac{1}{9}(p - 2a + b - 8)$. Similar calculations yield the following table for $9N_{i,j}$:

	0	1	2
0	$p - 2a + b - 8$	$p + a - 2b - 2$	$p + a + b - 2$
1	$p + a - 2b - 2$	$p + a + b - 2$	$p - 2a + b + 1$
2	$p + a + b - 2$	$p - 2a + b + 1$	$p + a - 2b - 2$

This gives

$$\frac{p-1}{3} \equiv S(1) - S(2) = (N_{1,1} - N_{1,0}) + (N_{1,2} - N_{1,0}) \equiv \frac{1-a-b}{3} \bmod 3,$$

hence

$$S(1) = N_{1,1} + 2N_{1,2} = N_{0,2} + 2(N_{0,0} + 1) = \frac{p-a+b}{3}$$
$$= \frac{p-1}{3} + \frac{1-a+b}{3} \equiv \frac{a-1}{3} \bmod 3.$$

In particular, we find $\chi(\rho) = \rho^{(1-a-b)/3}$ and $\chi(1-\rho) = \rho^{(a+1)/3}$ in agreement with our results in Chapter 7.

Remark. The case $\ell = 3$ is particularly simple because here the Jacobi sum and the corresponding primary prime coincide; in general, the Jacobi sum will be a product of conjugates of primes, and these primes need not even be principal. Nevertheless, it is possible (though certainly tedious) to derive the supplementary laws at least if ℓ is regular, that is, if it does not divide the class number of $\mathbb{Q}(\zeta_\ell)$. A necessary ingredient, however, is the prime ideal factorization of Jacobi sums – this will be the subject of Chapter 11.

It would be nice to know if Goldscheider (see Section 9.4) computed the supplementary laws of octic reciprocity in a similar way, or whether he had a better method.

10.3 Counting Points with Jacobi Sums

Consider the curve $C : x^c + y^d = 1$; in Section 10.1 we have counted – in the special case $c = 4$, $d = 2$ – the number of solutions over finite fields with p elements. Here we will solve this problem over any finite field $\mathbb{F}_q$, $q = p^s$.

To this end, observe that

$$N_{c,d} := N(x^c + y^d = 1) = \sum_{a+b=1} N(x^c = a)N(y^d = b).$$

Our first step will therefore be the computation of $N(x^n = a)$.

10.3 Counting Points with Jacobi Sums   327

Proposition 10.9. *Let χ be a character of order n over $\mathbb{F}_q^\times$. Then the equation $x^n = a$ has exactly $\sum_{j=0}^{n-1} \chi^j(a)$ solutions in $\mathbb{F}_q$.*

Proof. If $a = 0$, there is only one solution and $\sum_{j=0}^{n-1} \chi^j(a) = 1$ (because $\mathbb{1}(0) = 1$). If $a \neq 0$, then $x^n = a$ is solvable if and only if $\chi(a) = 1$; if it is solvable, there are n solutions (since χ has order n, the map $\chi : \mathbb{F}_q^\times \longrightarrow \mu_n = \langle \zeta_n \rangle$ is surjective; thus any preimage of ζ_n is an n-th root of unity in $\mathbb{F}_q$, and multiplying one solution with n-th roots of unity one gets all n solutions), and $\sum_{j=0}^{n-1} \chi^j(a) = n$. Finally, if $\chi(a) \neq 1$, then there is no solution, and $\sum_{j=0}^{n-1} \chi^j(a) = 0$, as is easily seen from the standard trick of multiplying this sum by $\chi(a) \neq 1$. □

We can use this result by taking multiplicative characters χ and ψ of order c and d, respectively; we find

$$N_{c,d} = \sum_{a+b=1} \sum_{j=0}^{c-1} \chi^j(a) \sum_{k=0}^{d-1} \psi^k(b) = \sum_{j=0}^{c-1} \sum_{k=0}^{d-1} \sum_{a+b=1} \chi^j(a)\psi^k(b)$$

$$= \sum_{j=0}^{c-1} \sum_{k=0}^{d-1} \sum_{a \in \mathbb{F}_q} \chi^j(a)\psi^k(1-a)$$

$$= -\sum_{j=0}^{c-1} \sum_{k=0}^{d-1} J(\chi^j, \psi^k) = q - \sum_{j=1}^{c-1} \sum_{k=1}^{d-1} J(\chi^j, \psi^k),$$

where we have used $J(\mathbb{1}, \mathbb{1}) = -q$ and $J(\mathbb{1}, \chi) = 0$ for $\chi \neq \mathbb{1}$ (see Exercise 4.20). This reduces the computation of the number of solutions of C over $\mathbb{F}_p$ to the computation of certain Jacobi sums. In Gauss's example of Section 10.1, taking a prime $p \equiv 1 \bmod 4$ and a character χ of order 4 on $\mathbb{F}_p$ we get

$$N(y^2 = 1 - x^4) = p - 1 - J(\chi^2, \chi) - J(\chi^2, \chi^3),$$

because, according to Section 4.6, $J(\chi^2, \chi^2) = J(\chi^2, \chi^{-2}) = \chi^2(-1) = 1$. If we normalize $\pi = a + bi$ by $\pi \equiv 1 \bmod 2 + 2i$, then $J(\chi^2, \chi) = a$, and we get Gauss's result $N(y^2 = 1 - x^4) = p - 1 - 2a$.

But we can do more: let N_s denote the number of solutions of $x^c + y^d = 1$ over the finite field $\mathbb{F} = \mathbb{F}_{p^s}$. If we want to use Proposition 10.9, we have to take a character χ_s of order c and a character ψ_s of order d, both defined on $\mathbb{F}$, and compute the Jacobi sums $J(\chi_s^i, \psi_s^k)$.

It doesn't matter which character χ_s of order c on $\mathbb{F}$ we take; so pick any character χ of order c on $\mathbb{F}_p$ and consider $\chi_s = \chi \circ N_{\mathbb{F}/\mathbb{F}_p}$. We claim that χ_s does indeed have order c. In fact, assume that $\chi(y) = \zeta_c$, where ζ_c is a primitive c-th root of unity. Since the relative norm in extensions of finite fields is surjective (Exercise 4.28), there is an $x \in \mathbb{F}$ such that $y = N_{\mathbb{F}/\mathbb{F}_p}(x)$. Thus $\chi_s(x) = \chi(N_{\mathbb{F}/\mathbb{F}_p}(x)) = \chi(y) = \zeta_c$, hence χ has order divisible by c. On the other hand, χ_s^c is clearly the trivial character, so our claim follows.

Choosing χ_s as above we may apply the theorem of Davenport-Hasse for Jacobi sums, that is, Corollary 4.33 (actually the Davenport-Hasse theorem was proved with this application in mind), and we get

$$J(\chi_s^i, \psi_s^k) = J(\chi^i, \psi^k)^s. \tag{10.9}$$

Applying this to our example $C_2 : y^2 = 1 - x^4$, we find $N_s = p^s - 1 - \pi^s - \overline{\pi}^s$.

Proposition 10.10. *Let $p \equiv 1 \bmod 4$ be prime, and let $\pi \equiv 1 \bmod 2 + 2i$ be a primitive prime divisor of p in $\mathbb{Z}[i]$. Then the equation $y^2 = 1 - x^4$ has exactly $p^s - 1 - \pi^s - \overline{\pi}^s$ solutions in the finite field with p^s elements.*

In his Disquisitiones Arithmeticae, long before he published his results in the quartic case, Gauss computed the number $N(x^3 + y^3 = 1)$ for primes $p \equiv 1 \bmod 3$ in terms of the representation $4p = L^2 + 27M^2$, where L is normalized by $L \equiv 1 \bmod 3$. Gauss used the method of Section 10.1; let us apply our formulas to the slightly more general case $N(x^m + y^m = 1)$ for odd $m \geq 3$; using the fact that $J(\chi, \chi^{-1}) = \chi(-1) = 1$ we get

$$N(x^m + y^m = 1) = p - (m-1) - \sum_{j+k \neq m} J(\chi^j, \psi^k),$$

and using the Davenport-Hasse theorem as above yields

$$N_s(x^m + y^m = 1) = p^s - (m-1) - \sum_{j+k \neq m} J(\chi^j, \psi^k)^s. \tag{10.10}$$

Specializing to the case $m = 3$ we finally get

Proposition 10.11. *Let $p \equiv 1 \bmod 3$ be prime, and let $\pi \equiv 1 \bmod 3$ be a primitive prime divisor of p in $\mathbb{Z}[\rho]$, where $\rho^2 + \rho + 1 = 0$. Then the equation $x^3 + y^3 = 1$ has exactly $p^s - 2 - \pi^s - \overline{\pi}^s$ solutions in the finite field with p^s elements.*

Proof. Exercise 10.8. □

10.4 The Classical Zeta Functions

We promised to explain some connections between Gauss's Last Entry and the Riemann Hypothesis. This is hard to do without ζ-functions, so here goes our review of the algebraic number theorist's analytic tool-kit. For a very nice introduction into this area, the reader should consult Stark [Sta4].

Riemann's ζ-function Let $a_1, a_2, \ldots$ be an infinite sequence of complex numbers. Then the complex function $f(s) = \sum_{n=1}^{\infty} a_n n^{-s}$ is called the *Dirichlet series* associated to the sequence $\{a_n\}_{n \in \mathbb{N}}$. The Dirichlet series $\zeta(s) = \sum_{n=1}^{\infty} n^{-s}$ converges in the half-plane $\operatorname{Re} s > 1$ and is called the Riemann ζ-function. Euler used the divergence of $\zeta(1)$ to give a new proof of the infinitude of primes; he also found that the unique factorization theorem in $\mathbb{Z}$ implies the extremely important representation of $\zeta(s)$ as an *Euler product*:

$$\zeta(s) = \prod_{p \in \mathbb{P}} \frac{1}{1 - p^{-s}}, \quad \operatorname{Re} s > 1.$$

In Exercise 10.12 you are requested to verify that $(1 - 2^{1-s})\zeta(s) = 1 - 2^{-s} + 3^{-s} - 4^{-s} \pm \ldots$; since this series converges for all $s \in \mathbb{C}$ with positive real part, this allows us to extend Riemann's ζ-function to the half plane $\operatorname{Re} s > 0$. Actually, much more is true:

Theorem 10.12. *The ζ-function $\zeta(s) = \sum_{n=1}^{\infty} n^{-s}$ converges to a holomorphic function in the half plane $\operatorname{Re} s > 1$ and can be extended to a holomorphic function on $\mathbb{C} \setminus \{1\}$, with a simple pole at $s = 1$. Moreover, $\lim_{s \to 1}(s-1)\zeta(s) = 1$.* □

Riemann conjectured that all the zeros of $\zeta(s)$ inside the *critical strip* $0 \leq \operatorname{Re} s \leq 1$ lie on the line $\operatorname{Re} s = \frac{1}{2}$. The Riemann conjecture still is one of the central problems in number theory. The special role played by the critical line is stressed by the

Theorem 10.13. Functional Equation of Riemann's ζ-function.
Let $\Gamma(s)$ denote the Gamma function and put $\xi(s) = \pi^{-s/2}\Gamma(\frac{s}{2})\zeta(s)$. Then $\xi(s) = \xi(1-s)$. □

In other words: the line $\operatorname{Re} s = \frac{1}{2}$ is a symmetry axis for the slightly twisted ζ-function. Moreover it can be shown that $\xi(s)$ takes only real values on the critical line: this allows a numerical verification of the Riemann conjecture for, say, the first million zeros (counted according to their imaginary part) by first finding these zeros (using Newton's method, for example) and then showing that there are no others (by integrating around the critical strip using Cauchy's integral formula).

The following proposition is a celebrated result of Euler on the values of Riemann's ζ-function at positive even integers; define the Bernoulli numbers B_n by

$$\frac{te^t}{e^t - 1} = \sum_{n=0}^{\infty} B_n \frac{t^n}{n!}.$$

Proposition 10.14. *The B_n are rational numbers, and we have*

$$2\zeta(2m) = (-1)^{m+1} \frac{(2\pi)^{2m}}{(2m)!} B_{2m}. \tag{10.11}$$

Note that $B_1 = \frac{1}{2}$ by this definition, in contrast to the more familiar $B_1 = -\frac{1}{2}$. Also observe that $B_{2n+1} = 0$ for all $n \geq 1$ (Exercise 10.15). The following tables give the Bernoulli numbers $B_{2n} = r_{2n}/s_{2n}$ for $2n \leq 50$.

$2n$	r_{2n}	s_{2n}
2	1	6
4	-1	30
6	1	42
8	-1	30
10	5	66
12	-691	2730
14	7	6
16	-3617	510

$2n$	r_{2n}	s_{2n}
18	43867	798
20	$-283 \cdot 617$	330
22	$11 \cdot 131 \cdot 593$	138
24	$-103 \cdot 2294797$	2730
26	$13 \cdot 657931$	6
28	$-7 \cdot 9349 \cdot 362903$	870
30	$5 \cdot 1721 \cdot 1001259881$	14322
32	$-37 \cdot 683 \cdot 305065927$	510

The functional equation relates (10.11) with the values of the ζ-function at negative odd integers:

Corollary 10.15. *For all integers $m < 0$ we have $\zeta(-2m+1) = -\frac{1}{2m}B_{2m}$.*

Proof. Exercise 10.13 □

Zeta Functions can be used to count things: Riemann's ζ-function, for example, counts primes. In fact, Euler used the divergence of $\zeta(1)$ to derive Euclid's theorem on the infinitude of primes. Later Hadamard and de la Vallée Poussin showed that $\zeta(s)$ does not possess zeros on the line $\operatorname{Re} s = 1$ and used this fact to prove the

Theorem 10.16. Prime Number Theorem. *Let $\pi(x)$ denote the number of primes in the interval $[1, x]$. Then $\pi(x) \sim \frac{x}{\log x}$.* □

Here $f \sim g$ stands for $\lim_{x \to \infty} f(x)/g(x) = 1$. There are other formulations of the prime number theorem, involving the function $\operatorname{Li}(x) = \int_2^x \frac{dt}{\log t}$. Since $\frac{x}{\log x} \sim \operatorname{Li}(x)$, the prime number theorem can be expressed by $\pi(x) \sim \operatorname{Li}(x)$. Actually, the difference $\pi(x) - \operatorname{Li}(x)$ can be bounded, and the error term depends on the behaviour of the zeros of Riemann's ζ-function inside the critical strip. If the Riemann Conjecture were true, one could show that

$$\pi(x) = \operatorname{Li}(x) + O(x^{\frac{1}{2}+\varepsilon})$$

for every $\varepsilon > 0$ (the constants involved in the big-O notation depend on the choice of ε).

The proof of the prime number theorem has been simplified considerably lately by Newman [New]; see Zagier [Zag] and Koch [Koc]. A previously unpublished "one-page proof" of the prime number theorem by Witt has recently surfaced in his collected works [Wit].

Dedekind's ζ-function In analogy to Riemann's ζ-function, one can define for every number field K a ζ-function named after Dedekind, namely

$$\zeta(K,s) = \sum_{\mathfrak{a} \neq (0)} N\mathfrak{a}^{-s},$$

where the sum is over all nonzero integral ideals in $\mathcal{O}_K$. Obviously, Riemann's ζ-function is just Dedekind's ζ-function for $K = \mathbb{Q}$. Moreover, the Unique Factorization Theorem for prime ideals in the Dedekind ring $\mathcal{O}_K$ implies that Dedekind's ζ-function is a product of Euler factors:

$$\zeta(K,s) = \prod_{\mathfrak{p} \in \mathbb{P}_K} \frac{1}{1 - N\mathfrak{p}^{-s}}.$$

The statement analogous to Theorem 10.12 is

Theorem 10.17. *Let K be a number field. Dedekind's ζ-function $\zeta(K,s)$ converges to a holomorphic function in the half plane $\operatorname{Re} s > 1$ and can be extended to a meromorphic function on $\mathbb{C}$, with a simple pole at $s = 1$. Moreover,*

$$\lim_{s \to 1}(s-1)\zeta(K,s) = \frac{2^{r_1}(2\pi)^{r_2}hR}{w\sqrt{d}},$$

where r_1 and r_2 denote the number of real and complex valuations of K, h the class number, R the regulator, and where d is the absolute value of the discriminant of K. □

The computation of the residue at $s = 1$ was done by Dedekind, the continuation of $\zeta(K,s)$ to a meromorphic function in $\mathbb{C}$ was conjectured by Weber (who could even prove it assuming the existence theorem of class field theory) and is due to Hecke (1917); Landau (1903) showed by elementary means that it can be extended to $\operatorname{Re} s > 1 - \frac{1}{(K:\mathbb{Q})}$.

Just as for Riemann's ζ-function, there is a functional equation in the general case. For its formulation, we introduce the complete ζ-function

$$\xi(K,s) = \left(\frac{\Gamma(s/2)}{\pi^{s/2}}\right)^{r_1} \left(\frac{\Gamma(s)}{\pi^s}\right)^{r_2} \sqrt{d}^s \zeta(K,s).$$

Theorem 10.18. Functional Equation of Dedekind's ζ-function. *The complete ζ-function $\xi(K,s)$ satisfies $\xi(K,s) = \xi(K, 1-s)$.* □

Using the functional equation, we can transform Theorem 10.17 into a statement about the ζ-function at $s = 0$ (see Exercise 10.7):

Corollary 10.19. *For number fields K, we have $\lim_{s \to 0} s^{-r}\zeta(K,s) = -\frac{hR}{w}$.*

Dedekind's (as well as Riemann's) ζ-functions have a set of 'trivial' zeros in the half plane $\operatorname{Re} s < 0$; in fact

Proposition 10.20. *Let K be a number field with r_1 real and r_2 non-real valuations. Then Dedekind's ζ-function $\zeta(K, s)$ has zeros of multiplicity $r_1 + r_2$ at the even negative integers, and of multiplicity r_2 at the odd negative integers.* □

Note that a zero of multiplicity 0 isn't a zero at all, i.e. ζ-functions of totally real fields (e.g. Riemann's ζ-function) do not vanish at the odd negative integers (see Proposition 10.22). Just as for Riemann's ζ-function, there is a conjecture on the non-trivial zeros of Dedekind's ζ-function: the statement that all the zeros of $\zeta(K, s)$ inside the critical strip $0 \leq \operatorname{Re} s \leq 1$ lie on the critical line $\operatorname{Re} s = \frac{1}{2}$ is called the Extended Riemann Hypothesis (ERH) or the Generalized Riemann Hypothesis (GRH).

The prime number theorem can be generalized to number fields; for its formulation, we need a prime ideal counting function. For a number field K, let $\pi_K(x)$ (and $\pi'_K(x)$) denote the number of prime ideals of norm $< x$ and degree 1 over $\mathbb{Q}$; then

Theorem 10.21. *We have $\pi_K(x) \sim \pi'_K(x) \sim \frac{x}{\log x}$.* □

Again, the difference between $\pi_K(x)$ and $\frac{x}{\log x}$ can be bounded by an error term whose quality depends on the behaviour of the zeros of $\zeta_K(s)$ inside the critical strip.

Hurwitz's ζ-function Hurwitz [Hu1] introduced the ζ-function

$$\zeta(s, b) = \sum_{n=0}^{\infty} (b + n)^{-s}, \quad \operatorname{Re} s > 1,\ 0 < b \leq 1. \tag{10.12}$$

It is related to the partial ζ-function

$$\zeta(\sigma_a, s) = \sum_{\substack{n \geq 1 \\ n \equiv a \bmod m}} n^{-s} \tag{10.13}$$

associated to the automorphism $\sigma_a : \zeta_m \longmapsto \zeta_m^a$ of $\mathbb{Q}(\zeta_m)/\mathbb{Q}$ via the obvious identity $\zeta(\sigma_a, s) = m^{-s}\zeta(s, \frac{a}{m})$ for $0 < a \leq m$. Although these ζ-functions in general cannot be written as Euler products and do not possess functional equations (some authors suggest that such functions shouldn't be called zeta functions), they can be analytically continued to $\mathbb{C} \setminus \{1\}$, with a simple pole in $s = 1$.

The importance of Hurwitz's ζ-functions $\zeta(s, b)$ stems from the fact that their values at negative integers are rational numbers whenever b is rational. Since we are going to need generalized Bernoulli numbers anyway, this is a good place to introduce them: let χ be a Dirichlet character defined modulo m and write

$$\sum_{a=1}^{m-1} \frac{\chi(a) t e^{at}}{e^{mt} - 1} = \sum_{n=0}^{\infty} B_{n,\chi} \frac{t^n}{n!}. \tag{10.14}$$

10.4 The Classical Zeta Functions

Clearly $B_{n,\chi} = B_n$ for $\chi = \mathbb{1}$. The table below contains some Bernoulli numbers $B_{n,\chi}$ for characters χ defined modulo 3 and modulo 4:

n	$3B_{n,\chi}$ $(f=3)$	$2B_{n,\chi}$ $(f=4)$
1	-1	-1
3	2	3
5	$-2 \cdot 5$	-5^2
7	$2 \cdot 7^2$	$7 \cdot 61$
9	$-2 \cdot 809$	$-3^2 \cdot 5 \cdot 277$
11	$2 \cdot 11 \cdot 1847$	$11 \cdot 19 \cdot 2659$
13	$-2 \cdot 7 \cdot 13^3 \cdot 47$	$-5 \cdot 13^2 \cdot 43 \cdot 967$
15	$2 \cdot 5 \cdot 419 \cdot 16519$	$3 \cdot 5 \cdot 47 \cdot 4241723$

For studying generalized Bernoulli numbers, the Bernoulli polynomials are an indispensable tool. They are defined by

$$\frac{te^{tX}}{e^t - 1} = \sum_{n=0}^{\infty} B_n(X) \frac{t^n}{n!}. \tag{10.15}$$

The following properties of Bernoulli polynomials are easy to verify:

- $B_n(X) \in \mathbb{Q}[X]$
- $B_n(1) = B_n$, $B_n(1-X) = (-1)^n B_n(X)$;
- $B_n(X) = \sum_{i=0}^{n} \binom{n}{i} B_i X^{n-i}$;
- $B_{n,\chi} = \frac{1}{m} \sum_{a=1}^{m-1} \chi(a) B_n(\frac{a}{m})$, where m is any multiple of the conductor of χ.

Since $B_1(X) = X - \frac{1}{2}$, the last property immediately implies that

$$B_{1,\chi} = \frac{1}{m} \sum_{a=1}^{m-1} \chi(a) a \tag{10.16}$$

for any $\chi \neq \mathbb{1}$.

Proposition 10.22. *For $0 < b \leq 1$, we have $\zeta(1-n, b) = -\frac{1}{n} B_n(b)$. In particular, $\zeta(\sigma_a, 0) = \frac{1}{2} - \frac{a}{m}$ for $0 < a \leq m$.*

Proof. Hurwitz [Hu1], Washington [Was, Thm. 4.2]. □

From these results of Hurwitz we can deduce some well known formulas in an amazingly simple way: the definition of Hurwitz's ζ-function gives

$$\sum_{n=0}^{N} n^k = \zeta(-k, 0) - \zeta(-k, N+1) = \frac{B_{k+1}(N+1) - B_{k+1}(0)}{k+1}.$$

In fact, the the left equality sign holds for integers $k \leq -2$, but from the uniqueness of the extension of analytic functions it must also hold for positive integers. Certainly a nice proof of the summation formula for k-th powers! See Stark [Sta2] for more.

Dirichlet's L-series Dirichlet's L-series are defined as follows:

$$L(s,\chi) = \sum_{n\geq 1}\chi(n)n^{-s} = \sum_{a=1}^{m}\chi(a)\zeta(\sigma_a,s).$$

They can be viewed as Riemann's ζ-function twisted with a Dirichlet character χ. This series coincides with $\zeta(s)$ for $\chi = 1$ and converges for $\operatorname{Re} s > 1$; if $\chi \neq 1$, it even converges in the half plane $\operatorname{Re} s > 0$. Since Dirichlet characters are multiplicative functions, L admits an Euler factorization

$$L(s,\chi) = \prod_{p\in\mathbb{P}} \frac{1}{1-\chi(p)p^{-s}}$$

for $\operatorname{Re} s > 1$, showing incidentally that $L(s,\chi) \neq 0$ whenever $\operatorname{Re} s > 1$. Dirichlet introduced his L-series in his proof that there are infinitely primes $\equiv a \bmod b$ if $(a,b) = 1$. The idea of the proof is quite simple: it is easy to show that

$$\log L(s,\chi) = \sum_{p}\chi(p)p^{-s} + g(s)$$

for some "error term" $g(s)$ which is bounded as $s \longrightarrow 1$ (in fact g is even holomorphic for $\operatorname{Re} s > \frac{1}{2}$). Now the orthogonality relations for characters imply

$$\sum_{\chi}\chi(a)^{-1}\log L(s,\chi) = \sum_{p\equiv a\bmod n}\phi(n)p^{-s} + \text{error}(s),$$

where the sum is over all characters of $(\mathbb{Z}/n\mathbb{Z})^{\times}$ and error(s) is bounded as $s \longrightarrow 1$. But $\log L(s, 1) \longrightarrow \infty$ for $s \longrightarrow 1$, and if it can be shown that $L(1,\chi) \neq 0$ for characters $\chi \neq 1$, then the left hand side (and, therefore, the right hand side as well) tends to ∞ as $s \longrightarrow 1$, and Dirichlet's theorem follows.

There are essentially two different ways of proving $L(1,\chi) \neq 0$ for $\chi \neq 1$; the simplest proof uses the fact that L-series are 'prime factors' of the ζ-functions of abelian extensions:

Theorem 10.23. *Let $K/\mathbb{Q}$ be an abelian extension, and let X be the character group belonging to K. Then*

$$\zeta(K,s) = \prod_{\chi\in X}L(s,\chi). \quad \square$$

This conjecture of Weber was proved by Takagi using class field theory. Assuming Theorem 10.23, the idea then is that a zero of $L(1,\chi)$ for $\chi \neq 1$ would cancel against the simple pole of $L(1, 1)$, making $\zeta(K, 1)$ convergent, which it isn't.

Dirichlet proved the non-vanishing of $L(1,\chi)$ in a different way: for characters such that $\chi^2 \neq 1$ he found a clever trick, then reduced the problem to real characters of prime conductor, that is, to Legendre symbols $\chi = (\frac{\cdot}{p})$, and finally computed $L(1,\chi)$ explicitly (note that he used the special case $(K : \mathbb{Q}) = 2$ of Theorem 10.23, that is the relation $\zeta(K,s) = \zeta(s)L(s,\chi)$, and that this is equivalent to the quadratic reciprocity law):

Theorem 10.24. *Let p be an odd prime and $\chi = (\frac{\cdot}{p})$; then*

$$L(1,\chi) = \begin{cases} \frac{\pi}{p\sqrt{p}}(\sum b - \sum a), & \text{if } p \equiv 3 \bmod 4, \\ \frac{1}{\sqrt{p}} \log \frac{\prod \sin(\frac{\pi b}{p})}{\prod \sin(\frac{\pi a}{p})} & \text{if } p \equiv 1 \bmod 4. \end{cases}$$

Here a and b run through the quadratic residues and non-residues modulo p, respectively. □

Comparing this with Theorem 10.17 one gets the class number formula for quadratic number fields; see Theorem 3.26 for the real case.

Dirichlet's L-functions also satisfy a functional equation. Put $\delta = 0$ if $\chi(-1) = 1$ and $\delta = 1$ if $\chi(-1) = -1$, let f denote the conductor of χ and $G(\chi)$ the corresponding Gauss sum (the Gauss sum without the minus sign is denoted as usual by $\tau(\chi)$), and define the root number

$$W(\chi) = i^\delta \frac{G(\chi)}{\sqrt{f}} = \frac{\tau(\chi)}{i^\delta \sqrt{f}}.$$

Then $\Lambda(s,\chi) = (\frac{f}{\pi})^{s/2} \Gamma(\frac{s+\delta}{2}) L(s,\chi)$ satisfies the functional equation

$$\Lambda(s,\chi) = W(\chi)\Lambda(1-s,\chi^{-1}).$$

Applying the functional equations of the ζ-function and Dirichlet's L-series to the equation in Theorem 10.23, we first get $\prod_\chi W(\chi) = 1$ (for a real quadratic character, this is equivalent to the determination of the sign of the quadratic Gauss sum), and then Theorem 4.22, the Conductor-Discriminant Formula. We finally mention that the values of Dirichlet's L-series at negative integers are basically generalized Bernoulli numbers: for positive integers $n \equiv \delta \bmod 2$, we have

$$L(1-n,\chi) = (-1)^{1+(n-\delta)/2} \frac{\tau(\chi)}{i^\delta} \left(\frac{2\pi}{f}\right)^n \frac{B_{n,\chi^{-1}}}{n!}.$$

10.5 Counting Points with Zeta Functions

In the classical situation, one takes the primes in $\mathbb{Z}$ (i.e. the object one wants to study), constructs the corresponding ζ-function, studies its properties, and uses the results to prove theorems about the number of primes below x, i.e. the prime number theorem. For curves over $\mathbb{F}_p$, the idea is the same: one takes the number N_s of solutions of C over $\mathbb{F}_{p^s}$ for each $s \geq 1$, constructs the corresponding ζ-function, studies its properties, and uses the results to prove theorems about the N_p, for example the Hasse-Weil bound (also called the Riemann hypothesis for curves).

Points at infinity Actually, counting the number N_p of solutions of C over $\mathbb{F}_p$ is not the best idea: the formulas only become smooth only if we take the infinite solutions into account. In order to explain what this means, let

$$f(X,Y) = \sum a_{i,j} X^i Y^j \in K[X,Y]$$

be a polynomial with coefficients $a_{i,j}$ in a field K. Let d denote the maximal degree $i+j$ of the monomials $X^i Y^j$ occurring in f. Then we can homogenize f by introducing a new variable Z and writing

$$F(X,Y,Z) = \sum a_{i,j} X^i Y^j Z^{d-i-j}.$$

For example, the polynomial $X^2 + Y^2 + X^2 Y^2 - 1$ has homogenization $X^2 Z^2 + Y^2 Z^2 + X^2 Y^2 - Z^4$. Let us look at the solutions of the homogenized equation $F(X,Y,Z) = 0$; we consider two solutions $(x,y,z), (x',y',z') \in K^3$ of $F = 0$ as equivalent if there exists an element $\lambda \in K^\times$ such that $x' = \lambda x$, $y' = \lambda y$, and $z' = \lambda z$. The equivalence class $[x : y : z]$ of such solutions corresponds to a solution of the original equation $f(x,y) = 0$ if and only if $z \neq 0$; the nontrivial solutions $[x : y : 0]$ (by nontrivial we mean that we do not count the trivial solution $(0,0,0)$ of the homogenized equation are called the *infinite solutions* of $f(x,y) = 0$.

We can say the same thing in a different (and better) way by introducing the projective plane $\mathbb{P}^2(K)$; this is the set of equivalence classes $[x : y : z] \neq [0 : 0 : 0]$ of points $(x,y,z) \in K \times K \times K$, with the equivalence relation defined as above. Although we cannot evaluate F at P for the point $P = [x : y : z]$ (since $[x : y : z] = [\lambda x : \lambda y : \lambda z]$ for $\lambda \in K^\times$ and $F(\lambda X, \lambda Y, \lambda Z) = \lambda^d F(X,Y,Z)$), the question whether $F(P) = 0$ or not has a well defined answer: thus it makes sense to speak of the zeros of F in $\mathbb{P}^2(K)$.

In order to determine the infinite solutions of $X^2 + Y^2 + X^2 Y^2 = 1$ over $K = \mathbb{F}_p$ we put $Z = 0$ in the homogenized equation $X^2 Z^2 + Y^2 Z^2 + X^2 Y^2 = Z^4$ and solve $X^2 Y^2 = 0$. There are exactly two solutions $[1 : 0 : 0]$ and $[0 : 1 : 0]$ (note that $[1 : 0 : 0] = [-1 : 0 : 0]$), showing that Gauss counted the infinite solutions twice (so did Herglotz, by the way).

Similarly, there are exactly m infinite solutions of the Fermat curve F_m : $X^m + Y^m + Z^m = 0$ over any field containing the $2m$-th roots of unity: in fact, putting $Z = 0$ gives $X^m = -Y^m$, i.e. $(XY^{-1})^m = -1$. This explains the curious expression $N_m + m$ in Equation (10.23): $N_m + m = N(X^m + Y^m + Z^m)$ is simply the number of solutions of F_m over $\mathbb{P}^2(\mathbb{F}_p)$.

Our next task is to define singular and non-singular points on curves. Geometrically, a point P on a curve C is nonsingular if C is "smooth" in a vicinity of P (i.e. C has no cusp in P, and there is a well defined tangent at P). More exactly: let C be a projective curve in $\mathbb{P}^2(K)$ given by a homogeneous polynomial $F(X,Y,Z)$ defined over an algebraically closed field K (that is, with coefficients in K). Then a point $P = [x : y : z]$ on C (i.e., $F(x,y,z) = 0$) is called *singular* if and only if all the partial derivatives vanish: $\frac{\partial F}{\partial X}(x,y,z) =$

$\frac{\partial F}{\partial Y}(x,y,z) = \frac{\partial F}{\partial Z}(x,y,z) = 0$. A curve C is called *nonsingular* if all points P on C are nonsingular. The reason for having to introduce the concept of nonsingularity is that only zeta functions of nonsingular curves are supposed to behave nicely.

The Fermat curve F_m is nonsingular: the partial derivatives vanish only for $X = Y = Z = 0$ which is not in $\mathbb{P}^2(K)$. The curve $C_2 : y^2 = 1 - x^4$, however, does have a singular point: its homogenization is $F(X, Y, Z) = Y^2 Z^2 + X^4 - Z^4 = 0$, and all the derivatives vanish in the infinite point $[0 : 1 : 0]$. The curve C_1 defined in Equation (10.1) is another example of a singular curve (see Exercise 10.17).

Zeta functions of curves over finite fields

Let C be a nonsingular projective curve defined by an irreducible polynomial $F \in \mathbb{F}_q[X, Y, Z]$, and let N_s denote the number of points on C over $\mathbb{F}_{q^s}$. Then we use these numbers to define the *congruence zeta function* by

$$Z_C(T) = \exp\left(\sum_{s=1}^{\infty} N_s \frac{T^s}{s}\right). \tag{10.17}$$

In order to motivate this definition, we will show

Proposition 10.25. *Let $A \in M_n(\mathbb{C})$ be a square matrix; then*

$$\det(I - At) = \exp\left(-\sum_{r=1}^{\infty} \mathrm{Tr}\,(A^r) \frac{t^r}{r}\right).$$

Proof. For $\lambda \in \mathbb{C}$, put $f_\lambda(t) = \exp\left(-\sum_{r=1}^{\infty} \lambda^r \frac{t^r}{r}\right)$; note that $f_\lambda(0) = 1$. Taking the logarithm on both sides we get $\log f_\lambda(t) = -\sum_{r=1}^{\infty} \lambda^r \frac{t^r}{r}$, and taking the derivative with respect to t we find

$$\frac{f'_\lambda(t)}{f_\lambda(t)} = -\lambda \sum_{r=1}^{\infty} (\lambda t)^{r-1} = -\frac{\lambda}{1 - \lambda t}.$$

Integrating (and using $f_\lambda(0) = 1$) yields $\log f_\lambda(t) = \log(1 - \lambda t)$, that is, $f_\lambda(t) = (1 - \lambda t)$. Let $\lambda_1, \ldots, \lambda_n$ denote the eigenvalues of A with multiplicity. Then $\mathrm{Tr}\, A^r = \lambda_1^r + \ldots + \lambda_n^r$, hence

$$\exp\left(-\sum_{r=1}^{\infty} \lambda^r \frac{t^r}{r}\right) = \prod_{j=1}^{n} f_{\lambda_j}(t) = \prod_{j=1}^{n} (1 - \lambda_j t) = \det(I - At).$$

This is what we wanted to prove. □

Thus if N_s is a "trace", then the congruence zeta function, originally defined as a formal power series in one variable T, is a polynomial in T! Though N_s is not a trace in general, it is almost one: in fact, in all our examples we will have $N_s = \lambda_1^{a_1} + \ldots + \lambda_r^{a_r} - \lambda_{r+1}^{a_{r+1}} - \ldots - \lambda_n^{a_n}$, and now the same proof as above shows that the corresponding zeta function is $Z(t) = \frac{(1-\lambda_1 t)\cdots(1-\lambda_r t)}{(1-\lambda_{r+1} t)\cdots(1-\lambda_n t)}$.

As an example, let us compute the zeta function for the (elliptic) curve $C: x^3 + y^3 = 1$ over $\mathbb{F}_p$, where $p \equiv 1 \bmod 3$ is a prime. Since there are exactly three points at infinity, namely $[1:-1:0]$, $[\rho:-1:0]$ and $[\rho^2:-1:0]$ (here ρ is a primitive cube root of 1 in $\mathbb{F}_p$), we have $N_s = p^s + 1 - \pi^s - \overline{\pi}^s$ by Proposition 10.11, hence $Z(T) = \frac{(1-\pi T)(1-\overline{\pi}T)}{(1-T)(1-pT)}$.

A simple computation shows that $Z(\frac{1}{pT}) = Z(T)$, that is, the zeta function of C over $\mathbb{F}_p$ satisfies a functional equation. This is no coincidence: in [SF2], F.K. Schmidt defined the *genus* of curves over finite fields (if C is a nonsingular projective curve defined by an irreducible homogeneous polynomial $F(X, Y, Z) \in K[X, Y, Z]$ of degree d, then there is the simple formula $g = (d-1)(d-2)/2$ for the genus), proved an analogue of the theorem of Riemann-Roch, and used it to show that such curves have rational ζ-functions satisfying a functional equation:

Theorem 10.26. *Let C be a nonsingular projective curve of genus g defined over a finite field $\mathbb{F}_q$. Then there exists a polynomial $P_1(T) \in \mathbb{Z}[T]$ of degree $2g$ and with $P_1(0) = 1$ such that*

$$Z_C(T) = \frac{P_1(T)}{(1-T)(1-qT)}. \tag{10.18}$$

Moreover, $Z_C(T)$ satisfies the functional equation

$$Z_C\left(\frac{1}{qT}\right) = (qT^2)^{1-g} Z_C(T). \tag{10.19}$$

□

Schmidt also conjectured that the analogue of the Riemann Hypothesis holds, which says that the roots of the polynomial $P_1(T)$ in (10.18) have absolute value $q^{-1/2}$. The connection with the classical Riemann Hypothesis becomes apparent after making the substitution $T = q^{-s}$: we find $\zeta(s) = \frac{P_1(q^{-s})}{(1-q^{-s})(1-q^{1-s})}$, hence the zeros of ζ_s are the zeros of $P_1(q^{-s})$. Write $P_1(T) = \prod_j(1-\alpha_j T)$; then $P_1(q^{-s}) = 0$ if and only if $q^s = \alpha_j$ for some j, and since $|q^s| = q^{\operatorname{Re} s}$, we see that $|\alpha_i| = q^{1/2}$ if and only if $\operatorname{Re} s = \frac{1}{2}$, in other words: the zeros of $\zeta(s)$ lie on the critical line if and only if the inverse roots of $P_1(T)$ have absolute value $\sqrt{q}$. Also, the functional equation looks more familiar when expressed with $\zeta(s)$: it is a simple exercise to transform (10.19) into

$$\zeta(1-s) = q^{(g-1)(2s-1)} \zeta(s). \tag{10.20}$$

10.5 Counting Points with Zeta Functions

Since we used the numbers N_s to construct our zeta function, it is hardly surprising that we can get them back: the most obvious way is of course by differentiating the logarithm of the zeta function; the following observation, however, based on the rationality of the zeta function, is more interesting:

Proposition 10.27. *Let N_s be the number of $\mathbb{F}_{q^s}$-rational points on a nonsingular projective curve C defined over $\mathbb{F}_q$, and let $Z_C(T)$ be the corresponding zeta function. Then $Z_C(T)$ is a rational polynomial in T if and only if there exist $\alpha_i, \beta_j \in \mathbb{C}$ such that $N_s = \sum_i \alpha_i^s - \sum_j \beta_j^s$ for all $s \geq 1$. In this case,*

$$Z_C(T) = \frac{\prod_i (1 - \alpha_i T)}{\prod_j (1 - \beta_j T)}.$$

Proof. If $N_s = \sum_i \alpha_i^s - \sum_j \beta_j^s$, then the zeta function clearly has the form $P(T)/Q(T)$ with $P(T) = \prod_j (1 - \beta_j T)$ and $Q(T) = \prod_i (1 - \alpha_i T)$. So assume conversely that $Z_C(T)$ is rational; since $Z(0) = 1$, we can write $Z_C(T) = P(T)/Q(T)$ with $P(0) = Q(0) = 1$. Since $\mathbb{C}$ is algebraically closed, there exist $\alpha_i, \beta_j \in \mathbb{C}$ such that $P(T) = \prod_j (1 - \beta_j T)$ and $Q(T) = \prod_i (1 - \alpha_i T)$. Forming the logarithmic derivative of $Z_C(T)$ we get

$$\frac{Z'_C(T)}{Z_C(T)} = \sum_i \frac{\alpha_i}{1 - \alpha_i T} - \sum_j \frac{\beta_j}{1 - \beta_j T}.$$

Multiplying through by T and using the geometric series we find

$$\frac{T Z'_C(T)}{Z_C(T)} = \sum_i \sum_{s \geq 1} (\alpha_i t)^s - \sum_j \sum_{s \geq 1} (\beta_j t)^s = \sum_{s \geq 1} \left(\sum_i \alpha_i^s - \sum_j \beta_j^s \right) T^s.$$

On the other hand, forming the logarithmic derivative of the zeta function directly from the definition yields

$$\frac{T Z'_C(T)}{Z_C(T)} = \sum_{s \geq 1} N_s T^s,$$

and comparing the coefficients proves our claim. □

The interesting thing is that the Riemann conjecture for curves C allows us to give bounds for the numbers N_s as follows: factorize the polynomial $P_1(T)$ in (10.18) over $\mathbb{C}$ as $P_1(T) = \prod_{i=1}^{2g} (1 - \alpha_i T)$; then $N_s = q^s + 1 - \sum_{i=1}^{2g} \alpha_i^s$, hence

$$|N_s - (q^s + 1)| = \left| \sum_{i=1}^{2g} \alpha_i^s \right| \leq \sum_{i=1}^{2g} |\alpha_i|^s = 2g \sqrt{q}^s.$$

This bound for the numbers N_s is called the Hasse-Weil bound.

Finally we remark that Theorem 10.26 is a powerful method for evaluating the number of points N_s of nonsingular curves C over finite fields $\mathbb{F}_{q^s}$: for example, if C is an elliptic curve (these have genus 1) defined over $\mathbb{F}_p$ then calculating the number of points over $\mathbb{F}_p$ gives N_1; from N_1 we can compute the inverse roots α and $\overline{\alpha}$ of $P_1(T)$ (in fact, we immediately get $P_1(T) = 1 - a_p T + pT^2$, where $a_p = p + 1 - N_1$), and then Theorem 10.26 says that $N_s = p + 1 - \alpha^s - \overline{\alpha}^s$ for all $s \geq 1$. In general, this method works once we know $N_1, \ldots, N_g$; examples are given in Exercises 10.19 and 10.20.

If we compare the formula $N_s = q^s + 1 - \sum_{i=1}^{2g} \alpha_i^s$ with Equation (10.10) after having added the m infinite solutions we might suspect that the zeros of the congruence zeta function of Fermat curves are Jacobi sums; this is actually true, as was shown by Davenport & Hasse [DaH].

The first verification of the Hasse-Weil bound for hyperelliptic curves is due to Eisenstein. He proved the special case $p \equiv 3 \bmod 8$ of the following theorem:

Theorem 10.28. *Let p be an odd prime; define integers $a, b, c, d \in \mathbb{N}$ by $p = a^2 + 4b^2$ if $p \equiv 1 \bmod 4$ and $p = c^2 + 2d^2$ if $p \equiv 1, 3 \bmod 8$. Then choose their signs in such a way that*

$$a \equiv 3, \ c \equiv (-1)^{(p+7)/8} \bmod 4 \ \text{if } p \equiv 1 \bmod 8,$$
$$c \equiv (-1)^{(p-3)/8} \bmod 4 \quad \text{if } p \equiv 3 \bmod 8,$$
$$a \equiv 1 \bmod 4 \quad \text{if } p \equiv 5 \bmod 8.$$

Then the number N_1 of points on $C : y^2 = x^8 - 28x^6 + 70x^4 - 28x^2 + 1$ with coordinates in $\mathbb{F}_p$ equals

$$N_1 = \begin{cases} p - 1 + 2a + 4c & \text{if } p \equiv 1 \bmod 8, \\ p - 1 + 4c & \text{if } p \equiv 3 \bmod 8, \\ p - 1 + 2a & \text{if } p \equiv 5 \bmod 8, \\ p - 1 & \text{if } p \equiv 7 \bmod 8. \end{cases}$$

This result can be proved (see Exercise 10.9) using Eisenstein sums. Curves of the form $y^2 = f(x)$, where f is a polynomial of degree ≥ 5 without multiple roots, are called *hyperelliptic curves*. Over fields of characteristic $\neq 2$, their affine part is nonsingular, but they have a singularity at the point at infinity. A nonsingular model for C has two points at infinity, hence the number of points on Eisenstein's curve C over $\mathbb{P}^2(\mathbb{F}_p)$ equals $N_1 + 2$. Since C has genus 3, Theorem 10.28 and the inequalities

$$|\#C(\mathbb{F}_p) - (p+1)| \leq \begin{cases} 2|a| + 4|c| < 6\sqrt{p} & \text{if } p \equiv 1 \bmod 8, \\ 4|c| < 4\sqrt{p} & \text{if } p \equiv 3 \bmod 8, \\ 2|a| < 2\sqrt{p} & \text{if } p \equiv 5 \bmod 8, \text{ and} \\ 0 & \text{if } p \equiv 7 \bmod 8 \end{cases}$$

show that the Riemann hypothesis for C is true.

The Weil Conjectures

Artin verified the Riemann conjecture in some special cases in his thesis; Hasse proved it 1933 for elliptic curves over finite fields (these are non-singular curves with a rational point and genus $g = 1$), and in 1940 Weil showed that the Riemann conjecture is true for curves of arbitrary genus. A few years later Weil considered a more general situation, that of projective varieties. We need a few definitions before we can explain what these are. First, the *projective space* $\mathbb{P}^n = \mathbb{P}^n(K)$ is defined as the set of equivalence classes of points $(x_0, \ldots, x_n) \in K^{n+1} \setminus \{(0, \ldots, 0)\}$ with respect to the equivalence relation $(x_0, \ldots, x_n) \sim (\lambda x_0, \ldots, \lambda x_n)$ for $\lambda \in K^\times$. A *projective algebraic set* is the set of zeros of a family of homogeneous polynomials in $K[X_0, \ldots, X_n]$. It is easy to see that the class of projective algebraic sets is closed under taking finite unions or arbitrary intersections; moreover, the empty set and the whole space $\mathbb{P}^n$ are projective algebraic sets. Thus we can define a topology on $\mathbb{P}^n$ (the *Zariski topology*) by calling a subset $X \subseteq \mathbb{P}^n$ *open* if X is the complement of a projective algebraic set; in particular, projective algebraic sets are *closed*. Moreover, X is called *irreducible* if it cannot be written as a disjoint union of projective algebraic sets. The *dimension* of $X \subseteq \mathbb{P}^n$ is the supremum over all $n \in \mathbb{N}$ for which there is a chain of irreducible closed subsets $Z_0 \subsetneq Z_1 \subsetneq \ldots \subsetneq Z_n = X$. Now a projective variety V is a nonsingular irreducible projective algebraic set in $\mathbb{P}^n$. The simplest example of a projective variety is the space $\mathbb{P}^n$ itself; it can be shown that $\dim \mathbb{P}^n = n$ as expected.

Weil made his conjectures in 1949, and they are still called Weil Conjectures despite the fact that their proof was completed by Deligne in 1974: let V be a nonsingular projective variety of dimension d. Replacing the number of points on a curve C by those on the variety V, the definition of $Z_C(T)$ applies to varieties as well, and it was conjectured to have the following properties:

1. Rationality: $Z_V(T)$ is a rational function of T, that is, $Z_V(T) \in \mathbb{Q}(T)$;
2. Functional Equation: There exists an integer E such that
$$Z_V\left(\frac{1}{q^d T}\right) = \pm (q^{d/2} T)^E Z_V(T);$$

3. Riemann Hypothesis: There exist polynomials $P_i(T) \in \mathbb{Z}[T]$, $0 \leq i \leq 2d$, with $P_0(T) = 1 - T$, $P_{2d}(T) = 1 - q^d T$, and
$$Z_V(T) = \frac{P_1(T) P_3(T) \cdots P_{2d-1}(T)}{P_0(T) P_2(T) \cdots P_{2d}(T)};$$
moreover,
$$P_i(T) = \prod_{j=1}^{B_i} (1 - \alpha_{ij} T),$$
where the α_{ij} are algebraic integers such that $|\alpha_{ij}| = q^{i/2}$.

Weil also conjectured that the integers B_i should have interpretations as the Betti numbers of a suitable cohomology theory.

Notes

Gauss's contributions

Gauss's diary was discovered in 1897 and became known to the mathematical world through its publication in the Mathematische Annalen in 1903 (a copy of the diary was also included in vol. X_1 of his collected works and published in 1917; see [274]). Dedekind verified the claim for all primes $p \equiv 1 \bmod 4$ smaller than 100, Bachmann remarked that the connection with the lemniscatic functions should be uncovered, and Fricke noticed that the congruence resembled the functional equation $\mathrm{sl}^2 x + \mathrm{cl}^2 x + \mathrm{sl}^2 x \, \mathrm{cl}^2 x = 1$ for the lemniscatic functions sl and cl. The first proof of Proposition 10.1 was given in 1921 by Herglotz [362]: he used elliptic functions, showing that Fricke's intuition was correct. Herglotz also noticed (as did Davenport & Hasse [DaH], as well as Deuring and Weil [We3]) that Gauss's conjectural Last Entry follows from results he himself had given in his first memoir on biquadratic residues.

Proofs for the cubic analogue of Proposition 10.3 can be built on results already found in Gauss's Disquisitiones Arithmeticae [263], as was shown by Libri [541]: he treated actually both the cubic and quartic case there. See also the articles of Stern [752] (who derives the cubic character of 2 from his results), Pépin [638], and Vaidyanathaswamy [807], or the books of Venkov [813], S. Chowla [Ch2], and Silverman & Tate [ST]. Kemnitz [Kem] used elliptic functions to give a proof in the spirit of Herglotz.

Gauss's last entry was proved using Jacobi sums by Morlaye [Mor], after Swinnerton-Dyer had claimed that a non-elementary proof was not known. Singh & Rajwade [SR] gave an elementary proof which is essentially Gauss's first proof via cyclotomy; yet another elementary proof was published by Monzingo [589].

The Fermat Congruence

Libri [Lib] observed that Fermat's Last Theorem for the exponent p would follow if there existed infinitely many primes q such that the congruence $x^p + y^p + 1 \equiv 0 \bmod q$ has only the trivial solution. Let $N(q,p) = \#\{(x,y) \in \mathbb{F}_q^\times \times \mathbb{F}_q^\times : x^p + y^p + 1 = 0\}$. Pellet [Pe1] showed that Libri's assumption is never satisfied because for every prime p, $N(q,p) > 0$ for q large enough. Another dismissal of a proposed proof of FLT by Lefébure is due to Pépin [Pep]. Bounds for $N(q,p)$ were derived by Pellet [Pe2], Cornacchia [Cor] (according to Dickson [158], his results contain an error) and Schur [Sch] gave integers q_p such that $N(q,p) > 0$ for all $q > q_p$. The best bound for q_p is due to Dickson [Di1], who proved the bound in Exercise 10.3 in a computational tour de force. Dickson's result was immediately generalized to equations $ax^e + by^e + cz^e = 0$ by Hurwitz [Hu3] in a paper that was even more technical since Hurwitz chose to give a proof from scratch without cyclotomy or Jacobi sums. See Washington [Was, §6.1] and Ribenboim [Rib, Ch. XII] for more.

Kummer and Cyclotomy

Mazur wrote a nice review [Maz] of Kummer's collected papers; see also the articles of Edwards [181, Ed1, Ed2] and Terjanian [Ter].

Section 10.2 on cyclotomic numbers is an attempt at understanding and correcting Kummer's formulas in [464]. He does not give any proofs there, and some of his formulas are definitely not true, but as we have shown, his basic idea is correct and leads very quickly to results in cyclotomy.

Cyclotomic numbers of small order were studied e.g. by Barkan [32], Dickson [158], Evans [221], E. Lehmer [497], Muskat [599], Muskat & Whiteman [603], Parnami, Agrawal & Rajwade [626], and Whiteman [841, 842, 844, 845, 846]. Applications of cyclotomy to power residues were given e.g. by Evans [225], E. Lehmer [500], Muskat [596, 597, 598, 601], and K.S. Williams [849, 850].

ζ-functions

As we have already remarked, Riemann's ζ-function was introduced into number theory by Euler; the reason why it is named after Riemann is that he was the first to extend it to the complex plane. In fact, even Dirichlet studied his L-series only over the reals. The careful investigation of the *complex* zeros of $\zeta(s)$ eventually led to the proof of the prime number theorem, which turned out to be equivalent to the nonvanishing of $\zeta(1 + it)$ and thus is intimately connected with the behaviour of $\zeta(s)$ as a complex function.

In the paper where Riemann introduced the ζ-function for complex variables, he stated five conjectures concerning the zeros of $\zeta(s)$ in the critical strip; two of them were proved by Hadamard in 1893, two others by von Mangoldt in 1894, the last one – the Riemann Conjecture – is still wide open. We know from a result of Conrey [Con] that at least 40% of all zeros (there are infinitely many) in the critical strip are on the line $\operatorname{Re} s = \frac{1}{2}$, and extensive computations by van de Lune, te Riele and Winter [LRW] have shown that the first 1,500,000,001 zeros in the critical strip (ordered by their imaginary part) do have real part $\frac{1}{2}$. Computations of the zeros of some Dedekind zeta functions were performed recently by Tollis [Tol].

Bernoulli numbers were discovered by Jakob Bernoulli, although Johann Faulhaber (1580–1635) already developed formulas for the sums of k-th powers of integers for all $k \leq 25$. Equation (10.11) shows that the Bernoulli numbers are, up to some well known factors, values of the zeta function at the even positive integers. Something similar is true for Hurwitz numbers defined in Chapter 8, as was apparently shown by Kluyver [Klu], whose paper was not accessible to me. Generalized Bernoulli numbers are usually attributed to Leopoldt; as a matter of fact, they already appear in the work of Sylvester [Syl].

The values $\zeta(2k-1)$ of the Riemann zeta function at odd positive integers ≥ 3 are still a mystery. Apéry could show that $\zeta(3)$ is irrational with a proof that Euler could have understood (see van der Poorten [vdP]), but it seems that his proof doesn't generalize too easily. Borel defined "higher regulators" $R_{2k} \in \mathbb{R}^\times/\mathbb{Q}^\times$ with the help of higher K-groups of $\mathbb{Z}$ and proved that $\zeta(2k-1) \equiv R_{2k} \mod \mathbb{Q}^\times$. His results are now part of the theory around the Beilinson and the Bloch-Kato conjectures; see Rapoport, Schappacher & Schneider [RSS] and Hulsbergen [Hu].

Weil Conjectures

The number of solutions of congruences of type $a_0 x_0^n + \ldots + a_m x_m^n \equiv 1 \mod p$ were the subject of many elementary investigations; Gauss dealt with the cases $m = 2$ and $n = 3, 4$, while Lebesgue studied $n = 2$ and $m \geq 2$. See also Unger [806] and Widmer [Wid].

The first zeta functions for fields other than number fields were introduced in the thesis [Kor] of Kornblum, a student of Landau who died in the first world war. Kornblum used L-series of type

$$L(s, \chi) = \sum_f \chi(f) q^{-s \deg f}; \qquad (10.21)$$

here the sum is over all monic polynomials $f \in \mathbb{F}_q[t]$, a finite field of q elements, $\deg f$ denotes the degree of f, and χ is a character on the residue class ring $\mathbb{F}_q/(m)$ of a fixed polynomial $m \in \mathbb{F}_q[t]$. Following Dirichlet's arguments, he succeeded in proving an analogue of Dirichlet's theorem for $\mathbb{F}_q[t]$, namely that there exist infinitely many irreducible polynomials $p \in \mathbb{F}_q[t]$ such that $p \equiv a \mod m$, if only the given polynomials $a, m \in \mathbb{F}_q[t]$ are coprime.

Next came Artin's thesis, where he transferred the theory of quadratic number fields to quadratic extensions of $\mathbb{F}_p(t)$, including the notion of an ideal class group and a proof of the quadratic reciprocity law for elements $f, g \in \mathbb{F}_q[t]$ (this last result is, as we have remarked in Chapter 1, due to Dedekind). Artin also saw the connection between the number of $\mathbb{F}_p$-rational points on hyperelliptic curves and the Riemann conjecture for these curves. His thesis discussed in some detail in the book of Thomas [Tho].

Artin's results were generalized to arbitrary finite extensions of $\mathbb{F}_q(t)$ (such fields are called function fields of one variable with a finite field of constants) in F.K. Schmidt's thesis [SF1]. Hasse [Has] succeeded in proving the Riemann conjecture for elliptic curves; Manin [Ma] rewrote Hasse's proof in a completely elementary language. He succeeded in replacing the isogenies of $E: y^2 = x^3 + ax + b$ in Hasse's proof by $K(t)$-rational points on the elliptic curve twisted by $\lambda = t^3 + at + b$; I owe this explanation to P. Roquette [Ro3]. Manin's proof has been simplified lately by Chahal [Cha]. See Zimmer [Zim] for another elementary proof.

It seems that Hasse's interest in these questions was sparked by a question that Davenport treated in his thesis supervised by Littlewood, namely: for odd primes p, let R_n denote the number of chains $a_1, \ldots, a_n$ of quadratic residues modulo p (thus e.g. $R_2 = RR$ as defined in Chapter 1). Then $|\#R_n - 2^{-n}p| = O(\sqrt{p})$ for $n = 1, 2, 3$, and the question is whether this generalizes. Hopf [Hop] had earlier got the bound $\frac{1}{\sqrt{6}}p$, and Davenport [Da1, Da2] could show that e.g. $|\#R_4 - 2^{-4}p| = O(p^{3/4})$. The complete history can be found in Roquette's article [Ro2], and Hasse's Vorlesungen [342] contain an elementary approach to these questions.

In 1948, Weil [We3] announced his conjectures on the zeta functions of projective varieties and proved them for plane curves of arbitrary genus g. The Weil bounds $|N_s - (q^s + 1)| \leq 2g\sqrt{q^s}$ can be improved trivially to $|N_s - (q^s + 1)| \leq \lfloor 2g\sqrt{q^s} \rfloor$; Serre [Se3] showed that actually $|N_s - (q^s + 1)| \leq g\lfloor 2\sqrt{q^s} \rfloor$.

After Dwork [Dwo] had proved the rationality of the congruence zeta functions (see Serre's exposition [Se1] or the obituary [KT] written by Katz & Tate), Grothendieck put the world of algebraic geometry upside down by inventing a completely new language and applying it to derive the rationality of the zeta function [Gr2], and finally Deligne [Del] used Grothendieck's results to prove the last of the Weil Conjectures, the Riemann hypothesis. The book of Monsky [Mon] deals with the situation in between Grothendieck and Deligne.

Since Weil's proof of the Riemann conjecture for curves was very deep (some simplifications were obtained by Mattuck & Tate [MT] and Grothendieck [Gr1]), Stepanov's elementary proof [Ste1, Ste2] for the case of hyperelliptic curves was quite welcome (see W.M. Schmidt [SW] for a detailed exposition and extension of Stepanov's ideas; they are also treated by Lidl & Niederreiter [LN]). More recently, the hyperelliptic case has been studied again by Stark [Sta1] and Zannier [Zan].

Bombieri refined Stepanov's ideas and succeeded in giving an elementary proof of the Riemann conjecture for curves in just a few pages, using the theorem of Riemann-Roch; see the original paper [Bo1] as well as the expositions by Bombieri himself [Bo2] and in the textbooks of Winnie Li [Li], Lorenzini [Lor] and Stichtenoth [Sti], or in the lecture notes of S.V. Hansen [Han]. In some special cases the Weil bounds may be improved; see Stöhr & Voloch [SV] for an approach using Weierstraß points.

Once more I can't help but advise the reader to take a look at the book of Ireland & Rosen [386] for an introduction to this kind of questions that is probably impossible to beat. Other beautiful introductions can be found in Joly [413], Koblitz [Ko1, Ko2, Ko3], Robba & Christol [RC], Small [Sma], Stepanov [Ste3] and Thomas [Tho]; if you can read Russian, you may also have a look at Glazunov & Postnikov [GP]. For historical details, check out Cartier [Car] and Houzel [Hou].

For "elementary" expositions of the cohomology behind the proof of the Weil Conjectures, see Dieudonné [Die], Hartshorne [Har], and Winnie Li [Li]; for a less modest survey, see Katz [Kat], or try the book of Freitag & Kiehl [FrKi]. There is some hope that the cohomological proof of the Riemann conjecture for projective varieties may carry over to the much deeper original

Riemann conjecture; see the lecture of Deninger in [Den]. Even the first step of such a transfer, namely the definition of the genus of a number field and a proof of a corresponding "Riemann-Roch theorem", is not obvious: for the most recent (and very promising) attempt, see the exposition of van der Geer & Schoof [GS].

Exponential Sums Zeta functions are not only used for estimating the number of solutions of equations over finite fields but also for bounding exponential sums. Let $\mathbb{F}$ be a finite field with $q = p^s$ elements, $\text{Tr} : \mathbb{F} \longrightarrow \mathbb{F}_p$ the trace, and $f \in \mathbb{F}[X]$ a polynomial. Define a character $\chi : \mathbb{F} \longrightarrow \mu_p : t \longmapsto \zeta_p^{\text{Tr}(t)}$; then expressions of the form

$$S_f = \sum_{t \in \mathbb{F}} \chi(f(t)) \tag{10.22}$$

are called *exponential sums*. The quadratic Gauss sum, for example, is the special case $\mathbb{F} = \mathbb{F}_p$ and $f(X) = X^2$. A typical result is

Theorem 10.29. *If $d = \deg f$ is not divisible by p, then*

$$|S_f| \leq (d-1)\sqrt{q}$$

for the exponential sum in (10.22). □

Similar estimates are due to Weil [We2] and are treated in Li [Li]. For a beautiful introduction to such questions, see Roquette [Ro1].

Exercises

10.1 (Gauss) Use Proposition 10.3 to show that $(2/p)_4 = (-1)^{b/4}$ for primes $p = a^2 + b^2 \equiv 1 \bmod 8$ with b even.
Hints: 2 is contained in A_0 or A_2. The map $f : a \longmapsto p - 1 - a$ is a bijection of $A_{00} = \{(a, a+1) : (a/p)_4 = (a+1/p)_4 = 1\}$ onto itself. Thus A_{00} has even cardinality unless f has a fixed point; this is the case if and only if $a = p-1-a$ has a solution in A_0, i.e. if and only if $a = \frac{1}{2}(p-1) \in A_0$. Now show that f has a unique fixed point if and only if $2 \in A_0$.

10.2 (Jordan [Jor, 156–161]) Let p be an odd prime, and r, s and A integers not divisible by p. Show that $rx^2 + sy^2 = A$ has exactly $p - (\frac{-rs}{p})$ solutions in $\mathbb{F}_p$.

10.3 Let N_m denote the cardinality of the solutions of the Fermat congruence $x^m + y^m \equiv 1 \bmod p$, where $p = mn+1$ is prime. Show that $N_m = m^2 N_{0,0} + 2m$ and use Proposition 10.6 to show that

$$|N_m + m - (p+1)| \leq (m-1)(m-2)\sqrt{p} \tag{10.23}$$

if n is even. Show that this implies that the Fermat congruence is always solvable when $p \geq (m-1)^2(m-2)^2 + 6m - 2$. Show that the same result holds for odd n with $N_m + m$ replaced by N_m. (Actually, a similar result holds for primes that are not necessarily $\equiv 1 \bmod m$: see Washington [Was].)

10.4 Put $f(x) = x^3 + a$ in Exercise 1.26 and compute $S_p(f)$. (Hints: if $p \equiv 2 \mod 3$, the result $S_p(f) = 0$ is almost trivial. For $p \equiv 1 \mod 3$, write $p = \pi\overline{\pi}$ for $\pi \equiv 1 \mod 3$ in $\mathbb{Z}[\rho]$ and show that $S_p(f) = -(4a/\pi)_6 \overline{\pi} - -(4a/\overline{\pi})_6 \pi$.)

10.5 We have expressed the solutions of $x^m + y^m = 1$ over finite fields using Jacobi sums $J(\chi, \psi)$. Generalize this to equations of type $a_1 x_1^m + \ldots + a_r x_r^m = 1$ by using generalized Jacobi sums

$$J(\chi_1, \ldots, \chi_r) = (-1)^{r-1} \sum_{t_1 + \ldots + t_r = 1} \chi_1(t_1) \cdots \chi_r(t_r)$$

and prove the analogues of the results in Chapter 4, in particular

$$G(\chi_1) \cdots G(\chi_r) = G(\chi_1 \cdots \chi_r) J(\chi_1, \ldots, \chi_r).$$

10.6 Show that $A \equiv -2 \mod \ell$ for the integer A defined in Proposition 10.7.

10.7 Prove Corollary 10.19.

10.8 Prove Proposition 10.11.

10.9 Prove Theorem 10.28. (Hint: The case $p \equiv 1 \mod 4$ can be handled with Jacobi sums, the case $p \equiv 7 \mod 8$ is almost trivial, and for $p \equiv 3 \mod 8$, you will have to use Eisenstein sums. See Adler [Adl].) Compute the ζ-function of Eisenstein's hyperelliptic curve by generalizing your proof.

10.10 Herglotz's proof of Gauss's Last Entry uses the following facts: write $\pi = a + bi$, $\gamma = i^{b/2} 2^{(p-1)/4}$ and $\mathrm{sl}(\pi z) = xW(x^4)/V(x^4)$ as in Theorem 8.9 (i.e. $V(X^4) = X^{1-p} W(X^{-4})$, $x = \mathrm{sl}\, z$, $y = \mathrm{sl}\, \pi z$ etc.); then there are polynomials $A, B, C, D \in \mathbb{Z}[i][X]$ such that

$$W(x^4) + V(x^4) = (1-i)A(x^4)C(x^4),$$
$$W(x^4) - V(x^4) = (1-i)(1-x^4)B(x^4)D(x^4),$$
$$(1+x^2)^{(p-1)/2} W(y^4) + V(y^4) = 2\gamma A(x^4)D(x^4),$$
$$(1+x^2)^{(p-1)/2} W(y^4) - V(y^4) = 2i\gamma x^2 B(x^4)C(x^4).$$

Defining $2n = p - 2a + 1$ and $2m = p + 2a + 1$, these polynomials have degrees $\deg A = m-2$, $\deg B = n-4$, $\deg C = n$ and $\deg D = m-2$. Can the existence of these polynomials be derived from the properties of W in Proposition 8.7? Herglotz uses Weierstraß' σ-functions for his proof.

10.11 (continued) Theorem 8.9 says that $W(x) \equiv x^{(p-1)/4} \mod \pi$ and $V(x) \equiv 1 \mod \pi$. Derive the following congruences modulo π:

$$(1-i)A(x^4)C(x^4) \equiv x^{p-1} - 1,$$
$$(1-i)(1-x^4)B(x^4)D(x^4) \equiv x^{p-1} - 1,$$
$$2\gamma A(x^4)D(x^4) \equiv (1-x^2)^{(p-1)/2} + (1+x^2)^{(p-1)/2}$$
$$2i\gamma x^2 B(x^4)C(x^4) \equiv (1-x^2)^{(p-1)/2} - (1+x^2)^{(p-1)/2}.$$

Now consider the $p - 5$ residues modulo π that are different from 0, ± 1, $\pm i$. The second congruence shows that they are the roots of the polynomials B and D of degree $n - 4$ and $m - 2$, respectively; congruences three and four show that, for a root x of B or D modulo π, the quotient $\frac{1-x^2}{1+x^2}$ is a quadratic

residue or nonresidue modulo π, respectively, that is, for such x the congruence $x^2y^2 + x^2 + y^2 \equiv 1 \bmod \pi$ is solvable or not.
Adding the four solutions corresponding to $x \equiv 0, \pm 1 \bmod \pi$ as well as the two infinite solutions, conclude that there are exactly $2(n-4) + 6 = 2n - 2 = p - 2a - 1$ solutions.

10.12 Show that $(1 - 2^{1-s})\zeta(s) = 1 - 2^{-s} + 3^{-s} - 4^{-s} \pm \ldots$ for all $s \in \mathbb{C}$ with real part > 1; use a similar formula for $(1 - 3^{1-s})\zeta(s)$ to conclude that $\zeta(s)$ has no pole on $\operatorname{Re} s = 1$ except in $s = 1$. Compute the limit $\lim_{s \to 1} \frac{1 - 2^{1-s}}{s-1}$ and verify that $\lim_{s \to 1}(s-1)\zeta(s) = 1$.

10.13 Derive Corollary 10.15 from Euler's computation of $\zeta(2n)$. Hints: you will need some properties of the Γ-function, namely

$$\Gamma(s+1) = s\Gamma(s), \quad \Gamma(s)\Gamma(1-s) = \frac{\pi}{\sin \pi s}, \quad \Gamma(s)\Gamma\left(s+\frac{1}{2}\right) = \frac{2\sqrt{\pi}}{2^{2s}}\Gamma(2s).$$

Use them to prove $\Gamma\left(\dfrac{1-s}{2}\right)\Gamma\left(\dfrac{1+s}{2}\right) = \dfrac{\pi}{\cos(\pi s/2)}$, as well as

$$\Gamma\left(\frac{s}{2}\right)\Gamma\left(\frac{1+s}{2}\right) = \frac{2\sqrt{\pi}}{2^s}\Gamma(s).$$

Then transform the functional equation of Riemann's ζ-function and derive

$$\zeta(1-s) = \frac{2}{(2\pi)^s}\Gamma(s)\cos\left(\frac{\pi s}{2}\right)\zeta(s).$$

Now put $s = 2m$ and use $\Gamma(2m) = (2m-1)!$.

10.14 Prove Proposition 10.5 by working directly with the definition of the $N_{i,j}$.

10.15 Show that $f(-x) = f(x) - x$ for $f(x) = \frac{xe^x}{e^x - 1}$ and conclude that $B_{2n+1} = 0$ for $n \geq 1$. Show similarly that, for any quadratic Dirichlet character, $B_{2n,\chi} = 0$ if χ is even $(\chi(-1) = 1)$ and $B_{2n+1,\chi} = 0$ if χ is odd. Prove that $B_{n,\chi} \in \mathbb{Q}(\chi)$, where $\mathbb{Q}(\chi)$ is the smallest extension of $\mathbb{Q}$ containing the values of χ.

10.16 Compute the congruence ζ-function for the curve $C_2 : Y^2Z^2 - Z^4 + X^4 = 0$. Why does the result not contradict the Weil conjectures (or Schmidt's Theorem 10.26)?

10.17 Show that the curve C_1 defined by Equation (10.1) is singular. Show that C_2 (and therefore C_1) is birational equivalent to the nonsingular curve $C_3 : Y^2 = X^3 + 4X$ (Hints: divide (10.2) by v^4 and put $t = w/v^2$, $u = 1/v$. Then $1 = u^4 - t^2 = (u^2 - t)(u^2 + t)$. Put $s = u^2 - t$ and conclude that $2u^2 = s + 1/s$. Now multiply by $8s^2$). Compute the number of $\mathbb{F}_p$-rational points on C_3, and find its zeta function.

10.18 Let $P(T) = \prod_{j=1}^{d}(1 - \omega_j T) \in \mathbb{Z}[X]$ be a polynomial of degree d, and let $q = p^f$ be a prime power. Show that the following two conditions are equivalent:
 1. $Z_C(T) = \frac{P(T)}{(1-T)(1-qT)}$ satisfies the functional equation (10.19);
 2. $d = 2g$, the map $\omega_j \longmapsto q/\omega_j$ is a permutation of the set $\{\omega_1, \ldots, \omega_d\}$, and $\prod_{j=1}^{2g} \omega_j = q^g$.

10.19 Consider the curve $C: x^3 + y^3 = z^3$ over $\mathbb{F}_p$ for primes $p \equiv 2 \bmod 3$. Show that $N_1 = p+1$, and deduce that the congruence ζ-function of C over $\mathbb{F}_p$ is given by $Z_C(T) = (1-pT^2)(1-T)^{-1}(1-pT)^{-1}$. Deduce that $N_s = 1 + p^s - (i\sqrt{p})^s - (-i\sqrt{p})^s$, and check that the Weil bounds are attained when s is even

10.20 (Hansen & Pedersen [HP], or Hansen [Han]) Let $q = p^n$ be a prime power and consider the projective curve $C: X_0^{q+1} = X_1^q X_2 + X_1 X_2^q$.
1. Show that C is nonsingular over any field of characteristic p and that it has genus $g = \frac{1}{2}q(q-1)$.
2. Show that C has q^3 points over $k = \mathbb{F}_{q^2}$ plus one infinite point. (Hint: for each $y = y_0$ in the affine equation $x^q + x = y^{q+1}$, there are q solutions in the algebraic closure of k; let x_0 be one of them and show that $x_0 \in k$ by verifying that $x_0^{q^2} + x_0^q = x_0^q + x_0$).
3. Observe that $N_2 = q^3 + 1 = q^2 + 1 + 2gq$; deduce that the inverse roots α_j ($1 \le j \le 2g$) of the polynomial $P_1(T)$ in the numerator of the ζ-function $Z_C(T)$ must all be equal to $\sqrt{-q}$. Compute $P_1(T)$ and check the functional equation.
4. Show that $N_s = 1 + q^s - (i\sqrt{q})^s - (-i\sqrt{q})^s$ for all $s \ge 1$, and check that the Weil bounds are attained when s is even.

10.21 Show that $\mathbb{P}^m(\mathbb{F}_q)$ is a nonsingular projective variety, compute its congruence ζ-function, and check the Weil conjectures.

10.22 Compute the congruence ζ-function of Eisenstein's hyperelliptic curve.

10.23 (Artin's Thesis) Let p be an odd prime, $k = \mathbb{F}_p(X)$, $f \in \mathbb{F}_p[X]$ a polynomial of degree ≥ 1, and $K = k(\sqrt{f})$ the splitting field of the irreducible polynomial $Y^2 - f(X) \in k[Y]$. Show that K/k is a quadratic extension, and that $\mathcal{O}_f = \mathbb{F}_p[X] \oplus \sqrt{f}\mathbb{F}_p[X]$ is a Dedekind ring.

10.24 (continued) Consider the field $\mathbb{F}_p(X)_\infty = \left\{ \sum_{n=-\infty}^{N} a_n X^n : a_n \in \mathbb{F}_p, N \in \mathbb{N} \right\}$ of finite Laurent series in X^{-1} (this field corresponds to the completion $\mathbb{R} = \mathbb{Q}_\infty$ of the rationals). The field K from the preceding exercise is called *real* if and only if f is a square in $\mathbb{F}_p(X)_\infty$, and *imaginary* otherwise. Show that K is real if and only if $f(x) = a_n x^n + \ldots + a_0$ satisfies $n = 2m$ and $a_n = a^2$ in $\mathbb{F}_p$. (Hint: write $f(X) = a^2 x^{2m}(1 + h(X))$ and show that the binomial expansion $\sqrt{1+h} = \sum_{r \ge 1} \binom{1/2}{r} h^r$ makes sense (and converges) in $\mathbb{F}_p(X)_\infty$.) Next show that $\mathcal{O}_f^\times \simeq \mathbb{F}_p^\times$ if K is imaginary, and $\mathcal{O}_f^\times \simeq \mathbb{F}_p^\times \times \mathbb{Z}$ if K is real.

10.25 (For this exercise and those below, see Ireland & Rosen [386, §11.4] or Moreno [Mor, 4.2, 4.3.2]). Show that $\mathbb{F}_q[X]$ is a Euclidean ring and deduce that it is a UFD. For nonzero $f \in \mathbb{F}_q[X]$ of degree $\deg f$, show that $(\mathbb{F}_q[X] : (f)) = q^{\deg f}$. In analogy with the Dedekind rings $\mathcal{O}_K$, this index is called the norm of f and will be denoted by Nf in the Exercises below.

10.26 Let $\lambda: \mathbb{F}_q[X] \longrightarrow \mu_n$ be a multiplicative character on $\mathbb{F}_q[X]$, that is a homomorphism from $\mathbb{F}_q(X)^\times$ into the n-th roots of unity μ_n. Show that

$$\sum_f \lambda(f) T^{-s \deg f} = \prod_p \frac{1}{1 - \lambda(p) T^{-s \deg p}} \qquad (10.24)$$

as an equality of formal power series, where f runs through the monic polynomials in $\mathbb{F}_q[X]$, while p runs through the monic irreducible polynomials (note that irreducible is equivalent to prime here), and deduce that

$$\sum_f \lambda(f) N f^{-s} = \prod_p \frac{1}{1 - \lambda(p) N p^{-s}}$$

converges for $s \in \mathbb{C}$ with $\operatorname{Re} s > 1$. (Hint: This is basically equivalent to the Unique Factorization property.)

10.27 Let χ be a character on $\mathbb{F}_q^\times$ and put $\psi(a) = \zeta_p^{\operatorname{Tr} a}$, where ζ_p is a primitive p-th root of unity, and where Tr denotes the trace in $\mathbb{F}_q/\mathbb{F}_p$. Show that

$$\lambda(x^n + a_{n-1}x^{n-1} + \ldots + a_0) = \chi((-1)^n a_0)\psi(a_{n-1})$$

defines a multiplicative character λ on $\mathbb{F}_q(X)^\times$. Now recall the definition of Kornblum's L-function 10.21 associated to λ

$$L(s; \lambda) = \sum_{f \text{ monic}} \lambda(f) N f^{-s}.$$

Show that $L(s; \lambda) = 1 + \sum_{d=1}^\infty q^{-ds} S_d$, where $S_d = \sum_{\deg f = d} \lambda(f)$, the sum being over all monic $f \in \mathbb{F}_q[X]$ with degree d. Show that $S_1 = -G(\chi, \psi)$ and $S_d = 0$ for all $d \geq 1$, and deduce

$$L(s; \lambda) = 1 - G(\chi, \psi) q^{-s}. \tag{10.25}$$

Observe that this gives an analytic continuation of $L(s; \lambda)$ to the whole complex plane. Show that $|G(\chi, \psi)| = \sqrt{q}$ implies that all zeros of $L(s; \lambda)$ lie on the line $\operatorname{Re} s = \frac{1}{2}$, and compute the zeros explicitly for the quadratic character on $\mathbb{F}_q$.

10.28 (Proof of the Davenport-Hasse Theorem 4.32) Put $F = \mathbb{F}_q$ and $E = \mathbb{F}_{q^s}$ and use the notation of Theorem 4.32. Let λ be the character on $F[X]$ associated to χ and ψ, and let λ' denote the one associated to the lifted characters χ' and ψ'. Prove the following claims:
   i) If $f \in F[X]$ is the (monic) minimal polynomial for $\alpha \in E^\times$, then $\lambda(f)^{s/\deg f} = \lambda'(f')$, where $f' = X + \alpha \in E[X]$.
   ii) $G(\chi', \psi') = -\sum_p (\deg p) \lambda(p)^{s/\deg p}$, where p runs through the monic irreducible polynomials of degree dividing s.
Now consider the formal identity

$$1 - G(\chi, \psi) T = \prod_p (1 - \lambda(p) T^{\deg p})^{-1},$$

where the product runs through the monic irreducible polynomials $p \in F[X]$ (which follows from combining (10.24) with (10.25)). Take the logarithmic derivative with respect to T, multiply through by T, and develop the denominators in a geometric series to get

$$-\sum_{i=1}^\infty G(\chi, \psi)^i T^i = \sum_{j=1}^\infty \sum_p (\deg p) \lambda(p)^j T^{j \deg p}.$$

The theorem now follows via ii) by comparing the coefficients of T^s.

10.29 For an integer $n \geq 1$, show that $F(z) = z^n - (1-z)^n$ satisfies the functional equation $F(z) = \overline{F(1-z)}$ and that its zeros lie on the line $\operatorname{Re} z = \frac{1}{2}$.

Additional References

[Adl] A. Adler, *Eisenstein and the Jacobian varieties of Fermat curves*, Rocky Mt. J. Math. **27** (1997), 1–60

[Art] E. Artin, *Quadratische Körper im Gebiete der höheren Kongruenzen*, Math. Z. **19** (1924), 153–246

[Bo1] E. Bombieri, *Counting points on curves over finite fields (d'après S. A. Stepanov)*, Sem. Bourbaki 1972/73, Exposé No. 430, Lecture Notes Math. **383** (1974), 234–241

[Bo2] E. Bombieri, *Hilbert's 8th problem: An analogue*, Proc. Symp. Pure Math. **28**, De Kalb 1974, 269–274 (1976)

[Car] P. Cartier, *Des nombres premiers a la géometrie algébrique (une brève histoire de la fonction zeta*, Chabert, Jean-Luc (ed.) et al., Cah. Semin. Hist. Math. **2**, Ser. 3 (1993), 51–77

[Cha] J.S. Chahal, *Manin's proof of the Hasse inequality revisited*, Nieuw Arch. Wiskd., IV. Ser. 13, No. **2** (1995), 219–232

[Ch1] S. Chowla, *The last entry in Gauss's diary*, Proc. Nat. Acad. Sci. USA **35** (1949), 244–246

[Ch2] S. Chowla, *The Riemann hypothesis and Hilbert's tenth problem*, London, Glasgow 1965

[Con] J.B. Conrey, *At least two fifths of the zeros of the Riemann zeta function are on the critical line*, Bull. Am. Math. Soc. **20** (1989), 79–81

[Cor] G. Cornacchia, *Sulla congruenza $x^n + y^n \equiv z^n \bmod p$*, Giornale di Mat. **47** (1909), 219–268

[Da1] H. Davenport, *On the distribution of quadratic residues (mod p)*, J. London Math. Soc. **6** (1931), 49–54

[Da2] H. Davenport, *On the distribution of the l-th power residues (mod p)*, J. London Math. Soc. **7** (1932), 117–121

[DaH] H. Davenport, H. Hasse, *Die Nullstellen der Kongruenzzetafunktionen in gewissen zyklischen Fällen*, J. Reine Angew. Math. **172** (1934), 151–182

[Del] P. Deligne, *La conjecture de Weil. I*, Publ. Math. IHES **43** (1974), 273–307

[Den] C. Deninger, *Evidence for a cohomological approach to analytic number theory*, First European congress of mathematics 1992, Vol. I; Prog. Math. **119**, 491–510, Birkhäuser 1994

[Deu] M. Deuring, *Die Typen der Multiplikatorenringe elliptischer Funktionenkörper*, Abh. Math. Sem. Hamburg **14** (1941), 197–272

[Di1] L.E. Dickson, *On the congruence $x^n + y^n + z^n \equiv 0 \bmod p$*, J. Reine Angew. Math. **135** (1909), 134–141

[Di2] L.E. Dickson, *Lower limit for the number of solutions of $x^e + y^e + z^e \equiv 0 \bmod p$*, J. Reine Angew. Math. **135** (1909), 181–188

[Die]   J.A. Dieudonné, *The Weil conjectures*, Math. Intell. **10** (1975), 7–21; the original is hard to find, but it is reprinted in [FrKi].

[Dwo]   B. Dwork, *The rationality of the zeta function of an algebraic variety*, Amer. J. Math. **82** (1960), 631–648

[Ed1]   H.M. Edwards, *The background of Kummer's proof of Fermat's last theorem for regular primes*, Arch. Hist. Exact Sci. **14** (1975), 219–236

[Ed2]   H.M. Edwards, *Postscript to "The background of Kummer's proof"*, Arch. Hist. Exact Sci. **18** (1977), 381–394

[FrKi] E. Freitag, R. Kiehl, *Etale cohomology and the Weil conjecture*, Erg. Math. Grenzgeb. **13**, Springer-Verlag 1988.

[FMS]   C. Friesen, J.B. Muskat, B.K. Spearman, K.S. Williams, *Cyclotomy of order 15 over $GF(p^2)$, $p \equiv 4, 11 \pmod{15}$*, Int. J. Math. Math. Sci. **9** (1986), 665–704

[GS]   G. van der Geer, R. Schoof, *Effectivity of Arakelov divisors and the theta divisor of a number field*, preprint 1998

[Gla]   J. C. Glashan, *Quinquisection of the cyclotomic equation*, Amer. J. Math. **21** (1899), 270–275; RSPM **8** (1900), 3

[GP]   N.M. Glazunov, A.G. Postnikov, *On existence of rational points on the curve $y^2 = f(x)$ over a simple finite field* (Russ.), Studies in number theory. Analytical number theory, Interuniv. Sci. Collect., Saratov **10** (1988), 4–8

[Gr1]   A. Grothendieck, *Sur une note de Mattuck-Tate*, J. Reine Angew. Math. **200** (1958), 208–215

[Gr2]   A. Grothendieck, *Formule de Lefschetz et rationalité des fonctions L*, Séminaire Bourbaki **17** (1964/65), No. 279, 15 pp. (1966).

[HP]   J.P. Hansen, J.P. Pedersen, *Automorphism groups of Ree type, Deligne-Lusztig curves and function fields*, J. Reine Angew. Math. **440** (1993), 99–109

[Han]   S.V. Hansen, *Rational Points on Curves over Finite Fields*, Lecture Notes Series No. **62**, University of Aarhus, 1996; URL: http://www.mi.aau.dk/~shave/

[Har]   R. Hartshorne, *Algebraic geometry*, GTM **52**, Springer Verlag, 1977; 3rd printing 1983

[Has]   H. Hasse, *Beweis des Analogons der Riemannschen Vermutung für die Artinschen und F. K. Schmidtschen Kongruenzzetafunktionen in gewissen elliptischen Fällen*, Nachr. Ges. Wiss. Göttingen No. **42** (1933), 253–262; Math. Abh. II, 85–94

[Hop]   H. Hopf, *Über die Verteilung quadratischer Reste*, Math. Z. **32** (1930), 222–231

[Hou]   C. Houzel, *La préhistoire des conjectures de Weil*, Development of mathematics 1900–1950, Birkhäuser 1994, 385–414

[Hu]   W.W.J. Hulsbergen, *Conjectures in arithmetic algebraic geometry. A survey*, Aspects of Mathematics; 2nd ed. 1994

[Hu1]   A. Hurwitz, *Einige Eigenschaften der Dirichlet'schen Funktionen*, Zeit. Math. Phys. **27** (1882), 86–101; Math. Werke I, 72–88

[Hu2]   A. Hurwitz, *Über die Entwicklungskoeffizienten der lemniskatischen Funktionen*, Math. Ann. **51** (1899), 196–226

[Hu3]   A. Hurwitz, *Über die Kongruenz $ax^e + by^e + cz^e \equiv 0 \mod p$*, J. Reine Angew. Math. **136** (1909), 272–292; Math. Werke II, Birkhäuser (1933), 430–445

[Jor]   C. Jordan, *Traité des substitutions et des équations algebraiques*, 1870

[Kat]   N. Katz, *An overview of Deligne's proof of the Riemann hypothesis for varieties over finite fields*, Math. Dev. Hilbert Probl., Proc. Symp. Pure Math. **28** (1976), 275–305

[KT]   N. Katz, J. Tate, *Bernard Dwork (1923–1998)*, Notices AMS **46** (1999), 338–343

[Kem]   T.M.H. Kemnitz, *Über die Lösungen der Kongruenz $x^3 + y^3 \equiv 1 \mod \pi$ und die Teilungsgleichungen der elliptischen äquianharmonischen Funktionen* (Dissertationsauszug), Jahrb. phil. Fak. Leipzig 1923, 119–122; FdM **48** (1921–22), 190

[Kle]   F. Klein, *Gauß' wissenschaftliches Tagebuch 1796–1814*, Math. Ann. **57**, 1–34

[Klu]   J.C. Kluyver, *De voortzetting van eene eenwardige functie, voorgesteld door eene dubbel oneinddige reeks*, Koninklijke Akad. Wetensch. Amsterdam **8** (1899/1900), 179–185

[Ko1]   N. Koblitz, *p-adic numbers, p-adic analysis, and zeta-functions*, Springer-Verlag 1977

[Ko2]   N. Koblitz, *The p-adic approach to solutions of equations over finite fields*, Amer. Math. Monthly **87** (1980), 115–118

[Ko3]   N. Koblitz, *Why study equations over finite fields?*, Math. Mag. **55** (1982), 144–149

[Koc]   H. Koch, *Zahlentheorie: algebraische Zahlen und Funktionen*, Vieweg 1997

[Kor]   H. Kornblum, *Über die Primfunktionen in einer arithmetischen Progression*, Math. Z. **5** (1919), 100–111

[Li]   W.C. Winnie Li, *Number theory with applications*, World Scientific, 1995

[Lib]   G. Libri, *Memoire sur la résolution de quelques équations indéterminées*, J. Reine Angew. Math. **9** (1832), 277–294

[LN]   R. Lidl, H. Niederreiter, *Finite Fields*, Encyclopedia of Mathematics and Its Applications **20**, 2nd ed., Cambridge Univ. Press 1996

[Lor]   D. Lorenzini, *An invitation to arithmetic geometry*, Grad. Stud. Math **9**, Amer. Math. Soc. 1995

[LRW]   J. van de Lune, H.J.J. te Riele, D.T. Winter, *On the zeros of the Riemann zeta-function in the critical strip. IV*, Math. Comput. **46** (1986), 667–681

[Ma] Yu.I. Manin, *On cubic congruences to a prime modulus* (Russ.), Izv. Akad. Nauk SSSR, Ser. Mat. **20** (1956), 672–678 Engl. transl. Am. Math. Soc., Transl., II. Ser. **13** (1960), 1–7

[MT] A. Mattuck, J. Tate, *On the inequality of Castelnuovo-Severi*, Abh. Math. Semin. Univ. Hamburg **22** (1958), 295–299

[Maz] B. Mazur, *Review of Ernst Eduard Kummer, Collected Papers*, Bull. Amer. Math. Soc. **83** (1977), 976–988

[Mon] P. Monsky, *p-adic analysis and zeta functions*, Lectures in Mathematics, Kyoto Univ. 4, 1970

[Mor] C.J. Moreno, *Algebraic curves over finite fields*, Cambridge Tracts in Mathematics **97**, CUP 1991

[Mor] B. Morlaye, *Démonstration élémentaire d'un théorème de Davenport et Hasse*, Ens. Math. **18** (1972), 269–276

[New] D.S. Newman, *Simple analytic proof of the prime number theorem*, Am. Math. Mon. **87** (1980), 693–696

[PAR] J. C. Parnami, M. K. Agrawal, A. R. Rajwade, *The number of points on the curve $y^2 = x^5 + a$ in $\mathbb{F}_q$ and application to local zeta-function*, Math. Stud. 48 (1984), 205–211 (1980)

[Pe1] A.E. Pellet, *Mémoire sur la théorie algébrique des équations*, Bull. Soc. Math. France **15** (1887), 61–102

[Pe2] A.E. Pellet, *Réponse à une question de E. Dubouis*, L'Interm. des Math. **18** (1911), 81–82

[Pep] Th. Pépin, *Sur diverses tentatives de démonstration du théorème de Fermat*, C. R. Acad. Sci. Paris **91** (1880), 366–368

[Ra1] A. R. Rajwade, *A note on the number of solutions N_p of the congruence $y^2 \equiv x^3 - Dx \pmod{p}$*, Proc. Cambridge Phil. Soc. **67** (1970), 603–605

[Ra2] A. R. Rajwade, *On the congruence $y^2 \equiv x^5 - a \pmod{p}$*, Proc. Cambridge Phil. Soc. **74** (1973), 473–475

[Ra3] A. R. Rajwade, *Notes on the congruence $y^2 \equiv x^5 - a \pmod{p}$*, Ens. Math. **21** (1975), 49–56

[RSS] M. Rapoport, N. Schappacher, P. Schneider, *Beilinson's conjectures on special values of L-functions*, Perspectives in Mathematics, 4 (1988)

[Rib] P. Ribenboim, *13 lectures on Fermat's Last Theorem*, Springer 1979

[RC] P. Robba, G. Christol, *Equations differentielles p-adiques. Applications aux sommes exponentielles*, Paris 1994

[Ro1] P. Roquette, *Exponential sums*, Lecture at the German-Russian mathematics seminar in Heidelberg 1995 and at the Escola de Algebra, Rio de Janeiro (1996); Matematica Contemporanea **14** (1998)

[Ro2] P. Roquette, *Zur Geschichte der Zahlentheorie in den dreißiger Jahren*, Math. Semesterber. **45** (1998), 1–38

[Ro3] P. Roquette, *Manins Beweis*, email from April 29, 1998

[SF1]   F.K. Schmidt, *Allgemeine Körper im Gebiet der höheren Kongruenzen*, Diss. Freiburg 1925
[SF2]   F.K. Schmidt, *Analytische Zahlentheorie in Körpern der Charakteristik p*, Math. Z. **33** (1931), 1–32
[SW]   W.M. Schmidt, *Equations over finite fields. An elementary approach*, Lecture Notes Math. 536, Springer-Verlag 1976
[Sch]   I. Schur, *Über die Kongruenz $x^m + y^m \equiv z^m$ mod p*, Jahresber. DMV **25** (1917), 114–117
[Se1]   J.-P. Serre, *Rationalité des fonctions ζ des variétés algébriques*, Semin. Bourbaki **12** (1959/60), No. 198, (1960)
[Se2]   J.-P. Serre, *Zeta and L functions*, Arithmetical algebraic Geom., Proc. Conf. Purdue Univ. 1963, 82–92 (1965)
[Se3]   J.-P. Serre, *Sur le nombre des points rationnels d'une courbe algébrique sur un corps fini*, C. R. Acad. Sci. Paris **296** (1983), 397–402; Œuvres III, 658–663
[ST]   J.H. Silverman, J. Tate, *Rational points on elliptic curves*, Undergraduate Texts in Mathematics, Springer-Verlag 1992
[SR]   S. Singh, A. R. Rajwade, *The number of solutions of the congruence $y^2 \equiv x^4 - a$ (mod p)*, L'Ens. Math. **20** (1974), 265–273
[Sma]   C. Small, *Arithmetic of finite fields*, New York 1991
[Sta1] H. Stark, *On the Riemann hypothesis in hyperelliptic function fields*, Proc. Sympos. Pure Math. **24**, Miss. 1972, 285–302 (1973)
[Sta2] H. Stark, *Values of zeta and L-functions*, Abh. Braunschw. Wiss. Ges. **33** (1982), 71–83
[Sta3] H. Stark, *Dirichlet's class-number formula revisited*, A tribute to Emil Grosswald, AMS, Contemp. Math. **143** (1993), 571–577
[Sta4] H. Stark, *Galois theory, algebraic number theory, and zeta functions*, From number theory to physics (Waldschmidt et al, eds.) Springer-Verlag 1992, 313–393
[Ste1] S.A. Stepanov, *On the number of points of a hyperelliptic curve over a finite prime field*, (Russian), Izv. Akad. Nauk SSSR, Ser. Mat. **33** (1969), 1171–1181; Engl. Transl.: Math. USSR, Izv. **3** (1971), 1103–1114 (1971)
[Ste2] S.A. Stepanov, *An elementary proof of the Hasse-Weil theorem for hyperelliptic curves*, J. Number Theory **4** (1972), 118–143
[Ste3] S.A. Stepanov, *Arithmetic of Algebraic Curves*, Monographs in Contemporary Mathematics, 1994
[Sti]   H. Stichtenoth, *Algebraic function fields and codes*, Springer 1993
[SV]   K.-O. Stöhr, J.F. Felipe, *Weierstrass points and curves over finite fields*, Proc. Lond. Math. Soc. **52** (1986), 1–19
[Syl]   J.J. Sylvester, *Sur une propriété des nombres premiers qui se rattache au théorème de Fermat*, C. R. Acad. Sci. Paris **52** (1861), 161–163

[Ter] G. Terjanian, *L'oeuvre arithmétique de Kummer*, Gaz. Math. Soc. Math. Fr. **66** (1995), 45–53

[Tho] A.D. Thomas, *Zeta-functions: An introduction to algebraic geometry*, Research Notes in Math. 12, 1977

[Tol] E. Tollis, *Zeros of Dedekind zeta functions in the critical strip*, Math. Comput. **66** (1997), 1295–1321

[vdP] A. van der Poorten, *A proof that Euler missed. Apéry's proof of the irrationality of $\zeta(3)$. An informal report*, Math. Intell. **1** (1979), 195–203

[Was] L. Washington, *Introduction to Cyclotomic Fields*, Graduate Texts in Math. **83**, Springer Verlag 1982; 2nd edition 1997

[We1] A. Weil, *Sur les courbes algébriques et les variétés qui s'en déduisent*, Hermann, Paris 1948

[We2] A. Weil, *On some exponential sums*, Proc. Nat. Acad. Sci. USA **34** (1948), 204–207

[We3] A. Weil, *Number of solutions of equations in finite fields*, Bull. Amer. Math. Soc. **55** (1949), 497–508;

[We4] A. Weil, *Prehistory of the Zeta-Function*, Number Theory, Trace Formulas and Discrete Groups (K. E. Aubert, E. Bombieri, D. Goldfeld, eds.) Academic Press 1989, 1–9

[Wid] A. Widmer, *Über die Anzahl der Lösungen gewisser Kongruenzen nach einem Primzahlmodul*, Diss. Zürich, 1919

[Wit] E. Witt, *Collected papers – Gesammelte Abhandlungen*, Springer Verlag 1998

[Zag] D. Zagier, *Newman's short proof of the prime number theorem*, Am. Math. Mon. **104** (1997), 705–708

[Zan] U. Zannier, *Polynomials modulo p whose values are squares (elementary improvements on some consequences of Weil's bounds)*, L'Ens. Math. **44** (1998), 95–102

[Zim] H.G. Zimmer, *An elementary proof of the Riemann hypothesis for an elliptic curve over a finite field*, Pac. J. Math. **36** (1971), 267–278

11. Eisenstein Reciprocity

In order to prove higher reciprocity laws, the methods known to Gauss were soon found to be inadequate. The most obvious obstacle, namely the fact that the unique factorization theorem fails to hold for the rings $\mathbb{Z}[\zeta_\ell]$, was overcome by Kummer through the invention of his ideal numbers. The direct generalization of the proofs for cubic and quartic reciprocity, however, did not yield the general reciprocity theorem for ℓ-th powers: indeed, the most general reciprocity law that could be proved within the cyclotomic framework is Eisenstein's reciprocity law. The key to its proof is the prime ideal factorization of Gauss sums; since we can express Gauss sums in terms of Jacobi sums and vice versa, the prime ideal factorization of Jacobi sums would do equally well.

Although Eisenstein's reciprocity law is only a very special case of more general reciprocity laws, it turned out to be an indispensable step for proving these general laws until Furtwängler [253] succeeded in finally giving a proof of the reciprocity law in $\mathbb{Q}(\zeta_\ell)$ without the help of Eisenstein's reciprocity law. It should also be noted that Eisenstein's reciprocity law holds for all primes ℓ, whereas Kummer had to assume that ℓ is regular, i.e. that ℓ does not divide the class number of $\mathbb{Q}(\zeta_\ell)$.

Using the prime ideal factorization of Gauss sums together with the trivial fact that the mth power of Gauss sums generate principal ideals in $\mathbb{Z}[\zeta_m]$, we will be able to deduce amazing properties of ideal class groups of abelian extensions of $\mathbb{Q}$. This idea goes back to work of Jacobi, Cauchy and Kummer, was extended by Stickelberger and revived by Iwasawa. Later refinements and generalizations due to Thaine, Kolyvagin and Rubin will be discussed only marginally.

11.1 Factorization of Gauss Sums

In the mathematical literature there exist many proofs for the *Stickelberger relation*, which gives the prime ideal factorization of Gauss sums. The simplest proof unfortunately works only for the primes $p \equiv 1 \bmod m$, and this is why we treat this case separately.

358   11. Eisenstein Reciprocity

First we will show that the Stickelberger relation follows almost from the fact that adjoining $G(\chi) = -\sum_{t \in \mathbb{F}_q^\times} \chi(t) \zeta_p^{\text{Tr}(t)}$ to $\mathbb{Q}(\zeta_m)$ generates an abelian extension. We will complete the proof by following Hilbert's Zahlbericht [368].

Let $\mathfrak{p}$ be a prime ideal in $K = \mathbb{Q}(\zeta_m)$ above $p \equiv 1 \bmod m$, and suppose that χ is a multiplicative character of order m on $\mathbb{F} = \mathcal{O}_K/\mathfrak{p}$. From Chapter 4 we know that $G(\chi)^m \in \mathbb{Z}[\zeta_m]$; moreover, K is the decomposition field of $\mathfrak{p}$ in $L = \mathbb{Q}(\zeta_{pm})$, because $p \equiv 1 \bmod m$ guarantees that p splits completely in $K/\mathbb{Q}$. We also know from $G(\chi)\overline{G(\chi)} = p$ that only prime ideals above p can occur in the prime ideal factorization of $\mu = G(\chi)^m$. Since $\Gamma = \text{Gal}(K/\mathbb{Q})$ acts transitively on the prime ideals above $\mathfrak{p}$, we can write $\mu \mathcal{O}_K = \mathfrak{p}^\gamma$ for some $\gamma = \sum_\sigma b_\sigma \sigma \in \mathbb{Z}[\Gamma]$, where $\mathbb{Z}[\Gamma]$ denotes the group ring of Γ, and where γ depends on the choice of the prime ideal $\mathfrak{p}$.

Remark. The group ring $\mathbb{Z}[G]$ of a finite group G is simply the set of formal sums $\{\sum_{\sigma \in G} a_\sigma \sigma\}$. If M is an abelian group on which G acts, then we can make $\mathbb{Z}[G]$ act on M via $(\sum_{\sigma \in G} a_\sigma \sigma) m = \sum_{\sigma \in G} a_\sigma \sigma(m)$. Actually, we have been doing this before without even noticing it when we used expressions like $\alpha^{\sigma-1}$ as an abbreviation for $\alpha^\sigma \alpha^{-1}$.

In order to determine γ we first take the absolute norm of μ^2 and find

$$N_{K/\mathbb{Q}} \mu^2 = N_{K/\mathbb{Q}}(\mu\overline{\mu}) = N_{K/\mathbb{Q}}(p^m) = p^{m(K:\mathbb{Q})} = p^{m\phi(m)}, \text{ and}$$
$$N_{K/\mathbb{Q}} \mu = N_{K/\mathbb{Q}} \mathfrak{p}^\gamma = p^S, \text{ where } S = \sum_{\sigma \in \Gamma} b_\sigma.$$

This implies $\sum_{\sigma \in \Gamma} b_\sigma = \frac{1}{2} m \phi(m)$. The proof we are about to give will proceed as follows:

1. First we observe that $0 \leq b_\sigma \leq m$: of course $b_\sigma \geq 0$ since $G(\chi)$ is integral; moreover, $b_\sigma \leq m$ follows from $\mu\overline{\mu} = p^m$.
2. Then we will show that $(b_\sigma, m) = 1$; this will follow from the fact that adjoining $G(\chi)$ to K gives an extension L of degree $(L:K) = m$.
3. The fact that L/K is abelian will allow us to derive that the b_σ form a complete system of coprime residues mod m;
4. Finally, a simple inequality will imply that the b_σ take the minimal positive coprime residues mod m, and Stickelberger's relation will follow.

Now we will prove that $(a, m) = 1$, where $\mathfrak{p}^a \parallel \mu$. To this end we will show that $K(\sqrt[m]{\mu})/K$ is an abelian extension of degree m; once we know this, the proof of $(a, m) = 1$ is immediate: suppose that $b = (a, m)$; then $\mathfrak{p}$ has ramification index $\frac{m}{b}$ in $K(\sqrt[m]{\mu})/K$ by the decomposition law in Kummer extensions. On the other hand $K(\sqrt[m]{\mu})/K$ is a sub-extension of $K(\zeta_p)/K$, which is completely ramified above $\mathfrak{p}$: this shows that $\mathfrak{p}$ has ramification index $(K(\sqrt[m]{\mu}):K) = m$. Comparing both expressions yields $b = 1$.

Lemma 11.1. *Let χ be a character of order m on $\mathbb{F}_p$, $p \equiv 1 \bmod m$, let $G(\chi)$ be the corresponding Gauss sum, and put $L = \mathbb{Q}(\zeta_m, G(\chi))$. Then $L \subseteq \mathbb{Q}(\zeta_{mp})$, and $(L : \mathbb{Q}(\zeta_m)) = m$.*

11.1 Factorization of Gauss Sums

Proof. Let g be a primitive root mod p and put $k = L \cap \mathbb{Q}(\zeta_p)$; define $\sigma \in \mathrm{Gal}\,(\mathbb{Q}(\zeta_{mp})/\mathbb{Q})$ by $\sigma : \zeta_p \longmapsto \zeta_p^g$, $\zeta_m \longmapsto \zeta_m$. Now $\chi(g)$ is a primitive m-th root of unity, hence the relation $G(\chi)^\sigma = \chi(g)^{-1}G(\chi)$ shows that σ^m is the smallest power of σ fixing L. This shows that $L/\mathbb{Q}(\zeta_m)$ is cyclic of degree m, and our claim follows. □

Now $L = K(\sqrt[m]{\mu})$ is an abelian extension of $\mathbb{Q}$; for each $\sigma \in \mathrm{Gal}\,(L/\mathbb{Q})$ define $a(\sigma) \in (\mathbb{Z}/m\mathbb{Z})^\times$ by $\sigma(\zeta_m) = \zeta_m^{a(\sigma)}$. From Corollary 4.17 we infer that $\mu^{\sigma-a(\sigma)} = \xi^m$ for some $\xi \in L^\times$. This shows that the exponent of $\mathfrak{p}$ in $\mu^{\sigma-a(\sigma)}$ is divisible by m, i.e., that $\sigma\gamma \equiv a(\sigma)\gamma \bmod m$. Let σ_a denote the automorphism of $\mathbb{Q}(\zeta_m)/\mathbb{Q}$ mapping $\zeta_m \longmapsto \zeta_m^a$ and write $\gamma = \sum_{(a,m)=1} b_a \sigma_a$. We find $\sigma_c \gamma = \sum b_a \sigma_a \sigma_c = \sum b_a \sigma_{ac}$ and $a(\sigma_c)\gamma = c\gamma = \sum a(\sigma) b_a \sigma_a = \sum a(\sigma) b_{ac} \sigma_{ac}$, hence $\sigma\gamma \equiv a(\sigma)\gamma \bmod m$ implies the congruence $b_{ac} \equiv c^{-1} b_a \bmod m$ for all $a, c \in (\mathbb{Z}/m\mathbb{Z})^\times$.

Now choose t such that b_t is minimal among the b_c; then

$$\sum_a b_a = \sum_c b_{ct} \equiv \sum_c c^{-1} b_t = b_t \sum_c c^{-1} \bmod m.$$

On the other hand, letting $M \subset \{1, \ldots, m-1\}$ denote the set of minimal positive coprime residues mod m, we find

$$\sum_{c \in M} c = \frac{1}{2}\left(\sum_{c \in M} c + \sum_{c \in M}(m-c)\right) = \frac{1}{2}m\phi(m).$$

Since the b_a are integers ≥ 1, this shows that $\sum_a b_a = \frac{1}{2}m\phi(m)$ can only hold if $b_t = 1$, that is if the b_a actually take all the values of M exactly once. Therefore $G(\chi)^m \mathcal{O}_m = \mathfrak{p}^\gamma$ for some suitable prime ideal $\mathfrak{p}$ in $\mathcal{O}_m = \mathbb{Z}[\zeta_m]$ above p, with $\gamma = \sum_t t^{-1} \sigma_t \in \mathbb{Z}[\Gamma]$, and where t^{-1} denotes the smallest positive integer such that $t^{-1}t \equiv 1 \bmod m$. If we denote the fractional part of a real number x by $\langle x \rangle$ (i.e., $\langle x \rangle = x - \lfloor x \rfloor$, where $\lfloor \cdot \rfloor$ is Gauss's floor function), then $G(\chi)^m \mathcal{O}_m = \mathfrak{p}^\gamma$, $\gamma = m\theta$, with $\theta = \sum_{(t,m)=1} \langle \frac{t}{m} \rangle \sigma_t^{-1}$. We have seen

Proposition 11.2. *Let χ be a character of order m on $\mathbb{F}_p$, $p \equiv 1 \bmod m$, and let $G(\chi)$ be the corresponding Gauss sum. Then there exists a prime ideal $\mathfrak{p} \mid p\mathcal{O}_m$ such that*

$$G(\chi)^m \mathcal{O}_m = \mathfrak{p}^{m\theta}, \quad \theta = \sum_{\substack{0 < t < m \\ (t,m)=1}} \langle \tfrac{t}{m} \rangle \sigma_t^{-1}. \tag{11.1}$$

360   11. Eisenstein Reciprocity

What we have proved so far about the prime ideal factorization of the Gauss sum would suffice to show that $\left(\frac{a}{\alpha}\right)_m = 1 \iff \left(\frac{\alpha}{a}\right)_m = 1$; if we want to prove that both expressions are always equal we need more information, i.e. we have to specify the prime ideal $\mathfrak{p}$ in the preceding proposition. In fact we claim that $\mathfrak{p} \parallel G(\chi)^m$ if $\chi = \left(\frac{\cdot}{\mathfrak{p}}\right)_m^{-1}$. Since $\mathfrak{p} = \mathfrak{P}^{p-1}$ in $\mathbb{Z}[\zeta_{mp}]$, we see that

$$\mathfrak{p} \parallel G(\chi)^m \iff \mathfrak{P}^{p-1} \parallel G(\chi)^m \iff \mathfrak{P}^{(p-1)/m} \parallel G(\chi).$$

Put $n = \frac{p-1}{m}$ and $\Pi = \zeta_p - 1$; we will compute $G(\chi)$ mod Π^{n+1}:

Lemma 11.3. *Let the notation be as above. Then*

$$G(\chi) \equiv \frac{\Pi^n}{n!} \mod \mathfrak{P}^{n+1}. \tag{11.2}$$

Proof. This is a slightly tricky computation:

$$-G(\chi) = \sum_{a=1}^{p-1} \chi(a)\zeta_p^a = \sum_{a=1}^{p-1} \chi(a)(1+\Pi)^a = \sum_{a=1}^{p-1} \chi(a) \sum_{j=0}^{a} \binom{a}{j} \Pi^j$$

$$\equiv^{1,2} \sum_{j=0}^{n} \sum_{a=1}^{p-1} a^{p-1-n} \binom{a}{j} \Pi^j =^{3,4} \sum_{a=1}^{p-1} a^{p-1-n} \binom{a}{n} \Pi^n$$

$$= \sum_{a=1}^{p-1} a^{p-1-n} \frac{a^n}{n!} \Pi^n \equiv (p-1)\frac{\Pi^n}{n!} \equiv^5 -\frac{\Pi^n}{n!} \mod \Pi^n \mathfrak{P}.$$

In this computation we have used the following facts:
1. $\chi(a) \equiv a^{p-1-n} \mod \mathfrak{p}$;
2. $\Pi^j \equiv 0 \mod \Pi^n$ for $j > n$;
3. $\binom{a}{j}$ is a polynomial of degree j in a; in particular, $a^{p-1-n}\binom{a}{j}$ contains a monomial of degree divisible by $p-1$ if and only if $j = n$;
4. $\sum_{a=1}^{p-1} a^k \equiv 0 \mod p$ if k is not divisible by $p-1$ (see Proposition 4.29);
5. $\mathfrak{P} \mid \Pi$; in particular, the congruence $G(\chi) \equiv \Pi^n/n!$ is valid modulo $\mathfrak{P}^n$. Observe the analogy to the computation in Section 8.7. □

The congruence $G(\chi) \equiv \Pi^n/n! \mod \mathfrak{P}^{n+1}$ implies, as we already have pointed out, that $\mathfrak{p} \parallel G(\chi)^m$, and we have proved:

Theorem 11.4. *Let $p \equiv 1 \mod m$ be prime, and let $\mathfrak{p}$ be a prime ideal above p in $K = \mathbb{Q}(\zeta_m)$. Then $\chi = \left(\frac{\cdot}{\mathfrak{p}}\right)_m^{-1}$ is a multiplicative character of order m on $\mathbb{F}_p$, and the corresponding Gauss sum $G(\chi)$ has the factorization $G(\chi)^m \mathcal{O}_m = \mathfrak{p}^{m\theta}$, where θ is defined in (11.1).*

This looks more complicated than it is; here are a few examples that illustrate the factorization of Gauss sums $G(\chi)^m$ for characters $\chi = (\cdot/\mathfrak{p})_m$ over $\mathbb{F}_p$, $p \equiv 1 \mod m$; here $\mathfrak{p}_i$ denotes the prime ideal $\sigma_i(\mathfrak{p})$ with $\mathfrak{p} = \mathfrak{p}_1$:

m	$G(\chi^{-1})^m$	$G(\chi)^m$	$J(\chi,\chi)$
2	$\mathfrak{p}$	$\mathfrak{p}$	
3	$\mathfrak{p}_1\mathfrak{p}_2^2$	$\mathfrak{p}_1^2\mathfrak{p}_2$	$\mathfrak{p}_1$
4	$\mathfrak{p}_1\mathfrak{p}_3^3$	$\mathfrak{p}_1^3\mathfrak{p}_3$	$\mathfrak{p}_1$
5	$\mathfrak{p}_1\mathfrak{p}_2^3\mathfrak{p}_3^2\mathfrak{p}_4^4$	$\mathfrak{p}_1^4\mathfrak{p}_2^2\mathfrak{p}_3^3\mathfrak{p}_4$	$\mathfrak{p}_1\mathfrak{p}_3$
7	$\mathfrak{p}_1\mathfrak{p}_2^4\mathfrak{p}_3^5\mathfrak{p}_4^2\mathfrak{p}_5^3\mathfrak{p}_6^6$	$\mathfrak{p}_1^6\mathfrak{p}_2^3\mathfrak{p}_3^2\mathfrak{p}_4^5\mathfrak{p}_5^4\mathfrak{p}_6$	$\mathfrak{p}_1\mathfrak{p}_4\mathfrak{p}_5$
8	$\mathfrak{p}_1\mathfrak{p}_3^3\mathfrak{p}_5^5\mathfrak{p}_7^7$	$\mathfrak{p}_1^7\mathfrak{p}_3^5\mathfrak{p}_5^3\mathfrak{p}_7$	$\mathfrak{p}_1\mathfrak{p}_5$

The factorization of $G(\chi)^m$ into prime ideals follows from the factorization of $G(\chi^{-1})$ given Theorem 11.4 and the relation $G(\chi)G(\chi^{-1}) = \pm p$. The corresponding results for the Jacobi sum can be derived from $J(\chi,\chi)^m = G(\chi)^{2m}G(\chi^2)^{-m}$ and $G(\chi^2)^m = \sigma_2(G(\chi)^m)$ (the last equality only holds for odd m; if m is even, $G(\chi^2)$ is known from the computations for $m/2$).

For odd prime values of m, one finds (see Exercise 11.2)

Corollary 11.5. *Let ℓ and $p \equiv 1 \bmod \ell$ be odd primes; assume that $\mathfrak{p}$ is a prime ideal above p in $\mathbb{Z}[\zeta_\ell]$, and put $\chi = (\,\cdot\,/\mathfrak{p})_\ell$. Then*

$$(J(\chi,\chi)) = \mathfrak{p}^s, \quad s = \sum_{t=1}^{\ell-1} \left\lfloor \frac{2t}{\ell} \right\rfloor \sigma_{-t}^{-1}.$$

Turning this procedure around, one can prove Proposition 11.2 by exploiting the fact that Jacobi sums are integral (see Exercise 11.5).

11.2 Eisenstein Reciprocity for ℓ-th Powers

Now that we know the prime ideal factorization of the Gauss sum, we will use the special case $m = \ell$ prime to prove Eisenstein's reciprocity law. This law will take its simplest form if we restrict it to numbers $\equiv 1$ modulo a high power of $\lambda = 1 - \zeta_\ell$ (compare the special case of cubic and quartic residues). We will call $\alpha \in \mathbb{Z}[\zeta_\ell]$ *semi-primary* if $(\alpha, \ell) = 1$ and $\alpha \equiv a \bmod (1 - \zeta_\ell)^2$ for some $a \in \mathbb{Z}$.

Lemma 11.6. *Let ℓ be an odd prime, and suppose that $(\alpha, \ell) = (\beta, \ell) = 1$ for some $\alpha, \beta \in \mathbb{Z}[\zeta_\ell]$. Then*

i) *there is a unique $c \in \mathbb{Z}/\ell\mathbb{Z}$ such that $\zeta_\ell^c \alpha$ is semi-primary;*
ii) *if α, β are semi-primary, then so are $\alpha \pm \beta$ and $\alpha\beta$;*
iii) *α^ℓ is semi-primary;*
iv) *if $\alpha \in \mathbb{Z}[\zeta + \zeta^{-1}]$, then α is semi-primary;*
v) *if α is a semi-primary unit, then $\alpha \in \mathbb{Z}[\zeta + \zeta^{-1}]$;*
vi) *Jacobi sums are semi-primary; more exactly: if $\chi, \psi \neq \mathbb{1}$ are characters of order ℓ on $\mathbb{F}_q$, then $J(\chi, \psi) \equiv 1 \bmod (1 - \zeta_\ell)^2$.*

Proof. These are straightforward computations: i) Let $\lambda = \zeta_\ell - 1$; then $\mathfrak{l} = (\lambda)$ is the prime ideal above ℓ in $\mathbb{Q}(\zeta_\ell)$, and we find

$$\alpha = \sum_{j=0}^{\ell-1} a_j \zeta_\ell^j = \sum_{j=0}^{\ell-1} a_j(1+\lambda)^j \equiv \sum_{j=0}^{\ell-1} a_j(1+j\lambda) = a + b\lambda \bmod \mathfrak{l}^2$$

for some $a, b \in \mathbb{Z}$. Observe that $\ell \nmid a$ since $(\alpha, \ell) = 1$, and define $c \in \mathbb{Z}$ by $ac \equiv b \bmod \ell$; then $\zeta_\ell^c = (1-\lambda)^c \equiv 1 - c\lambda \bmod \mathfrak{l}^2$, hence $\zeta_\ell^c \alpha \equiv (a+b\lambda)(1-c\lambda) \equiv a + (ac-b)\lambda \equiv a \bmod \mathfrak{l}^2$.
ii) is clear;
iii) Let $\alpha \equiv \beta \bmod \lambda$; then $\alpha^\ell - \beta^\ell = \prod_{j=0}^{\ell-1}(\alpha - \zeta^j \beta) \equiv 0 \bmod \lambda^\ell$. Thus $\alpha \equiv a \bmod \lambda$ implies immediately $\alpha^\ell \equiv a^\ell \equiv a \bmod (1-\zeta)^2$, and α^ℓ is semi-primary as claimed.
iv) Assume that $\alpha \equiv a + b\lambda \bmod \lambda^2$ for integers $a, b \in \mathbb{Z}$; then $\alpha \equiv a + b - b\zeta \bmod \lambda^2$ and $\overline{\alpha} \equiv a + b - b\zeta^{-1} \bmod \lambda^2$ imply that $0 = \alpha - \overline{\alpha} \equiv b(\zeta - \zeta^{-1}) \bmod \lambda^2$, since α is real. But $\lambda \,\|\, (\zeta - \zeta^{-1})$ shows $b \equiv 0 \bmod \lambda$, hence we have $b \equiv 0 \bmod \ell$ and $\alpha \equiv a \bmod \lambda^2$.
v) We can write $\alpha = \pm\zeta^j \alpha_0$, where $\alpha_0 \in \mathbb{Z}[\zeta + \zeta^{-1}]$ is a real unit. Since α is semi-primary, we have $\alpha \equiv a \bmod \lambda^2$. Now $a \equiv \overline{a} \equiv \pm \zeta^{-j} \alpha_0 \bmod \lambda^2$, together with the fact that $\lambda \nmid \alpha_0$ (since α_0 is a unit) implies $\lambda^2 \mid (\zeta^j - \zeta^{-j})$; this in turn is only possible if $j \equiv 0 \bmod \ell$, i.e., if $\alpha = \alpha_0$ is real.
vi) using the congruences $(\chi(t) - 1)\psi(1 - t) \equiv \chi(t) - 1 \bmod (1 - \zeta_\ell)^2$ and $q \equiv p^f \equiv 1 \bmod \ell$ we find

$$\begin{aligned} J(\chi, \psi) &= -\sum_{t \neq 0,1} \chi(t)\psi(1-t) \\ &= -\sum_{t \neq 0,1}(\chi(t) - 1)\psi(1-t) - \sum_{t \neq 0,1} \psi(1-t) \\ &\equiv -\sum_{t \neq 0,1}(\chi(t) - 1) - \sum_{t \neq 0,1} \psi(1-t) \\ &\equiv \chi(1) + \psi(1) + (q - 2) = q \equiv 1 \bmod (1-\zeta_\ell)^2. \end{aligned}$$

This completes the proof. $\square$

Our next result concerns the power character of $\mu = G(\chi)^m$; since the proof does not depend on $m = \ell$ being prime, we treat the general case of arbitrary $m \geq 2$ (observe that the lemma below contains the quadratic reciprocity law as the special case $m = 2$!):

Lemma 11.7. *Let $p \equiv 1 \bmod m$ be prime, let $\mathfrak{p}$ be a prime ideal above p in $K = \mathbb{Q}(\zeta_m)$, and let $\mu = G(\chi)^m$, where $G(\chi)$ is the Gauss sum corresponding to $\chi = \left(\frac{\cdot}{\mathfrak{p}}\right)^{-1}$. Then for all prime ideals $\mathfrak{q}$ in $\mathcal{O}_k$ such that $\mathfrak{q} \nmid pm$ we have*

$$\left(\frac{\mu}{\mathfrak{q}}\right)_m = \left(\frac{N\mathfrak{q}}{\mathfrak{p}}\right)_m,$$

where $N\mathfrak{q} = q^f$ is the absolute norm of $\mathfrak{q}$.

11.2 Eisenstein Reciprocity for ℓ-th Powers

Proof. The decomposition law in cyclotomic fields implies the congruence $q^f \equiv 1 \bmod m$; hence

$$(-G(\chi))^{q^f} \equiv \sum_t \chi(t)^{q^f} \zeta_m^{tq^f} = \sum_t \chi(t) \zeta_m^{tq^f}$$
$$= -\chi(q^f)^{-1} G(\chi) = \left(\tfrac{N\mathfrak{q}}{\mathfrak{p}}\right)_m (-G(\chi)) \bmod q\mathcal{O}_k.$$

This implies $(-G(\chi))^{q^f-1} \equiv \left(\tfrac{N\mathfrak{q}}{\mathfrak{p}}\right)_m \bmod q$. On the other hand we have

$$(-G(\chi))^{q^f-1} = ((-1)^m \mu)^{(q^f-1)/m} \equiv \left(\tfrac{\mu}{\mathfrak{q}}\right)_m \bmod \mathfrak{q},$$

and comparing both expressions we get the claimed equality. $\square$

For each prime ideal $\mathfrak{p} \nmid m$ in $\mathbb{Z}[\zeta_m]$ define $\Phi(\mathfrak{p}) = G(\chi_\mathfrak{p})^m$ with $\chi_\mathfrak{p} = \left(\tfrac{\cdot}{\mathfrak{p}}\right)_m^{-1}$; we extend Φ multiplicatively to all ideals prime to m, and from the multiplicativity of Φ, the norm N, and of the power residue symbol $\left(\tfrac{\cdot}{\cdot}\right)_m$ we deduce that

$$\left(\frac{\Phi(\mathfrak{a})}{\mathfrak{q}}\right)_m = \left(\frac{N\mathfrak{q}}{\mathfrak{a}}\right)_m \tag{11.3}$$

for all ideals $\mathfrak{a}$ which are products of prime ideals of degree 1 not dividing m. Observe that we did not use the Stickelberger relation for deriving (11.3). It comes in now: if $\mathfrak{a} = \alpha \mathcal{O}_k$ is principal, then there exists a unit $\varepsilon(\alpha) \in \mathcal{O}_k^\times$ such that

$$\Phi(\mathfrak{a}) = \varepsilon(\alpha) \alpha^\gamma, \tag{11.4}$$

where $\gamma = m\theta \in \mathbb{Z}[G]$ as in (11.1). We want to compute the residue symbol $(\alpha^\gamma/\mathfrak{q})$: first note that, for $m = \ell$ prime,

$$\left(\frac{\sigma_t^{-1}(\alpha^t)}{\mathfrak{q}}\right)_\ell = \left(\frac{\sigma_t^{-1}(\alpha)}{\mathfrak{q}}\right)_\ell^t = \left(\frac{\sigma_t^{-1}(\alpha)}{\mathfrak{q}}\right)_\ell^{\sigma_t} = \left(\frac{\alpha}{\mathfrak{q}^{\sigma_t}}\right)_\ell;$$

this shows immediately that

$$\left(\frac{\alpha^\gamma}{\mathfrak{q}}\right)_\ell = \prod_t \left(\frac{\alpha}{\mathfrak{q}^{\sigma_t}}\right)_\ell = \left(\frac{\alpha}{N\mathfrak{q}}\right)_\ell,$$

where $N\mathfrak{q} = p^f$ denotes the absolute norm of $\mathfrak{q}$. Since we have proved Stickelberger's relation only for prime ideals of degree 1, the reciprocity formula just proved is only valid for such α which are products of prime ideals of degree 1. Before we will see how Hilbert dealt with this difficulty, we take care of the unit $\varepsilon(\alpha)$ defined in Equation (11.4): if we want a simple formula like $\left(\tfrac{\alpha}{N\mathfrak{q}}\right)_\ell = \left(\tfrac{N\mathfrak{q}}{\alpha}\right)_\ell$ to hold, we must make sure that $\varepsilon(\alpha)$ is an ℓ-th power residue modulo all prime ideals $\mathfrak{q}$: the only way to do this is to show that $\varepsilon(\alpha)$ is an actual ℓ-th power if α is semi-primary:

Lemma 11.8. *The unit $\varepsilon(\alpha)$ defined in Equation (11.4) is a root of unity, and if α is semi-primary and $m = \ell$ is an odd prime, then $\varepsilon(\alpha) = \pm 1$.*

Proof. We begin by showing that $\varepsilon(\alpha)$ is an m-th root of unity. Since $\mathbb{Q}(\zeta_m)/\mathbb{Q}$ is abelian it is sufficient to show that $|\varepsilon(\alpha)| = 1$, because this implies that $|\varepsilon(\alpha)^\sigma| = 1$ for all $\sigma \in \mathrm{Gal}\,(\mathbb{Q}(\zeta_m)/\mathbb{Q})$, and a well known result due to Kronecker asserts that the only algebraic integers with this property are roots of unity.

The fact that $|\Phi(\mathfrak{p})|^2 = p = N\mathfrak{p}$ for all prime ideals of degree 1 implies at once that $|\Phi(\alpha)|^2 = |N(\alpha)|^m$. On the other hand, letting $\sigma = \sigma_{-1}$ denote complex conjugation we have $|\alpha^\gamma|^2 = \alpha^\gamma \alpha^{\gamma\sigma}$ and

$$\gamma(1+\sigma) = \sum t^{-1}\sigma_t + \sum t^{-1}\sigma_t\sigma_{-1} = \sum t^{-1}\sigma_t + \sum t^{-1}\sigma_{-t}$$
$$= \sum t^{-1}\sigma_t + \sum (m-t)^{-1}\sigma_t = m\sum \sigma_t,$$

hence $\alpha^\gamma \alpha^{\gamma\sigma} = |N(\alpha)|^m$. This yields our first claim that $|\varepsilon(\alpha)| = 1$.

This much is true without m being prime or α being semi-primary – now we suppose that $\alpha \equiv z \bmod \mathfrak{l}^2$, where $\mathfrak{l} = (1 - \zeta_\ell)\mathcal{O}_k$ is the prime ideal above the rational prime $m = \ell$. Applying $\sigma \in \mathrm{Gal}\,(k/\mathbb{Q})$ yields $\alpha^\sigma \equiv z \bmod \mathfrak{l}^2$, since $\mathfrak{l}$ is an ambiguous ideal, i.e. $\mathfrak{l}^\sigma = \mathfrak{l}$. This shows that

$$\alpha^\gamma \equiv z^\gamma \equiv z^{1+2+\ldots+(\ell-1)} \equiv z^{\ell(\ell-1)/2} \equiv \left(\frac{z}{\ell}\right)^\ell \equiv \pm 1 \bmod \mathfrak{l}^2.$$

Now look at $\Phi(\alpha) = \varepsilon(\alpha)\alpha^\gamma$: if we can show that $\Phi(\alpha) \equiv \pm 1 \bmod \mathfrak{l}^2$, then we can conclude that, for semi-primary α, we have $\varepsilon(\alpha) \equiv \pm 1 \bmod \mathfrak{l}^2$. But the only semi-primary roots of unity are ± 1, and this proves our claim.

The proof of the congruence $\Phi(\alpha) \equiv 1 \bmod \mathfrak{l}^2$ is straightforward:

$$\Phi(\alpha) = G(\chi_\mathfrak{p})^\ell = \left(-\sum_{t\neq 0} \chi_\mathfrak{p}(t)\psi(t)\right)^\ell \equiv -\sum_{t\neq 0} \psi(\ell t) = \psi(0) = 1 \bmod \ell,$$

and this suffices because $\mathfrak{l}^2 \mid \ell$ for $\ell > 2$. $\square$

Now we will remove the condition that (α) be a product of prime ideals of degree 1. To this end let $\alpha \in \mathcal{O}_k$ be a semi-primary integer, assume that $\sigma : \zeta_\ell \longmapsto \zeta_\ell^r$ generates $\mathrm{Gal}\,(k/\mathbb{Q})$, and define

$$\beta = \alpha^S, \quad \text{where } S = \prod(1 - \sigma^e);$$

here the product is over all integers $e \neq \ell - 1$ which divide $\ell - 1$. We claim that only prime ideals of degree 1 occur in the prime ideal factorization of $\beta\mathcal{O}_k$. In fact, suppose that $\mathfrak{p}$ is a prime ideal of degree $f > 1$ dividing β. Put $ef = \ell - 1$: then $(1 - \sigma^e)$ occurs in the product S above, and we can write $\beta = (\alpha^{h(\sigma)})^{1-\sigma^e}$, where h is some polynomial in $\mathbb{Z}[x]$. But σ^e fixes $\mathfrak{p}$, hence $\mathfrak{p}$ divides the numerator and the denominator of β equally often, and this shows that it cannot occur in the prime ideal factorization of $\beta\mathcal{O}_k$.

Since α is semi-primary, so is β, and from what we have proved we know that $(\beta/q) = (q/\beta)$. We also know that $(\alpha^\sigma/q)_\ell = (\alpha/q)_\ell^\sigma = (\alpha/q)_\ell^r$, hence we find
$$\left(\frac{\alpha}{q}\right)_\ell^{\prod(1-r^e)} = \left(\frac{\alpha}{q}\right)_\ell^{\prod(1-\sigma^e)} = \left(\frac{\beta}{q}\right)_\ell = \left(\frac{q}{\beta}\right)_\ell$$
$$= \left(\frac{q}{\alpha}\right)_\ell^{\prod(1-\sigma^e)} = \left(\frac{q}{\alpha}\right)_\ell^{\prod(1-r^e)}.$$

But since the product of the numbers $1 - r^e$ is not divisible by ℓ, we conclude that $(\alpha/q) = (q/\alpha)$. At this point we know that $(\alpha/q)_\ell = (q/\alpha)_\ell$ holds for all semi-primary $\alpha \in \mathcal{O}_k$ and all primes $q \neq \ell$. Since $(\cdot/\cdot)_\ell$ is multiplicative in the denominator, we have proved $(\alpha/a)_\ell = (a/\alpha)_\ell$ for all semi-primary $\alpha \in \mathcal{O}_k$ and all $a \in \mathbb{Z}$ not divisible by ℓ.

Theorem 11.9. (Eisenstein's Reciprocity Law for ℓ-th Powers) *Let $\ell \in \mathbb{N}$ be prime and suppose that $a \in \mathbb{Z}$ and $\alpha \in \mathbb{Z}[\zeta_\ell]$ are relatively prime and semi-primary; then*
$$\left(\frac{a}{\alpha}\right)_\ell = \left(\frac{\alpha}{a}\right)_\ell.$$

Moreover, we have

i) $\left(\frac{\alpha}{a}\right) = 1$ *if* $(\alpha, a) = 1$ *and* $\alpha \in \mathbb{Z}[\zeta_\ell + \zeta_\ell^{-1}]$ *is real;*
ii) $\left(\frac{a}{b}\right) = 1$ *for all* $a, b \in \mathbb{Z}$ *such that* $(a, b) = (b, \ell) = 1$.
iii) *The first supplementary law:* $\left(\frac{\zeta}{a}\right) = \zeta^{(a^{\ell-1}-1)/\ell}$.
iv) *The second supplementary law:* $\left(\frac{1-\zeta}{a}\right) = \left(\frac{\zeta}{a}\right)^{\frac{\ell-1}{2}}$.

Proof. Only the assertions i) – iv) are left to prove:
i) Let $G = \text{Gal}(k/\mathbb{Q})$ denote the Galois group of $k = \mathbb{Q}(\zeta_\ell)$; then complex conjugation τ generates a subgroup $H = \langle \tau \rangle$ of order 2 in G. For a prime p let $\mathfrak{p}$ denote a prime ideal in $\mathcal{O}_k$ above p. Then
$$\left(\frac{\alpha}{\mathfrak{p}^\tau}\right) = \left(\frac{\alpha^\tau}{\mathfrak{p}^\tau}\right) = \left(\frac{\alpha}{\mathfrak{p}}\right)^\tau = \left(\frac{\alpha}{\mathfrak{p}}\right)^{-1}$$

implies that
$$\left(\frac{\alpha}{p}\right) = \prod_{\sigma \in G/H} \left(\frac{\alpha}{\mathfrak{p}^\sigma \mathfrak{p}^{\sigma\tau}}\right) = 1.$$

ii) If $\ell \nmid a$ this is a special case of i); but now $\left(\frac{\ell}{b}\right) = \left(\frac{\ell-b}{b}\right) = 1$.
iii) Let $p\mathcal{O}_k = \mathfrak{p}_1 \ldots \mathfrak{p}_g$; then
$$\left(\frac{\zeta}{p}\right) = \prod_{j=1}^{g} \left(\frac{\zeta}{\mathfrak{p}_j}\right) = \prod_{j=1}^{g} \zeta^{\frac{p^f-1}{\ell}} = \zeta^{g\frac{p^f-1}{\ell}}.$$

The observation

$$\frac{p^{fg}-1}{\ell} = \frac{p^f-1}{\ell} \cdot (p^{f(g-1)} + \ldots + p^f + 1) \equiv g \cdot \frac{p^f-1}{\ell} \mod \ell$$

shows that the claim holds for prime $a = p$. Now

$$\frac{(mn)^{\ell-1}-1}{\ell} = \frac{m^{\ell-1}-1}{\ell} n^{\ell-1} + \frac{n^{\ell-1}-1}{\ell}$$
$$\equiv \frac{m^{\ell-1}-1}{\ell} + \frac{n^{\ell-1}-1}{\ell} \mod \ell$$

proves the assertion by induction on the primes dividing a.

iv) This follows immediately from i), iii), and the fact that $(1-\zeta)^2\zeta^{-1}$ is real. □

11.3 The Stickelberger Congruence

If p is a prime which does not split completely in $\mathbb{Q}(\zeta_m)$, then the computation of the prime ideal factorization of the Gauss sum corresponding to a character on $\mathbb{Z}[\zeta_m]/\mathfrak{p} \simeq \mathbb{F}_{p^f}$ becomes more difficult: historically, the first obstacle was overcome by Galois through his construction of finite fields; the first character sums over finite fields of order p^2 and p^3 were studied by Eisenstein [Eis], Kummer [465, §2] gave the prime ideal factorization of Jacobi sums in the general case, and Stickelberger [759] gave the corresponding result for Gauss sums that will be discussed below. The main idea will be to study the prime ideal decomposition of Gauss sums for characters χ over $\mathbb{F}_q$ first in the field $\mathbb{Q}(\zeta_{p(q-1)})$, and then take norms down to $\mathbb{Q}(\zeta_m)$, where m is the order of χ.

We start by introducing some notation. Let $q = p^f$ be the power of a prime p which will remain fixed throughout this section. For an integer $a \in \mathbb{Z}$, let $\overline{a}$ denote the unique integer satisfying $0 \leq \overline{a} < q-1$ and $a \equiv \overline{a} \mod q-1$. Write it in the form

$$\overline{a} = a_0 + a_1 p + \ldots + a_{f-1} p^{f-1}. \qquad (11.5)$$

Then we define $s(a) = a_0 + a_1 + \ldots + a_{f-1}$ and $\gamma(a) = a_0! a_1! \cdots a_{f-1}!$.

Theorem 11.10. *Let $\mathfrak{P}$ be a prime ideal above p in $\mathbb{Q}(\zeta_{q-1})$, and let $\omega = (\,\cdot\,/\mathfrak{P})^{-1}$. Then the corresponding Gauss sums $G(\omega^a)$ satisfy the Stickelberger congruence*

$$\frac{G(\omega^a)}{\pi^{s(a)}} \equiv \frac{1}{\gamma(a)} \mod \mathcal{P} \qquad (11.6)$$

for all $a \in \mathbb{N}$, where $\pi = \zeta_p - 1$ and $\mathcal{P} = (\mathfrak{P}, \pi)$. Since $\mathcal{P} \parallel \pi$ and $\gamma(a)$ is a $\mathcal{P}$-adic unit, this implies in particular that $\mathcal{P}^{s(a)} \parallel G(\omega^a)$.

11.3 The Stickelberger Congruence   367

This is the main theorem on Gauss sums – among its corollaries are Theorems 4.31 and 4.32 of Davenport-Hasse (see [DaH] and Exercise 11.9) as well as a host of amazing results on class groups of abelian extensions of $\mathbb{Q}$, some of which we will discuss below. Also note that the choice $a = \frac{p-1}{2}$ turns (11.6) into (8.35), since $\omega^{(p-1)/2} = (\frac{\cdot}{p})$ is the quadratic residue character. The Hasse diagram for the fields and ideals occurring in the proofs below are displayed in Figure 11.1.

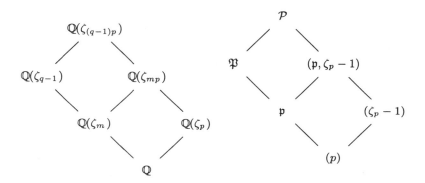

Fig. 11.1. Some subfields of $\mathbb{Q}(\zeta_{(q-1)p})$

Proof of Theorem 11.10. We will prove Stickelberger's congruence by induction on $s(a)$. For $s(a) = 0$, the claim is trivial (recall that $G(\mathbf{1}) = 1$). If $s(a) = 1$, then $a = p^r$ for some $r \geq 1$. Now

$$G(\omega^p) = G(\omega), \quad s(ap) = s(a) \quad \text{and} \quad \gamma(ap) = \gamma(a). \qquad (11.7)$$

The last two equations are obvious, and the first is proved easily: since $G(\omega^p) = -\sum_{t \in \mathbb{F}_q^\times} \chi(t^p)\zeta^{\text{Tr}(t)}$, it is sufficient to prove that $\text{Tr}(t) = \text{Tr}(t^p)$. But this is clear in light of $\text{Tr}(t) = t + t^p + \ldots + t^{p^{f-1}}$.

Therefore it is sufficient to prove the claim for $a = 1$. This is done as follows: first we notice that

$$-G(\omega) = \sum_{t \in \mathbb{F}_q^\times} \omega(t)\zeta^{\text{Tr}(t)} = \sum_{t \in \mathbb{F}_q^\times} \omega(t)(\zeta^{\text{Tr}(t)} - 1)$$

since $\sum_t \omega(t) = 0$. The last sum has the advantage that all summands are divisible by $\zeta - 1$; moreover,

$$\frac{\zeta^m - 1}{\zeta - 1} = 1 + \zeta + \zeta^2 + \ldots + \zeta^{m-1} \equiv m \bmod \pi$$

since $\zeta^r \equiv 1 \bmod \pi$ for all $r \in \mathbb{Z}$. This shows $-\frac{G(\omega)}{\pi} \equiv \sum_t \omega(t) \text{Tr}(t) \bmod \pi$. Since we are summing over roots of unity t, we have $\omega(t) = t^{-1}$ and $\text{Tr}(t) = t + t^p + \ldots + t^{p^{f-1}}$. Thus

$$-\frac{G(\omega)}{\pi} \equiv \sum_{t \in \mathbb{F}_q^\times} t^{-1}(t + t^p + \ldots + t^{p^{f-1}})$$

$$= \sum_{t \in \mathbb{F}_q^\times} (1 + t + \ldots + t^{p^{f-2}}) = (q-1) \equiv -1 \bmod \mathcal{P}.$$

This proves Stickelberger's congruence for $a = 1$, thus for all $a < q - 1$ such that $s(a) = 1$.

Now we do the induction step, so assume that (11.6) is proved for all $0 < a < q-1$ with $s(a) \leq r$, $r \geq 1$. Suppose $s(a) = r + 1$ and write $a = a_i p^i + \ldots a_{f-1} p^{f-1}$ with $a_i > 0$. Using (11.7) we may assume that $i = 0$, i.e. that $a = a_0 + a_1 p + \ldots a_{f-1} p^{f-1}$ with $a_0 > 0$. Then $a - 1 = a_0 - 1 + a_1 p + \ldots a_{f-1} p^{f-1}$, hence $s(a-1) = s(a) - 1$, and

$$\frac{G(\omega^{a-1})}{\pi^{s(a-1)}} \equiv \frac{1}{\gamma(a-1)} \bmod \mathcal{P}$$

from the induction assumption. Next $G(\omega^{a-1})G(\omega) = G(\omega^a)J(\omega, \omega^{a-1})$, and writing $b = q - 1 - (a-1) = q - a$ we find

$$-J(\omega, \omega^{a-1}) \equiv \sum_t t^{-1}(1-t)^b \equiv \sum_t t^{-1} \sum_{j=0}^b (-1)^j \binom{b}{j} t^j$$

$$= \sum_{j=0}^{q-1} (-1)^j \binom{b}{j} \sum_t t^{j-1} \bmod \mathfrak{P}.$$

Since the inner sum vanishes modulo $\mathfrak{p}$ unless $j - 1$ is divisible by $q - 1$ (which happens if and only if $j = 1$), we get

$$J(\omega, \omega^{a-1}) \equiv -\binom{b}{1} = -b = a - q \equiv a \equiv a_0 \bmod \mathfrak{P}.$$

In particular, $J(\omega, \omega^{a-1})$ is a $\mathfrak{P}$-unit, and we conclude that

$$\frac{G(\omega^a)}{\pi^{s(a)}} = \frac{G(\omega^{a-1})}{\pi^{s(a-1)}} \frac{G(\omega)}{\pi^{s(1)}} J(\omega, \omega^{a-1})^{-1} \equiv \frac{1}{\gamma(a-1)} \frac{1}{a_0} = \frac{1}{\gamma(a)} \bmod \mathcal{P}.$$

This proves our claims. □

Since $\gamma(a)$ is a $\mathcal{P}$-adic unit and $\mathcal{P} \| \pi$, the Stickelberger congruence implies that $\mathcal{P}^{s(a)} \| G(\omega^a)$. In order to find the complete prime ideal factorization of $G(\omega^a)$, let $\sigma_b \in \Gamma = \text{Gal}(\mathbb{Q}(\zeta_{p(q-1)})/\mathbb{Q})$ be defined by $\sigma_b : \zeta_{q-1} \mapsto \zeta_{q-1}^b$, $\zeta_p \mapsto \zeta_p$. Assume that $\mathcal{P}^{r\sigma_b^{-1}} \| G(\omega^a)$; applying σ_b gives $\mathcal{P}^r \| G(\omega^a)^{\sigma_b} = G(\omega^{ab})$, and we see that $r = s(ab)$.

11.3 The Stickelberger Congruence

Corollary 11.11. *We have* $(G(\omega^a)) = \mathcal{P}^\Theta$, *where* $\Theta = \sum_{\sigma_b \in \Gamma/Z} s(ab)\sigma_b^{-1}$. *Here Z denotes the decomposition group of $\mathfrak{p}$.*

This corollary gives the complete prime ideal factorization of Gauss sums; we will now show that the formulation of Theorem 11.4 carries over to the general case:

Theorem 11.12. *Let $p \nmid m$ be prime and let $\mathfrak{p}$ be a prime ideal above p in $K = \mathbb{Q}(\zeta_m)$; then $\chi = \left(\frac{\cdot}{\mathfrak{p}}\right)_m^{-1}$ is a multiplicative character of order m on $\mathbb{F} = \mathbb{F}_q$, where $q = p^f = N\mathfrak{p}$ is the absolute norm of $\mathfrak{p}$. The corresponding Gauss sum $G(\chi) = -\sum_{t \in \mathbb{F}^\times} \chi(t)\zeta_m^{Tr(t)}$ has the factorization $G(\chi)^m = \mathfrak{p}^{m\theta}$ for θ as in (11.1).*

Proof. We start by recalling that the decomposition group Z of $\mathcal{P}$ is generated by σ_p (see Proposition 3.1.iv), hence we have $Z = \{1, \sigma_p, \ldots, \sigma_p^{f-1}\}$. Now we need

$$s(a) = (p-1)\sum_{i=0}^{f-1}\left\langle \frac{ap^i}{q-1}\right\rangle. \tag{11.8}$$

For a proof, consider the following set of congruences modulo $q-1$:

$$\begin{aligned}
a &= a_0 + a_1 p + \ldots + a_{f-1} p^{f-1} \\
ap &\equiv a_{f-1} + a_0 p + \ldots + a_{f-2} p^{f-1} \mod(q-1) \\
&\ldots \\
ap^{f-1} &\equiv a_1 + a_2 p + \ldots + a_0 p^{f-1} \mod(q-1)
\end{aligned}$$

The right hand side of the i-th congruence equals $(q-1)\langle\frac{ap^{i-1}}{q-1}\rangle$; summing up we find

$$\sum_{i=0}^{f-1}\left\langle\frac{ap^i}{q-1}\right\rangle = s(a)\frac{1+p+\ldots+p^{f-1}}{q-1} = \frac{s(a)}{p-1},$$

and this proves (11.8). Using (11.8) we get

$$s(ab)\sigma_b^{-1} = (p-1)\sum_{i=0}^{f-1}\left\langle\frac{abp^i}{q-1}\sigma_b^{-1}\right\rangle \equiv (p-1)\sum_{i=0}^{f-1}\left\langle\frac{abp^i}{q-1}\sigma_{bp^i}^{-1}\right\rangle \mod Z,$$

hence

$$\sum_{\sigma_b \in \Gamma/Z} s(ab)\sigma_b^{-1} \equiv (p-1)\sum_{t \bmod q-1}\left\langle\frac{at}{q-1}\right\rangle\sigma_t^{-1} \mod Z.$$

Using $\mathcal{P}^{p-1} = \mathfrak{P}$, we find

$$(G(\omega^a)) = \mathfrak{P}^T, \quad \text{where } T = \sum_{t \bmod q-1} \left\langle \frac{at}{q-1} \right\rangle \sigma_t^{-1}. \tag{11.9}$$

Now we are in a position to complete the proof by using Kummer's trick (cf. Section 10.2) of going up to $\mathbb{Q}(\zeta_{q-1}, \zeta_p)$, where $q = p^f = N\mathfrak{p}$: every ideal prime to m in $\mathbb{Z}[\zeta_m]$ has norm $\equiv 1 \bmod m$, hence, in particular, $q \equiv 1 \bmod m$. Let $\mathfrak{P}$ denote the prime ideal above $\mathfrak{p}$ in $\mathbb{Q}(\zeta_{q-1})$; then $\mathfrak{P}$ still has inertia degree f, hence $\mathcal{O}_{q-1}/\mathfrak{P} \simeq \mathcal{O}_m/\mathfrak{p} \simeq \mathbb{F}_q$. The advantage of working in $\mathcal{O}_{q-1}$ is, as we have seen, that $(\mathcal{O}_{q-1}/\mathfrak{P})^\times$ has μ_{q-1} as a set of representatives.

We have $\chi = \omega^a$ with $a = \frac{q-1}{m}$. From (11.9) we get

$$G(\chi) = \mathfrak{P}^T, \quad \text{with } T = \sum_{t \bmod q-1} \left\langle \frac{t}{m} \right\rangle \sigma_t^{-1}. \tag{11.10}$$

Next let us look at the automorphisms of $\mathbb{Q}(\zeta_{q-1})/\mathbb{Q}(\zeta_m)$. Write $q - 1 = mn$; then $\sigma_t : \zeta_{q-1} \mapsto \zeta_{q-1}^t$ fixes $\zeta_m = \zeta_{q-1}^n$ if and only if $nt \equiv 1 \bmod q - 1$, i.e. if and only if $t \equiv 1 \bmod m$. We conclude that the relative norm of $\mathbb{Q}(\zeta_{q-1})/\mathbb{Q}(\zeta_m)$ is simply $\nu = \sigma_1 + \sigma_{1+m} + \ldots + \sigma_{1+m(n-1)}$. Since $\left\langle \frac{t}{m} \right\rangle$ in (11.10) only depends on $t \bmod m$, we can write

$$\sum_{t \bmod q-1} \left\langle \frac{t}{m} \right\rangle \sigma_t^{-1} = \sum_{t \bmod m} \left\langle \frac{t}{m} \right\rangle \sum_{j=0}^{n-1} \sigma_{t+jm}^{-1}$$

$$= \sum_{j=0}^{n-1} \sigma_{1+jm} \sum_{t \bmod m} \left\langle \frac{t}{m} \right\rangle \sigma_t^{-1}$$

Since the first sum is just the relative norm ν that occurred before, we find, using $\mathfrak{P}^\nu = \mathfrak{p}$, that

$$G(\chi)^m = \mathfrak{p}^{m\theta}, \quad \text{with } \theta = \sum_{(t,m)=1} \left\langle \frac{t}{m} \right\rangle \sigma_t^{-1}. \tag{11.11}$$

This concludes our proof. □

Congruences for Jacobi Sums

Equation (11.6) also generalizes many of the congruences for cubic and quartic Jacobi sums that we have proved. In fact, assume for the sake of simplicity that $f = 1$, that is, let $p = mn + 1$ be prime, $K = \mathbb{Q}(\zeta_m)$, let $\mathfrak{p}$ denote a prime ideal in $\mathcal{O}_K$ above p, and let $\chi = (\cdot/\mathfrak{p})_m^{-1}$. Then $s(a) = a$ for $0 < a < m$, and we can identify χ with $\omega^{(p-1)/m} = \omega^n$ via the isomorphism $(\mathcal{O}_K/\mathfrak{p})^\times \simeq (\mathbb{Z}[\zeta_{p(q-1)}]/\mathcal{P})^\times$. Now Stickelberger's congruence gives $G(\chi^a)/\pi^a \equiv 1/(an)! \bmod \mathcal{P}$. This implies

$$J(\chi^a, \chi^b) = \frac{G(\chi^a)G(\chi^b)}{G(\chi^{a+b})} \equiv \frac{(an+bn)!}{(an)!(bn)!} = \binom{an+bn}{an} \bmod \mathcal{P} \tag{11.12}$$

whenever $0 < a, b, a+b < m$. Since the left hand side is an element in $\mathcal{O}_K = \mathbb{Z}[\zeta_m]$, the congruence is even valid modulo $\mathfrak{P} = \mathcal{P} \cap \mathcal{O}_K$. Observe that this contains the results of Corollaries 6.6, 7.6 and Exercise 9.1 as special cases. Also observe that this congruence is compatible with Proposition 4.28 since the binomial coefficient on the right hand side of (11.12) is divisible by p if and only if $a+b = m-1$ (under the restrictions $0 < a, b, a+b < m$).

11.4 Class Groups of Abelian Number Fields

Our first application will exploit the fact that the Stickelberger relation can be regarded as a statement about the class group: for every prime ideal $\mathfrak{p}$ in $\mathcal{O} = \mathbb{Z}[\zeta_m]$ prime to m, the ideal $\mathfrak{p}^{m\theta}$ is principal, because it is generated by $G(\chi)^m$, where χ is a character of order m on $(\mathcal{O}/\mathfrak{p})^\times$.

Stickelberger's Theorem

Let $K/\mathbb{Q}$ be a finite abelian extension with Galois group G and conductor m. Let σ_a denote the restriction of the automorphism $\zeta_p \longmapsto \zeta_p^a$ of $M = \mathbb{Q}(\zeta_m)$ to K. Then

$$\theta = \theta(K) = \frac{1}{m} \sum_{\substack{0 < a < m \\ (a,m)=1}} a\sigma_a^{-1} \in \mathbb{Q}[G]$$

is called the *Stickelberger element* corresponding to K. Clearly $\theta(K)$ is the restriction of $\theta(M)$ to K.

Now we claim that $(b-\sigma_b)\theta \in \mathbb{Z}[G]$ for integers $b \in \mathbb{Z}$ such that $(b, m) = 1$: in fact, from $\sigma_b \sigma_{ab}^{-1} = \sigma_a^{-1}$ we deduce that

$$(b - \sigma_b)\theta = \sum_{(a,m)=1} \left(b \left\langle \frac{a}{m} \right\rangle - \left\langle \frac{ba}{m} \right\rangle \right) \sigma_a^{-1},$$

and this element of $\mathbb{Q}[G]$ has integral coefficients. This allows us to define the *Stickelberger ideal* $I_0(K)$ as the ideal in $\mathbb{Z}[G]$ generated by elements of the form $(b - \sigma_b)\theta$. We also use the name Stickelberger ideal for $I(K) = \mathbb{Z}[G] \cap \theta\mathbb{Z}[G]$. We have already seen that $I_0(K) \subseteq I(K)$; unfortunately, these ideals are different in general (see Exercise 11.22). Nevertheless we have

Lemma 11.13. *If $K = \mathbb{Q}(\zeta_m)$ is a full cyclotomic field, then $I(K) = I_0(K)$.*

Proof. We already know that $I_0(K) \subseteq I(K)$ for any abelian field K, so it is sufficient to prove the converse. To this end, take any $\beta = \sum_t b_t \sigma_t \in \mathbb{Z}[G]$ with the property $\beta\theta \in \mathbb{Z}[G]$; we have to show that $\beta\theta \in I_0(K)$. The familiar trick of substituting $a = ct$ gives

$$\beta\theta = \left(\sum_t b_t \sigma_t \right) \left(\sum_a a\sigma_a^{-1} \right) = \sum_c \left(\sum_t \left\langle \frac{ct}{m} \right\rangle b_t \right) \sigma_c^{-1}.$$

372   11. Eisenstein Reciprocity

Since K is the full cyclotomic field, the automorphisms σ_c are all different, and we can deduce that the coefficient of σ_1 must be an integer, i.e. that $\sum_t tb_t \equiv 0 \bmod m$. But now $\beta\theta = \left(\sum_t b_t\sigma_t\right)\theta = \sum_t b_t(t - \sigma_t)\theta + \sum_t tb_t\theta$; the first sum is clearly in $I_0(K)$, the second is an integral multiple of $m\theta = (m + 1 - \sigma_{m+1})\theta$ and therefore also lies in $I_0(K)$.   □

Stickelberger used Gauss sums to construct annihilators of the ideal class groups of cyclotomic fields; his result looks quite innocent:

Theorem 11.14. *(Stickelberger's Theorem) Let $K/\mathbb{Q}$ be an abelian extension. Then the Stickelberger ideal $I(K)$ annihilates $\mathrm{Cl}\,(K)$.*

Thus if $\alpha \in \mathbb{Z}[G]$, where $G = \mathrm{Gal}\,(K/\mathbb{Q})$, is such that $\alpha\theta \in \mathbb{Z}[G]$, then $c^\alpha = 1$ for any ideal class $c \in \mathrm{Cl}\,(K)$. In other words: any element of $\mathbb{Z}[G]$ that kills the denominator of the Stickelberger elementattached to $K/\mathbb{Q}$ also kills the class group of K. Let us admit right away that Theorem 11.14 is useless for real abelian fields, because then $\sigma_a = \sigma_{-a}$, thus $\theta = \frac{1}{m}\sum a\sigma_a^{-1} = \frac{1}{m}\sum_{a<m/2}[a + (m - a)]\sigma_a^{-1} = \sum a < m/2 \sigma_a^{-1}$, and this last expression is the relative norm from $\mathbb{Q}(\zeta_m + \zeta_m^{-1})$: this clearly kills the ideal class of any intermediate field since $\mathbb{Q}$ has class number 1.

Proof of Theorem 11.14. Let m be the conductor of K; then we have $K \subseteq M = \mathbb{Q}(\zeta_m)$. For a prime ideal $\mathfrak{p} \nmid m$ in M, let $\chi = \chi_\mathfrak{p} = (\frac{\cdot}{\mathfrak{p}})_m^{-1}$ be the corresponding m-th power character. Then we know that $G(\chi)^m = \mathfrak{p}^{m\theta}$ is principal in M, and since the class group is generated by classes of prime ideals coprime to m (any ideal class contains an ideal coprime to any given ideal), we conclude that $m\theta$ annihilates $\mathrm{Cl}\,(M)$. Our first aim is to show that this holds for any $\beta\theta \in I(M)$ (not just for $\beta = m$), and then we will have to pull everything back to $K \subseteq M$.

So assume that we are given a $\beta \in \mathbb{Z}[\Gamma]$ such that $\beta\theta \in \mathbb{Z}[\Gamma]$, where $\Gamma = \mathrm{Gal}\,(M/\mathbb{Q})$. We want to show that $\mathfrak{a}^{\beta\theta}$ is principal for any integral ideal $\mathfrak{a}$ in M; clearly it is sufficient to prove this for ideals coprime to m. To this end, write $\mathfrak{a} = \prod_\mathfrak{p} \mathfrak{p}$ and put $\gamma = \prod_\mathfrak{p} G(\chi_\mathfrak{p})$. Then $\mathfrak{a}^{m\beta\theta} = (\gamma^{m\beta})$ is a principal ideal in M. Put $\alpha = \gamma^{m\beta}$ and $L = M(\sqrt[m]{\alpha})$. Then L/M is a Kummer extension; moreover, $(\gamma^\beta) = \mathfrak{a}^{\beta\theta}$ is an integral ideal in M since $\beta\theta \in \mathbb{Z}[\Gamma]$, hence (α) is an m-th ideal power and therefore L/M is unramified outside m. On the other hand, $L \subseteq M(\zeta_P)$, where P is the product of primes below the prime ideals $\mathfrak{p}$ dividing $\mathfrak{a}$, and every subextension Λ/M of $M(\zeta_P)/M$ ramifies at some p. Since $p \nmid m$, this implies that $L = M$, hence $\gamma^\beta \in M$, and we have shown that $\beta\theta$ annihilates $\mathrm{Cl}\,(M)$.

Now assume that $\mathfrak{a}$ is an integral ideal in K; then $\mathfrak{a}^\sigma = \mathfrak{a}$ for any $\sigma \in \mathrm{Gal}\,(M/K)$. In particular, σ permutes the prime ideals $\mathfrak{p}$ that divide $\mathfrak{a}$. Let s be an automorphism of $M(\zeta_p)/\mathbb{Q}(\zeta_p)$ whose restriction to M is σ (note that, in particular, $\zeta_p^s = \zeta_p$); this is possible since $p \nmid m$. Then right from the definition of a Gauss sum we deduce that $G(\chi_\mathfrak{p})^s = G(\chi_{\mathfrak{p}^\sigma})$. But this implies that γ^β is fixed by σ, hence $\gamma^\beta \in K$ and $\beta\theta(K)$ kills $\mathrm{Cl}\,(K)$.   □

11.4 Class Groups of Abelian Number Fields

A) Quadratic Fields It might seem that this result is not what we wanted, because we were looking for an *integer* annihilating $\text{Cl}(K)$, not some element $\beta\theta$ in the group ring $\mathbb{Z}[G]$. But consider an imaginary quadratic number field $K = \mathbb{Q}(\sqrt{d})$ with discriminant $d \neq -3, -4, -8$. If we put

$$R = \sum_{\substack{(d/r)=+1 \\ 1 \leq r < d}} r, \quad N = \sum_{\substack{(d/n)=-1 \\ 1 \leq n < d}} n,$$

then $\theta(K) = \frac{1}{|d|}(R + \sigma N)$, where σ is the nontrivial automorphism of $K/\mathbb{Q}$. The definition of θ implies that $|d|\theta \in \mathbb{Z}[G]$; actually, much more is true:

Lemma 11.15. *For any discriminant d of a complex quadratic number field, R and N are divisible by d unless $d \in \{-3, -4, -8\}$.*

Proof. If $d = -\ell$ with $\ell > 3$ prime, this is trivial: choose $a \not\equiv 1 \bmod \ell$ such that $(d/a) = +1$ and observe that $aR \equiv R \bmod \ell$.

Now assume that $d = d'm$ for a prime discriminant d' and some $m \neq \pm 1$. Let $C = \{a + d\mathbb{Z} : (d/a) = +1\}$ be the group of quadratic residues modulo d and consider the homomorphism $\pi : C \longrightarrow (\mathbb{Z}/d'\mathbb{Z})^\times$. Since π is onto, $\#\ker\pi = \frac{1}{2}\phi(m')$, so among the $\frac{1}{2}\phi(d)$ summands in R, exactly $\frac{1}{2}\phi(m')$ reduce modulo d' to a given element in $(\mathbb{Z}/d'\mathbb{Z})^\times$. Thus we see $R \equiv \frac{1}{2}\phi(m')(\sum_{(a,d')=1} a) \bmod d'$. But if $d' = \pm\ell$ is an odd prime, the last sum is $1 + 2 + \ldots + \ell - 1 = \frac{1}{2}\ell(\ell-1) \equiv 0 \bmod \ell$, if $d' = -4$, it is $1 + 3 \equiv 0 \bmod 4$, and if $d' = \pm 8$, it is $1 + 3 + 5 + 7 \equiv 0 \bmod 8$. This proves our claims. □

Thus $\theta(K) \in \mathbb{Z}[G]$ for these d, hence Stickelberger's theorem says that $\theta(K) = (R + \sigma N)/d$ annihilates $\text{Cl}(K)$. But so does $1 + \sigma$, hence $\text{Cl}(K)$ is also killed by $h = \frac{1}{|d|}|R - N|$:

Proposition 11.16. *Let $d < -4$ be a discriminant of an imaginary quadratic number field. Then $h = \frac{N-R}{d}$ annihilates the ideal class group of $k = \mathbb{Q}(\sqrt{d})$, i.e., the h-th power of any ideal in $\mathcal{O}_k$ is principal.*

Note that we have proved that the imaginary quadratic number fields with discriminant $d = -3$ or $d = -4$ have class number 1 in Chapters 6 and 7 by using Jacobi sums, and that the corresponding result for $d = -8$ was shown to hold in Chapter 9 using Eisenstein sums.

B) Quartic Fields Proposition 11.16 is just the tip of an iceberg; in order to show what can be done and to get a feeling for the problems yet to solve, let us look at the complex cyclic quartic fields K of conductor f. Recall that K is a CM-field, that is a totally complex quadratic extension of a totally real number field K^+. For CM-fields K, the *minus* or *relative class group* $\text{Cl}^-(K)$ is defined as the kernel of the norm map $N_{K/K^+} : \text{Cl}(K) \longrightarrow \text{Cl}(K^+)$. If we let σ denote a generator of $\text{Gal}(K/\mathbb{Q})$, then $1 + \sigma^2 = j_{K^+ \hookrightarrow K} \circ N_{K/K^+}$, where $j_{K^+ \hookrightarrow K} : \text{Cl}(K^+) \longrightarrow \text{Cl}(K)$ is the canonical transfer of ideal classes. In particular, $1 + \sigma^2$ kills $\text{Cl}^-(K)$ since N_{K/K^+} does.

Instead of just two sums N and R as in the quadratic case, here we have four of them: let ψ be an odd character (that is, $\psi(-1) = -1$; recall that K is complex) of order 4 on $(\mathbb{Z}/f\mathbb{Z})^\times$; then $C_j = \sum_{\psi(a)=i^j} a$. We claim that $C_j \equiv 0 \bmod f$ unless $f \in \{5, 16\}$. This is easily checked for prime power conductors, and if f is not a prime power, we use the same argument as in the quadratic case. Thus we can write $C_j = fD_j$ for integers D_j. Then Stickelberger's theorem says that $\theta = D_0 + D_1\sigma + D_2\sigma^2 + D_3\sigma^3$ kills $\mathrm{Cl}\,(K)$ since $\theta \in \mathbb{Z}[G]$. On the other hand, $1+\sigma^2$ kills $\mathrm{Cl}^-(K)$, hence so does $(D_0-D_2)+(D_1-D_3)\sigma$ (this follows from $\theta \equiv (D_0 - D_2) + (D_1 - D_3)\sigma \bmod (1 + \sigma^2)$). Now we use the following lemma:

Lemma 11.17. *Let $G = \langle\sigma\rangle$ be a cyclic group of order 4 and assume that the G-module R is annihilated by $1+\sigma^2$ and $a+b\sigma \in \mathbb{Z}[G]$. Then R is annihilated by $a^2 + b^2$.*

Proof. Applying σ to the relation $c^a c^{b\sigma} = 1$ and using the fact that $c^{\sigma^2} = c^{-1}$ we get $1 = c^{a\sigma} c^{b\sigma^2} = c^{a\sigma} c^{-b}$. Raising this to the b-th power we find $1 = c^{-b^2} c^{ab\sigma} = c^{-b^2} c^{-a^2}$, and this proves our claim that $a^2 + b^2$ kills G. □

Using this lemma in the case at hand we find that $(D_0-D_2)^2 +(D_1-D_3)^2$ kills $\mathrm{Cl}^-(K)$. Finally we claim that $h^- = \frac{1}{2}[(D_0 - D_2)^2 + (D_1 - D_3)^2]$ is an integer. But $C_0 + C_1 + C_2 + C_3 = \sum a + \sum(f-a)$, where the sums are over all $1 \leq a < f/2$ with $(a, f) = 1$ (the summand $a = f/2$ does never occur: either f is odd, or f is even and $f/2 \notin (\mathbb{Z}/f\mathbb{Z})^\times$). Since there are $\phi(f)/2$ summands, we find $C_0 + C_1 + C_2 + C_3 = f\phi(f)/2$, hence $D_0 + D_1 + D_2 + D_3 = \phi(f)/2$. But $\phi(f)/2$ is even, since f is either divisible by a prime $\equiv 1 \bmod 4$ or divisible by 16. This implies that $D_0 - D_2$ and $D_1 - D_3$ have the same cardinality, and therefore h^- is an integer. We have proved:

Proposition 11.18. *Let $K/\mathbb{Q}$ be a cyclic quartic field with conductor $f \neq 5, 16$ and Galois group $G = \mathrm{Gal}\,(K/\mathbb{Q})$; let $\psi \in \widehat{G}$ be an odd character of order 4 on $(\mathbb{Z}/f\mathbb{Z})^\times$, define C_j as the sum of all $t \in (\mathbb{Z}/f\mathbb{Z})^\times$ with $\psi(t) = i^j$ and put $C_j = fD_j$. Then $2h^-$ annihilates the minus class group of K, where*

$$h^- = \frac{1}{2}\left((D_0 - D_2)^2 + (D_1 - D_3)^2\right).$$

It is not too hard to prove more general versions; unfortunately, it seems, the method we have presented only shows that some power of 2 times h^- kills $\mathrm{Cl}^-(K)$ for general abelian extensions $K/\mathbb{Q}$. It seems that even in the case of full cyclotomic fields of prime power conductor, the best result that can be achieved with algebraic methods is that $2h^-$ kills the minus class group. For prime power conductors, one can use genus theory to show that, in Proposition 11.18, the class number $h(K)$ is odd, hence that h^- annihilates the minus class group of K.

11.4 Class Groups of Abelian Number Fields

Herbrand's Theorem

Herbrand's Theorem is a result on the fine structure of the p-class group of $\mathbb{Q}(\zeta_p)$, strengthening previous classical results in a very beautiful way. The story starts, predictably, with Kummer.

A) Kummer and Hecke In his letter to Kronecker (May 17, 1847), Kummer conjectured that the class number h of $K = \mathbb{Q}(\zeta_p)$ is divisible by p (such primes he called *irregular*; primes not dividing h were called *regular*) if p divides the numerator of one of the Bernoulli numbers $B_2, B_4, \ldots, B_{p-3}$. He also conjectured that units congruent to a rational integer modulo p must be p-th powers if $p \nmid h$ (this would later become known as *Kummer's Lemma*) and expressed his hope that the first conjecture would imply the second. In a letter to Dirichlet (Sept. 16, 1847) he not only sketched a proof of these conjectures but also introduced the class number h^+ of the maximal real subfield $K^+ = \mathbb{Q}(\zeta_p + \zeta_p^{-1})$ of K and showed that $h^+ \mid h$:

Theorem 11.19. *A prime p is irregular if and only if p divides the numerator of one of the Bernoulli numbers $B_2, B_4, \ldots, B_{p-3}$. Moreover, the quotient $h^- = h/h^+$ is integral, and $p \mid h^+$ only if $p \mid h^-$.* □

In another letter to Kronecker (Dec. 28, 1849), Kummer announced that he had found a unit $\varepsilon \in \mathbb{Z}[\zeta_{37}]$ that is congruent to a rational integer modulo p but not an p-th power (in modern terminology: $K(\sqrt[37]{\varepsilon})/K$ is an unramified cyclic extension of $K = \mathbb{Q}(\zeta_{37})$, or, since $h_{37} = 37$: the Kummer extension $K(\sqrt[37]{\varepsilon})$ is the Hilbert class field of K).

The number $i(p)$ of indices $i \leq \frac{p-3}{2}$ such that $p \mid B_{2i}$ is called the *index of irregularity*. It is very hard to find good bounds on $i(p)$; see Metsänkylä [Met] for more.

Kummer also introduced the relative class number h_p^- of $K = \mathbb{Q}(\zeta_p)$: let h_p and h_p^+ denote the class numbers of K and its maximal real subfield $K^+ = \mathbb{Q}(\zeta_p + \zeta_p^{-1})$, respectively. Then $h_p = h_p^+ h_p^-$, and both factors are integers. Kummer's observation that $p \mid h_p^+$ implies $p \mid h_p^-$ was later refined by Hecke [Hec], who used the class field theory of Furtwängler to show

Proposition 11.20. *We have* rank $\text{Cl}_p(K^+) \leq$ rank $\text{Cl}_p^-(K)$. □

As a corollary of the two preceding theorems we note that $p \mid h_p$ if and only if $p \mid B_2 B_4 \cdots B_{p-3}$. This prompts the question whether the p-part of the minus class group of $\mathbb{Q}(\zeta_p)$ can be broken into smaller pieces such that the nontriviality of such a piece is controlled by the p-divisibility of a corresponding Bernoulli number.

B) Idempotents Let K be a totally complex abelian number field with maximal real subfield K^+; then the restriction J of complex conjugation to K generates $H = \text{Gal}(K/K^+)$. Moreover, J acts on the class group, and for each odd prime p there is a decomposition of $\text{Cl}_p(K)$ into a plus and a minus part.

376   11. Eisenstein Reciprocity

This decomposition can be put into a quite general framework: let R be a commutative ring with 1, and M an R-module. An element $e \in R$ is called an *idempotent* if $e^2 = e$. If e is an idempotent, then so is $1 - e$: in fact, $(1-e)^2 = 1 - 2e + e^2 = 1 - e$ as desired. The existence of idempotents allows us to decompose R-modules M into smaller pieces: for $m \in M$, we get $m = 1m = (e + 1 - e)m = em + (1-e)m$, so M is the sum of the submodules (!) eM and $(1-e)M$. Moreover, this sum is direct: this is due to the fact that e and $1-e$ are orthogonal idempotents, that is, we have $e(1-e) = 0$. Thus if $m \in eM \cap (1-e)M$, then $m = em_1 = (1-e)m_2$, hence $m = em_1 = e^2 m_1 = e(1-e)m_2 = 0m_2 = 0$.

As a simple example, assume that a group $H = \{1, J\}$ acts on M; define $M^+ = \{m \in M : Jm = m\}$ and $M^- = \{m \in M : Jm = -m\}$. If M has odd order n, then M is a $\mathbb{Z}/n\mathbb{Z}$-module, and since 2 has a multiplicative inverse in $\mathbb{Z}/n\mathbb{Z}$, the element $e = \frac{1+J}{2}$ is in $R = (\mathbb{Z}/p^n\mathbb{Z})[H]$. Next $J^2 = 1$ implies that $e^2 = e$, that is, e is an idempotent in R. As we have seen, this implies $M = eM \oplus (1-e)M$; moreover, $eM \subseteq M^+$ since $Je = e$, and $(1-e)M \subseteq M^-$ since $J(1-e) = -(1-e)$, and this implies that we actually have $M^+ = eM$ and $M^- = (1-e)M$. In the case $p = 2$, there is a weak substitute in form of Exercise 11.11.

The prototype for such considerations is the decomposition of the class group $\mathrm{Cl}_p(K)$, p an odd prime, into plus and minus parts. Now $\mathrm{Cl}_p(K)$ is not only acted upon by H but by the whole Galois group $G = \mathrm{Gal}\,(K/\mathbb{Q})$. Since this is an abelian group, it contains many idempotents:

Proposition 11.21. *Let G be a finite abelian group with character group $G\hat{\ } = \mathrm{Hom}\,(G, \mathbb{C}^\times)$. Let R be an integral domain containing $\chi(\sigma)$ for all $\sigma \in G$, and suppose that $\#G$ is a unit in R. Then the elements*

$$\varepsilon_\chi = \frac{1}{\#G} \sum_{\sigma \in G} \chi(\sigma) \sigma^{-1} \in R[G]$$

form a complete set of orthogonal idempotents of $R[G]$, that is we have $\varepsilon_\chi^2 = \varepsilon_\chi$ (idempotent), $\varepsilon_\chi \varepsilon_\psi = 0$ for $\chi \neq \psi$ (orthogonal), and $\sum_{\chi \in \hat{G}} \varepsilon_\chi = 1$ (complete). Moreover, $\tau \varepsilon_\chi = \chi(\tau) \varepsilon_\chi$ for every $\tau \in G$.

Proof. This is a straightforward verification. □

For understanding Herbrand's Theorem, it is sufficient to look at the quotient group $\mathcal{C} = \mathrm{Cl}\,(K)/\mathrm{Cl}\,(K)^p$ of the class group. This is clearly an $\mathbb{F}_p[G]$-module, and the character group $G\hat{\ }$ is generated by the character ω that maps $\sigma_a \in G$ to $a \bmod p$ (this becomes a character in the above sense upon identifying $\mathbb{F}_p^\times$ with μ_{p-1}). Writing $\varepsilon_i := \varepsilon_{\omega^i}$ for $1 \leq i \leq p-1$, we find that

$$\varepsilon_i = \frac{1}{p-1} \sum_{t=1}^{p-1} t^i \sigma_t^{-1}$$

are the idempotents constructed above. For any $\mathbb{F}_p[G]$-module M we can therefore define $M_i = \varepsilon_i M$. It is a formal consequence of the properties of complete sets of orthogonal idempotents that we have $M = M_1 \oplus \ldots \oplus M_{p-1}$: the existence of the sum is deduced from $1 = \sum e_\chi$, the fact that the sum is direct follows from the orthogonality relations in the same way as in the special case of the idempotents e and $1 - e$ above.

It is an easy exercise to show that $M^- = M_1 \oplus \ldots \oplus M_{p-2}$ and $M^+ = M_2 \oplus \ldots \oplus M_{p-1}$, or, equivalently, that $\frac{1-J}{2} = \sum_{\chi \text{ odd}} e_\chi$ and $\frac{1+J}{2} = \sum_{\chi \text{ even}} e_\chi$ (recall that $J = \sigma_{-1}$). In particular, the decomposition of M into eigenspaces is finer than the one into a plus and minus part, at least for $p > 3$.

C) Pollaczek, Takagi, Herbrand and Ribet Let us recall the situation: We have an odd prime p, the cyclotomic field $K = \mathbb{Q}(\zeta_p)$, and its Galois group $G = \text{Gal}(K/\mathbb{Q})$; we want to study the action of G on the p-class group $\text{Cl}_p(K)$. Using the idempotents of $\mathbb{F}_p[G]$ introduced above we get the decomposition

$$\mathcal{C} = \text{Cl}(K)/\text{Cl}(K)^p = \mathcal{C}_1 \oplus \ldots \oplus \mathcal{C}_{p-1},$$

where the submodules $\mathcal{C}_i = \varepsilon_i \mathcal{C}$ can be described more concretely by

$$\mathcal{C}_i = \{c \in \mathcal{C} : \sigma_t(c) = c^{t^i} \text{ for } 1 \leq t \leq p-1\}.$$

Thus Hecke's result (Proposition 11.20) can be expressed by saying that $\text{rank}\,\mathcal{C}^+ \leq \text{rank}\,\mathcal{C}^-$. By studying the interplay between Kummer theory and class field theory, Leopoldt [Leo] was able to refine this inequality considerably:

Theorem 11.22. *We have* $\text{rank}\,\mathcal{C}_{2n} \leq \text{rank}\,\mathcal{C}_{p-2n} \leq 1 + \text{rank}\,\mathcal{C}_{2n}$ *for all* $1 \leq n \leq \frac{p-3}{2}$. □

Actually this is only a special case of Leopoldt's famous 'Spiegelungssatz' (reflection theorem).

In general it is hard to tell which of the subspaces $\mathcal{C}_i$ are trivial and which are not; the following theorem due to Pollaczek [Pol], Takagi [Tak] and Herbrand [Her] refines Kummer's Theorem 11.19:

Theorem 11.23. *We have* $\mathcal{C}_1 = 1$, *and for odd integers* $3 \leq i \leq p-2$, $p \nmid B_{p-i}$ *implies* $\mathcal{C}_i = 1$.

This shows that the p-divisibility of certain Bernoulli numbers controls the minus class group $\text{Cl}_p^-(K)$ in a very precise way. The proof that $\mathcal{C}_1 = 1$ is actually quite easy: if $c \in \mathcal{C}_1$, then $c^{\sigma_t} = c^t$, hence, by Stickelberger's theorem, $1 = c^{\sum t\sigma_t^{-1}} = c^{\sum tt^{-1}} = c^{p-1} = c^{-1}$, and the claim follows. The general case is hardly more difficult: we know from Stickelberger's Theorem that $(b - \sigma_b)\theta$ kills the class group $\text{Cl}(K)$; in particular it kills $\mathcal{C}$. Thus $\mathcal{C}_i$ is annihilated by $(b - \sigma_b)\theta e_i$. Now $\sigma e_i = \omega^i e_i$ implies that

$$\theta e_i = \Big(\frac{1}{p}\sum_{a=1}^{p-1} a\omega^{-i}(a)\Big)e_i = B_{1,\omega^{-i}}e_i, \tag{11.13}$$

hence mC_i is killed by $(b - \omega(b))B_{1,\omega^{-i}}e_i$. Choosing b in such a way that $b - \omega(b)$ is not divisible by p and observing that e_i is an automorphism on C_i we see that mC_i is killed by $B_{1,\omega^{-i}}$. Finally, putting $n = p - i - 1$ in the congruence

$$\frac{1}{n+1}B_{n+1} \equiv B_{1,\omega^n} \bmod p. \tag{11.14}$$

(see Washington [Was1, Corollary 5.13]) and observing that $\omega^{p-1} = 1$ gives $B_{1,\omega^{-i}} \equiv \frac{1}{p-i}B_{p-i} \bmod p$. This shows that mC_i is killed by B_{p-i}, and Herbrand's theorem follows.

Takagi [Tak] and Herbrand [Her] showed moreover that if one assumes the truth of Vandiver's conjecture that p does not divide the class number of $\mathbb{Q}(\zeta_p + \zeta_p^{-1})$, then the converse also holds: this is done by writing down the corresponding p-class fields explicitly as Kummer extensions generated by p-th roots of certain cyclotomic units; Vandiver's conjecture is needed for securing that these units aren't p-th powers.

Example. For $p = 37$ we find that B_{32} is the only Bernoulli number with index ≤ 34 divisible by 37. Since $h_{37}^- = 37$, we conclude that the minus class group of $K = \mathbb{Q}(\zeta_{37})$ consists only of C_5. In particular, we see that $\sigma_2(c) = c^{32}$ for each ideal class $c \in \text{Cl}^-(K)$.

Using deep properties of modular curves, Ribet [Rb1] succeeded in removing Vandiver's conjecture from the converse of Herbrand's theorem:

Theorem 11.24. *For odd integers $3 \leq i \leq p - 2$, the relation $p \mid B_{p-i}$ implies $C_i \neq 1$.* □

For an exposition of Ribet's proof without the technical details, see Tamme [Ta]. The explicit construction of the class fields corresponding to nontrivial C_i was studied by Harder & Pink [HP] as well as by Harder's students Lippert [Lip] and Kleinjung [Klj]. Generalizations to cyclotomic fields of conductor pq were given by Kamienny [Ka].

Gut [Gut] studied a similar situation in 1951: consider the fields $L = \mathbb{Q}(\zeta_{4n})$ and $K = \mathbb{Q}(\zeta_n)$; the relative class group $\text{Cl}(L/K)$ is defined to be the kernel of the norm map $N_{L/K} : \text{Cl}(L) \longrightarrow \text{Cl}(K)$; let $C = \text{Cl}_p(L/K)$ be its p-Sylow subgroup. Next define the *Euler numbers E_n* by

$$\sum_{n=0}^\infty E_n \frac{x^n}{n!} = \frac{2}{e^x + e^{-x}};$$

actually Euler numbers are essentially generalized Bernoulli numbers since $E_n = \frac{2}{n+1}B_{n+1,\chi}$, where χ is the nontrivial Dirichlet character modulo 4.

Then Gut showed that $\#C$ is divisible by p if and only if one of the Euler numbers $E_2, \ldots, E_{p-3}$ is divisible by p. Kleboth [Kle] proved the analogous result over $\mathbb{Q}(\zeta_3)$. The natural question whether this result can be improved in the direction of Herbrand's theorem was only studied after Mazur and Wiles had proved the main conjecture of Iwasawa theory which contained such an extension of Gut's result as a very special case.[1] Recently, Ernvall [Er2] has proved such a generalization of Herbrand's theorem using the elementary techniques of Herbrand.

The Stickelberger Ideal

The most attractive results about the structure of class groups of abelian fields are known only for cyclotomic fields of prime power conductor, and in this case the desired result follows from the computation of the index of the Stickelberger ideal:

Theorem 11.25. *(The Index of the Stickelberger Ideal) Let $m = p^n$ be a prime power, $K = \mathbb{Q}(\zeta_m)$, $G = \text{Gal}(K/\mathbb{Q})$, $R = \mathbb{Z}[G]$, $\theta = \theta(K)$ the corresponding Stickelberger element, and $I = R \cap R\theta$ the Stickelberger ideal. Moreover, let $J = \sigma_{-1}$ denote complex conjugation, and define*

$$R^- = \{x \in R \mid Jx = -x\}, \quad I^- = I \cap R^-.$$

Then

$$(R^- : I^-) = h^-(K) := Qw \prod_{\chi \text{ odd}} \left\{ -\frac{1}{2f_\chi} \sum_{t=1}^{f_\chi - 1} \chi(t) t \right\}. \tag{11.15}$$

The analytic class number formula says that the number $h^-(K)$ defined above coincides with the minus class number $h^-(K) = \#\text{Cl}^-(K)$ for any abelian extension of $\mathbb{Q}$. Here $Q = (E_K : W_K E_{K^+})$ denotes Hasse's unit index, which is known to be trivial for cyclotomic fields of prime power conductor, and in general takes only the values 1 or 2 (see Exercise 11.10). The product is over all odd characters of the character group $X(K/\mathbb{Q})$ associated to abelian extensions of $\mathbb{Q}$ in Section 4.5. Using the generalized Bernoulli numbers defined in Equation (10.16) we can also express h^- in the form

$$h^- = Qw \prod_{\chi \text{ odd}} \left(-\frac{1}{2} B_{1,\chi}\right).$$

Note that these h^- coincide with the integers defined in Propositions 11.16 and 11.18 above (see Exercise 11.23).

Let us give a few examples. Assume that p is an odd prime and let $K = \mathbb{Q}(\zeta_{p^n})$. We claim that Theorem 11.25 provides us with an algebraic proof

---

[1] Karl Rubin kindly explained that to me in an email from July 29, 1998.

that $h^-(K)$ annihilates the odd part of $\mathrm{Cl}^-(K)$. In fact, we know $\mathrm{Cl}_p^-(K) = \mathrm{Cl}_p(K)^{1-J} = \mathrm{Cl}_p(K)^{R^-}$, so for any $c \in \mathrm{Cl}_p^-(K)$, we find $c^{h^-} \in \mathrm{Cl}(K)^{I^-}$, because $h^- = (R^- : I^-)$; but I^- annihilates the ideal class group of K, hence h^- annihilates $\mathrm{Cl}_p^-(K)$ for every odd prime p. For $p = 2$ the result is not as strong: from Exercise 11.11 we know that $\mathrm{Cl}_2^-(K)^2 \subseteq \mathrm{Cl}_2(K)^{(1-J)}$, so the above reasoning only shows that $2h^-$ annihilates $\mathrm{Cl}_2^-(K)$.

The proof that $h^-(K) = (R^- : I^-)$ also shows that $h^-(K)$ is an integer: a direct integrality proof for general abelian extensions was given by Hasse [Has1].

For the proof of Theorem 11.25 we need a few concepts. Exercise 11.13 generalizes the notion of an index of free abelian groups (lattices, to be exact). From Exercise 11.14 we will also borrow the fact that if V is a $\mathbb{Q}$-vector space and $T : V \longrightarrow V$ an invertible endomorphism, then $(A : TA) = |\det T|$ for any lattice A in V.

Proof of Theorem 11.25. We will split up the index $(R^- : I^-)$ into more manageable parts. To this end, we put $\mathcal{S} = \{\alpha \in R : \alpha\theta \in R\}$, so $I = R \cap R\theta = \theta\mathcal{S}$. We also introduce the $\mathbb{Q}$-vector space $V = \mathbb{Q}[G]^- = 2e^-\mathbb{Q}[G] = \{\alpha \in \mathbb{Q}[G] : (1+J)\alpha = 0\}$, where $2e^- = 1 - J \in \mathbb{Z}[G]$, and the linear map $T : V \longrightarrow V$ defined by $T\alpha = \theta\alpha$. Clearly V is a $\mathbb{Q}$-vector space of dimension $r = \frac{1}{2}\#G$, and R^- is a submodule of V of full rank r. Now

$$(R^- : I^-) = \frac{(R^- : 2e^-\mathcal{S})(2e^-\mathcal{S} : T(2e^-\mathcal{S}))(T(2e^-\mathcal{S}) : 2I^-)}{(I^- : 2I^-)}.$$

Since I^- has rank r, we see $(I^- : 2I^-) = 2^r$. Moreover, $(2e^-\mathcal{S} : T(2e^-\mathcal{S})) = |\det T|$; in order to compute this determinant, we extend T to a map on $\mathbb{C}[G]^- = \{\alpha \in \mathbb{C}[G] : (1+J)\alpha = 0\}$ and observe that the idempotents e_χ, χ odd, form a basis of $\mathbb{C}[G]^-$ over $\mathbb{C}$. The relation (11.13), when generalized from primes p to prime powers, says that $e_\chi \theta = B_{1,\chi^{-1}}\theta$. Thus the e_χ are eigenvectors of T with eigenvalues $B_{1,\chi^{-1}}$, hence $\det T = \prod_{\chi \text{ odd}} B_{1,\chi^{-1}}$.

All that is left to do now is to compute $(R^- : 2e^-\mathcal{S})$ and $(T(2e^-\mathcal{S}) : 2I^-)$. We define a homomorphism $\phi : R \longrightarrow \mathbb{Z}/p^n\mathbb{Z}$ by putting $\phi(\sigma_a) = a + p^n\mathbb{Z}$ and claim that

$$0 \longrightarrow \mathcal{S} \longrightarrow R \xrightarrow{\phi} \mathbb{Z}/p^n\mathbb{Z} \longrightarrow 0$$

is an exact sequence. Surjectivity of ϕ being clear, we have to show that $\ker \phi = \mathcal{S}$. Take an $\alpha = \sum x_b \sigma_b \in R$; here and below, the sums are over all $1 \leq b < p^n$ with $p \nmid b$. Then

$$p^n \alpha \theta = \sum_a \sum_b a x_b \sigma_a^{-1} \sigma_b = \sum_c \sigma_c \sum_a a x_{ac}.$$

Now $\alpha \in \mathcal{S}$ means $\alpha\theta \in R$, and this implies that the coefficient of σ_1 in $p^n \alpha\theta$ is divisible by p^n; but then $\sum_a a x_a \equiv 0 \bmod p^n$, hence $\phi(\alpha) = 0$ and

$\alpha \in \ker \phi$. Conversely, assume that $\phi(\alpha) = 0$. Since ϕ is a homomorphism, this implies $\phi(\alpha\sigma_c^{-1}) = 0$, hence the coefficient of σ_1 in $\alpha\sigma_c^{-1}\theta$ is an integer; now this coefficient coincides with the coefficient of σ_c in $\alpha\theta$, therefore $\alpha\theta \in R$ and $\alpha \in S$ as claimed.

Note that $R^- = 2e^-R$, although this is not obvious as 2 is not invertible in $\mathbb{Z}$. In fact it is sufficient to show that $R^- \subseteq 2e^-R$. But $\alpha = \sum x_a\sigma_a \in R^-$, the sum being over all $1 \leq a < m$ with $p \nmid a$, implies that $x_a = -x_{m-a}$, hence $\alpha = \sum_a x_a\sigma_a = (1-J)\sum_b x_b\sigma_b$, where b runs over the integers $1 \leq b < \frac{m}{2}$ not divisible by p.

Now we use the simple fact that $(A:B) = (A^f : B^f)(A_f + B : B)$, where $f: A \longrightarrow A$ is a group endomorphism and where $B \subseteq A$ is a subgroup of the abelian group A (see Exercise 11.12). Applying this to the situation $A = R$, $B = S$ and $f = 1 - J$, we find $(R:S) = (R^- : 2e^-S)$ because the kernel of $1 - J : R \longrightarrow R$ is $(1+J)R \subseteq S$. Thus $(R^- : 2e^-S) = p^n$.

Finally we claim that $(T(2e^-S) : 2I^-) = \delta$ with $\delta = 2$ if p is odd and $\delta = 1$ if $p = 2$; in particular, $\delta \cdot p^n = w$ gives the number of roots of unity in K. Taking this for granted and putting everything together we get

$$(R^- : I^-) = \frac{(R^- : 2e^-S)(2e^-S : T(2e^-S))(T(2e^-S) : 2I^-)}{(I^- : 2I^-)}$$

$$= \frac{p^n \cdot |\prod B_{1,\chi^{-1}}| \cdot \delta}{2^r} = w \prod_{\chi \text{ odd}} \left(-\frac{1}{2}B_{1,\chi}\right),$$

where $w = \#W_K$ denotes the number of roots of unity in K. The fact that the product $\prod_{\chi \text{ odd}}(-\frac{1}{2}B_{1,\chi})$ is positive follows from the analytic class number formula; if you don't want to invoke analytic machinery, take the absolute value on both sides.

In order to study the index $(T(2e^-S) : 2I^-)$ we introduce a homomorphism $\psi : T(2e^-S) \longrightarrow \mathbb{Z}/2\mathbb{Z}$ by putting $\psi(2e^-\gamma\theta) = \omega(\gamma) + 2\mathbb{Z}$, where $\omega : R \longrightarrow \mathbb{Z}$ induced by mapping σ_a to 1. The identity $2 = 2e^- + 2e^+$ (with $e^+ = 1 + J$) shows that $2e^-\gamma\theta = 2\gamma\theta - 2e^+\gamma\theta$; since $2e^+\theta = N$ is the norm, we have $2e^+ \in S$, $N \in I$, and $2e^+\gamma\theta = \gamma N = \omega(\gamma)N$. If $\omega(\gamma)$ is even, then this implies that $2e^-\gamma\theta \in 2I \cap R^- = I^-$. Conversely, if $2e^-\gamma\theta \in 2I^-$, then $\omega(\gamma)N \in 2R$, thus $2 \mid \omega(\gamma)$. This proves that $\ker \psi = 2I^-$.

If there is a $\gamma \in S$ such that $\psi(\gamma)$ is odd, then similarly $2e^-\gamma\theta - N \in 2I$, and then $N \in R \setminus 2R$ implies that $2e^-\gamma\theta \notin 2I^-$, which in turn means that ψ is onto. Now if p is odd, then $p^n \in S$, so $\gamma = p^n$ does it. If $p = 2$, however, then we claim that ψ is the trivial map. To this end we observe that $(\sum x_b\sigma_b)\theta = (\sum x_b b)\theta$ for $\sum x_b\sigma_b \in R$; since the b's are all odd if $p = 2$, we see that $\sum x_b b \equiv \sum x_b = \omega(\sum x_b\sigma_b) \bmod 2$. In particular, the existence of a $\gamma \in S$ with odd $\omega(\gamma)$ implies that the odd integer $\omega(\gamma)$ is in S: but S also contains $m = 2^n$, and since S is an ideal, it must be equal to R. But this is a contradiction because it would imply $\theta \in R$. This completes our proof. $\square$

382   11. Eisenstein Reciprocity

Sinnott [Sin] defined a Stickelberger ideal $I(K)$ for general abelian extensions $K/\mathbb{Q}$ in such a way that Theorem 11.25 essentially remains valid; more exactly he showed that $(R^- : S^-) = c^- h^-$ for certain 'dirt factors' c^-. Similar class number formulas hold for the plus part when Stickelberger ideals are replaced by cyclotomic units. Recently, a unified approach combining the plus and minus side was discovered by Anderson [An]. Anderson also inspired a new proof of Sinnott's formulas by Ouyang [Ou] in which Sinnott's quite technical calculations are replaced by arguments using spectral sequences. For computational aspects involving Stickelberger ideals, see Schoof [Sf3].

The fact that the index $(R^- : I^-)$ coincides with the minus class number $h^- = \# \operatorname{Cl}^-(K)$ prompts the question whether there is an isomorphism $R^-/I^- \simeq \operatorname{Cl}^-(K)$ as abelian groups (or even as $\operatorname{Gal}(K/\mathbb{Q})$-modules). The answer to the second question is no (see Washington [Was1]), and the first question can be answered negatively using the following result due to Jha [Jha, p. 78]:

Proposition 11.26. *Let $p \equiv 3 \bmod 4$ be a prime, $K = \mathbb{Q}(\zeta_p)$, and let R^- and I^- be as above. Moreover, let h be defined as in Proposition 11.16. Then h divides the exponent t of R^-/I^-, and, in particular, $h \mid h^-(K)$.*

Proof. Consider the homomorphism $\lambda : R \longrightarrow \mathbb{Z}$ induced by $\sigma_a \longmapsto (\frac{a}{p})$. Observe that $\lambda(\theta) = h$. Since $1 - J \in R^-$ and t kills R^-/I^-, we have $(1 - J)t \in I^- = R^- \cap R\theta$. Thus $(1 - J)t = \gamma \theta$ for some $\gamma \in R$, hence $2t = \lambda((1 - J)t) = \lambda(\gamma)\lambda(\theta) = \lambda(\gamma) \cdot h$. Now $\lambda(\gamma)$ is an integer, hence h divides $2t$; but h is odd by genus theory, and the claim follows.   □

We need another ingredient:

Proposition 11.27. *Let ℓ, q and $p = 2q+1$ be odd primes, and assume that ℓ is a primitive root modulo q. Put $K = \mathbb{Q}(\zeta_p)$ and let h be defined as in Proposition 11.16. Then ℓ does not divide $h^-(K)/h$.*

Proof. We have already seen that $h \mid h^-(K)$, hence $h^-(K)/h$ is an integer. From the definition of $h^-(K)$ and h we find immediately that $h^-(K)/h = 2p\prod_\chi(-\frac{1}{2}B_{1,\chi})$, where the product is over all characters of $\mathbb{F}_p^\times$ with exact order $p - 1 = 2q$ (the class number h corresponds to the single odd character of order 2 in the product (11.15)). Now $B_{1,\chi} \in \mathbb{Q}(\zeta_q)$, and the product over all characters of order $2q$ is simply the norm of $B_{1,\psi}$, where ψ is the character defined by $\psi(\ell) = \zeta_{2q}$. But since ℓ is inert in $\mathbb{Q}(\zeta_q)$, the norm of $B_{1,\psi}$ is divisible by ℓ if and only $B_{1,\psi}$ is. But this is easily seen not to be the case: if we write $B_{1,\psi} = \frac{1}{p}\sum_{a=1}^{p-1} a\psi(a)^{-1} = \sum_{j=0}^{q-1} a_j \zeta_q^j$, then $a_0 = \psi(1) + (p-1)\psi(p-1) = 2-p$ and (observe that $\zeta_{2q} = \zeta_q^{(q+1)/2}$) $a_{(q+1)/2} = \ell\psi(\ell) + (p-\ell)\psi(p-\ell) = 2\ell - p$. But a sum $\sum_{j=0}^{q-1} a_j \zeta_q^j$ is divisible by an integer n if and only if n divides all the differences $a_j - a_0$; since $a_{(q+1)/2} - a_0 = 2\ell - p - (2-p) = 2\ell - 2$ and since ℓ is odd, $B_{1,\psi}$ is not divisible by ℓ.   □

For a proof that R^-/I^- is not isomorphic to $\mathrm{Cl}^-(K)$ as an abelian group we use the analytic class number formula which says that $h^-(K)$ is the order of $\mathrm{Cl}^-(K)$, and that h is the class number of $k = \mathbb{Q}(\sqrt{-p})$. Now take $\ell = 3$ and let p and q be as above. Then $(K : k) = q$ is not divisible by ℓ, consequently the transfer of ideal classes $j : \mathrm{Cl}(k) = \mathrm{Cl}^-(k) \longrightarrow \mathrm{Cl}^-(K)$ is injective (see Exercise 11.25). If we can find a p such that the 3-class group $\mathrm{Cl}_3(k)$ of k is non-cyclic, then so is $\mathrm{Cl}^-(K)$. If 3 is a primitive root modulo q, then the fact that $3 \nmid h^-/h$ implies that $\mathrm{Cl}_3^-(K) \simeq \mathrm{Cl}_3(k)$. Thus $h/3$ kills $\mathrm{Cl}_3^-(K)$ while the exponent of R^-/I^- is divisible by h: hence R^-/I^- is not isomorphic to $\mathrm{Cl}^-(K)$.

Finding such q is easy: $q = 30689$ (here $\mathbb{Q}(\sqrt{-p})$ has class group $(\mathbb{Z}/3\mathbb{Z} \oplus \mathbb{Z}/9\mathbb{Z} \oplus \mathbb{Z}/5\mathbb{Z})$ and $q = 38333$ (here $\mathbb{Q}(\sqrt{-p})$ has class group $(\mathbb{Z}/3\mathbb{Z} \oplus \mathbb{Z}/27\mathbb{Z})$ are the two smallest examples.[2]

Brumer and Stark

To some degree, the results of Stickelberger on the annihilation of class groups can be generalized (at least conjecturally) from abelian extensions of $\mathbb{Q}$ to those of arbitrary number fields. So let K/k be an abelian extension of number fields with Galois group $G = \mathrm{Gal}(K/k)$ and conductor $\mathfrak{f} = \mathrm{cond}(K/k)$. Let S denote a finite set of places containing all ramified and all archimedean places, and write $(\mathfrak{a}, S) = 1$ if an integral ideal $\mathfrak{a}$ is not divisible by any finite prime in S. Then define the partial ζ-function

$$\zeta_S(\sigma, s) = \sum_{\substack{(\mathfrak{a},S)=1 \\ (\mathfrak{a},K/k)=\sigma}} N\mathfrak{a}^{-s}. \tag{11.16}$$

Here $(\mathfrak{a}, K/k)$ denotes the Artin symbol. Siegel has shown that the values $\zeta_S(\sigma, 0)$ are rational (this is a deep result!), hence the *Brumer-Stark elements*

$$\theta_{S,K/k} = \theta_{S,G} = \theta_S := \sum_{\sigma \in G} \zeta_S(\sigma, 0) \sigma^{-1} \tag{11.17}$$

are elements of the group ring $\mathbb{Q}[G]$. It follows from work of Deligne & Ribet [DeR] (as well as from Shintani's formulas [Shi]) that the denominator of θ_S is bounded by the number of roots of unity in K; in fact, if we denote the group of roots of unity in K by W_K and put $w = \#W_K$, then

$$w \, \theta_{S,K/k} \in \mathbb{Z}[G]. \tag{11.18}$$

More generally, if some $\xi \in \mathbb{Z}[G]$ kills W_K, then $\xi \theta_{S,K/k} \in \mathbb{Z}[G]$.

Let us see what happens when $k = \mathbb{Q}$ and $K = \mathbb{Q}(\zeta_m)$ for some $m \geq 3$. In this case, the Artin symbol $(a, K/\mathbb{Q})$ maps an ideal (a) generated by

---

[2] Thanks go to René Schoof and Larry Washington for communicating the ideas that led to these examples (emails from Nov. 13, 1999).

a positive integer a coprime to m to the element $\sigma_a : \zeta_m \mapsto \zeta_m^a$ of the Galois group $G = \text{Gal}(K/\mathbb{Q})$. Thus the elements $a \in \mathbb{N}$ whose Artin symbols coincide form a residue class modulo m, and the partial ζ-function defined in (11.16) coincides with the partial ζ-function (10.13) studied in Chapter 10. Here clearly $S = \{\infty\} \cup \{p : p \mid m\}$. Recall that

$$\zeta(\sigma_a, 0) = \frac{1}{2} - \left\langle \frac{a}{m} \right\rangle$$

by Theorem 10.22. This shows that the corresponding Brumer-Stark elements are $\theta_{S,K/k} = \frac{1}{2}\nu - \theta$, where θ is the Stickelberger element defined in (11.11). Note that $\#W_K = 2m$, and that $2m\theta_{S,K/k} \in \mathbb{Z}[G]$ as predicted by (11.18).

Back to the general case. Let $\mathbb{P}$ denote the set of all places in K, and let $\mathbb{P}_\infty$ be its subset of infinite places. Then the subgroup

$$K^\circ = \{\alpha \in K^\times : |\alpha|_v = 1 \text{ for all } v \in \mathbb{P}_\infty\} \qquad (11.19)$$

is called the group of *anti-units* (the group of units E_K is defined by replacing $\mathbb{P}_\infty$ in (11.19) with $\mathbb{P}\setminus\mathbb{P}_\infty$; this explains the 'anti'). It follows from Kronecker's Lemma (see Exercise 11.1) that, for anti-units α, an ideal (α) determines α up to a root of unity in W_K. This observation guarantees that the following conjecture makes sense:

The Brumer-Stark Conjecture
Let K/k be an abelian extension of number fields. Then for each ideal $\mathfrak{a}$ in K there is an $\alpha \in K^\circ$ such that $\mathfrak{a}^{w \cdot \theta_{S,K/k}} = (\alpha)$, and such that the extension $K(\sqrt[w]{\alpha})/k$ is abelian.

In fact, α is determined up to a factor $\zeta \in W_K$, and $K(\sqrt[w]{\alpha})/k$ is abelian if and only if $K(\sqrt[w]{\zeta\alpha})/k$ is (since $K(\sqrt[w]{\zeta})/k$ is abelian, $K(\sqrt[w]{\alpha})/k$ is abelian if and only if $K(\sqrt[w]{\alpha}, \sqrt[w]{\zeta})/k$ is).

If K is a cyclotomic extension of $\mathbb{Q}$, the Brumer-Stark conjecture follows from Stickelberger's theorem (see Exercise 11.18). The fact that the Brumer-Stark elements give rise to abelian extensions of number fields k made Stark look more closely at what is happening here; see Tate's book [Ta1]. Recently, the Stark conjectures have been used to find explicit generators for certain Hilbert class fields by e.g. H. Bauer [Bau], Dummit & Hayes [DuH], Dummit, Sands & Tangedal [DST] and Roblot [Ro1, Ro2].

Amazingly, the Brumer-Stark theory can be generalized yet further: for abelian extensions K/k, define elements

$$\theta_n(K/k) = \sum_{\sigma \in G} \zeta_K(\sigma, -n)\sigma^{-1}.$$

Clearly $\theta_0(K/k)$ is the Brumer-Stark element for K/k. Let $w_n(K)$ be the maximal integer m such that $\text{Gal}(K(\zeta_m)/K)$ has exponent dividing n; in particular, we have $w_1(K) = w = \#W_K$. Again it follows from the work of Deligne and Ribet that

$$w_n(K)\theta_n(K/k) \in \mathbb{Z}[G].$$

Coates [Co2] has shown that their results even imply that the elements

$$(N\mathfrak{a}^{n+1} - (\mathfrak{a}, K/k))\theta_n(K/k) \tag{11.20}$$

are integral. This allows us to define the n-th Stickelberger ideal $I_n(K/k)$ as the $\mathbb{Z}[G]$-ideal generated by elements in (11.20), as was suggested by Brumer (see Rideout's thesis [Rid]). Stickelberger's theorem (combined with Lemma 11.13) says that the ideal $I_0(K/\mathbb{Q})$ annihilates $\mathrm{Cl}(K)$. What do the $I_n(K/\mathbb{Q})$ annihilate?

Note that $\mathrm{Cl}(K)$ can be interpreted as the reduced K-group $\widetilde{K}_0(\mathcal{O}_K)$; it is also known that Milnor's $K_2(\mathcal{O}_K)$ is a finite group, and it can be shown that $I_1(K/\mathbb{Q})$ kills $K_2(\mathcal{O}_K)$, except perhaps for the 2-part. The conjecture of Birch & Tate (see Birch [54] for its origin) predicts that, for K totally real, $\#K_2(\mathcal{O}_K)$ equals $w_2(K) \cdot |\zeta_K(-1)|$. In general, it is expected that $I_n(K/\mathbb{Q})$ annihilates Quillen's K-groups $K_{2n}(\mathcal{O}_K)$.

Iwasawa Theory

One of the origins of Iwasawa theory is the construction of functions that interpolate zeta functions p-adically. In fact, since Hurwitz's ζ-function assumes rational values at the negative integers, it is tempting to ask whether there exists a continuous function defined on $\mathbb{Z}_p$ that takes the same values there, at least up to some trivial factors. The answer is yes, as was shown by Leopoldt and Kubota (who constructed p-adic L-functions this way) as well as Iwasawa (who used p-adic integration). These p-adic L-functions can be used to show that, for large enough $n \in \mathbb{N}$, the p-class number h_n of $\mathbb{Q}(\zeta_{p^n})$ is given by

$$h_n = \#A_n = p^t, \quad t = \mu p^n + \lambda n + \nu; \tag{11.21}$$

here $\mu, \lambda, \nu \in \mathbb{N}_0$ are integers depending only on the prime p.

Later Iwasawa could prove a similar formula for p-class numbers in arbitrary $\mathbb{Z}_p$-extensions; these are infinite abelian extensions k_∞/k of a number field k with Galois group $\mathrm{Gal}(k_\infty/k) \simeq \mathbb{Z}_p$. A famous result of Ferrero and Washington says that $\mu = 0$ for every abelian extension $F/\mathbb{Q}$. A notoriously difficult question is whether $\mu = \lambda = 0$ for all totally real number fields F; this conjecture of Greenberg has only been verified in special cases. The number of independent $\mathbb{Z}_p$-extensions of a given number field k is at most $r_2 + 1$ (where r_2 denotes the number of complex primes of k), with equality if Leopoldt's conjecture on p-adic regulators is true.

Iwasawa also found a way to reinterpret p-adic L-functions in terms of his theory of $\mathbb{Z}_p$-extensions; this led to a very natural conjecture on the nature of these functions: the "Main Conjecture" of Iwasawa theory. See Nekovar [Nek].

Mazur and Wiles

This Main Conjecture soon occupied a central part of the research in Iwasawa theory, and when Mazur and Wiles eventually proved it in 1984, it was already known that it had quite a few important corollaries. For example, R. Greenberg had by then deduced the following conjecture of G. Gras from the Main Conjecture: Let K be an abelian extension of $\mathbb{Q}$ with conductor m, $L = \mathbb{Q}(\zeta_m)$, and put $G = \mathrm{Gal}\,(K/\mathbb{Q})$. The units in $E = E_K$ that can be written as products or quotients of elements of the form $\pm N_{L/K} \prod_a (1 - \zeta_m^a)$ form a group

$$C = C_K = \langle \pm N_{L/K}(\zeta_m^j \prod_a (1 - \zeta_m^a)) \rangle \cap E_K \qquad (11.22)$$

called the group of *cyclotomic* (sometimes also called *circular*) units of K; note, however, that there are even more definitions of cyclotomic units than of Stickelberger ideals floating around.

Gras conjectured that, for even characters $\chi \in G\widehat{}$, the components $\mathrm{Cl}_p(\chi)$ and $(E/C)_p(\chi)$ not only should have the same order, but that they are isomorphic as $\mathbb{Z}_p[G]$-modules. Thanks to Mazur and Wiles, this is now a theorem.

Another corollary of the Main conjecture is a class number formula conjectured by Iwasawa and Leopoldt. In order to formulate it, let us write $a \sim b$ for p-adic integers $a, b \in \mathbb{Z}_p$ when a and b are divisible by the same p-power.

Theorem 11.28. *Let $F/\mathbb{Q}$ be an abelian extension with Galois group G, and assume that p is an odd prime not dividing $(F : \mathbb{Q})$. Let $\chi \neq \omega$ be an odd character of $\mathrm{Gal}\,(F/\mathbb{Q})$; then*

$$\#\mathrm{Cl}_p(F)(\chi) \sim B_{1,\chi^{-1}}^g. \qquad (11.23)$$

Here $g = (\mathbb{Q}_p(\chi) : \mathbb{Q}_p)$, where $\mathbb{Q}_p(\chi)$ is the smallest extension of $\mathbb{Q}_p$ containing the values of χ. □

The condition that $p \nmid (F : \mathbb{Q})$ was removed later by D.R. Solomon [Sol]. As an example, take $F = \mathbb{Q}(\zeta_p)$; here $G \simeq \mathbb{Z}/(p-1)\mathbb{Z}$, hence χ takes values in $\mu_{p-1} \subset \mathbb{Q}_p$, and we have $g = 1$. Theorem 11.28 says that if i is odd then $\#\mathrm{Cl}_p(F)(\omega^i) \sim B_{1,\omega^{-i}}$, congruence (11.14) says that $B_{1,\omega^{-i}} \equiv \frac{1}{p-i}B_{p-i} \bmod p$, hence we find that $\mathrm{Cl}_p(F)(\omega^i) \neq 1$ if and only if $p \mid B_{p-i}$; this is of course the theorem of Herbrand–Ribet.

In a similar way we can explain (and improve on) the results of Gut and Kleboth: take $F = \mathbb{Q}(\zeta_{4p})$; its characters are either characters χ belonging to $K = \mathbb{Q}(\zeta_p)$, or they have the form $\psi\chi$, where ψ is the nontrivial Dirichlet character modulo 4. The relative class group $\mathrm{Cl}_p(F/K)$ corresponds to characters $\psi\chi$, that is, $\mathrm{Cl}_p(F/K) \simeq \bigoplus_\chi \mathrm{Cl}_p(F)(\psi\chi)$, and $\mathrm{Cl}_p(K) \simeq \bigoplus_\chi \mathrm{Cl}_p(F)(\chi)$. Since ψ is an odd character, $\psi\omega^i$ is odd if and only if i is even; for such values of i we find that $\#\mathrm{Cl}_p(F)(\psi\omega^i) \sim B_{1,\psi\omega^{-i}}$ by (11.23), and $B_{1,\psi\omega^{-i}} \equiv \frac{1}{p-i}B_{p-i}(\psi) \bmod p$ by (11.14). Finally $\frac{1}{p-i}B_{p-i}(\psi) \sim E_{p-1-i}$ by the definition of Euler numbers, hence we find that for even integers $i \leq p$, the component $\mathrm{Cl}_p(F)(\psi\omega^i)$ is nontrivial if and only if $p \mid E_{p-1-i}$.

Yet another consequence of the Main Conjecture ([MW, Thm. 5]) due to Coates [Co1] is the formula $\#K_2(\mathcal{O}_K) = w_2(K) \cdot |\zeta_K(-1)| \cdot 2^a$ for real abelian extensions $K/\mathbb{Q}$ and some 2-power 2^a, that is, the truth of the Birch-Tate conjecture up to 2-powers. The problem with the 2-part came from the fact that Iwasawa's Main Conjecture was usually formulated only for odd primes p; the analogous conjecture for $p = 2$ was stated by Iwasawa (see Federer [Fe]), Kolster [Kol] showed that it would imply the 2-primary part of the Birch-Tate conjecture, and finally Wiles [Wil] proved the Main Conjecture also for $p = 2$. For a proof using the simpler methods described in the next subsection, see Greither [Gre].

The connection between K-groups and the class groups of cyclotomic fields is much stronger than indicated by this last result. In fact, Kurihara [Kur] showed that Vandiver's conjecture would follow from conjectures about the structure of Quillen's K-groups $K_n(\mathbb{Z})$; he exploits this relationship to construct a surjection $K_4(\mathbb{Z}) \otimes \mathbb{Z}/p\mathbb{Z} \longrightarrow \mathcal{C}_{p-3}$, where $\mathcal{C}_{p-3} = \mathrm{Cl}_p(F)(\omega^{p-3})$, and then shows that $K_4(\mathbb{Z})$ is small enough to enforce $\mathcal{C}_{p-3} = 0$ (in fact, today we know that $K_4(\mathbb{Z}) = 0$). A more general result in this direction is due to Soulé [Sou]; see Ghate [Gha] for an introduction.

Thaine, Kolyvagin and Rubin

Stickelberger's relation contains no information on the class group of real abelian fields (see Exercise 11.18, for example). In [Tha], Thaine used cyclotomic units to construct annihilators of ideal class groups of real abelian fields F: let $m = \mathrm{cond}\, F$ be the conductor of F, $G = \mathrm{Gal}\,(F/\mathbb{Q})$ its Galois group, put $K = \mathbb{Q}(\zeta_m)$, and define the subgroup $C_F \subseteq E_F$ of cyclotomic units as in (11.22) above. Then Thaine proved that for any prime p not dividing $(F : \mathbb{Q})$, 2θ kills $\mathrm{Cl}_p(F)$ whenever $\theta \in \mathbb{Z}[G]$ kills the p-Sylow subgroup of E_F/C_F.

By refining Thaine's construction, Kolyvagin could not only give an elementary proof of Gras' conjecture but also of Ribet's converse of Herbrand's theorem. Rubin finally showed how Kolyvagin's theory of Euler systems could be applied to prove the main conjecture of Iwasawa theory; see e.g. his appendix in Lang's [La2], [Ru1], or the survey [PR] by Perrin-Riou. All this is explained beautifully in the second edition of Washington's book [Was1]. Rubin has written a book on Euler systems that will be published soon. His CIME lectures [CGR] on this topic have just appeared.

NOTES

Normal Integral Bases

It was not completely accurate when we said that our proof of Eisenstein's reciprocity law would follow Hilbert's arguments as laid out in his Zahlbericht

[368]: in fact we left out all of his results on normal integral bases of number fields, because they are not needed for deriving Eisenstein's reciprocity law. We cannot disregard them completely, however, because these results moved to the center of mathematical interest during the 1970's. This was due to the completely unexpected connections with Artin's L-series; in order to see what has happened we have to go back to Hilbert's Zahlbericht.

In Chapter 3 we have seen that, for odd primes p, all subfields of $\mathbb{Q}(\zeta_p)$ have a normal integral basis. More generally, Hilbert showed

Theorem 11.29. (Satz 132) *Let $K/\mathbb{Q}$ be an abelian extension of $\mathbb{Q}$ such that $(\operatorname{disc} K, (K:\mathbb{Q})) = 1$. Then $\mathcal{O}_K$ has a NIB.* □

Hilbert's proof was quite simple: he used the theorem of Kronecker and Weber to embed K in some $L = \mathbb{Q}(\zeta_m)$ and applied Proposition 3.6 to reduce the problem to finding a NIB of $\mathcal{O}_L$, which is easy.

Now where are the Gauss sums? Take odd primes p, ℓ with $p \equiv 1 \bmod \ell$, and let k denote the subfield of degree ℓ in $\mathbb{Q}(\zeta_p)$. Since $\operatorname{disc} k = p^{\ell-1}$, $k/\mathbb{Q}$ satisfies the hypothesis of Satz 132 and thus has a NIB generated by $\nu \in \mathcal{O}_k$ (this means that a NIB is given by $\{\nu, \nu^\sigma, \ldots, \nu^{\sigma^{\ell-1}}\}$, where σ is a generator of $G = \operatorname{Gal}(k/\mathbb{Q})$. In this situation, the element

$$\Omega = \nu + \zeta_\ell \nu^\sigma + \zeta_\ell^2 \nu^{\sigma^2} + \ldots + \zeta_\ell^{\ell-1} \nu^{\sigma^{\ell-1}} \in \mathbb{Z}[\zeta_p]$$

is called a *root number* (Wurzelzahl) of k by Hilbert. The principal properties of root numbers are given by

Theorem 11.30. (Satz 133) *Let σ be the $\mathbb{Q}$-automorphism of $K = \mathbb{Q}(\zeta_\ell)$ defined by $\zeta_\ell \longmapsto \zeta_\ell^r$, where r is a primitive root modulo ℓ, and let k be as above. Then root numbers of k have the following properties:*

i) $\omega := \Omega^\ell \in \mathcal{O}_K$, and $\omega^{\sigma-r}$ is an ℓ-th power in $K^\times$.
ii) $\Omega \equiv \pm 1 \bmod (1 - \zeta_\ell)$, and $\omega \equiv \pm 1 \bmod (1 - \zeta_\ell)^\ell$.
iii) $N_{K/\mathbb{Q}}\omega = p^{\ell(\ell-1)/2}$.
iv) $\omega \in K^\times \setminus K^{\times \ell}$. □

As a matter of fact, Hilbert neither states nor proves property iv), but when he claims that properties i), ii) and iii) suffice to characterize root numbers, he makes use of it:

Theorem 11.31. (Satz 134) *If $\omega \in \mathcal{O}_K$ satisfies the properties i) – iv) in Theorem 11.30, then $\Omega = \sqrt[\ell]{\omega}$ is a root number of k.* □

Next Hilbert studies the prime ideal decompositions of root numbers with the techniques discussed in Section 11.1 (this is no problem: the essential property is that root numbers as well as Gauss sums generate *abelian* extensions).

We still haven't seen any Gauss sums, but now they enter the picture: since $\mathbb{Q}(\zeta_p)$ has a NIB, so does k (its subfield of degree ℓ), and the proof of Proposition 3.6 shows that we can take

$$\lambda_0 = \zeta_p + \zeta_p^{R^\ell} + \ldots + \zeta_p^{R^{(m-1)\ell}}$$
$$\lambda_1 = \zeta_p^R + \zeta_p^{R^\ell+1} + \ldots + \zeta_p^{R^{(m-1)\ell+\ell}}$$
$$\ldots$$
$$\lambda_{\ell-1} = \zeta_p^{R^\ell-1} + \zeta_p^{R^{2\ell}-1} + \ldots + \zeta_p^{R^{m\ell}-1}$$

as a NIB for k, where R is a primitive root modulo p. Hilbert calls the corresponding root number $\Lambda = \Omega$ a Lagrangian root number *(Lagrange'sche Wurzelzahl)*; but $\Lambda = \sum_{j=0}^{p-2} \zeta_\ell^j \zeta_p^{R^j}$ is nothing but the Gauss sum for the character χ over $\mathbb{F}_p$ that maps R^j mod p to ζ_ℓ^j. In particular, Hilbert's results on root numbers apply to Gauss sums.

Back to normal integral bases. Speiser [Sp] later observed that Hilbert's condition (disc $K, (K : \mathbb{Q})) = 1$ could be weakened; he found

Theorem 11.32. *If $K/\mathbb{Q}$ is a normal extension such that $\mathcal{O}_K$ has a NIB, then the ramification index of each prime does not divide $(K : \mathbb{Q})$ i.e., $K/\mathbb{Q}$ is tamely ramified. Moreover, if $K/\mathbb{Q}$ is abelian, this condition is sufficient.*

This is not so hard to prove: recall from Hilbert's work that if K has a NIB generated by α, then k has a NIB generated by Tr$_{K/k}\alpha$. In particular, the existence of a NIB for K implies that Tr$_{K/k}\mathcal{O}_K = \mathcal{O}_k$. Now it is easy to see that a prime ideal $\mathfrak{p}$ in $\mathcal{O}_k$ divides Tr$_{K/k}\mathcal{O}_K$ if and only if $\mathfrak{p}$ is wildly ramified in K/k.

In 1932, E. Noether [Noe] looked at this problem from the viewpoint of the local-global principle and showed that Speiser's condition $e(p) \nmid (K : \mathbb{Q})$ was equivalent to the existence of a local NIB, i.e. if $K/\mathbb{Q}$ is normal, then p is tamely ramified in $K/\mathbb{Q}$ if and only if $\mathcal{O}_\mathfrak{p}$ has a NIB, where $\mathcal{O}_\mathfrak{p}$ is the integral closure of the completion $K_\mathfrak{p}$ of K at some prime ideal $\mathfrak{p}$ above p (see Chapman [Ch1] for a simple proof).

There are two obvious ways to generalize these results of Hilbert, Speiser and Noether: one can replace $\mathbb{Q}$ by a general number field (the naive way of doing this does not work at all: see Exercise 11.31 for a simple counterexample. In fact, Greither, Replogle, Rubin & Srivastav [GRR] have recently shown that $\mathbb{Q}$ is the only number field such that all tame abelian extensions have a normal integral basis), and one can look at non-abelian extensions of $\mathbb{Q}$. In the last direction we have the following result of Martinet [Ma1]:

Proposition 11.33. *If $K/\mathbb{Q}$ is a tame normal extension with Galois group $Gal(K/\mathbb{Q}) \simeq D_p$ (dihedral group of order $2p$), then $K/\mathbb{Q}$ has a NIB.*

The next simplest non-abelian groups are the quaternion groups. Here Martinet [Ma2] found:

Proposition 11.34. *There exist tame normal extensions $K/\mathbb{Q}$ with Galois group isomorphic to H_8 (the quaternion group of order 8) which do (or do not) possess a NIB.*

Now H_8 has a unique irreducible character χ of degree 2. Artin showed how to attach an L-series $L(s,\chi)$ to the pair K and χ; the corresponding function $\Lambda(s,\chi)$ (obtained by multiplying $L(s,\chi)$ by appropriate Γ-factors as in Section 10.4) satisfies a functional equation of type $\Lambda(s,\chi) = W(\chi)\Lambda(1-s,\overline{\chi})$, where $W(\chi)$ is a root of unity called the *Artin root number* of χ; it follows from the functional equation that $W(\chi) = \pm 1$ for real-valued characters χ. Fröhlich & Queyrut [FQ] showed that $W(\chi) = +1$ whenever χ is the character of a real representation. On the other hand, Armitage showed that, for the irreducible 2-dimensional character of H_8, the root number does assume negative values. This led Serre to the 'crazy idea' that the value of $W(\chi)$ might be connected with the existence of a NIB, and in fact, Fröhlich eventually managed to prove

Theorem 11.35. *Let $L/\mathbb{Q}$ be a tame normal extension with $\mathrm{Gal}(K/\mathbb{Q}) \simeq H_8$. Then $L/\mathbb{Q}$ has a NIB if and only if $W(\chi) = 1$.*

It can be shown that $W(\chi) = 1$ is also equivalent to $L(1/2,\chi) = 0$. See the Notes in Narkiewicz [Nar, Chapter 4] for more.

There is (in some sense) a final answer which was given by M. J. Taylor (by now, of course, the whole area has been generalized – at least conjecturally – almost beyond recognition by Chinburg, Fröhlich, M.J. Taylor, and others):

Theorem 11.36. *Let $L/\mathbb{Q}$ be a tame normal extension with $\mathrm{Gal}(K/\mathbb{Q}) = G$. If G has no symplectic characters, then $L/\mathbb{Q}$ has a NIB. If G has a symplectic character, then $\mathcal{O}_L$ or $\mathcal{O}_L \oplus \mathcal{O}_L$ is a free $\mathbb{Z}[G]$-module.*

A symplectic character of a finite group G is a character corresponding to a representation of G that factorizes through the symplectic group $\mathrm{Sp}_{2n}(\mathbb{C})$. Note that abelian groups or groups of odd order do not possess symplectic characters, but that H_8 does. For a leisurely introduction to this area, see Erez [Ere]; the real stuff is in Fröhlich [Fr1]. For connections between NIB's, Gauss sums and Leopoldt's Spiegelungssatz see Brinkhuis [Br2, Br3].

The Stickelberger Relation

Theorem 11.4 was already known to Cauchy, Jacobi, Eisenstein, and Kummer. Of course, they had to use a different language since ideals had not yet been invented then. Jacobi's substitute for Gauss sums were the polynomials

$$F(x,\alpha) = x + \alpha x^{g_1} + \ldots + \alpha^{p-2} x^{g_{p-2}},$$

where x is an indeterminate, α a complex number with $\alpha^{p-1} = 1$, and where the exponents g_j are defined by $g_j \equiv g^j \bmod p$, $0 \leq g_j \leq p-1$, with g a primitive root modulo p. It is clear that substituting $x = \zeta_p$ gives $F(\alpha) := F(\zeta_p, \alpha) = -G(\chi)$, where χ is the character modulo p of order $p-1$ defined by $\chi(g) = \alpha$. Thus the relation $G(\chi)G(\chi^{-1}) = \chi(-1)p$ translates into $F(x,\alpha)F(x,\alpha^{-1}) = \alpha^{(p-1)/2}(p-1-x-\ldots-x^{p-1})$. Jacobi also shows that

$F(\alpha^a)F(\alpha^b) = \psi_{a,b}(\alpha)F(\alpha^{a+b})$ as long as α^a, α^b and α^{a+b} are different from 1, and that $\psi_{a,b}(\alpha) \in \mathbb{Z}[\alpha]$: of course $\psi_{a,b}(\alpha) = -J(\chi^a, \chi^b)$ is a Jacobi sum. He then replaces α by g and shows that the congruence $\psi_{a,b}(g) \equiv -\frac{(a+b)!}{a!b!} \bmod p$ holds (this is how Cauchy and Jacobi could determine the "prime ideal factorization" of Jacobi sums without having the notion of ideal numbers, let alone ideals). This is of course just the congruence (11.12): in fact, we have $n = 1$ since $m = p - 1$, and if we write $\mathcal{P} = (1 - \zeta_p, g - \alpha)$, then the congruence is valid modulo $\mathcal{P}$. Replacing α by g turns the Jacobi sum on the left hand side into $\psi_{a,b}(g)$, and the resulting congruence is not only valid modulo $(1 - \zeta_p)$ but modulo p since both sides are elements of $\mathbb{Z}$.

Next Jacobi compares the equality $\psi_{a,b}(\alpha) = \frac{F(\alpha^a)F(\alpha^b)}{F(\alpha^{a+b})}$ with the congruence $\psi_{a,b}(g) \equiv -\frac{(a+b)!}{a!b!} \bmod p$ and concludes that $F(\alpha^a)$ seems to behave very much like $-\frac{1}{a!} \bmod p$. He then goes on to prove a special case of Stickelberger's congruence.

Cauchy's work is somewhat hard to read (the motto 'more Landau, less Goethe!' would have stood him in good stead). His main work [123] on cyclotomy has about as many pages as this book. The basic relations for Gauss sums are all there, but scattered throughout his treatise. The multiplication formula $G(\chi_1)G(\chi_2) = J(\chi_1, \chi_2)G(\chi_1\chi_2)$ for Gauss sums occurs as (9) on p. 7, and the relations $G(\chi)G(\chi^{-1}) = \chi(-1)p$, $J(\chi, \chi)\overline{J(\chi, \chi)} = p$ as well as $G(\chi)^2 = p^*$ for the quadratic character χ on $\mathbb{F}_p^\times$ can be found on pp. 92–93. On p. 15, he gives $4p^\mu = x^2 + ny^2$, where $n \equiv 3 \bmod 4$, and on p. 18 he shows that μ is congruent to the smallest integer $\equiv \pm 2B_{(n+1)/4} \bmod p$. On p. 106, he discusses a similar result for $n \equiv 1 \bmod 4$. The last chapters are dedicated to congruences for binomial coefficients: on p. 410, he gives Gauss's result that $x \equiv \binom{2n}{n} \bmod p$ for $p = 3n + 1 = x^2 + 3y^2$, and on the next 15 pages he discusses analogous results with 3 replaced by other primes $\equiv 3 \bmod 4$ up to $p = 43$.

The general congruence 11.10 is due to Stickelberger [759]; for different proofs, see Brinkhuis [Br1], Coates [Co2], Conrad [Con], Gillard [Gil], Gras [Gra], Joly [413], and Mertens [581]. There are also various textbooks containing proofs of Stickelberger's relation: see e.g. Ireland & Rosen [386], Lang [La1, La2], Moreno [Mo1], and Washington [Was1]. The very simple proof we have given is taken from Gras [Gra] (it coincides essentially with Lang's presentations). The proof of Davenport & Hasse [DaH] is presented in the book [386] of Ireland & Rosen. See also Fröhlich [Fr2], Ibrahimoglu [Ibr] and Washington [Was2] for proofs of Stickelberger's theorem.

Why Hilbert did not mention Stickelberger's general relation in his Zahlbericht is quite mysterious; Davenport rediscovered Stickelberger's contribution in 1934 after he and Hasse had given a new proof of the relation. Three months after Davenport's discovery, Hasse [Has2] writes

> I found this proof very nice indeed, and much simpler than I expected from my first scanning of Stickelberger's paper.

It is conceivable that Hilbert's first impression was similar.

Schwering [Sch] proved that Jacobi sums for characters of odd prime order $\ell > 3$ are congruent to 1 mod $(1 - \zeta_\ell)^3$; this is sharper than the congruence in Lemma 11.6.vi).

A drastic improvement of the Stickelberger congruence is due to Gross & Koblitz [GK], who gave precise p-adic expressions for Gauss sums that contain Stickelberger's result as a very special case. Washio, Shimaura, & Shiratani derive a congruence following from the Gross-Koblitz formula from Stickelberger's congruence. See also Koblitz [Kob] (an excellent book providing a lot of insight, but requiring quite some background at various places), and Lang [La2] for a more elementary treatment.

A completely new approach to the Stickelberger relation using the arithmetic of the Jacobian variety of the curve $y^2 = 1 - x^l$, where l is an odd prime, was presented by Shimura and Taniyama in [ST, p. 129]; see also Kubota [Kub].

Eisenstein's Reciprocity Law

Eisenstein's reciprocity law for residues of ℓ-th powers is due to Eisenstein [204] himself. He published his proof in 1850, using Kummer's language of ideal numbers. Jacobi had claimed in 1839 (see [403]) to be in possession of this law in the special cases $n = 5, 8$ and 12, but never published anything on them. In [400, p. 263], he writes

> Mit den Resten der 8ten und 5ten Potenzen, welche ganz neue Principien nöthig machen, bin ich ziemlich weit vorgerückt; sobald ich den betreffenden Reciprocitätsgesetzen die wünschenswerthe Vollendung gegeben habe, werde ich sie der Akademie mittheilen.[3]

Whether Jacobi knew that the corresponding rings $\mathbb{Z}[\zeta_5]$ and $\mathbb{Z}[\zeta_8]$ are Euclidean is questionable: in his lectures, the Euclidean algorithm is not used to prove unique factorization (in fact, this problem is not addressed at all) but to the problem of computing power residue symbols using reciprocity! In a letter to Jacobi, Hermite [Hrt] showed in 1845 that $\mathbb{Z}[\zeta_p]$ is a principal ideal ring for $p = 5$ and $p = 7$ by a different method. Whatever the reasons, Jacobi did not publish anything on this. Even when Reuschle wrote to Jacobi on Nov. 11, 1846 and asked him for criteria for $(10/p)_n$ for $n = 5, 7, 8, 9$ (he was computing the period length of decimal fractions for a table he was compiling: see Hertzer [365]), Jacobi's answer from Dec. 13, 1846 (published by Lampe [475]) contains criteria for $(10/p)_8$ plus the rather shallow remark that criteria for $(10/p)_5$ would depend on the factorization of p in $\mathbb{Z}[\zeta_5]$.

Eisenstein seems to have rediscovered these special cases in 1844, as his letter to Stern (probably July 1844) shows:

---

[3] I have advanced considerably the theory of the 8th and 5th power residues which require completely new principles. As soon as I have given these reciprocity laws the desired perfection, I will communicate them to the academy.

Die Reste der 8ten, 12ten und auch der 5ten Potenzen, welche fertig sind, arbeite ich jetzt aus. Das ist ein Feld, auf dem ich mich ganz frei bewegen kann, denn hier hat selbst Jacobi nichts, wie er mir gesteht.[4]

At that time, Eisenstein visited Jacobi weekly, and Jacobi's accusation of plagiarism lay two years ahead.

Apart from the allusions by Jacobi and Eisenstein, the first contribution to quintic residuacity is due to Pépin [635], who used an approach via Jacobi sums. Later, L. Tanner hit upon results on quintic power residues without recognizing them as such (he was studying the coefficients of quintic Jacobi sums); Tanner's results were explained by E. Lehmer in [500].

Hilbert's proof of Eisenstein's reciprocity law in Section 11.2 can be simplified somewhat by using Theorem 11.12; see Ireland & Rosen [386]. For other (but similar) proofs, see Landau [Lan], Spearman & Williams [745], as well as Weil's beautiful paper [826].

The generalization to ℓ^2-th powers was sketched by Furtwängler in [255], and the law for ℓ^ν-th powers was proved by Hasse. In the case $\ell = 2$, Hasse could only prove the reciprocity law for $2^{\nu-2}$-th powers in $\mathbb{Q}(\zeta_{\ell^\nu})$. Here is his result for odd prime powers:

Theorem 11.37. *Let $m = \ell^n$ be an odd prime power. Assume that $a \in \mathbb{Z}$ and $\alpha \in \mathbb{Z}[\zeta_m]$ be relatively prime; if $\alpha \equiv \xi^\ell \bmod (1 - \zeta_m)^2$ and $a^{\ell-1} \equiv 1 \bmod m$, then*

$$\left(\frac{a}{\alpha}\right)_m = \left(\frac{\alpha}{a}\right)_m.$$

Apparently Hasse, at the time of writing [336], was not aware of the papers of Western [836, 837] which contain stronger results and simpler proofs; Western's discussion of primary elements is not very clear, but fortunately Berndt, Evans & K.S. Williams gave a readable and simplified account of Western's paper in their excellent book [48]. Nevertheless, an explicit definition of primariness in Western's sense is still a desideratum. Bohnicek claims in [63] that [62] contains a proof for Hilbert's n-th power reciprocity law for number fields K in which Eisenstein's n-th power reciprocity holds; unfortunately, [62] was unaccessible to me. Takagi [790] proves Eisenstein's reciprocity law for ℓ-th powers in arbitrary number fields containing ζ_ℓ. Wojcik [872] gives a version of Eisenstein's reciprocity law for n-th powers based on yet another definition of primary integers; since the special case $n = 2$ of his law is an incorrect formulation of the quadratic reciprocity law, his proof (which is based on his results from [873]) needs to be checked. It seems that a definitive treatment of Eisenstein's reciprocity law for n-th powers is still lacking. For an application of Eisenstein's reciprocity law to n-th powers of

---

[4] I am now elaborating the residues of the 8th, 12th and also the 5th powers, which are completed. This is an area where I can move freely, as even Jacobi admits not to have anything on them.

integers (this problem was discussed in the Notes of Chapter 4) see Kraft & Rosen [436]. Hayes [Hay2] uses Eisenstein's reciprocity law for computing conductors of what he calls Eisenstein characters; in [Hay1] he proved an analogue of Eisenstein's reciprocity law in function fields.

Class Numbers

The integrality of $h = \frac{R-N}{d}$ (Lemma 11.15) was proved by Cauchy [Cau] and Stickelberger [759]. The proof given here is Stickelberger's, which is much simpler than the one in Hasse's book [342]. Hasse also proved the integrality of the minus class number h_p^- in Equation (11.15) (see [Has1]). Cauchy noticed the connections with Bernoulli numbers; see also Voronoi [Vor]. For some slick proofs of congruences between Bernoulli numbers (originally due to Kummer and Voronoi), see Johnson's article [Joh]. For generalizations of many results about Bernoulli numbers to generalized Bernoulli numbers, see Ernvall [Er1].

Computation of cyclotomic invariants (that is, irregular primes, irregularity index, Iwasawa invariants etc.) continues despite the proof of Fermat's last theorem; for the latest results, see Buhler, Crandall, Ernvall, Metsänkylä, & Shokrollahi [BC1, BC2].

Special cases of Proposition 11.16 were already known to Cauchy and Jacobi; since they only had Gauss sums over $\mathbb{F}_p$ at their disposal, all they could treat were primes $p \equiv 1 \bmod d$. Jacobi even restricted to prime values $d = -\ell$, but conjectured that $\frac{N-R}{\ell}$ always equals the class number of $\mathbb{Q}(\sqrt{-\ell})$ [Jac]. Cauchy and Jacobi published their results at about the same time (shortly after Jacobi's visit in Paris in 1829), but apparently they have been written independently (the same remark applies to Cauchy's and Jacobi's versions of Gauss's sixth proof of quadratic reciprocity that we were talking about in the Notes of Chapter 8). The extension of these results to primes not necessarily of the form $p \equiv 1 \bmod \ell$ was accomplished by Stickelberger [759]. Hilbert's Zahlbericht [368] only gives the part due to Cauchy and Jacobi, as does e.g. the exposition in Ireland & Rosen [386]. A proof of the general result along these lines borrowing ideas from the paper of Coates [Co2] is given in Exercise 11.8. Mitchell [Mi1, Mi2] showed, using Jacobi sums, that the minus class number of the subfield $K \subseteq \mathbb{Q}(\zeta_p)$ of degree e annihilate certain parts of the minus class group of K. MacKenzie [McK] derives relations in the class group of $\mathbb{Q}(\zeta_n)$ that seem to come from the fact that Jacobi sums are principal; his proof, however, uses Fourier transforms, and it would be desirable to see if his method can be used to find the prime ideal factorization of Jacobi sums.

Euler numbers were first studied by Euler in 1755; they satisfy the relation

$$\sum_{\nu=0}^{m} \binom{2m}{2\nu} E_{2\nu} = 0$$

for $m \geq 1$, and this implies that Euler numbers are integral; the first few values are $E_2 = -1$, $E_4 = 5$, $E_6 = -61$, $E_8 = 1385$. Their connection with class groups of $\mathbb{Q}(\zeta_{4m})$ was studied by Gut [Gut] and Ernvall & Metsänkylä [EM]. For a survey of known results see Salié [Sal].

The index of the Stickelberger ideal (Theorem 11.25) was computed by Iwasawa, who also seems responsible for introducing the Stickelberger ideal itself (of course Kummer and Stickelberger never talked about ideals in group rings). Our calculation of $(R^- : I^-)$ is based on an unpublished (but weblished) manuscript by Robin Chapman [Ch2] and is close in spirit to the one given by Lang [La2]. The treatment in Washington [Was1] is closer to the original computation by Iwasawa. Jha [Jha] wrote a survey on class number formulas and Stickelberger ideals, and so did Kimura [Kim]; Kimura's book seems to be the better choice but unfortunately it is written in Japanese.

For surveys on Iwasawa theory, the main conjecture, Euler systems etc. we refer the reader to Coates [Co3] (he also discusses relevant work of Kubert & Lang on the occurrence of the Stickelberger ideal in the theory of cusps of modular forms), Lang [La3], Nekovar [Nek], Rubin [Ru1, Ru2] and Tamme [Ta] as well as to the books on cyclotomic fields by Lang [La2] and Washington [Was1].

Fermat's Last Theorem

The claim that the equation $x^3 + y^3 = z^3$ has only trivial solutions in integers was first claimed (with a completely inadequate proof) by al-Hogendi more than six centuries before Fermat: see Rashed [Ras] for more on this, as well as for other details about the contributions of Arabic mathematics to number theory.

Legendre included his results on Fermat's Last Theorem as a second supplement to his book on number theory; the first supplement was added in 1816, and the book was brought into its final form for the third edition in 1830. Legendre also studied the equation $x^3 + y^3 = az^3$ for $a \in \mathbb{N}$ and claimed that there are no non-trivial solutions if $a = 1, 2, 3, 4, 5, 6, 8, 16\ldots$; Pépin noticed, however, that $17^3 + 37^3 = 6 \cdot 21^3$, as did Lucas in a letter to Sylvester as well as Dudeney in his booklet "The Canterbury Puzzles".

The results on Fermat's Last Theorem in Exercises 11.32 – 11.37 can all be found in Hasse's Zahlbericht [340] as well as in the third volume of Landau's Vorlesungen [Lan]. Frobenius [Fro] showed how the criteria of Wieferich and Mirimanoff could be extended to primes $q > 3$; using quite complicated computations, this has been done up to $q = 89$ by Granville & Monagan [GM] and then to $q = 113$ by J. Suzuki [Su]. Wieferich derived his result from a congruence due to Kummer; a simple proof of this congruence using Herbrand's theorem was given recently by Granville [Gr2].

For a proof of a result containing Exercise 11.35 see Wendt [Wen]; his method was taken up again by Fee & Granville [FG], as well as Lenstra & Stevenhagen [LeSt]; see also Helou [Hel]. Attempts at attacking the case of prime pairs p, $6p + 1$ are due to Granville [Gr1]. For other connections between reciprocity and Fermat's Last Theorem, see Bachmann [Ba], Delcour

[154], Edwards [181], Furtwängler [254], Holzer [373], Noguès [Nog], Terjanian [Ter, 798], and Vandiver [811], as well as Ribenboim's excellent pre-Wiles classic [Ri1] and his article [Ri2].

After centuries of research on certain types of diophantine equations, it was eventually noticed that equations like $x^3+y^3 = az^3$ or $z^2 = x^4+y^4$ belong to the family of elliptic curves; in fact, Fermat's proof of FLT for $n = 4$ via infinite descent has been developed into an algorithm that allows us to compute the group of rational points for a large class of elliptic curves (unfortunately, the non-triviality of the Tate-Shafarevich group $\text{III}(E/\mathbb{Q})$ complicates things considerably; Fermat and Euler were simply lucky that their curves had trivial III). Only the cases $n = 3, 4$ and 7 of Fermat's equation are known to lead to elliptic curves: $x^3 + y^3 = z^3$ is already elliptic and has the Weierstraß form $y^2 = x^3 - 432$, the quartic Fermat equation $x^4 + y^4 = z^4$ leads to the elliptic curve $z^2 = x^4 + y^4$ with Weierstraß form $y^2 = x^3 - 4x$ (see Exercise 10.17 for the analogous problem of $z^2 = x^4 - y^4$), and Lamé's solution of $x^7 + y^7 = z^7$ boils down to solving $u^2 = s^4 + 6s^2t^2 - \frac{1}{7}t^4$, which can also be written as $y^2 = x(x^2 - 3 \cdot 7^2 x + 2^4 \cdot 7^3)$; this is an elliptic curve of conductor 7^2 whose only rational points are its two torsion points. Since there are no elliptic curves of 5-power conductor, a similar proof for the case $n = 5$ of FLT probably doesn't exist. In this connection it is interesting to note that Chowla [Cho] has shown that the Fermat curve $x^p + y^p + z^p = 0$ has a nontrivial rational point if and only if the hyperelliptic curve $y^2 = 4x^p + 1$ does.

Hellegouarch associated the elliptic curve $E_{a,b,c} : y^2 = x(x-a^p)(x+b^p)$ to any solution a, b, c of $A^p + B^p = C^p$ in order to study torsion points on elliptic curves; Frey was the first to suggest that $E_{a,b,c}$ should have properties that are so weird that the curve cannot exist. After contributions of Serre, Ribet succeeded in proving that the conjecture of Taniyama-Shimura-Weil would imply FLT. Wiles, with a little help from R. Taylor, eventually managed to prove enough of this conjecture to be able to derive Fermat's Last Theorem. For an exposition of his proof plus an explanation of the terms used above, see the Boston Proceedings edited by Cornell, Silverman, & Stevens [CSS]. Remarkably, Stickelberger's congruence is still present there: look up Theorem 4.4.1. in Tate's contribution, where these congruences play a role in the classification of certain finite flat group schemes.

Other expositions of the proof of Fermat's Last Theorem (or, rather, of a large part of the Taniyama-Shimura conjecture) ordered approximately by level of difficulty are Cox [Cox], van der Poorten [vdP], Hellegouarch [Hll], J. Kramer [Kr1, Kr2], Moreno [Mo2], K. Murty [Mu1, Mu2], R. Murty [Mu], Schoof [Sf1, Sf2], Bertolini & Canuto [BC], Darmon [Dar], Ribet [Rb2], and Darmon, Diamond & R. Taylor [DDT]. Note that some of these surveys were written before the gap in Wiles' first proof was filled.

Exercises

11.1 Prove Kronecker's assertion that any algebraic integer $\alpha \in \mathcal{O}_K$ such that $|\alpha^\sigma| = 1$ for every embedding $\sigma : K \hookrightarrow \mathbb{C}$ is a root of unity. Give a counterexample in the case where α is not integral.

11.2 Prove Corollary 11.5.

11.3 (cf. Sharifi [731]) Generalize Exercises 6.5 and 7.12 to ℓth powers: for primes $p = \Phi_\ell(\ell x)$, show that any divisor a of x is an ℓ-th power residue modulo p. (Hint: observe that $p = N(1 - \ell x \zeta_\ell)$ and use Eisenstein's reciprocity law).

11.4 Let $p \equiv 1 \bmod 5$ be a prime, and $\pi \in \mathbb{Z}[\zeta_5]$ a semi-primary element of norm p. Let $\chi = (\cdot/\pi)$ be the quintic power residue character; then show that $J(\chi,\chi) = \pi\sigma_3(\pi)$, where σ_3 is the automorphism of $\mathbb{Q}(\zeta_5)/\mathbb{Q}$ mapping ζ_5 to ζ_5^3. How does multiplying π by the primary unit ε^2, where $\varepsilon = \frac{1}{2}(1+\sqrt{5})$, influence the product $\pi\sigma_3(\pi)$? Show also that $J(\chi^2,\chi^2) = \pi\sigma_2(\pi)$, and use Proposition 4.27 to deduce that $(\pi/2) = (\pi/2)$.

11.5 Assume the notation of Proposition 11.2. Write $G(\chi^m) = \mathfrak{p}^\theta$, where $\theta = \sum_a b_a \sigma_a$. Use the facts that $\sum_a b_a = \frac{1}{2}\phi(m)m$ and that the Jacobi sums $J(\chi,\chi^t)$ are integral for $t = 1, 2, \ldots, m-2$ to give a new proof of Proposition 11.2.

11.6 (Conrad [Con]) Let $1 \le a_1, \ldots, a_r < q-1$ be integers; generalize the congruence (11.12) to
$$J(\omega_1^{a_1}, \ldots, \omega_r^{a_r}) \equiv \frac{(a_1 + \ldots + a_r)!}{a_1! \cdots a_r!} \bmod \mathcal{P}$$
for any $r \ge 2$. Here $J(\chi_1, \ldots, \chi_r)$ is the generalized Jacobi sum defined by
$$J(\chi_1, \ldots, \chi_r) = \sum_{\substack{t_1,\ldots,t_r \in \mathbb{F}_q \\ t_1+\cdots+t_r=1}} \chi_1(t_1) \cdots \chi_r(t_r).$$

11.7 Here we sketch the proof of Stickelberger's Relation as given in Davenport & Hasse [DaH]. Define a function $S(a)$ by $\mathcal{P}^{S(a)} \| G(\omega^a)$; we have to show that $S(a) = s(a)$.

$$S(\alpha) \ge 0 \tag{11.24}$$
$$S(\alpha+\beta) \le S(\alpha) + S(\beta) \tag{11.25}$$
$$S(\alpha+\beta) \equiv S(\alpha) + S(\beta) \bmod p - 1 \tag{11.26}$$
$$S(1) = 1 \tag{11.27}$$
$$S(\alpha p) = S(\alpha) \tag{11.28}$$
$$\sum_{\alpha \bmod q-1} S(\alpha) = \frac{f(p-1)(q-1)}{2} \tag{11.29}$$

Once we have proved these claims we can complete the proof as follows: from (11.24), (11.26) and (11.27) we deduce that $S(\alpha) \ge \alpha$ for $0 \le \alpha \le p-1$. From (11.25) and (11.27) we get $S(\alpha) \le \alpha$, and we conclude that $S(\alpha) = \alpha$ for

$0 \le \alpha \le p-1$. Now (11.25) and (11.28) imply $S(\alpha) \le \alpha_0 + \ldots + \alpha_{f-1} = s(\alpha)$, where $\alpha = \alpha_0 + \alpha_1 p + \ldots + \alpha_{f-1} p^{f-1}$. But now (11.29) gives

$$\sum_{\alpha \bmod q - 1} S(\alpha) = \frac{1}{2}f(p-1)(q-2) = \sum_{\alpha \bmod q - 1} s(\alpha),$$

and this implies the claim $S(\alpha) = s(\alpha)$. Now

- (11.24) follows directly from the fact that the Gauss sum $G(\chi)$ is an algebraic integer;
- (11.25) is also a direct consequence of the integrality of the Jacobi sums $J(\chi, \psi)$;
- (11.26) can be deduced from $J(\chi, \psi) \in \mathbb{Z}[\zeta_{q-1}]$;
- (11.27): redo the calculation we did in our proof of (11.2);
- (11.28): follows from $G(\omega^{ap}) = G(\omega^a)$;
- (11.29): note that $G(\omega^a)\overline{G(\omega^a)} = q$ implies $S(\alpha) + S(q-1-\alpha) = (p-1)f$ and form the sum over all $1 \le \alpha \le q-2$.

11.8 Prove Proposition 11.16 directly, that is, without using Stickelberger's Theorem 11.14.

Hints: 1. Show that it is sufficient to show that $\mathfrak{p}^h$ is principal for all prime ideals that split in k by using the fact that every ideal class contains an ideal prime to any given ideal (the only problem is to get around the ramified primes; an alternative solution is to show that ramified primes are principal if d is a prime discriminant, and that h is even otherwise).

2. Put $L = \mathbb{Q}(\zeta_m)$, where $m = |d|$, and observe that k is contained in the decomposition field K of p. Let $\mathfrak{P}$ denote a prime ideal above $\mathfrak{p}$ in $\mathcal{O}_K$; the following Hasse diagram (where $\widetilde{F} = F(\zeta_p)$) shows what's going on:

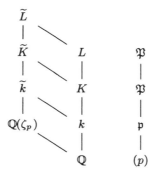

Let $\chi = (\,\cdot\,/\mathfrak{P})_m^{-1}$ be the inverse of the m-th power character in $(\mathcal{O}_L/\mathfrak{P})^\times$, and let $G(\chi)$ denote the corresponding Gauss sum. Then $\mu = G(\chi)^m \in \mathcal{O}_K$ by Proposition 4.25. Show that $\mathfrak{p}^{(R-N)/m}$ is principal in $\widetilde{k}$.

3. Put $\gamma = N_{\widetilde{K}/\widetilde{k}} G(\chi)$; use an argument about ramification to show that $k(\gamma) = k$.

11.9 Use Stickelberger's congruence to prove the Davenport-Hasse theorem 4.32 (compare Exercise 10.28).

Hints: consider the algebraic number $\eta = G(\chi')/G(\chi)^{(E:F)}$.

1. Show that $\eta \in \mathbb{Q}(\zeta_m)$, where m denotes the order of χ (which equals the order of χ');
2. show that the prime ideal factorization of η contains only prime ideals above p;
3. show that η is a unit in $\mathbb{Z}[\zeta_m]$ (use the prime ideal factorization of the Gauss sum);
4. show that $|\eta| = 1$ and deduce that η must be a root of unity;
5. use Stickelberger's congruence to show that $\eta \equiv 1 \bmod \mathfrak{p}$ for any prime ideal $\mathfrak{p}$ above p in $\mathbb{Z}[\zeta_m]$ and conclude that $\eta = 1$.

Use the same idea to give a proof of the Davenport-Hasse relation in Theorem 4.31.

11.10 Let k be a totally real number field and K a totally complex quadratic extension. Show that $Q = (E_K : W_K E_k)$ divides 2.

11.11 Let M be a finite additive group on which a group $H = \{1, J\}$ of order 2 acts. Put $M^- = \{m \in M : Jm = -m\}$ and show that $(1-J)M \subseteq M^- \subseteq M$. Using $(1-J)M^- = 2M^-$, deduce that $2M^- \subseteq (1-J)M$.

11.12 Let $B \subseteq A$ be abelian groups and $f : A \longrightarrow A$ a group homomorphism. Then $(A : B) = (A^f : B^f)(A_f + B : B)$ whenever these indices exist, where $A^f = f(A)$, $B^f = f(B)$, and $A_f = \ker f$. Hint: show that the epimorphism $A/B \longrightarrow A^f/B^f$ has kernel $(A_f + B)/B$.

11.13 Let V be a $\mathbb{Q}$-vector space. An abelian group $A \subseteq V$ is called a lattice in V if $A = v_1\mathbb{Z} \oplus \ldots \oplus v_n\mathbb{Z}$, where $\{v_1, \ldots, v_n\}$ is a basis of V/F. Show that, given lattices A and B in V, there exists a lattice C in V containing A and B (can you find a counter example for vector spaces V over, say, $\mathbb{Q}(\sqrt{2})$?). For any such lattice, define
$$(A : B) = \frac{(C : A)}{(C : B)},$$
where the indices on the right hand side are the usual indices of abelian groups, and show that this definition does not depend on the choice of C. Show that this index has the following properties:
i) $(A : B)$ coincides with the usual index if $B \subseteq A$;
ii) $(A : B) = (B : A)^{-1}$;
iii) $(A : B)(B : C) = (A : C)$.

11.14 (continued) Let A be a lattice in V, and assume that $T : V \longrightarrow V$ is a linear map with the property that there is an integer $m \in \mathbb{N}$ such that $mTA \subseteq A$. Then $(A : TA) = |\det T|$. Hints: $(A : TA) = (A : mTA)(mTA : TA) = (A : mTA)(TA : mTA)^{-1}$; clearly $(TA : mTA) = m^n$, so it is sufficient to show that $(A : TA) = |\det T|$ for any linear map $T : V \longrightarrow V$ such that $TA \subseteq A$. For help, cf. Cohn [Coh, IV.8, Lemma 7]. Alternatively, consider lattices $B \subseteq A$ and define $\text{vol}(A)$ to be the volume of the parallelepiped spanned by the basis vectors of A. Show that $(A : B) = \text{vol}(B)/\text{vol}(A)$, and deduce our claim from $\text{vol}(TA) = |\det T| \cdot \text{vol}(A)$.

11.15 Verify the following table containing information about the subgroups of R occurring for $m = 3$ and $m = 4$ in our computation of the index of the Stickelberger ideal:

m	θ	S	S^-	$2e^-S$	$I = S\theta$	I^-
3	$\frac{1}{3}(1+2J)$	$(1+J, 3)$	$3R^-$	$3R^-$	R	R^-
4	$\frac{1}{4}(1+3J)$	$(1+J, 2-2J)$	$2R^-$	$4R^-$	$(1+J, 1-J)$	R^-

Note that $I^- = (1-J)I$ for $m = 3$ while $(1-J)I = 2I^-$ for $m = 4$. Can you generalize?

11.16 Recall our proof that $2h^-$ kills the minus class group $\mathrm{Cl}\,(K^-)$ of $K = \mathbb{Q}(\zeta_p)$ and show that we have proved more, namely that twice the exponent of R^-/I^- annihilates $\mathrm{Cl}\,(K^-)$.

11.17 Let $\mathfrak{P}$ be a prime ideal in $K = \mathbb{Q}(\zeta_m)$ and put $\chi = (\cdot/\mathfrak{P})_m$. Show that $G(\chi)^m \mathcal{O}_K = \mathfrak{P}^{m\theta'}$, where $\theta' = m\sum_{(a,m)=1}\left(1 - \langle\frac{a}{m}\rangle\right)\sigma_a^{-1}$. Show that this implies $G(\chi)^m \mathcal{O}_K = \mathfrak{P}^{m(\nu/2+\theta_S)}$, where θ_S is the Brumer-Stark element for $K/\mathbb{Q}$ as defined in (11.17).

11.18 Verify the Brumer-Stark conjecture for cyclotomic extensions of $\mathbb{Q}$. Also show that $\theta_S = 0$ for real abelian extensions $K/\mathbb{Q}$.

11.19 Check what goes wrong in the proof of Proposition 11.16 if $d = -8$.

11.20 In the proof of Proposition 11.16, can you show that $\gamma \in k$ by Galois theory?

11.21 Prove Proposition 11.16 for ramified prime ideals. (Hint: if d is a prime discriminant, the claim is trivial since $\mathfrak{p}$ is principal in this case. If d is composite, $\mathfrak{p}^2 = (p)$ is principal, and it is sufficient to show that h is even).

11.22 (Washington [Was1, §6.2]) Let $K = \mathbb{Q}(\zeta_{12})$. Show that $\theta(K) = 1 + \sigma \in I(K) \setminus I_0(K)$.

11.23 Show that the integer defined in (11.15) gives $h^- = (N-R)/|d|$ for complex quadratic number fields of conductor d, and $h^- = \frac{1}{2}[(D_0 - D_2)^2 + (D_1 - D_3)^2]$ for complex cyclic quartic fields, where N, R and the D_j are defined in Propositions 11.16 and 11.18.

11.24 Show that the integer h^- defined in Proposition 11.18 is odd when the conductor f is an odd prime. Show that $h^- = 1$ for $f = 16$, and verify this by showing that the class number of $\mathbb{Q}(\sqrt{-2+\sqrt{2}})$ is 1.

11.25 Let K/k be an extension of number fields. Define a map $j : \mathrm{Cl}\,(k) \longrightarrow \mathrm{Cl}\,(K)$ by mapping an ideal class $c = [\mathfrak{a}]$ to $[\mathfrak{a}\mathcal{O}_K] \in \mathrm{Cl}\,(K)$ and show that $\ker j$ is killed by $(K : k)$ (Hint: take the relative norm). Show that $\mathrm{Cl}^-(k)$ gets mapped to $\mathrm{Cl}^-(K)$.

11.26 For primes p, define a function $\Gamma_p : \mathbb{N} \longrightarrow \mathbb{Z}$ by $\Gamma_p(n) := (-1)^n \prod j$, where the product is over all $1 \leq j \leq n-1$ such that $p \nmid j$. Prove that $\Gamma_p(m+n) \equiv \Gamma_p(m) \bmod p^{v_p(n)}$ unless $p = 2$ and $n \equiv 4 \bmod 8$. Show that this congruence allows us to extend the function Γ_p continuously to a function $\Gamma_p : \mathbb{Z}_p \longrightarrow \mathbb{Z}_p^\times$ by putting $\Gamma_p(x) = \lim_{n \to x} \Gamma_p(n)$, where the n tend p-adically to $x \in \mathbb{Z}_p$. Verify that this p-adic Gamma function satisfies $\Gamma_p(0) = 1$, and that $\Gamma_p(x+1)/\Gamma_p(x) = -x$ or $= -1$ according as $x \in \mathbb{Z}_p^\times$ or $x \in p\mathbb{Z}_p$.

11.27 A non-empty set I is called *partially ordered* if there is an order relation $<$ defined on I such that
 1. $i < j$ and $j < k \Longrightarrow i < k$;
 2. $i < j$ and $j < i \Longrightarrow i = j$;
 3. $i < i$ for all $i \in I$.

The set I is called *directed* if, in addition, it has the property
 1. for all $i, j \in I$ there is a $k \in I$ such that $i < k$ and $j < k$.

Now consider a family X_i $(i \in I)$ of compact topological spaces indexed by a directed set I, and assume that for each pair $(i,j) \in I \times I$ there exists a continuous epimorphism $\pi_{ij} : X_j \longrightarrow X_i$ such that
i) $\pi_{ii} = \mathrm{id}$;
ii) if $i < j < k$, then $\pi_{ij} \circ \pi_{jk} = \pi_{ik}$;
then the triple $(X_i; \pi_{ij}; I)$ is called an *projective system*.
Given such a projective system, we can form the direct product $\widetilde{X} = \prod_{i \in I} X_i$ and make it into a topological space by giving it the product topology; this ensures that the projection maps $\pi_i : \widetilde{X} \longrightarrow X_i$ are continuous. Now define the *projective limit* of this projective system by

$$\varprojlim X_i = \{x \in \widetilde{X} : \pi_{ij} \circ \pi_j(x) = \pi_i(x) \text{ for all } i < j\}.$$

1. Use the axiom of choice to show that $\varprojlim X_i$ is non-empty;
2. Use Tychonov's theorem to show that $\varprojlim X_i$ is compact;
3. Show that X is a group if each X_i is a group;
4. For a ring R, show that X is an R-module if each X_i is.

11.28 Let ℓ be a prime and consider the finite groups $X_n = \mathbb{Z}/\ell^n\mathbb{Z}$ together with the projections $\pi_{mn} : \mathbb{Z}/\ell^n\mathbb{Z} \longrightarrow \mathbb{Z}/\ell^m\mathbb{Z}$ for $0 < m < n$. Endow the X_n with the discrete topology and show that the triple $(X_n; \pi_{mn}; \mathbb{N})$ is a projective system. Show that $\varprojlim X_n \simeq \mathbb{Z}_\ell$ as topological groups, where the topology on $\mathbb{Z}_\ell$ is induced by the ℓ-adic valuation.

11.29 Let $K_0 \subseteq K_1 \subseteq \ldots$ be a tower of normal number fields, and put $G_n = \mathrm{Gal}\,(K_n/K_0)$. Define epimorphisms $\pi_{mn} : G_n \longrightarrow G_m$ for $m < n$ such that $(G_n; \pi_{mn}; \mathbb{N})$ becomes a projective system, and show that $\Gamma = \varprojlim G_n$ is topologically isomorphic to the Galois group of $K_\infty = \bigcup_n K_n$, endowed with the Krull topology.

11.30 Let $K = K_0 \subset K_1 \subset \ldots \subset K_n \subset \ldots$ be a $\mathbb{Z}_p$-extension of a number field K, that is, a family of number fields such that $\mathrm{Gal}\,(K_n/K) \simeq \mathbb{Z}/p^n\mathbb{Z}$ (the preceding exercise shows that $\mathrm{Gal}\,(K_\infty/K) \simeq \mathbb{Z}_p$). Put $A_n = \mathrm{Cl}\,_p(K_n)$ and show that the relative norms $N_{mn} = N_{K_n/K_m}$ make $(A_n; N_{mn}; \mathbb{N})$ into a projective system.

11.31 Put $K = \mathbb{Q}(\sqrt{-5})$ and $L = K(i)$. Show that the result of Hilbert and Speiser is not valid for general number fields by showing that L/K is unramified (hence tame), that $\mathcal{O}_L$ has an integral basis over $\mathcal{O}_K$, but does not have a NIB over $\mathcal{O}_K$.

11.32 (Furtwängler 1912) Let p be an odd prime, and assume that $x^p + y^p + z^p = 0$ for pairwise coprime integers $x, y, z \in \mathbb{Z}$ with $p \nmid xyz$. Use the unique factorization theorem for prime ideals to deduce that $(x + y\zeta^i) = \mathfrak{A}_i^p$ for ideals $\mathfrak{A}_i$, $i = 0, 1, \ldots, p-1$. Show that $\alpha = \zeta^y x + \zeta^{-x} y$ is semi-primary. Now use Eisenstein's reciprocity law to deduce that $\left(\frac{\alpha}{r}\right)_p = \left(\frac{r}{\alpha}\right)_p = \left(\frac{r}{\mathfrak{A}_j}\right)_p^p = 1$ for each prime $r \mid x$, and deduce that $r^{p-1} \equiv 1 \bmod p^2$.

11.33 (Wieferich 1909) Suppose that $x^p + y^p + z^p = 0$ for some odd prime $p \nmid xyz$; then $2^{p-1} \equiv 1 \bmod p^2$. (Hint: Use the preceding exercise).
Remark. Primes p satisfying $2^{p-1} \equiv 1 \bmod p^2$ are called *Wieferich primes*. The only Wieferich primes below $4 \cdot 10^{12}$ are 1093 and 3511 (see Crandall, Dilcher & Pomerance [CDP]).

11.34 (S. Germain 1823) Suppose that $x^p + y^p + z^p = 0$ for some odd prime $p \nmid xyz$; then $\ell = 2p + 1$ is not prime.

11.35 (Legendre 1823) Suppose that $x^p + y^p + z^p = 0$ for some odd prime $p \nmid xyz$; then the numbers $2p + 1, 4p + 1$ and $8p + 1$ are not prime.

11.36 (Furtwängler 1912) Suppose that $x^p + y^p + z^p = 0$ for some odd prime $p \nmid xyz$, and that $(x,y) = (y,z) = (x,z) = 1$; assume moreover that $p \nmid (x^2 - y^2)$; then $r^{p-1} \equiv 1 \bmod p^2$ for every prime $r \mid (x - y)$.

11.37 (Mirimanoff 1911) Suppose that $x^p + y^p + z^p = 0$ for some prime $p \nmid xyz$, $p > 3$; then $3^{p-1} \equiv 1 \bmod p^2$.

11.38 Transform the Fermat curve $x^3 + y^3 = 1$ into Weierstraß form. (Hint: put $x = u + v$, $y = u - v$).

11.39 Transform the Fermat curve $w^4 + 1 = z^2$ into Weierstraß form. (Hint: write it as $1 = (w^2 - z)(w^2 + z)$ and put $x = w^2 + z$. Then $2w^2 = x + \frac{1}{x}$; multiply by x^2 and put $wx = y$).

11.40 Ribenboim [Ri1] sketches a proof for FLT in the case $n = 7$; fill in the details and transform the resulting curve $u^2 = s^4 + 6s^2t^2 - \frac{1}{7}t^4$ into the form $E : y^2 = x(x^2 - 3 \cdot 7^2 x + 16 \cdot 7^3)$. Use simple 2-descent to show that $E(\mathbb{Q}) = E(\mathbb{Q})_{\text{tors}} \simeq \mathbb{Z}/2\mathbb{Z}$, and conclude that $x^7 + y^7 = z^7$ has only trivial solutions in $\mathbb{Z}$.

Additional References

[An] G.W. Anderson, *Another look at the index formulas of cyclotomic number theory*, J. Number Theory **60** (1996), 142–164

[Ba] P. Bachmann, *Das Fermatproblem in seiner bisherigen Entwicklung*, Reprint Springer-Verlag 1976

[Bau] H. Bauer, *Zur Berechnung von Hilbertschen Klassenkörpern mit Hilfe von Stark-Einheiten*, Diss. TU Berlin, 1998

[BC] M. Bertolini, G. Canuto, *The Shimura-Taniyama-Weil conjecture* (Italian), Boll. Unione Mat. Ital. (VII) **10** (1996), 213-247

[Br1] J. Brinkhuis, *Gauss sums and their prime factorization*, Ens. Math. **36** (1990), 39–51

[Br2] J. Brinkhuis, *On a comparison of Gauss sums with products of Lagrange resolvents*, Compos. Math. **93** (1994), 155–170

[Br3] J. Brinkhuis, *Normal integral bases and the Spiegelungssatz of Scholz*, Acta Arith. **69** (1995), 1–9

[BC1] J. Buhler, R. Crandall, R. Ernvall, T. Metsänkylä, *Irregular primes and cyclotomic invariants to four million*, Math. Comp. **61** (1993), 151–153

[BC2] J. Buhler, R. Crandall, R. Ernvall, T. Metsänkylä, M.A. Shokrollahi, *Irregular primes and cyclotomic invariants to eight million*, J. Symbolic Computation **11** (1998)

[CT] Ph. Cassou-Noguès, M.J. Taylor, *Un élément de Stickelberger quadratique*, J. Number Theory **37** (1991), 307–342

[Cau] A.L. Cauchy, *Mémoire sur la théorie des nombres; Note VIII*, Mém. Inst. France **17** (1840), 525–588 Œuvres (1) III, 265–292

[Ch1] R.J. Chapman, *A simple proof of Noether's theorem*, Glasg. Math. J. **38** (1996), 49–51

[Ch2] R.J. Chapman, http://www.maths.ex.ac.uk/~rjc/rjc.html

[Cho] S. Chowla, *L-series and elliptic curves*, Lecture Notes Math. **626**, Springer 1977, 1–42

[Co1] J. Coates, *On K_2 and some classical conjectures in algebraic number theory*, Ann. of Math. **95** (1972), 99–116

[Co2] J. Coates, *p-adic L-functions and Iwasawa's theory*, Algebraic Number Fields, Durham 1975, (1977), 269–353

[Co3] J. Coates, *The work of Mazur and Wiles on cyclotomic fields*, Semin. Bourbaki 1980/81, Exp. 12, Lect. Notes Math. **901** (1981), 220–241

[CGR] J. Coates, R. Greenberg, K. Ribet, K. Rubin, C. Viola, *Arithmetic Theory of Elliptic Curves*, CIME Lectures 1997, Springer LNM 1716, 1999

[CoP] J. Coates, G. Poitou, *Du nouveau sur les racines de l'unité*, Gaz. Math., Soc. Math. Fr. **15** (1980), 5–26

[CoS] J. Coates, W. Sinnott, *An analogue of Stickelberger's theorem for the higher K-groups*, Invent. Math. **24** (1974), 149–161

[Coh] H. Cohn, *Advanced Number Theory*, Dover 1980

[Con] K. Conrad, *Jacobi sums and Stickelberger's congruence*, L'Enseign. Math. **41** (1995), 141–153

[CSS] G. Cornell, J.H. Silverman, G. Stevens (eds.), *Modular Forms and Fermat's Last Theorem*, Springer 1997

[Cox] D.A. Cox, *Introduction to Fermat's Last Theorem*, Amer. Math. Mon. **101** (1994), 3–14

[CDP] R. Crandall, K. Dilcher, C. Pomerance, *A search for Wieferich and Wilson primes*, Math. Comp. **66** (1997), 433–449

[Dar] H. Darmon, *The Shimura-Taniyama conjecture (d'apres Wiles)*, Russ. Math. Surv. **50** (1995), 503–548; translation from Usp. Mat. Nauk **50** (1995), 33–82

[DDT] H. Darmon, F. Diamond, R. Taylor, *Fermat's Last Theorem*, Current developments in mathematics, Cambridge International Press (1995), 1–107

[DaH] H. Davenport, H. Hasse, *Die Nullstellen der Kongruenzzetafunktionen in gewissen zyklischen Fällen*, J. Reine Angew. Math. **172** (1934), 151–182

[DeR] P. Deligne, K.A. Ribet, *Values of Abelian L-functions at negative integers over totally real fields*, Invent. Math. **59** (1980), 227–286

[DuH] D.S. Dummit, D.R. Hayes, *Checking the p-adic Stark Conjecture*, ANTS II, Lecture Notes Comput. Sci. **1122** (1996), 91–97

[DST] D.S. Dummit, J.W. Sands, B.A. Tangedal, *Computing Stark units for totally real cubic fields* Math. Comput. **66** (1997), 1239–1267

[Eis] G. Eisenstein, *Zur Theorie der quadratischen Zerfällung der Primzahlen $8n+3$, $7n+2$ und $7n+4$*, J. Reine Angew. Math. **37** (1848), 97–126; Werke II, 506–535

[Ere] B. Erez, *Representations of groups in algebraic number theory. An introduction* (Ital.), Proc. Colloq., Locarno/Italy 1988-1989, Note Mat. Fis. (3) (1990), 41–65

[Er1] R. Ernvall, *Generalized Bernoulli numbers, generalized irregular primes, and class number*, Ann. Univ. Turku **178** (1979), 72 pp.

[Er2] R. Ernvall, *A generalization of Herbrand's theorem*, Ann. Univ. Turku, Ser. A I **193** (1989), 15 pp.

[EM] R. Ernvall, T. Metsänkylä, *Cyclotomic invariants and E-irregular primes* Math. Comp. **32** (1978), 617–629; Corr.: ibid. **33** (1979), 433

[Fe] L.J. Federer, *Regulators, Iwasawa modules, and the main conjecture for $p=2$*, Number theory related to Fermat's last theorem, Prog. Math. **26** (1982), 287–296

[FG] G.J. Fee, A. Granville, *The prime factors of Wendt's Binomial Circulant determinant*, Math. Comp. **57** (1991), 839–848

[Fro] G. Frobenius, *Über den Fermat'schen Satz III*, Sitzungsber. Akad. Wiss. Berlin (1914), 653–681

[Fr1] A. Fröhlich, *Galois module structure of algebraic integers*, Springer-Verlag 1983

[Fr2] A. Fröhlich, *Stickelberger without Gauss sums*, Algebraic Number fields, Univ. Durham 1975, 589–607 (1977)

[FQ] A. Fröhlich, J. Queyrut, *On the functional equation of the Artin L-function for characters of real representations*, Invent. Math. **20** (1973), 125–138

[Gha] E. Ghate, *Vandiver's conjectue via K-theory*, Summer School on Cyclotomic Fields, June 1999; preprint, see http://www.math.tifr.res.in/~eghate/

[Gil] R. Gillard, *Relations de Stickelberger*, Sém. Théor. Nombres Grenoble, 1974

[Gr1] A. Granville, *Sophie Germain's Theorem for prime pairs p, $6p+1$*, J. Number Theory **27** (1987), 63–72

[Gr2] A. Granville, *The Kummer-Wieferich-Skula approach to the First Case of Fermat's Last Theorem*, Advances in Number Theory (F.Q. Gouvea, N. Yui, eds.), Oxford University Press 1993, 479–498

[GM] A. Granville, M.B. Monagan, *The first case of Fermat's last theorem is true for all prime exponents up to $714,591,416,091,389$*, Trans. Am. Math. Soc. **306** (1988), 329–359

[Gra] G. Gras, *Sommes de Gauss sur les corps finis*, Publ. Math. Besançon (1977/78), 71 pp

[Gre]   C. Greither, *Class groups of abelian fields, and the main conjecture*, Ann. Inst. Fourier **42** (1991), 449–500

[GRR]   C. Greither, D.R. Replogle, K. Rubin, A. Srivastav, *Swan modules and Hilbert-Speiser number fields*, preprint 1999

[GK]   B.H. Gross, N. Koblitz, *Gauss sums and the p-adic Γ-function*, Annals of Math. **109** (1979), 569–581

[Gut]   M. Gut, *Euler'sche Zahlen und Klassenzahl des Körpers der 4ℓ-ten Einheitswurzeln*, Commentarii Math. Helvet. **25** (1951), 43–63

[HP]   G. Harder, R. Pink, *Modular konstruierte unverzweigte abelsche p-Erweiterungen von $\mathbb{Q}(\zeta_p)$ und die Struktur ihrer Galoisgruppen*, Math. Nachr. **159** (1992), 83–99

[Has1] H. Hasse, *Über die Klassenzahl abelscher Zahlkörper*, Akademie-Verlag Berlin 1952; Reprint Akademie-Verlag Berlin 1985

[Has2] H. Hasse, *Letter to Davenport, May 20, 1934*, Trinity College, Cambridge, UK.

[Hay1] D.R. Hayes, *Hecke characters and Eisenstein reciprocity in function fields*, J. Number Theory **43** (1993), 251–292

[Hay2] D.R. Hayes, *The conductors of Eisenstein characters in cyclotomic number fields*, Finite Fields Appl. **1** (1995), 278–296

[Hec]   E. Hecke, *Über nicht-reguläre Primzahlen und den Fermatschen Satz*, Gött. Nachr. (1910), 420–424

[Hll]   Y. Hellegouarch, *Invitation aux mathematiques de Fermat-Wiles*, Paris: Masson, 397 pp, 1997

[Hel]   C. Helou, *On Wendt's determinant*, Math. Comput. **66** (1997), 1341–1346

[Her]   J. Herbrand, *Sur les classes des corps circulaires*, J. Math. Pures Appl. **11** (1932), 417–441

[Hrt]   C. Hermite, *Lettre à Jacobi, Aug. 06, 1845*, Œuvres de Charles Hermite I, 100–121

[Ibr]   I. Ibrahimoglu, *A proof of Stickelberger's theorem*, Hacettepe Bull. Nat. Sci. Eng. **12** (1983), 279–287

[Jac]   C. G. J. Jacobi, *Observatio arithmetica de numero classium divisorum quadraticorum formae $yy+Azz$, designante A numerum primum formae $4n+3$*, J. Reine Angew. Math. **9** (1832), 189–192; Ges. Werke **6**, 240–244, Berlin 1891

[Jha]   V. Jha, *The Stickelberger ideal in the spirit of Kummer with application to the first case of Fermat's last theorem*, Queen's Papers in Pure and Applied Mathematics **93**, 181 pp. (1993)

[Joh]   W. Johnson, *p-adic proofs of congruences for the Bernoulli numbers*, J. Number Theory **7** (1975), 251–265

[Ka]   S. Kamienny, *Modular curves and unramified extensions of number fields*, Compositio Math. **47** (1982), 223–235

[KM] I. Kersten, J. Michalicek, *On Vandiver's conjecture and $\mathbb{Z}_p$-extensions of $\mathbb{Q}(\zeta_{p^n})$*, J. Number Theory **32** (1989), 371–386

[Kim] T. Kimura, Algebraic class number formulae for cyclotomic fields (Japanese), Sophia University, Department of Mathematics. IX, 281 pp. (1985)

[Kle] H. Kleboth, *Untersuchung über Klassenzahl und Reziprozitätsgesetz im Körper der 6ℓ-ten Einheitswurzeln und die Diophantische Gleichung $x^{2\ell} + 31y^{2\ell} = z^{2\ell}$ für eine Primzahl ℓ grösser als 3*, Diss. Univ. Zürich 1955, 37 pp.

[Klj] T. Kleinjung, *Konstruktion unverzweigter Erweiterungen von Zahlkörpern durch Wurzelziehen aus zyklotomischen Einheiten*, Diplomarbeit Univ. Bonn, 1994

[Kob] N. Koblitz, *p-adic analysis: a short course on recent work*, London Math. Soc. LNS **46**, 1980

[Kol] M. Kolster, *A relation between the 2-primary parts of the main conjecture and the Birch-Tate-conjecture*, Can. Math. Bull. **32** (1989), 248–251

[Kr1] J. Kramer, *Über die Fermat-Vermutung*, Elem. Math. **50** (1995), 12–25

[Kr2] J. Kramer, *Über den Beweis der Fermat-Vermutung. II*, Elem. Math. **53** (1998), 45–60

[Kub] T. Kubota, *An application of the power residue theory to some abelian functions*, Nagoya Math. J. **27** (1966), 51–54

[Kur] M. Kurihara, *Some remarks on conjectures about cyclotomic fields and K-groups of $\mathbb{Z}$*, Compos. Math. **81** (1992), 223–236

[La1] S. Lang, *Algebraic Number Theory*, Addison-Wesley 1970; reissued as Graduate Texts in Mathematics **110**, Springer-Verlag 1986; 2nd. ed. 1994

[La2] S. Lang, *Cyclotomic fields. I and II*, Graduate Texts in Mathematics **121**, Springer-Verlag 1990

[La3] S. Lang, *Classes d'idéaux et classes de diviseurs*, Semin. Delange-Pisot-Poitou 1976/77, Exp. 28, 9 pp. (1977)

[Lan] E. Landau, *Vorlesungen über Zahlentheorie*, Leipzig 1927; Chelsea 1969

[Leg] A. M. Legendre, , Mem. Acad. Sci. Inst. France **6** (1823), 1–60; see also Théorie des nombres, 2nd ed. (1808), second supplement (1825), 1–40.

[LeSt] H. W. Lenstra, P. Stevenhagen, *Class field theory and the first case of Fermat's Last Theorem*, see [CSS], 499–503 (1997).

[Leo] H. W. Leopoldt, *Zur Struktur der ℓ-Klassengruppe galoisscher Zahlkörper*, J. Reine Angew. Math. **199** (1958), 165–174

[Lip]   M. Lippert, *Konstruktion unverzweigter Erweiterungen von Zahlkörpern durch Wurzelziehen aus Elementen eines Eulersystems*, Diplomarbeit Univ. Bonn, 1996

[McK]   R.E. MacKenzie, *Class group relations in cyclotomic fields*, Amer. J. Math. **74** (1952), 759–763

[Ma1]   J. Martinet, *Bases normales et constante de l'équation fonctionelle des fonctions L d'Artin*, Sém. Bourbaki 1973/74, Exposé 450, Lecture Notes Math. **431** (1975), 273–294

[Ma2]   J. Martinet, H_8, Algebr. Number Fields, Proc. Symp. London Math. Soc., Univ. Durham 1975, 525–538 (1977)

[MW]   B. Mazur, A. Wiles, *Class fields of abelian extensions of* $\mathbb{Q}$, Invent. Math. **76** (1984), 179–330

[Met]   T. Metsänkylä, *The index of irregularity of primes*, Expo. Math. **5** (1987), 143–156

[Mir1] M. D. Mirimanoff, *Sur le dernier théorème de Fermat*, C. R. Acad. Sci. Paris **150** (1910), 204–206

[Mir2] M. D. Mirimanoff, *Sur le dernier théorème de Fermat*, J. Reine Angew. Math. **139** (1911), 309–324

[Mi1]   H.H. Mitchell, *On the generalized Jacobi-Kummer cyclotomic function*, Amer. Math. Soc. Trans. **17** (1916), 165–177; FdM **46** (1916–18), 255

[Mi2]   H.H. Mitchell, *Proof that certain ideals in a cyclotomic realm are principal ideals*, Trans. Amer. Math. Soc. **19** (1918), 119–126

[Mo1]   C.J. Moreno, *Algebraic curves over finite fields*, Cambridge Tracts in Mathematics **97**, CUP 1991

[Mo2]   C.J. Moreno, *Fermat's Last Theorem: From Fermat to Wiles*, Rev. Colomb. Mat. **29** (1995), 49–88

[Mu1]   V.K. Murty, *Fermat's last theorem*, Analysis, geometry and probability, Texts Read. Math. **10** (1996), 125–139

[Mu2]   V.K. Murty, *Modular elliptic curves*, Seminar on Fermat's last theorem, CMS Conf. Proc. **17** (1995), 1–38

[Mu]   M.R. Murty, *Topics in Number Theory*, Lecture Notes 1993

[Nar]   W. Narkiewicz, *Elementary and analytic theory of algebraic numbers*, Warszawa 1974; 2nd ed. Springer-Verlag 1990

[Nek]   J. Nekovar, *Iwasawa's main conjecture. (A survey)*, Acta Math. Univ. Comenianae **50/51** (1987), 203–215

[Noe]   E. Noether, *Normalbasis bei Körpern ohne höhere Verzweigung*, J. Reine Angew. Math. **167** (1932), 147–152

[Nog]   R. Nogues, *Théorème de Fermat: son histoire*, Paris 1932; 2nd ed. 1966; reprint of 1st ed. Sceaux 1992

[Ou]   Yi Ouyang, *Spectral sequence of universal distribution and Sinnott's index formula*, preprint 1999; http://front.math.ucdavis.edu/math.NT/9911031

[PR] B. Perrin-Riou, *Travaux de Kolyvagin et Rubin*, Semin. Bourbaki, Vol. 1989/90, Astérisque **189-190**, Exp. No. 717 (1990), 69–106

[Pol] F. Pollaczek, *Über die irregulären Kreiskörper der ℓ-ten und ℓ^2-ten Einheitswurzeln*, Math. Z. **21** (1924), 1–37

[Ras] R. Rashed, *The development of Arabic mathematics: between arithmetic and algebra*, Kluwer 1994

[Ri1] P. Ribenboim, *13 lectures on Fermat's Last Theorem*, Springer-Verlag 1979

[Ri2] P. Ribenboim, *Fermat's Last Theorem, before June 23, 1993*, Number theory (K. Dilcher, ed.), Mathematical Society, CMS Conf. Proc. **15** (1995), 279–294

[Rb1] K. Ribet, *A modular construction of unramified p-extensions of $\mathbb{Q}(\mu_p)$*, Invent. Math. **34** (1976), 151–162

[Rb2] K. Ribet, *Galois representations and modular forms*, Bull. Am. Math. Soc. **32** (1995), 375–402

[Rid] D. Rideout, *A generalization of Stickelberger's theorem*, thesis, McGill University, Montreal 1970

[Ro1] X.-F. Roblot, *Unités de Stark et corps de classes de Hilbert*, C. R. Acad. Sci., Paris **323** (1996), 1165–1168

[Ro2] X.-F. Roblot, *Algorithmes de factorisation dans les extensions relatives et applications de la conjecture de Stark à la construction des corps de classes de rayon*, Diss. Univ. Bordeaux, 1997

[Ru1] K. Rubin, *Kolyvagin's system of Gauss sums*, Arithmetic algebraic geometry, Prog. Math. **89** (1991), 309–324

[Ru2] K. Rubin, *Euler systems and exact formulas in number theory*, Jahresber. DMV **98** (1996), 30–39

[Sal] H. Salié, *Eulersche Zahlen*, Sammelband Leonhard Euler, Deutsche Akad. Wiss. Berlin 293–310 (1959)

[San] J.W. Sands, *Abelian fields and the Brumer-Stark conjecture*, Compositio Math. **53** (1984), 337–346

[Sf1] R. Schoof, *Fermat's Last Theorem*, Jahrbuch Überblicke Math. 1995 (Beutelspacher, ed.), Vieweg 1995, 193–211

[Sf2] R. Schoof, *Wiles' proof of Taniyama-Weil conjecture for semi-stable elliptic curves over $\mathbb{Q}$*, Gaz. Math., Soc. Math. Fr. **66** (1995), 7–24

[Sf3] R. Schoof, *Minus class groups of the fields of the lth roots of unity*, Math. Comp. **67** (1998), 1225–1245

[Sch] K. Schwering, *Zur Theorie der arithmetischen Funktionen, welche von Jacobi $\psi(\alpha)$ genannt werden*, J. Reine Angew. Math. **93** (1882), 334–337

[ST] G. Shimura, Y. Taniyama, *Complex multiplication of abelian varieties and its applications to number theory*, Tokyo 1961

[Shi] T. Shintani, *On evaluation of zeta functions of totally real algebraic number fields at non-positive integers*, J. Fac. Sci. Univ. Tokyo **23** (1976), 393–417

[Sin] W. Sinnott, *On the Stickelberger ideal and the circular units of an abelian field*, Invent. Math. **62** (1980), 181–234

[Sol] D.R. Solomon, *On the class groups of imaginary abelian fields*, Ann. Inst. Fourier **40** (1990), 467–492

[Sou] C. Soulé, *Perfect forms and the Vandiver conjecture*, J. Reine Angew. Math. **517** (1999), 209–221

[Sp] A. Speiser, *Gruppendeterminante und Körperdiskriminante*, Math. Annalen **77** (1916), 546–562

[Su] J. Suzuki, *On the generalized Wieferich criteria*, Proc. Japan Acad. **70** (1994), 230–234

[Tak] T. Takagi, *Zur Theorie des Kreiskörpers*, J. Reine Angew. Math. **157** (1927), 230–238; Coll. Papers 246–255

[Ta] G. Tamme, *Über die p-Klassengruppe des p-ten Kreisteilungskörpers*, Ber. Math.-Stat. Sekt. Joanneum (1988), 48 pp.

[Ta1] J. Tate, *Les conjectures de Stark sur les fonctions L d'Artin en $s = 0$*, Progress in Math. (J. Coates, S. Helgason, eds), Birkhäuser, 1984

[Ta2] J. Tate, *Brumer-Stark-Stickelberger*, Sémin. Theor. Nombres 1980–1981, Exp. No. **24** (1981), 16 pp.

[Ter] G. Terjanian, *Sur l'équation $x^{2p} + y^{2p} = z^{2p}$*, C. R. Acad. Sci. Paris **285** (1977), 973–975

[Tha] F. Thaine, *On the ideal class groups of real abelian number fields*, Ann. Math. **128** (1988), 1–18

[vdP] A. van der Poorten, *Notes on Fermat's last theorem*, Wiley, New York 1996

[Vor] G.F. Voronoi, Über die Summe der quadratischen Reste einer Primzahl $p = 4m + 3$ (Russ.), St. Petersb. Math. Ges. **5**; FdM **30** (1899), 184

[Was1] L. Washington, *Introduction to Cyclotomic Fields*, Graduate Texts in Math. **83**, Springer Verlag 1982; 2nd edition 1997

[Was2] L. C. Washington, *Stickelberger's theorem for cyclotomic fields, in the spirit of Kummer and Thaine*, Number Theory (eds.: J.-M. de Koninck, C. Levesque), Proceedings of the Intern. Number Theory Conference at Laval (1987) W. Gruyter, Berlin, New York, 990–993

[WSS] T. Washio, A. Shimaura, K. Shiratani, *On certain congruences for Gauss sums*, Sci. Bull. Fac. Educ., Nagasaki Univ. **55** (1996), 1–8

[Wen] E. Wendt, *Arithmetische Studien über den "letzten" Fermatschen Satz, welcher aussagt, daß die Gleichung $a^n = b^n + c^n$ für $n \geq 2$ in ganzen Zahlen nicht auflösbar ist*, J. Reine Angew. Math. **113** (1894), 335–347

[Wie] A. Wieferich, *Zum letzten Fermatschen Theorem*, J. Reine Angew. Math. **136** (1909), 293–302

[Wil] A. Wiles, *The Iwasawa conjecture for totally real fields*, Ann. of Math. **131** (1990), 493–540

[WHS] K.S. Williams, K. Hardy, B.K. Spearman, *Explicit evaluation of certain Eisenstein sums*, Number theory, Banff/Alberta 1988, 553–626 (1990)

Appendix A. Dramatis Personae

Pierre de Fermat	1607	1665
Giulio Carlo Fagnano	1682	1766
Christian Goldbach	1690	1764
Leonhard Euler	1707	1783
Joseph Louis Lagrange	1736	1813
Adrien Marie Legendre	1752	1833
Carl Friedrich Gauss	1777	1855
Augustin-Louis Cauchy	1789	1857
Victor Amédée Lebesgue	1791	1875
Karl Georg Christian von Staudt	1798	1867
Niels Henrik Abel	1802	1829
Carl Gustav Jacob Jacobi	1804	1851
Viktor Jakovlevic Bouniakowsky	1804	1889
Gustav Peter Lejeune Dirichlet	1805	1859
Moritz Abraham Stern	1807	1894
Joseph Liouville	1809	1882
Ernst Eduard Kummer	1810	1893
Evariste Galois	1811	1832
Theodor Schönemann	1812	1868
Angelo Genocchi	1817	1889
Mathieu Schaar	1817	1867
Christian Julius Johannes Zeller	1822	1899
Ferdinand Gotthold Eisenstein	1823	1852
Leopold Kronecker	1823	1891
Bernhard Riemann	1826	1866
Henry John Stanley Smith	1826	1883
Théophile Pépin	1826	1904
Julius Wilhelm Richard Dedekind	1831	1916
Emile Léonard Mathieu	1835	1890
Paul Gustav Heinrich Bachmann	1837	1920
Franz Mertens	1840	1927

HEINRICH WEBER .... 1842 – 1913
ALLAN JOSEPH CHAMPNEYS CUNNINGHAM .... 1842 – 1928
JEGOR IVANOVICH ZOLOTAREV .... 1847 – 1878
LEOPOLD BERNHARD GEGENBAUER .... 1849 – 1903
GEORG FERDINAND FROBENIUS .... 1849 – 1917
LUDWIG STICKELBERGER .... 1850 – 1936
FRANZ GOLDSCHEIDER .... 1852 – 1926
ADOLF HURWITZ .... 1859 – 1919
KURT HENSEL .... 1861 – 1941
DAVID HILBERT .... 1862 – 1943
PHILIPP FRIEDRICH PIUS FURTWÄNGLER .... 1869 – 1940
STJEPAN BOHNICEK .... 1872 – 1956
ALFRED EDWARD WESTERN .... 1873 – 1961
LEONARD EUGENE DICKSON .... 1874 – 1954
TEIJI TAKAGI .... 1875 – 1960
WILLIAM SEALEY GOSSET .... 1876 – 1937
KARL RUDOLF FUETER .... 1880 – 1950
WALTER KARL JULIUS LIETZMANN .... 1880 – 1959
GUSTAV HERGLOTZ .... 1881 – 1953
EMMY NOETHER .... 1882 – 1935
ERNST JACOBSTHAL .... 1882 – 1965
ERICH HECKE .... 1887 – 1947
ALBERT THORALF SKOLEM .... 1887 – 1963
CARL LUDWIG SIEGEL .... 1896 – 1981
NIKOLAI GRIGORIEVICH CHEBOTARYOV .... 1894 – 1947
EMIL ARTIN .... 1898 – 1962
HELMUT HASSE .... 1898 – 1979
RICHARD DAGOBERT BRAUER .... 1901 – 1977
ARNOLD SCHOLZ .... 1904 – 1942
ABRAHAM ADRIAN ALBERT .... 1905 – 1972
ANDRÉ WEIL .... 1906 – 1998
HAROLD DAVENPORT .... 1907 – 1969
JACQUES HERBRAND .... 1908 – 1931
HANS REICHARDT .... 1908 – 1991
ALEXANDER GOTTFRIED WALTER ANTON AIGNER .... 1909 – 1988
CLAUDE CHEVALLEY .... 1909 –
ALBERT LEON WHITEMAN .... 1915 – 1995
KENKICHI IWASAWA .... 1917 – 1998
IGOR ROTISLAVOVIC SHAFAREVICH .... 1923 –
JOHN TORRENCE TATE .... 1925 –
JEAN-PIERRE SERRE .... 1926 –
YUTAKA TANIYAMA .... 1927 – 1958
EMMA LEHMER .... 1928 –

Appendix B. Chronology of Proofs

#	proof	year	comments
1.	Legendre	1788	Quadratic forms; incomplete
2.	Gauss 1 [264]	1801	Induction; April 8, 1796
3.	Gauss 2 [265]	1801	Quadratic forms; June 27, 1796
4.	Gauss 3 [266]	1808	Gauss' Lemma; May 6, 1807
5.	Gauss 4 [267]	1811	Cyclotomy; May 1801
6.	Gauss 5 [268]	1818	Gauss's Lemma; 1807/08
7.	Gauss 6 [269]	1818	Gauss sums; 1807/08
8.	Cauchy [122]	1829	Gauss 6
9.	Jacobi [400]	1830	Gauss 6
10.	Dirichlet [165]	1835	Gauss 4
11.	Lebesgue 1 [482]	1838	$N(x_1^2 + \ldots + x_q^2 \equiv 1 \bmod p)$
12.	Schönemann [722]	1839	quadratic period equation
13.	Eisenstein 1 [187]	1844	generalized Jacobi sums
14.	Eisenstein 2 [189]	1844	Gauss 6
15.	Eisenstein 3 [192]	1844	Gauss's Lemma
16.	Eisenstein 4 [197]	1845	Sine
17.	Eisenstein 5 [198]	1845	infinite products
18.	Liouville [549]	1847	Cyclotomy
19.	Lebesgue 2 [485]	1847	Lebesgue 1
20.	Schaar [702]	1847	Gauss's Lemma
21.	Genocchi [295]	1852	Gauss's Lemma
22.	Dirichlet [170]	1854	Gauss 1
23.	Lebesgue 3 [487]	1860	Gauss 7, 8
24.	Kummer 1 [469]	1862	Quadratic forms
25.	Kummer 2 [469]	1862	Quadratic forms
26.	Dedekind 1 [174, §154]	1863	quadratic forms
27.	Gauss 7 [270]	1863	quadratic periods; Sept. 1796
28.	Gauss 8 [271]	1863	quadratic periods; Sept. 1796
29.	Mathieu [567]	1867	Cyclotomy
30.	von Staudt [749]	1867	Cyclotomy
31.	Bouniakowski [70]	1869	Gauss's Lemma
32.	Stern [756]	1870	Gauss's Lemma
33.	Zeller [881]	1872	Gauss's Lemma
34.	Zolotarev [883]	1872	Permutations
35.	Kronecker 1 [438]	1872	Zeller

#	proof	year	comments
36.	Schering [709]	1875	Gauss 3
37.	Kronecker 2 [440]	1876	Induction
38.	Mansion [564, 565]	1876	Gauss's Lemma
39.	Dedekind 2 [149]	1877	Gauss 6
40.	Dedekind 3 [153]	1877	Dedekind Sums
41.	Pellet 1 [631]	1878	Stickelberger-Voronoi
42.	Pépin 1 [640]	1878	Cyclotomy
43.	Schering [711, 712]	1879	Gauss's Lemma
44.	Petersen [651]	1879	Gauss's Lemma
45.	Genocchi [297, 300]	1880	Gauss's Lemma
46.	Kronecker 3 [443]	1880	Gauss 4
47.	Kronecker 4 [442]	1880	quadratic period
48.	Voigt [815]	1881	Gauss's Lemma
49.	Pellet 2 [632]	1882	Mathieu 1867
50.	Busche 1 [106]	1883	Gauss's Lemma
51.	Gegenbauer 1 [285]	1884	Gauss's Lemma
52.	Kronecker 5 [444]	1884	Gauss's Lemma
53.	Kronecker 6 [446, 447, 448]	1885	Gauss 3
54.	Kronecker 7 [449]	1885	Gauss's Lemma
55.	Bock [58]	1886	Gauss's Lemma
56.	Lerch [527]	1887	Gauss 3
57.	Busche 2 [108]	1888	Gauss's Lemma
58.	Hacks [322]	1889	Schering
59.	Hermes [364]	1889	Induction
60.	Kronecker 8 [450]	1889	Gauss's Lemma
61.	Tafelmacher 1 [787]	1889	Stern
62.	Tafelmacher 2 [787]	1889	Stern/Schering
63.	Tafelmacher 3 [787]	1889	Schering
64.	Busche 3 [110]	1890	Gauss's Lemma
65.	Franklin [236]	1890	Gauss's Lemma
66.	Lucas [558, 559]	1890	Gauss's Lemma
67.	Pépin 2 [643]	1890	Gauss 2
68.	Fields [233]	1891	Gauss's Lemma
69.	Gegenbauer 2 [290]	1891	Gauss's Lemma
70.	Gegenbauer 3 [292]	1893	Gauss's Lemma
71.	Schmidt 1 [719]	1893	Gauss's Lemma
72.	Schmidt 2 [719]	1893	Gauss's Lemma
73.	Schmidt 3 [719]	1893	Induction
74.	Gegenbauer 4 [293]	1894	Gauss's Lemma
75.	Bang [29]	1894	Induction

#	proof	year	comments
76.	Mertens 1 [578]	1894	Gauss's Lemma
77.	Mertens 2 [578]	1894	Gauss sums
78.	Busche 4 [111]	1896	Gauss's Lemma
79.	Lange 1 [476]	1896	Gauss's Lemma
80.	de la Vallée Poussin [808]	1896	Gauss 2
81.	Lange 2 [477]	1897	Gauss's Lemma
82.	Hilbert [368]	1897	Cyclotomy
83.	Alexejewsky [12]	1898	Schering
84.	Pépin 3 [645]	1898	Legendre
85.	Pépin 4 [645]	1898	Gauss 5
86.	König [432]	1899	Induction
87.	Fischer [234]	1900	Resultants
88.	Takagi [788]	1903	Zeller
89.	Lerch [531, 532]	1903	Gauss 5
90.	Mertens 3 [580]	1904	Eisenstein 4
91.	Mirimanoff & Hensel [588]	1905	Stickelberger-Voronoi
92.	Busche 5 [112]	1909	Zeller
93.	Busche 6 [113]	1909	Eisenstein
94.	Aubry [23]	1910	= Eisenstein 3
95.	Aubry [23]	1910	= Voigt
96.	Aubry [23]	1910	= Kronecker
97.	Pépin [649]	1911	Gauss 2
98.	Petr 1 [653]	1911	Mertens 3
99.	Pocklington [663]	1911	Gauss 3
100.	Dedekind 4 [152]	1912	Zeller
101.	Heawood [345, 346]	1913	= Eisenstein 3
102.	Frobenius 1 [242]	1914	Zeller
103.	Frobenius 2 [243]	1914	Eisenstein 3
104.	Lasker [478]	1916	Stickelberger-Voronoi
105.	Cerone [125, 126]	1917	Eisenstein 4
106.	Bartelds & Schuh [37]	1918	Gauss's Lemma
107.	Stieltjes [767]	1918	Lattice points
108.	Teege 1 [795]	1920	Legendre
109.	Teege 2 [796]	1921	Cyclotomy
110.	Arwin [18]	1924	Quadratic forms
111.	Rédei 1 [669]	1925	Gauss's Lemma
112.	Rédei 2 [670]	1926	Gauss's Lemma
113.	Whitehead [839]	1927	Genus theory (Kummer)
114.	Petr 2 [654]	1927	theta functions
115.	Skolem 1 [736]	1928	Genus theory

#	proof	year	comments
116.	Petr 3 [656]	1934	Kronecker (signs)
117.	van Veen [812]	1934	Eisenstein 3
118.	Fueter [251]	1935	quaternion algebras
119.	Whiteman [840]	1935	Gauss's Lemma
120.	Dockeray [177]	1938	Eisenstein 3
121.	Dörge [178]	1942	Gauss's Lemma
122.	Rédei 3 [673]	1944	Gauss 5
123.	Lewy [538]	1946	Cyclotomy
124.	Petr 4 [657]	1946	Cyclotomy
125.	Skolem 2 [738]	1948	Gauss 2
126.	Barbilian [30]	1950	Eisenstein 1
127.	Rédei 4 [675]	1951	Gauss 3
128.	Brandt 1 [76]	1951	Gauss 2
129.	Brandt 2 [77]	1951	Gauss sums
130.	Brewer [84]	1951	Mathieu, Pellet
131.	Furquim de Almeida [256]	1951	Finite fields
132.	Zassenhaus [879]	1952	Finite fields
133.	Riesz [684]	1953	Permutations
134.	Fröhlich [245]	1954	Class Field Theory
135.	Ankeny [14]	1955	Cyclotomy
136.	D. H. Lehmer [492]	1957	Gauss's Lemma
137.	C. Meyer [583]	1957	Dedekind sums
138.	Holzer [374]	1958	Gauss sums
139.	Rédei 5 [676]	1958	Cyclotomic polynomial
140.	Reichardt [677]	1958	Gauss 3
141.	Carlitz [117]	1960	Gauss 1
142.	Kubota 1 [454]	1961	Cyclotomy
143.	Kubota 2 [455]	1961	Gauss sums (sign)
144.	Skolem 3 [740]	1961	Cyclotomy
145.	Skolem 4 [741]	1961	finite fields
146.	Hausner [343]	1961	Gauss sums
147.	Swan 1 [778]	1962	Stickelberger-Voronoi
148.	Koschmieder [435]	1963	Eisenstein, sine
149.	Gerstenhaber [305]	1963	Eisenstein, sine
150.	Rademacher [665]	1964	Finite Fourier analysis
151.	Weil [829]	1964	Theta functions
152.	Kloosterman [431]	1965	Holzer
153.	Chowla [129]	1966	Finite fields
154.	Burde [97, 98]	1967	Gauss's Lemma
155.	Kaplan 1 [418]	1969	Eisenstein

#	proof	year	comments
156.	Kaplan 2 [419]	1969	quadratic congruences
157.	Birch [54]	1971	K-theory (Tate)
158.	Reshetukha [680]	1971	Gauss sums
159.	Agou [2]	1972	finite fields
160.	Brenner [83]	1073	Zolotarev
161.	Honda [375]	1973	Gauss sums
162.	Milnor & Husemöller [587]	1973	Weil 1964
163.	Allander [13]	1974	Gauss's Lemma
164.	Berndt & Evans [47]	1974	Gauss's Lemma
165.	Hirzebruch & Zagier [372]	1974	Dedekind Sums
166.	Rogers [686]	1974	Legendre
167.	Castaldo [121]	1976	Gauss's Lemma
168.	Frame [235]	1978	Kronecker (signs)
169.	Hurrelbrink [385]	1978	K-theory
170.	Auslander & Tolimieri [25]	1979	Fourier transform
171.	Brown [89]	1981	Gauss 1
172.	Goldschmidt [316]	1981	cyclotomy
173.	Kac [415]	1981	Eisenstein, sine
174.	Barcanescu [31]	1983	Zolotarev
175.	Zantema [878]	1983	Brauer groups
176.	Ely [205]	1984	Lebesgue 1
177.	Eichler [184]	1985	Theta function
178.	Barrucand & Laubie [36]	1987	Stickelberger-Voronoi
179.	Peklar [630]	1989	Gauss's Lemma
180.	Barnes [33]	1990	Zolotarev
181.	Swan 2 [779]	1990	Cyclotomy
182.	Rousseau 1 [688]	1990	Exterior algebras
183.	Rousseau 2 [689]	1991	Permutations
184.	Keune [428]	1991	Finite fields
185.	Kubota 3 [458]	1992	geometry
186.	Russinoff [692]	1992	Gauss's Lemma
187.	Garrett [261]	1992	Weil 1964
188.	Motose [593]	1993	group algebras
189.	Rousseau[690]	1994	Zolotarev
190.	Young [877]	1995	Gauss sums
191.	Brylinski [90]	1997	group actions
192.	Merindol [577]	1997	Eisenstein, sine
193.	Watanabe [823]	1997	Zolotarev
194.	Ishii [387]	1998	Gauss 4
195.	Motose [594]	1999	group algebras
196.	Lemmermeyer [515]	2000	Lebesgue 1, Ely

Appendix C. Some Open Problems

We'll put this empty page to use by collecting a number of interesting projects and open problems.

- Check out the unsolved problems concerning quadratic residues (F5, F6, F8) in Guy's "Unsolved Problems in Number Theory".
- Generalize Nakhash's proof of $(5/p) = (p/5)$ from Exercise 2.26 to arbitrary primes, thereby giving a new proof of the quadratic reciprocity law.
- Reconstruct Gauss's computation of $(3/p)_3$ announced in his entry from Jan. 06, 1809 in his diary. More generally, find a *simple* cyclotomic proof for the supplementary laws for n-th power residues, at least for $n \in \{3, 4, 8\}$.
- Let p be a prime inert in the real quadratic number field $\mathbb{Q}(\sqrt{m})$, and let ε_m denote its fundamental unit. For units with positive norm, generalize Exercises 6.30 – 6.32, that is, determine $\varepsilon^{(p+1)/4} \bmod p$ in terms of appropriate binary quadratic forms.
- Generalize the test of Berrizbeitia & Berry (Exercise 7.24) by replacing the cubic reciprocity law with Eisenstein's. The results of H.C. Williams (*A class of primality tests for trinomials which includes the Lucas-Lehmer test*, Pac. J. Math. **98** (1982), 477–494) suggest that this should not be too hard. Can Lenstra's Primality Test be generalized so as to include primality tests based on elliptic curves?
- Eliminate the analytic class number formula from the proof of Scholz's octic reciprocity law. Is there a proof analogous to that of Schönemann-Lehmer for the quartic case?
- Develop the theory of Eisenstein sums, and give a proof for the octic reciprocity law that is a simple as that for the quartic case. Can the Eisenstein sums be generalized to the Brumer-Stark situation, i.e. can the division values be replaced by the units predicted by Brumer-Stark?
- Try to derive Herglotz's relations in Exercise 10.10 from the properties of sl(x) discussed in Chapter 8.
- In Chapter 11 we have seen an algebraic proof for the fact that $2h^-$ kills the minus class group of cyclotomic number fields. Eliminate the factor 2 in this statement. Moreover, given any factor n of h^-, prove by algebraic methods that n divides the order of the minus class group (e.g. by constructing a corresponding class field).

References

1. V. W. Adkisson, *Note on the law of biquadratic reciprocity*, Bull. Amer. Math. Soc. **38** (1932), 529–530; FdM **58 - I** (1932), 176; cf. p. 201
2. S. Agou, *Une démonstration de la loi de réciprocité quadratique*, Publ. Dép. Math. (Lyon) **9**, fasc. 3 (1972), 55–57; Zbl 308.10001; MR 48 # 10959; cf. p. 417
3. A. Aigner, *Kriterien zum 8. und 16. Potenzcharakter der Reste 2 und −2*, Deutsche Math. **4** (1939), 44–52; FdM **65** I (1939), 112; cf. p. 310
4. A. Aigner, *A generalization of Gauss' lemma*, Amer. Math. Monthly **57** (1950), 408–410; Zbl 38.18100; MR 12 # 80b; cf. p. 23
5. A. Aigner, *Einige handliche Regeln für biquadratische Reste*, Math. Nachr. **17** (1959), 219–223; Zbl 92.04302; MR 22 # 4697; cf. p. 171, 310
6. A. Aigner, *Bemerkung zur Lösung zum Problem Nr. 29*, Elemente Math. **15** (1960), 66–67; Zbl 93.25801; cf. p. 313
7. A. Aigner, *Quadratische und kubische Restkriterien für das Auftreten einer Fibonacci-Primitivwurzel*, J. Reine Angew. Math. **274/275** (1975), 139–140; Zbl 302.12002; MR 51 # 10286; cf. p. 70, 174, 228
8. A. Aigner, *Zahlentheorie*, Berlin - New York, Walter de Gruyter, 1975; Zbl 289.10001; MR 56 #2901; cf. p. 227
9. A. Aigner, H. Reichardt, *Stufenreihen im Potenzrestcharakter*, J. Reine Angew. Math. **184** (1942), 158–160; FdM **68 - I** (1942), 56; Zbl 27.01102; MR 5, 91c; cf. p. 173, 310
10. H. P. Alderson, *On the quintic character of* 2, Mathematika **11** (1964), 125–130; Zbl 127.26802; MR 30 # 4718; cf. p. 141, 141
11. H. P. Alderson, *On the septimic character of 2 and 3*, Proc. Cambridge Phil. Soc. **74** (1973), 421–433; Zbl 265.10003; MR 48 # 2119; cf. p. 141
12. W. P. Alexejevsky, *Über das Reciprocitätsgesetz der Primzahlen* (Russ.), Samml. Mitt. Math. Ges. Charkov (2) **6** (1898) 200–202; FdM **30** (1899), 184; RSPM **7** (1899), 132; cf. p. 415
13. C. G. Allander, *Gauss's law of reciprocity - a lucid proof*, (Swed.), Nordisk. Mat. Tidskr. **22** (1974), 23–25, 40; Zbl 285.10001; MR 49 # 8923; cf. p. 417
14. N. C. Ankeny, *The law of quadratic reciprocity*, Norske Vid. Selsk. Forh., Trondheim **28** (1956), 145–146; Zbl 72.03303; MR 17 # 1185i; cf. p. 416

15. N. C. Ankeny, *Criterion for r-th power residuacity*, Pac. J. Math. **10** (1960), 1115–1124; Zbl 113.26801; MR 22 # 9479; cf. p. 141
16. F. Arndt, *Bemerkungen über die Verwandlung der irrationalen Quadratwurzel in einen Kettenbruch*, J. Reine Angew. Math. **31** (1846), 343–358; cf. p. 69
17. E. Artin, J. Tate, *Class field Theory*, Benjamin New York Amsterdam 1967; cf. p. 136
18. A. Arwin, *A contribution to the theory of closed chains*, Annals of Math. (2) **25** (1924), 91–117; FdM **50** (1924), 87; cf. p. 415
19. A. Arwin, *Einige periodische Kettenbruchentwicklungen*, J. Reine Angew. Math. **155** (1926), 111–128; FdM **52** (1926), 187; cf. p. 170
20. A. Aubry, *Théorie élémentaire des résidus quadratiques*, Ens. Math. **9** (1907), 24–36; FdM **38** (1907), 224; cf. p. 26
21. A. Aubry, *Le lemme fondamentale de la théorie des nombres*, Ens. Math. **9** (1907), 286–305; FdM **38** (1907), 224; cf. p. 23
22. A. Aubry, *Étude élémentaire sur le théorème de Fermat*, L'Ens. Math. **9** (1907), 417–460; FdM **38** (1907), 224; cf. p. 25
23. A. Aubry, *Exposition élémentaire de la loi de réciprocité dans la théorie des nombres*, Ens. Math. **12** (1910), 457–475; FdM **41** (1910), 230; cf. p. 23, 170, 415
24. A. Aubry, *Étude élémentaire sur les résidus quadratiques*, Mathesis (3), **10** (1910), 8–11, 33–35; FdM **41** (1910), 230–231; cf. p. 23
25. L. Auslander, R. Tolimieri, *Is computing with the finite Fourier transform pure or applied mathematics?*, Bull. Am. Math. Soc., N. S. **1** (1979), 847–897; Zbl 475.42014; MR 81e:42020; cf. p. 139, 417
26. P. Bachmann, *Niedere Zahlentheorie* **I**, Leipzig 1902; FdM **33** (1902), 192–183; cf. p. xiv, 24, 101, 227
27. P. Bachmann, *Die Lehre von der Kreistheilung*, Leipzig 1872, 1927; FdM **4** (1872), 78–79; cf. p. 101
28. C. Banderier, *Résidus quadratiques. Lois de réciprocité quadratique*, Univ. Rouen, 1997; cf. p. 24
29. A. S. Bang, *Nyt Bevis for Reciprocitetsaetninger*, Nyt. Tidss. for Math. **V B** (1894), 92–96; FdM **25** (1893/94), 281; cf. p. 414
30. D. Barbilian, *Das Reziprozitätsgesetz mit Anwendung auf die Galoissche Theorie* (Roum., French and Russian summaries), Acad. Republ. popul. Romane, Bul. Sti. A **2** (1950), 731–736; Zbl 40.161; cf. p. 416
31. S. Barcanescu, *On Zolotarev's proof of the quadratic reciprocity law*, Institutul National Pentru Creatie Stiintifica Si Tehnica, Preprint Series in Mathematics no. **15**, 1983; cf. p. 417
32. Ph. Barkan, *Partitions quadratiques et cyclotomie*, Sémin. Delange-Pisot-Poitou **1974/75** (1975), exp. 13, 12 pp; Zbl 324.10037; MR 53, 5459; cf. p. 343
33. F. W. Barnes, *A permutation reciprocity law*, Ars. Comb. **29 A** (1990), 155–159; Zbl 716.11004; MR 97f:11006; cf. p. 417

34. P. Barrucand, H. Cohn, *Note on primes of type $x^2 + 32y^2$, class number, and residuacity*, J. Reine Angew. Math. **238** (1969), 67–70; Zbl 207.36202; MR 40 # 2641; cf. p. 172, 173
35. P. Barrucand, F. Laubie, *Ramification modérée dans les corps de nombres de degré premier*, Sém. Théor. Nombres, Bordeaux 1981/82, exposé 13; Zbl 503.12001; MR 84k:12002; cf. p. 118, 137
36. P. Barrucand, F. Laubie, *Sur les symboles des restes quadratiques des discriminants*, Acta Arith. **48** (1987), 81–88; Zbl 609.12004; MR 88h:11080; cf. p. 118, 417
37. A. L. Bartelds, F. Schuh, *Elementair bewijs der uitgebreide wederkeeringheidswet van Legendre*, Nieuw Arch. Wisk. **12** (1918), 420–438; FdM **46** (1916–18), 1445; cf. p. 415
38. O. Baumgart, *Ueber das quadratische Reciprocitätsgesetz. Eine vergleichende Darstellung der Beweise*, Diss. Göttingen 1885; Zeitschrift Math. Phys. 30 (1885); FdM **17**(1885), 26–28, cf. p. xiv, 14, 24, 101
39. A. Bayad, *Loi de réciprocité quadratique dans les corps quadratiques imaginaires*, Ann. Inst. Fourier **45** (1995), 1223–1237; Zbl 843.11047; MR 96j:11139; cf. p. 277
40. B. D. Beach, H. C. Williams, *A numerical investigation of the Diophantine equation $x^2 - dy^2 = -1$*, Proc. 3rd Southeastern Conf. on Combinatorics, Graph Theory and Computing, Boca Raton 1972, 37–66; Zbl 261.10015; MR 50, 231; cf. p. 175
41. N.G.W.H. Beeger, *Sur le caractère de 2 comme résidu de degré huitième*, Nieuw Arch. Wiskunde **12** (1917), 188–189; FdM **46** (1916–18), 191; cf. p. 310
42. N.G.W.H. Beeger, *A problem in the theory of numbers and it's history*, Nieuw Arch. Wiskunde (2), **22** (1948), 306–309; Zbl 29.25202; MR 9, 336f; cf. p. 312
43. D. Bernardi, *Résidus de puissances*, Sém. Delange-Pisot-Poitou 1977/78 **2**, no. 28 (1978), 12 pp; Zbl 403.12003; MR 80h:10005; cf. p. 141, 175
44. D. Bernardi, *Résidus de puissances*, Publ. Math. d'Orsay, Diss. Univ. Paris XI 1979; Zbl 416.10001; MR 80i:10006; cf. p. 141, 175
45. D. Bernardi, *Résidus de puissances et formes quadratiques*, Ann. Inst. Fourier **30** (1980), 7–17; Zbl 425.10003; MR 82e:10006; cf. p. 141, 175
46. B.C. Berndt, *A generalization of a theorem of Gauss on sums involving $[x]$*, Amer. Math. Monthly **82** (1975), 44–51; MR 50 #4460; cf. p. 23
47. B.C. Berndt, R.J. Evans, *Least positive residues and the law of quadratic reciprocity*, Delta **4** (1974), 65–69; Zbl 291.10004; MR 50 # 12881; cf. p. 417
48. B.C. Berndt, R.J. Evans, K.S. Williams, *Gauss and Jacobi sums*, John Wiley & Sons 1998; MR 99d:11092; cf. p. 139, 201, 293, 393
49. B.C. Berndt, U. Dieter, *Sums involving the greatest integer function and Riemann–Stieltjes integration*, J. Reine Angew. Math. **337** (1982), 208–220; MR 84c:10006; cf. p. 23

50. L. Bianchi, *Osservazioni circa il carattere quadratico dei numeri in un corpo quadratico*, Rom. Acc. L. Rend. (5), **29** (1920), 223–230; Opere I, 116–124; FdM **47** (1919/20), 143; cf. p. 170
51. L. Bianchi, *Lezioni sulla Teoria dei numeri algebrici*, Pisa 1923; cf. p. 170
52. C. E. Bickmore, *On the numerical factors of $a^n - 1$*, Mess. Math. **25** (1895), 1–44; FdM **26** (1895), 202; cf. p. 29, 70, 179, 205, 229, 310
53. C. E. Bickmore, *On the numerical factors of $a^n - 1$*, Mess. Math. **26** (1896), 1–38; FdM **27** (1896), 139; cf. p. 70, 179, 310
54. B. J. Birch, *K_2 of global fields*, Institute on Number theory, Proc. Symp. Pure Math. XX, Amer. Math. Soc. (1971), 87–95; Zbl 218.12010; MR 48 11052; cf. p. 21, 385, 417
55. A. Blanchard, C. Blanchard (eds.), *Séminaire d'arithmétique. Formes quadratiques sur $\mathbb{Z}$; Formes quadratiques sur $\mathbb{Q}_p$; Théorème de Minkowsi–Hasse; Loi de réciprocité de Hilbert*, Faculté des Sciences de Marseille, 1970; cf. p. 70
56. F. van der Blij, *Méthodes algébriques et analytiques dans la théorie des nombres*, Bull. Soc. Math. Belg. **15** (1963), 3–17; Zbl 112.26804; MR 27 # 2488; cf. p. 23
57. W. Bock, *Zusatz zu dem Artikel 129 der Disquisitiones arithmeticae von Gauß*, Hamb. Mitt. **5** (1920), 307–309; FdM **47** (1919/20), 108
58. W. Bock, *Ueber eine neue zahlentheoretische Funktion*, Hamb. Mitt. **6** (1886), 187–194; FdM **18** (1886), 146; cf. p. 414
59. S. Bohnicek, *Das Reziprozitätsgesetz der biquadratischen Potenzreste aus dem Gebiet der imaginären Zahlen* (Croat.), Agram Ak. **154** (1904), 7–79; FdM **35** (1904), 229; cf. p. 201
60. S. Bohnicek, *Über das Reziprozitätsgesetz für achte Potenzreste im Körper der achten Einheitswurzeln* (Croat.), Sitzungsberichte der südslawischen Akademie **165** (1906), 1–49; FdM **40** (1909), 267; cf. p. 312
61. S. Bohnicek, *Zur Theorie des relativbiquadratischen Zahlkörpers*, Math. Ann. **63** (1907), 85–144; transl. from Agram. Ak. (Croat.) **163**, 41–112; FdM **37** (1906), 243; cf. p. 201
62. S. Bohnicek, *Theorie der Potenzreste in algebraischen Zahlkörpern*, (Croat.) Agram Ak. **177** (1909), 1–96; FdM **41** (1910), 247; cf. p. 312, 393
63. S. Bohnicek, *Zur Theorie der achten Einheitswurzeln*, Wiener Sitzungsber. Abt. IIa **120** (1911), 25–47; FdM **42** (1911), 233–234; cf. p. 312, 393
64. S. Bohnicek, *Anwendungen der Lemniskatenteilung*, Wiener Sitzungsber. Abt. IIa **121** (1912), 679–717; FdM **43** (1912), 273–274; cf. p. 277
65. P. Bonaventura, *Il teorema di reciprocità nei numeri interi complessi e le funzioni lemniscatiche*, Giornale di Matematichi di Battaglini **30** (1892), 300–310; FdM **24** (1892), 183; cf. p. 277

66. P. Bonaventura, *Sul teoremo di reciprocità delle teorie dei residui quadratici nei numeri interi del campo* $(1, i\sqrt{2})$, Giornale di Matematichi di Battaglini **30** (1892), 221–234; FdM **24** (1892), 183; cf. p. 170
67. H. Bork, *Untersuchungen über das Verhalten zweier Primzahlen in Bezug auf ihren quadratischen Restcharakter*, Diss. Halle Pr. Askem. Gymn. Berlin, 21 pp; FdM **17** (1885), 152; cf. p. 29
68. W.J. Bouniakowski, *Sur quelques formules, qui résultent de la combinaison des résidus quadratiques et non quadratiques des nombres premiers*, Bull. Acad. St. Pétersbourg **13** (1868), 25–32; FdM **2** (1869/70), 89–90; BSMA **1** (1870), 240; cf. p. 23
69. W.J. Bouniakowski, *Sur les congruences binômes exponentielles à base 3 et sur plusiers nouveaux théorèmes relatifs aux résidus et aux racines primitives*, Bull. Acad. St. Pétersbourg **14** (1869), 356–381; FdM **2** (1869/70), 89; BSMA **2** (1871), 299; cf. p. 25
70. W.J. Bouniakowski, *Sur un théorème relatif à la théorie des résidus et son application à la démonstration de la loi de réciprocité de deux nombres premiers*, Bull. Acad. St. Pétersbourg **14** (1869), 432–447; BSMA **2** (1871), 300; FdM **2** (1869/70), 92–93; cf. p. 413
71. W.J. Bouniakowski, *Sur le symbole de Legendre* $\left(\frac{a}{p}\right)$, Bull. Acad. St. Pétersbourg **14** (1869); FdM **2** (1869/70), 93; cf. p. 23
72. W.J. Bouniakowski, *Sur quelques propositions nouvelles relatives au symbole de Legendre* $\left(\frac{a}{p}\right)$, Bull. Acad. St. Pétersbourg **22** (1876); FdM **8** (1876), 95–96; cf. p. 23
73. J. Brandler, *Residuacity properties of real quadratic units*, thesis, Univ. of Arizona, Tucson, 1970; cf. p. 173
74. J. Brandler, *Residuacity properties of real quadratic units*, J. Number Theory **5** (1973), 271–287; Zbl 272.12002; MR 49, 2657; cf. p. 172, 173
75. J. Brandler, *On a theorem of Barrucand*, Bolletino U.M.I. (4), **12** (1975), 50–55; Zbl 334.12007; MR 53, 342; cf. p. 173
76. H. Brandt, *Über das quadratische Reziprozitätsgesetz*, Ber. Verh. Sächs. Akad. Wiss. Leipzig Math.-Nat. Kl. **99** (1951), 17 pp; Zbl 44.03205; MR 13 # 537b; cf. p. 416
77. H. Brandt, *Über das quadratische Reziprozitätsgesetz im rationalen Zahlkörper*, Math. Nachr. **6** (1951), 125–128; Zbl 43.27301; MR 13 # 537c; cf. p. 416
78. H. Brandt, *Das quadratische Reziprozitätsgesetz im Gauss'schen Zahlkörper*, Comment. Math. Helvet. **26** (1952), 42–54; Zbl 46.26801; MR 13 # 726b; cf. p. 170
79. H. Brandt, *Über das quadratische Reziprozitätsgesetz im Körper der dritten Einheitswurzeln*, Nova Acta Leopoldina (N.F.), **15** (1952), 163–188; Zbl 48.27303; MR 15 # 14e; cf. p. 226
80. H. Brandt, *Binäre quadratische Formen im Gaußschen Zahlkörper*, Math. Nachr. **7** (1952), 151–158; Zbl 46.27302; MR 14, 358b; cf. p. 170

81. I. Braun, *Das quadratische Reziprozitätsgesetz aus der Sicht der allgemeinen Klassenkörpertheorie*, Hausarbeit Göttingen 1976; cf. p. 13
82. J. L. Brenner, *A new property of the Jacobi symbol*, Duke Math. J. **29** (1962), 29–32; Zbl 102.27906; MR 24 #A3127; cf. p. 23
83. J. L. Brenner, *Zolotarev's theorem on the Legendre symbol*, Pac. J. Math. **45** (1973), 413–414; Zbl 253.10005; MR 48 # 2038; cf. p. 23, 417
84. B. W. Brewer, *On the quadratic reciprocity law*, Amer. Math. Monthly **58** (1951), 177–179; Zbl 42.03910; MR 12 # 589j; cf. p. 416
85. R. Bricard, *Sur le caractère quadratique du nombre 3 par rapport à un nombre premiere quelconque*, Nouv. Ann. (3) **16** (1897), 546–549; FdM **28** (1897), 170; cf. p. 25
86. E. Brown, *A theorem on biquadratic reciprocity*, Proc. Amer. Math. Soc. **30** (1971), 220–222; Zbl 207.05201, p. 52 and 221.10006; MR 43 # 6182; cf. p. 176
87. E. Brown, *Quadratic forms and biquadratic reciprocity*, J. Reine Angew. Math. **253** (1972), 214–220; Zbl 229.10002; MR 45 # 5080; cf. p. 172
88. E. Brown, *Biquadratic reciprocity laws*, Proc. Amer. Math. Soc. **37** (1973), 374–376; Zbl 257.10005; MR 47 # 1727; cf. p. 173
89. E. Brown, *The first proof of the quadratic reciprocity law, revisited*, Amer. Math. Monthly **88** (1981), 257–264; Zbl 459.01006; MR 82d:10008; cf. p. 4, 21, 33, 70, 417
90. J.-L. Brylinski, *Central extensions and reciprocity laws*, Cah. Topologie Geom. Differ. Categoriques **38** (1997), 193–215; Zbl 886.18003; cf. p. 417
91. J. Bucher, *Neues über die Pellsche Gleichung*, Mitt. d. Naturforsch. Ges. Luzern **14** (1943), 3–18; Zbl 63.00652; MR 9, 78e; cf. p. 312
92. D. A. Buell, *Binary quadratic forms. Classical theory and modern computations*, Springer 1989; Zbl 698.10013; MR 92b:11021; cf. p. 175
93. D. A. Buell, R. H. Hudson, K. S. Williams, *Extension of a theorem of Cauchy and Jacobi*, J. Number Theory **19** (1984), 309–340; Zbl 551.12014; MR 86i:11002; cf. p. 313
94. D. A. Buell, Ph. A. Leonard, K. S. Williams, *Note on the quadratic character of a quadratic unit*, Pac. J. Math. **92** (1981), 35–38; Zbl 475.10023; MR 83j:12005; cf. p. 174
95. D. A. Buell, K. S. Williams, *Is there an octic reciprocity law of Scholz type?*, Amer. Math. Monthly **85** (1978), 483–484; Zbl 383.10004; cf. p. 312
96. D. A. Buell, K. S. Williams, *An octic reciprocity law of Scholz type*, Proc. Amer. Math. Soc. **77** (1979), 315–318; Zbl 417.10002; MR 81a:10006; cf. p. 312
97. K. Burde, *Reziprozitätsgesetze für Gitterpunktsummen*, Diss. Univ. Göttingen, 1964; MR 31 # 114; cf. p. 416
98. K. Burde, *Reziprozitätsgesetze für Gitterpunktsummen*, J. Reine Angew. Math. **226** (1967), 159–174; Zbl 173.04302; MR 35 #111; cf. p. 416

99. K. Burde, *Ein rationales biquadratisches Reziprozitätsgesetz*, J. Reine Angew. Math. **235** (1969), 175–184; Zbl 169.36902; MR 39 # 2694; cf. p. 166, 172
100. K. Burde, *Zur Herleitung von Reziprozitätsgesetzen unter Benutzung von endlichen Körpern*, J. Reine Angew. Math. **293/294** (1977), 418–427; Zbl 349.10003; MR 57 # 16178; cf. p. 173, 201, 227
101. K. Burde, *Pythagoräische Tripel und Reziprozität in Galoisfeldern*, J. Number Theory **12** (1980), 278–282; Zbl 435.12008; MR 81i:10005; cf. p. 171
102. K. Burde, *Ein Reziprozitätsgesetz in Galoisfeldern*, J. Number Theory **13** (1981), 66–87; Zbl 447.10007; MR 82e:12022; cf. p. 310
103. K. Burde, *Ergänzungsgesetze*, Math. Nachr. **163** (1993), 15–26; Zbl 814.11003; MR 94g:11004; cf. p. 310
104. W. Burnside, *On cyclotomic quinquisection*, Proc. London Math. Soc. **14** (1915), 251–259; FdM **45** (1914/15), 1252; cf. p. 141
105. W. Burnside, *On cyclotomic octosection*, Trans. Cambr. Phil. Soc. **22** (1919), 405–411; FdM **47** (1919/20), 150; cf. p. 313
106. E. Busche, *Ueber eine Beweismethode in der Zahlentheorie und einige Anwendungen derselben, insbesondere auf das Reziprozitätsgesetz in der Theorie der quadratischen Reste*, Diss. Göttingen 1883; cf. p. 414
107. E. Busche, *Arithmetischer Beweis des Reciprocitätsgesetzes für die biquadratischen Reste*, J. Reine Angew. Math. **99** (1886), 261–274; FdM **18** (1886), 147; cf. p. 201
108. E. Busche, *Ueber grösste Ganze*, J. Reine Angew. Math. **103** (1888), 118–125; FdM **20** (1888), 183–184; cf. p. 23, 414
109. E. Busche, *Beweis des quadratischen Reciprocitätsgesetzes in der Theorie der aus den vierten Wurzeln der Einheit gebildeten complexen Zahlen*, Hamb. Mitt. **2** (1890), 80–92; FdM **22** (1890), 207; cf. p. 170
110. E. Busche, *Ueber die Function $\sum_{x=1}^{(q-1)/2}[\frac{px}{q}]$*, J. Reine Angew. Math. **106** (1890), 65–80; FdM **22** (1890), 208; cf. p. 23, 414
111. E. Busche, *Beweis des quadratischen Reciprocitätsgesetzes*, Hamburger Mitt. **3** (1896), 233–234; FdM **27** (1896), 142; cf. p. 415
112. E. Busche, *Eine geometrische Veranschaulichung des quadratischen Restcharakters*, Hamb. Mitt. **4** (1909), 403–409; FdM **40** (1909), 246–247; cf. p. 415
113. E. Busche, *Zur Theorie der Funktion [x]*, J. Reine Angew. Math. **136** (1909), 39–57; FdM **40** (1909), 247; cf. p. 201
114. E. Busche, *Über die Theorie der biquadratischen Reste*, J. Reine Angew. Math. **141** (1912), 146–161; FdM **43** (1912), 246; cf. p. 201
115. F. S. Carey, *Notes on the division of the circle*, Quart. J. Math. **26** (1893), 322–371; FdM **25** (1893/94), 287; cf. p. 228
116. L. Carlitz, *A theorem of Stickelberger*, Math. Scand. **1** (1953), 82–84; Zbl 50.26701; MR 15, 13f; cf. p. 137

117. L. Carlitz, *A note on Gauss' first proof of the quadratic reciprocity theorem*, Proc. Amer. Math. Soc. **11** (1960), 563–565; Zbl 99.03003; MR 22 # 7997; cf. p. 416
118. L. Carlitz, *Some sums involving fractional parts*, Amer. Math. Monthly **82** (1975), 264–269; MR 51 #5465; cf. p. 23
119. P. Cartier, *Sur une généralization des symboles de Legendre-Jacobi*, Enseign. Math. **16** (1970), 31–48; Zbl 195.05802; cf. p. 23
120. J. W. S. Cassels, *A note on the diophantine equation $x^3 + y^3 + z^3 = 3$*, Math. Comp. **44** (1985), 265–266; Zbl 556.10007; MR 86d:11021; cf. p. 229
121. P. Castaldo, *I grafi e la legge di reciprocità dei numeri primi*, Archimede **28** (1976), 114–123; Zbl 365.10001; MR 58 # 27712; cf. p. 417
122. M. Cauchy, *Sur la théorie des nombres*, Bull. de Férussac **12** (1829), 205–221 Œuvres S. 2, II, 88–107; cf. p. 138, 270, 413
123. M. Cauchy, *Mémoire sur la theorie des nombres*, Œuvres de Cauchy, Sér. I, **III**, 5–??; cf. p. 202, 270, 313, 391
124. A. Cayley, *Eisenstein's geometrical proof of the fundamental theorem for quadratic residues*, Quart. Math. J. **1** (1857), 186–191; Collected Math. Papers III, 39–43
125. A. Cerone, *Sulle formole di moltiplicazione delle funzioni circolari e teorema di reciprocità pei residui quadratici*, Periodico di Mat. **31** (1916), 171–175; FdM **46** (1916–18), 1449; cf. p. 415
126. A. Cerone, *Sulla moltiplicazione e divisione dell'argomento nelle funzioni circolari*, Periodico di Mat. **32** (1917), 49–60; FdM **46** (1916–18), 1449; cf. p. 415
127. S. Chowla, *The cubic character of 2 mod p*, Proc. Lahore Phil. Soc. **6** (1944), 12; Zbl 63.00873; MR 7, 243i; ibid. **7** (1945); Zbl 63.00878; MR 7, 243h; cf. p. 225
128. S. Chowla, *A property of biquadratic residues*, Proc. Nat. Acad. Sci. India **14** (1944), 45–46; Zbl 63.00871; MR 7, 243f; cf. p. 201
129. S. Chowla, *An algebraic proof of the law of quadratic reciprocity*, Norske Vid. Selsk. Forh. (Trondheim), **39** (1966), 59; Zbl 149.28601; MR 34 # 7443; cf. p. 416
130. K. S. Chua, *A note on Gauss's lemma*, Bull. Number Theory Relat. Top. **15**, No. 1-3 (1991), 24–27; Zbl 836.11003; cf. p. 23
131. Y. Chuman, N. Ishii, *On the quartic residue of quadratic units of negative norm*, Math. Japonica **32** (1987), 389–420; Zbl 628.10004; MR 89a:11106; cf. p. 174
132. M. J. Collison, *The forgotten reciprocity laws*, Diss. New York Univ. 1976; cf. p. 25, 201, 225
133. M. J. Collison, *The origins of the cubic and biquadratic reciprocity laws*, Arch. History Exact Sci. **17** (1977), 63–69; Zbl 358.01011; MR 56, 52; cf. p. 25, 201, 274

134. G. Cooke, *Notes on an application of algebraic number theory to a study of cubic residues*, Lecture Notes 1972; cf. p. xv, 224, 228
135. G. Cooke, *Lectures on the power reciprocity laws of algebraic number theory*, Cornell University, 1974; cf. p. xv
136. A. E. Cooper, *A topical history of the theory of quadratic residues*, Diss. Chicago, 1926; cf. p. xiii, 23,
137. K. Corrádi, I. Kátai, *On the theory of quadratic residues* (Hungar.), Mat. Lapok **18** (1967), 75–81; Zbl 149.28504; MR 37 # 2669
138. M.J. Cowles, *A reciprocity law for cubic polynomials*, Diss. Pennsylv. State Univ. 1976
139. D. A. Cox, *Quadratic reciprocity: the conjecture and application*, Amer. Math. Monthly **95** (1988), 442–448; Zbl 652.10002; cf. p. 23
140. D. A. Cox, *Primes of the form $x^2 + ny^2$; Fermat, Class Field Theory, and Complex Multiplication*, Wiley 1989; Zbl 701.11001; MR 90m:11016; cf. p. 11, 275
141. R. Cuculière, *Histoire de la loi de réciprocité quadratique: Gauss et Tate*, Study group on ultrametric analysis 1979–1981 **36** (1979), 14 pp; Zbl 464.10002; MR 82k:01022; cf. p. 24
142. R. Cuculière, *Histoire d'un théorème d'Arithmetique: la loi de réciprocité*, Paris 1980, 71 pp.; cf. p. xv, 8, 24
143. A. Cunningham, *On 2 as a 16-ic residue*, Proc. London Math. Soc. (1), **27** (1895/96), 85–122; FdM **27** (1896), 146; cf. p. 310
144. A. Cunningham, *On 4-tic residuacity and reciprocity*, Proc. London Math. Soc. (2), **1** (1904), 132–150; FdM **34** (1903), 228; cf. p. 171
145. A. Cunningham, *On 8-vic, 16-ic, etc. residuacity*, Lond. Math. Soc. Proc. (2), **9** (1910), 1–14; FdM **41** (1910), 227–228; cf. p. 171
146. A. Cunningham, Th. Gosset, *4-tic & 3-bic residuacity tables*, Mess. Math. (2), **50** (1920), 1–30; FdM **47** (1919/20), 125; cf. p. 201, 226, 310
147. K. Dalen, *On a theorem of Stickelberger*, Math. Scand. **3** (1955), 124–126; Zbl 68.03102; MR 17, 130a; cf. p. 137
148. V. Dantscher, *Bemerkungen zum analytischen Beweise des cubischen Reciprocitätsgesetzes*, Math. Ann. **12** (1877), 241–253; FdM **9** (1877), 132–133; cf. p. 225, 277
149. R. Dedekind, *Sur la théorie des nombres entiers algébriques*, Darboux Bull. **XI** (1877); Engl. Transl. by John Stillwell: *Theory of algebraic integers*, Cambridge Univ. Press. 1996; cf. p. 95, 414
150. R. Dedekind, *Réponse à une remarque de M. Sylvester concernant les "Leçons sur la théorie des nombres de Dirichlet"*, C. R. Acad. Sci. Paris **91** (1880), 154–156; Werke I, 236–237; FdM **12** (1880), 124–125; cf. p. 26
151. R. Dedekind, *Über die Anzahl der Idealklassen in reinen kubischen Zahlkörpern*, J. Reine Angew. Math. **121** (1900), 40–123; FdM **30** (1899), 198–200; Ges. Math. Werke II, 148–233; cf. p. 225, 228

152. R. Dedekind, *Über den Zellerschen Beweis des quadratischen Reziprozitätsgesetzes*, H. Weber Festschrift Leipzig (1912), 23–36; Ges. Math. Werke II, 340–353; FdM **43** (1912), 246; cf. p. 415
153. R. Dedekind, *Schreiben an Herrn Borchardt über die Theorie der elliptischen Modulfunktionen*, J. Reine Angew. Math. **83** (1877), 265–292; Ges. Werke I, 174–201; cf. p. 414
154. A. Delcour, *Dernier théorème de Fermat et réciprocité quadratique*, Bull Soc. Math. Belg., Ser. B **34** (1982), 205–232; Zbl 496.10007; MR 84d:10021; cf. p. 396
155. P. Delezoide, *La loi de réciprocité quadratique*, Revue de Math. Spéciales **7** (1989/90), 339–342; cf. p. 23
156. J.B. Dence, T.P. Dence, *Cubic and quartic residues modulo a prime*, Missouri J. Math. Sci. **7** (1995), 24–31; MR 95m:11009
157. L.E. Dickson, *Historical note on the proof of the quadratic reciprocity law in a posthumous paper by Gauß*, Bull. Amer. Math. Soc. (2) **21** (1915), 441; FdM **45** (1914/15), 3334; cf. p. 21
158. L.E. Dickson, *Cyclotomy, higher congruences, and Waring's problem. I*, Amer. J. Math. **57** (1935), 391–424; FdM **61** (1935), 175; Zbl 12.01203; cf. p. 225, 343
159. E. Dintzl, *Ueber den zweiten Ergänzungssatz des biquadratischen Reciprocitätsgesetzes*, Monatsh. Math. **10** (1899), 88–96; FdM **30** (1899), 184; RSPM **7** (1899), 128; cf. p. 201
160. E. Dintzl, *Der zweite Ergänzungssatz des kubischen Reciprocitätsgesetzes*, Monatsh. Math. **10** (1899), 303–306; FdM **30** (1899), 184–185; cf. p. 225
161. E. Dintzl, *Über die Legendreschen Symbole für quadratische Reste in einem imaginären quadratischen Zahlkörper mit der Klassenzahl 1*, Wiener Ber. **116** (1907), 785–800; FdM **38** (1907), 254; cf. p. 277
162. L. Dirichlet, *Recherches sur les diviseurs premiers d'une classe de formules du quatrième degré*, J. Reine Angew. Math. **3** (1828), 35-69; Werke I, 63–69; cf. p. 24, 171
163. L. Dirichlet, *Letter to Gauss*, 8. 4. 1828, Werke II, 376–378; cf. p. 174
164. L. Dirichlet, *Démonstration nouvelles de quelques théorèmes relatifs aux nombres*, J. Reine Angew. Math. **3** (1828), 390–393; Werke I, 99–104; cf. p. 4
165. L. Dirichlet, *Über eine neue Anwendung bestimmter Integrale auf die Summation endlicher oder unendlicher Reihen*, Abh. Preuss. Akad. Wiss. 1835, 649–664; Werke I, 237–256; cf. p. 102, 413
166. L. Dirichlet, *Démonstration d'une propriété analogue à la loi de Réciprocité qui existe entre deux nombres premiers quelconques*, J. Reine Angew. Math. **9** (1832), 379–389; Werke I, 173–188; cf. p. 154, 170
167. L. Dirichlet, *Untersuchungen über die Theorie der quadratischen Formen*, Abh. Königl. Preuss. Akad. Wiss. 1833, 101–121; Werke I, 195–218; cf. p. 171

168. L. Dirichlet, *Recherches sur diverses applications de l'analyse infinitésimale à la théorie des nombres*, J. Reine Angew. Math. **19** (1839), 324–369; ibid. **21** (1840), 1–12, 134–155; Werke I, 411–496

169. L. Dirichlet, *Recherches sur les formes quadratiques à coefficients et à indéterminées complexes*, J. Reine Angew. Math. **24** (1842), 291–371; Werke I, 533–618; cf. p. 170

170. L. Dirichlet, *Über den ersten der von Gauss gegebenen Beweise des Reciprocitätsgesetzes in der Theorie der quadratischen Reste*, J. Reine Angew. Math. **47** (1854), 139–150; Werke II, 121–138; cf. p. 21, 413

171. L. Dirichlet, *Sur la première demonstration donnée par Gauss de la loi de réciprocité dans la théorie des residus quadratiques*, French transl. of [170]; J. Math. pure appl. (II), **4** (1854), 401–420; cf. p. 21

172. L. Dirichlet, *Über den biquadratischen Charakter der Zahl "Zwei"*, J. Reine Angew. Math. **57** (1860), 187–188; Werke II, 261–262; cf. p. 174

173. L. Dirichlet, *Sur le caractère biquadratique du nombre 2; extrait d'une lettre adressée à M. Stern*, French transl. of [172]; J. Math. pure appl. II, **4** (1860), 367–368; cf. p. 174

174. L. Dirichlet, R. Dedekind, *Vorlesungen über Zahlentheorie*; Braunschweig 1863; 2nd ed. 1871, 3rd ed. 1879/80, 4th ed. 1894; Ital. transl. 1881; English transl. AMS 1999; cf. p. 22, 26, 69, 70, 413

175. A. Dirks, *Übersicht über die Beweise des quadratischen Reziprozitätsgesetzes*, Hausarbeit Göttingen, 1971; cf. p. 24

176. Dittmar, *Zur Theorie der Reste, insbesondere derer vom dritten Grade, nebst einer Tafel der cubischen Reste aller Primzahlen von der Form $6n + 1$ zwischen den Grenzen 1 bis 100*, Pr. Berlin 1873; FdM **5** (1873), 100; cf. p. 225

177. N. R. C. Dockeray, *The law of quadratic reciprocity*, Math. Gaz. **22** (1938), 440–453; Zbl 19.39307; cf. p. 416

178. K. Dörge, *Beweis des Reziprozitätsgesetzes für quadratische Reste*, Mathem. Ann. **118** (1942), 310–311; FdM **68 - I** (1942), 65; Zbl 26.20104; MR 5, 91b; cf. p. 416

179. H. Dörrie, *Das quadratische Reciprocitätsgesetz in quadratischen Zahlkörpern mit der Classenzahl 1*, Diss. Göttingen, 1898, 75 pp.; FdM **30** (1899), 201; cf. p. 170, 226

180. R. Dressler, E. E. Shult, *A simple proof of the Zolotareff-Frobenius theorem*, Proc. Amer. Math. Soc. **54** (1976), 53–54; Zbl 317.10005; MR 52 # 10563; cf. p. 24

181. H.M. Edwards, *Kummer, Eisenstein, and higher reciprocity laws*, Number theory related to Fermat's last theorem, Progress in Math. **26** (ed.: N. Koblitz), 1983, 31–43; Zbl 498.12003; MR 85c:01023; cf. p. 274, 343, 396

182. H.M. Edwards, *Euler and quadratic reciprocity*, Math. Mag. **56** (1983), 285–291; Zbl 523.01010; MR 86a:01017; cf. p. 13

183. S. Eichenberg, *Über das quadratische Reciprocitätsgesetz und einige quadratische Zerfällungen der Primzahlen*, Diss. Gött., 1886; cf. p. 23
184. M. Eichler, *The quadratic reciprocity law and the elementary theta function*, Glasgow J. Math. **27** (1985), 19–30; Zbl 581.10011; MR 87f:11004; cf. p. 417
185. G. Eisenstein, *Beweis des Reciprocitätssatzes für die cubischen Reste in der Theorie der aus den dritten Wurzeln der Einheit zusammengesetzten Zahlen*, J. Reine Angew. Math. **27** (1844), 289–310; Math. Werke I, 59–80; cf. p. 225
186. G. Eisenstein, *Einfacher Algorithmus zur Bestimmung des Wertes von $(\frac{a}{b})$*, J. Reine Angew. Math. **27** (1844), 317–318; Math. Werke I, 95–96; cf. p. 26
187. G. Eisenstein, *Neuer und elementarer Beweis des Legendre'schen Reciprocitäts-Gesetzes*, J. Reine Angew. Math. **27** (1844), 322–329; Math. Werke I, 100–107; cf. p. 413
188. G. Eisenstein, *Nachtrag zum cubischen Reciprocitätssatze für die aus den dritten Wurzeln der Einheit zusammengesetzten Zahlen, Criterien des cubischen Characters der Zahl 3 und ihrer Teiler*, J. Reine Angew. Math. **28** (1844), 28–35; Math. Werke I, 81–88; cf. p. 225, 229
189. G. Eisenstein, *La loi de réciprocité tirée des formules de Mr. Gauss, sans avoir déterminée préalablement la signe du radical*, J. Reine Angew. Math. **28** (1844), 41–43; Math. Werke I, 114–116; cf. p. 413
190. G. Eisenstein, *Lois de réciprocité*, J. Reine Angew. Math. **28** (1844), 53–67; Math. Werke I, 126–140; cf. p. 201,
191. G. Eisenstein, *Einfacher Beweis und Verallgemeinerung des Fundamentaltheorems für die biquadratischen Reste*, J. Reine Angew. Math. **28** (1844), 223–245; Math. Werke I, 141–163; cf. p. 201
192. G. Eisenstein, *Geometrischer Beweis des Fundamentaltheorems für die quadratischen Reste*, J. Reine Angew. Math. **28** (1844), 246–248; Math. Werke I, 164–166; Engl. Transl. Quart. J. Math. **1** (1857), 186–191, or A. Cayley: Coll. Math. Papers III, 39–43; cf. p. 31, 413
193. G. Eisenstein, *Allgemeine Untersuchungen über die Formen dritten Grades mit drei Variablen, welche der Kreistheilung ihre Entstehung verdanken*, J. Reine Angew. Math. **28** (1844), 289–374; cf. p. 226
194. G. Eisenstein, *Allgemeine Untersuchungen über die Formen dritten Grades mit drei Variablen, welche der Kreistheilung ihre Entstehung verdanken*, J. Reine Angew. Math. **29** (1845), 19–53; cf. p. 226
195. G. Eisenstein, *Letter to Gauss*, 25.01.1845 Math. Werke II, 825–831; cf. p. 275
196. G. Eisenstein, *Letter to Gauss*, 18.02.1845, Math. Werke II, 832–837; cf. p. 275
197. G. Eisenstein, *Application de l'algèbre à l'arithmétique transcendante*, J. Reine Angew. Math. **29** (1845), 177–184; Math. Werke I, 291–298; cf. p. 201, 275, 413

198. G. Eisenstein, *Beiträge zur Theorie der elliptischen Funktionen I: Ableitung des biquadratischen Fundamentaltheorems aus der Theorie der Lemniskatenfunctionen, nebst Bemerkungen zu den Multiplications- und Transformationsformeln*, J. Reine Angew. Math. **30** (1846), 185–210; Math. Werke I, 299–324; cf. p. 201, 277, 413
199. G. Eisenstein, *Beiträge zur Theorie der elliptischen Funktionen VI, Genaue Untersuchung der unendlichen Doppelprodukte, aus welchen die elliptischen Functionen als Quotienten zusammengesetzt sind*, J. Reine Angew. Math. **35** (1847), 153–274; Math. Werke I, 457–478; cf. p. 217
200. G. Eisenstein, *Letter to Gauss*, 18.08.1847, Math. Werke II, 845–855; cf. p. 313
201. G. Eisenstein, *Letter to Gauss*, 09.03.1848, Math. Werke II, 856–859; cf. p. 299, 313
202. G. Eisenstein, *Über einige allgemeine Eigenschaften der Gleichung, von welcher die Teilung der ganzen Lemniskate abhängt, nebst Anwendungen derselben auf die Zahlentheorie*, J. Reine Angew. Math. **39** (1850), 224–287; Math. Werke II, 556–619; cf. p. 254, 281, 300
203. G. Eisenstein, *Über ein einfaches Mittel zur Auffindung der höheren Reciprocitätsgesetze und der mit ihnen zu verbindenen Ergänzungssätze*, J. Reine Angew. Math. **39** (1850), 351–364; Math. Werke II, 623–636; cf. p. 274
204. G. Eisenstein, *Beweis der allgemeinsten Reciprocitätsgesetze zwischen reellen und complexen Zahlen*, J. Reine Angew. Math. **39** (1850), 712–721; Math. Werke II, 189–198; cf. p. 392
205. J. S. Ely, *A geometric approach to the quadratic reciprocity law*, Comm. Algebra **12** (1984), 1533–1544; Zbl 539.10005; MR 86f:11008; cf. p. 30, 417
206. P. Epstein, *Theorie der Potenzreste für zusammengesetzte Moduln*, Archiv Math. Phys. (III) **12** (1907), 134–150; FdM **38** (1907), 237; cf. p. 136
207. P. Epstein, *Zur Auflösbarkeit der Gleichung $x^2 - Dy^2 = -1$*, J. Reine Angew. Math. **171** (1934), 243–252; FdM **60**-I (1934), 119; cf. p. 176, 180
208. E. Esrafilian, M. Sangani-Monfared, *Reciprocity laws from Euler to Langlands*, Proc. 3rd Intern. Coll. on numerical analysis, SCT Publishing, 39-48 (1995); Zbl 912.11001; cf. p. 25
209. D. R. Estes, G. Pall, *Spinor genera of binary quadratic forms*, J. Number Theory **5** (1973), 421–432; Zbl 268.10010; MR 48 # 10979; cf. p. 172
210. D. R. Estes, G. Pall, *A reconsideration of Legendre-Jacobi-Symbols*, J. Number Theory **5** (1973), 433–434; Zbl 268.10004; MR 48 # 10960; cf. p. 26
211. L. Euler, *Theoremata circa divisores numerorum in hac forma $pa^2 \pm qb^2$ contentorum*, Comm. Acad. Sc. Petersburg **14** (1744/46), 151–181; Opera Omnia I-2, p. 194–222; cf. p. 4, 70

212. L. Euler, *Theorematum quorundam ad numeros primos spectantium demonstratio*, Novi Comm. Acad. Sci. Petropol. **8** (1736), 1741, 141–146 Opera Omnia I–2, p. 33–37; cf. p. 4
213. L. Euler, *Theoremata circa divisores numerorum*, Opera Omnia I–2, p. 62–85; cf. p. 4
214. L. Euler, *Theoremata circa residua ex divisione potestatum relicta*, Opera Omnia I–2, p. 493–518; cf. p. 4
215. L. Euler, *Observationes circa divisionem quadratorum per numeros primes*, Opera Omnia **I - 3** (1783), 477–512, in particular p. 511, 512; cf. p. 4
216. L. Euler, *De criteriis aequationis $fxx + gyy = hzz$ utrum ea resolutionem admittet necne*, Opuscula anal. **1** (1783), p. 211–241; Opera Omnia **I - 3**, 1–24; cf. p. 5
217. L. Euler, *Tractatus de numerorum doctrina capita sedecim quae supersunt*, Comment. Arithm. **2** (1849), 503–575; Opera Omnia V, 182–283; cf. p. 13, 229
218. L. Euler, *Leonhard Euler und Christian Goldbach: Briefwechsel 1729 – 1764* (eds.: A. P. Juskevic, E. Winter), Abh. Deutsche Akad. Wiss. Berlin, Akademie-Verlag 1965; cf. p. 3
219. L. Euler, *Letter to Goldbach, 28. 8. 1742*, see [218]; cf. p. 4
220. R. Evans, *Unambiguous evaluations of bidecic Jacobi and Jacobsthal sums*, J. Aust. Math. Soc. **28** (1979), 235–240; Zbl 417.10034; MR 81c:10042; cf. p. 141
221. R. Evans, *The cyclotomic numbers of order sixteen*, Math. Comp. **146** (1979), 827–835; Zbl 403.10003; MR 80b:10053; cf. p. 343
222. R. Evans, *The 2^r-th power character of 2*, J. Reine Angew. Math. **315** (1980), 174–189; Zbl 419.10003; MR 81f:10006; cf. p. 310
223. R. Evans, *Rational reciprocity laws*, Acta Arithmetica **39** (1981), 281–294; Zbl 472.10006; MR 83h:10006; cf. p. 173, 310
224. R. Evans, *The octic period polynomial*, Proc. Amer. Math. Soc. **87** (1983), 389–393; Zbl 503.10025; MR 84b:10055; cf. p. 313
225. R. Evans, *The octic and bioctic character of certain quadratic units*, Utilitas Math. **25** (1984), 153–157; Zbl 558.10003; MR 86j:11109; cf. p. 174, 343
226. R. Evans, *Residuacity of primes*, Rocky Mt. J. Math. **19** (1989), 1069–1081; Zbl 699.10012; MR 90m:11008; cf. p. 299
227. D. K. Faddeev, *Zum neunten Hilbertschen Problem*, in: Die Hilbertschen Probleme, Ostwalds Klassiker der exakten Wissenschaften 252, Akad. Verlagsges., Leipzig 1979, p. 164–176; Russian original: Moscow 1969; MR 40 # 5573; cf. p. viii
228. E. T. Federighi, R. G. Roll, *A letter to the Editor*, Fibonacci Quart. **4** (1966), 85–88; cf. p. 70
229. U. Felgner, *Reciprocitätsgesetze*, 1983, unpublished lecture notes; cf. p. 201, 226

230. U. Felgner, *On Bachet's diophantine equation $x^3 = y^2 + k$*, Monatsh. Math. **98** (1984), 185–191; Zbl 545.10010; MR 86c:11015; cf. p. 230
231. D.D. Fenster, *Why Dickson Left Quadratic Reciprocity out of his History of the Theory of Numbers*, Amer. Math. Monthly Aug.-Sept. (1999); cf. p. xiii
232. P. de Fermat, *Œuvres*; cf. p. 1, 3, 12
233. J. C. Fields, *A simple statement of proof of reciprocal theorem*, American J. Math. **13** (1891), 189–190; FdM **22** (1890), 207; cf. p. 414
234. E. Fischer, *Ueber Eisenstein's Beweis des quadratischen Reciprocitätsgesetzes*, Monatsh. f. Math. **11** (1900), 176–182; FdM **31** (1900), 189–190; cf. p. 415
235. J. S. Frame, *A short proof of quadratic reciprocity*, Amer. Math. Monthly **85** (1978), 818–819; Zbl 401.10006; cf. p. 417
236. F. Franklin, *A proof of the theorem of reciprocity for quadratic residues*, Mess. Math. (2), **19** (1890), 176–177; FdM **22** (1890), 207; cf. p. 414
237. G. Frattini, *Carattere quadratico di 2 come conseguenza prossima del teorema di Wilson*, Rom. Ist. Tecn. Ann. **8** (1883), 89–94; FdM **15** (1883), 152; cf. p. 25
238. G. Frei, *The reciprocity law from Euler to Eisenstein*, in: The intersection of history and mathematics; Papers presented at the history of mathematics symposium, held in Tokyo, Japan, August 31 - September 1, 1990. Basel: Birkhäuser, Sci. Networks, Hist. Stud. **15** (1994), 67–90; Zbl 818.01002; MR 95k:01016; cf. p. 200, 226
239. G. Frei (ed.), *Die Briefe von E. Artin an H. Hasse, 1923 – 1953*, Univ. Laval, Québec, and ETH Zürich, Preprint 1981; cf. p. 200, 204
240. J. B. Friedlander, K. H. Rosen, *Refinements of a congruence of Gauss*, Elemente Math. **32** (1977), 62–64; Zbl 355.10004; MR 56 # 8472; cf. p. 23
241. C. Friesen, B. K. Spearman, K. S. Williams, *Another proof of Eisenstein's law of cubic reciprocity and its supplement*, Rocky Mt. J. Math. **16** (1986), 395–402; Zbl 601.10002; MR 87g:11011; cf. p. 225
242. G. Frobenius, *Über das quadratische Reziprozitätsgesetz I*, Sitzungsberichte Berliner Akad. (1914), 335–349; FdM **44** (1913), 225; Ges. Abhandl., 628–642; cf. p. 21, 415
243. G. Frobenius, *Über das quadratische Reziprozitätsgesetz II*, Sitzungsberichte Berliner Akad. (1914), 484–488; FdM **45** (1914/15), 308; Ges. Abhandl., 643–647; cf. p. 415
244. A. Fröhlich, *The restricted biquadratic residue symbol*, Proc. London Math. Soc. (3), **9** (1959), 189–207; Zbl 87.03301; MR 21 # 5616; cf. p. 173
245. A. Fröhlich, *On fields of class two*, Proc. London Math. Soc. (3) **4** (1954), 235–256; Zbl 55.03301; MR 16, 116b; cf. p. 416
246. R. Fueter, *Synthetische Zahlentheorie*, 1916; FdM **46** (1916–18), 244–245; 3rd edition de Gruyter 1950; MR 13, 207c; cf. p. 101, 226

247. R. Fueter, *Reziprozitätsgesetze in quadratisch-imaginären Körpern I*, Gött. Nachr. (1927), 336–346; FdM **54** (1928), 192–193; cf. p. 277
248. R. Fueter, *Reziprozitätsgesetze in quadratisch-imaginären Körpern II*, Gött. Nachr. (1927), 427–445; FdM **54** (1928), 192–193; cf. p. 277
249. R. Fueter, *Les lois de réciprocité dans un corps quadratique imaginaire*, Enseignement Math. **26** (1928), 316–317; FdM **54** (1928), 193; cf. p. 277
250. R. Fueter, *Die Klassenkörper der komplexen Multiplikation und ihr Einfluß auf die Entwicklung der Zahlentheorie*, Jahresber. DMV **20** (1911), 1–47; FdM **42** (1911), 228; cf. p. 277
251. R. Fueter, *Zur Theorie der Brandtschen Quaternionenalgebren*, Math. Ann. **110** (1935), 650–661; Zbl 10.29104; cf. p. 416
252. Ph. Furtwängler, *Über die Reziprozitätsgesetze zwischen l-ten Potenzresten in algebraischen Zahlkörpern, wenn l eine ungerade Primzahl bedeutet*, Math. Ann. **58** (1904), 1–50; FdM **34** (1903), 236; cf. p. 136
253. Ph. Furtwängler, *Die Reziprozitätsgesetze für Potenzreste mit Primzahlexponenten in algebraischen Zahlkörpern I*, Math. Ann. **67** (1909), 1–31; FdM **40** (1909), 265–266; cf. p. 357
254. Ph. Furtwängler, *Letzter Fermatscher Satz und Eisensteinsches Reziprozitätsprinzip*, Wiener Akad. Ber., Abt. IIa **121** (1912), 589–592; FdM **43** (1912), 272; cf. p. 396
255. Ph. Furtwängler, *Über die Reziprozitätsgesetze für Primzahlpotenzexponenten*, J. Reine Angew. Math. **157** (1926), 15–25; FdM **52** (1926), 156; cf. p. 393
256. F. Furquim de Almeida, *The law of quadratic reciprocity*, Bol. Soc. Mat. Sao Paulo **3** (1951), 3–8; MR 13 # 437d; cf. p. 416
257. Y. Furuta, *A reciprocity law of the power reciprocity symbol*, J. Math. Soc. Japan **10** (1958), 46–54; Zbl 83.26002; MR 21 # 2646; cf. p. 310
258. Y. Furuta, *Norms of units of quadratic fields*, J. Math. Soc. Japan **11** (1959), 139–145; Zbl 202.33004; MR 22 # 3716; cf. p. 174, 177
259. Y. Furuta, P. Kaplan, *On quadratic and quartic characters of quadratic units*, Sci. Rep. Kanazawa Univ. **26** (1981), 27-30; Zbl 487.12005; MR 83j:12003; cf. p. 174
260. Y. Furuta, T. Kubota, *Central extensions and rational quadratic forms*, Nagoya Math. J. **130** (1993), 177–182; Zbl 771.11040; MR 94j:11107; cf. p. 175
261. P. Garrett, *Quadratic reciprocity (following Weil?)*, online manuscript; URL: http://www.math.umn.edu/~garrett/index.shtml; cf. p. 417
262. C.F. Gauss, *Ostwald's Klassiker der exakten Wissenschaften* **122**, *Sechs Beweise des Fundamentaltheorems über quadratische Reste von Carl Friedrich Gauss*, (E. Netto, ed.) Leipzig, 1901; FdM **32** (1901), 193; cf. p. 24
263. C.F. Gauss, *Disquisitiones Arithmeticae*, Werke I; French transl. 1807; Paris 1910, FdM **42** (1911), 236; 1953, Zbl 51.03003; German transl. Springer 1889, reprint Chelsea 1965, MR 32 #5488; Russian transl.

Moscow 1959, MR 23 #A2352; English transl. New Haven, London 1966, Zbl 136.32301; reprint Springer Verlag 1986, Zbl 585.10001; Spanish transl. Bogota 1995, Zbl 899.01034; cf. p. 6, 12, 14, 22, 100, 101, 137, 223, 227, 268, 342

264. C.F. Gauss, *1st Proof*, Disquisitiones arithmeticae, Art. 125–145 (1801), 94–145; Werke I, p. 73–111; cf. p. 413
265. C.F. Gauss, *2nd Proof*, Disquisitiones arithmeticae, Art. 262 (1801), 262; cf. p. 21, 413
266. C.F. Gauss, *(3rd Proof) Theorematis arithmetici demonstratio nova*, Comment. Soc. regiae sci. Göttingen **XVI** (1808), 69; Werke II, p. 1–8; cf. p. 21, 23, 413
267. C.F. Gauss, *(4th Proof) Summatio serierum quarundam singularium*, Comment. Soc. regiae sci. Göttingen 1811; Werke II, p. 9–45; cf. p. 21, 137, 413
268. C.F. Gauss, *(5th Proof) Theorematis fundamentalis in doctrina de residuis quadraticis demonstrationes et amplicationes novae*, 1818; Werke II, 47–64, in particular p. 51; cf. p. 21, 200, 413
269. C.F. Gauss, *(6th Proof) Theorematis fundamentalis in doctrina de residuis quadraticis demonstrationes et amplicationes novae*, 1818; Werke II, 47–64, in particular p. 55; cf. 21, 413
270. C.F. Gauss, *7th Proof*, Werke II, p. 233; cf. p. 413
271. C.F. Gauss, *8th Proof*, Werke II, p. 234; cf. p. 413
272. C.F. Gauss, *Theoria residuorum biquadraticorum, Commentatio prima*, Comment. Soc. regiae sci. Göttingen **6** (1828), 27 ff; Werke II, 65–92; cf. p. 15, 101, 154, 170, 174, 202, 200
273. C.F. Gauss, *Theoria residuorum biquadraticorum, Commentatio secunda*, Comment. Soc. regiae sci. Göttingen **7** (1832), 93–148; Werke II, 93–148; cf. p. 15, 203, 200
274. C.F. Gauss, *Zwei Notizen über die Auflösung der Kongruenz $xx + yy + zz \equiv 0 \pmod{p}$*, Nachlaß, Werke **VIII** (1900), 3–4; cf. p. 342
275. C.F. Gauss, *Notizen über cubische und biquadratische Reste*, Nachlaß, Werke **VIII** (1900), 5–14; cf. p. 224
276. C.F. Gauss, *Zur Theorie der kubischen Reste*, Nachlaß, Werke **VIII** (1900), 15–19; cf. p. 225
277. C.F. Gauss, *Fragmente zur Theorie der aus einer Cubikwurzel zu bildenden ganzen algebraischen Zahlen*, Werke **VIII** (1900), 20ff; cf. p. 226
278. C.F. Gauss, *Zum Reziprozitätsgesetz der quadratischen und der biquadratischen Reste*, Werke **X, 1**, 53–55; cf. p. 199, 204
279. C.F. Gauss, *Hauptmomente des Beweises für die biquadratischen Reste*, Werke **X, 1**, 56–57; cf. p. 201
280. C.F. Gauss, *Beweis des Reziprozitätssatzes für die biquadratischen Reste, der auf die Kreisteilung gegründet ist*, Werke **X, 1**, 65–69; cf. p. 200

281. C.F. Gauss, *Gauss an Sophie Germain*, letter from April 30, 1807; Werke **X, 1**, 70–74; cf. p. 199
282. P. Gazzaniga, *Sui residui ordine qualunque rispetto i moduli primi*, Ven. Ist. Att. (6), **10** (1886), 1271–1280; FdM **18** (1886), 145–146; cf. p. 23
283. L. Gegenbauer, *Ueber das cubische Reciprocitätsgesetz*, Wiener Ber. **81** (1880), 1–5; FdM **12** (1880), 124; cf. p. 225
284. L. Gegenbauer, *Algorithmen zur Bestimmung des verallgemeinerten Legendre'schen Symbols*, Wiener Ber. **82** (1880), 931; FdM **12** (1880), 125; cf. p. 26
285. L. Gegenbauer, *Quadratisches Reciprocitätsgesetz*, Wiener Ber. **90** (1884), 10 pp.; cf. p. 414
286. L. Gegenbauer, *Über das verallgemeinerte Legendre'sche Symbol*, Wiener Ber. **91** (1885), 1089–1101; FdM **13** (1881), 138; cf. p. 26
287. L. Gegenbauer, *Über das Legendre-Jacobi'sche Symbol*, Wiener Berichte **91** (1885), 11–33; FdM **17** (1885), 151; cf. p. 26
288. L. Gegenbauer, *Über das Symbol (m/n)*, Wiener Ber. **92** (1885), 876–892; FdM **17** (1885), 151; cf. p. 23
289. L. Gegenbauer, *Note über das quadratische Reciprocitätsgesetz*, Wiener Ber. **97** (1888), 427–431; FdM **20** (1888), 179
290. L. Gegenbauer, *Note über das Legendre-Jacobi'sche Symbol*, Wiener Ber. **100** (1891), 855–864; FdM **23** (1891), 188; cf. p. 26, 414
291. L. Gegenbauer, *Über den quadratischen Restcharakter*, Wiener Ber. **100** (1891), 1072–1087; FdM **23** (1891), 188; cf. p. 26
292. L. Gegenbauer, *Beweis des quadratischen Reciprocitätsgesetzes*, Monatsh. f. Math. **4** (1893), 190–192; FdM **25** (1893/94), 281; cf. p. 414
293. L. Gegenbauer, *Einige Bemerkungen zum quadratischen Reciprocitätsgesetz*, Wiener Ber. **103** (1894), 285–294; FdM **25** (1893/94), 281; cf. p. 414
294. L. Gegenbauer, *Zur Theorie der biquadratischen Reste*, Akad. Versl. Amsterdam **10** (1901), 195–207; FdM **32** (1901), 193; cf. p. 201
295. A. Genocchi, *Note sur la théorie des residus quadratiques*, Mém. cour. et mém. des savants étrangers Acad. Roy Sci. Lettres Belgique **25** (1851/53), 54 pp; cf. p. 413
296. A. Genocchi, *Sulla formula sommatoria di Eulero, e sulla teoria de residui quadratici*, Ann. Sci. Mat. Fis. **3** (1852), 406–436; cf. p. 102
297. A. Genocchi, *Sur la loi de réciprocité de Legendre étendue aux nombres non premiers*, C. R. Acad. Sci. Paris **90** (1880), 300–302; FdM **12** (1880), 123–124; cf. p. 414
298. A. Genocchi, *Teoremi di Sofia Germain intorno ai residui biquadratici*, Bull. bibliogr. storia sci. mat. fis. **17** (1884), 248–252; FdM **16** (1884), 163; cf. p. 201
299. A. Genocchi, *Ancora un cenno dei residui cubici e biquadratici*, Bull. bibliogr. storia sci. mat. fis. **18** (1885), 231–234; FdM **17** (1885), 151; cf. p. 201, 225

300. A. Genocchi, *Sur la loi de réciprocité de Legendre étendue aux nombres non premiers*, Bull. bibliogr. storia sci. mat. fis. **18** (1885), 235–237 (reprint of [297]); FdM **17** (1885), 152; cf. p. 414
301. A. Genocchi, *Sur quelques théorèmes qui peuvent conduire à la loi de réciprocité de Legendre*, Bull. bibliogr. storia sci. mat. fis. **18** (1885), 238–243; FdM **17** (1885), 152
302. A. Genocchi, *Remarques sur une demonstration de la loi de réciprocité*, C. R. Acad. Sci. Paris **101** (1885), 425–427; FdM **17** (1885), 152–153
303. A. Genocchi, *Intorno all' ampliazione d'un lemma del Gauss*, Bull. bibliogr. storia sci. mat. fis. **18** (1885), 650–651; FdM **18** (1886), 26; cf. p. 23
304. A. Genocchi, *Première partie du chapitre XIII de la "Note sur la théorie des résidus quadratiques"*, J. Reine Angew. Math. **104** (1889), 345–347; reprint of [295]; FdM **21** (1889), 177
305. M. Gerstenhaber, *The 152nd proof of the law of quadratic reciprocity*, Am. Math. Month. **70** (1963), 397–398; Zbl 113.26705; MR 27 # 100; cf. p. 416
306. M. Gerstenhaber, *Email from 7 Apr 1995*; cf. p. v
307. R.E. Giudici, *Residui quadratici in un campo de Galois*, Atti Accad. Naz. Lincei Rend. Cl. Sci. Fis. Mat. Natur. (8), **52** (1972), 461–466; Zbl 253.12020; MR 48 # 8455; cf. p. 136
308. R.E. Giudici, *Quadratic residues in $GF(p^2)$*, Math. Mag. **44** (1971), 153–157; Zbl 214.30201; MR 51 # 5563; cf. p. 136
309. H. Glause, *Geometrischer Beweis der Ergänzungssätze zum bikubischen Reziprozitätsgesetz*, Diss. Kiel, 1918; cf. p. 226
310. P.J.A. Gmeiner, *Eine neue Darstellung des biquadratischen Charakters*, Wiener Ber. **100** (1891), 1093–1100; FdM **23** (1891), 189; cf. p. 137
311. P.J.A. Gmeiner, *Die Ergänzungssätze zum bikubischen Reciprocitätsgesetz*, Wiener Ber. **100** (1891), 1330–1361; FdM **23** (1891), 189–190; cf. p. 226
312. P.J.A. Gmeiner, *Die bicubische Reciprocität zwischen einer reellen und einer zweigliedrigen regulären Zahl*, Monatsheft Math. Phys. **3** (1892), 179–192; FdM **24** (1892), 180; cf. p. 226
313. P.J.A. Gmeiner, *Die bicubische Reciprocität zwischen einer reellen und einer zweigliedrigen regulären Zahl*, Monatsheft Math. Phys. **3** (1892), 199–210; FdM **24** (1892), 180; cf. p. 226
314. P.J.A. Gmeiner, *Das allgemeine bicubische Reciprocitätsgesetz*, Wiener Ber. **101** (1892), 562–584; FdM **24** (1892), 180; cf. p. 226
315. F. Goldscheider, *Das Reciprocitätsgesetz der achten Potenzreste*, Wissensch. Beil. z. Progr. d. Luisenstädt. Realgymn. Berlin 1889, 30 pp; FdM **21** (1889), 178; cf. p. 173, 202, 205, 310
316. D. M. Goldschmidt, *Some amazing properties of the function $f(x) = x^2$*, Math. Medley **9** (1981), 29–36; Zbl 498.10003; MR 83k:10006; cf. p. 13, 23, 417

317. T. Gosset, *On the quartic residuacity of* $1+i$, Mess. Math. (2), **40** (1910), 165–169; FdM **42** (1911), 213–214; cf. p. 202
318. T. Gosset, *On the law of quartic reciprocity*, Mess. Math. (2), **41** (1911), 65–90; FdM **42** (1911), 212–213; cf. p. 172, 201
319. L. v. Grosschmid, *Bemerkungen über den quadratischen Charakter der Zahl* 2 (Hungar.), Math. és phys. lapok **27** (1918), 80–90; FdM **46** (1916–18), 217–218; cf. p. 25
320. H. Gupta, *The Legendre and Jacobi-Symbols for k-ic residues*, J. Number Theory **4** (1972), 219–222; Zbl 231.10005; MR 45 # 6745
321. W. Habicht, *Ein elementarer Beweis des kubischen Reciprocitätsgesetzes*, Math. Ann. **139** (1960), 343–365; Zbl 104.26501; MR 22 # 4698; cf. p. 225
322. J. Hacks, *Schering's Beweis des Reciprocitäts-Satzes für die quadratischen Reste dargestellt mit Hülfe des Zeichens* $[x]$, Acta Math. **12** (1889), 109–111; FdM **20** (1888), 179; cf. p. 414
323. E. Halberstadt, *Sommes de Gauss, symbole de Jacobi*, Revue de Mathématiques Spéciales Oct. 1992, 137–143
324. Halphén, *Sur le caractère biquadratique du nombre* 2, C. R. Acad. Sci. Paris **66** (1868), 190–193; FdM **1** (1868), 48; cf. p. 174
325. F. Halter-Koch, *Kriterien zum 8. Potenzcharakter der Reste 3, 5 und 7*, Math. Nachr. **44** (1970), 129–144; Zbl 208.06201; MR 43 # 1916; cf. p. 141, 310
326. F. Halter-Koch, *Zerlegungsgesetze und binäre quadratische Formen*, Tagungsber. Oberwolfach **35** 1984; cf. p. 175
327. F. Halter-Koch, *An Artin character and representations of primes by binary quadratic forms III*, Manuscr. Math. **51** (1985), 163–169; Zbl 568.12002; MR 86f:11077; cf. p. 175
328. F. Halter-Koch, *On the quartic character of certain quadratic units and the representation of primes by binary quadratic forms*, Rocky Mt. J. Math. **16** (1986), 95–102; Zbl 599.12003; MR 87f:11088; cf. p. 175
329. F. Halter-Koch, *Quadratische Einheiten als 8. Potenzreste*, Proc. Int. Conf. on class numbers and fundamental units, Katata, Japan (1986), 1–15; Zbl 617.12010; MR 88k:11070; cf. p. 174
330. F. Halter-Koch, *Binäre quadratische Formen und Diederkörper*, Acta Arith. **51** (1988), 141–172; Zbl 683.10021; MR 90g:11046; cf. p. 172, 175
331. F. Halter-Koch, *Representation of primes by binary quadratic forms of discriminant* $-256q$ *and* $-128q$, Glasgow Math. J. **35** (1993), 261–268; Zbl 783.11018; MR 94b:11034; cf. p. 175
332. F. Halter-Koch, N. Ishii, *Ring class fields modulo 8 of* $\mathbb{Q}(\sqrt{-m})$ *and the quartic character of units of* $\mathbb{Q}(\sqrt{m})$ *for* $m \equiv 1 \bmod 8$, Osaka J. Math. **26** (1989), 625–646; Zbl 705.11062; MR 90j:11123; cf. p. 174
333. F. Halter-Koch, P. Kaplan, K.S. Williams, *An Artin character and representations of primes by binary quadratic forms II*, Manuscr. Math. **37** (1982), 357–381; Zbl 485.12003; MR 83i:12003; cf. p. 175

334. F. Hajir, F. R. Villegas, *Explicit elliptic units I*, Duke Math. J. **90** (1997), 495–521; Zbl 898.11025; MR 98k:11075; cf. p. 277

335. H. Hasse, *Das allgemeine Reziprozitätsgesetz und seine Ergänzungssätze in beliebigen algebraischen Zahlkörpern für gewisse nicht-primäre Zahlen*, J. Reine Angew. Math. **153** (1924), 192–207; FdM **50** (1924), 105; Math. Abh. I, 217–232; cf. p. 201

336. H. Hasse, *Das Eisenstein'sche Reziprozitätsgesetz der n-ten Potenzreste*, Math. Ann. **97** (1927), 599–623; Math. Abh. I, 269–293; FdM **53** (1927), 145; cf. p. 393

337. H. Hasse, *Zum expliziten Reziprozitätsgesetz*, Abh. Math. Sem. Hamburg **7** (1929), 52–63; FdM **55-I** (1929), 700; cf. p. 273

338. H. Hasse, *Der 2^n-te Potenzcharakter von 2 im Körper der 2^n-ten Einheitswurzeln*, Rend. Circ. Matem. Palermo (2), **7** (1958), 185–243; Zbl 126.06903; MR 21 # 4143; cf. p. 310

339. H. Hasse, *Der 2^n-te Potenzcharakter von 2 im Körper der 2^n-ten Einheitswurzeln*, Ber. Dirichlet-Tagung (1963), 59–69; Zbl 283.12002; cf. p. 310

340. H. Hasse, *Bericht über neuere Untersuchungen und Probleme aus der Theorie der algebraischen Zahlkörper*, Jahresber. DMV, Ergänzungsband **6** (1930), 204 pp., Teil II: Reziprozitätsgesetz. Reprint: Physica Verlag, Würzburg 1965; Zbl 138.03202; MR 33 #4045a, #4045b; MR 42 #1795; cf. p. 395

341. H. Hasse, *Zahlentheorie*, Akademie-Verlag 1949; Zbl 35.02002; Engl. Transl. *Number Theory*, Springer Verlag 1980; Zbl 423.12002; cf. p. 60, 122, 137

342. H. Hasse, *Vorlesungen über Zahlentheorie*, Springer Verlag 1950; MR 14, 534c; 2nd. ed. 1964; Zbl 38.17703; MR 32 #5569; cf. p. 24, 26, 345, 394

343. A. Hausner, *On the quadratic reciprocity theorem*, Arch. Math. **12** (1961), 182–183; Zbl 99.26501; MR 24 # A1866; cf. p. 416

344. H. Hayashi, *On a simple proof of Eisenstein's reciprocity law*, Mem. Fac. Sci. Kyushu Univ. Ser. A **28** (1974), 93–99; Zbl 335.10008; MR 50 # 7000; cf. p. 225

345. P.J. Heawood, *On a graphical demonstration of the fundamental properties of quadratic residues*, London Math. Soc. Proc. (2), **12** (1913), 373–376; FdM **44**, 225; cf. p. 415

346. P.J. Heawood, *The law of quadratic reciprocity*, Math. Gaz. London **23**, 198–200; FdM **65** II (1939), 1143; Zbl 21.00803; cf. p. 415

347. E. Hecke, *Reziprozitätsgesetz und Gauss'sche Summen in quadratischen Zahlkörpern*, Göttinger Nachr. (1919), 265–278; Mathem. Werke (1959), 235–248; FdM **47** (1919/20), 145; cf. p. 103, 137

348. E. Hecke, *Vorlesung über die Theorie der algebraischen Zahlen*, Akadem. Verlagsges. Leipzig, 1923; FdM **49** (1923), 106–108; cf. p. v, 119

349. E. Hecke, *Lectures on the Theory of Algebraic Numbers* (Engl. Transl. of [348]), Springer Verlag 1981; Zbl 504.12001; MR 83m:12001; cf. p. 103, 137
350. G. Heinitz, *Eine neue Bestimmung des quadratischen Restcharakters*, Wiss. Beil. z. Pr. Realsch. Seesen a. Harz **694**, Göttingen 1893, 45 pp; cf. p. 26
351. G. Heinitz, *Elementare Berechnung der Zahl µ, welche den quadratischen Restcharakter bestimmt*, Diss. Göttingen 1893; cf. p. 26
352. Ch. Helou, *An explicit 2^nth reciprocity law*, Diss. Cornell Univ. 1984; cf. p. 312
353. Ch. Helou, *An explicit 2^nth reciprocity law*, J. Reine Angew. Math. **389** (1988), 64–89; Zbl 638.12008; MR 89e:11072; cf. p. 312
354. Ch. Helou, *Classical explicit reciprocity*, Théorie des nombres, Proc. Int. Number Theory Conf., Laval 1987 (eds.: J. M. de Koninck, C. Levesque) (1989), 359–370; Zbl 714.11083; MR 91g:11140; cf. p. 173
355. Ch. Helou, *On rational reciprocity*, Proc. Amer. Math. Soc. **108** (1990), 861–866; Zbl 701.11002; MR 90g:11007; cf. p. 310
356. Ch. Helou, *Power reciprocity for binomial cyclotomic integers*, J. Number Theory **71** (1998), 245–256; MR 99j:11126
357. Ch. Helou, *A note on the power residue symbol*, Proc. Conf. Turku 1999, preprint
358. Ch. Helou, R. Roll, L. Washington, *Power residue character of rational primes*, preprint 1999; cf. p. 141
359. K. Hensel, *Arithmetische Untersuchungen über Diskriminanten und ihre außerwesentlichen Theiler*, Diss. Univ. Berlin 1884; cf. p. 174
360. K. Hensel, *Über die zu einem algebraischen Zahlkörper gehörigen Invarianten*, J. Reine Angew. Math. **129** (1905), 68–85; FdM **36** (1905), 286–287; cf. p. 137
361. K. Hensel, *Die Verallgemeinerung des Legendreschen Symboles für allgemeine algebraische Körper*, J. Reine Angew. Math. **147** (1917), 233–248; FdM **46** (1916–18), 253; cf. p. 136
362. G. Herglotz, *Zur letzten Eintragung im Gauss'schen Tagebuch*, Ber. Verh. Sächs. Akad. Wiss. Leipzig Math.-Nat. Kl. **73** (1921), 271–276; Ges. Werke 415–420; cf. p. 201, 342
363. G. Herglotz, *Über das quadratische Reziprozitätsgesetz in imaginären quadratischen Zahlkörpern*, Berichte über die Verhandlungen der sächsischen Akad. der Wiss. zu Leipzig, Math.-Phys. Klasse **73** (1921), 303–310; Ges. Werke 396–410; FdM **48** (1921/22), 170; cf. p. 137, 256, 277
364. J. Hermes, *Beweis des quadratischen Reciprocitätsgesetzes durch Umkehrung*, Arch. Math. Phys. (2), **5** (1889), 190–198; FdM **19** (1887), 180; cf. p. 414
365. H. Hertzer, *Periode des Dezimalbruches für $1/p$, wo p eine Primzahl*, Archiv Math. Phys. (3) **2** (1902), 249–252; cf. p. 392

366. W. Hettkamp, *Quadratischer Restcharakter von Grundeinheiten und 2-Klassengruppe quadratischer Zahlkörper*, Diss. Münster, 1981; cf. p. 175
367. D. Hilbert, *Über den Dirichlet'schen biquadratischen Zahlkörper*, Math. Ann. **45** (1894), 309–340; FdM **25** (1893/94), 303–305; cf. p. 170
368. D. Hilbert, *Die Theorie der algebraischen Zahlen* (Zahlbericht), Jahresber. DMV **4** (1897), 175–546; FdM **28** (1897), 157–162; French transl.: Toulouse Ann. (3) **1** (1905), 257–328; FdM **41** (1911), 244; English transl.: Springer Verlag 1998; Roumanian transl: Bukarest 1998; cf. p. 86, 101, 103, 358
369. D. Hilbert, *Über die Theorie der relativ-quadratischen Zahlkörper*, Jber. DMV **6** (1899), 88–94; cf. p. 70
370. K. S. Hilbert, *Das allgemeine quadratische Reciprocitätsgesetz in ausgewählten Kreiskörpern der 2^n-ten Einheitswurzeln*, Diss. Göttingen, 1900, 355 pp; FdM **31** (1900), 204; cf. p. 136
371. T. Hiramatsu, *Introduction to Higher Reciprocity Law* (Japan.), Makino Shoten 1998, ISBN 4-7952-0120-X C3341; cf. p. xv
372. F. Hirzebruch, D. Zagier, *The Atiyah-Singer Theorem and Elementary Number Theory*, Math. Lecture Series **3**, Publish & Perish 1974; Zbl 288.10001; MR 58 # 31291; cf. p. 417
373. L. Holzer, *Takagische Klassenkörpertheorie, Hassesche Reziprozitätsformel und Fermatsche Vermutung*, J. Reine Angew. Math. **173** (1935), 114–124; Zbl 11.38902; cf. p. 396
374. L. Holzer, *Zahlentheorie I*, Teubner, Leipzig, 1958; Zbl 81.03702; MR 23 # A827; cf. p. 416
375. T. Honda, *Invariant differentials and L-functions; Reciprocity law for quadratic fields and elliptic curves over* $\mathbb{Q}$, Rend. Sem. Mat. Univ. Padova **49** (1973), 323–325; Zbl 283.14006; MR 50 # 13041; cf. p. 417
376. C.M. Huber, *On the prime divisors of the cyclotomic functions*, Trans. Amer. Math. Soc. **27** (1925), 43–48; FdM **51** (1925), 125; cf. p. 102
377. R. H. Hudson, *Diophantine determinations of $3^{(p-1)/8}$ and $5^{(p-1)/4}$*, Pac. J. Math. **111** (1984), 49–55; Zbl 531.10003; MR 85c:11005; cf. p. 310
378. R. H. Hudson, K. S. Williams, *Some new residuacity criteria*, Pac. J. Math. **91** (1980), 135–143; Zbl 456.10001; MR 82f:10004; cf. p. 201, 310
379. R. H. Hudson, K. S. Williams, *A new criterion for 7 to be a fourth power (mod p)*, Israel J. Math. **38** (1981), 221–230; Zbl 466.10002; MR 82e:10007; cf. p. 201
380. R. H. Hudson, K. S. Williams, *A new formulation of the law of octic reciprocity for primes $\equiv 3 \bmod 8$ and its consequences*, Intern. J. Math. Math. Sci. **5** (1982), 565–584; Zbl 488.10002; MR 83m:10005; cf. p. 310
381. R. H. Hudson, K. S. Williams, *An application of a formula of Western to the evaluation of certain Jacobsthal sums*, Acta Arith. **41** (1982), 261–276; Zbl 488.10035; MR 84a:10041; cf. p. 313

382. R. H. Hudson, K. S. Williams, *Extensions of theorems of Cunningham–Aigner and Hasse–Evans*, Pac. J. Math. **104** (1983), 111–132; Zbl 467.10001; MR 84e:10005; cf. p. 310
383. H. P. Hübler, *Über die kubischen Reste*, Diss. Jena, 1871; cf. p. 225, 277
384. S. Humble, *A proof of the law of reciprocity for Jacobi symbols*, Math. Gaz. **49** (1965), 169–170; Zbl 131.28505; cf. p. 25
385. J. Hurrelbrink, *The elements of $K_2(\mathbb{Z}_S)$*, Manuscripta Math. **24** (1978), 173–177; Zbl 374.13009; MR 57 #12463; cf. p. 21, 417
386. K. Ireland, M. Rosen, *A classical introduction to modern number theory*, Springer Verlag, 2nd. ed. 1990; cf. p. 139, 391
387. H. Ishii, *Functional equations and the law of quadratic reciprocity*, Mem. Inst. Sci. Engrg. Ritsumeikan Univ. No. **57** (1998), 1–3 (1999); cf. p. 417
388. N. Ishii, *Cusp forms of weight one, quartic reciprocity and elliptic curves*, Nagoya Math. J. **98** (1985), 117–137; Zbl 556.10019; MR 86j:11047; cf. p. 175
389. N. Ishii, *On the quartic residue symbol of totally positive quadratic units*, Tokyo J. Math. **9** (1986), 53–65; Zbl 604.12003; MR 87i:11151; cf. p. 175
390. N. Ishii, *On the eighth power residue of totally positive quadratic units*, Investigations in number theory, Adv. Stud. Pure Math. **13** (1988), 413–431; Zbl 675.12007; MR 89j:11106; cf. p. 175
391. N. Ishii, *Quadratic units and congruences between Hilbert modular forms*, Investigations in number theory, Adv. Stud. Pure Math. **13** (1988), 261–275; Zbl 661.10037; MR 90b:11047; cf. p. 175
392. S. Ishimura, *On Gaussian sums associated with a character of order 5 and a rational prime number $p \equiv 1 \bmod 5$*, J. Tsuda College **8** (1976), 27–35; MR 57, 3086
393. H. Ito, *A note on Dedekind sums*, Number Theory Conf. Banff 1988, (1990), 239–248; Zbl 697.10024; MR 92d:11043; cf. p. 225
394. I. I. Ivanov, *Zur Theorie der quadratischen Reste* (Russ.), Ann. Inst. Polytechn. **28** (1921), 185–189; FdM **48** (1921/22), 1161–1162; cf. p. 20
395. S. Iyanaga, *Sur un lemme d'arithmétique élémentaire dans la démonstration de la loi générale de réciprocité*, C. R. Acad. Sci. Paris **197** (1933), 728–730; Zbl 7.33702
396. C.G.J. Jacobi, *De residuis cubicis commentatio numerosa*, J. Reine Angew. Math. **2** (1827), 66–69; Werke VI, 233–237; cf. p. 224, 227, 313
397. C.G.J. Jacobi, *Letter to Gauss*, 27. 10. 1826, Gesammelte Werke VII, 391–392; cf. p. 171
398. C.G.J. Jacobi, *Letter to Gauss*, 08. 02. 1827; Gesammelte Werke VII, 393–400; cf. p. 138, 144, 313
399. C.G.J. Jacobi, *Lettre à Legendre*, 5. 8. 1827, Gesammelte Werke I, 390–396 (ed.: C. W. Borchardt), Berlin 1881
400. C.G.J. Jacobi, *Über die Kreistheilung und ihre Anwendung auf die Zahlentheorie*, Berliner Akad. Ber. 1837, 127–136; Werke VI, 245–274; cf. p. 10, 138, 138, 224, 225, 270, 313, 413

401. C.G.J. Jacobi, *Über die Kreistheilung und ihre Anwendung auf die Zahlentheorie*, J. Reine Angew. Math. **30** (1846), 166–182; cf. p. 270
402. C.G.J. Jacobi, *Lecture Notes*, Königsberg 1837 (H. Pieper, ed.), in preparation; cf. p. vii, xv, 12, 24, 170, 200, 227, 270
403. C.G.J. Jacobi, *Über die complexen Primzahlen, welche in der Theorie der Reste der 5ten, 8ten und 12ten Potenzen zu betrachten sind*, Berliner Akad. Ber. 1839, 86–91; J. Reine Angew. Math. **19** (1839), 314–318; cf. p. 179, 392
404. C.G.J. Jacobi, *Sur les nombres premiers complexes que l'on doit considerer dans la théorie des résidus de cinquième, huitième et douzième puissance*, J. Math. pures appl. **8** (1843), 268–272
405. E. Jacobsthal, *Anwendung einer Formel aus der Theorie der quadratischen Reste*, Diss. Berlin 1906, 39 pp; FdM **37** (1906), 226; cf. p. 26, 140
406. S. Jakubec, *Power residues*, Number theory, Summer School 1983 Chlebske / Czech. (1985), 19–21
407. S. Jakubec, *Note on the Jacobi sum $J(\chi,\chi)$*, J. Théor. Nombres Bordeaux **7** (1995), 461–471; Zbl 853.11068; MR 97a:11124; cf. p. 141
408. S. Jakubec, *Criterion for 3 to be eleventh power*, Acta Math. Inform. Univ. Ostrav. **3** (1995), 37–43; Zbl 876.11002; MR 98f:11088; cf. p. 141
409. W. Jänichen, *Zur Theorie der quadratischen Reste*, Arch. Math. Phys. (3), **26** (1917), 201–204; FdM **46** (1916/18), 195; cf. p. 26
410. W. Jänichen, *Über einige zahlentheoretischen Relationen aus der Theorie der quadratischen Reste*, Arch. Math. Phys. (3), **28** (1919/20), 85–89; FdM **47** (1919/20), 108–109; cf. p. 26
411. M. Jenkins, *Proof of an Arithmetical Theorem leading, by means of Gauss' fourth demonstration of Legendre's law of reciprocity, to the extension of that law*, Proc. London Math. Soc. **2** (1867), 29–32; FdM **2** (1869/70), 92–93; cf. p. 22, 23
412. Ch. U. Jensen, *Remark on a characterization of certain ring class fields by their absolute Galois group*, Proc. Amer. Math. Soc. **14** (1963), 738–741; Zbl 128.03503; MR 27 # 3622; cf. p. 225
413. J. R. Joly, *Démonstration cyclotomique de la loi de réciprocité cubique*, Bull. Sci. Math. (2), **96** (1972), 273–278; Zbl 256.10002; MR 48 # 2109; cf. p. 225, 345
414. J. R. Joly, *Équations et variétés algébriques sur un corps fini*, Ens. Math. (2), **19** (1973), 1–117; Zbl 282.14005; MR 48 # 6065; cf. p. 138, 391
415. V.G. Kac, *Simple Lie groups and the Legendre symbol*, in: Algebra, Carbondale 1980, LNM **848** (1981), 110–123; Zbl 498.22013; MR 82f:20072; cf. p. 417
416. J.-M. Kantor, *Les problèmes de Hilbert et leur devenir*, Analyse diophantienne et géométrie algébrique, Cahiers Sém. Hist. Math. (2) **3** (1993), 95–112; MR 94i:01021; cf. p. viii

417. G. Kantz, *Über einen Satz aus der Theorie der biquadratischen Reste*, Deutsche Math. **5** (1940), 269–272; FdM **66** - **I** (1940), 138–139; Zbl 25.24904; MR 7 # 145e; cf. p. 201
418. P. Kaplan, *Une démonstration géométrique de la loi de réciprocité quadratique*, Proc. Japan Acad. **45** (1969), 779–780; Zbl 216.03703; MR 42 # 4474; cf. p. 416
419. P. Kaplan, *Démonstration des lois de réciprocité quadratique et biquadratique*, J. Fac. Sci. Tokyo **16** (1969), 115–145; Zbl 216.03704; MR 41 # 1683; cf. p. 200, 225, 417
420. P. Kaplan, *Divisibilité par 8 du nombre de classes des corps quadratiques dont le 2-groupe des classes est cyclique, et réciprocité biquadratique*, J. Math. Soc. Japan **25** (1973), 596–608; Zbl 276.12006; MR 48 # 2113
421. P. Kaplan, *Sur le 2-groupe des classes d'idéaux des corps quadratiques*, J. Reine Angew. Math. **283/284** (1976), 313–363; Zbl 337.12003; MR 53 # 8009; cf. p. 173
422. P. Kaplan, *Cours d'Arithmétique III*, Lecture Notes
423. P. Kaplan, K. S. Williams, *An Artin character and representation of primes by binary quadratic forms*, Manuscripta Math. **33** (1981), 339–356; Zbl 464.12003; MR 82d:10036; cf. p. 175
424. P. Kaplan, K. S. Williams, Y. Yamamoto *An application of dihedral fields to representations of primes by binary quadratic forms*, Acta Arith. **44** (1984), 407–413; Zbl 553.10018: MR 86e:11025; cf. p. 172, 175
425. I. Kaplansky, *Quadratic reciprocity in Dickson's History*, Notices Amer. Math. Soc. **40** (1993), no. 9, 1155; cf. p. xiii
426. E. Karst, *Tables on fifth power residuacity*, BIT, Nordisk Tidskr. Inform.-Behandl. **7** (1967), 114–122; Zbl 163.03901; cf. p. 141
427. S. A. Katre, A. R. Rajwade, *Euler's criterion for quintic nonresidues*, Canad. J. Math. **37** (1985), 1008–1024; Zbl 712.11004; MR 87c:11002; cf. p. 141
428. F. Keune, *Quadratic reciprocity and finite fields*, Nieuw Arch. Wisk. (4), **9** (1991), 263–266; Zbl 765.11003; MR 93g:11006; cf. p. 417
429. S. Kirchhoff, *Bemerkungen zum quadratischen Reziprozitätsgesetz*, Hausarbeit Göttingen, 1977; cf. p. 24
430. H.D. Kloosterman, *De complexe geheele getallen von Gauss I*, Math. Tijds. v. Studeerenden **1**, 53–61, 125–134, 190–196; ibid. **2**, 6–13
431. H.D. Kloosterman, *The law of quadratic reciprocity*, Indag. Math. **27** (1965), 163–164; Zbl 123.04206; MR 30 # 1974; cf. p. 416
432. J. König, *Das Reciprocitätsgesetz in der Theorie der quadratischen Reste*, Acta Math. **22** (1899), 181–192; RSPM **7** 1898, 147; FdM **29** (1898), p. 154; cf. p. 201, 415
433. J. König, *Über 3. und 4. Potenzreste*, Berichte über die Verhandlungen der königl. sächs. Gesellschaft der Wiss. Leipzig, Math. Phys. Kl. **64**, 237–256; FdM **43** (1912), 270–271

434. L. Koschmieder, *Über eine besondere äquianharmonische elliptische Funktion und ihre Verwendung in der Zahlentheorie*, Sitzungsber. Berlin. Math. Ges. **21** (1922), 12–15; cf. p. 225, 277
435. L. Koschmieder, *Zu Eisensteins transzendentem Beweis des quadratischen Reziprozitätsgesetzes*, Comment. Math. Helv. **37** (1962/63), 235–239; Zbl 115.04003; MR 26 # 4967; cf. p. 277, 416
436. J. Kraft, M. Rosen, *Eisenstein reciprocity and n-th power residues*, Amer. Math. Monthly **88** (1981), 269–270; Zbl 458.10002; MR 82g:10008; cf. p. 136, 394
437. K. Kramer, *Residue properties of certain quadratic units*, J. Number Theory **21** (1985), 204–213; Zbl 567.12003; MR 87h:11106; cf. p. 174
438. L. Kronecker, *Über Zeller's Beweis des quadratischen Reciprocitätsgesetzes*, Berl. Monatsber. (1872), 846–848; Werke V, 445–448; cf. p. 413
439. L. Kronecker, *Bemerkungen zur Geschichte des Reciprocitätsgesetzes*, Monatsber. Berlin (1875), 267–275; Werke II, 1–10; FdM **7** (1875), 18–19; Ital. translation in Bull. bibliogr. storia sci. mat. fis. **18**, 244–249; cf. p. 4, 22
440. L. Kronecker, *Ueber das Reciprocitätsgesetz*, Monatsber. Berlin (1876), 331–341; Werke II, 11–23; FdM **8** (1876), 93–95; cf. p. 23, 414
441. L. Kronecker, *Sur la loi de réciprocité*, Bull. Sci. Math. (2), **4** (1880), 182–192; Werke II, 25–36; FdM **12** (1880), 123; French translation of [440]
442. L. Kronecker, *Ueber die Potenzreste gewisser complexer Zahlen*, Monatsber. Berlin (1880), 404–407; Werke II, 95–101; cf. BSMA (2) **9** (1885), 121–122; FdM **12** (1880), 125; cf. p. 101, 414
443. L. Kronecker, *Ueber den vierten Gauss'schen Beweis des Reciprocitätsgesetzes für die quadratischen Reste*, Monatsber. Berlin (1880), 686–698; Werke IV, 275–294; FdM **12** (1880), 123; BSMA (2) **9** (1885), 122–124; cf. p. 22, 26, 414
444. L. Kronecker, *Beweis des Reciprocitätsgesetzes für die quadratischen Reste*, Monatsber. Berlin (1884), 519–539; FdM **16** (1884), 156–158; Werke II, 497–522; cf. p. 414
445. L. Kronecker, *Beweis des Reciprocitätsgesetzes für die quadratischen Reste*, J. Reine Angew. Math. **96** (1884), 348–350; FdM **16** (1884), 156–158; Werke II, 523–526; extract from [444]
446. L. Kronecker, *Ueber den dritten Gauss'schen Beweis des Reciprocitätsgesetzes für die quadratischen Reste*, Monatsber. Berlin 1884, 645–649; FdM **16** (1884), 156–158; Werke II, 527–532; cf. p. 21, 414
447. L. Kronecker, *Der dritte Gauss'sche Beweis des Reciprocitätsgesetzes für die quadratischen Reste, in vereinfachter Darstellung*, J. Reine Angew. Math. **97** (1885), 93–94; FdM **16** (1884), 156–158; Werke II, 533–536; cf. p. 21, 414
448. L. Kronecker, *Zum dritten Gauss'schen Beweis des Reciprocitätsgesetzes für die quadratischen Reste (Bemerkungen zu Herrn Ernst Schering's*

Mitteilung), Monatsber. Berlin (1885), 117–118; Werke II, 537–540; cf. p. 21, 414
449. L. Kronecker, *Die absolut kleinsten Reste reeller Grössen*, Monatsber. Berlin (1885), 383–386; FdM **17** (1885), 150–151; cf. p. 23, 414
450. L. Kronecker, *Beweis des Reciprocitätsgesetzes für die quadratischen Reste*, J. Reine Angew. Math. **104** (1889), 348–351; Werke III, 137–144; FdM **21** (1889), 177; cf. p. 414
451. P. B. Kronheimer, M. J. Larsen, J. Scherk, *Casson's invariant and quadratic reciprocity*, Topology **30** (1991), 335–338; Zbl 729.57007; MR 92g:57028; cf. p. 141
452. D. S. Kubert, *Product formulae on elliptic curves*, Invent. Math. **117** (1994), 227–273; Zbl 834.14016; MR 95d:11075; cf. p. 225
453. T. Kubota, *Anwendung Jacobischer Thetafunktionen auf die Potenzreste*, Nagoya Math. J. **19** (1961), 1–13; Zbl 104.03801; MR 25 # 3926; cf. p. 200, 204, 275
454. T. Kubota, *Reciprocities in Gauss' and Eisenstein's number fields*, J. Reine Angew. Math. **208** (1961), 35–50; Zbl 202.33302; MR 25 # 3030; cf. p. 201, 266, 275, 416
455. T. Kubota, *Über quadratische Charaktersummen*, Nagoya Math. J. **19** (1961), 15–25; MR 25 #3927; cf. p. 416
456. T. Kubota, *Notes on analytic theory of numbers*, Univ. Chicago, 1963; Zbl 344.12001; cf. p. 201
457. T. Kubota, *Some arithmetical applications of an elliptic function*, J. Reine Angew. Math. **214/215** (1964), 141–145; Zbl 146.27901; MR 29 # 2240; cf. p. 201, 266, 275
458. T. Kubota, *The foundation of class field theory based on the principles of space diagrams* (Japan.), Sugaku **44** (1992), 1–12; Engl. Transl. Sugaku Expo. **8** (1995), 1–12; Zbl 839.11054; MR 94k:11123; cf. p. 417
459. L. Kuipers, *A simple proof of the "Eisenstein" theorem*, Tamkang J. Math. **5** (1974), 5–7; MR 51 # 10206; cf. p. 23
460. L. Kuipers, *An extension of a theorem of Eisenstein*, Proc. Amer. Math. Soc. **43** (1974), 69–70; MR 48 #8368; cf. p. 23
461. E.E. Kummer, *Eine Aufgabe, betreffend die Theorie der kubischen Reste*, J. Reine Angew. Math. **23** (1842), 285–286; Coll. Papers I, 143–144; cf. p. 278
462. E.E. Kummer, *Über die Divisoren gewisser Formen der Zahlen, welche aus der Theorie der Kreistheilung entstehen*, J. Reine Angew. Math. **30** (1846), 107–116; Coll. Papers I, 103–116; cf. p. 101
463. E.E. Kummer, *De residuis cubicis disquisitiones nonnullae analyticae*, J. Reine Angew. Math. **32** (1846), 341–359; Coll. Papers I, 145–164; cf. p. 278
464. E.E. Kummer, *Allgemeine Reziprozitätsgesetze für beliebig hohe Potenzreste*, Berliner Akad. Ber. (1850), 154–165; Coll. Papers I, 345–357; cf. p. 321, 343

465. E.E. Kummer, *Über die Ergänzungssätze zu den allgemeinen Reziprozitätsgesetzen*, J. Reine Angew. Math. **44** (1852), 93–146; Coll. Papers I, 485–538; cf. p. 138, 272, 366
466. E.E. Kummer, *Über die allgemeinen Reziprozitätsgesetze der Potenzreste*, Berliner Akad. Ber. (1858), 158–171; Coll. Papers I, 673–687; cf. p. 18
467. E.E. Kummer, *Über die Ergänzungssätze zu den allgemeinen Reziprozitätsgesetzen*, J. Reine Angew. Math. **56** (1858b), 270–279; Coll. Papers I, 688–698
468. E.E. Kummer, *Über die allgemeinen Reziprozitätsgesetze unter den Resten und Nichtresten der Potenzen, deren Grad eine Primzahl ist*, Berliner Akad. Abh. (1859), 19–159; Coll. Papers I, 699–839; cf. p. 19, 22, 235, 273
469. E.E. Kummer, *Zwei neue Beweise der allgemeinen Reziprozitätsgesetze unter den Resten und Nichtresten der Potenzen, deren Grad eine Primzahl ist*, Berliner Akad. Abh. 1861; J. Reine Angew. Math. **100** (1887), 10–50; FdM **18** (1886), 146; Coll. Papers I, 842–882; cf. p. 69, 413
470. E.E. Kummer, *Letter to Kronecker*, January 16, 1842; Werke I, 46–48; cf. p. 227
471. R. Kurata, *The law of quadratic reciprocity* (Japan.), Nihon Hyoronsha, Tokyo 1992
472. S. Kuroda, *Über die Zerlegung rationaler Primzahlen in gewissen nichtabelschen Körpern*, J. Math. Soc. Japan **3** (1951), 148–156; Zbl 45.01401; MR 13 # 442a; cf. p. 175
473. Ladrasch, *Von den kubischen Resten und Nichtresten*, Pr. Dortmund 1870; FdM **2** (1869), 90
474. J. L. Lagrange, *Recherches d'Arithmetique, 2nde partie*, Nouv. Mém. de l'Acad. de Berlin (1775), 349–352; Œuvres III, 759–795; cf. p. 4, 13, 25
475. E. Lampe, *Zwei Briefe von C. G. Jacobi, die in den gesammelten Werken desselben nicht abgedruckt sind*, Arch. Math. Phys. (3) **2** (1902), 253–256; FdM **33** (1902), 201; cf. p. 392
476. E.J. Lange, *Ein elementarer Beweis des Reciprocitätsgesetzes*, Ber. Verh. Sächs. Akad. Wiss. Leipzig Math.-Nat. Kl. **48** (1896), 629–633; FdM **27** (1896), 143; cf. p. 415
477. E.J. Lange, *Ein elementarer Beweis des Reciprocitätsgesetzes*, Ber. Verh. Sächs. Akad. Wiss. Leipzig Math.-Nat. Kl. **49** (1897), 607–610; FdM **28** (1897), 169; cf. p. 415
478. E. Lasker, *Über eine Eigenschaft der Diskriminante*, Sitzungsber. Berl. Math. Ges. **15** (1916), 176–178; FdM **46** (1916–18), 240; cf. p. 137, 415
479. R. Laubenbacher, D. Pengelley, *Eisenstein's misunderstood geometric proof of the quadratic reciprocity theorem*, College Mathematics Journal **25** (1994), 29–34; cf. p. 10, 21

480. R. C. Laubenbacher, D. J. Pengelley, *Gauss, Eisenstein, and the "third" proof of the quadratic reciprocity theorem: Ein kleines Schauspiel*, Math. Intell. **16**, No. 2 (1994) 67–72; Zbl 815.01005; MR 95c:11006; cf. p. 10
481. V.A. Lebesgue, *Recherches sur les nombres*, J. math. pure appl. **2** (1837), 253–292; cf. p. 201, 202
482. V.A. Lebesgue, *Recherches sur les nombres*, J. math. pure appl. **3** (1838), 113–144; cf. p. 25, 228, 413
483. V.A. Lebesgue, *Suite des recherches sur les nombres*, J. math. pure appl. **4** (1839), 9–59; cf. p. 174
484. V.A. Lebesgue, *Démonstration de quelques théorèmes relatifs aux résidus et aux non-résidus quadratiques*, J. math. pures appl. **7** (1842), 137–159
485. V.A. Lebesgue, *Démonstration nouvelle élémentaire de la loi de réciprocité de Legendre, par. M. Eisenstein, précédée et suivie de remarques sur d'autres démonstrations, que peuvent être tirées du même principe*, J. math. pures appl. **12** (1847), 457–473; cf. p. 30, 413
486. V.A. Lebesgue, *Sur le symbole $\left(\frac{a}{b}\right)$ et quelques-unes de ses applications*, J. math. pures appl. **12** (1847), 497–517; cf. p. 26
487. V.A. Lebesgue, *Note sur les congruences*, C. R. Acad. Sci. Paris **51** (1860), 9–13; cf. p. 21, 413
488. V. A. Lebesgue, *Détermination de la valeur du symbole $\left(\frac{b}{a}\right)$, dû à Jacobi*, C. R. Acad. Sci. Paris **59** (1864), 940–944; cf. p. 22, 26
489. A.M. Legendre, *Recherches d'analyse indéterminée*, Histoire de l'Academie Royale des Sciences de Paris (1785), 465–559, Paris 1788; cf. p. 6, 7
490. A.M. Legendre, *Essai sur la théorie des nombres*, 1st ed. Paris 1798, 2nd ed. Paris 1808, 3rd ed. Paris 1830; German transl. (*Zahlentheorie*) of the 3rd. ed., Leipzig 1886; Quatrième édition conforme à la troisième, nouveau tirage corrigé, Paris 1955; cf. p. 6, 18
491. D.N. Lehmer, *Certain theorems in the theory of quadratic residues*, Amer. Math. Monthly **20** (1913), 151–157; cf. p. 26
492. D.H. Lehmer, *A low energy proof of the reciprocity law*, Amer. Math. Monthly **64** (1957), 103–106; MR 18 # 641c; cf. p. 416
493. D.H. Lehmer, *Gauss' first proof of the law of quadratic reciprocity*, in: D. E. Smith, A source book in mathematics (1959), 112–118; cf. p. 21
494. D.H. Lehmer, E. Lehmer, *Cyclotomic resultants*, Math. Comp. **48** (1987), 211–216; MR 88c:11075; cf. p. 102
495. E. Lehmer, *The quintic character of 2 and 3*, Duke Math. J. **18** (1951), 11–18; Zbl 45.02002; MR 12, 677a; cf. p. 141
496. E. Lehmer, *Period equations applied to difference sets*, Proc. Amer. Math. Soc. **6** (1955), 433–442; MR 16,904a; cf. p. 102
497. E. Lehmer, *On cyclotomic numbers of order sixteen*, Can. J. Math. **6** (1954), 449–454; Zbl 57.03603; MR 16, 115b; cf. p. 343

498. E. Lehmer, *Criteria for cubic and quartic residuacity*, Mathematika **5** (1958), 20–29; Zbl 102.28002; MR 20, 1668; cf. p. 101, 102, 167, 171, 212, 227
499. E. Lehmer, *On Euler's criterion*, J. Australian Math. Soc. **1** (1959/61), 64–70; Zbl 90.25904; MR 21,7191; cf. p. 140, 227
500. E. Lehmer, *Artiads characterized*, J. Math. Analysis and Appl. **15** (1966), 118–131; Zbl 151.02704; MR 34 # 1261; cf. p. 343, 393
501. E. Lehmer, *On the quadratic character of the Fibonacci root*, Fibonacci Quart. **4** (1966), 135–138; Zbl 144.27405; MR 39 # 160; cf. p. 174, 228
502. E. Lehmer, *On the divisors of the discriminant of the period equation*, Amer. J. Math. **90** (1968), 375–379; Zbl 169.06401; MR 37 # 2718; cf. p. 102
503. E. Lehmer, *On the quadratic character of some quadratic surds*, J. Reine Angew. Math. **250** (1971), 42–48; Zbl 222.12007; MR 44 # 3986; cf. p. 173
504. E. Lehmer, *On some special quartic reciprocity laws*, Acta Arith. **21** (1972), 367–377; Zbl 241.10003; MR 46 # 1747; cf. p. 166, 172
505. E. Lehmer, *Power characters of quadratic units*, Proc. 1972 Number Theory Conf. Boulder, Colorado (1972), 128–132; Zbl 334.12006; MR 53 # 10756; cf. p. 175, 228
506. E. Lehmer, *On the cubic character of quadratic units*, J. Number Theory **5** (1973), 385–389; Zbl 274.12006; MR 48 # 271; cf. p. 175, 228
507. E. Lehmer, *On the quartic character of quadratic units*, J. Reine Angew. Math. **268/269** (1974), 294–301; Zbl 289.12007; MR 49 # 10662; cf. p. 70, 175
508. E. Lehmer, *Generalizations of Gauss' Lemma*, Number Theory and Algebra, Academic Press, New York 1977, 187–194; Zbl 382.10003; MR 57 # 9639; cf. p. 137, 162
509. E. Lehmer, *Rational reciprocity laws*, Amer. Math. Month. **85** (1978), 467–472; Zbl 383.10003; MR 58 # 16482; cf. p. v, 166, 174
510. E. Lehmer, *On special primes*, Pac. J. Math. **118** (1985), 471–478; MR 86f:11062; cf. p. 203
511. E. Lehmer, *An indeterminate in number theory*, J. Austral. Math. Soc. **46** (1989), 469–472; Zbl 676.10006; MR 90d:11008; cf. p. 141, 141
512. F. Lemmermeyer, *Rational quartic reciprocity*, Acta Arithmetica **67** (1994), 387–390; Zbl 833.11049; MR 95m:11010; cf. p. 173
513. F. Lemmermeyer, *On the 4-rank of real quadratic number fields*, submitted 1996; cf. p. 176
514. F. Lemmermeyer, *Rational quartic reciprocity II*, Acta Arithmetica **80**, No. 3 (1997), 273–276; Zbl 878.11040; MR 98f:11004; cf. p. 166, 173
515. F. Lemmermeyer, *Kreise und Quadrate modulo p*, Math. Sem.Ber. **47** (2000); cf. p. 30, 417
516. H.W. Lenstra, P. Stevenhagen, *Artin reciprocity and Mersenne primes*, preprint 1999; cf. p. 70

517. Ph.A. Leonard, B.C. Mortimer, K.S. Williams, *The eleventh power character of 2*, J. Reine Angew. Math. **286/287** (1976), 213–222; Zbl 332.10003; MR 54 # 10115; cf. p. 142
518. Ph.A. Leonard, K.S. Williams, *The septimic character of 2, 3, 5 and 7*, Pac. J. Math. **52** (1974), 143–147; Zbl 285.10002; MR 51 # 319; cf. p. 142
519. Ph.A. Leonard, K.S. Williams, *A diophantine system of Dickson*, Rendiconti Atti Accad. naz. Lincei **56** (1974), 145–150; Zbl 299.12017; MR 51 # 10308; cf. p. 141
520. Ph.A. Leonard, K.S. Williams, *Forms representable by integral binary quadratic forms*, Acta Arithmetica **26** (1974), 1–9; Zbl 284.10006; MR 50 # 7028; cf. p. 175
521. Ph.A. Leonard, K.S. Williams, *The quadratic and quartic character of certain quadratic units I*, Pac. J. Math. **71** (1977), 101–106; Zbl 354.12003; cf. p. 175
522. Ph.A. Leonard, K.S. Williams, *A rational sixteenth power reciprocity law*, Acta Arith. **33** (1977), 365–377; Zbl 363.10003; MR 57 # 219; cf. p. 310
523. Ph.A. Leonard, K.S. Williams, *The quadratic and quartic character of certain quadratic units II*, Rocky Mt. J. Math. **9** (1979), 683–691; Zbl 413.12004; MR 81b:12006; cf. p. 174
524. Ph.A. Leonard, K.S. Williams, *The quartic characters of certain quadratic units*, J. Number Theory **12** (1980), 106–109; Zbl 427.12002; MR 81k:12011; cf. p. 175
525. Ph.A. Leonard, K.S. Williams, *A representation problem involving binary quadratic forms*, Arch. Math. **36** (1981), 53–56; Zbl 449.10016; MR 83b:12006; cf. p. 175
526. Ph.A. Leonard, K.S. Williams, *An observation on binary quadratic forms of discrimiant $-32q$*, Abh. Math. Semin. Univ. Hamb. **53** (1983), 39–40; Zbl 517.10019; MR 85g:11100; cf. p. 175
527. M. Lerch, *Modification de la troisième démonstration donnée par Gauss de la loi de reciprocité de Legendre*, J. Sciencias Matem. Astron. **8** (1887), 137–146; FdM **19** (1887), 180; cf. p. 21, 414
528. M. Lerch, *Sur un théorème arithmetique de Zolotarev*, Cesk. Akad. Prague, Bull. Int. Cl. Math. **3** (1896), 34–37; cf. p. 24
529. M. Lerch, *Sur un théorème de Zolotarev*, Bull. intern. de l'Ac. François Joseph (1896), 4 pp.; FdM **27** (1896), 141; cf. p. 24
530. M. Lerch, *Über einen arithmetischen Satz von Zolotarev* (Bohem.), Rozpravy **5** (1896), 8 pp.; FdM **27** (1896), 141; cf. p. 24
531. M. Lerch, *Über den fünften Gaußschen Beweis des Reziprozitätsgesetzes für die quadratischen Reste*, Sep.-Abdr. Sitzungsber. Kgl. Böhm. Ges. d. Wiss. 1903, Prag, 12 pp; FdM **34** (1903), 227; cf. p. 22, 415

532. M. Lerch, *Sur la cinquième démonstration de Gauß de la loi de réciprocité de Legendre*, J. Sciencias Matem. Astron. **15** (1904), 97–104; FdM **34** (1903), 227; cf. p. 415
533. M. Lerch, *Sur quelques applications d'un théorème arithmétique de Jacobi*, Krakau Anz. 1904, 57–70; FdM **35** (1904), 211–212
534. R. Levavasseur, *Quelques démonstrations relatives à la théorie des nombres entiers complexes cubiques*, Ann. Univ. Lyon **21** (1908), 1–22; FdM **41** (1910), 248; cf. p. 225
535. R. Levavasseur, *Quelques démonstrations relatives à la théorie des nombres entiers complexes cubiques*, Lyon Ann. (2), **24**, 66 pp; FdM **39** (1908), 271; cf. p. 225
536. C. A. Levin, *En faktor uppdelning av binomet $x^{p-1} - 1$ (mod p)*, Elementär Mat. Fys. Kemi, Stockholm **20**, 81–89; FdM **63** II (1937), 909
537. J. F. Lewandowski, *Über die äquianharmonische Funktion*, Monatsh. Math. Phys. **21** (1910), 155–170; FdM **41** (1910), 513; cf. p. 225, 277
538. H. Lewy, *Waves on beaches*, Bull. Amer. Math. Soc. **52** (1946), 737–775; Zbl 61.46101; cf. p. 416
539. M. Libri, *Mémoire sur la théorie des nombres I*, J. Reine Angew. Math. **9** (1832), 54–80; cf. p. 25
540. M. Libri, *Mémoire sur la théorie des nombres II*, J. Reine Angew. Math. **9** (1832), 169–188; cf. p. 25
541. M. Libri, *Mémoire sur la théorie des nombres III*, J. Reine Angew. Math. **9** (1832), 261–276; cf. p. 25, 342
542. H. von Lienen, *Regeln für kubische und höhere Potenzreste*, Diss. Bochum 1970, 93pp.; cf. p. 310
543. H. von Lienen, *Kriterien für n-te Potenzreste*, Math. Z. **129** (1972), 185–193; Zbl 243.10004; MR 47 # 1728; cf. p. 310
544. H. von Lienen, *Primzahlen als achte Potenzreste*, J. Reine Angew. Math. **266** (1974), 107–117; Zbl 279.10003; MR 49 # 4916; cf. p. 310
545. H. von Lienen, *Reelle Reziprozitätsgesetze für kubische und biquadratische Potenzreste*, Habilitationsschrift, Techn. Univ. Braunschweig, 1976, 166 pp.; Zbl 447.10006;
546. H. von Lienen, *Reelle kubische und biquadratische Legendre-Symbole*, J. Reine Angew. Math. **305** (1979), 140-154; Zbl 392.10005; MR 80d:10004; cf. p. 227
547. W. Lietzmann, *Über das biquadratische Reziprozitätsgesetz in algebraischen Zahlkörpern*, Diss. Göttingen, 1904, vi+93 pp; FdM **35** (1904), 229; cf. p. 201
548. C.-E. Lind, *Untersuchungen über die rationalen Punkte der ebenen kubischen Kurven vom Geschlecht Eins*, Diss. Uppsala, 1940; cf. p. 175
549. J. Liouville, *Sur la loi de réciprocité dans la théorie des résidus quadratiques*, J. math. pure appl. (I), **12** (1847), 95–96; cf. p. 413
550. J. Liouville, *Sur la loi de réciprocité dans la théorie des résidus quadratiques*, C. R. Acad. Sci. Paris **24** (1847), 577–578; cf. p. 275

551. R. Lipschitz, *Sur un théorème arithmétique*, C. R. Acad. Sci. Paris **108** (1889), 489–492; FdM **21** (1889), 177–178; cf. p. 23
552. E. Liverance, *A formula for the root number of a family of elliptic curves*, J. Number Theory **51** (1995), 288–305; Zbl 831.14012; MR 96e:11086; cf. p. 228
553. P. Llorente, *On k-th power residuacity*, Rev. Un. Mat. Argentina **28** (1977), 60–67; Zbl 409.10003; MR 56 #8526; cf. p. 142
554. C. Longo, *Sui residui n^{ici} d'un modulo primo p*, Atti Accad. Ligure **2** (1942), 83–86; MR 11 # 81g; cf. p. 136
555. J. J. Durán Loriga, *Sobre los residuos cuadraticos*, Revista R. Acad. Madrid **5**, 211–220; RSPM **16** (1907), 45
556. S. Louboutin, *Norme relative de l'unité fondamentale et 2-rang du groupe des classes de certains corps biquadratiques*, Acta Arith. **58** (1991), 273–288; Zbl 726.11067; MR 93a:11090; cf. p. 170
557. S. Lubelski, *Zur Reduzibilität von Polynomen in der Kongruenztheorie*, Acta Arith. **1** (1936), 169–183; see also Prace Mat.-Fiz. **43** (1936), 207–221; cf. p. 137
558. E. Lucas, *Nouvelle démonstration de la loi de réciprocité*, Assoc. Franç. Limoges **19$_1$** (1890), 147; FdM **22** (1890), 211; cf. p. 414
559. E. Lucas, *Sur la loi de réciprocité des résidus quadratiques*, St. Petersbourg, Mélanges math. et astr. **8**, 65–66; FdM **23** (1891), 188; cf. p. 414
560. D. Lucon, *La factorisation quadratique des polynomes cyclotomiques*, Revue de Mathématiques Spéciales **3** (1990/91), 167–171
561. K. Ludwig, *Studien über den 5. Gauss'schen Beweis fr das quadratische Reziprozitätsgesetz und die analogen Entwicklungen im Gebiet der komplexen Zahlen*, Diss. Univ. Breslau, 1924; cf. p. 22
562. M. Mandl, *Ueber die Verallgemeinerung eines Gaussischen Algorithmus*, Monatsh. f. Math. **1** (1890/91), 465–472; FdM **22** (1890), 206; cf. p. 23
563. M. Mandl, *On the generalization of a theorem by Gauss and its application*, Quart. J. **25** (1891), 227–236; FdM **23** (1891), 187–188; cf. p. 23
564. P. Mansion, *On the law of reciprocity of quadratic residues*, Mess. Math. (2) **5** (1876), 140–143; FdM **8** (1876), 93; cf. p. 414
565. P. Mansion, *On the law of reciprocity of quadratic residues*, Nouv. Corresp. Math. **2** (1876), 233–239, 266–272; FdM **8** (1876), 93; cf. p. 414
566. J. Martinet, *Apropos de classes d'idéaux*, Sém. Théor. Nombres Bordeaux, 1971/72, exp. no. 5, 10 pp.; Zbl 285.12006; cf. p. 229
567. E. Mathieu, *Mémoire sur la théorie des résidus biquadratiques*, J. de Math. Pures Appl. (2) **12** (1867), 377–438; cf. p. 26, 101, 413
568. E. Mathieu, *Extrait d'un mémoire sur la théorie des résidus biquadratiques*, C. R. Acad. Sci. Paris **64** (1867), 568–571

569. P.B. Massell, *Ray class field extensions of real quadratic fields and solvability of congruences*, J. Number Theory **20** (1985), 262–272; MR 87b:11104; Zbl 579.12007; cf. p. 174
570. A. Matrot, *Sur les residus quadratiques*, Assoc. Franç. Limoges **19** (1890), 82–88; FdM **22** (1890), 211; cf. p. 25
571. May, *Die Quadratreste und Nichtreste, 1. Theil*, Pr. Dillingen, 1872; BSMA **10** (1876), 292; FdM **4** (1872), 74–75
572. J. Mayer, *Ueber n-te Potenzreste und binomische Congruenzen dritten Grades*, Diss. Univ. München 1895, Progr. d. Königl. Humanist. Gymnasiums Freising; FdM **26** (1895), 211; cf. p. 136
573. J. McDonnell, *On quadratic residues*, Amer. Math. Soc. Bull. (2) **19** (1913), 457 (Abstract)
574. J. McDonnell, *On quadratic residues*, Amer. Math. Soc. Trans. **14** (1913), 477–480; FdM **44** (1913), 225
575. G.I. Mel'nikov, *Two proofs of the cubic law of reciprocity with the aid of the theory of elliptic functions*, Uchenje Zap. Leningrad (Fiz. Matem. Fak.), **14** (1955); cf. p. 277
576. G.I. Mel'nikov, I.Sh. Slavutskij, *Über zwei vergessene Beweise des quadratischen Reziprozitätsgesetzes* (Russian), Tr. Inst. Istor. Estest. Tekh. 28 (1959), 201–218; Zbl 101.27901; cf. p. 24
577. J.Y. Merindol, *Symbole de Legendre et résultant*, Revue de Math. Spéciales **107** (1996/97), 721–730; cf. p. 417
578. F. Mertens, *Ueber den quadratischen Reciprocitätssatz und die Summen von Gauss*, Wiener Ber. **103** (1894), 1005–1022; FdM **25** (1893/94), 282; cf. p. 415
579. F. Mertens, *Ueber die Gaussischen Summen*, Berl. Ber. (1896), 217–219; FdM **27** (1896), 144; cf. p. 103
580. F. Mertens, *Über eine Darstellung des Legendreschen Zeichens*, Wiener Ber. **113** (1904), 905–910; FdM **35** (1904), 208–209; cf. p. 415
581. F. Mertens, *Die Kummersche Zerfällung der Kreisteilungsresolvente*, Wiener Ber. **114** (1905), 1359–1375; FdM **36** (1905), 124–125; cf. p. 391
582. F. Mertens, *Über die Darstellung der Legendreschen Symbole der biquadratischen, kubischen und bikubischen Reste durch Thetareihen*, Wiener Ber. Abt. IIa **115** (1906), 1339–1360; FdM **37** (1906), 234; cf. p. 201, 226, 277
583. C. Meyer, *Ueber einige Anwendungen Dedekindscher Summen*, J. Reine Angew. Math **198** (1957), 143–203; Zbl 79.10303; MR 21 # 3396; cf. p. 416
584. G. Meyer, *Zur Theorie der quadratischen und kubischen Reste*, Diss. Göttingen, Grunert Arch. **63** (1879), 50 pp; FdM **11** (1879), 129–130
585. A. Micali, *Quadratic residues and the quadratic reciprocity law* (Portug.), Notas. Mat. **35** (1966), 1–35; Zbl 161.04406

586. J. W. Milnor, *Introduction to Algebraic K theory*, Annals of Math. Studies, v. **72**, Princeton University Press (1971), p. 101–106; Zbl 237.18005; MR 50 # 2304; cf. p. 21, 70
587. J. Milnor, D. Husemöller, *Symmetric bilinear forms*, Springer-Verlag 1973; MR 58 #22129; cf. p. 417
588. D. Mirimanoff, K. Hensel, *Sur la relation $(\frac{D}{p}) = (-1)^{n-h}$ et la loi de réciprocité*, J. Reine Angew. Math. **129** (1905), 86–87; FdM **36** (1905), 286–287; cf. p. 137, 415
589. M. G. Monzingo, *On the distribution of consecutive triples of quadratic residues and quadratic nonresidues and related topics*, Fibonacci Q. **23** (1985), 133–138; Zbl 564.10002; cf. p. 26, 342
590. L. J. Mordell, *The diophantine equation $x^4 + my^4 = z^2$*, Quart. J. Math. **18** (1967), 1–6; Zbl 154.29702; cf. p. 175, 174
591. P. Morton, *A generalization of Zolotarev's theorem*, Am. Math. Mon. **86** (1979), 374–375; Zbl 418.10003; MR 80d:10005; cf. p. 24
592. P. Morton, *Governing fields for the 2-classgroup of $\mathbb{Q}(\sqrt{-qq_1p_2})$ and a related reciprocity law*, Acta Arith. **55** (1990), 267–290; Zbl 709.11061; MR 91k:11101; cf. p. 175
593. K. Motose, *On commutative group algebras*, Sci. Rep. Hirosaki Univ. **40** (1993), 127–131; Zbl 819.11054; MR 95b:11009; cf. p. 417
594. K. Motose, *On commutative group algebras. III*, Bull. Fac. Sci. Technol. Hirosaki Univ. **1** (1999), no. 2, 93–97; cf. p. 417
595. A. Movahhedi, M. Zahidi, *Symboles des restes quadratiques des discriminants dans les extensions modérément ramifiées*, Acta Arith. **92** (2000), 239–250; cf. p. 137
596. J.B. Muskat, *On certain prime power congruences*, Abh. Math. Sem. Hamburg **26** (1963), 102–110; Zbl 107.03502; cf. p. 343
597. J.B. Muskat, *Criteria for solvability of certain congruences*, Canad. J. Math. **16** (1964), 343–352; Zbl 126.07601; MR 29 # 1170; cf. p. 343
598. J.B. Muskat, *On the solvability of $x^e \equiv e \pmod{p}$*, Pac. J. Math. **14** (1964), 257–260; Zbl 117.27701; MR 28 # 2997; cf. p. 343
599. J.B. Muskat, *The cyclotomic numbers of order 14*, Acta Arith. **11** (1965/66), 263–279; Zbl 139.28101; MR 33 # 1302; cf. p. 343
600. J.B. Muskat, *Reciprocity and Jacobi sums*, Pac. J. Math. **20** (1967), 275–280; Zbl 204.06402; MR 35 # 1543; cf. p. 343
601. J.B. Muskat, *On Jacobi sums of certain composite orders*, Trans. Amer. Math. Soc. **134** (1968), 483–502; Zbl 174.07701; MR 38 # 1075; cf. p. 228
602. J.B. Muskat, *On simultaneous representations of primes by binary quadratic forms*, J. Number Theory **19** (1984), 263–282; cf. p. 175
603. J.B. Muskat, A. L. Whiteman, *The cyclotomic numbers of order twenty*, Acta Arith. **17** (1970), 185–216; Zbl 216.30801; MR 42 # 3050; cf. p. 343

604. J.B. Muskat, Y. C. Zee, *Sign ambiguities of Jacobi sums*, Duke Math. J. **40** (1973), 313–334; Zbl 268.10019; MR 47, 3326;
605. T. Nagell, *Zahlentheoretische Notizen*, Skrifter Norske Videnskaps Akad. Oslo **13** 1923/24; cf. p. 227
606. T. Nagell, *Einige Sätze über kubische und biquadratische Reste*, Skrifter Oslo (1924), 16–19; cf. p. 202, 227
607. T. Nagell, *Sur les restes et nonrestes cubiques*, Arkiv. Math. **1** (1952), 579–586; Zbl 46.26705
608. T. Nagell, *Sur quelques problèmes dans la théorie des restes quadratiques et cubiques*, Arkiv f. Mat. **3** (1956), 211–222; Zbl 74.27104
609. A. Nakhash, *Connections between the Fibonacci and Lucas progressions and the binomial coefficients,*; cf. p. 73 preprint 1999;
610. P.S. Nasimoff, *Anwendung der Theorie der elliptischen Funktionen auf die Theorie der Zahlen*, Moskau 1885; FdM **17** (1885); cf. p. 277
611. P.S. Nasimoff, *Extrait d'une lettre*, Bull. Sci. Math. (2), **9** (1885), 15–20; cf. p. 26
612. P.S. Nasimoff, *Application de la théorie des fonctions elliptiques à la théorie des nombres*, Ann. Ec. Norm. Paris **5** (1888), 23–48, 147–176; cf. p. 277
613. P.S. Nasimoff, *Application of the Theory of Elliptic Functions to the Theory of Numbers*, transl. from the Russian, mimeographed; see Bull. Amer. Math. Soc. **35**, p. 735; cf. p. 277
614. O. Neumann, *Bemerkungen aus heutiger Sicht über Gauß' Beiträge zu Zahlentheorie, Algebra und Funktionentheorie*, NTM Schr. Geschichte Natur. Tech. Medizin **16** (1979), no. 2, 22–39; Zbl 439.01007; MR 81c:01024; cf. p. 22, 274
615. O. Neumann, *Zur Genesis der algebraischen Zahlentheorie. II*, NTM Schr. Geschichte Natur. Tech. Medizin **17** (1980), no. 1, 32–48; Zbl 496.01004; MR 82i:01032a; cf. p. 22, 100, 274
616. O. Neumann, *Zur Genesis der algebraischen Zahlentheorie III*, NTM Schr. Geschichte Natur. Tech. Medizin **17** (1980), no 2, 38–58; Zbl 496.01005; MR 82i:01032b; cf. p. xii, 274
617. N. Nielsen, *Note sur les résidus quadratiques*, Vidensk. Selsk. Overs. (1916), 191–201; FdM **46** (1916–18), 196
618. N. Nielsen, *Recherches sur les résidus quadratiques et sur les quotients de Fermat*, Ann. Ecole Norm. Sup. (3) **31** (1914), 161–294; FdM **45** (1914/15), 323–324; BSMA (2) **44** (1920), 35–36
619. N. Nielsen, *Note sur le nombre premier 2*, Nyt Tidsskr. Mat. **28** (1917), 7–9; RSPM **26** (1918), 15; cf. p. 25
620. H. Niemeyer, *Bernoullische Zahlen in imaginär-quadratischen Zahlkörpern*, Staatsexamensarbeit Hamburg 1966 (1976); cf. p. 276, 277
621. R. Odoni, *On Gauss sums (mod p^n)*, Bull. London Math. Soc. **5** (1973), 325–327; Zbl 269.10020
622. R. Okazaki, *Quartic residue character of quadratic units*, preprint 1994

623. G. Oltramare, *Considérations générales sur les racines des nombres premiers*, J. Reine Angew Math. **45** (1853), 303–344; cf. p. 25
624. J.S. Park, D.Y. Kim, *An application of the cubic reciprocity law*, Bull. Honam. Math. Soc. **13** (1996), 57–66; MR 97k:11006; cf. p. 225
625. J. C. Parnami, M. K. Agrawal, S. Pall, A. R. Rajwade, *A new proof of the Leonard and Williams criterion for 3 to be a 7th power*, J. Indian Math. Soc., New Ser. **45** (1984), 129–134; Zbl 633.10004; cf. p. 142
626. J. C. Parnami, M. K. Agrawal, A. R. Rajwade, *Jacobi sums and cyclotomic numbers for a finite field*, Acta Arith. **41** (1982), 1–13; Zbl 491.12019; cf. p. 343
627. J. C. Parnami, M. K. Agrawal, A. R. Rajwade, *Criterion for 2 to be ℓ-th power*, Acta Arith. **43** (1984), 361–364; Zbl 539.10006; MR 86i:11004; cf. p. 142
628. S.J. Patterson, *Erich Hecke und die Rolle der L-Reihen in der Zahlentheorie*, in: Ein Jahrhundert Mathematik 1890–1990, Dokumente Gesch. Math. **6**, 629–655; Vieweg, 1990; MR 91m:01020; cf. p. 18
629. K.C. Peitz, J.H. Jordan, *Quadratic reciprocity in $\mathbb{Z}(\sqrt{-3})$*, Washington State Univ. Technical Report **8** (1966); cf. p. 226
630. T. Peklar, *A proof of the quadratic reciprocity law* (Slowen.), Obt. Mat. Fiz. **36** (1989), 129–133; Zbl 679.10004; cf. p. 417
631. A.E. Pellet, *Sur la décomposition d'une fonction entière en facteurs irréducibles suivant un module permier*, Comptes Rendus Paris **86** (1878), 1071–1072; FdM **10** (1878), 295–296; cf. p. 137, 414
632. A.E. Pellet, *Sur les résidus cubiques et biquadratiques suivant un module premier*, Bull. Soc. Math. France **10** (1882), 157–162; FdM **14** (1882), 123; cf. p. 73, 202, 225, 227, 414
633. A.E. Pellet, *Sur les fonctions réduites suivant un module premier*, Bull. Soc. Math. France **17** (1889), 156–167; FdM **21** (1889), 179; cf. p. 202, 225
634. A.E. Pellet, *Sur les caractères cubiques et biquadratiques*, C. R. Acad. Sci. Paris **108** (1889), 609–610; FdM **21** (1889), 178; cf. p. 202, 227,
635. T. Pépin, *Sur les résidus de cinquième puissance*, C. R. Acad. Sci. Paris **76** (1873), 151–156; BSMA (2) **5** (1873), 122; FdM **5** (1873), 100–101; cf. p. 393
636. T. Pépin, *Sur les résidus cubiques*, C. R. Acad. Sci. Paris **79** (1874), 1403–1407; BSMA (2) **8** (1875), 42; FdM **6** (1874), 114; cf. p. 225
637. T. Pépin, *Sur les résidus de septième puissance*, C. R. Acad. Sci. Paris **80** (1875), 811–815; FdM **7** (1875), 91
638. T. Pépin, *Étude sur la théorie des résidus cubiques*, J. Math. pures appl. (3) **2** (1876), 313–325; FdM **8** (1876), 96; cf. p. 225, 342
639. T. Pépin, *Sur les lois de réciprocité dans la théorie des résidus des puissances*, C. R. Acad. Sci. Paris **84** (1877), 762–765; FdM **9** (1877), 132; cf. p. 141, 142

640. T. Pépin, *Mémoire sur les lois de réciprocité relatives aux résidus des puissances*, Atti della Accademia Pontificia dei Nuovi Lincei Roma **31** (1878), 40–149; FdM **12** (1880), 122–123; BSMA 1880, p. 181; cf. p. 224, 414

641. T. Pépin, *Démonstration d'un théorème de M. Sylvester sur les diviseurs d'une fonction cyclotomique*, C. R. Acad. Sci. Paris **90** (1880), 526–528; cf. p. 102

642. T. Pépin, *Sur trois théorèmes de Gauss*, Rom. Acc. P. d. N. L. **38** (1885), 197–200; FdM **18** (1886), 146; cf. p. 174

643. T. Pépin, *Nouvelle démonstration de la loi de réciprocité de Legendre*, Rom. Acc. P. d. N. L. **43** (1890), 192–198; FdM **22** (1890), 206; cf. p. 414

644. T. Pépin, *Démonstration d'un théorème de Liouville*, Rom. Acc. P. d. N. L. Mem. **5** (1890), 131–151; FdM **22** (1890), 212; cf. p. 174

645. T. Pépin, *Dissertation sur deux démonstrations du théorème de réciprocité de Legendre*, Rom. Acc. P. d. N. L. **51** (1898), 123–144; FdM **29** (1898), 154; RSPM **7** (1898), 113; cf. p. 20, 415

646. T. Pépin, *Nouvelle formule relative aux résidus quadratiques*, C. R. Acad. Sci. Paris **128** (1899), 1553–1556; FdM **30** (1899), 183–184; RSPM **8** (1900), 67; BSMA (2) **25** (1901), 108; cf. p. 23

647. T. Pépin, *Etude historique sur la théorie des résidus quadratiques*, Memorie della Accademia Pontificia dei Nuovi Lincei Roma **16** (1900), 229–276; FdM **31** (1900), 188–189; cf. p. 22

648. T. Pépin, *Théorie des Nombres*, Memorie della Accademia Pontificia dei Nuovi Lincei Roma **23** (1905), 109–177; FdM **36** (1906), 256–257

649. T. Pépin, *Théorie des Nombres; Suite et fin*, Memorie della Accademia Pontificia dei Nuovi Lincei Roma **29** (1911), 319–339; FdM **42** (1911), 202; RSPM **21** (1913), 86; cf. p. 415

650. J. Perott, *Sur l'équation $t^2 - Du^2 = -1$*, J. Reine Angew. Math. **102** (1888), 185–223; FdM **19** (1887), 183–184; cf. p. 175, 310, 313

651. J. Petersen, *A new proof of the theorem of reciprocity*, Amer. J. Math. pure and appl. **2** (1879), 285–286; FdM **11** (1879), 132; cf. p. 414

652. J. Petersen, *Reciprocitetssätningen*, Tidssk. Mat. udgived of Zeuthen (4) **III** (1879), 86–90; FdM **11** (1879), 132; BSMA (2) **8** (1884), 175

653. K. Petr, *Poznámka o Legendre-Jacobiove symbolu $(\frac{P}{Q})$* (Eine Bemerkung über das Legendre-Jacobische Symbol (P/Q), Bohem.), Casopis **40** (1911), 162–165; FdM **42** (1911), 212; RSPM **21** (1913), 104; cf. p. 415

654. K. Petr, *Über die lineare Transformation der Thetafunktionen* (Czech.), Rozpravy Ceske Akad. ved. (2) **36**, No. 1 (1927), 10 pp; FdM **53** (1927), 345; cf. p. 415

655. K. Petr, *Über die Eisensteinschen Beweise des Reziprozitätsgesetzes bei den biquadratischen und bikubischen Resten* (Czech.), C. R. Congrès Math. Pays slaves 1929 (1930), 119–128; FdM **56-I** (1930), 158; cf. p. 201, 225, 226, 277

656. K. Petr, *Poznámka k důkazu zákona reciprocity pro kvadratiské zbytky* (Remarque concernant la loi de réciprocité des residus quadratiques, Czech., French summary), Časopis **62** (1934), 228–230; FdM **59** - II (1933), 935; cf. p. 416
657. K. Petr, *On alternating functions in a cyclotomic field*, Rozpravy Il. Tridy Ceske Akad. **56** (1946), 12 pp; MR 10, 15c; cf. p. 416
658. J. V. Pexider, *Über Potenzreste*, Archiv Math. Phys. (III) **14** (1909), 71–93; FdM **39** (1908); cf. p. 26
659. G. Pickert, *Von einer Siebenerprobe zu den quadratischen Resten*, Math. Semesterber. **37** (1990), 59–67; cf. p. 23
660. J. Piehler, *Bemerkungen zur Verteilung der kubischen Reste*, Math. Ann. **134** (1957), 50–52; Zbl 78.03701
661. H. Pieper, *Variationen über ein zahlentheoretisches Thema von Carl Friedrich Gauss*, Birkhäuser Verlag Basel-Stuttgart, 1978, 183 pp; Zbl 373.10001; MR 81b:10001; cf. p. 24, 139
662. H. Pieper, *Über Legendres Versuche, das quadratische Reziprozitätsgesetz zu beweisen*, Acta hist. Leopoldina **27** (1997), 223–237; MR 98k:01011; cf. p. 5, 18, 20
663. H.C. Pocklington, *The determination of the exponent to which a number belongs, the practical solution of certain congruences, and the law of quadratic reciprocity*, Math. Proc. Cambr. Phil. Soc. **16**, (1911), 1–5; FdM **42** (1911), 209; cf. p. 415
664. P. Porcelli, G. Pall, *A property of Farey sequences, with applications to qth power residues*, Can. J. Math. **3** (1951), 52–53; Zbl 42.26802; MR 12:804f; cf. p. 136
665. H. Rademacher, *Lectures on elementary number theory*, New York-Toronto-London 1964; MR 30 # 1079; cf. p. 416
666. H. Rademacher, E. Grosswald, *Dedekind sums*, Carus Math. Monographs No. 16, 1972; MR 50 # 9767
667. A.R. Rajwade, *Cyclotomy – a survey article*, Math. Stud. **48** (1985), 70–115 (1980); Zbl 587.12012; MR 84b:12031; cf. p. 101
668. Rao D. Rameswar, *A few results on congruences and quadratic residues*, Booklinks Corp. 1973, 24 pp; MR 49:198; cf. p. 23
669. L. Rédei, *Ein neuer Beweis des quadratischen Reciprocitätssatzes*, Acta Sci. Math. Szeged **2** (1925), 134–138; FdM **51** (1925), 129; cf. p. 415
670. L. Rédei, *Ein neuer Beweis des quadratischen Reciprocitätssatzes*, J. Reine Angew. Math. **155** (1926), 103–106; FdM **52** (1926), 143; cf. p. 415
671. L. Rédei, *Über die Grundeinheit und die durch 8 teilbaren Invarianten der absoluten Klassengruppe im quadratischen Zahlkörper*, J. Reine Angew. Math. **171** (1934), 131–148; cf. p. 172, 175
672. L. Rédei, *Ein neues zahlentheoretisches Symbol mit Anwendung in der Klassenkörpertheorie I*, J. Reine Angew. Math. **180** (1938), 1–43; FdM **65** I (1939), 106–107; Zbl 21.00701; cf. p. 175

673. L. Rédei, *Kurze Darstellung des fünften Gauss'schen Beweises für den quadratischen Reziprozitätssatz*, Comment. Math. Helv. **16** (1944), 264–265; Zbl 60.08607; MR 6:256a; cf. p. 22, 416
674. L. Rédei, *Die Primfaktoren der Zahlenfolge* 1, 3, 4, 7, 11, 18, ..., Portugaliae Math. **8** (1949), 59–61; Zbl 39.03501; cf. p. 70, 174
675. L. Rédei, *Einfacher Beweis des quadratischen Reziprozitätssatzes*, Mathemat. Z. **54** (1951), 25–26; Zbl 42.03911; MR 12:675g; cf. p. 416
676. L. Rédei, *Zur Theorie der Polynomideale über kommutativen nullteilerfreien Hauptidealringen*, Math. Nachr. **18** (1958), 313–332; MR 20 #2334; cf. p. 416
677. H. Reichardt, *Eine Bemerkung zur Theorie des Jacobischen Symbols*, Math. Nachr. **19** (1958), 171–175; Zbl 88.25501; MR 21:3394; cf. p. 23, 416
678. C. G. Reuschle, *Mathematische Abhandlungen, enthaltend neue zahlentheoretische Tabellen*, Progr. Königl. Gymn. Stuttgart (1856), 61 pp.; cf. p. 141, 205, 310
679. I.V. Reshetukha, *A question in the theory of cubic residues*, Matem. Zametki **7** (1970), 469–476; Zbl 194.350; cf. p. 278
680. I.V. Reshetukha, *Generalized sums for characters and their applications to the reciprocity laws* (Russ.), Ukrain. Mat. Z. **23** (1971), 270–276; Zbl 261.10027; MR 47 # 3328; cf. p. 225, 278, 417
681. I.V. Reshetukha, *An analytic definition of a product of cubic character*, Ukrainian Math. J. **27** (1975), 152–158; Zbl 313.10037; cf. p. 225, 278
682. I.V. Reshetukha, *Application of a cyclic determinant in the theory of cubic residues*, Mat. Zametki **34** (1983), 783–795; MR 85k:11037; engl. transl.: Math. Notes **34** (1983), 884–891; Zbl 532.10023; cf. p. 278
683. I.V. Reshetukha, *A product related to cubic Gaussian sums*, Ukrainian Math. J. **37** (1985), 611–616; Zbl 599.10027; MR 87d:11009; cf. p. 278
684. M. Riesz, *Sur le lemme de Zolotareff et sur la loi de réciprocité des restes quadratiques*, Math. Scand. **1** (1953), 159–169; Zbl 51.27504; MR 15:200d; cf. p. 24, 416
685. J. B. Roberts, *Integral power residues as permutations*, Amer. Math. Monthly **76** (1969), 379–385; Zbl 179.06603; MR 39 # 1403; cf. p. 24
686. K. Rogers, *Legendre's theorem and quadratic reciprocity*, J. Number Theory **6** (1974), 339–344; Zbl 292.10005; MR 52:10564; cf. p. 417
687. K. H. Rosen, *Least positive residues and the quadratic character of 2*, Canad. Math. Bull. **23** (1980), 355–358; Zbl 438.10003; MR 82b:10005; cf. p. 25
688. G. Rousseau, *Exterior algebras and the quadratic reciprocity law*, L'enseignement Math. **36** (1990), 303–308; Zbl 729.11003; MR 92e:11006; cf. p. 417
689. G. Rousseau, *On the quadratic reciprocity law*, J. Austral. Math. Soc. **51** (1991), 423–425; Zbl 748.11002; MR 93b:11005; cf. p. 31, 417

690. G. Rousseau, *On the Jacobi symbol*, J. Number Theory **48** (1994), 109–111; Zbl 814.11002; MR 93b:11005; cf. p. 24, 417
691. D. E. Rowe, *Gauss, Dirichlet, and the law of biquadratic reciprocity*, Math. Intell. **10** (1988), 13–25; Zbl 644.10001; MR 89e:01031; cf. p. 201
692. D. M. Russinoff, *A mechanical proof of quadratic reciprocity*, J. Autom. Reasoning **8** (1992), 3–21; Zbl 767.68087; MR 93b:68078; cf. p. 25, 417
693. L. Saalschütz, *Zur Lehre von den quadratischen Resten*, Archiv Math. Phys. (III) **9** (1905), 220–230; cf. p. 26
694. S. Saidi, *Semicrosses and quadratic forms*, Europ. J. Combinatorics **16** (1995), 191–196; Zbl 822.05020; cf. p. 228
695. G. Sansone, *Sulle equazioni indeterminate delle unità di norma negativa dei corpi quadratici reali*, Rend. Acad. d. L. Roma (6) **2** (1925), 479–484; FdM **51** (1925), 132–133; cf. p. 175, 180
696. G. Sansone, *Ancora sulle equazioni indeterminate delle unità di norma negativa dei corpi quadratici reali*, Rend. Acad. d. L. Roma (6) **2**, 548–554; FdM **51** (1925), 132–133; cf. p. 175, 180
697. Ph. Satgé, *Lois de réciprocité et lois de décomposition*, Sém. Théorie des Nombres Bordeaux, exp. 9 (1974/75), 10 pp; Zbl 334.12018; MR 52:13742; cf. p. 175
698. Ph. Satgé, *Décomposition des nombres premiers dans des extensions non abéliennes*, Ann. Inst. Fourier **27** (1977), 1–8; MR 57 #16257; cf. p. 175
699. K. Sato, *A remark on the criteria for 3 to be a ninth power* (mod p), Math. Scand. **60** (1987), 148–150; Zbl 634.10005; cf. p. 142
700. K. Sato, T. Karakisawa, *Explicit criteria in L, M ($4p = L^2 + 27M^2$) for cubic residuacity*, (Japan.), J. Coll. Eng. Nikon Univ. Ser. B **30** (1989), 117–119; Zbl 676.10005; MR 89m:11005; cf. p. 225
701. U. Sbrana, *Alcune proprietà dell'equazione per la divisione dei periodi di una funziona equianarmonica*, Battaglini giorn. **42** (1904), 297–311; RSPM **13** (1905), 116; cf. p. 277
702. M. Schaar, *Nouvelles démonstrations de la loi de réciprocité pour les residus quadratiques*, Bulletin de l'Academie de Belgique **14** (1847), 79–83; cf. p. 413
703. M. Schaar, *Mémoire sur la théorie des residus quadratiques*, Acad. Roy. Sci. Lettres Beaux Arts Belgique **24** (1852), 14 pp; cf. p. 103
704. M. Schaar, *Recherches sur la théorie des residus quadratiques*, Acad. Roy. Sci. Lettres Beaux Arts Belgique **25** (1854), 20 pp; cf. p. 103
705. W. Scharlau, *Quadratic reciprocity laws*, J. Number Theory **4** (1972), 78–97; Zbl 241.12005; MR 45 # 1835; cf. p. 21
706. W. Scharlau, *Quadratic and Hermitian Forms*, Grundlehren der math. Wiss., Springer Verlag 1985; cf. p. 21
707. W. Scheibner, *Zur Theorie des Legendre-Jacobi'schen Symbols* $(\frac{n}{m})$, Leipz. Abh. **24** (1899), 369–410; FdM **31** (1900), 190; cf. p. 26, 201

708. W. Scheibner, *Zur Theorie des Legendre-Jacobi'schen Symbols* $\left(\frac{n}{m}\right)$, Leipz. Abh. **27** (1902), 653–752; FdM **33** (1902), 204; RSPM **13** (1905), 40; cf. p. 26, 170, 201, 226
709. E. Schering, *Zum dritten Gauss'schen Beweis des Reciprocitätssatzes für die quadratischen Reste*, Berl. Ber. (1885), 113–117; FdM **17** (1885), 153; Werke II, 103–106; cf. p. 21, 414
710. E. Schering, *Verallgemeinerung des Gauss'schen Criteriums für den quadratischen Restcharakter einer Zahl in Bezug auf eine andere*, Berl. Ber. (1876), 330–331; Werke I, 285–286; FdM **8** (1876), 93–95; cf. p. 23
711. E. Schering, *Neuer Beweis des Reciprocitäts-Satzes für die quadratischen Reste*, Gött. Nachr. (1879), 217–224; FdM **11** (1879), 130–131; Werke I, 331–336; cf. p. 414
712. E. Schering, *Nouvelle démonstration de la loi de réciprocité dans la théorie des résidus quadratique*, C. R. Acad. Sci. Paris **88** (1879), 1073–1075; FdM **11** (1879), 130–131; Werke I, 337–340; cf. p. 414
713. E. Schering, *Bestimmung des quadratischen Rest-Charakters*, Göttinger Abh. **24** (1879), 1–47; Werke I, 341–386; cf. p. 26
714. E. Schering, *Zur Theorie der quadratischen Reste*, Acta Math. **1** (1882), 153–170; FdM **15** (1883), 141; Werke II, 69–86; cf. p. 23
715. A. Schiappa-Monteiro, *Sur un théorème relatif à la théorie des nombres*, Revista Scientifica **1** (1884); FdM **16** (1884), 139; cf. p. 25
716. A. Schinzel, *A refinement of a theorem of Gerst on power residues*, Acta Arith. **17** (1970), 161–168; Zbl 233.10003; cf. p. 136
717. A. Schinzel, *Abelian binomials, power residues and exponential congruences*, Acta Arith. **32** (1977), 245–274; Zbl 409.12029; MR 55 # 2829; Addendum: ibid. **36** (1980), 101–104; Zbl 438.12014; cf. p. 136
718. A. Schinzel, *Les residus de puissances et des congruences exponentielles*, Journ. Arithm. Caen, Astérisque **41/42** (1977), 203–209; Zbl 364.12001; MR 56 # 5488; cf. p. 136
719. H. Schmidt, *Drei neue Beweise des Reciprocitätssatzes in der Theorie der quadratischen Reste*, J. Reine Angew. Math. **111** (1893), 107–120; FdM **25** (1893/94), 279–280; cf. p. 32, 414
720. G. Schmitt, *Verschiedene Beweise des quadratischen Reziprozitätsgesetzes*, Staatsexamensarbeit Heidelberg; cf. p. 24
721. Th. Schönemann, *Ueber die Congruenz $x^2 + y^2 \equiv 1 \pmod{p}$*, J. Reine Angew. Math. **19** (1839), 93–112; cf. p. 25
722. Th. Schönemann, *Theorie der symmetrischen Functionen der Wurzeln einer Gleichung. Allgemeine Sätze über Congruenzen nebst einigen Anwendungen derselben*, J. Reine Angew. Math. **19** (1839), 289–308; cf. p. 101, 160, 174, 413
723. A. Scholz, *Über die Lösbarkeit der Gleichung $t^2 - Du^2 = -4$*, Math. Z. **39** (1934), 95–111; Zbl 9.29402; cf. p. 166
724. A. Scholz, *Einführung in die Zahlentheorie*, Göschen **5131** (eds.: B. Schoeneberg), 1973; Zbl 248.10001; cf. p. 23

725. F. Schuh, *Theorie der hoogere-machtscongruenties en der hoogere-machtsresten*, 272 pp, Leyden, van der Hoek 1928; RSPM **34** (1929), 132; cf. p. 136
726. V. Schulze, *Potenzreste*, Acta Arith. **33** (1977), 379–404; Zbl 355.10005; MR 57 # 220; cf. p. 136
727. V. Schulze, *Erweiterung eines Satzes von Schinzel über Potenzreste*, Acta Arith. **41** (1982), 383–394; Zbl 486.12001; cf. p. 136
728. K. Schwering, *Untersuchung über die fünften Potenzreste und die aus den fünften Einheitswurzeln gebildeten ganzen Zahlen*, Z. Math. u. Phys. **27** (1882), 102–119; FdM **14** (1882); BSMA (2) **8** (1884), 196–197
729. K. Schwering, *Multiplication der lemniscatischen Funktion sin am u*, J. Reine Angew. Math. **107** (1891), 196–240; FdM **23** (1894), 476–480; cf. p. 277
730. J. Schwermer, *Über Reziprozitätsgesetze in der Zahlentheorie*, Mathematische Miniaturen 3, Arithmetik und Geometrie (eds: Knörrer et al.), Birkhäuser 1986, 29–69; MR 88c:11064; cf. p. 25
731. R.T. Sharifi, *On cyclotomic polynomials, power residues, and reciprocity laws*, L'Ens. Math. **43** (1997), 319–336; Zbl 904.11003; cf. p. 102, 203, 229, 397
732. K. Shiratani, *Die Diskriminante der Weierstraß'schen elliptischen Funktionen und das Reziprozitätsgesetz in besonderen imaginär-quadratischen Zahlkörpern*, Abh. Math. Sem. Univ. Hamburg **31** (1967), 51–61; Zbl 163.05101; MR 35 # 6647; cf. p. 277
733. K. Shiratani, *Über eine Anwendung elliptischer Funktionen auf das biquadratische Reziprozitätsgesetz*, J. Reine Angew. Math. **268/269** (1974), 203–208; Zbl 289.12001; MR 50 # 2050; cf. p. 201, 277
734. C. L. Siegel, *Über das quadratische Reziprozitätsgesetz in algebraischen Zahlkörpern*, Nachr. Akad. Wiss. Göttingen (1960), 1–16; Ges. Abh. 3 (1966), 334–349; Zbl 109.26604; MR 23:A 878; cf. p. 103
735. W. Sierpinsky, *Contribution à l'étude des restes cubiques*, Ann. Soc. Polon. Math. **22** (1949), 269–272; Zbl 38.18102; MR 11:641a; cf. p. 29
736. Th. Skolem, *Geschlechter und Reziprozitätsgesetze*, Norsk. Mat. Forenings Skrifter (1), **18** (1928), 38 pp; FdM **54** (1928), 192; cf. p. 415
737. Th. Skolem, *Ein einfacher Beweis der sogenannten Zählertransformationsformel der Jacobischen Symbole*, Avhandlinger Oslo **11** (1932) 7pp; FdM **58** - I (1932), 177; Zbl 5.19402
738. Th. Skolem, *A property of ternary quadratic forms and its connection with the quadratic reciprocity theorem* (Norw.), Norsk Mat. Tidsskr. **30** (1948), 1–10; Zbl 30.34303; MR 10 # 15d; cf. p. 416
739. Th. Skolem, *On a certain connection beteeen the discriminant of a polynomial and the number of its irreducible factors mod p*, Norske Mat. Tidsskr. **34** (1952), 81–85; Zbl 47.25403; MR 14:251a; cf. p. 137

740. Th. Skolem, *Remarks on proofs by cyclotomic formulas of reciprocity laws for power residues*, Math. Scand. **9** (1961), 229–242; Zbl 115.26403; MR 24:A1243; cf. p. 102, 225, 416
741. Th. Skolem, *A proof of the quadratic law of reciprocity with proofs of two so-called "Ergänzungssätze"*, Norske Vid. Selsk. Forh. Trondheim **34** (1961), 18–24; Zbl 106.03401; MR 30:4720; cf. p. 416
742. I.Sh. Slavutskij, *Ein verallgemeinertes Lemma von Zolotarev* (Russian), Acad. Republ. Popul. Roum., Rev. Math. Pur. Appl. **8** (1963), 455–457; Zbl 135.09203; cf. p. 24
743. H.J. Smith, *Report on the theory of numbers*, Coll. Math. Papers **I**, Chelsea (1894), 55–92; 2nd ed. 1965; cf. p. 22, 101, 201, 275
744. H.J. Smith, *On complex binary quadratic forms*, Proc. Royal Soc. **13**, 278–298: Coll. Math. Papers **I**, Chelsea (1894), 418–442; cf. p. 170
745. B. K. Spearman, K. S. Williams, *A simple proof of Eisenstein's reciprocity law from Stickelberger's theorem*, Indian J. Pure Appl. Math. **17** (1986), 169–174; Zbl 594.12004; cf. p. 393
746. B. Spies, *Der Gauss'sche Beweis des biquadratischen Reziprozitätsgesetzes*, Diss. T.U. Braunschweig, 1983; cf. p. 201
747. T. A. Springer, *The Theory of Quadratic Forms in the Disquisitiones Arithmeticae*, in: Carl Friedrich Gauss, Four Lectures on his Life and Work (A. F. Monna, ed.), Communications of the Math. Inst. Rijksuniversiteit Utrecht 1978, 42–53; cf. p. 21
748. J. Stalker, *Complex Analysis. Fundamentals of the Classical Theory of Functions*, Birkhäuser, 1998; cf. p. 204
749. C. von Staudt, *Ueber die Functionen Y und Z, welche der Gleichung $\frac{4(x^p-1)}{x-1} = Y^2 \mp pZ^2$ Genüge leisten, wo p eine Primzahl der Form $4k \pm 1$ ist*, J. Reine Angew. Math. **67** (1867), 205–217; cf. p. 26, 413
750. L.M. Steenvoorden, *Prime numbers and quadratic residues, a trove in the garden of numbers* (Dutch), Nieuw Tijdschr. Wisk. **75** (1988), 153–158; MR 90a:11009; cf. p. 13
751. C. Stengel, *Über quadratische Nichtreste von der Form $8h + 1$*, J. Reine Angew. Math. **153** (1924), 208–214; FdM **50** (1924), 87–88; cf. p. 101
752. M.A. Stern, *Bemerkungen über höhere Arithmetik*, J. Reine Angew. Math. **6** (1830), 147–158; cf. p. 225
753. M.A. Stern, *Démonstration de quelques théorèmes sur les nombres*, J. Reine Angew. Math. **12** (1834), 288–290; cf. p. 25
754. M.A. Stern, *Recherches sur la théorie des résidus quadratiques*, Acad. Royale Bruxelles **15** (1841), 1–38; cf. p. 25, 26
755. M.A. Stern, *Ueber eine der Theilung der Zahlen ähnliche Untersuchung und deren Anwendung auf die Theorie der quadratischen Reste*, J. Reine Angew. Math. **61** (1863), 66–94; cf. p. 25
756. M.A. Stern, *Über einen einfachen Beweis des quadratischen Reciprocitätsgesetzes und einige damit zusammenhängende Sätze*, Gött. Nachr (1870), 237–253; FdM **2** (1869/70), 93–94; cf. p. 413

757. M.A. Stern, *Über quadratische, trigonale und bitrigonale Reste*, J. Reine Angew. Math. **71** (1870), 137–163; cf. p. 26
758. P. Stevenhagen, *Divisibility by 2-powers of certain quadratic class numbers*, Report 91-12, Univ. Amsterdam; J. Number Theory **43** (1993), 1–19; Zbl 767.11054; MR 94b:11113; cf. p. 312
759. L. Stickelberger, *Über eine Verallgemeinerung der Kreistheilung*, Math. Ann. **37** (1890), 321–367; cf. p. 137, 145, 225, 310, 313, 366, 391, 394
760. L. Stickelberger, *Über eine neue Eigenschaft der Diskriminanten algebraischer Zahlkörper*, Verhandl. I. Internat. Math. Kongress 1897, Zürich, 183–193; FdM **29** (1898), 172–173; cf. p. 145
761. T.J. Stieltjes, *Over het quadratische rest-karakter van het getal 2*, Nieuw Arch. Wiskunde **9** (1882), 193–195; FdM **14** (1882), 122; Œuvres Complètes I, 137–140; cf. p. 25
762. T.J. Stieltjes, *Le nombre 2 comme résidu quadratique*, Œuvres Complètes **I** (1882), 141–144; cf. p. 25
763. T.J. Stieltjes, *Bijdrage tot de theorie der derde-en vierde-macht resten*, Amst. Versl. en Meded. **17** (1882), 338–417; FdM **14** (1882), 123; Œuvres Complètes I, 145–209; French translation: *Contributions à la théorie des résidus cubiques et biquadratiques*, Toulouse Ann. **11** (1883), 1–65; Arch. Neerl. 18, 358–436, FdM **15** (1883), 141; Œuvres Complètes I, 210–275; FdM **28** (1897), 170–171; cf. p. 202, 225
764. T.J. Stieltjes, *Sur la théorie des résidus biquadratiques* (Lettre à Hermite), Bull. Sci. Math. Paris (2), **7** (1883), 139–142; FdM **15** (1883), 141–142; Œuvres Complètes I, 308–310; cf. p. 202
765. T.J. Stieltjes, *Sur le caractère quadratique du nombre 2 comme résidue ou non-résidue quadratique*, Bull. Sci. Math. Paris (2), **8** (1884), 175–176; FdM **16** (1884), 156; Œuvres Complètes I, 361–363; cf. p. 25
766. T.J. Stieltjes, *Sur le caractère quadratique du nombre 2*, Toulouse Ann. **11** (1884), 5–8; FdM **28** (1897), 169; cf. p. 25
767. T.J. Stieltjes, *Sur la loi de réciprocité de Legendre*, Œuvres Complètes **II** (1918), 567–573; FdM **46** (1916–18), 188–189; cf. p. 415
768. E. Storchi, *Nuova dimostrazione di un teorema sui numeri primi*, Periodico Mat. **18**, 247–276; FdM 65 I (1939), 127; cf. p. 25, 70
769. E. Storchi, *Alcuni criteri di divisibilità per i numeri di Mersenne e il carattere $6^{co}, 12^{mo}, 48^{mo}$, dell'intero 2*, Bull. Unione Mat. Ital. (3) **10** (1955), 363–375; Zbl 64.27703; MR 17,127h; cf. p. 70
770. E. Storchi, *Un metodo per la fattorizzazione dei numeri della forma $a^n \pm 1$*, Ist. Lomb. Sci. Lett. **88** (III) (1955), 405–441; Zbl 66.29103; MR 17,585c; cf. p. 70
771. X. Stouff, *Sur les lois de réciprocité et les sous-groupes du groupe arithmetique*, C. R. Acad. Sci. Paris **116** (1893), 308–309; BSMA **19** (1895), 157; cf. p. 202

772. X. Stouff, *Les lois de réciprocité et les sous-groupes du groupe arithmetique*, Ann. de l'Ec. Norm. (3), **10** (1893), 295–314; BSMA **20** (1896), 31–32; FdM **25** (1893/94), 281–282; cf. p. 202
773. M. V. Subbarao, *The algebra of biquadratic residues*, J. Madras Univ. Sect. B **30** (1960), 123–131; Zbl 104.26803; MR 25:43
774. Zh. Sun, *Notes on quadratic residue symbols and rational reciprocity laws*, Nanjing Daxue Xuebao Shuxue Bannian Kan **9** (1992), 92–103; Zbl 791.11001; MR 94b:11007; cf. p. 173
775. Zh. Sun, *Combinatorial sum $\sum_{k\equiv r\ (m)} \binom{n}{k}$ and its applications in number theory II* (Chinese), J. Nanjing Univ. Biquarterly **5** (1993), 105–118; Zbl 808.11016; MR 94j:11021
776. Zh. Sun, *On the theory of cubic residues and nonresidues*, Acta Arith. **84** (1998), 291–335; MR 99c:11005; cf. p. 226, 228
777. Zh. Sun, *Supplements to the theory of biquadratic residues*, Acta Arith., to appear
778. R.G. Swan, *Factorization of polynomials over finite fields*, Pac. J. Math. **12** (1962), 1099–1106; Zbl 113.01701; MR 26 #2432; cf. p. 137, 416
779. R.G. Swan, *Another proof of the quadratic reciprocity law?*, Amer. Math. Month. **97** (1990), 138–139; Zbl 748.11003; MR 91b:11003; cf. p. 80, 417
780. J.J. Sylvester, *Sur la fonction $E(x)$*, C. R. Acad. Sci. Paris **50** (1860), 732–734; cf. p. 23
781. J.J. Sylvester, *Instantaneous proof of a theorem of Lagrange on the divisors of the form $Ax^2+By^2+Cz^2 = 0$, with a postscript on the divisors of the functions which multisect the primitive roots of unity*, Amer. J. Math. **3** (1880), 390–392; FdM **13** (1881), 137; Math. Papers III, 446–448; cf. p. 102
782. J.J. Sylvester, *Sur les diviseurs des fonctions cyclotomiques*, C. R. Acad. Sci. Paris **90** (1880), 287–290, 345–347; FdM **12** (1880), 129; cf. p. 102
783. J.J. Sylvester, *Sur la loi de réciprocité dans la théorie des nombres*, C. R. Acad. Sci. Paris **90** (1880), 1053–1057; FdM **12** (1880), 124; cf. p. 26
784. J.J. Sylvester, *Sur la loi de réciprocité dans la théorie des nombres*, C. R. Acad. Sci. Paris **90** (1880), 1104–1106; FdM **12** (1880), 124; cf. p. 26
785. J.J. Sylvester, *On the multisection of the roots of unity*, John Hopkins Univ. Circulars **1** (1881), 150–152; Math. Papers III, 477–478; cf. p. 102
786. J.J. Sylvester, *Sur les diviseurs des fonctions des périodes des racines primitives de l'unité*, C. R. Acad. Sci. Paris **92** (1881), 1084–1086; FdM **13** (1881), p. 136; Math. Papers III, 479–480; cf. p. 102
787. A. Tafelmacher, *Zu dem dritten Gauss'schen Beweise des Reciprocitäts-Satzes für die quadratischen Reste gehörende Untersuchungen*, Diss. Göttingen 1889, Pr. Gymn. Osnabrück 1890, 1–24; FdM **22** (1890), 206; cf. p. 21, 414
788. T. Takagi, *A simple proof of the law of quadratic reciprocity for quadratic residues*, Proc. Phys.-Math. Soc. Japan, Ser. II **2** (1903), 74–78;

FdM **35** (1904), 210; Coll. Papers, 2nd ed. Springer Verlag 1990, 10–12; Zbl 707.01017; cf. p. 415

789. T. Takagi, *Über eine Eigenschaft des Potenzcharacters*, Proc. Phys.-Math. Soc. Japan **9** (1917), 166-169; Coll. Papers, 68–70; cf. p. 136

790. T. Takagi, *Über das Reciprocitätsgesetz in einem beliebigen algebraischen Zahlkörper*, J. of the College of Science Tokyo **4** (1922); Coll. Papers, 179–216; FdM **48** (1921–22), 169; cf. p. 393

791. F. Tano, *Sur quelques points de la théorie des nombres*, Darboux Bull. (2), **14**, 215–218

792. J. Tate, *Problem 9: The general reciprocity law*, Proc. Symp. Pure Math. **28** (1976), 311–322; Zbl 348.12013; MR 55:2849; cf. p. viii

793. L. Taylor, *The general law of quadratic reciprocity*, Fibonacci Quart. **13** (1975), 318, 321, 324, 328, 330, 333, 336, 339, 342, 344, 349, 384; Zbl 316.10002; MR 52:3029; cf. p. 26

794. L. Taylor, *A conjecture relating quartic reciprocity and quartic residuacity to primitive Pythagorean triples*, Fibonacci Quart. **14** (1976), 180–181; Zbl 353.10002; cf. p. 182

795. H. Teege, *Über den Legendreschen Beweis des sogenannten Reziprozitätsgesetzes in der Lehre von den quadratischen Resten und seine Vervollständigung durch den Nachweis, dass jede Primzahl von der Form $8n+1$ quadratischer Nichtrest unendlich vieler Primzahlen von der Form $4n+3$ ist*, Mitt. Math. Ges. Hamburg **5** (1920), 6–19; cf. p. 20, 415

796. H. Teege, *Ein Kreisteilungsbeweis für das quadratische Reziprozitätsgesetz*, Mitt. Math. Ges. Hamburg **6** (1921), 136–138; FdM **51** (1925), 128; cf. p. 415

797. H. Teege, *Über den Zusammenhang von $f \equiv 1 \cdot 2 \cdot \ldots \cdot \frac{p-1}{2} \pmod{p}$ mit der Klassenzahl der binären quadratischen Formen von positiver Diskriminante $+p$*, Mitt. Math. Ges. Hamburg **6** (1921), 87–100; FdM **50** (1924), 94; cf. p. 101

798. G. Terjanian, *Sur la loi de réciprocité des puissances l-èmes*, Acta Arith. **54** (1989), 87–125; Zbl 679.12006; cf. p. 396

799. J. Thomae, see A. Voigt [815]; cf. p. 414

800. N. Tihanyi, *Kennzeichen kubischer Reste* (Hungar.), Math. és phys. lapok **27** (1918), 67–79; FdM **46** (1916–18), 217; cf. p. 226

801. N. Tschebotaröw, *Eine Aufgabe aus der algebraischen Zahlentheorie*, Acta Arith. **2** (1937), 221–229; Zbl 18.05101; cf. p. 170

802. M. Tsunekawa, *Relation entre $\left(\frac{a+\sqrt{m}}{p}\right)$ et la loi des résidus quadratiques dans le champ de nombres quadratiques $R(\sqrt{m})$* (japan.), Bull. Nagoya Inst. Technol. **7** (1955), 253–255 (engl. summary); B.S. **19** 1 (1958), p. 19; cf. p. 170

803. M. Tsunekawa, *Elementary proof of the law of quadratic reciprocity in $R(i)$* (Japan.), Sugaku **7** (1955), 23–24; cf. p. 170

804. V. M. Tsvetkov, *Theorem of Stickelberger–Voronoi*, J. Sov. Math. **29** (1985), 1352–1355; Zbl 563.12007; cf. p. 137

805. T. Uehara, *On a congruence relation between Jacobi sums and cyclotomic units*, J. Reine Angew. Math. **382** (1987), 199–214; Zbl 646.12002; MR 89a:11113
806. W. Unger, *Ueber einige Summen cubischer und biquadratischer Charaktere*, Diss. Univ. Bonn, 1921; cf. p. 27, 140, 344
807. R. Vaidyanathaswamy, *The algebra of cubic residues*, J. Indian Math. Soc. (N.S.), **21** (1957), 57–66; Zbl 87.03802; MR 20:7010; cf. p. 226, 342
808. C. de la Vallée Poussin, *Recherches arithmétiques sur la composition des formes binaires quadratiques*, Mém. Acad. Belgique **53** (1895/86). no. 3, 59 pp.; FdM **27** (1896), 164–165; cf. p. 69, 415
809. H.S. Vandiver, *Problem 152*, Amer. Math. Monthly **15** (1908), 46, 235; cf. p. 174
810. H.S. Vandiver, *On sets of three consecutive integers which are quadratic or cubic residues of primes*, Bull. Amer. Math. Soc. **31**, 33–38; FdM **51** (1925), 128; cf. p. 26, 174
811. H.S. Vandiver, *Laws of reciprocity and the first case of Fermat's last theorem*, Proc. Nat. Acad. Sci. **11** (1925), 292–298; FdM **51** (1925), 144; cf. p. 396
812. S. C. van Veen, *De wederkkerigsheidswet der kwadraatresten*, Math. Tijds. v. Studeerenden **1** (1934), 148–153; FdM **59 - II** (1933), 935; Zbl 7.00402; cf. p. 416
813. B. A. Venkov, Elementary number theory (Russ.), Moskow, Engl. Transl. 1970; Zbl 204.37101; cf. p. 171, 342
814. S. Vögeli-Fandel, *Rationale Reziprozitätsgesetze*, Diplomarbeit Univ. Heidelberg, 1997; cf. p. 174
815. A. Voigt, *Abkürzung des dritten Gauss'schen Reciprocitätsbeweises*, Z. Math. Phys. **26** (1881), 134; FdM **13** (1881), 138; cf. p. 21, 414
816. G. Voronoi, *Sur les nombres entiers algébriques dépendant des racines de l'equation cubique* (Russ.), Pétersbourg 1894, 155 pp; FdM **25** (1893/94), 302–303; cf. p. 137
817. G. Voronoi, *Sur une propriété du discriminant des fonctions entières*, Verhandl. III. Internat. Math. Kongress Heidelberg 1905, 186–189; FdM **36** (1905), 238–239; cf. p. 137
818. M. Ward, *Conditions for the solubility of the diophantine equation $x^2 - My^2 = -1$*, Trans. Am. Math. Soc. **33** (1931), 712–718; cf. p. 70, 174
819. M. Ward, *The prime divisors of Fibonacci numbers*, Pac. J. Math. **11** (1961), 379–386; Zbl 112.26904; MR 25 # 2029; cf. p. 70
820. M. Watabe, *An arithmetical application of elliptic functions to the theory of cubic residues*, Proc. Japan Acad. Ser. A **53** (1977), 178–181; Zbl 391.12001; MR 58:5479; cf. p. 225, 277
821. M. Watabe, *An arithmetical application of elliptic functions to the theory of biquadratic residues*, Abh. Math. Sem. Univ. Hamburg **49** (1979), 118–125; Zbl 405.12001; MR 80k:12018; cf. p. 201

822. K. Watanabe, Y. Miyagawa, T. Higuchi, *A remark on the analytic proof of the law of biquadratic reciprocity*, Sci. Rep. Yokohama Nat. Univ., Sect. I Math. Phys. Chem. **43** (1996), 55–72; MR 97m:11006; cf. p. 275
823. T. Watanabe, *Random walks on* $SL(2, \mathbb{F}_2)$ *and Jacobi symbols of quadratic residues*, Advances in combinatorial methods and applications to probability and statistics (N. Balakrishnan, ed.), Birkhäuser, Statistics for Industry and Technology. 125–134 (1997); Zbl 887.11021; MR 98h:11005; cf. p. 417
824. W. Waterhouse, *A tiny note on Gauss' lemma*, J. Number Theory **30** (1988), 105–107; Zbl 663.12002; MR 89i:11113; cf. p. 23
825. H. Weber, *Elliptische Funktionen*, 1891; cf. p. 23
826. A. Weil, *La cyclotomie, jadis et naguère*, Enseign. math. **20** (1974), 247–263; Sem. Bourbaki 1973/74, Exp. 452, Lecture Notes Math. **431** (1975), 318–338; Œuvres III (1980), 311–328; Zbl 362.12003; cf. p. 393
827. A. Weil, *Review of "Mathematische Werke, by Gotthold Eisenstein"*, Bull. Amer. Math. Soc. **82** (1976), 658–663; Œuvres III (1980), 398–401; cf. p. 278
828. A. Weil, *Une lettre et un extrait de lettre à Simone Weil*, Œuvres I (1926–1951), 244–255; cf. p. xi, 22
829. A. Weil, *Sur certains groupes d'opérateurs unitaires*, Acta Math. **111** (1964), 143–211; Œuvres III (1980), 1–69; Zbl 203.03305; MR 29 # 2324; cf. p. 416
830. A. Weiler, G. Röhrle, *Stickelberger-Relation und Eisenstein-Reziprozität*, Manuskript, Tübingen 1985
831. P. Weinberger, *The cubic character of quadratic units*, Proc. of the Number Theory Conf. Boulder, Colorado, 1972, 241–242; MR 52:10673; Zbl 321.12002; cf. p. 175, 228
832. W.P. Welmin, *Zur Theorie der Reste achten Grades in den algebraischen Zahlkörpern*, (Russ.), Warsaw, 1912, 229 pp; FdM **43** (1912), 271; cf. p. 312
833. W.P. Welmin, *Über die Theorie der Reste achten Grades im algebraischen Körper*, Ann. Univ. Warschau I-IX (1912), 1–56; cf. p. 312
834. W.P. Welmin, *Das quadratische Reziprozitätsgesetz in beliebigen quadratischen Zahlkörpern*, J. Reine Angew. Math. **149** (1919), 147–173; FdM **47** (1919/20), 143–144; cf. p. 170
835. A.E. Western, *Certain systems of quadratic complex numbers*, Trans. Cambridge Phil. Soc. **17** (1897/98), 109–148; FdM **30** (1899), p. 201; cf. p. 170
836. A.E. Western, *An extension of Eisenstein's law of reciprocity*, London Math. Soc. Proc. (2), **6** (1907/08), 16–28; FdM **39** (1908), 254; cf. p. 393
837. A.E. Western, *An extension of Eisenstein's law of reciprocity*, London Math. Soc. Proc. (2), **6** (1907/08), 265–297 FdM **39** (1908), 254; cf. p. 393

838. A.E. Western, *Some criteria for the residues of eighth and other powers*, London Math. Soc. Proc. (2), **9** (1910/11), 244–272; FdM **42** (1911), 233; cf. p. 310
839. R. F. Whitehead, *A proof of the law of quadratic reciprocity*, J. London Math. Soc. **2** (1927), 51–55; FdM **53** (1927), 125; cf. p. 415
840. A.L. Whiteman, *On the law of quadratic reciprocity*, Bull. Am. Math. Soc. **41** (1935), 359–360; Zbl 11.38804; cf. p. 416
841. A.L. Whiteman, *Finite Fourier series and cyclotomy*, Proc Nat. Acad. Sci. USA **37** (1951), 373–378; Zbl 42.27203; MR 13, 113f; cf. p. 343
842. A.L. Whiteman, *Cyclotomy and Jacobsthal sums*, Amer. J. Math. **74** (1952), 89–99; Zbl 46.26803; MR 13, 626j; cf. p. 343
843. A.L. Whiteman, *The sixteenth power residue character of* 2, Can. J. Math. **6** (1954), 364–373; Zbl 55.27102; MR 16:14a; cf. p. 310
844. A.L. Whiteman, *The cyclotomic numbers of order sixteen*, Trans. Amer. Math. Soc. **86** (1957), 401–413; Zbl 78.03702; MR 19, 1160h; cf. p. 343
845. A.L. Whiteman, *The cyclotomic numbers of order ten*, Proc. Sympos. Appl. Math., vol. **10**, Amer. Math. Soc., Providence, RI, 1960, pp. 95–111; Zbl 96.02604; MR 22 # 4682; cf. p. 343
846. A.L. Whiteman, *The cyclotomic numbers of order twelve*, Acta Arith. **6** (1960), 53–76; Zbl 91.04302; MR 22 # 9480; cf. p. 343
847. H. C. Williams, *The quadratic character of a certain quadratic surd*, Utilitas Math. **5** (1974), 49–55; Zbl 278.10015; MR 51 # 345; cf. p. 177
848. H. C. Williams, E. R. Zarnke, *Computer solution of the Diophantine equation* $x^2 - dy^4 = -1$, Proc. 2nd Manitoba Conf. on Numerical Math., Winnipeg (1972), 405–415; Zbl 311.10024; MR 51 # 3051
849. K.S. Williams, *A quadratic partition of primes* $\equiv 1$ *(mod 7)*, Math. Comp. **28** (1974), 1133–1136; Zbl 296.10012; MR 49 # 10632; cf. p. 142, 343
850. K.S. Williams, *Elementary treatment of a quadratic partition of primes* $p \equiv 1$ *(mod 7)*, Illinois J. Math. **18** (1974), 608–621; Zbl 292.12022; MR 50 # 4549; cf. p. 141, 343
851. K.S. Williams, 3 *as a ninth power (mod p)*, Math. Scand. **35** (1974), 309–317; Zbl 297.10002; MR 51:3039; cf. p. 142
852. K.S. Williams, *On Eulers criterion for cubic non-residues*, Proc. Amer. Math. Soc. **47** (1975), 277–283; Zbl 298.10001; MR 51:3038
853. K.S. Williams, *On Eulers criterion for quintic non-residues*, Pac. J. Math. **61** (1975), 543–550; Zbl 317.10007; MR 53:7917; cf. p. 141
854. K.S. Williams, 2 *as a ninth power (mod p)*, J. Indian Math. Soc. (N.S.), **39** (1975), 167–172; Zbl 415.10002; MR 53:13089; cf. p. 142
855. K.S. Williams, *Note on the supplement to the law of cubic reciprocity*, Proc. Amer. Math. Soc. **47** (1975), 333–334; Zbl 296.10003; MR 50:9761; cf. p. 225
856. K.S. Williams, *Cubic nonresidues (mod p)*, Delta **6** (1976), 23–28; Zbl 326.10004; MR 54:5095; cf. p. 226

857. K.S. Williams, *A rational octic reciprocity law*, Pac. J. Math. **63** (1976), 563–570; Zbl 327.10006; MR 54:2568; cf. p. 310
858. K.S. Williams, *The quadratic character of 2 (mod p)*, Math. Mag. **49** (1976), 89–90; Zbl 317.10006; MR 52:13604; cf. p. 25, 30
859. K.S. Williams, *Explicit criteria for quintic residuacity*, Math. Comp. **30** (1976), 847–853; Zbl 341.10004; MR 54:218; cf. p. 101, 141
860. K.S. Williams, *Note on a result of Barrucand and Cohn*, J. Reine Angew. Math. **285** (1976), 218–220; Zbl 325.10002; MR 54, 5182; cf. p. 174
861. K.S. Williams, *Explicit forms of Kummer's complementary theorems to his law of quintic reciprocity*, J. Reine Angew. Math. **288** (1976), 207–210; Zbl 335.12014; MR 54:12611; cf. p. 141
862. K.S. Williams, *On the supplement to the law of biquadratic reciprocity*, Proc. Amer. Math. Soc **59** (1976), 19–22; Zbl 339.10003; MR 54:2569; cf. p. 201
863. K.S. Williams, *On Eisenstein's supplement to the law of cubic reciprocity*, Bull. Calcutta Math. Soc. **69** (1977), 311–314; Zbl 423.12004; MR 81f:10007; cf. p. 226
864. K.S. Williams, *On Scholz's reciprocity law*, Proc. Amer. Math. Soc. **64** (1977), 45–46; Zbl 372.12004; MR 56:11880; cf. p. 174
865. K.S. Williams, *Note on Burde's rational biquadratic reciprocity law*, Canad. Math. Bull. **20** (1977), 145–146; Zbl 363.10002; MR 56:5400; cf. p. 173
866. K.S. Williams, *On the evaluation of* $(\varepsilon_{q_1 q_2}/p)$, Rocky Mt. J. Math. **10** (1980), 559–566; Zbl 476.12003; MR 81m:12010; cf. p. 175
867. K.S. Williams, *On Yamamoto's reciprocity law*, Proc. Amer. Math. Soc. **111** (1991), 607–609; Zbl 724.11003; MR 91f:11003; cf. p. 175
868. K.S. Williams, J. D. Currie, *Class numbers and biquadratic reciprocity*, Can. J. Math. **34** (1982), 969–988; Zbl 488.12004; MR 84b:12014; cf. p. 202
869. K.S. Williams, C. Friesen, L. J. Howe, *Criteria for biquadratic residuacity modulo a prime p involving quaternary representations of p*, Can. J. Math. **37** (1985), 337–370; Zbl 573.10003; MR 86f:11078; cf. p. 142
870. K.S. Williams, K. Hardy, C. Friesen, *On the evaluation of the Legendre-symbol* $(\frac{A+B\sqrt{m}}{p})$, Acta Arith. **45** (1985), 255–272; Zbl 565.10004; MR 87b:11006; cf. p. 166, 167, 173
871. J. Wojcik, *Some remarks about the power residue symbol*, Acta Arith. **66** (1994), 351–358; Zbl 827.11061; MR 95i:11121
872. J. Wojcik, *An extension of the Eisenstein reciprocity law*, Comment. Math. Prace Mat. **35** (1995), 277–285; Zbl 857.11054; MR 97a:11171; cf. p. 393
873. J. Wojcik, *Reciprocity law for powers of cyclotomic integers*, Comment. Math. Prace Mat. **35** (1995), 287–300; Zbl 857.11053; MR 97a:11172; cf. p. 393

874. P. J. Wu, *A rational reciprocity law*, Ph. D. thesis, Univ. South Calif, 1976; cf. p. 310
875. B. F. Wyman, *What is a reciprocity law*, Amer. Math. Month. **79** (1972), 571–586; Zbl 244.12010; MR 46:7199; Corr.: ibid. **80** (1973), 281; Zbl 272.12009; cf. p. 174
876. Y. Yamamoto, *Congruences mod 2^i ($i = 3, 4$) for the class number of quadratic fields*, Proc. International Conf. on Class numbers and Fundamental Units of Algebraic Number Fields, Katata, Japan 1986, 205–215; Zbl 622.12006; MR 88k:11074
877. P. T. Young, *Quadratic reciprocity via Lucas sequences*, Fib. Quarterly **33** (1995), 78–81; Zbl 826.11004; MR 95m:11011; cf. p. 417
878. H. Zantema, *Global restrictions on ramification in number fields*, Manuscr. Math. **43** (1983), 87–106; Zbl 519.12005; MR 85g:11101; cf. p. 417
879. H. J. Zassenhaus, *The quadratic law of reciprocity and the theory of Galois fields*, Proc. Glasgow Math. Assoc. **1** (1952), 64–71; Zbl 52.27802; MR 14:450c; cf. p. 416
880. Y. Ch. Zee, *Some sufficient conditions for quintic residuacity*, Proc. Amer. Math. Soc. **54** (1976), 8–10; Zbl 323.10003; MR 52:10562; cf. p. 141, 142
881. Ch. Zeller, *Beweis des Reciprocitätsgesetzes für die quadratischen Reste*, Berl. Monatsber. (1872), 846–847; FdM **4** (1872), 76; cf. p. 413
882. Ch. Zeller, *Bestimmung des quadratischen Restcharakters durch Kettenbruchdivision; Versuch einer Ergänzung zum dritten und fünften Beweise des Gaussschen Fundamentaltheorems*, Gött. Nachr. (1879), 197–216; FdM **11** (1879), 129; cf. p. 21, 26
883. G. Zolotarev, *Nouvelle démonstration de la loi de réciprocité de Legendre*, Nouv. Ann. Math (2), **11** (1872), 354–362; FdM **4** (1872), 75–76; cf. p. 23, 413
884. A.M. Zuravskij, *Das kubische Reziprozitätsgesetz* (Russ.), Leningrad Journ. Soc. Phys. Math. **1** (1927), 204–232; FdM **53** (1927), 145; cf. p. 226
885. A.M. Zuravskij, *Über den kubischen Charakter der Zahl $1 - \rho$* (Russ., Engl. summary), Ann. Inst. Mines **7** (1928), 15–25; FdM **55-II** (1929), 701; cf. p. 226

Remark. We have used abbreviations for the following review journals:

FdM: Fortschritte der Mathematik
BSMA: Bulletin des Sciences de Mathematique et Astronomie
B.S. Bulletin Signalétique
MR: Mathematical Reviews
RSPM: Revue Semestrielle des Publications Mathématiques
Zbl: Zentralblatt

Author Index

Abel, 239, 269, 275
Adkisson, 201
Adleman, 26, 139
Adler, 140, 275, 276, 347
Agargün, 20
Agou, 417
Agrawal, 142, 343
Ahlborn, 23
Aigner, A., 23, 70, 171, 173, 174, 227, 228, 310, 313
Aigner, M., 12
Aladov, 26
Alderson, 141
Alexejewsky, 415
Allander, 417
Alling, 277
Anderson, 382
Andrews, 103
Ankeny, 141, 416
Anshel, 228
Antropov, 11
Apéry, 343
Armitage, 390
Arndt, 69
Artin, ix, xiii, 25, 136, 201, 204, 341, 344, 349
Arwin, 170, 415
Aubry, 11, 23, 25–27, 170, 415
Auslander, 139, 417
Ayoub, 267

Bach, 25, 119
Bachet, 1, 11
Bachmann, xii, xiv, 11, 18, 100–102, 179, 200, 227, 277, 342, 395
Backlund, 20
Bambah, 103
Banderier, 24
Bang, 414
Barbilian, 416
Barcanescu, 417

Barkan, 278, 343
Barner, 11
Barnes, 417
Barrucand, 118, 137, 173, 417
Bartelds, 415
Bass, 21, 70
Bauer, 384
Baumgart, xiv, 14, 24, 101
Bayad, 277
Beach, 175
Beeger, 141, 310, 312
Bennett, 26
Bernardi, 141, 175
Berndt, B., 23, 138–140, 146, 201, 293, 310, 393, 417
Bernoulli, Jak., 267, 343
Bernoulli, Joh., 267
Berrizbeitia, 228, 231
Berry, 228, 231
Bertolini, 396
Bessel, 15
Beuker, 202
Bhaskaracharya, 11
Bianchi, 170
Bickmore, 70, 179, 205, 229, 310
Biermann, 34, 272
Bierstedt, 27
Billing, 175
Bindschedler, 277
Birch, xii, 21, 385, 417
Birkhoff, 27
Blanchard, A., 70
Blanchard, C., 70
van der Blij, 23
Bock, 70, 414
Böge, xv
Bölling, 175
Bohnicek, 201, 277, 312, 393
Bombieri, 345
Bonaventura, 170, 277
Borchardt, 271
Borel, 344

Borevich, xii, 60, 70
Bork, 29
Borwein, J.M., 269
Borwein, P.B., 269
Bouniakowski, 23, 25, 413
Brahmagupta, 11
Brandis, xv
Brandler, 172, 173, 176
Brandt, 170, 226, 416
Brauer, A., 20, 26, 27, 310
Braun, H., 141
Braun, I., 13
Bray, 73
Brenner, 23, 417
Bressoud, 103
Brewer, 313, 416
Bricard, 25, 270
Brillhart, 12, 27, 70, 185
Brinkhuis, 390, 391
Brown, 4, 20, 21, 33, 70, 172, 173, 175, 176, 417
Brumer, 383, 385
Brylinski, 417
Bucher, 312
Bühler, 11
Buell, 174, 175, 312, 313
Bugaieff, 26, 277
Buhler, 394
Burde, 26, 166, 171, 201, 227, 310, 416
Burkhardt, 11
Burnside, 141, 313
Busche, 23, 170, 201, 414, 415
Bussey, 13

Cabillon, 269
Cailler, 228
Canuto, 396
Carcavi, 12
Carey, 228
Caris, 26
Carlitz, 23, 27, 103, 137, 141, 416
Carrega, 101
Cartier, 23, 345
Cassels, 60, 175, 229

Cassou-Noguès, 311
Castaldo, 417
Cauchy, 15, 31, 102, 138, 202, 269, 270, 313, 357, 390, 394, 413
Cayley, 100, 141, 201, 228
Cerone, 415
Chahal, 344
Chang, 20, 26
Chapman, xv, 389, 395
Chebotarev, x, 103, 170
Chebyshev, 22
Chevalley, x
Chinburg, 390
Chowla, P., 25
Chowla, S., 25, 103, 138, 140, 201, 202, 225, 228, 342, 396, 416
Christol, 345
Chua, 23
Chuman, 174
Cipolla, 26
Clarke, 18
Clausen, 227
Coates, xii, 387, 391, 394, 395
Cochrane, 16
Cohen, E., 140
Cohen, H., 12, 26, 139, 142
Cohen, S.D., 141
Cohn, 69, 74, 173, 277, 399
Collins, 26
Collison, 25, 201, 225
Conrad, 391, 397
Conrey, 343
Cooke, xv, 224, 228
Cooper, xiii, 23, 26
Cornacchia, 12, 342
Cornaros, 25
Cornell, 396
Corrales-Rodriganez, 137
Cosgrave, 231
Cowles, J., 228
Cowles, M., 228
Cox, 11, 23, 263, 396

Craig, 229
Crandall, 394, 401
Crelle, 269
Cremona, 17
Crstici, 27
Cuculière, xv, 1, 8, 24
Cunningham, 171, 201, 226, 310
Currie, 202

Dalen, 137
Dantscher, 225, 277
Darmon, 396
Davenport, 16, 23, 132, 138, 340, 344, 391, 397
Dedekind, 22, 25, 26, 69, 81, 84, 95, 101, 125, 225, 228, 229, 274, 276, 342, 344, 413–415
Delaunay, 103
Delcour, 395
Delezoide, 23
Deligne, xii, 345, 383
Deloup, 141
Delsarte, 23
Deninger, 346
Dennis, xv
Deuring, 342
Diamond, 396
Dickson, xiii, 12, 14, 21, 141, 145, 175, 225, 342, 343
Dieter, 23
Dieudonné, 345
Digby, 3, 12
Dilcher, 401
Dintzl, 201, 225, 276, 277
Diophantus, 1, 11, 14
Dirichlet, xiii, 4, 7, 19–21, 26, 27, 70, 102, 137, 153, 167, 170, 171, 174, 175, 199, 209, 256, 270, 343, 375, 413
Dirks, 24
Dittmar, 225
Dockeray, 416
Dörge, 26, 416
Dörrie, 170, 226

Dressler, 24
Dudeney, 395
Dummit, 384
Dunton, 225, 228
Dupré, 20
Dwork, 202, 345
Dym, 139

Edwards, 13, 16, 274, 343, 396
Eichenberg, 23
Eichler, 417
Eisenstein, vii, 10, 16, 24, 26, 31, 114, 136, 137, 139, 201, 203, 209, 225, 236, 270, 281, 310, 313, 340, 366, 390, 392, 413
El Hassani, 278
Ely, 30, 417
Enneper, 269, 277
Ennola, 173
Epstein, 136, 175, 176
Erchinger, 100
Erez, 390
Ernvall, 379, 394, 395
Errera, 20
Esrafilian, 25
Estermann, 103
Estes, 26, 172
Euler, vi, 3, 11, 13, 70, 100, 131, 175, 201, 222, 229, 268, 276, 329, 343, 394
Evans, 138–141, 146, 173, 174, 201, 202, 215, 293, 299, 310, 313, 343, 393, 417
Eymard, 36

Faddeev, viii
Fagnano, 267
Faulhaber, 343
Federer, 387
Federighi, 70
Fee, 395
Felgner, xv, 201, 226, 230
Feng, 25
Fenster, xiii

Fermat, P., 1, 11
Fermat, S., 11
Ferrero, 385
Fields, 414
Filaseta, 254
Fischer, 415
Flanders, 136
Fletcher, 20
Fontené, 100
Fouche, 25
Fowler, 20
Frame, 417
Franklin, 414
Frattini, 25
Frei, 25, 70, 200, 226
Freitag, 345
Frenicle, 13
Frey, 70, 396
Fricke, 342
Friedlander, 23
Friedmann, 26
Friesen, 25, 142, 164, 173, 225
Frobenius, 21, 395, 415
Fröhlich, 70, 173, 390, 416
Froemke, 23
Fuchs, 22
Fueter, xii, 101, 225, 226, 228, 277, 416
Furquim de Almeida, 416
Furtwängler, ix, xiii, 136, 143, 357, 375, 393, 396, 401
Furuta, 174, 175, 177, 310

Galois, 366
Garrett, 417
Gauss, vi, xiii, 9, 11, 15, 20, 68, 85, 100, 137, 154, 170, 174, 179, 202–204, 223–226, 239, 268, 344
Gazzaniga, 23
Gebel, 13
Gebre-Egziabher, 70
Gee, 265
van der Geer, 346
Gegenbauer, 23, 26, 201, 225, 414

Genocchi, 23, 102, 201, 225, 413
Germain, 199, 201, 225, 402
Gerst, 136
Gerstenhaber, v, 416
Ghate, 387
Gillard, 278, 391
Girard, 1
Giudici, 26, 136
Glaisher, 85, 101
Glashan, 141
Glause, 226
Glazunov, 345
Gmeiner, 137, 226
Götting, 26
Goldbach, 3, 12, 13
Goldfeld, 228
Goldscheider, 173, 174, 205, 310, 311, 326
Goldschmidt, 13, 23, 417
Goldstein, C., 11, 12, 20
Golomb, 70, 227, 231
Gosset, 172, 201, 202, 226, 310
Gottlieb, 100
Granville, 395
Gras, 138, 386, 391
Grassmann, 16
Gray, 22
Greenberg, M.J., 103
Greenberg, R., 385, 386
Greene, 138
Greither, 387, 389
Grimson, 140
Gross, 392
von Grosschmid, 25, 26
Grossmann, 23
Grosswald, 1
Grothendieck, 345
Grytczuk, 25
Guinot, 11
Gundelfinger, 16
Gupta, 26
Gut, 378, 386, 395
Guthmann, 228
Guy, 58

Habicht, 225
Hacks, 414
Hadamard, 330, 343
Hagge, 100
Hahn, 140
Hajir, 277, 278
Hall, 16, 143
Halphen, 174
Halter-Koch, 141, 172, 174, 175, 310
Hansen, J.P., 349
Hansen, S.V., 345, 349
Harder, 378
Hardman, 136
Hardy, G.H., xii
Hardy, K., 140, 164, 173
Hartshorne, 345
Hasse, vi, ix, xiii, 16, 24, 26, 60, 86, 122, 132, 137, 138, 173, 201, 310, 340, 341, 344, 380, 391, 393–395, 397
Hausner, 416
Hayashi, 225
Hayes, 384, 394
Heath, 11
Heath-Brown, 278
Heawood, 415
Hecke, v, xii, 103, 137, 331, 375
Heegner, 265
Heinitz, 26
Hellegouarch, 25, 396
Helou, 141, 173, 310, 312, 395
Hensel, 15, 16, 136, 137, 174, 415
Herbrand, 375, 377, 378
Herglotz, 137, 201, 209, 256, 276, 277, 336, 342
Hermes, 100, 414
Hermite, 202, 392
Hertzer, 392
Hettkamp, 175
van den Heuvel, xv
Higuchi, 275
Hilbert, D., viii, xiii, 16, 18, 20, 60, 70, 84, 101, 103, 104, 119, 170, 201, 209, 358, 394, 415
Hilbert, K.S., 136
Hiramatsu, xv, 275
Hirzebruch, 417
Hofmann, 1, 11
al-Hogendi, 395
Holzer, 16, 396, 416
Honda, 417
Hopf, 26, 345
Houzel, 345
Howe, 142
Huard, 138
Huber, 102
Hudson, 140, 201, 310, 313
Hübler, 225, 277
Hull, 141
Hulsbergen, 344
Humble, 25
Hurrelbrink, 21, 174, 417
Hurwitz, 275, 276, 332, 342, 343
Husemöller, 417
Hwang, 141

Ibrahimoglu, 391
Inaba, 70
Ireland, xii, 30, 79, 104, 139, 345, 349, 391, 393, 394
Ishida, 70
Ishii, H., 417
Ishii, N., 174
Ito, 225, 278
Ivanov, 20
Iwasawa, xii, 357, 385, 386, 395

Jacobi, vii, xiii, xv, 10, 24, 31, 135, 136, 138, 144, 166, 170, 171, 179, 180, 202, 203, 224, 227, 268–270, 275, 313, 357, 390, 392, 394, 413
Jacobsthal, 26, 140, 193
Jänichen, 26
Jakubec, 141, 227
Jatem, 26

Jenkins, 22, 23
Jensen, 225
Jha, 382, 395
Johnson, K.R., 23
Johnson, W., 394
Joly, 138, 225, 345, 391
Jones, 26
Jordan, 136, 226, 346
Joshi, 11
Joux, 141

Kac, 417
Kamienny, 378
Kantor, viii
Kantz, 201
Kaplan, 172–175, 200, 225, 416
Kaplansky, xiii, xv
Karakisawa, 225
Karamate, 103
Karst, 141
Katre, 140, 141
Katz, 276, 345
Kawada, 270
Kedlaya, 277
Kemnitz, 342
Keune, 417
Kida, 139
Kiehl, 345
Kiepert, 270
Kim, 225
Kimura, T., 395
King, 277
Kirchhoff, 24
Kirwan, 100
Kleboth, 379, 386
Kleinjung, 378
Kloosterman, 416
Kluyver, 343
Knapp, 263
Knorr, 20
Koblitz, 263, 345, 392
Koch, 330
König, 12, 201, 415
Kolster, 174, 387
Kolyvagin, 357, 387

Kornblum, 344, 350
Koschmieder, 225, 277, 416
Koutsky, 26
Kozuka, 103
Kraft, 136, 394
Kramer, J., 396
Kramer, K., 174, 175
Krazer, 141, 277
Kronecker, xv, 21–23, 26, 69, 101–103, 201, 227, 275, 276, 375, 413
Kronheimer, 141
Krull, 275
Krusemarck, 275
Kubert, 138, 225, 395
Kubota, 100, 175, 200, 201, 266, 275, 385, 392, 416, 417
Kühne, 25
Kuipers, 23
Kummer, vii, xiii, xv, 15, 18, 19, 22, 69, 85, 101, 102, 119, 136, 138, 227, 270, 272, 321, 343, 357, 366, 370, 375, 390, 395, 413
Kurihara, 387
Kuroda, 175

Lafon, 36
Lagrange, J., 175
Lagrange, J.-L., 4, 13, 14, 25, 26, 29, 137, 223, 268
Lambert, 4
Lamé, 15, 396
Lampe, 392
Landau, xii, 18, 102, 331, 344, 393, 395
Landsberg, 102
Lang, 139, 387, 391, 395
Lange, 415
Langevin, 139
Langlands, xi
Larsen, 141
Larson, 12
Lasker, 137, 415
Laubenbacher, 10, 11, 21

Laubie, 118, 137, 417
Lawden, 281
Lazarus, 228
Le Vavasseur, 225
Lebesgue, H., 100
Lebesgue, V.A., 21, 22, 24–26, 30, 102, 174, 201, 202, 228, 273, 278, 344, 413
Lee, 140
Lefébure, 342
Legendre, vi, 5, 6, 11, 12, 15, 16, 26, 224, 268, 270, 402
Lehmer, D.H., 21, 26, 27, 56, 58, 102, 228, 416
Lehmer, D.N., 26
Lehmer, E., v, 27, 70, 85, 101, 102, 137, 140, 141, 162, 166, 167, 171, 173, 174, 203, 212, 228, 343
Leibniz, 12
Lemmermeyer, 30, 166, 173, 176, 181, 417
Lenstra, A.K., 139
Lenstra, H.W., 26, 59, 70, 73, 89, 102, 137, 139, 395
Leonard, 140–142, 174, 175, 310
Leopoldt, 70, 137, 343, 377, 385, 386
Leprevost, 141
Lerch, 21, 24, 102, 414, 415
Lewandowski, 225, 277
Lewy, 416
Li, 138, 139, 345, 346
Libri, 25, 269, 342
Lichtenbaum, 138
Lidl, 139, 140, 345
von Lienen, 175, 227, 310
Lietzmann, 201
Lind, 175
Linden, 26
Liouville, 15, 269, 275, 413
Lippert, 378
Lipschitz, 23, 271
Littlewood, 344

Liverance, 228
Llorente, 142
Loewy, 25, 100
Longo, 136
Loos, 26
Lorenzini, 345
Louboutin, xv, 170
Loxton, 278
Lubelski, 103, 137
Lubin, 103
Lucas, 56, 72, 395, 414
Ludwig, 22
Lüneburg, 275
van de Lune, 343

MacKenzie, 394
Mahnke, 12
Mahoney, 11
Manders, 26
Mandl, 23
von Mangoldt, 343
Manin, 22, 344
Mansion, 414
Mantel, 25
Markuschewitsch, 269
Martinet, xv, 229, 389
Masahito, 277
Maser, 18
Massell, 174
Mathews, 270, 277
Mathieu, 26, 101, 413
Matrot, 25
Matter, 276
Matthews, C., 267
Matthews, K., xv
Mattuck, 345
Mayer, 136
Mazur, 343, 379, 386
McClintock, 26
McEliece, 138
McGettrick, 278
McGuinness, 175
McKean, 139
Mel'nikov, 24, 270, 277
Merindol, 417

Merrill, 25
Mersenne, 56
Mertens, 18, 69, 103, 201, 226, 277, 391, 415
Metsänkylä, 375, 394, 395
Meyer, C., 278, 416
Meyer, P., 270
Meyer, S., 26
Mihailescu, 92, 139
Miller, 26
Mills, 27
Milnor, 21, 70, 385, 417
Minding, 28
Minkowski, 16, 20
Minnigerode, 170
Mirimanoff, 137, 402, 415
Mitchell, H., 394
Mitchell, P., 16
Mitrinovic, 27
Mitzscherling, 100
Miyagawa, 275
Mollin, xv, 25, 69, 215
Monagan, 395
Monsky, 345
Monzingo, 26, 140, 342
Morain, 92, 141, 146
Mordell, 16, 28, 103, 174, 175
Moreau, 20
Moree, 70
Morehead, 70
Moreno, 278, 349, 391, 396
Morlaye, 342
Mortimer, 142
Morton, 24, 175
Motose, 417
Movahhedi, 137
Murty, K., 396
Murty, R., 396
Muskat, 175, 228, 343

Nagaev, 270
Nagell, 202, 227
Nakamula, 278
Nakhash, xv, 73, 418
Narkiewicz, 390

Naryskina, 276
Nashier, 141
Nasimoff, 26, 277
Naur, 25
Nekovar, 385, 395
Nemenzo, 69
Neukirch, 248
Neumann, xii, 16, 22, 34, 100, 103, 274
Newman, 330
Niederreiter, 139, 140, 345
Nielsen, 25
Niemeyer, 276, 277, 280
Noether, 389
Noguès, 396

Olbers, 15, 96
Oltramare, 25
Ono, 70, 136, 139
Onuki, 103
Opolka, 11, 16, 18, 265
Ore, 25
Ostwald, 24
Ouyang, 382

Padma, 141
Pall, 26, 136, 142, 172
Panaitopol, 275
Park, 225
Parnami, 141, 142, 343
Pascal, 12
Patterson, 18, 278
Paucker, 100
Pedersen, 349
Peitz, 226
Peklar, 417
Pell, 11
Pellet, 73, 116, 137, 202, 225, 227–229, 342, 414
Pengelley, 10, 11, 21
Pépin, 13, 15, 20, 22, 23, 59, 72, 102, 141, 142, 174, 175, 181, 224, 225, 269, 342, 393, 395, 414, 415
Perott, 175, 310, 313

Perrin-Riou, 387
Petersen, 100, 414
Pethö, 13
Petin, 138
Petr, 103, 201, 225, 226, 277, 415, 416
Pexider, 26
Pfleiderer, 100
Phong, 26
Pickert, 23
Pieper, xv, 5, 18, 20, 24, 139, 200
Pierpont, 100
Piltz, 20
Pinch, 311
Pink, 378
Pocklington, 25, 26, 90, 175, 415
Pollaczek, 377
Polya, 105
Pomerance, 139, 401
van der Poorten, 25, 344, 396
Popovici, 23
Porcelli, 136
Postnikov, 140, 345
Poulakis, 141
Prasolov, 270, 275, 277
Propp, xv

Quellet, 275
Queyrut, 390
Quillen, 385, 387

Rademacher, 416
Rajwade, 101, 140–142, 313, 342, 343
Rameswar, 23
Rankin, 141
Rapoport, 344
Rashed, 11, 395
Rédei, 22, 26, 70, 172, 174, 175, 181, 204, 415
Reich, 26
Reichardt, 23, 137, 173, 175, 310, 416
Replogle, 389
Reshetukha, 225, 278, 417

Reuschle, 141, 205, 310, 392
Reynolds, 27
Ribenboim, 342, 396, 402
Ribet, 377, 378, 383, 396
Richelot, 100
Rideout, 385
Rieger, 22
te Riele, 343
Riemann, 200, 329, 343
Riese, 103
Riesz, 24, 416
Rishi, 141
Robba, 345
Roberts, 24
Robinson, D., 202
Robinson, R.M., 70
Roblot, 384
Rocci, 26
Rogers, 20, 417
Roll, 70, 141
Roquette, xv, 344–346
Rose, 175
Rosen, K.H., 23, 25
Rosen, M., xii, 30, 79, 103, 104, 136, 139, 269, 270, 345, 349, 391, 393, 394
Rosenberg, 21
Rousseau, 24, 31, 417
Rowe, 201
Rubin, 357, 387, 389, 395
Rumely, 139
Rumsey, 138
Rusin, 17
Russinoff, 25, 417
Ryan, 25

Saalschütz, 26
Saidi, 228
Salié, 395
Salvadori, 103, 139
Sandor, 27
Sands, 384
Sangani-Monfared, 25
Sansone, 175, 180
Satgé, 175

Sato, 142, 225, 227
Sbrana, 225, 277
Schaar, 103, 413
Schappacher, 103, 261, 276, 277, 344
Scharlau, W., 11, 16, 18, 21, 25, 272
Scheffler, 20, 140
Scheibner, 26, 170, 201, 226, 275
Schering, 21, 23, 26, 200, 414
Scherk, 141
Schertz, 265
Schiappa-Monteiro, 25
Schinzel, 25, 136
Schmid, H.L., 138
Schmidt, F.K., x, 25, 338, 344
Schmidt, H., 32, 414
Schmidt, W.M., 345
Schmitt, 24
Schneider, 344
Schönemann, 25, 101, 160, 162, 174, 254, 274, 413
Schönheim, 26
Scholz, 23, 27, 160, 166, 183
Schoof, 26, 137, 346, 382, 383, 396
Schröter, 100
Schuh, 136, 415
Schuhmann, 34
Schulte, 136
Schumacher, 100
Schur, 103, 342
Schwarz, 25
Schwering, 277, 280, 392
Schwermer, 25
Scott, 141
de Séguier, 103
Seidel, xv
Selberg, 141
Selmer, 228
Senior, 143
Serre, 263, 345, 390, 396
Serret, 12
Sesiano, 11
Shafarevich, xii, 60, 70

Shallit, 26, 59, 102, 119
Shanks, 26, 103, 228
Shapiro, 21
Sharifi, 102, 203, 229, 397
Shimaura, 392
Shimura, xi, 71, 187, 276, 392
Shintani, 383
Shirai, 227
Shiratani, 201, 277, 392
Shokrollahi, 394
Shult, 24
Siegel, 18, 103, 267, 383
Sierpinski, 310
Silverman, 342, 396
Silvester, 70
Singh, 26, 342
Sinnott, 382
Skolem, 16, 102, 137, 225, 415, 416
Slavutskij, 24
Small, 345
Smith, H.J., 21–23, 101, 170, 201, 275
Smith, S., 25
Solomon, 386
Solovyev, 270, 275, 277
Sommer, xii
Sorenson, 26
Soulé, 387
Spearman, 138, 140, 144, 225, 393
Speckmann, 26
Speiser, 69, 103, 389
Spies, 201
Springer, 21
Srivastav, 389
Stöhr, 345
Stalker, 204, 275
Stankowitsch, 26
Stanton, 138
Stark, 141, 328, 333, 345, 383
von Staudt, 26, 100, 101, 413
Steenvoorden, 13
Stefanescu, 275
Steinbacher, 103
Stengel, 101

Stepanov, 27, 105, 138, 140, 345
Stern, 25, 26, 225, 227, 313, 342, 392, 413
Sterneck, 26
Stevenhagen, 59, 70, 265, 312, 395
Stevens, 396
Stevin, 1
Stichtenoth, 345
Stickelberger, 116, 131, 137, 140, 145, 146, 225, 310, 313, 357, 358, 366, 391, 394, 395
Stieltjes, 25, 202, 225, 415
Stillwell, 32
Storchi, 25, 70
Stouff, 202
Strnad, 100
Stroeker, 175
Sun, Q., 229
Sun, Z., 173, 174, 226, 228
Suzuki, J., 395
Swan, 80, 137, 416, 417
Swinnerton-Dyer, xii
Sylvester, 23, 26, 102, 228, 343, 395

Tafelmacher, 21, 100, 414
Takagi, ix, xiii, 136, 261, 276, 277, 334, 377, 378, 393, 415
Takenouchi, 277
Tamarkin, 26
Tamme, 378, 395
Tangedal, 384
Taniyama, 187, 276, 392
Tanner, 141, 393
Tardy, 203
Tate, viii, x, 21, 136, 342, 345, 384, 385, 396
Taylor, L., 26, 181
Taylor, M.J., 311, 390
Taylor, R., 396
Teege, 20, 101, 103, 415
Terasoma, 137
Terjanian, 343, 396
Terras, 139

Thaine, 137, 357, 387
Thakur, 139
Thomas, 344
Thue, 27
Tihanyi, 226
Tolimieri, 139, 417
Tollis, 343
Tomic, 103
Tonelli, 26
Top, 175
Torelli, 26
Trost, 136
Tschebotaröw, 170
Tsunekawa, 170
Tsvetkov, 137
Tunnell, 119
Turaev, 141
Tychonov, 401

Unger, 27, 140, 344
Upadhyaya, 141

Vaidyanathaswamy, 25, 226, 342
de la Vallée Poussin, 69, 330, 415
Vandermonde, 100, 137
Vandiver, 26, 27, 174, 396
Vaughan, 143
Vazzana, 174
van Veen, 416
Venedem, xv
Venkataraman, 141
Venkov, 171, 342
Villegas, 277
Vinogradov, 27, 105, 278
Vladut, 277
Vögeli–Fandel, xv, 174
Voigt, 21, 414
Voloch, 345
Voronoi, 116, 137, 394
Vuillemin, 11

Wada, 69
Wagon, 12
Wagstaff, 23
Wall, 70

Walling, 25
Wallis, 267
Walsh, 25
Walum, 140
Wantzel, 100
Ward, 70, 174
Warren, 73
Washington, 79, 124, 137, 141, 342, 346, 382, 383, 385, 387, 391
Washio, 392
Watabe, 201, 225, 277
Watanabe, K., 275
Watanabe, T., 417
Waterhouse, 23, 103, 137, 143
Weber, 23, 26, 69, 103, 137, 265, 276, 277, 331
Weierstraß, 248
Weil, A., v, xi, 11, 22, 138, 274–276, 278, 341, 345, 346, 393, 416
Weil, S., 22
Weinberger, 175, 228
Welmin, 170, 312
Wendt, 395
Western, 170, 291, 296, 310, 393
Whitehead, 415
Whiteman, 25, 140, 310, 313, 343, 416
Whyburn, 26, 136
Wichert, 270
Widmer, 140, 344
Wieferich, 401
Wiles, xii, 379, 386, 387, 396
Williams, H.C., 59, 102, 139, 175, 177, 228, 232
Williams, K.S., 16, 25, 30, 101, 138–141, 144, 164, 172–176, 201, 202, 225, 226, 293, 310, 312, 313, 343, 393

Wilson, 70
Winter, 343
Witt, 330
Wojcik, 393
Woltman, 58, 70
Wong, 137
Wright, xii
Wu, 310
Wussing, 34
Wyman, 174

Xu, 13

Yamamoto, K., 103, 140, 146, 278, 313
Yamamoto, Y., 172, 175
Ying, 25
Yoon, 141
Yorinaga, 70
Young, J., 231
Young, P.T., 417

Zagier, 12, 69, 330, 417
Zahidi, 137
Zaldivar, xv
Zantema, 417
Zarnke, 175, 228
Zassenhaus, 103, 416
Zee, 141, 142
Zeitz, 20
Zeller, 21, 26, 413
Ziegler, 12
Zignago, 20, 33
Zimmer, 13, 344
Zolotarev, 23, 32, 413
Zuravskij, 226

Subject Index

addition formula, 241, 279
algebraic integer, 43
ambiguous
 ideal, 47
 ideal class, 47
anti-units, 384
Artin root number, 390
Artin symbol, 86, 87, 114, 130, 383

Beilinson conjecture, 344
Bernoulli numbers, 276, 329, 343, 375, 394
 generalized, 332
Bernoulli polynomials, 333
Beta function, 139
Binet's formula, 73
Birch-Tate conjecture, 385, 387
Bloch-Kato conjecture, 344

character
 additive, 127
 Dirichlet, 121
 multiplicative, 127
 primitive, 71, 105, 122
 real, 124
class group
 genus, 47
 narrow, 46, 47
 wide, 46
class number formula, 100, 105, 277, 307, 308, 312
complex multiplication, 187, 245, 249, 263
conductor, 122
 -discriminant formula, 125, 335
congruence
 Eisenstein's, 255
 Euler's, 238, 255
 Gauss's, 192, 202, 203
conjugate, 43
constructible, 104
convolution, 134

cusp form, 263
cyclotomic numbers, 101, 106, 145, 223, 230, 321, 322
cyclotomic polynomial, 81, 83, 101
cyclotomic units, 97, 386

Davenport-Hasse relation, 128, 132, 138
Davenport-Hasse theorem, 133, 328, 350, 398
decomposition field, 75
decomposition group, 75, 79, 125
defining modulus, 122
degree, 248
difference set, 201
dimension, 341
direct limit, 66
directed set, 400
discriminant, 43, 116
 prime, 45
divisor, 247
 principal, 248
duplication formula, 241, 268

E-primary, 217
Eisenstein series, 249, 263
Eisenstein sum, 133, 139, 373
elliptic curve, 396
elliptic function, 247
elliptic integral, 267
embedding, 43
Euler numbers, 378, 387, 394
Euler product, 329
Euler system, 387
Euler's criterion, 4
exponential sum, 97, 346

Fermat numbers, 4, 70, 72, 73, 203, 231, 314
Fermat prime, 59, 104
Fermat's Last Theorem, 11, 15, 16, 395

Subject Index 485

Fibonacci numbers, 70, 73, 168
finite field, 127
Fourier transform, 134
Frobenius automorphism, 127, 131, 238, 255, 261
Frobenius symbol, 85
function field, 344
functional equation, 329, 331, 335, 338, 341, 342
fundamental unit, 44, 160, 167, 175

Gamma function, 139, 240
    p-adic, 400
Gauss sum, 93, 126, 238, 278, 357
    cubic, 209, 212, 223, 278
    elliptic, 289, 299, 311
    octic, 289
    quadratic, 27, 265, 335
    quartic, 190, 200, 266
Gauss's Last Entry, 190, 317
Gauss's Lemma, 9, 23, 114, 161, 200, 236, 302
genus, 338
    character, 52
    field, 54
    principal, 47
    spinor, 172
    theory, 47
grössencharacter, 276
Gross-Koblitz formula, 140

half system, 9
Hasse-Weil bound, 335, 339
Hensel's Lemma, 60, 62, 63, 68
Hilbert class field, 169, 264, 375
holomorphic, 243
Hurwitz numbers, 276, 343
hyperelliptic curve, 340

idèle class group, ix
idempotent, 376
    orthogonal, 376
inertia field, 75
inertia group, 75, 125
irreducible, 341

irregularity index, 375
Iwasawa theory, 385

Jacobi sum, 128, 326, 340, 391, 393
    cubic, 209
    octic, 289
    quartic, 190
Jacobi symbol, 10
Jacobsthal sum, 140, 193

K-groups, 21, 60, 65, 174, 344, 385
Kronecker delta, 322
Kronecker symbol, 44
Kronecker's Jugendtraum, 190, 260, 277
Krull topology, 401
Kummer extension, 119
Kummer sum, 278
Kummer's Lemma, 375

L-series
    Artin, 388
    Dirichlet, 334
Lagrange resolvants, 137
Langlands' program, 25
Laurent series, 247
Legendre symbol, 6, 45
lemniscate, 239
lemniscatic sine, 239, 300
Leopoldt's Conjecture, 385
local-global principle, 389
Lucas numbers, 73
Lucas-Lehmer test, 70, 91, 92

Main Conjecture, 379, 385
meromorphic, 243
Mersenne numbers, 4, 70, 73, 315
Mersenne prime, 168, 206
Minkowski bound, 55
minus class group, 373
modular function, 263

nonsingular, 337
norm, 43, 127, 349
norm residue, viii
norm residue symbol, 60

normal integral basis, 83, 387

partially ordered, 400
Pell's equation, 11, 44
period equation
    e-th, 102
    cubic, 210, 223, 228, 278
    general, 84
    octic, 313
    quadratic, 21, 95
    quartic, 201
    quintic, 141
    sextic, 228
period lattice, 247
Primality Test, 85, 88
    Lenstra, 92, 232
    Lucas-Lehmer, 56
    Pocklington, 90
primary, 191, 209, 221, 289
prime
    inert, 44
    irregular, 375
    ramified, 44
    regular, 357, 375
    split, 44
product formula, 53, 54, 64
projective
    algebraic set, 341
    limit, 401
    space, 341
    system, 401
    variety, 341

quadratic form, 9, 21
quadratic period, 95

ramification field, 75
ramification group, 75
ray class field, 261
Reciprocity law
    Artin's, ix, 119
    Burde's, 159, 167, 188, 204
        octic, 290
    cubic, 209–232
        in $\mathbb{Z}[\zeta_{12}]$, 220

Eisenstein's, vii, 136, 296–299, 357, 365, 392
Hilbert's, viii, 64, 70, 201
Kummer's, viii
Lehmer's, 167
octic, 289, 304
quadratic, vi, 4, 10, 23, 24, 28, 30, 31, 43, 47, 50–52, 60, 64, 71, 80, 81, 85, 88, 95, 96, 100, 101, 103, 200, 223, 335
    general, viii
    in $\mathbb{Z}[\rho]$, 209
    in $\mathbb{Z}[i]$, 154, 188
quartic, vii, 194
    in $\mathbb{Z}[\zeta_8]$, 198
rational, xiii, 153
    cubic, 227
Scholz's, 102, 160, 162, 167
    octic, 307
sextic, 216–219
relative class group, 373
residue symbol
    general, 111
    rational, 153
Riemann hypothesis, 132, 329, 335, 341
ring class field, 225
root number, 335

similar, 47, 50
singular, 336
snake lemma, 49, 72
Spiegelungssatz, 377, 390
split sequence, 75
Steinberg symbol, 65
Stickelberger ideal, 371, 379, 395
Stickelberger index, 379
Stickelberger relation, 357, 391
Stickelberger's congruence, 128, 366, 391, 396, 398
Stickelberger's theorem, 371

Tate-Shafarevich group, 136, 175
Theorem

90 (Hilbert's), 47, 209
Abel's, 250
Dirichlet's, 19
Herbrand's, 375
Kronecker-Weber, 80, 103, 104
Legendre's, 7, 14, 16, 28, 54
Liouville's, 248
Principal Genus, 53, 69
Stickelberger's, 371, 372
Stickelberger-Voronoi, 116, 137
Wilson's, 203
totally
    negative, 45
    positive, 45
trace, 43, 127

unit group, 44

Vandiver's conjecture, 378, 387

Weierstraß
    σ-function, 250
    $\wp$-function, 248, 276
    ζ-function, 250
weight, 263
Weil Conjectures, xi, 341, 344, 345
Wieferich prime, 401
wreath product, 186, 202
Wurzelzahl, 388

Zariski topology, 341
zeta function
    congruence, 337
    Dedekind's, 331
    Hurwitz's, 332
    partial, 332
    Riemann's, 248, 263, 329

Springer Monographs in Mathematics

Abhyankar, **Resolution of Singularities of Embedded Algebraic Surfaces,** 2nd Edition, 1998
Arnold, **Random Dynamical Systems,** 1998
Aubin, **Some Nonlinear Problems in Riemannian Geometry,** 1998
Baues, **Combinatorial Foundation of Homology and Homotopy,** 1999
Brown, **Buildings,** 1st Edition 1989, Reprint 1996, 1998
Crabb/James, **Fibrewise Homotopy Theory,** 1998
Dineen, **Complex Analysis on Infinite Dimensional Spaces,** 1999
Elstrodt/Grunewald/Mennicke, **Groups Acting on Hyperbolic Space,** 1997
Fadell/Husseini, **Geometry and Topology of Configuration Spaces,** 2000
Fedorov/Kozlov, **A Memoir on Integrable Systems,** 1998
Flenner/O'Carroll/Vogel, **Joins and Intersections,** 1999
Griess, **Twelve Sporadic Groups,** 1998
Ivrii, **Microlocal Analysis and Precise Spectral Asymptotics,** 1998
Kozlov/Maz'ya, **Differential Equations with Operator Coefficients,** 1999
Lemmermeyer, **Reciprocity Laws,** 2000
Landsman, **Mathematical Topics Between Classical and Quantum Mechanics,** 1998
Malle/Matzat, **Inverse Galois Theory,** 1999
Mardešić, **Strong Shape and Homology,** 1999
Narkiewicz, **The Development of Prime Number Theory,** 2000
Ranicki, **High-dimensional Knot Theory,** 1998
Ribenboim, **The Theory of Classical Valuations,** 1999
Rudyak, **Thom Spectra, Orientability and Cobordism,** 1998
Serre, **Local Algebra - Multiplicities,** 2000
Springer/Veldkamp, **Octonions, Jordan Algebras and Exceptional Groups,** 2000
Sznitman, **Brownian Motion, Obstacles and Random Media,** 1998
Üstünel/Zakai, **Transformation of Measure on Wiener Space,** 1999

Printing: Weihert-Druck GmbH, Darmstadt
Binding: Buchbinderei Schäffer, Grünstadt